Liquid Crystals and Ordered Fluids

Volume 4

Liquid Crystals and Ordered Fluids

A Continuation Order Plan is available for this series. A continuation order will bring delivery of each new volume immediately upon publication. Volumes are billed only upon actual shipment. For further information please contact the publisher.

Liquid Crystals and Ordered Fluids

Volume 4

Edited by
ANSELM C. GRIFFIN
University of Southern Mississippi
Hattiesburg, Mississippi

and

JULIAN F. JOHNSON
University of Connecticut
Storrs, Connecticut

PLENUM PRESS • NEW YORK-LONDON

Library of Congress cataloged the second volume of this title as follows:

Main entry under title:

Liquid crystals and ordered fluids.

Papers, from a symposium of the Division of Colloid and Surface Chemistry held in Chicago during the national meeting of the American Chemical Society, August 1973, of the 3d of a series of meetings; papers of the 1st are entered under the title: Ordered fluids and liquid crystals; papers of the 2d are entered under: Symposium on Ordered Fluids and Liquid Crystals, 2d, New York, 1969.

1. Liquid crystals—Congresses. I. Johnson, Julian Frank, 1923–   ed. II. Porter, Roger Stephen, 1928–   ed. III. American Chemical Society. Division of Colloid and Surface Chemistry.

QD923.L56   548'.9   74-1269

Library of Congress Card Catalog Number 74-1269

ISBN-13: 978-1-4612-9658-4   e-ISBN-13: 978-1-4613-2661-8
DOI: 10.1007/978-1-4613-2661-8

Proceedings of the American Chemical Society Symposium on Liquid Crystals and Ordered Fluids, held March 29–April 1, 1982, in Las Vegas, Nevada

© 1984 Plenum Press, New York
Softcover reprint of the hardcover 1st edition 1984

A Division of Plenum Publishing Corporation
233 Spring Street, New York, N.Y. 10013

PREFACE

        This volume represents a collection of selected papers
presented at a symposium of the same name sponsored by the Division
of Colloid and Surface Chemistry held at the national Spring meeting
of the American Chemical Society in Las Vegas, Nevada, March 29 -
April 1, 1982. Also included are invited papers from a number of
outstanding overseas liquid crystal scientists who were unable to
attend the symposium. The attendance at the symposium itself and
the number of papers contained herein is reflective of the high
level of current interest in (and maturity of) the field of liquid
crystal research.

        Included in this volume are papers mainly derived from
the fields of chemistry and physics ranging in content from the
design and synthesis of new mesogenic materials to theoretical
physical treatments of anisotropic liquids. One of the significant
aspects of current liquid crystal research is the increasing col-
laboration between chemist and physicist. The overlap of these two
areas has been growing over the last several years and many contributions to this volume involve a molecular approach to the chemical
physics of liquid crystalline materials.

        In addition to the classical areas of liquid crystalline
investigation which are well represented here, there are several
contributions involving new developments/concepts which are in the
emerging stage, examples of which follow: pleochroic dyes, discotics,
ferroelectrics, incommensurate smectics, adiabatic colarimetry,
mesogenic polymers and polymer models, non-aqueous lyotropics and
ordered biological arrays, spectroscopic determination of detailed
molecular dynamics in mesogens, surface alignment for device applications, blue phases in cholesterics, improved theoretical interpretation of mesophase behavior, and liquid chromatography/mass
spectrometry analysis of liquid crystalline materials.

                                        Julian F. Johnson

                                        Anselm C. Griffin

CONTENTS

CONTENTS

SOME NOVEL FERROELECTRIC SMECTIC LIQUID CRYSTALS

J. W. Goodby and T. M. Leslie

Bell Laboratories
Murray Hill, New Jersey 07974

ABSTRACT

Ferroelectric smectic liquid crystals show potential for becoming increasingly important in electrooptic display device applications. However, the numbers of materials available for these applications are limited, particularly those which exhibit ferroelectric, chiral smectic C phases. In this paper we report a number of new systems which exhibit smectic C phases at temperatures close to room temperature making it possible to provide a eutectic mixture which exhibits a stable chiral smectic C phase at room temperature.

INTRODUCTION

Clark and Lagerwall[1] recently demonstrated an electrooptic effect which utilized the properties of a ferroelectric smectic phase for its switching mechanism. This new configuration also has memory properties. The device employs two tilted configurations of the molecules within their layers as the two stable states; switching from one state to the other involves the movement of a boundary between the two domains by applying an electric field across the

1

molecular layers. Switching speeds are extremely fast (in the
microsecond range) and coupled with the fact that the device is
inherently bistable makes for a potentially very interesting display concept.

Bistable electrooptic effects have been shown to occur for
chiral smectic C, chiral smectic I and chiral smectic F phases, but
not for the more ordered tilted phases G and H. The response times
are greatly effected by the phase employed, the more ordered its
structure, the slower the effect; thus making the C phase the most
desirable phase for use in a device of this type.

Although the chiral smectic C phase was discovered recently[2]
it has been the subject of more structural studies than of its
device properties. Hence, it was regarded generally as the optically active analogue of the C phase with a similar novelty to that
of the cholesteric phase. Thus concerted materials research in
this area has been limited by the requirements of structural and
physical studies.

However, the synthesis of novel materials provides the first
step in making this new device concept viable. In this present
work we discuss the behavior of a number of homologous series which
exhibit smectic C properties, and the problems encountered with
blending these in order to provide a suitable eutectic mixture for
further examination of the properties of this device.

RESULTS AND DISCUSSION

The Effects of the Position of the Chiral Center on Smectic C
Properties.

The 4-n-alkoxyphenyl 4'-n-alkoxybenzoates[3]

$$C_nH_{2n+1}O-\langle\!\bigcirc\!\rangle-COO-\langle\!\bigcirc\!\rangle-OC_mH_{2m+1}$$

are well known achiral materials which strongly favor smectic C
properties. Typically, as n and m increase in value (to n = 8,
m = 4) smectic C properties are injected into the homologous series.
Similarly, by incorporating a (+)-2-methylbutyloxy function in this
molecular configuration it is possible to produce chiral smectic
properties. The 4-(2'-methylbutyl)phenyl 4'-n-alkoxybenzoates[4] for
example

$$C_nH_{2n+1}O-\langle\!\!\bigcirc\!\!\rangle-COO-\langle\!\!\bigcirc\!\!\rangle-O(CH_2)_y-\overset{*}{C}HC_2H_5 \;\; (y{=}1) \qquad (I)$$
$$\underset{\displaystyle CH_3}{\vert}$$

exhibit ferroelectric smectic C phases for values of n $\geq$ 8. These
compounds also usually exhibit cholesteric and smectic A phases,
with the C phases occurring at approximately 30-35°, thus making
them useful materials for further studies. If the value of y is
increased, thus making the chiral centre further away from the core
section, then the temperature range of the C phase increases. Typically, the A-C transition temperatures rise between 10 and 15° for
each increase in the value of y by one. This has two advantages,
firstly it gives the device a wider operating temperature range and
secondly it increases the pitch length of the helicoidal structure
of the C phase. However, it has a disadvantage in reducing the
strength of the polarisable nature of the material; the better integrated the chiral centre is with the conjugated core system, the
greater will be the strength of the resulting polarisable ferroelectric dipole. Therefore, by moving the branch away from the
core the strength of the ferroelectric dipole will be reduced.

Reversal of the central linkage has interesting effects; the
smectic A phase is eliminated and a direct cholesteric - smectic C
transition is observed. Table 1 shows the transition temperatures
obtained for the (+)-4-n-alkoxyphenyl 4'-(4"-methylhexyloxy)benzo-
ates. All of the esters studied melt to the smectic C phase below
35° (except n=6) and have smectic C to cholesteric transition tem-

TABLE 1

$$C_2H_5-\overset{*}{\underset{\underset{CH_3}{|}}{C}H}(CH_2)_3O-\text{⟨○⟩}-COO-\text{⟨○⟩}-OC_nH_{2n+1}$$

n	Iso-Ch	Ch-S_C^*	m.p.
6	63.8	(42.7)	44.5
7	63.7	46.4	30.0
8	62.2	47.2	35.0
9	57.2	45.4	32.2
10	62.2	47.7	33.0

( ) Monotropic transition temperature

peratures at approximately 50°, eutectic mixtures of these materials
provide smectic C properties at room temperature to 50°. However,
they are not directly useful for devices that require fabrication
in the A phase, but they provide strong C behavior which can advan-
tageously be employed in mixtures with other different smectic C
materials which do exhibit A phases. An alternative method of
increasing the pitch of the chiral C phase but without moving the
asymmetric centre further away from the core can be made by the
inclusion of a spacer group between the core and the branch. This
spacer group has to be partially conjugated with the central core
in order to retain the polarizability qualities of the core. For
example, materials of the type

$$RO-\text{⟨○⟩}-COO-\text{⟨○⟩}-COOR'$$

and

$$RO-\text{⟨○⟩}-COO-\text{⟨○⟩}-CH=CHCOOR'$$

fall into this category, where the extra spacer group is (COO) and
(CH=CHCOO) respectively and R' is an optically active group.

TABLE 2

$$C_nH_{2n+1}O-\bigcirc-COO-\bigcirc-COOCH_2\overset{*}{C}HC_2H_5$$
$$CH_3$$

n	Iso-S_A	S_A-S_C^*	m.p.
5	(48.0)	–	63.0
6	55.0	–	39.0
7	53.0	(19.0)	35.0
8	58.0	(32.0)	33.0
9	57.0	(32.0)	56.0
10	59.5	(35.0)	52.0
12	60.0	(35.5)	55.0
14	60.0	(35.0)	40.0
16	60.2	(34.5)	49.0

( ) Monotropic Transition Temperature

The (−)-4-(2-methylbutyl)benzoate esters of the 4-n-alkoxyben-
zoic acids

$$C_nH_{2n+1}O-\bigcirc-COO-\bigcirc-COO(CH_2)_y\overset{CH_3}{\underset{*}{-CHC_2H_5}} \quad (y=1) \quad (II)$$

exhibit smectic A and chiral smectic C phases. Unlike the 4-(2'-
methylbutyl)phenyl 4'-n-alkoxybenzoates (I) these materials do not
exhibit cholesteric phases. The diester configuration favors
smectic A phases more strongly, as demonstrated by eutectic studies
(see later). The smectic A to smectic C transition temperatures
fall into the 30 to 35° temperature range as shown in Table 2. The
C phase is injected at the n-heptyloxy member, with the A-C transition temperatures rising rapidly and then leveling off at the
n-decyloxy homologue and then slowly falling as the series is
ascended. This pattern of behavior is very typical for smectic C
phases exhibited by esters. The variation in the phase behavior
for this series is shown in Fig. 1 as a function of increasing
terminal n-alkoxy chain length.

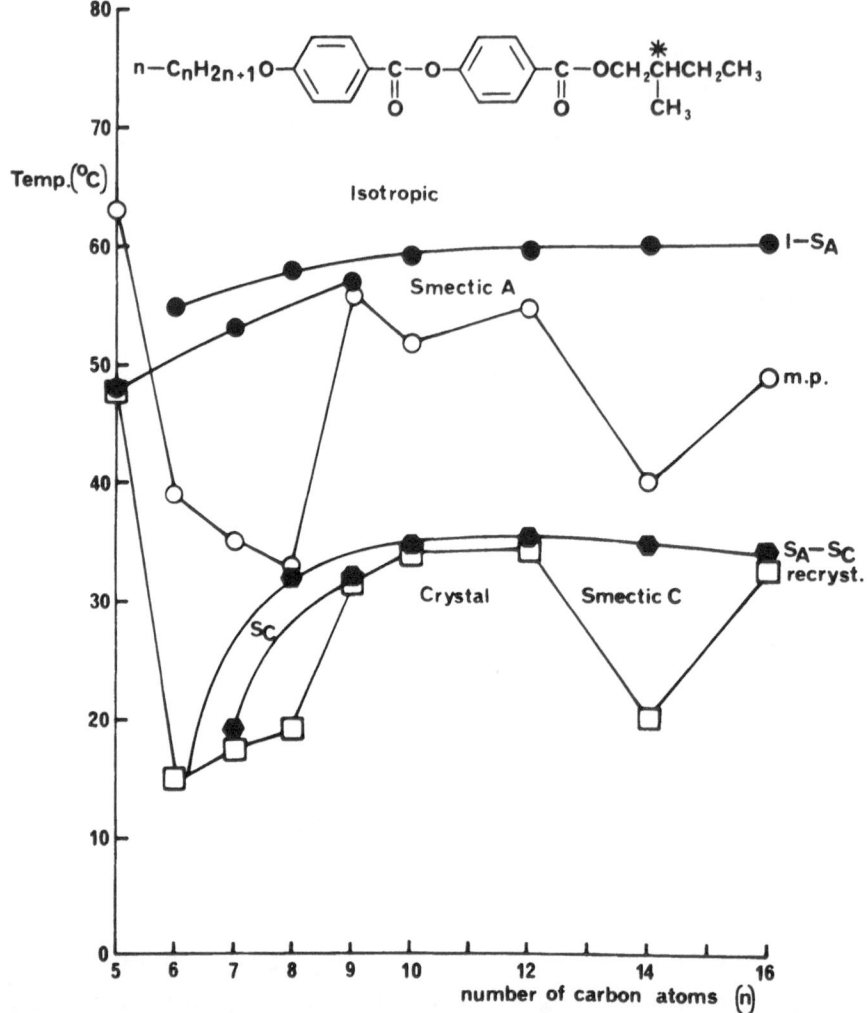

Fig. 1.   Plot of the transition temperatures against the number of
           carbon atoms (n) in the n-alkoxy chain of the (-)-4-(2-
           methylbutyl) benzoate esters of 4-n-alkoxybenzoic acids.
           Key: ●, I-S$_A$; ◑ , S$_A$-S$_C$, ○, crystal-mesophase; □ ,
           mesophase-crystal.

     The melting points of these materials are relatively high
except for the n-octyloxy and the n-tetradecyloxy compounds which
exhibit monotropic C phases only a few degrees below their melting
points. The transition temperatures for the n-octyloxy member are

$$\text{cryst} \rightarrow S_A \rightarrow \text{Iso} \rightarrow S_A \rightarrow S_C \rightarrow \text{cryst}$$
$$\phantom{\text{cryst}} \; 33 \quad 58 \quad 58 \quad 32 \quad 18$$

making it useful for mixture investigations.

The microscopic textures of the chiral smectic C phase for this compound are shown in Fig. 2. The material exhibits a banded focal-conic fan textures; the width of the banding is proportional to the pitch length of the phase. At the transition to the C phase a normal Schlieren texture is formed, this is quickly supplanted by a pseudohomeotropic texture. In this textural form the helical axis of the phase is vertical (perpendicular) to the glass supports, and often some residual Schlieren texture is retained around the focal-conic domains where the molecular layers are curved.

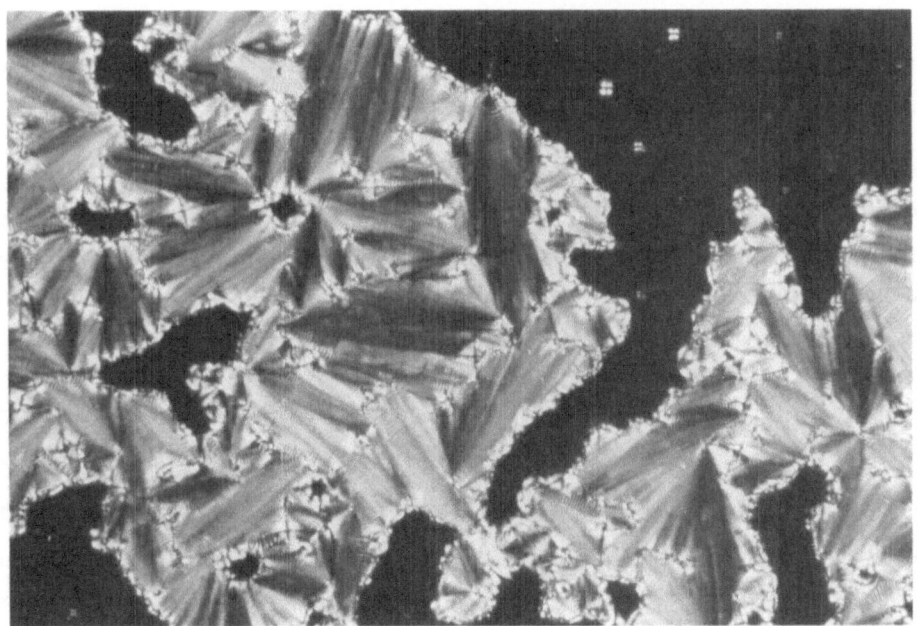

Fig. 2. The striated focal-conic fan and pseudohomeotropic texture of the chiral smectic C phase of the (-)-4-(2'-methylbutyl) benzoate ester of 4-n-octyloxybenzoic acid. (X100).

TABLE 3

$$C_nH_{2n+1}O-\bigcirc-COO-\bigcirc-CH=\underset{\underset{CN}{|}}{C}COOCH_2\overset{*}{\underset{\underset{CH_3}{|}}{C}H}C_2H_5$$

n	Iso-Ch	Ch-S_A	Iso-S_A	S_A-S_C^*	m.p.
5	(71.0)+	–	–	–	92.7
6	(82.0)	(67.0)	–	–	82.2
7	(82.6)	(74.1)	–	(58.0)	74.9
8	(85.6)	(82.5)	–	(60.6)	86.0
9	(86.6)	(85.5)	–	–	86.7
10	–	–	87.7	–	86.1
12	–	–	89.0	(71.7)	81.4
14	–	–	91.0	–	84.0
16	–	–	92.3	–	87.5

( ) Monotropic transition temperature

+ 'virtual' transition temperature

If the value of y in this series is increased the smectic C phase is stabilized and this is reflected by an increase in the A-C transition temperatures.

Similar behavior can be observed for the derivatives of 4-hydroxycinnamic acid. In this case the strength of the polarisable dipole of the materials under examination can be influenced by the inclusion of an electron withdrawing group in the spacer function. The derivatives of α-cyano-4-hydroxycinnamic acid

$$C_nH_{2n+1}O-\bigcirc-COO-\bigcirc-CH=\underset{\underset{CN}{|}}{C}-COOCH_2\overset{*}{\underset{\underset{CH_3}{|}}{C}H}C_2H_5 \qquad (III)$$

were synthesized and shown to exhibit chiral smectic C phases. The transition temperatures of these materials are given in Table 3 and shown as a function of increasing n-alkoxy chain length (n) in Fig. 3.

The earlier members of this series exhibit cholesteric phases, unlike series (II) which only exhibit smectic phases. This is

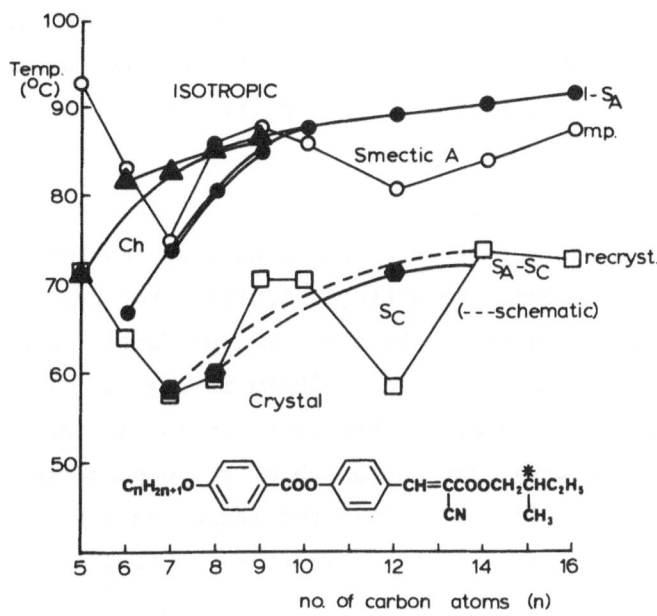

Fig. 3.   Plot of the transition temperatures against the number of
          carbon atoms (n) in the n-alkoxy chain of the (-)-α-cyano-
          4-(2'-methylbutyl)cinnamate esters of 4-n-alkoxybenzoic
          acids. Key: ▲, I-Ch; ●, I or CH-S_A; ◗ , S_A-S_C ○,
          crystal-mesophase, □ , mesophase-crystal.

probably due to the inclusion of the α-cyano group. The later

members, n-decyloxy onwards, exhibit direct Iso-A transitions. The

melting points of this series (III) are particularly high, forming

a peak at the n-octyloxy and n-nonyloxy homologues which exhibit

direct crystal to isotropic liquid melting behavior. However, both

of these materials exhibit three monotropic phases on cooling (choles-

teric S_A and S_C).

     The chiral smectic C phase is injected into the series at the

n-heptyloxy member and the A-C transition temperatures rise as the

series is ascended. These temperatures effectively level off at

the n-tetradecyloxy compound at a value of approximately 70°.

Again this sudden injection of C properties at the n-heptyloxy homo-

logue is typical of achiral systéms. However, it is interesting to

note that the Iso-Ch and A -Iso or Ch transition temperatures have
the correct sense of alternation, but the A-C temperatures are
opposite in alternation to that which is usually expected for this
transition.

Microscopic studies of the textures exhibited by these materials show typical focal-conic and plane textures for the cholesteric
phase, and focal conic homeotropic textures for the A phase. The
chiral smectic C phase exhibits the banded focal-conic, pseudohomeotropic and Schlieren textures as shown in Fig. 4. The arcs in the
fan texture are wider apart than those exhibited for compounds of
type (II) indicating that the pitch of the C phase is greater in
series (III) than in (II). Thus the larger spacer group of this
series increases the distance between the chiral centre and the
aromatic core, which probably in turn increases the pitch of the
phase even though the same chiral group is utilized. The transition temperatures for the A-C phase change are greater in series
(III) than in (II) or (I) indicating that the lateral cyano group
and the inclusion of a larger spacer group aids C formation. However, this also detrimentally raises the melting points.

Biphenyl derivatives also provide a useful alternative source
for smectic C phases. The directly analogous (-)-2-methylbutyl
4'-n-alkoxybiphenyl-4-carboxylates

$$C_nH_{2n+1}O-\langle\!\!\!\bigcirc\!\!\!\rangle-\langle\!\!\!\bigcirc\!\!\!\rangle-COO(CH_2)_{\overline{y}|}\overset{*}{C}HC_2H_5 \quad (y = 1)$$
$$\qquad\qquad\qquad\qquad\qquad\qquad CH_3$$
(IV)

exhibit both smectic A and chiral smectic C phases, with the A-C
transitions occurring in the 40-50° temperature range. Full details
of the transition temperatures for the n-pentyloxy to the n-decyloxy
members are given in Table 4.

The Iso-A transition temperatures alternate in the correct
manner (even members higher than odd for the n-alkoxy substitution),
but the A-C phase changes again alternate in the opposite sense to

TABLE 4

$$C_nH_{2n+1}O-\bigcirc\!\!-\!\!\bigcirc-COOCH_2\overset{*}{\underset{CH_3}{C}}HC_2H_5$$

n	Iso-S_A	S_A-S_C^*	m.p.
5	64.0	43.9	42.7
6	66.0	–	48.0
7	64.7	–	57.5
8	65.9	(44)	49.2
10	66.2	(41.2)	48.2

( ) Monotropic Transition Temperature

Fig. 4.   The striated fan, pseudohomeotropic and Schlieren textures of the chiral smectic C phase of the (-)-α-cyano-
4-(2'-methyl-butyl)cinnamate ester of 4-n-dodecyloxy-
benzoic acid (X100).

the Iso-A ones and therefore opposite to that which is usually
observed for A-C transitions. Most of the C phases observed are
monotropic in nature (except for the early members of the series).
If the methylene spacer chain between the branch and the core is
increased in value to y = 2; then for the n-octyloxy homologue

$$n\text{-}C_8H_{17}O\text{-}\langle\!\!\langle\;\;\rangle\!\!\rangle\text{-}\langle\!\!\langle\;\;\rangle\!\!\rangle\text{-}COO(CH_2)_2\overset{*}{C}HC_2H_5$$
$$\overset{|}{C}H_3$$

the following phase sequence is observed:

$$Iso \rightarrow S_A \rightarrow S_C^{\,*} \rightarrow recryst \qquad m.p.$$
$$\;\;\;\;72 \;\;\;\; 58 \;\;\;\; 41.5 \qquad\qquad\;\; 59$$

If y is increased again to a value of y = 3; then for the n-octyloxy
homologue

$$n\text{-}C_8H_{17}O\text{-}\langle\!\!\langle\;\;\rangle\!\!\rangle\text{-}\langle\!\!\langle\;\;\rangle\!\!\rangle\text{-}COO(CH_2)_3\overset{*}{C}H\;C_2H_5$$
$$\overset{|}{C}H_3$$

the following phase sequence is obtained:

$$Iso \rightarrow S_A \rightarrow S_G^{\,*} \rightarrow recryst \qquad m.p.$$
$$\;\;\;\;68 \;\;\;\; 62.2 \;\;\;\; 26 \qquad\qquad\;\; 38$$

Therefore, as the branch of the chiral centre is moved further away
from the core the temperature of the orthogonal smectic phase to
tilted smectic phase change increases, i.e. the formation of tilted
properties is enhanced. However, in this case the stability of the
G phase rises more rapidly by this action than does the C phase,
hence G properties predominate at longer methylene spacer chain
lengths.

This type of behavior is also observed in the analogous
straight chain derivatives[6]

$$C_nH_{2n+1}O\text{-}\langle\!\!\langle\;\;\rangle\!\!\rangle\text{-}\langle\!\!\langle\;\;\rangle\!\!\rangle\text{-}COOC_mH_{2m+1}$$

when m is low in value (1-3) only orthogonal phases (A, B and E) are observed.   Increases, in the value of m to 4-7 with n between 8-14 produces tilted C phases, and for higher values of n smectic G phases are obtained.   Thus, the action of including a methyl branch in the ester terminal alkyl chain has the property of <u>compacting</u> and <u>moving</u> this observed phase sequence pattern <u>towards shorter</u> ester chain lengths for these homologous series.

Effects of the Dipolar Properties of the Prime or Central Linkage on Smectic C Formation

The smectic C phase, like the hexatic B phase can be stabilized by stronger lateral dipolar forces.   For example, if the ester linkage in series (IV) is reversed the ketone function is no longer conjugated with the central polarisable core; then C properties give way to the formation of phases more crystalline in nature.   The materials

$$C_2H_5\overset{*}{\underset{\underset{CH_3}{|}}{C}}HCH_2O-\bigcirc\!\!-\!\!\bigcirc-OOCC_9H_{19}$$

(+)-4-n-decanoyloxy-4'-(2"-methylbutyloxy)biphenyl

$$Iso \rightarrow S_B$$
117

$$C_8H_{17}O-\bigcirc\!\!-\!\!\bigcirc-OOC-CH_2\overset{*}{\underset{\underset{CH_3}{|}}{C}}H\ C_2H_5$$

(+)-4-(3"-methylpentanoyloxy)-4'-n-octyloxybiphenyl

$$Iso \rightarrow S_G$$
82

and

$$C_8H_{17}O-\bigcirc\!\!-\!\!\bigcirc-OOC-(CH_2)_2\overset{*}{\underset{\underset{CH_3}{|}}{C}}HC_2H_5$$

(+)-4-(4"-methylhexanoyloxy)-4'-n-octyloxybiphenyl

$$\text{Iso} \rightarrow \underset{93}{S_A} \rightarrow \underset{92.5}{S_G}$$

strongly favor tilted smectic properties, but prefer to exhibit
crystal smectic phases rather than smectic C phases, i.e., they
exhibit the G phase which is the tilted analogue of the crystal B
phase. Fig. 5 shows the smectic G phase of (+)-4-(3"-methylpent-
anoyloxy)-4'-n-octyloxybiphenyl separating from the isotropic
liquid in its natural texture. The dendritic growth patterns
coalesce to produce a normal mosaic texture.

Similarly, the reverse is true when the chiral centre is
closely linked to the conjugated system of the core, as in homologous series (I). The orthogonal smectic A phase is suppressed

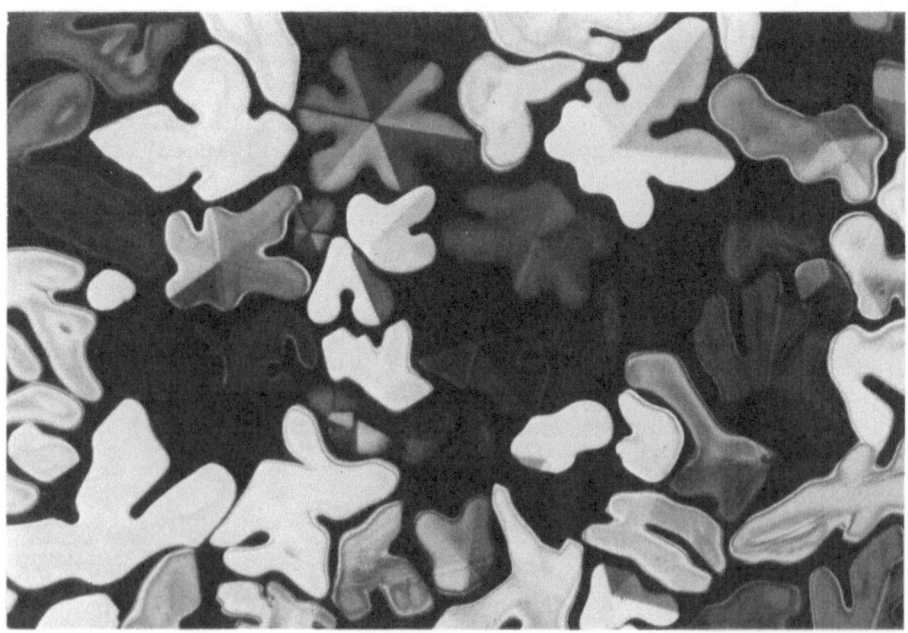

Fig. 5.   The smectic G phase separating from the isotropic liquid
           of (+)-4-(3"-methylpentanoyloxy)-4'-n-octyloxybiphenyl
           (X100).

and direct cholesteric to smectic C transitions are observed. In
this case, the alkoxy oxygen atom of the chiral moiety is linked to
the ester ketone function by mesomeric relay through the first
phenyl ring, as shown below

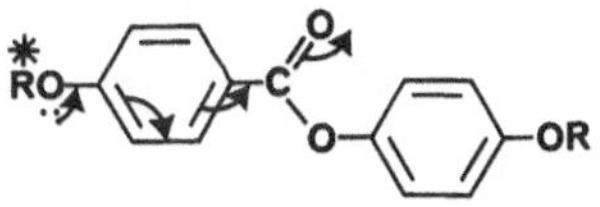

whereas this is not the case for the reversed system. The chiral
centre is influenced by this polarisation and this in turn advan-
tageously affects the stability of the C phase as opposed to the A
phase.

Thus, these results provide an insight into the delicate
balance between dipolar forces which stablize the C phase.

Effect of the Strength of the Dipole Associated with the Chiral Centre

So far only the properties of materials which possess a chiral
centre carrying a methyl branch have been discussed. This is
simply because optically pure S-(-)-2-methylbutan-1-ol is readily
available. Studies involving different branching groups are more
difficult to perform because of the need to synthesize and resolve
optical isomers. However, some preliminary investigations of
chloro and trifluoromethyl branched systems have been made.

Initial studies were made on derivatives of 2-chlorobutyric
acid; the acid was resolved with the aid of optically pure brucine.
The homologous series,

(+)-4-(2''-chlorobutanoyloxy)-4'-n-alkoxybiphenyls

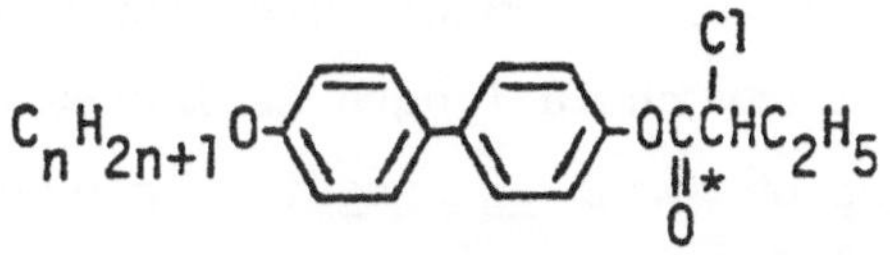

(V)

(for values of n = 5 to 10 inclusive) was synthesized and its properties examined. The transition temperatures for these materials

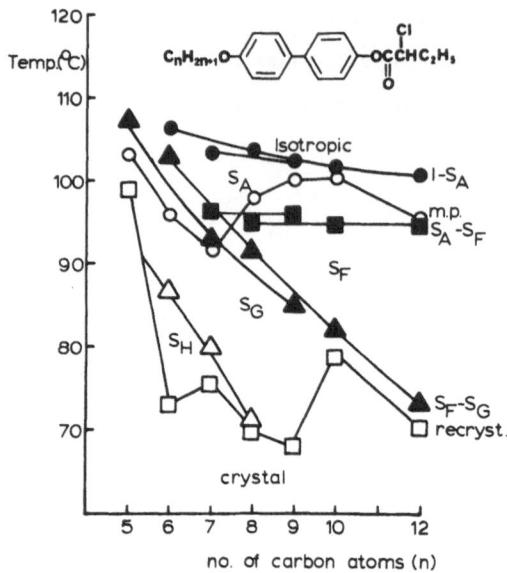

Fig. 6.   Plot of the transition temperatures against the number of
          carbon atoms (n) in the n-alkoxy chain of the (+)-4-(2"-
          chlorobutanoyloxy)-4'-n-alkoxybiphenyls. Key: ●, I-S$_A$;
          ■ , S$_A$-S$_F$; ▲ I, S$_A$ or S$_F$-S$_G$; △, S$_G$-S$_H$; ○, crystal-meso-
          phase: □ , mesophase-crystal.

are shown in Table 5 and as a function of increasing terminal
n-alkoxy chain length (n) in Fig. 6.

     All of the members studied exhibited Iso-A transitions except
for the n-pentyloxy homologue which showed an Iso-G phase change,
the subsequent members all possessed G phases except for the
n-decyloxy homologue which recrystallised before this determination
could be made. The n-octyloxy members onwards exhibit smectic F
phases, the upper transition temperatures for which remain fairly
constant as the n-alkoxy chain length is increased. The F phase is
therefore <u>unaffected</u> by small <u>changes</u> in the overall <u>molecular</u>
<u>structure</u>. The n-hexyloxy to n-octyloxy members inclusive exhibit
H phases on cooling the G phase. The G-H transition temperatures
fall away sharply on increasing the n-alkoxy chain length, and the

n-nonyloxy members onwards recrystallise before the H phase is observed.

Initial classification of the phase types was made by micro-scopic observations of the textures exhibited by these materials. Typically for the n-octyloxy homologue the A phase separates from the isotropic liquid on cooling in bâtonnets, these coalesce and form a focal-conic texture (Fig. 7). Cooling of this texture produces a transition to the F phase; this change is characterised by transition bars (Fig. 8) which expand and join together to produce a broken focal-conic texture (Fig. 9). Subsequent cooling of this texture produces a transition to another tilted phase, smectic G, which is characterised by its broken focal-conic texture (Fig. 10). The breakages in this case are less in number than those of the F phase giving the texture a more "chunky" appearance.

Fig. 7.   The focal-conic fan texture of the smectic A phase of (+)-4-(2"-chlorobutanoyloxy)-4¹n-octyloxybiphenyl (X100).

Fig. 8.   The transition bars formed at the A to F transition for
           the focal-conic texture of (+)-4-(2"-chlorobutanoyloxy)-
           4'-n-octyloxybiphenyl (X100).

Fig. 9.   The broken focal-conic fan texture of the smectic F phase
           of (+)-4-(2"-chlorobutanoyloxy)-4'-n-octyloxybiphenyl
           (X100).

Fig. 10.   The broken focal-conic fan texture of the smectic G phase
           of (+)-4-(2"-chlorobutanoyloxy)-4'-n-octyloxybiphenyl
           (X100).

Fig. 11.   The broken focal-conic fan texture at the point of
recrystallization of the smectic H phase of (+)-4-(2"-
chlorobutanoyloxy)-4¹n-octyloxybiphenyl (X100).

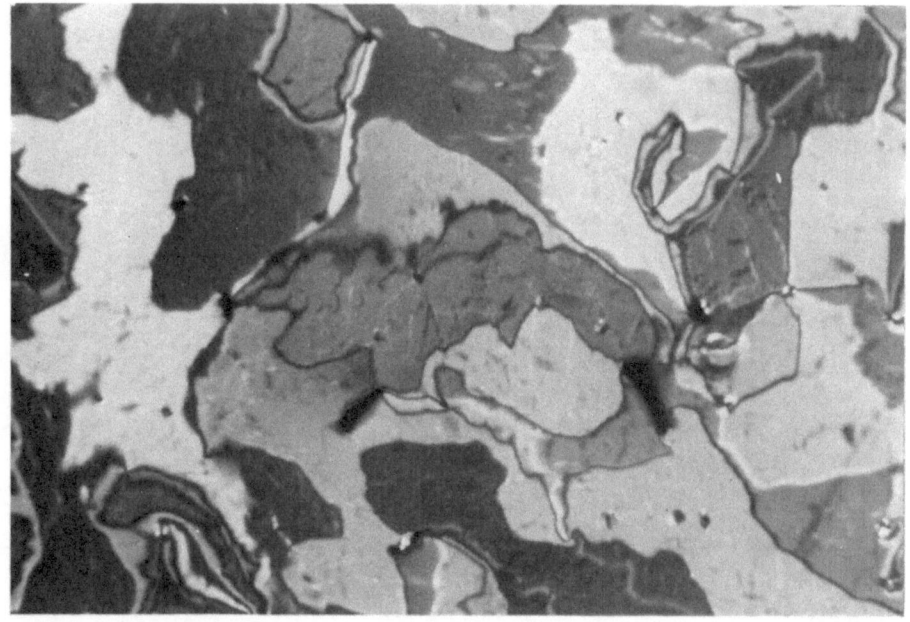

Fig. 12.   The Schlieren-mosaic texture of the smectic F phase of
           (+)-4-(2"-chlorobutanoyloxy)-4'-n-octyloxybiphenyl
           (X100).

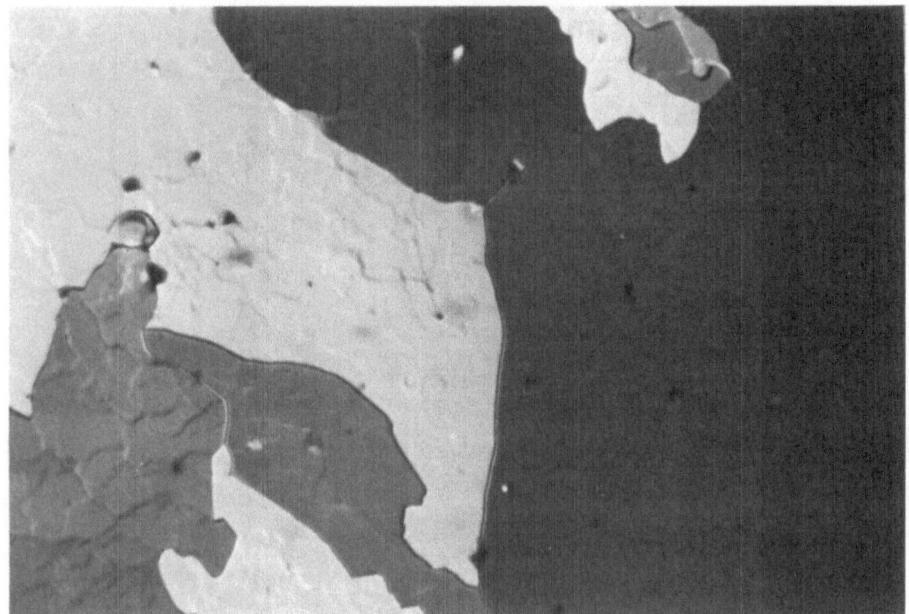

Fig. 13.   The mosaic texture of the smectic G phase (+)-4-(2"chloro-
           butanoyloxy)-4'-n-octyloxybiphenyl (X100).

On further cooling, the G phase gives way to the formation of the H
phase. This occurs on the point of recrystallization and the fans
do not produce a final texture before they are shattered by crystal
formation (Fig. 11).

    Alternatively the A phase can form a homeotropic texture and
the modifications formed on cooling this phase exhibit paramorphotic
textures based on the original texture. The F phase exhibits a
Schlieren-mosaic texture (Fig. 12) and the G phase exhibits a
mosaic texture texture (Fig. 13).

    Confirmation of the classification of the A, F, and G phases
of the n-octyloxy member was obtained by co-miscibility studies
with the standard material N-(4-n-nonyloxybenzylidene)-4'-n-butyl-
aniline (90.4) (A, F and G phases).[7] The miscibility diagram of
state for binary mixtures of these two materials is shown in

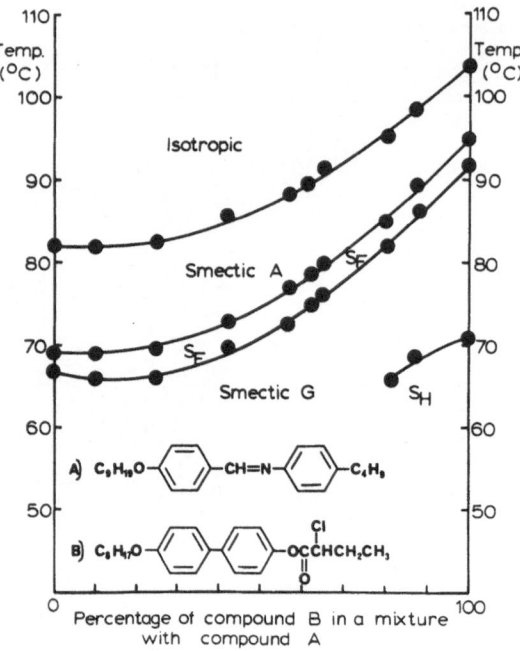

Fig. 14.   Diagram of state for mixtures (wt%) between N-(4-n-nonyl-
           oxybenzylidene)-4'-n-butylaniline (90.4) (A) and (+)-4-
           (2"-chlorobutanoyloxy)-4'-n-octyloxybiphenyl (B).

Fig. 14. The phases identified as A, F and G in the test ester
were found to be separately co-miscible with the A, F and G phases
of the standard compound.

     Unlike, the analogous methyl branched esters which exhibit A
and G phases, the chloro derivatives tend to exhibit more fluid
tilted smectic phases. This indicates that the stronger dipolar
nature of the optical centre may favour the formation of fluid (C)
and hexatic (F and I) tilted phases over that of crystal phases (G
and H).

     Preliminary studies on branches of the trifluoromethyl type
show unusual results. The material

$$C_8H_{17}O-\langle\bigcirc\rangle-\langle\bigcirc\rangle-COO\overset{*}{C}HC_6H_{13}$$
$$\underset{CF_3}{|}$$

Iso →	S_A →	S_E	m.p.
74	69		45.5

exhibits orthogonal phases, but this might be due to the branch being too close to the core structure. Thus, it is too early yet to draw any conclusions about the properties of this system.

However, the results obtained on these chloro and other known chloro materials indicate that the more dipolar the nature of the optical centre, the greater is the tendency toward formation of tilted phases, particularly of the C type.

<u>Higher Temperature Chiral Smectic C Phases</u>

The blending of smectic materials in order to produce eutectic mixtures with wide chiral smectic C temperature ranges requires some of the components to possess high A or Ch-C transition temperatures.

Increases in the values of the upper transition temperatures of the C phase can be obtained by the introduction of an extra phenyl group into the molecular structure of the smectogen, i.e., by producing a three phenyl ring system. They can be divided into two groups by the general structures

$$RO-\langle\bigcirc\rangle-X-\langle\bigcirc\rangle-X-\langle\bigcirc\rangle-OR'$$

and

$$RO-\langle\bigcirc\rangle-\langle\bigcirc\rangle-X-\langle\bigcirc\rangle-OR'$$

where R or R' is a terminal alkyl chain which contains a chiral centre. The following table gives some typical examples:

a) $C_2H_5\overset{*}{C}HCH_2O$—⟨benzene⟩—OOC—⟨benzene⟩—COO—⟨benzene⟩—$OCH_2\overset{*}{C}HC_2H_5$
    with CH_3 substituents

Iso → Ch → S_C^* → cryst   m.p.
     149.1   128.9   113.2   135

b) $C_2H_5\overset{*}{C}HCH_2OOC$—⟨benzene⟩—OOC—⟨benzene⟩—COO—⟨benzene⟩—$COOCH_2\overset{*}{C}HC_2H_5$
    with CH_3 substituents

Iso → Ch → S_A → S_C^* → cryst   m.p.
     160   148   112   105.5   135

c) $C_2H_5\overset{*}{C}HCH_2O$—⟨benzene⟩—COO—⟨benzene⟩—COO—⟨benzene⟩—OC_6H_{13}
    with CH_3 substituent

Iso → Ch → S_C^* → cryst   m.p.
     172.5 69.9 56.4   88.4

d) $C_2H_5\overset{*}{C}H(CH_2)_3O$—⟨benzene⟩—COO—⟨benzene⟩—COO—⟨benzene⟩—OC_6H_{13}
    with CH_3 substituent

Iso → Ch → S_C^* → cryst   m.p.
     175.3 115   52.3   78.5

e) $C_{10}H_{21}O$—⟨benzene⟩—COO—⟨benzene (Cl)⟩—COO—⟨benzene⟩—$OCH_2\overset{*}{C}HC_2H_5$
    with Cl and CH_3 substituents

Iso → S_A → S_C^* → cryst.   m.p.
     62.3   49.7 24.2   43.6

f) $C_8H_{17}O$—⬡⬡—OOC—⬡—$OCH_2\overset{*}{C}HC_2H_5$
                                                 |
                                                CH_3

Iso → Ch → S_C^* → cryst.   m.p.

146.3   121.5 92   103.8

g) $C_8H_{17}O$—⬡⬡—OOC—⬡—$OCH_2\overset{*}{C}HC_6H_{13}$
                                                 |
                                                CH_3

Iso → Ch → S_C^* → cryst   m.p.
141.3   125.3 93.9   108.2

These materials are only a few examples from a large number of relatively unexplored homologous series which will probably yield compounds with better device properties than the ones shown.

Eutectic Mixtures

Binary mixtures of some of these materials can provide useful eutectic mixtures, for example, a miscibility diagram of state for mixtures of the compounds

$C_{14}H_{29}O$—⬡—COO—⬡—$COOCH_2\overset{*}{C}HC_2H_5$
                                              |
                                             CH_3

and

$C_{12}H_{25}O$—⬡—COO—⬡—$CH=CCOOCH_2\overset{*}{C}HC_2H_5$
                                          | |
                                         CN CH_3

is shown in Fig. 15. In the region of 75% by weight of the first material and 25% by weight of the second, a suppression in the recrystallization temperatures is found. Thus, the chiral C phase exists from approximately 38° down to -10° on cooling. After the mixture is left to recrystallise over time, the melting point was

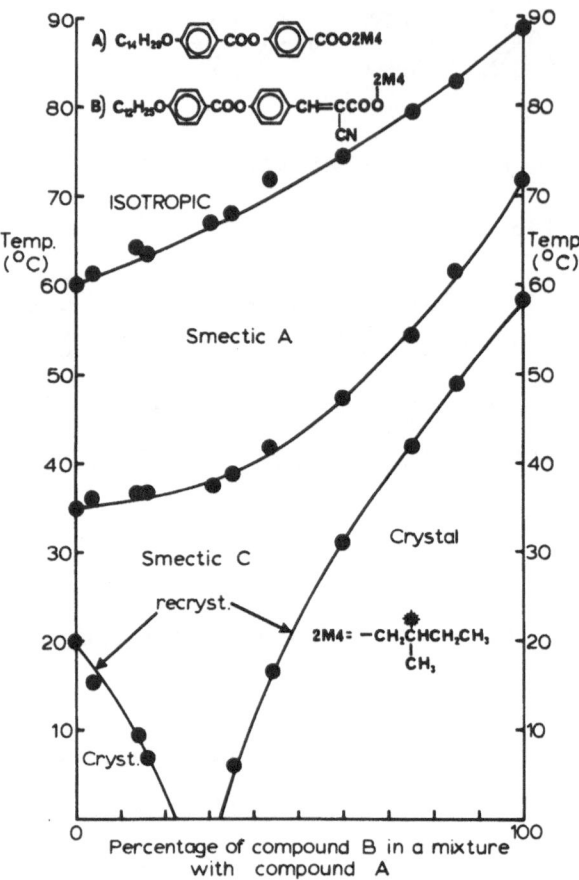

Fig. 15.   Diagram of state for mixtures (wt%) between the (-)-4-
(2'-methylbutyl)benzoate ester of 4-n-tetradecyloxybenzoic
acid (A) and the (-)-α-cyano-4-(2'-methylbutyl)cinnamate
ester of 4-n-dodecyloxybenzoic acid (B).

determined and found to be 28°. The A to C transition temperature
curve appears fairly normal, which is not always the case for this
type of curve, as many exhibit strong depressions when the two
materials are not totally compatible. The eutectic mixture from
the 3:1 blend of these two materials is useful for further physical
studies because the mix exhibits both A and C phases. However,
more advantageous properties can be obtained in multicomponent
mixtures, for example;

Mixture 1.

4M6O—⬡—COO—⬡—COO—⬡—OC$_6$H$_{13}$ 24.96%

2M4O—⬡—COO—⬡—COO—⬡—OC$_6$H$_{13}$ 23.66%

C$_{10}$H$_{21}$O—⬡—⬡—COO2M4   45.80%

C$_{10}$H$_{21}$O—⬡—COO—⬡—COO—⬡—O2M4   5.58%
                          Cl

$$cryst \rightarrow S_C^* \rightarrow Ch \rightarrow Iso$$
$$38.8 \quad 80.5 \quad 95.5$$

Mixture 2.

4M6O—⬡—COO—⬡—COO—⬡—OC$_6$H$_{13}$ 22.1%

C$_{12}$H$_{25}$O—⬡—COO—⬡—CH=CCOO2M4   33.1%
                                     CN

2M4O—⬡—COO—⬡—COO—⬡—OC$_8$H$_{13}$ 8.4%

C$_{10}$H$_{21}$O—⬡—COO—⬡—COO—⬡—O2M4   36.4%
                          Cl

$$Cryst \rightarrow S_C^* \rightarrow S_A \rightarrow Ch \rightarrow Iso$$
$$29 \quad 51.4 \quad 78.9 \quad 86.4$$

Mixture 3.

4M6O—⟨◯⟩—COO—⟨◯⟩—COO—⟨◯⟩—OC_6H_{13} 25.8%

4M6O—⟨◯⟩—COO—⟨◯⟩—OC_7H_{15} 22.7%

$C_8H_{17}O$—⟨◯⟩—COO—⟨◯⟩—COO2M4   51.5%

$$S_C^* \rightarrow S_A \rightarrow Ch \rightarrow Iso$$
$$\quad 5 \quad\quad 70.3 \quad 83.6$$

Mixture 4.

4M6O—⟨◯⟩—COO—⟨◯⟩—COO—⟨◯⟩—OC_6H_{13} 26%

4M6O—⟨◯⟩—COO—⟨◯⟩—OC_7H_{15} 30.8%

4M6O—⟨◯⟩—COO—⟨◯⟩—$OC_{10}H_{21}$ 11.9%

$C_8H_{17}O$—⟨◯⟩—COO—⟨◯⟩—COO2M4   25.1%

$C_{10}H_{21}O$—⟨◯⟩—COO—⟨◯⟩—COO—⟨◯⟩—O2M4   6.2%

$$S_C^* \rightarrow S_A \rightarrow Ch \rightarrow Iso$$
$$\quad 48 \quad\quad 59 \quad\quad 83$$

where the percentages given are weight percents and 2M4 is

$-CH_2\overset{*}{C}HC_2H_5$ and 4M6 is $-(CH_2)_3\overset{*}{C}HC_2H_5$.
$\quad\;\; CH_3 \qquad\qquad\qquad\qquad\quad CH_3$

   These mixtures are only some of the examples prepared and were
chosen to illustrate the following points:

   1) Mixtures containing materials which have either a biphenyl
or a 4-benzoyloxy-4'-alkylbenzoate in their molecular structures
invariably prefer to exhibit smectic A phases, as shown by mixture
3. In this case the large amount of the n-octyloxy ester actually
suppresses C formation and promotes A characteristics. This makes

diesters of this type useful for injecting smectic A phases into mixtures which only exhibit cholesteric and C phases.

2) The esters of group (I), the 4-n-alkoxyphenyl 4'-(4"-methyl-hexyloxy)benzoates, greatly reduce the melting points of mixtures to which they are added, as shown by mixtures 3 and 4.

3) Laterally substituted materials (e.g. CN or Cl) aid the formation of A phases in some mixtures, see mixture 2, (the A phase is currently required in device fabrication).

4) Mixtures also provide the opportunity to control the pitch length of the phase by mixing materials of opposite twist direction.

EXPERIMENTAL

The materials were synthesized by standard procedures[8-10] and purified by column chromatography over silica-gel (60-200 mesh, 3x30 cm) using hexane-dichloromethane, (1:1), as eluant. The final products were recrystallized from either ethanol or light-petrol (40-60°). Their structures were elucidated by elemental analysis and infra-red spectroscopy and their purities were checked by thin-layer chromatography and HPLC.

The transition temperatures of the final products and of the mixtures obtained on blending them were determined by optical micro-scopy using a Zeiss Universal polarising microscope in conjunction with a Mettler FP52 hot-stage and control unit.

CONCLUSION

A number of related systems have been studied and have been shown to exhibit chiral smectic C phases. The mixing of certain of these materials provide eutectic mixtures which exhibit C phases at room temperature, with the C phase being succeeded by the A phase on heating. The pitch of the C phase can be controlled by mixing materials of opposite twist direction. A full account of the physical properties of these materials will be published elsewhere.

REFERENCES

1. N. A. Clark and S. T. Lagerwall, App. Phys. Lett. <u>36(1)</u>, 899
       (1980).

2. R. B. Meyer, L. Liebert, L. Strzelecki and P. Keller, J. de
       Physique Lett. <u>36</u>, L69, 1475.

3. D. Marzotko Math. Nat. Dissertatin Halle, 1973, D. Demus, H.
       Demus, and H. Zaschke, Flussige Kristalle in Tabellen,
       VEB Deutscher Verlag Leipzig 1974, p. 65.

4. M. V. Loseva, N. I. Chernova, and N. I. Doroshina, Abstracts
       of the proceedings of the Third Liquid Crystal Conference,
       Budapest, 1979, Abstract G3.

5. R. B. Meyer, Mol. Cryst. Liq. Cryst. <u>40</u>, 33 (1977).

6. J. W. Goodby and G. W. Gray, J. de Physique <u>37</u>, 17 (1976).

7. J. W. Goodby and G. W. Gray, Mol. Cryst. Liq. Cryst. Lett. <u>56</u>,
       43 (1979).

8. J. W. Goodby and G. W. Gray, J. de Physique <u>40</u>, 27 (1979).

9. G. W. Gray and D. G. McDonnell, Mol. Cryst. Liq. Cryst. <u>37</u>,
       189 (1976).

10. Beilstein, B1^{3}, 1671.

LIQUID CRYSTALLINE ESTERS OF PHENYLHYDROQUINONE AND 3-PHENYL 4-HYDROXYBENZOIC ACID

R. J. Cox, W. Volksen and B. L. Dawson

IBM Research Laboratory
5600 Cottle Road
San Jose, California 95193

ABSTRACT

Two new series of diester liquid crystals were prepared. The central benzene ring in each case was laterally substituted with a bulky phenyl group. The transition temperatures were determined by DSC along with the values of ΔS and ΔH for one series. The thermodynamic parameters were compared with those of known series in which the lateral substituents were less bulky. It was concluded that the mechanism resulting in the large decrease in transition temperature in the subject compounds is quite different from that found in the other series which has been studied.

Liquid crystals of the diester type are well known, and have been used by several authors as probes to investigate the role of molecular structure of mesophase stability.[1,2,3] Dewar and Goldberg[1] demonstrated the effect of inversion of the central carbonyls on the mesophase stability of such a system. Dewar and Griffin[2] prepared a series of diesters from substituted hydroquinones in which the substituents were fluorine chlorine bromine iodine and methyl. They measured the enthalpy of the nematic to

isotropic transition and found a smooth relation between this and
the size of the substituents. Dubois and Beguin[4] also studied the
effect of lateral substitution on a similar series in which
p-hydroxybenzoic acid was diesterified. They did not measure the
enthalpy change for the transitions but they did observe a depression
in the transition temperature as the size of the substituent
increased.

In order to further elucidate the effect of molecular
structure on these compounds we have prepared two series of
diesters in which the central benzene ring is substituted with a
bulky phenyl group. The first series was prepared starting from
phenylhydroquinone and gave a symmetrically substituted compound,
while the second was from 3-phenyl-4-hydroxybenzoic acid giving the
unsymmetrical series. The synthesis of the compounds was straightforward and is outlined in Figures 1 and 2. The synthesis of the
unsymmetrical compounds obviously allows for the preparation of
materials which contain different alkoxy groups, however we have
so far only prepared those in which both are the same. The
compounds were all purified by high pressure liquid chromatography
and subsequent recrystallization. The structures were confirmed

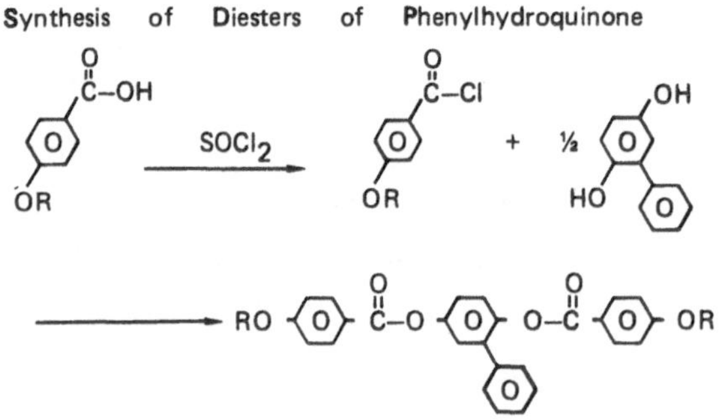

Figure 1

Synthesis   of   Diesters   of   3-Phenyl-4Hydroxy Benzoic   Acid

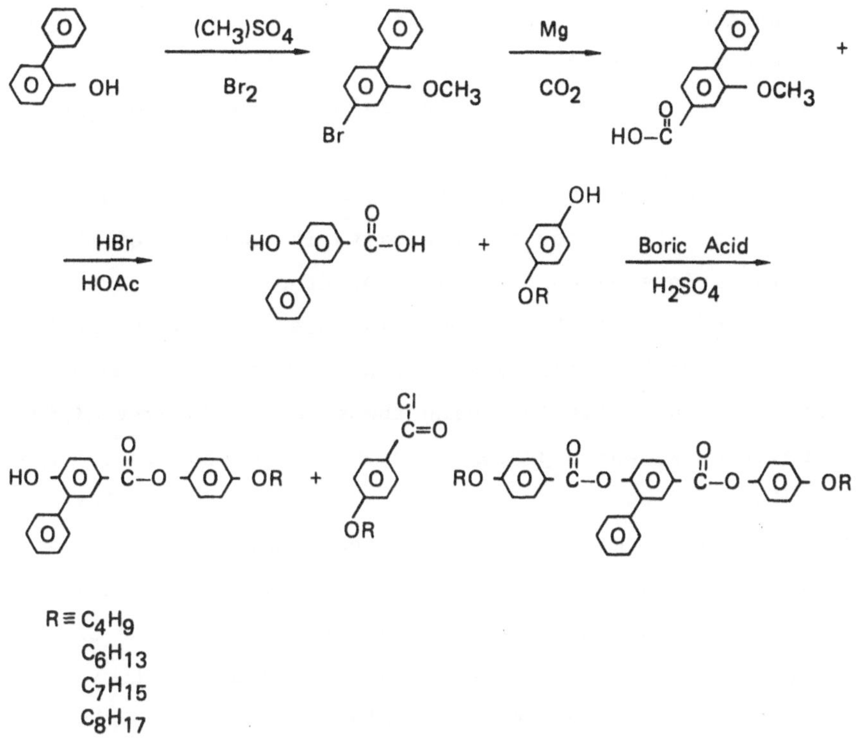

R ≡ C_4H_9
C_6H_{13}
C_7H_{15}
C_8H_{17}

Figure 2

by IR and NMR spectrometry and the purities determined by DSC
techniques and combustion analysis. The thermal properties were
measured on either a Dupont 990 or a Perkin-Elmer DSC-2 differential
scanning calorimeter. The enthalpy changes were measured on a
DSC-2 which had been computer interfaced. This system has been
described previously by Doelmann and co-workers.[5] The identification of the textures, and conformation of the transition temperature was made using a polarizing microscope equipped with a
Mettler PF-5 hot stage.

It was initially a surprise to us that these compounds should
show any liquid crystallinity at all, and although we have made no
accurate determination of the length to breadth ratio, a molecular
model, shown in Figure 3, has a geometry much different from the
lathe like shape of most liquid crystals. A summary of the
transition temperatures and the changes in the enthalpy and
entropy are given in Fig. 4 for the symmetrical compounds. The
first thing to notice, not surprisingly, is that all of the liquid
crystal transitions except one, are monotropic. The next is the
multiplicity of crystal forms found in these compounds, particularly
in those with short chain alkoxy groups. This makes any calculation
of binary phase diagrams involving these compounds very difficult,
as one never knows which phase has crystallized out in the mixture
or if in fact several have. In several cases the enthalpy changes
for the crystal transitions were not measured. This was due to the
instability of these phases, which would form and then immediately
revert to a more stable crystal. The formation of some of these
crystal phases was quite dependent on the thermal history of the
sample and often could be observed only by annealing or by

Figure 3

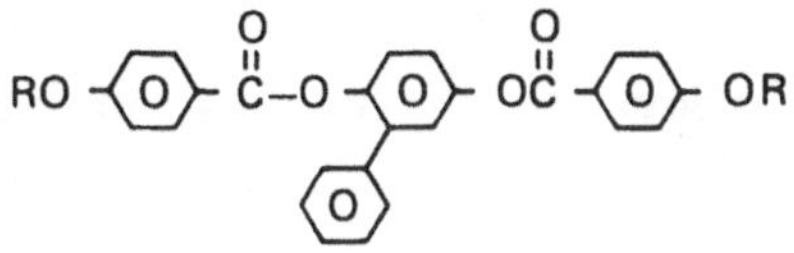

$$RO-⬡-\overset{\overset{O}{\|}}{C}-O-⬡-O\overset{\overset{O}{\|}}{C}-⬡-OR$$

(with a central benzene ring substituent)

R	Transition		Temp (°C)	ΔH (KCal/Mole)	ΔS (Cal/Mole/°K)
CH_3	C_3 ⟶	I	144	9.1	21.8
	C_1 ⟶	I	129	—	—
	C_2 ⟶	I	135	—	—
	N ⟶	I	(79)	0.27	0.77
	S ⟶	I	(41)	1.9	6.1
C_2H_5	C_1 ⟶	I	133	8.6	21.2
	C_2 ⟶	I	152	9.6	22.6
	N ⟶	I	(98)	0.43	1.2
C_3H_7	C_2 ⟶	I	117	8.2	21.0
	C_1 ⟶	I	135	9.9	24.3
	N ⟶	I	(66)	0.36	1.1
C_4H_9	C_2 ⟶	I	101	9.0	24.1
	C_1 ⟶	I	94	—	—
	N ⟶	I	(76)	0.45	1.3
C_5H_{11}	C_1 ⟶	C_2	100	—	—
	C_2 ⟶	N	90	—	—
	N ⟶	I	(70)	—	—
C_6H_{13}	C ⟶	N	65	5.1	15.1
	N ⟶	I	70	0.42	1.2
C_7H_{15}	C ⟶	I	67	9.4	27.6
	N ⟶	I	(56)	0.34	1.1
C_8H_{17}	C ⟶	I	80.7	14.2	40.1
	N ⟶	I	(77.1)	0.39	1.1

Figure 4

approaching the phase at a specific temperature rate. An interesting observation was made on the methoxy member of this series. On
super cooling the nematic phase, which was observed at 79°C, a
smectic phase appeared at 41°C. That it was smectic was confirmed
both by the texture observed under cross polarizers and also by
the large entropy change associated with the transition. We felt
that the appearance of a smectic in a compound in which the alkoxy

groups are as small as methoxy was unusual and we were interested
to see if the other members of the series, particularly those with
longer alkoxy chains, showed the same smectic phase. We were
surprised that even though in many cases the compounds could be
super cooled to 0°C before crystallization began, there were no
other smectic phases observed. This does not mean of course that
if it were possible to further super cool or if some binary mixtures
were prepared, virtual smectic transitions might not be observed
in the other members of the series. The compounds in the unsymmetrical series also showed only monotropic transitions. A
summary of these compounds and their transitions is given in Fig.
5. In these cases the propensity to crystallize was so great that
the enthalpy could not be measured by DSC and only the temperatures
could be recorded. The nematic structure was confirmed by observation of the texture using polarizing light microscopy. In Fig. 6
is shown a plot of the solid transition temperatures and the nematic
transition temperatures for the symmetric series, as a function of
the number of carbon atoms in the tail, it has the usual odd-even
effect with the even numbered plot forming a smooth curve and lying

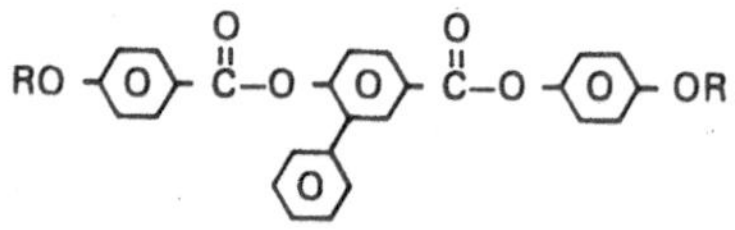

R	Transition		Temp (°C)
C_4H_9	C ⟶ I		108
	N ⟶ I		(87)
C_6H_{13}	C ⟶ I		80
	N ⟶ I		(70)
C_7H_{15}	C ⟶ I		72
	N ⟶ I		(63)
C_8H_{17}	C ⟶ I		69
	N ⟶ I		(64)

Figure 5

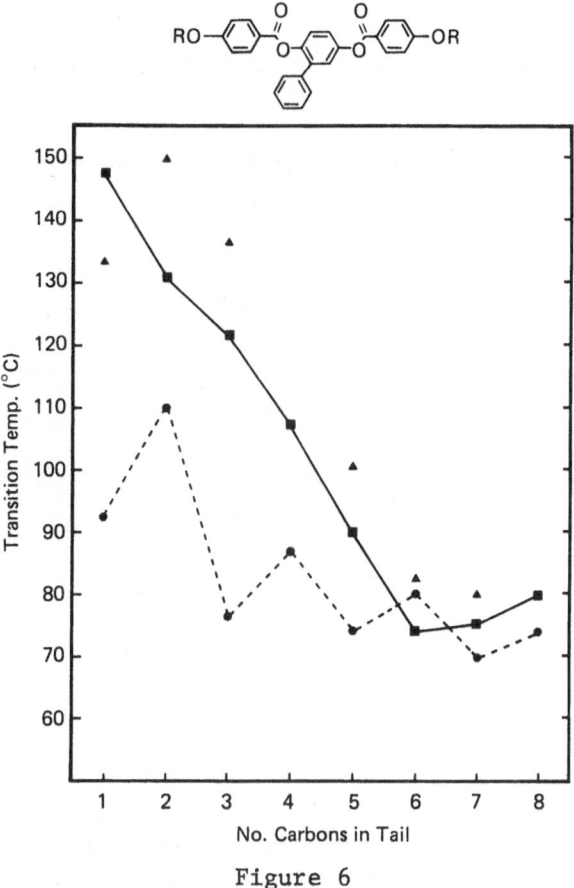

Figure 6

just above the curve for the odd numbered compounds. These curves
tend to merge as the tail length increases.

Figure 7 compares the transition temperatures of the symmetrically substituted dimethoxy compound with the transition
temperatures of the series studied by Dewar and Griffin in 1975[2]
in which the central ring was substituted with increasingly larger
groups. They noted the decrease in the nematic to isotropic
temperatures as well as an increase in the enthalpy and entropy
changes as the size of the substituent increased. Their conclusions

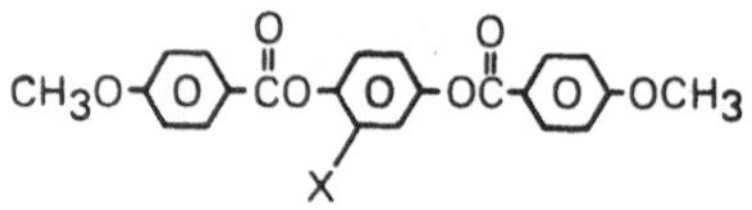

$$CH_3O-\bigcirc-\overset{\overset{O}{\parallel}}{C}O-\bigcirc-O\overset{\overset{O}{\parallel}}{C}-\bigcirc-OCH_3$$

X

X	$T_{N \to I}$	$\Delta H_{N \to I}$	$\Delta S_{N \to I}$
H*	301	0.408	0.711
CH₃*	252	0.529	1.01
Cl*	252	0.463	0.881
F*	278	0.404	0.732
Br*	241	0.487	0.947
I*	223	0.493	0.994
C₆H₅	79	0.27	0.77

*M. J. S. Dewar and A. C. Griffin, J. Am. Chem. Soc. 97 6662 (1975)

Figure 7

from this study were that the decrease in the nematic to isotropic temperature was not due to a decrease in the enthalpy change, as one would expect if the larger groups were keeping the molecules further apart in the nematic phase, but was in fact due to a large increase in ΔS as the size of the substituent grew, and that this increase in ΔS was due to an interlocking of the substituents in adjacent molecules, thus in effect increasing the order in the nematic phase. If this explanation is indeed true we would have expected to see a further increase in ΔS in the benzene substituted compound through overlapping of Pi bonds of the pendent benzene rings. Instead we see that the ΔS for this nematic to isotropic transition is only slightly larger than for the unsubstituted compound and that the value for ΔH is considerably smaller than

for this derivative, and that in this case the large decrease in
the transition temperature is in fact due to the large decrease in
ΔH. It would seem then that the large phenyl rings are not mutually
attractive substituents which hinder motion parallel to the long
axis of the molecules, but instead simply force the molecules
further apart than they would be in the unsubstituted case. This
decreases the value of ΔH and thus the transition temperature.

In other alkoxy substituted series[6] in which any significant
number of lateral substitutions have been prepared and studied,
only the values for the transition temperatures were recorded so
that one cannot make the same analysis as with the methoxy
compounds. The transition temperatures nevertheless do decrease
in the expected manner as the size of the substituent increases,
and for the benzene substituent the decrease is greater than for
the others studied.

Concluding, we note several interesting things about this
series of liquid crystals.
1.   The appearance of liquid crystallinity in compounds which are
so far from the classical rod like geometry.
2.   The multiplicity of crystal phases present in this series and
the appearance of the smectic phase in the methoxy derivative not
found in the higher homologes.
3.   The different mechanism which apparently is responsible for
the lowering of the nematic to isotropic temperature from that of
other lateral substituted compounds in this series.

REFERENCES

1.   M. J. S. Dewar and R. S. Goldberg, J. Org. Chem., 35, 2711
        (1970).

2.   M. J. S. Dewar and A. C. Griffin, J. Am. Chem. Soc., 97, 6662
        (1975).

3.   M. J. S. Dewar and R. M. Riddle, J. Am. Chem. Soc., $\underline{97}$, 6658
        (1975).

4.   J. C. Dubois and A. Beguin, Mol. Cryst. Liq. Cryst., $\underline{77}$, 193
        (1978).

5.   A. Dollman, A. R. Greggs and E. M. Barrall, <u>Analytical</u>
        <u>Calorimetry</u>, <u>Vol</u>. <u>4</u>, R. S. Porter and J. F. Johnson, Eds.,
        Plenum Press, New York, 1977, p. 4.

6.   H. Schubert and W. Weissflog, unpublished results.

THE SYNTHESIS AND CHARACTERIZATION OF SOME PERYLENE AND ANTHRA-
QUINOIC DICHROIC DYES FOR LIQUID CRYSTAL DISPLAY APPLICATIONS

T. M. Leslie, J. W. Goodby and R. W. Filas

Bell Laboratories
Murray Hill, New Jersey 07974

ABSTRACT

A series of dyes utilizing perylene as a fluorescent dichroic
chomophore have been synthesized and their order parameters measured.
For the N,N'-di-n-alkylperylene -3,4,9,10 - diimides, order parameters in the commercially available liquid crystal mixture E7,
ranging from 0.723 to 0.745 (corresponding to dichroic ratios of
8.7 to 13.8) were obtained. The dichroic ratio was obtained by measuring the polarized absorption spectra in both the parallel and
perpendicular directions to the nematic director and dividing the
parallel component by the perpendicular one ($A_{||}/A_{\perp}$). Further non-
fluorescent derivatives of perylene tetracarboxylic-3,4,9,10-
dianhydride have been synthesized and their properties characterized.
Alternative dyes employing the 2,6-diaminoanthraquinone chromophore
have been synthesized and their order parameters also measured.
Order parameters in E7 ranging from 0.66 to 0.69 (dichroic ratio
5.5 to 5.8) have been obtained for diazo and diazoxybenzene derivatives of this chromophore under the same conditions.

INTRODUCTION

The guest-host liquid crystal display concept has been known for a number of years.[1,2] However, until the discovery of the anthraquinone-based dichroic dyes,[3,4] this class of display was unsuitable for commercial display purposes. For example, azo dichroic dyes exhibit high order parameters, often ranging up to 0.8[2,5] (dichroic ratio 13) but because of photochemical instability they degrade rapidly.

Anthraquinone-based dichroic dyes offer high photostability, adequate solubility for display purposes, and are now commercially available. These dyes have reported order parameters of the order of 0.7 (dichroic ratio 8) in E7, however, higher values are desirable in order to improve the contrast of guest-host displays.

Four 2,6-disubstituted anthraquinone and several N,N' - disubstituted perylene diimides have been chosen in order to study the effect of the overall molecular shape, on the order parameter and solubility in a nematic host, E7.

RESULTS AND DISCUSSIONS

    1)   2,6-bis-diazo and 2,6-bis-diazoxyanthraquinones

The 4-N,N-dimethylaniline and 4-N,N-diethylaniline derivatives of 2,6-bis-diazoanthraquinone (I) and their oxidized counterparts (II) were synthesized.

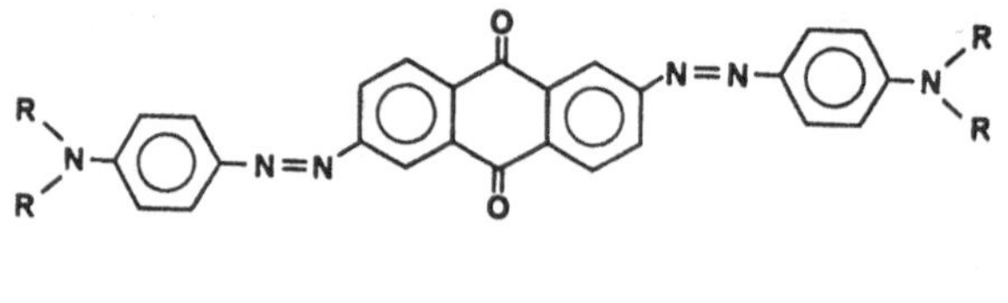

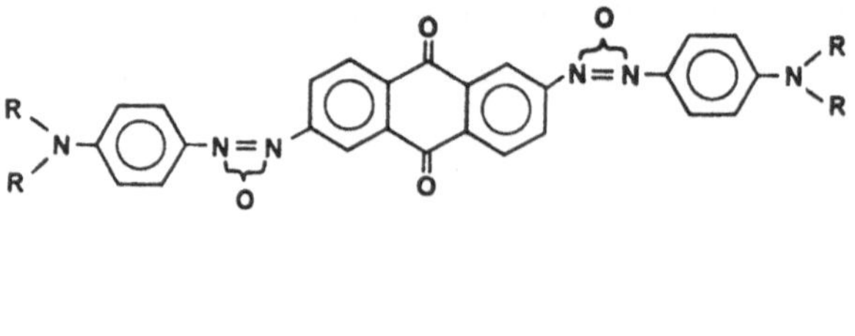

Both the azo (AAPNX2) and azoxy (AAOPNX2) compounds were found to have good solubilities and reasonable order parameters in E7.

On moving the substitution pattern from the 1,5 position[1,2] to the 2,6 position, the color of the dichroic dye changes from purple or blue to red without a reduction in order parameters as compared to the commercially available dyes.

These compounds have order parameters ranging from 0.657 to 0.688 (dichroic ratio 6.75 to 7.62) in E7 (shown in Table 1) which are comparable with some of the anthraquinone dyes with high order parameters which are now commercially available. It has been suggested that dyes substituted only in the β positions relative to the carbonyl function of the anthraquinone base "are not suited for display applications".[6] These dyes do, however, have order parameters in the range that are presently used in guest-host displays.

The azo derivates do suffer from a similar photochemical instability as previously mentioned for other azo dyes. The axoxy derivatives, however, are more soluble and offer better photochemical stabilities than the azo analogues.

TABLE 1

	DICHROIC RATIO*	ORDER PARAMETER*	λ_{max} ($\nu = 0$)
AAPNM2	6.73 ± 0.38	0.657 ± 0.014	519.2
AAPNE2	7.62 ± 0.14	0.688 ± 0.004	540.2
AAOPNM2	7.59 ± 0.10	0.687 ± 0.003	522.9
AAOPNE2	6.81 ± 0.10	0.659 ± 0.004	526.3

---

* Error limits were determiend by taking the maximum possible absorbance error, adding it to $A_{||}$ and substracting it from $A_{\perp}$. The dichroic ratio and order parameter were recalculated using $A_{||}$ + error and $A_{\perp}$ - error to determine the maximum possible error.

2) N,N'-dialkyl perylene-3,4,9,10-diimides

The perylene chromophore has been chosen because of its high photochemical stability and that it is possible to synthesize symmetric molecules in which the molecular long axis and the transition moment are almost coincident.

$$C_n H_{2n+1} - N \overset{\displaystyle}{} N - C_n H_{2n+1}$$

The effect of increasing the n-alkyl chain length on the order parameter (dichroic ratio) for the N,N' dialkyl perylene-3,4,9,10-diimide (PIN-n) series, is shown in Figure 1. The order parameter was found to increase rapidly on increasing the terminal alkyl chain length from n = 4 to 8. This is followed by a slight decrease in the observed order parameter for longer alkyl chain lengths of 10 and 18 as shown in Table 2. The solubility although poor ($\sim$ 0.005%) of the PIN-n series follows the same behavior found for the order parameters. The solubility increases rapidly by extending the alkyl chains from n = 4 to 8, then decreases for the higher n-alkyl homologues of 10 to 18 carbon atoms. At smaller values of n (n < 8) it

TABLE 2

	DICHROIC RATIO*	ORDER PARAMETER*	λ_{max} (ν = 0)
PIN-4	8.84 ± 0.22	0.723 ± 0.005	535.0
PIN-6	9.35 ± 0.28	0.736 ± 0.006	535.4
PIN-8	9.77 ± 0.24	0.745 ± 0.005	535.5
PIN-10	9.69 ± 0.28	0.734 ± 0.006	535.5
PIN-18	9.55 ± 0.62	0.740 ± 0.013	535.5

* See footnote, Table 1

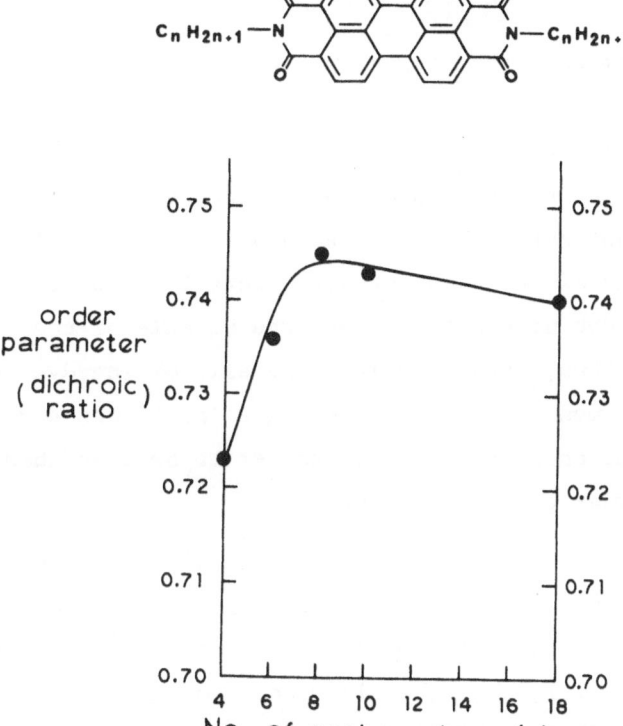

Fig. 1. Plot of the order parameter versus the number of carbon atoms in the n-alkyl chain for the N,N'-dialkyl
perylene-3,4,9,10-diimides.

is believed that the terminal alkyl chains are not totally fluid
and effectively extend the molecular long axis of the dye molecule.
Extending the terminal hydrocarbon chains to n = 18 has no benefi-
cial effect on either the solubility or the order parameter. For
the PIN-n series, as the n-alkyl chains become more fluid, there is
little distinction caused by the increase in chain length. At n = 8
both the solubility and the observed order parameter have been
maximized.

It is important to note that the observed order parameter
calculated using Equation (1)

$$S = \frac{D-1}{D+2} \qquad\qquad (1)$$

may differ from the true microscopic order parameter of the dye. This has been noted by Osman, Pietroneio, Scheffer, and Zeller[6] to be the case for certain anthraquinone dyes. This occurs when the angle between the transition moment of the dye and its molecular long axis is not zero. This angle is defined as β. The same may be true for the PIN-n series of dyes since the two terminal alkyl chains of the dye need not be along the C_2 axis of the perylene chromophore. Thus, the molecular long axis of the dye molecule and the transition moment may be at an angle (β) to one another. This would cause the observed order parameter to be less than the true microscopic order parameter of the dye.

For example, the true value of the order parameter for PIN-8 may be higher than 0.745. By multiplying the observed order parameter of PIN-8 by the factor $[1/2 \ (3 \cos^2 \beta-1)]^{-1}$ which for $\beta = 20^{\circ}$, corresponds to 1.21, the true microscopic order parameter for PIN-8 is found to be (1.21 x 0.745), i.e., 0.901. These dyes are fluorescent in phenylcyclohexane mixtures such as 1132 (Merck). Their fluorescence is quenched in biphenyls such as E7 (Merck) and phenyl-pyrimidines for example RO-TN-605 (Roche).

3) N,N'-bis-4n-alkyl and alkoxyphenyl perylene-3,4,9,10-dimides

Three dyes of the general structure shown below were synthesized in order to extend the rigid aromatic core structure.

The 4-n-butyl (PIP-4), 4-n-octyl (PIP-8), and 4-n-octlyloxy (PIP-O-8) dyes were found to be more soluble in E7 than the PIN-n dyes. Increasing the size of the aromatic core was found to have two effects on the perylene dyes. Firstly, the observed order para-

TABLE 3

	DICHROIC RATIO*	ORDER PARAMETER*	λ_{max} ($\nu = 0$)
PIN-8	9.77 ± 0.24	0.745 ± .005	535.5
PIP-4	11.9 ± 1.3	0.785 ± 0.018	537.0
PIP-8	10.3 ± 0.4	0.757 ± 0.007	537.0
PIP-0-8	14.0 ± 3.3	0.812 ± 0.032	537.0

---

* See footnote, Table 1

meters for all of the three dyes was found to be greater than those observed for the PIN-n series as shown in Table 3. The observed order parameter for PIP-8 is slightly higher than the order parameter observed for PIN-8. This indicates that for n = 8 in either of the systems, the transition moment of the dye may not lie along the molecular long axis to approximately the same extent. Secondly, the solubility of PIP-8 is greater than the solubility of the analogous PIN-8. This indicates that extending the aromatic core of the molecule coupled with the hydrogen bonding interacts between the imide ketone functions and the ortho-hydrogen atoms of the phenyl rings of the substituents in the PIP compounds play an important role in the solubility of the dye.

For the earlier homologues as the n-alkyl chain is shortened as in the case of PIP-4, the solubility of the dye decreases. This is similar to that which was found for the PIN-n series. However, for PIP-4 with the extended aromatic core, the observed order parameter increases to 0.785. Thus indicating the molecular long axis and the transition moment of the dye molecule appear to be becoming coincident. The strength of the hydrogen bonding interaction between the ortho-hydrogen atoms and the imide ketones can be increased by the substitution of an electron withdrawing group in the para position of the ring.

However, in PIP-O-8 where the alkoxy function donates electrons into
the aromatic core by mesomeric relay, the order parameter is found
to be significantly higher than the order parameter for PIN-8 and
PIP-8. Altering the interaction between the imide ketones and the
substituent on the imide nitrogen may, therefore, play an important
role in determining the order parameter for the perylene dye system.

The solubility of PIP-O-8 was found to be intermediate to PIN-8
and PIP-8. Thus the order of solubility is found to be PIN-8 < PIP-
O-8 < PIP-8. The oxygen atom in the para position has the advantage
of increasing the observed order parameter but the disadvantage of
reducing the solubility of the dye.

4)   4,4'-N,N dialkylaminophenyl perylene-3,4,9,10-diimides

Both the 4-N,N-dimethyl and the 4-N,N-diethyl derivatives of

this symmetric chromophore were synthesized. The use of the 4-N,N-
dialkyl aniline substitutent on the perylenediimide base provides
two important structural features. Firstly, it provides an electron
donating substituent in the para position similar to the para-alkoxy
derivate PIP-O-8. Secondly, it provides two alkyl substituent at
each end of the dye molecule. Thus, the molecular long axis and the
transition moment of the dye are effectively colinear especially for
the dimethyl derivative.   The observed order parameter (Table 4)
for the dimethyl derivative PIPNM2 is not unexpected for this compound with very short terminal alkyl chains. The observed order

TABLE 4

	DICHROIC RATIO*	ORDER PARAMETER*	λ_{max} $(\nu = 0)$
PIPNM2	14.4 ± 4.6	0.817 ± 0.071	535.5
PIPNE2	10.9 ± 1.2	0.767 ± 0.019	535.6

*See footnote, Table 1

parameter for the dialkyl derivative (PIPNE2) is lower but can be construed as the same within experimental error. By extending the four terminal alkyl chains by only one methylene group, the observed order parameter drops from 0.817 for the dimethyl derivative to 0.767 for the diethyl derivative. This result is unexpected and may simply be related to the relative solubilities of the dyes which reflects the error associated with the observed order parameters. It is very difficult to draw conclusions for this class of dye with only two examples of very short terminal alkyl chain length. Thus, higher homologues will have to be synthesized with n = 4,5 and 6 before detailed conclusions can be made.

CONCLUSIONS

It has been found that the highly symmetric perylene diimide chromophore can be used to produce dichroic dyes that exhibit high order parameters (dichroic ratios) of 0.8 and above for certain substitution patterns and are photochemically stable. Their solubilities preclude their immediate use in commercial displays. However, only a limited number of perylene derivatives have been investigated so far, and it is possible that alternatively substituted perylene dyes may prove useful in the future.

2,6 disubstituted anthraquinone dyes have lower observed order parameters (dichroic ratios) than the perylene dyes but are equivalent to some of the commerically available 1,5 disubstituted anthraquinoic dyes.

The stability of azoxy dichroic dyes has been found to be much improved over their azo analogues without an appreciable change in the order parameter (dichroic ratio).

EXPERIMENTAL

The dichroic ratio ($A_{\parallel}$ /$A_{\perp}$) of the dyes was measured in the nematic host E7 (Merck) at 20.0° using a Varian Cary 219 spectro-photometer equipped with prism polarizers and thermostatted sample holders. The sample temperature was measured ($\pm 0.1^{\circ}$) with a ther-mocouple embedded in the sample holder using a Fluke 2190A digital thermometer. Reusable sample cells were constructed from "Suprasil" quartz slides which had been treated for homogeneous orientation by thermal evaporation of silicon monoxide at an angle of 30° to the plane of the slide. The cells were glued together with an epoxy resin along two opposite edges so that they could be easily filled by capillary action or cleaned for reuse. Quartz fibers provided a spacing of 50 μm as determined by a Zeiss light section microscope. Dye solutions were filtered through a fine porosity sintered glass funnel prior to filling a cell. Once filled, a cell was placed in a Mettler FP52 hot-stage and observed microscopically between crossed polarizers. Heating samples to just below the nematic-isotropic transition facilitated the removal of all defects in the homogeneous texture. The baseline was recorded for each polarization (either parallel or perpendicular to the director) using cells filled only with E7, then the cell holder in the sample compartment was exchanged for an identical one containing the cell with the dye solution. The pair of cells used in the sample compartment were matched to ± 0.001 absorbance unit over the visible spectrum. Order parameters were calculated from the dichroic ratio, D, using the expression

$$S = \frac{D-1}{D+2}$$

which assumes that the transition moment is parallel to the dye's long axis. Obviously this assumption is not valid for molecules

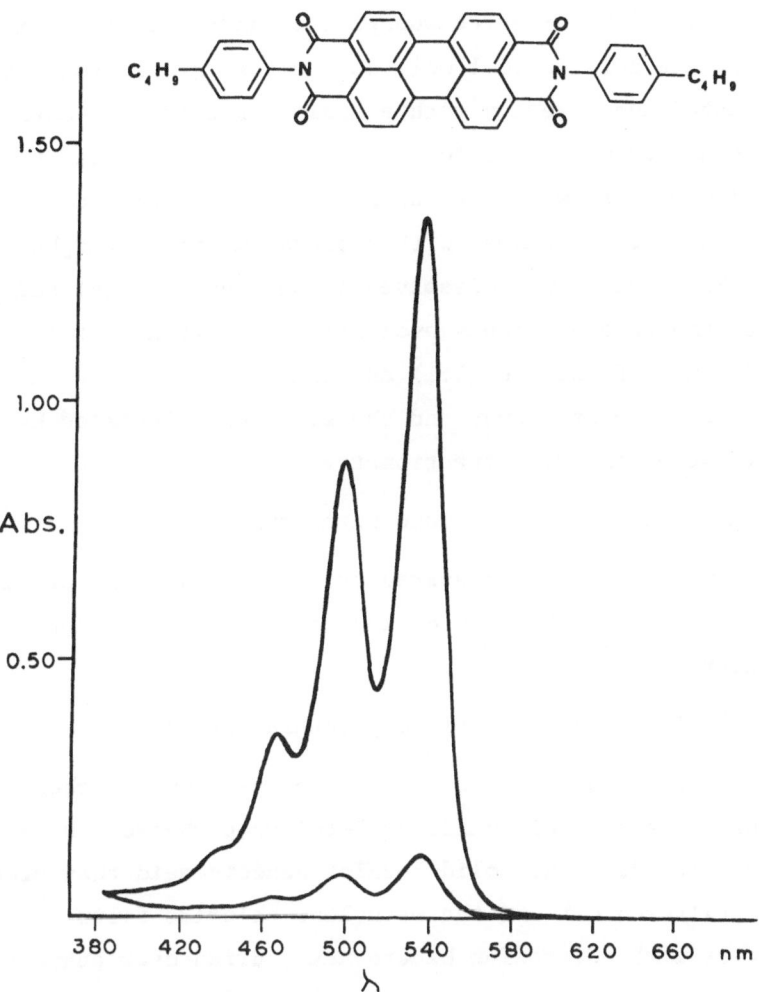

Fig. 2.   A typical set of spectra obtained for PIP-4 to be used in
          the calculation of the dichroic ratio and the observed
          order parameter.

with lower symmetry like AAOPNE2 and other known anthraquinones.[6]
A typical set of spectra used for the determination of the order
parameter are shown in Figure 2.

Preparation of N,N'-disubstituted-perylene-3,4,9,10-diimides

Perylene-3,4,9,10-tetracarboxylic dianhdydride (0.05 mol) and
the appropriate aniline (0.1 mol) were either heated together to
200° or heated under reflux with a minimal amount of N-methyl-2-
pyrrolidinone (20 ml) for 8 to 16 hours with continuous stirring.
The reaction mixture was taken up in ether (100 ml) and filtered.
The solid residue was washed with copious amounts of light-petrol
(35-65). The residue was dissolved in acetone (slurry) and purified by column chromatography over silica-gel (8 x 50 cm) using
acetone-dimethyl formamide (1:1) as eluant. The purity was checked
by thin-layer chromatography and the structure eluciated by infra-
red spectroscopy and mass spectrometry.

Preparation of 2.5 bis-diazo anthraquinones

These materials were prepared by standard methods of diazotiza-
tion of 2,5 diaminoanthraquinone and coupled with the appropriate
4-N,N-dialkylanilines.

Preparation of 2,5-bis-diazoxy anthraquinones

The diazo compounds were allowed to react undisturbed at room
temperature in a 2:1 mole ratio of 3-chloro-perbenzoic acid in
methylene chloride. The solid 3-chlorobenzoic acid that precipitated was filtered off and the methylene chloride filtrate was
washed twice with 10% sodium bicarbonate, dried over anhydrous
magnesium and evaporated to dryness under reduced pressure.

REFERENCES

1.   G. H. Heilmeier, L. A. Zanoni, Appl. Phys. Lett. 13, 91, 1968;
         G. H. Heilmeier, J. A. Castalleno, L. A. Zanoni, Mol.
         Cryst. Liq. Cryst. 8, 293, 1969.
2.   D. L. White, G. N. Taylor, J. Appl. Phys. 45, 4718, 1974.
3.   J. Constant, M. G. Pellantt, I. H. C. Roe, Seventh Inter-
         national Liquid Crystal Conference, Bordeaux, July 1978.

4.   M. G. Pellatt, I. H. C. Roe, Mol. Cryst. Liq. Cryst. <u>59</u>,
         299-316, 1980.

5.   R. J. Cox, Mol. Cryst. Liq. Cryst. <u>55</u>, 1979.

6.   M. A. Osman, L. Pietroneio, T. J. Scheffer, H. R. Zeller, J.
         Chem. Phys. <u>74</u>, 5377, 1981.

SOME NEW THERMOTROPIC DISCOGENS

Pierre Le Barny[*], Jean Billard[**] and
Jean-Claude Dubois[*]

* Laboratoire Central de Recherches, Thomson-CSF
Domaine de Corbeville, 91401 Orsay Cedex, France

** Laboratoire de Physique de La Matière Condensée
(Equipe Associée au C.N.R.S.), College de France
75231 Paris Cedex 05, France

ABSTRACT

New thermotropic discogens are synthesized and characterized by microcalorimetric measurements and microscopic examinations. The mesophases are identified by the miscibility method. The rufigallol hexa-n-nonanoate exhibits a stable D_c columnar discophase. Some chemical possibilities for new side chains are explored. The hexa-p-n-heptyloxycinnamoyloxytriphenylene exhibits a fluid discophase stable from 146.7 to 242°C. The discogenic potentialities of nine aromatic cores are evaluated, good candidates are found. Five binary phase diagrams with non ideal behavior are given. One enhanced intermediate columnar discophase with mosaic texture is observed. A binary mixture exhibiting a twisted fluid discophase with small pitch is proposed.

INTRODUCTION

One of the problems to advance the knowledges about the discotic states of the matter is to elaborate more diversified compounds[1] exhibiting different stable discophases. In this way this paper reports a new compound exhibiting a stable columnar D_c phase. To offer a larger choice of chemical possibilities we have tested new side chains and some aromatic central cores.

With p-n-alkyloxycinnamoyloxy side groups can be obtained a triphenylene derivative with the widest fluid discotic phase temperature range until now. Some aromatic cores with good discogenic potentialities are studied to stimulate new synthesis.

Many of the phase diagrams established for these studies exhibit non ideal behavior. In one case an intermediate column discophase is enchanced. A binary mixture exhibiting a twisted fluid discophase with small pitch is described.

STABLE DISCOPHASE OF TYPE C

One compound exhibiting a stable D_B and a monotropic D_C discophases (the rufigallol hexa-n-octanoate) have been previously described.[2] The structures of these two columnar mesophases have been determined by X-ray diffraction.[3] To obtain a derivative exhibiting a stable D_C phase, four rufigallol hexa-n-alkanoates are synthesized following the same way[2].

The transition temperatures and their molar enthalpy changes are determined with microcalorimetric measurements (Du Pont Instruments, 990 Thermal-analyser) and microscopic examinations (Leitz, Panphot with Mettler FP 52 heating stage). The identifications of the mesophases are deduced from the total miscibilities. The binary isobaric phase diagrams are built from microscopic observations of contact preparations.[4]

TABLE 1

Discogenic rufigallol hexa-n-alkanoates

In this table and in the following the temperatures (°C) and molar enthalpy changes (underlined in kcal/mole) correspond to the transitions between the phases indicated by crosses. Data between parenthesis correspond to monotropic or virtual transitions.

$$R = C_mH_{2m+1}-\underset{O}{\overset{O}{\|}}{C}-O-$$

m	K_2	K_1	D_C	D_B	L
6		x	98	x 127	x
			5.87	3.86	
7		x	107.5	x 127.5	x [2]
			11.3	3.3	
			x $\left(\begin{array}{c}98.8\\ \sim0\end{array}\right)$ x		
8	x 61	x 80	x 106	x 123	x
		8.0	0.02	3.0	
9		x	106	x 127.8	x
			13.3	3.0	

The textures of all the observed discophases are similar to the ones of the rufigallol hexa-n-octanoate.[2] The rufigallol heptanoate and octanoate are totally miscible in the D_B state (Fig. 1). From the Fig. 2 are deduced the existences of thermodynamically stable D_B and D_C phases for the rufigallol nonanoate. The last compound having the lowest melting point in this series is the first example for discogenic material exhibiting a stable D_C phase. With the rufigallol nonanoate for reference compound, the D_B phase of the rufigallol decanoate is identified.

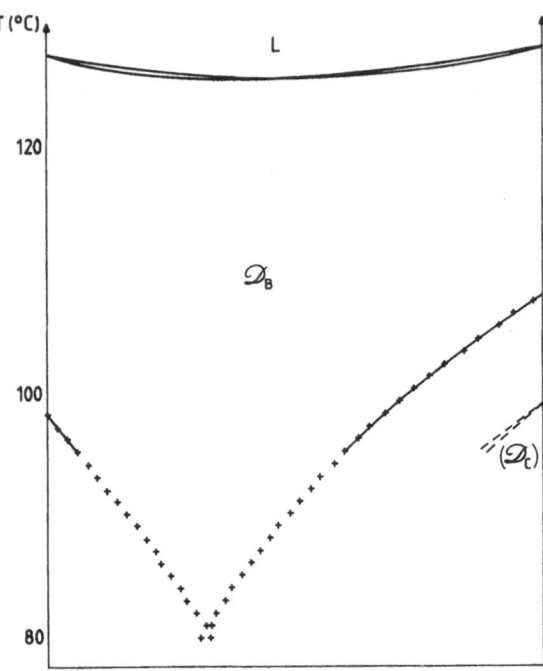

Fig. 1.   Phase diagram for the mixtures of rufigallol hexa-n-octan-
           oate (on the right) and rufigallol hexa-n-heptanoate (on
           the left).   In this figure and in the following the
           dashed lines correspond to the observed monotropic transi-
           tion.   Crosses are calculated values from the Le Chatelier
           Schröder relations[25,26]

NEW SIDE CHAINS

Until now fluid discophase (D_F) with nematic-like order have
been obtained with n alkanoates side chains for the truxene derivatives[5] and with p-n-alkyl or p-n-alkyloxybenzoyloxy side chains in
the triphenylene series.[6]

Two tentatives to modify slightly these last side chains are
not successful: 2, 3, 6, 7, 10, 11, (4' nonyl transcyclohexyl)-
carbonyl oxy triphenylene (melting at 105°C; 13.2 kcal/mole) and
2,3,6,7,10,11 hexa(4'n octyloxy tetra 2',3',5',6' fluoro) benzoyloxy triphenylene (melting 112°C; over 2.1 kcal/mole) exhibit only
columnar discophases.[7]

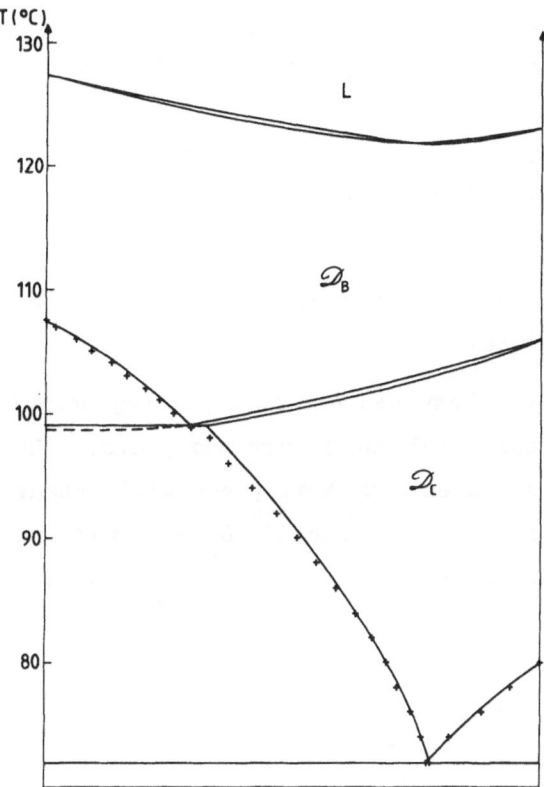

Fig. 2.   Phase diagram for the mixtures of rufigallol hexa–n–
            nonanoate (on the right) and rufigallol hexa–n–octanoate
            (on the left).

Because the sterical encumbrance the synthesis of hexaphenol
or rufigallol benzoates is difficult. Are elaborated benzoates of
phloroglucinol but using a 3–5 alkyloxy disubstituted benzoic acid
to obtain six side chains. The phloroglucinol (tri(3–5–dioctyloxy)
benzoate is prepared from 3–5 dihydroxybenzoic acid:

$$HO \text{—} \langle O \rangle \text{—} COOH + CH_2N_2 \longrightarrow HO \text{—} \langle O \rangle \text{—} COOCH_3 \qquad [8]$$

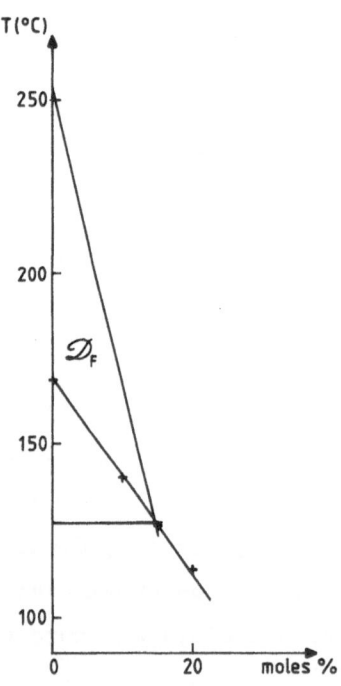

Triester is synthesized by reacting the acid chloride in pyridine at room temperature.

The 3-5-dioctyloxy benzoic acid phloroglucinol triester melts at 32°C (16,4 kcal/mole) and is not mesogenic. But the phase diagram for their mixtures with hexa p-n-heptyloxybenzoyloxy triphenylene.[10] (Fig. 3) gives a virtual[11] D_F - liquid transition at very low temperature. A similar result is obtained for a columnar -

Fig. 3.   The interesting part of the phase diagram for the mixtures of phloroglucinol tri (3-5-dioctyloxy) benzoate (on the right) and 2,3,6,7,10,11 hexa-4'-n-heptyloxy benzoyloxy triphenylene (on the left).

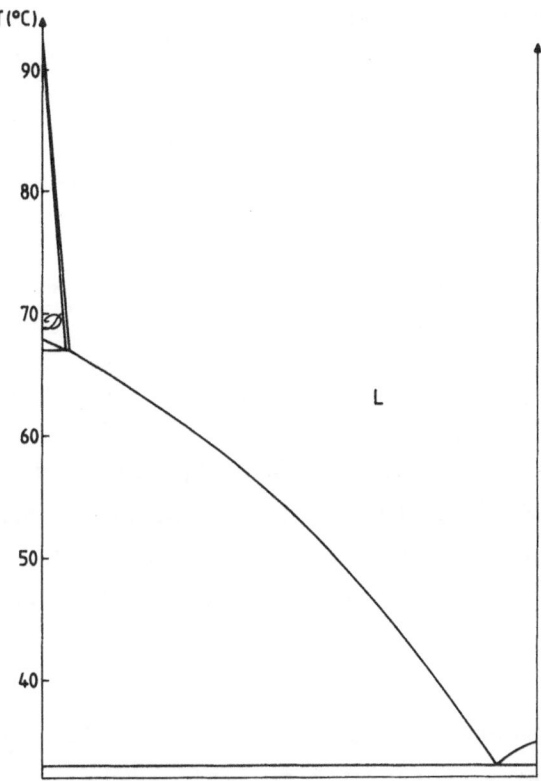

Fig. 4.   Phase diagram for the mixtures of phloroglucinol tri
           (3-5-dioctyloxy) benzoate (on the right) with 2,3,6,7,
           10,11 hexa-n-heptyloxytriphenylene (on the left).

liquid virtual transitions's temperature from the study of their
mixtures with the hexa n-heptyloxytriphenylene[12]: Fig. 4.

    Another way to solve the sterical encumbrance problem is to
use p-substituted cinnamoyloxy side chains. This conjugated radical
stabilize the cholesteric mesophase.[13] The comparisons between the
known p-substituted cinnamic acids[14] give the chain length the most
favourable to the molecular parallelism: pentyloxy or longer p-
alkyloxy substituent. For the 2,3,6,7,10,11 hexa alkyl and alkyloxy benzoyloxytriphenylene the larger D_F ranges are obtained[10] for
pentyloxy to heptyloxy derivatives. To verify these forecasts

4-n-pentyloxy trans cinnamic acid esters are synthesized.
The trans p-alkyloxy cinnamic acids are prepared from the trans
hydroxy cinnamic acid:

$$HO-\langle O \rangle-CH=CH-C\underset{OH}{\overset{O}{\backslash}} + CH_2N_2 \longrightarrow HO-\langle O \rangle-CH=CH-COOCH_3 \quad [8]$$

$$HO-\langle O \rangle-CH=CH-COOCH_3 \xrightarrow[\text{2) RBr}]{\text{1) } C_2H_5ONa/C_2H_5OH} R-O-\langle O \rangle-CH=CH-COOCH_3$$
$$[9]$$

$$R-O-\langle O \rangle-CH=CH-COOCH_3 \xrightarrow[\text{2) HCl}]{\text{1) } KOH/C_2H_5OH} R-O-\langle O \rangle-CH=CH-COOH$$

Hexa esters are synthesized by reacting the acid chlorides in
pyridine at room temperature. The 2,3,6,7,10,11-hexa 4'-n-hepty-
loxy trans cinnamoyloxy triphenylene melts at 146.5°C (3.8 kcal/
mole) and exhibits a fluid discophase with schlieren texture to
~ 242°C. The clearing transition is first order but with a small
enthalpy change. A monotropic crystalline form (melting point:
130°C) is also observed. A good homeotropic alignment is obtained
over glass surfaces coated with mellitic acid.[7] The discophase is
identified D_F from the phase diagram of their mixtures with the
hexa p-n-nonyloxybenzoyloxytriphenylene[6,7] (totally miscible in the
D_F state with the heptyloxy homolog): Fig. 5. The two triphenylene derivatives of the Fig. 5 have, practically, the same size,
this one with the new side chains have a more larger D_F range
beginning at lower temperature. This derivative have, until now,
the widest D_F temperature range.

    The hexahydroxybenzene is obtained by reduction of the tetra-
hydroxyquinone.[15] Their 4-n-pentyloxy trans cinnamic acid hexa
ester can be easily obtained and melts at 148°C (14.8 kcal/mole)
and is not mesogenic. The phase diagram of their mixtures with the
preceding compound (Fig. 6) gives the virtual D_F - liquid transition temperature: 122°C, not far from the melting point.

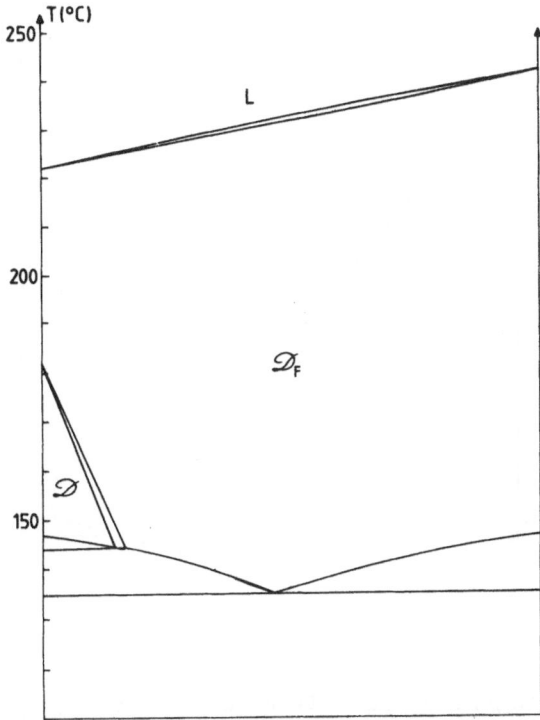

Fig. 5.   Phase diagram for the mixtures of 2,3,6,7,10,11 hexa-4'-
           n-heptyloxy trans cinnamoyloxytriphenylene (on the right)
           and 2,3,6,7,10,11-hexa 4'-n-nonyloxybenzoyloxytriphenylene
           (on the left).

NEW CENTRAL CORES

The number of central cores used for discogenic molecules is
small to day.[16]   To enlarge the choice is used a previously reported method[17] to evaluate the aptitude of discoid cores to give
discogenic derivatives.   The efficiency of this method has been
established with the rufigallol alkanoates series.[2]   The phase
diagrams (Figs. 7,8, and 9) for the mixtures of some cores with
ternary, binary or lower symmetry (Table II) with the 2,3,6,7,10,11
hexa-4'-n-nonyloxybenzoyloxy triphenylene give, by extrapolation,
the virtual D_F - liquid, columnar discophase - liquid or columnar -
fluid discophases transition's temperatures.   These values are

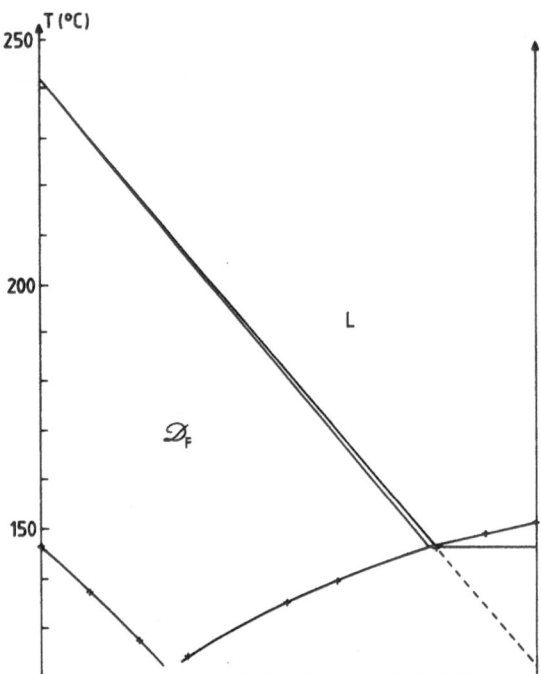

Fig. 6. Phase diagram of the mixtures of hexa 4-n-pentyloxy trans
         cinnamoyloxy benzene (on the right) and 2,3,6,7,10,11-hexa-4'
         -n-heptyloxy trans cinnamoyloxy triphenylene (on the left).

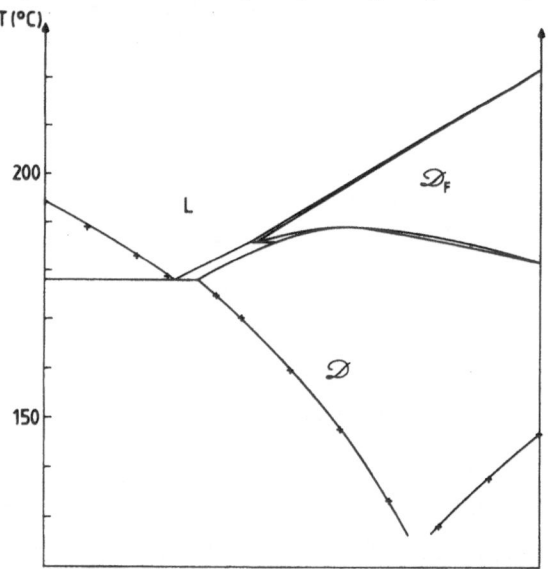

Fig. 7. Phase diagram for the mixtures of tetraphenylphenylene
         diamine (on the left) with 2,3,6,7,10,11-hexa-4'-n-nonyloxy-
         benzoyloxytriphenylene (on the right).

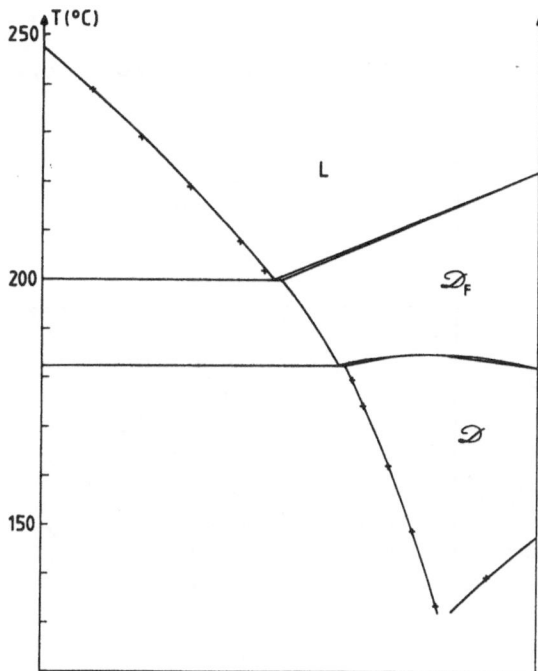

Fig. 8.   Phase diagram for the mixtures of 2,4,6-tri (2-pyridyl)-
s-triazine (on the left) and the same reference compound
used for fig. 7.

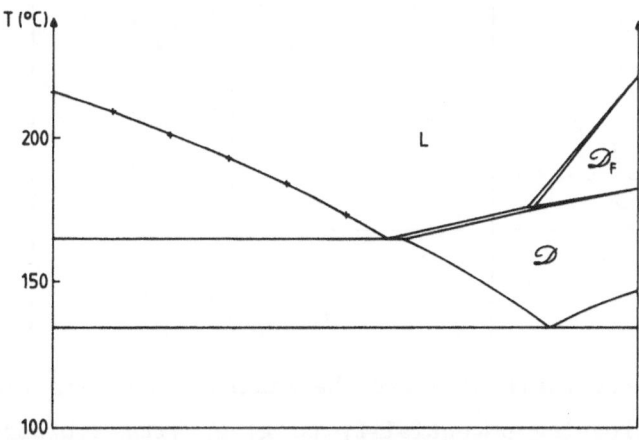

Fig. 9.   Phase diagram for the mixtures of anthracene (on the
left) and the same reference compound used for fig. 7.

TABLE II

Discogenic potentialities of some cores

Studied compound	K	$\mathcal{D}$	$\mathcal{D}_F$		L	Source
benzene core with two N substituents (φ groups)	x		194 ; <u>8.15</u>		x	[18]
			x	(160)	x	
		x	(160)		x	
1,3,5-trisubstituted benzene (φ groups)	x		172 ; <u>6.5</u>		x	[19]
			x	(110)	x	
		x	(110)		x	
triazine, R = pyridyl	x		247.5 ; <u>6.4</u>		x	[19]
		x (160)	x	(180)	x	
cyclopentadienone core (φ groups)	x		216.5 ; <u>7.5</u>		x	[19]
			x	(140)	x	
		x	(150)		x	
cyclopentadiene core (φ groups)	x		178 ; <u>8.4</u>		x	[19]
			x	(95)	x	
		x	(140)		x	
triazine, R = phenyl-O-	x		230 ; <u>10.6</u>		x	[19]
			x	(140)	x	
		x	(150)		x	
perylene core	x		275 ; <u>5.5</u>		x	[19]
		x (120)	x	(160)	x	
fluoranthene/acenaphthylene core	x		105 ; <u>4.2</u>		x	[19]
			x	(very low)	x	
		x	(60)		x	
anthracene core	x		216		x	[19]
			x (very low)		x	
		x	(140)		x	

reported in the Table II where the central cores are ranged in the order of decreasing potentiality to give, with appropriate side chains, fluid discophases. No relationship between the core's symmetry or size appears. From these datas (excepted the perylene) these cores are most favourable to give columnar discophases as the triphenylene (T_F = 199°C and virtual columnar discotic - liquid temperature 55°C.[17]) But are studied here discogenic potentiality for only one columnar mesophase.

BINARY MIXTURES

Some previously reported binary phase diagrams exhibit co-existing curves between non solid phases with minimum (Figs. 1 and 2) or maximum (Figs. 7 and 8) establishing a non ideal behavior for at the last one of the phases in equilibrium, frequent for the dis-cophases.[1,20] The non ideal behavior enhances an intermediate col-umnar discophase appeared in the mixtures of 1,3,5 triphenyl benzene (Table II) and hexa heptyloxybenzoyloxytriphenylene: Fig. 10. This mesophase has a mosaic texture, after pressing over the cover slip, smaller areas with slightly non uniform extinctions are observed. The extrapolation to the pure discogen of the columnar - fluid discophases equilibrium curves conducts to a virtual transition at 151°C. This conclusion is confirmed by the phase diagram for the mixtures with the hexa nonyloxybenzoyloxytriphenylene and by micro-scopic examination of the pure supercooled material.

The twisted fluid discophase[21] $(D_F*)^7$ can be obtained by mixing a chiral compound and a non chiral discogen exhibiting D_F phase[21]. To have a sufficiently elevated D_F - liquid virtual transition temperature we have used a chiral compound of a dis-cogenic series[22] the hexa-(S-4-methyl hexanoyloxy) benzene. This hexaester is prepared by refluxing in a mixture of toluene and pyridine S-4-methyl hexanoyl chloride with hexaphenol. The S-4-methyl hexanoic acid is synthesized from S (-) methyl 2 butanol.

[23] CH_3—⟨O⟩—SO_2Cl + $HO-CH_2-\overset{::}{C}H-C_2H_5$ → CH_3—⟨O⟩—$SO_2-O-CH_2-\overset{::}{C}H-C_2H_5$
$\qquad\qquad\qquad\qquad\quad \underset{CH_3}{|}\qquad\qquad\qquad\qquad\qquad\qquad\qquad\qquad \underset{CH_3}{|}$

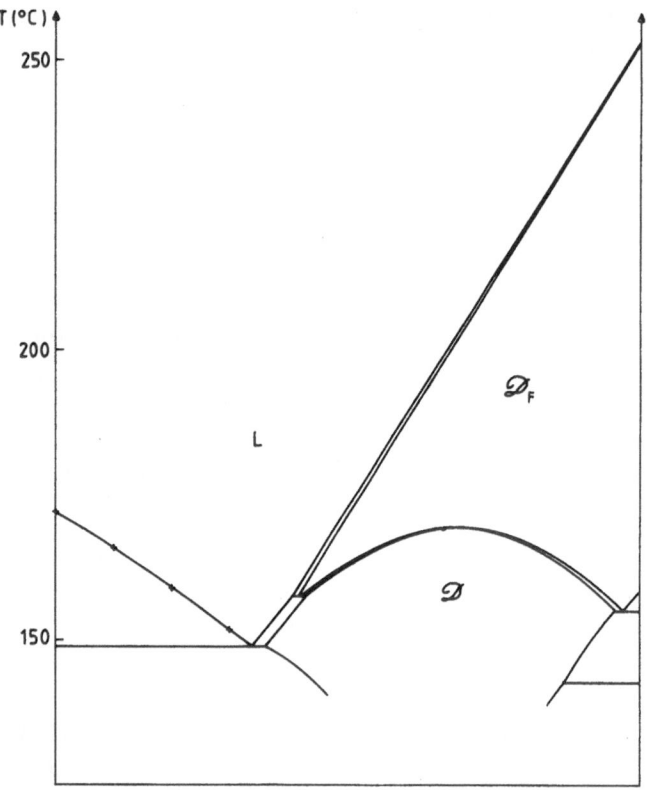

Fig. 10. Phase diagram for the mixtures of 1,3,5-triphenylbenzene (on the left) and 2,3,6,7,10,11-hexa-4'-n-heptyloxybenz-oyloxytriphenylene (on the right).

This non discogenic compound melts at 125.5°C (<u>7.8</u> kcal/mole). The stable D_F^* mixture containing the biggest proportion of this chiral compound we have observed is obtained with the hexa 4'-n heptyloxy

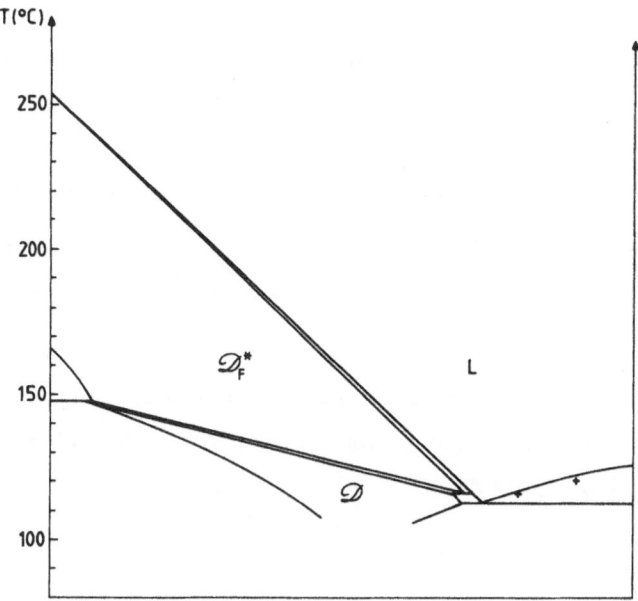

Fig. 11. Phase diagram for the mixtures of hexa-(S-4-methylhexanoy-
          loxy) benzene (on the right) and 2,3,6,7,10,11 hexa-4'-n-
          heptyloxybenzoyloxy triphenylene (on the left).

benzoyloxytriphenylene: Fig. 11. In the mixture containing approximately 70 mole % of the chiral component at 118°C the colours of
rotatory power dispersion can be observed in a contact preparation.
Consequently the rotatory power is big, indicating a small pitch.

CONCLUSIONS

    To increase the diversity of thermotropic discogenic materials
some directions are explored:

    (I) A compond exhibiting a stable D_C discophase is obtained:
the rufigallol hexa-n-nonanoate.

    (II) The possibility to obtain D_F phase over a wide tempera-
ture range (146.7 to 242°C) with p-n-alkyloxy trans cinnamoyloxy
side chains is established.

(III) The discogenic potentialities of some aromatic cores (Table II) are studied. We hope these datas will stimulate new synthesis of discogenic materials to discover new mesophases and increase the knowledges about the relations between the molecular structure and the thermodynamical stability of the discophases.

(IV) Numerous binary isobaric phase diagrams establish non ideal behavior. In one case (Fig. 10) an intermediate columnar discophase is observed.

(V) A D_F binary mixture (Fig. 11) containing sufficient proportion of chiral compound has a small pitch and exhibits an important rotatory power.

ACKNOWLEDGEMENTS

We thank Madame A. Beguin and Dr. P. Keller and C. Vauchier for the synthesis of used compounds and calorimetric measurements.

We thank the French D.G.R.S.T. who financially supported this work.

REFERENCES

1. J. Billard, Liquid Crystals of One- and -Two Dimensional Orders (edited by W. Helfrich and G. Heppke) Springer, Berlin, 383 (1980).

2. A. Queguiner, A. Zann and J. C. Dubois, Liquid Crystals (edited by S. Chandrasekhar) Heyden, London 35 (1980).

3. J. Billard, J. C. Dubois, C. Vauchier and A. M. Levelut, Mol. Cryst. Liq. Cryst., 66, 115 (1981).

4. L. Kofler and A. Kofler, Thermomikromethoden, Verlag Chemie, Weinheim (1956).

5. C. Destrade, J. Malthete, Nguyen Huu Tinh and H. Gasparoux, Phys. Lett., 78A, 82 (1980).

6. Nguyen Huu Tinh, C. Destrade and H. Gasparoux, Phys. Lett.
        72A, 251 (1979).

7. C. Vauchier, A. Zann, P. Le Barny, J. C. Dubois and J.
        Billard, Mol. Cryst. Liq. Cryst., 66, 103 (1981).

8. A. I. Vogel's, Textbook of Practical Organic Chemistry
        Including Qualitative Organic Analysis, Longman,
        London, 291 (1978).

9. W. J. Hickinbottom, Reaction of Organic Compounds, Longman,
        London 110 (1962).

10. Nguyen Huu Tinh, H. Gasparoux and C. Destrade, Mol. Cryst.
        Liq. Cryst. 68, 101 (1981).

11. M. Domon and J. Billard, Pramana, suppl.n° 1, 131 (1975).

12. J. Billard, J. C. Dubois, Nguyen Huu Tinh and A. Zann, Nouv.
        J. de Chi. 2, 535 (1978).

13. R. M. Cherkashina, A. V. Tolmachev and V. G. Tischenko,
        Zhurnal Fizicheskoi Khimii, 54, 2381 (1980); Russ.
        J. Phys. Chem., 54, 1359 (1980).

14. D. Demus, H. Demus and H. Zaschke, Flüssige Kristalle in
        Tabellen, Verlag f. Grundstoffindustrie, Leipzig,
        57 (1974).

15. A. J. Fitiadi and W. F. Sager, Organic Syntheses, Collective
        volume 5, John Wiley, NY, 595 (1973).

16. J. C. Dubois and J. Billard, these Proceedings.

17. A. Beguin, J. Billard, J. C. Dubois, Nguyen Huu Tinh and
        A. Zann, J. de Phys., 40C3, 15 (1979).

18. Eastman Kodak Co., Rochester, N.Y.

19. Aldrich Chemical Co., Milwaukee, WI

20. M. Dvolaitzky and J. Billard, Mol. Cryst. Liq. Cryst. Let.,
        64, 247 (1981).

21. C. Destrade, Nguyen Huu Tinh, J. Malthete and J. Jacques,
        Phys. Let., 79A, 189 (1980).

22. S. Chandrasekhar, B. K. Sadashiva and K. A. Suresh, Pramana,
        9, 471 (1977).

23. D. Dolphin, Z. Muljiani, J. Cheng and R. B. Meyer, J. Chem. Phys. 58, 413 (1973).

24. H. W. Gibson, Mol. Cryst. Liq. Cryst. 27, 43 (1974).

25. J. Mathete, M. Leclercq, J. Gabard, J. Billard and J. Jacques, C.R. Acad. Sci. Paris, 273C, 265 (1971).

26. J. Malthete, M. Leclercq, M. Dvolaitzky, J. Gabard, J. Billard, V. Pontikis and J. Jacques, Mol. Cryst. Liq. Cryst., 23, 233 (1973).

SYNTHESES AND PROPERTIES OF LIQUID CRYSTALLINE HETEROCYCLOALKANES

H. Zaschke, A. Isenberg and H.-M. Vorbrodt

Martin-Luther-Universität Halle-Wittenberg
Sektion Chemie
DDR-4020 Halle/Saale, Weinbergweg 16

Over the past few years many new substances having liquid crystalline properties were synthesized. The presence of chemical compounds having long extended rod-like molecular shape is still a fundamental premise and guarantee for the occurrence of thermotropic liquid crystalline phases.

Most of the synthesized liquid crystals contain aromatic structures the 1,4-disubstituted benzene system representing a perfect rigid group.[1,2]

Over the past few years our research team systematically replaced the benzene ring by heteroaromatic or alicyclic systems.[3,4] The insertion of heteroatoms strongly influences the formation of mesomorphic range. Melting and clearing temperatures, polymorphism of liquid crystalline phases and especially physical properties considerably change. The effect of heteroatoms, varying polarity, polarizability and transmission of effects of substituents over the molecule, can be demonstrated clearly.

Intensive studies of the effect of the cyclohexane ring system on the formation of mesophases were carried out. In typical liquid-crystalline compounds having aromatic structure benzene rings were

substituted by cyclohexane systems systematically. The higher
clearing temperatures of the perhydrated compounds related to
those of the analogous aromatic compounds were surprising.

Following results were received during our studies of mesomorphic behaviour of alicyclic compounds:

1.   Aromatic state is not needed for the formation of mesophase.

2.   The trans-configurated cyclohexane-system is sufficiently
      rigid and able to replace the aromatic 6-membered ring-system.

3.   Compared with aromatic analogous, the trans-4-alkyl-cyclo-
      hexanecarbonyloxy-group increases the clearing temperatures,
      that means, it favours the formation of mesophase.

4.   The thermal stability of liquid crystalline phases of the per-
      hydrated compounds is higher than that of the corresponding
      aromatic substances.

5.   Alicyclic compounds show useful physical properties with respect
      to a technical application, such as low values of viscosity
      and of elastic constants.

These conclusions on the influence of heteroaromatic and alicyclic rings on the formation of mesophase caused us to combine both
groups within one molecule. These structure should be a cyclohexane
system, which carbon-atoms are replaced by heteroatoms in (Fig. 1).
Developing this synthese conception we supposed the effect of cyclo-
hexanering on the melting behaviour and viscosity and the influence
of the heteroatoms on the dipole moment and dielectric properties
being kept for heteroalicyclic structure.

This manuscript shall give a report on some results of our
investigations on liquid crystalline 1,3-dioxanes, 1.3-dithianes
and piperazines.

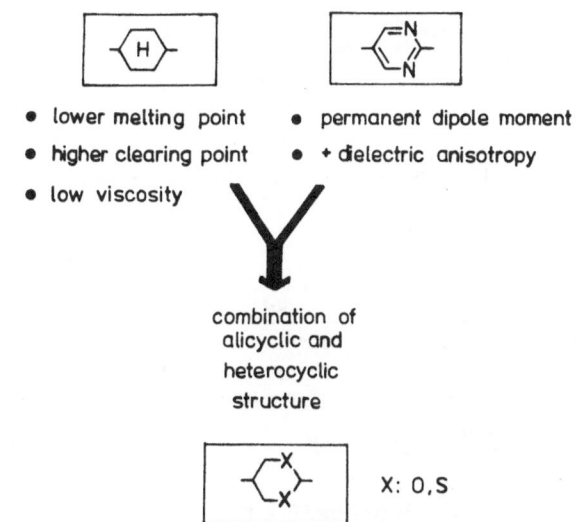

- lower melting point   • permanent dipole moment
- higher clearing point   • + dielectric anisotropy
- low viscosity

combination of
alicyclic and
heterocyclic
structure

X: O,S

Fig. 1 Influence of structure elements on the liquid crystalline properties.

1.   Liquid-crystalline 2,5-disubstituted 1,3-dioxanes

Well known 1,3-dioxanes like cyclic acetales can be obtained
by acid catalyzed condensation of substituted propan-1,3-diols
with aldehydes in good yields. By this reaction we obtained a
mixture consisting of the cis- and the trans-isomeric compounds.

The isomers can be separated by recrystallization of the raw
material from methanol or n-hexane easily. The isomers are classi-
fiable exactly by ^{1}H-NMR spectroscopy.

Figure 2 shows examples of variation of substituents of bi-
phenyl- and terphenyl-analogous 1,3-dioxanes.

Hitherto investigations carried out show the trans-1,3-
dioxanes being able to form mesophases only.[5-9] The liquid
crystalline melting properties of homologous series of the alkyl-
and alkoxysubstituted 1,3-dioxanes do not exhibit any deviation
from the typical behaviour of homologous series. They tend to form

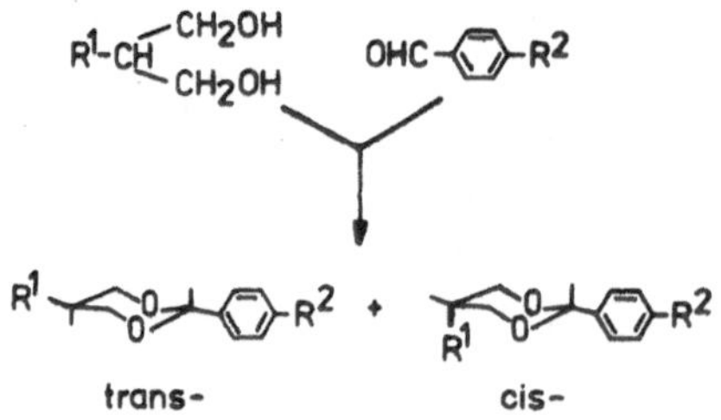

R¹	R²
alkyl	alkyl
alkyl	alkoxy
alkyl	acyloxy, halogen, CN
alkyl / alkoxy } $-C_6H_4-$	CN

Fig. 2 Synthesis of 1,3-dioxanes

thermic stable nematic and smectic phases. Bisalkylsubstituted 1,3-dioxanes melt at low temperatures. They have a distinctly marked tendency to form smectic phases. In this series smectic B is dominating. 5-alkyl-2-[4-alkoxyphenyl]-1,3-dioxanes show nematic and smectic A phases.

Considering the homologous series of the 5-n-hexyl-2-[4-alkoxyphenyl]-1,3-dioxanes an irregularity (anomaly) in the course of phase transition temperatures is remarkable. Clearing temperatures characteristically alternate, but S_A-N-phase-transition shows inversed alternation related to clearing temperature. Compounds bearing an even number of atoms in the side-chain have higher smectic A- to nematic transition temperatures than substances with odd number of atoms. We already established this phenomenon considering the S_A-N phase transition of 5-n-heptyl-2-[4-alkoxyphenyl]-pyrimidines.[10]

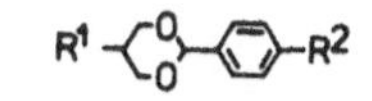

$$R^1 \text{—} \langle {}^O_O \rangle \text{—} \langle \rangle \text{—} R^2$$

R¹	R²	K	S_B	S_A	N	I
C_6H_{13}	C_3H_7	• 35	• 44	—	—	•
C_6H_{13}	C_6H_{13}	• 36,5	• 38,5	—	—	•
C_6H_{13}	Br	• 56	—	(• 39)	—	•
C_6H_{13}	CN	• 48	—	—	(• 42)	•
C_6H_{13}	OC_6H_{13}	• 34	—	• 45	• 53	•
$C_{16}H_{33}O$	OC_7H_{15}	• 51	• 68,5	—	—	•

Fig. 3 Mesomorphic ranges of some biphenyl analogous 1,3-dioxanes

Like in almost all cases of cyano-substituted compounds the nematic mesophase is also dominating in 5-alkyl-2[4-cyanophenyl]-1,3-dioxanes, occurring as metastable phase.

Contrary to those of 4-alkyl-4'-cyanobiphenyles the clearing temperatures are increased, compared to trans-4-alkyl-1[4-cyanophenyl]-cyclohexanes clearing points are decreased.[11,12]

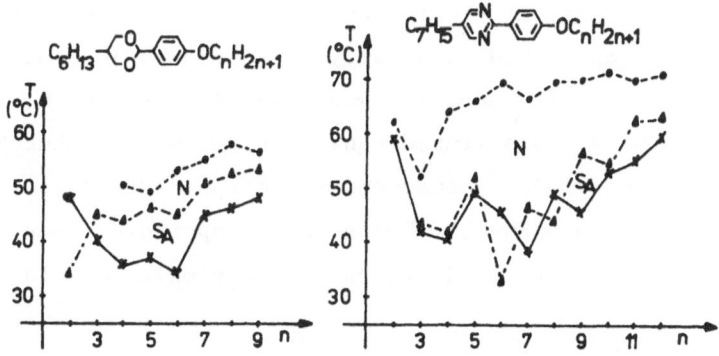

Fig. 4 Anomaly in S_A-N-phase transition

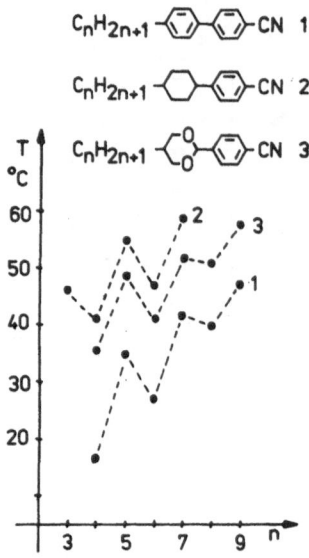

Fig. 5   Comparison of clearing temperatures of the homologous
         series 1, 2 and 3.

    Although the 5-n-alkyl-2-[4-cyanophenyl]-1,3-dioxanes form
nematic phase in supercooled state only, stable mesomorphic phases
over wide ranges exist in binary and multicomponent systems. This
behaviour can be explained especially by the low values of melting
enthalpies.[5,13] In terphenylanalogous 1,3-dioxanes which can be prepared by reaction of substituted benzaldehydes with 2-phenylpropan-
1,3-dioles in good yields pronounced mesomorphic ranges occur. The
phenyl-substituted propan-1,3-dioles were prepared by reduction of
the corresponding substituted esters of phenylmalonic acid with
LiAlH$_4$.

    A comparison of clearing temperatures of derivatives of 1,3-
dioxanes, cyclohexanes and benzene clearly demonstrates that both

the dioxane system and the cyclohexane ring influence the formation of the mesophase (Fig. 6).

Two aromatic rings or polar substitutents connected with the
cyclohexane – respectively 1,3-dioxanesystem unfavourable change
the liquid crystalline range and cause a decrease of clearing
temperature. But if there is only one of such groups attached to
the cyclus, no markable differences to analogous aromatic substances
concerning the clearing temperature are observed. This statement
is also confirmed by the esters bearing only one phenyl group at
the dioxane system. Their clearing temperatures have the same
magnitude as those of comparable aromatic esters. Higher melting
points are typical for dioxane derivatives. Installation of an
ethylene group into the molecule unfavourably influences the clearing temperature, because this flexible fragment strongly diminishes
the rigidity of the molecule.

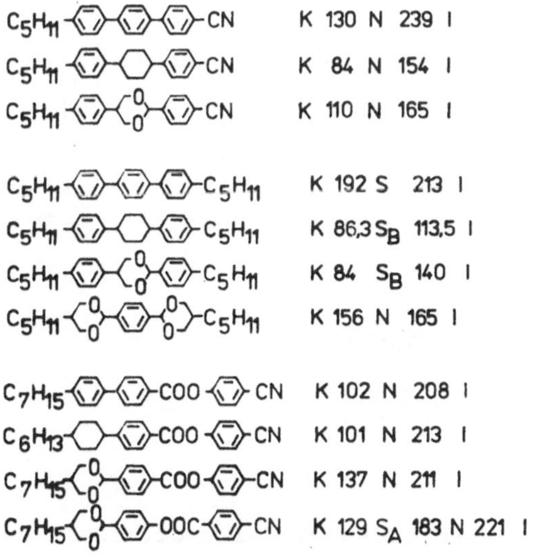

C_5H_{11} —〈〉—〈〉—〈〉—CN   K 130 N 239 I

C_5H_{11} —〈〉—〈〉—〈〉—CN   K 84 N 154 I

C_5H_{11} —〈〉—〈O/O〉—〈〉—CN   K 110 N 165 I

C_5H_{11} —〈〉—〈〉—〈〉—C_5H_{11} K 192 S 213 I

C_5H_{11} —〈〉—〈〉—〈〉—C_5H_{11} K 86,3 S_B 113,5 I

C_5H_{11} —〈〉—〈O/O〉—〈〉—C_5H_{11} K 84 S_B 140 I

C_5H_{11} —〈O/O〉—〈〉—〈O/O〉—C_5H_{11} K 156 N 165 I

C_7H_{15} —〈〉—〈〉—COO—〈〉—CN   K 102 N 208 I

C_6H_{13} —〈〉—〈〉—COO—〈〉—CN   K 101 N 213 I

C_7H_{15} —〈O/O〉—〈〉—COO—〈〉—CN   K 137 N 211 I

C_7H_{15} —〈O/O〉—〈〉—OOC—〈〉—CN   K 129 S_A 183 N 221 I

Fig. 6 Comparison of liquid crystalline behaviour of analogous
1,3-dioxane, cyclohexane and benzene derivatives.

$$C_6H_{13}-\bigcirc-CH_2-CH_2-\overset{\displaystyle O}{\underset{\displaystyle O}{\bigcirc}}-\bigcirc-CN$$

K 102 N 108 I

2. Liquid Crystalline 1,3-dithianes

Variation of heteroatoms causes markable changes of properties, as we could demonstrate in diazine series and in 5-membered hetero-aromatic ring containing systems.[3] The exchange of oxygen atoms in dioxane for sulphur atoms should be interesting. 2,5-disubstituted 1,3-dithianes can be synthesized according to the method of preparating the 1,3-dioxanes.

Substituted 1,3-dithianes can be prepared by condensation of 1,3-dimercaptopropanes with aldehydes in the presence of p-toluene sulphonic acid, the equilibrium of reaction being shifted to the side of reaction products. They are stable compounds, both in acid and in alkaline medias. The needed 1,3-dimercaptopropanes are prepared by treatment of 1,3-dibrompropane with thiourea.

$$R^1-CH(CH_2Br)_2 + 2\ SC(NH_2)_2 \longrightarrow R^1-CH(CH_2SH)_2$$

$$\downarrow + R^2-CHO$$

$$R^1-\bigcirc\overset{S}{\underset{S}{\diagup}}-R^2$$

$$R^1 = C_nH_{2n+1}$$

$$R^2 = R^3-\bigcirc-$$

$$R^3 = CN,\ OH,\ R^4-\bigcirc-COO-$$

$$R^4 = C_nH_{2n+1},\ C_nH_{2n+1}O,\ CN$$

The 5-alkyl-2-[4-cyanophenyl]-1,3-dithianes have high melting temperatures. Even in supercooled melt no liquid crystalline behaviour could be detected.

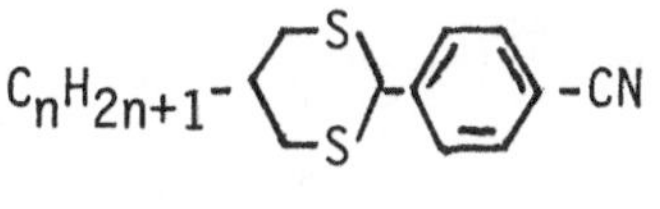

$$C_nH_{2n+1} - \left\langle \begin{array}{c} S \\ S \end{array} \right\rangle \bigcirc - CN \qquad \begin{array}{ccc} n = 6 & K\ 84\ I \\ n = 7 & K\ 98\ I \end{array}$$

A lengthening of molecules results in compounds forming meso-morphic properties.

The mesomorphic ranges of 1,3-dithianes of this type of esters correspond to those of the analogous dioxanes. They can be synthesized easily by reaction of 2-alkyl-1,3-dimercaptopropanes with 4-hydroxybenzaldehyd, followed by conversion of the obtained phenols into the esters via Einhorn by means of substituted benzoylchlorides. According to our investigations till now formation of nematic mesophase is preferred. These substances also have high melting temperatures. (See Fig. 7 below.)

3.   Liquid Crystalline Piperazines

Our group was engaged in syntheses of piperazines over a long time. We already reported on biphenyl- and terphenyl-analogous piperazines preferable forming smectic phases.[1,4,14] The piperazine system exists in chair conformation, substituents being arranged in equatorial position.

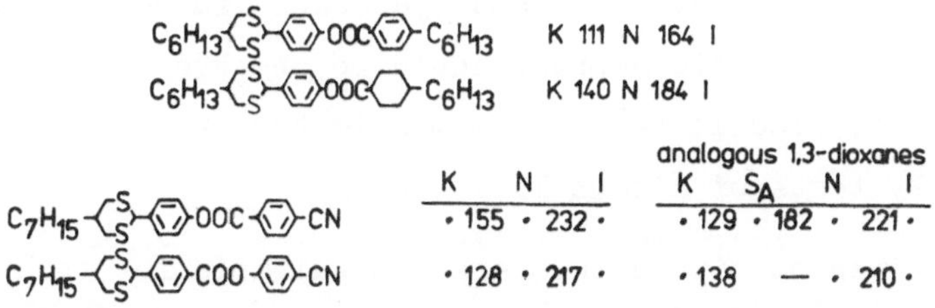

$$C_6H_{13}\left\langle\begin{array}{c}S\\S\end{array}\right\rangle\bigcirc OOC\bigcirc C_6H_{13} \qquad K\ 111\ N\ 164\ I$$

$$C_6H_{13}\left\langle\begin{array}{c}S\\S\end{array}\right\rangle\bigcirc OOC\bigcirc C_6H_{13} \qquad K\ 140\ N\ 184\ I$$

	K	N	I	analogous 1,3-dioxanes K	S_A	N	I
$C_7H_{15}\left\langle\begin{array}{c}S\\S\end{array}\right\rangle\bigcirc OOC\bigcirc CN$	· 155	· 232 ·		· 129	· 182	· 221 ·	
$C_7H_{15}\left\langle\begin{array}{c}S\\S\end{array}\right\rangle\bigcirc COO\bigcirc CN$	· 128	· 217 ·		· 138	—	· 210 ·	

Fig. 7   Comparison of the liquid crystalline behaviour of 1,3-dithianes and 1,3-dioxanes.

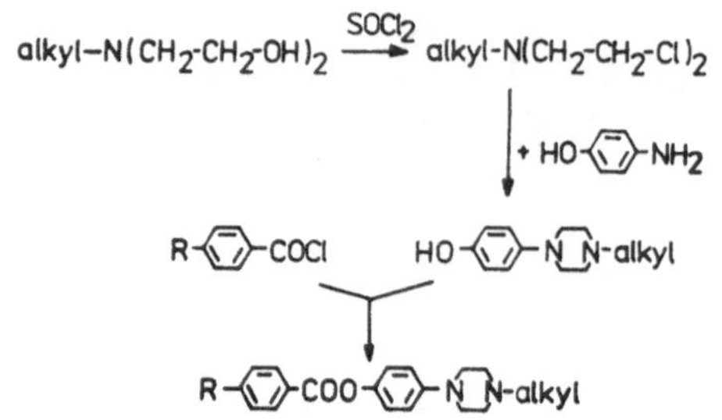

Fig. 8 Way of synthesis of unsymmetrical substituted piperazines.

Figure 8 explains the way of preparation of further piperazine substances. From a great number of known syntheses of 1-phenyl-piperazines we selected the following method: Alkylation of diethanolamin in the presence of soda without solvent led to N-alkyldiethanolamines in 50-85% yield. By reaction with thionylchloride they were converted to the very reactive N-losts, being stable over some months as hydrochlorides. Bis-(β-chlorethyl)-alkylamines react with 4-aminophenol in aceton/water or dimethylsulfoxide solvent to give 1-alkyl-4-(4-substituted-phenyl)-piperazines, whose sodium salts were used to react with substituted benzoylchlorides in benzene.

Figure 9 presents the liquid crystalline behaviour of some piperazine derivatives. At clearing temperatures all substances decompose, especial markedly detectable after thermic treatment over a long time.

Alkyl- and alkoxysubstituted esters form nematic phase only. The halogen-, trifluoromethyl- and cyanosubstituted compounds have smectic phases. Smectic A-phase could be observed as typical fan-shaped texture, smectic B-phase as mosaic texture. The phases were classified by tests of miscibility exactly.

Derivatives of piperazines are not suitable for technical application because of their high melting temperatures and decomposibility. Wide mesomorphic ranges can be observed on azomethine

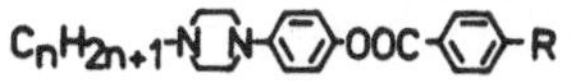

$$C_nH_{2n+1} - N \bigcirc N - \langle \bigcirc \rangle - OOC - \langle \bigcirc \rangle - R$$

n	R	K		S_B		S_A		N		I
3	C_6H_{13}	• 90		–		–		• 157		•
5	C_4H_9	• 98		–		–		• 198		•
3	OC_4H_9	• 118		–		–		• 211		•
3	Cl	• 134	(• 131		• 134	)		• 197		•
3	Br	• 142	(• 133		• 141	)		• 195		•
3	CF_3	• 141	• 165		• 188		–			•
3	NO_2	• 137	–		–			• 198		•
3	CN	• 153	–		(• 129	)		• 234		•

Fig. 9 Mesomorphic ranges of 1-alkyl-4-[4-(4-subst.-phenyl-
carbonyloxy)phenyl]-piperazines.

derivatives of piperazine. Like all mesomorphic azomethines these
compounds exhibit polymorphism in smectic mesophase, smectic A, B,
C and E phases dominating.

Our investigations on liquid crystals bearing heteroalicyclic
systems in the molecule permit following statements:

1. Heteroalicyclic 6-membered ring systems can be used as material
    for liquid crystal molecules. In liquid crystalline phase chair
    conformation dominates. The substituents in 1,4-positions are
    trans configurated.

2. The mesomorphic ranges of these compounds can be compared with
    those of the corresponding cyclohexane derivatives. The com-
    parison of clearing temperatures of analogous derivatives with
    aromatic respectively heteroalicyclic systems confirms the
    heteroalicyclic system contributing to formation of thermo-
    dynamic stable mesophases.

3. Liquid crystals with heterocyclic groups in the molecules
    markedly tend to form smectic phases, pointing out strong
    intermolecular interaction forces.

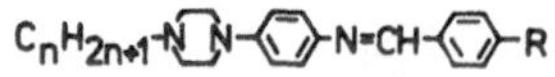

$$C_nH_{2n+1}\text{-N}\underset{\text{N}}{\bigcirc}\text{-}\bigcirc\text{-N=CH-}\bigcirc\text{-R}$$

n	R	K		S_E		S_B		S_A		N		I
5	Cl	•	101	•	159	•	201	•	228	•	233	•
3	C_4H_9	•	52	•	126	•	180	—		•	212	•
5	C_4H_9	•	71	•	111	•	191	•	193	•	203	•
6	C_4H_9	•	59	•	106	•	190	—		•	194	•
3	OC_7H_{15}	•	85 S 104 S_G 112		•	177	•	180	•	220	•
5	OC_7H_{15}	•	85 S_E 108		•	193	•	202	•	216	•

Fig. 10   Liquid crystalline azomethines with a piperazine group.

4.   The heteroatoms and their position in molecule strongly in-
     fluence the physical properties of substances.

Over the last years further teams worked at syntheses and investigation of mesophases of heterocycloalkanes. The paramagnetic
liquid crystals of oxazolidine-2-N-oxide type,[15] the existence
of thermotropic mesophases of l-n-alkyl-α-D-glucopyranosides,[16]
dibenzo-18-crown-6 derivatives,[17] piperidines,[18] new thiapyranes as
well as their sulfoxides and sulfones[18] are remarkable.

Considering all hitherto existing investigations of heterocycloalkanes one must state the syntheses of these compounds -
excepted the simple 1,3-dioxanes and piperazines - mostly demand-
ing a large preparation expense. Favourable liquid crystalline and
physical properties of these types of substances give weight to
their scientific treatment and technical use, for instance in
electrooptic field.

REFERENCES

1.   D. Demus, H. Demus, H. Zaschke, Flüssige Kristalle in Tabellen,
         Leipzig, VEB Deutscher Verlag für Grundstoffindustrie
         (1974).

2.   H. Kelker, R. Hatz, Handbook of Liquid Crystals, Verlag Chemie,
         Weinheim (1980).

3. H. Zaschke, Wiss. Z. Martin-Luther-Universität Halle-Wittenberg,
      Math.-naturwiss. R. 29, 35 (1980).

4. H.-J. Deutscher, H.-M. Vorbrodt, H. Zaschke, Z. Chem. 21, 9
      (1981).

5. H.-M. Vorbrodt, S. Deresch, H. Kresse, A. Wiegeleben, D. Demus,
      H. Zaschke, J. prakt. Chem. 323, 902 (1981).

6. H. Zaschke, H.-M. Vorbrodt, D. Demus, W. Weißflog, DDR-WP 139
      852 (13.12.1978); CA: 93, 213 415w.

7. H. Zaschke, H.-M. Vorbrodt, D. Demus, H. Kresse, DDR-WP 139
      867 (13.12.1978); CA: 93, 123 719u.

8. H. Sorkin, Mol. Cryst. Liq. Cryst. 56, 279 (1980).

9. Y. Y. Hsu, US 4200580 (5.5.1979); CA: 93, 150016u.

10. H. Zaschke, J. prakt. Chem. 317, 617 (1975).

11. G. W. Gray, J. K. Harrison, J. A. Nash, Electron. Letters
      9, 130 (1973).

12. R. Eidenschink, Kontakte (Merck) 1, 15 (1979);
      R. Eidenschink, D. Erdmann, J. Krause, L. Pohl, Angew. Chem.
      90, 133 (1978).

13. D. Demus, H. Zaschke, Mol. Cryst. Liq. Cryst. 63, 129 (1981).

14. K. Kinashi, S. Takenaka, S. Kusabayashi, Mol. Cryst. Liq.
      Cryst. 67, 49 (1981).

15. M. Dvolaitzky, J. Billard, F. Poldy, Tetrahedron 32, 1835
      (1976).

16. E. Barrall, B. Grand, M. Oxsen, E. T. Samulski, P. C. Moews,
      J. R. Knox, R. R. Gaskill, J. L. Haberfeld, Org. Coat.
      Plast. Chem. 40, 67 (1979).

17. N. G. Lukyanenko, A. V. Bogatskii, V. A. Pastushok, E. Y.
      Kulygina, L. R. Berdnikova, M. U. Mamina, Khim. Geterot-
      sikl. Soedin. 599, (1981).

18. L. A. Karamysheva, K. V. Roitman, S. I. Torgova, E. I. Kovshev,
      Advances in Liquid Crystal Research and Applications,
      edited by L. Bata, Pergamon Press, Oxford - Akademiai Kiado,
      Budapest 997, (1980).

CHARACTERIZATION OF TWO ISOMERIC CYANO-ESTER HOMOLOGOUS SERIES - I

J. W. Goodby, T. M. Leslie, P. E. Cladis
and P. L. Finn

Bell Laboratories
Murray Hill, New Jersey 07974

ABSTRACT

Current theories concerning the incidence of the reentrant nematic phase in liquid crystals generally incorporate the extent to which the molecules are paired in the nematic phase as a basis for the occurrence of reentrant phenomenon. In this study we sought to prepare materials in which the dipolar forces within the constituent molecules could be altered and hence, the degree of pairing in the nematic phase could be changed. Thus, a comparison of the properties of the 4-cyanophenyl 4'-n-alkoxybenzoates with those of the 4-n-alkoxyphenyl 4'-cyanobenzoates was made. It was found that the latter series exhibited unusual properties; an injection of smectic C phases and dual crystalline forms which melted at different temperatures. Indeed, the injection of smectic C properties into the series appears to indicate that the pairing is so strong that the reentrant phase gives way to smectic C formation; this formation in itself is unusual when observed for cyano compounds.

INTRODUCTION

Although the reentrant nematic phase was discovered only recently,[1] there have been a large number of investigations into its
true nature.[2-8] However, most of these studies have involved macroscopic theories concerning its structure and properties; yet very
few have probed the extent to which the phase is affected by
changes in the molecular structure of its constituent molecules.

The reentrant nematic phase has so far only been observed at
one atmosphere in single components which possess three phenyl rings
and at least one central-linkage group (COO, CH=N, etc.).[4] Reentrant phenomenon has also been observed in binary mixtures in which
the two components possessed only two phenyl rings each.[1,7] For
example, certain binary mixtures of 4 -n-hexyl-4'-cyanobiphenyl and
4 -n-octyl-4'-cyanobiphenyl exhibit reentrant properties.[7,8] It
should be noted however, that all the single materials and components in mixtures that exhibit this behavior have a terminal cyano
or nitro group.[9] These terminal groups produce a very strong dipole
which is directed along the long axis of the molecule. Thus in
most cases these materials have a strong positive dielectric anisotropy associated with them.

In this present study we chose to investigate the properties
of two closely related systems, firstly the 4-n-alkoxyphenyl
4'-cyanobenzoates

$C_nH_{2n+1}O$ —⟨ ⟩— OC —⟨ ⟩— CN (I)

and secondly the 4-cyanophenyl 4'-n-alkoxybenzoates

$C_nH_{2n+1}O$ —⟨ ⟩— $C-O$ —⟨ ⟩— CN (II)

which are known materials.[10] Although, these two sets of compounds
have almost identical molecular structures, electronically they are

very different. Materials of structure (I) have the central ketonic
oxygen atom (C=O) of the central linkage linked by mesomeric relay
through the second phenyl ring to the strongly electronegative
cyano group, whereas this is not the case in the second set of
materials which have another oxygen atom separating and disrupting
this relay. Hence, the strength of the central dipole of the ester
function in I is reduced in comparison to that of II.

The properties of each individual series and mixtures (both
within and between the series) were examined in order to ascertain
the effects of changing the strength of the dipoles on reentrant
behavior.

RESULTS

Homologous Series

The 4-n-alkoxyphenyl 4'-cyanobenzoates (I) (nOPCB) were synthesized by standard methods and pruified by chromatography. The
transition temperatures for this homologous series are presented in
Table 1 and are shown as a function of increasing terminal alkoxy

TABLE 1

C_nH_{2n+1}O—⟨phenyl⟩—OCO—⟨phenyl⟩—CN

n	I-N	N-S_A	I-S_A	S_A-S_C	M.P.[*]
6	91	-	-	-	83
7	93	78.5	-	-	72.5
8	93	87	-	(51)	69
9	-	-	95	(54)	69
10	-	-	98	(56)	67
12	-	-	104	(62)	76

( ) Monotropic Transition Temperature.

chain length in Fig. 1. Similarly, the results for the 4-cyano-
phenyl 4'-n-alkoxybenzoates (II) (CPnOB) are given in Table 2 and
depicted in Fig. 2.

A number of comparisons can be drawn from Figs. 1 and 2, firstly
series (I) exhibits greater smectogenic properties than does series
(II). The first smectic phase is observed for the n-decyl homologue
(CP10OB) in series (II) and for the n-heptyl homologue (7OPCB) in
series (I). Furthermore, the first series also exhibits a second
smectic phase as well. This second phase was shown to be of the
smectic C type. The smectic C phase is not usually observed in
materials which contain a terminal cyano group linked directly to
the central core, in fact the homologous series[11]

R O—⟨phenyl⟩—COO—⟨phenyl⟩—CH=N—⟨phenyl⟩—CN

is the only one reported besides this novel series which exhibits
smectic C properties. Thus, the 4-n-alkoxyphenyl 4'-cyanobenzoate
series (I) is the only known system with two phenyl rings and a
terminal cyano group (attached directly to the core) which exhibits
smectic C behavior. Typically the smectic C phase is injected into
the series at longer n-alkoxy chain lengths, i.e., the n-octyloxy
homologue. All of the C phases are monotropic often occurring at
the point of recrystallisation. However, enantiotropic C phases
can be obtained for study in some binary eutectic mixtures. After
injection of the C phase at the n-octyloxy homologue the transition
temperatures for the later members rise and then level off.

Correspondingly the 4-cyanophenyl 4'-n-alkoxybenzoates (II)
contrast quite strongly with series (I). The first smectic phase
(S_A) is injected only at the n-decyloxy member with the earlier members only exhibiting nematic phases. This series does not exhibit
any smectic C properties nor do its members show any tendency
towards exhibiting C phases in miscibility studies involving
compounds of type I.

TABLE 2

$$C_nH_{2n+1}O-\!\!\!\bigcirc\!\!\!-CO.O-\!\!\!\bigcirc\!\!\!-CN$$

n	I-N	N-S_A	M.P.*
6	82.5	-	67.8
8	84.2	-	73.4
10	86.5	79.0	79.0

* The melting points were determined by optical microscopy.
A detailed description of the melting behavior of these
materials by X-ray analysis and differential scanning
calorimetry, is given elsewhere.

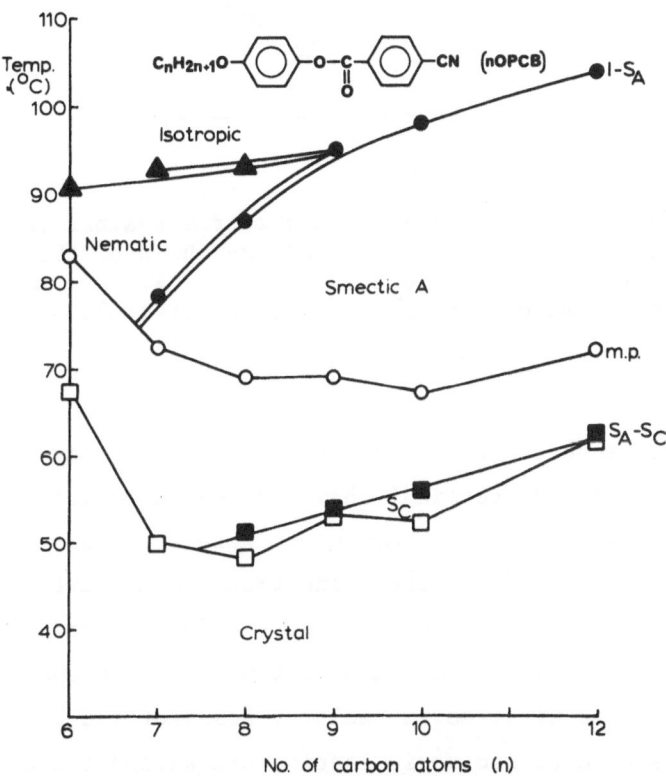

Fig. 1. Plot of the transition temperatures against the number of
carbon atoms (n) in the n-alkoxy chain of the 4-n-alkoxy-
phenyl 4'-cyanobenzoates. Key: ▲, N-I; ●, S_A-N or I;
■ , S_C-S_A; ○, crystal-mesophase; □ mesophase-crystal.

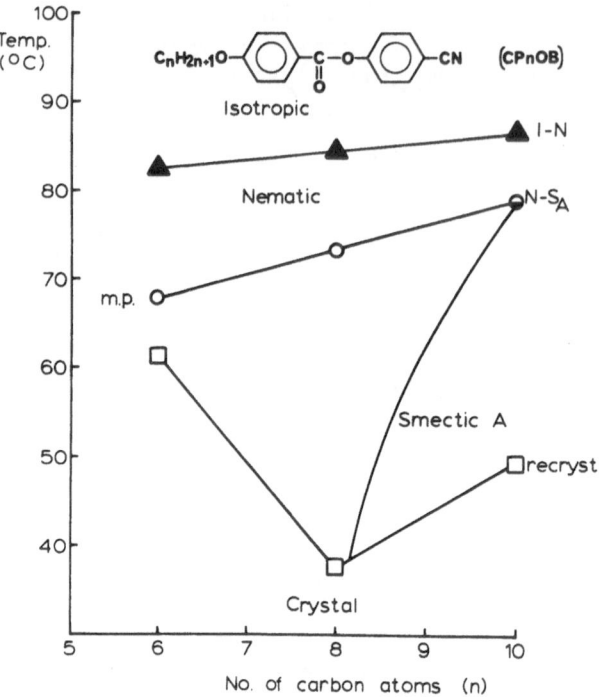

Fig. 2.   Plot of the transition temperatures against the number of
           carbon atoms (n) in the n-alkoxy chain of the 4-cyano-
           phenyl 4'-n-alkoxybenzoates. Key: ▲, N-I; ●, S$_A$-N; ○,
           crystal-mesophase; □ , mesophase-crystal.

        A further investigation into these series was made by adding
some rigidity into the terminal chain by the inclusion of a double
bond.   Thus, derivatives of bromobutene were synthesised analogous
to both of series I and II, the transition temperatures for these
materials are reported in Table 3.   As the double bond is moved
nearer to the central core structure, thus becoming conjugated with
the first aromatic ring, so the melting point of the compound rises
whilst the nematic to isotropic liquid, transition temperature
falls.   Therefore, the more rigid the terminal chain (and the whole
molecule) becomes the greater is the supression of liquid crystal
properties.   It is interesting to note however, that 4-(1"-butenoxy)

TABLE 3

a) 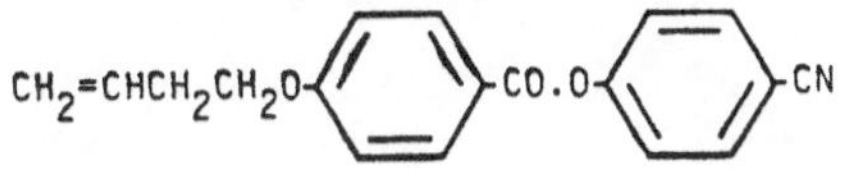 $CH_2=CHCH_2CH_2O$-⟨⟩-$CO.O$-⟨⟩-CN

| I ⟷ N | mp | Recryst |
| 62 | 86.5 | 50.9 |

b) 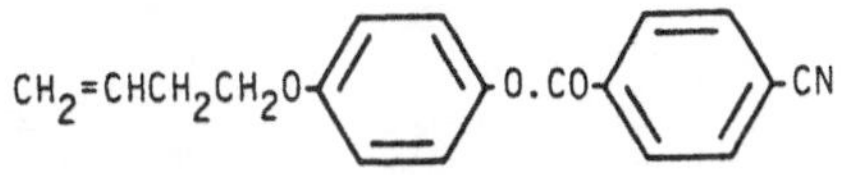 $CH_2=CHCH_2CH_2O$-⟨⟩-$O.CO$-⟨⟩-CN

| I ⟷ N | mp | Recryst |
| 59.6 | 89.3 | 53.4 |

c) 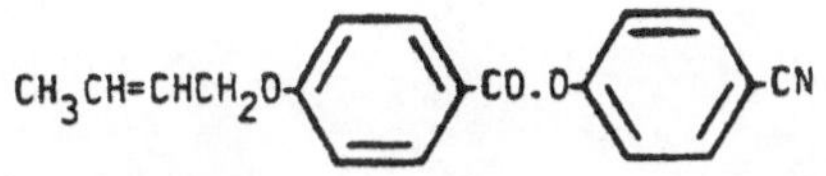 $CH_3CH=CHCH_2O$-⟨⟩-$CO.O$-⟨⟩-CN

| I ⟷ N | mp | Recryst |
| 100.2 | 127.2 | 100 |

d) 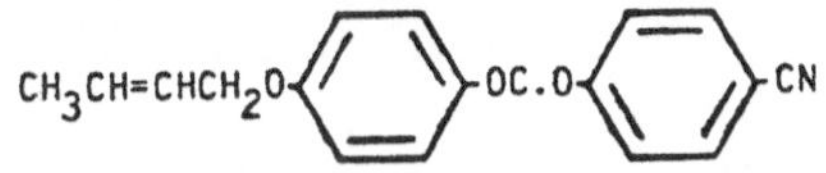 $CH_3CH=CHCH_2O$-⟨⟩-$OC.O$-⟨⟩-CN

| | mp | Virtual N-I |
| | 173 | 15° |

phenyl 4'-cyanobenzoate (compound (b)) has a tendency to exhibit smectic C properties just on the point of recrystallisation.

Phase Characterisation

The phases exhibited by series (I) and (II) were classified by thermal optical microscopy. The second homologous series proved relatively straightforward as all the members studied exhibit nematic phases except for the n-decyloxy homologue which also showed a smectic A phase characterised by its focal-conic and homeotropic textures.

The 4-n-alkoxyphenyl 4'-cyanobenzoates were found to exhibit novel dimorphism in that the later homologues (n-heptyloxy onwards) possessed both smectic A and smectic C phases. These phases were characterised by their microscopic textures; typically for the

Fig. 3.   The focal-conic fan and homeotropic textures of the
           smectic A phase of 4-n-dodecyloxyphenyl r'-cyanobenzoate
           (X100).

n-dodecyloxy member the smectic A phase separated from the isotropic liquid in the form of batonnets these coalesced to produce an
unbroken focal conic fan texture.   The phase also exhibited the
homeotropic texture; both textures are shown together in Fig. 3.
On cooling the A phase, a transition to the C phase took place with
the fans becoming broken and mottled in appearance and the homeotropic areas becoming birefringent and exhibiting a schlieren texture.   Both of these forms of texture are typical of the smectic C
phase and are shown together in Fig. 4 for the n-dodecyloxy compound.

Miscibility Studies

A large variety of miscibility studies was undertaken in order
to investigate the properties of these compounds.   Firstly, a miscibility involving 4-cyanophenyl 4'-n-hexyloxybenzoate (CP60B) and

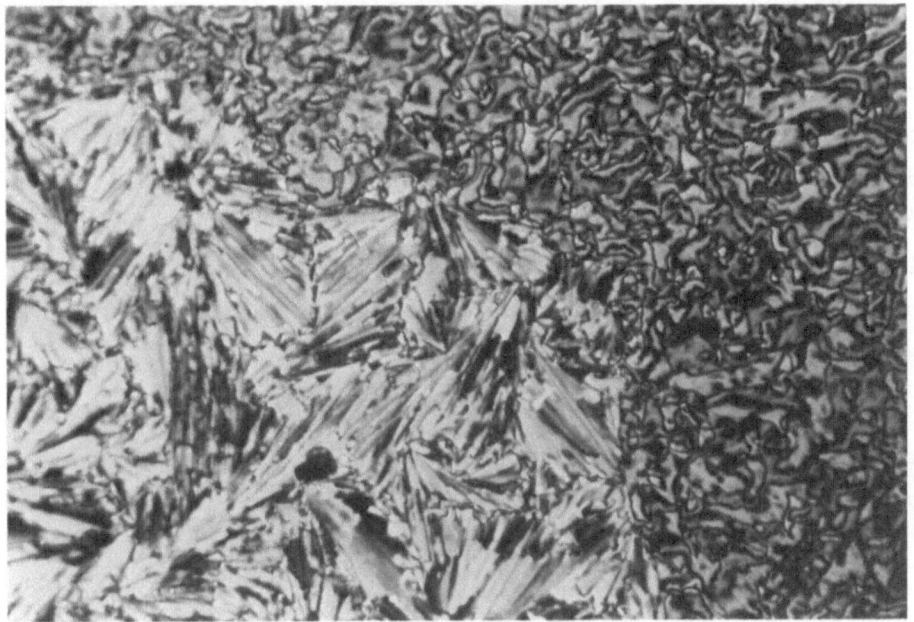

Fig. 4.   The broken focal-conic and Schlieren textures of the
           smectic C phase of 4-n-dodecyloxyphenyl 4'-cyanobenzoate
           (X100).

4-(2"-butenoxy)phenyl 4'-cyanobenzoate was made in order to determine the "virtual" N-I of the second compound; this value was found
to be 15°. Thus showing a great decrease in N-I values as the
terminal chain becomes more rigid.

Secondly, an investigation into the co-miscibility properties
of the smectic C phases of the 4-n-alkoxyphenyl 4'-cyanobenzoates
was made. Miscibility studies involving standard smectic C compounds provided interesting results; the smectic C phases of the
test materials could not be shown to be co-miscible with those of
the standard materials. However, there were large rises in the stability of the A phase particularly in the region of 50% by weight
of each component. For example, a miscibility study involving
4-n-heptyloxyphenyl 4'-n-octyloxybenzoate (N and S_C phases) as the
standard material and 4-n-octyloxyphenyl 4'-cyanobenzoate, as the

test compound, gave an increased stability of 8° over the N-S$_A$ transition temperature of the test compound in a mixture of 60% by
weight of the standard material and 40% by weight of the test compound. This is rather startling because the standard compound does
not exhibit a smectic A phase at all and only shows a N-S$_C$ transition
at 61.5°. The two smectic C phases could not be shown to be co-
miscible in mixtures cooled down to room temperature and in fact
most mixtures recrystallised before this determination could be
made.

This surprising elevation of smectic A thermal stability is
probably due to the splitting of the paired molecules of the cyano
compound by the standard material.[8] Thus, if the formation of the
C phase is dependent on this pairing then the separation of the
pairs by the unpaired standard will suppress C formation.

In order to try and overcome this barrier a miscibility study
of one member of the homologous series (I) with the standard
material,[12]

C$_7$H$_{15}$O—⟨ ⟩—COO— ... H —OOC—⟨ ⟩—OC$_7$H$_{15}$

Crystal ⟶ S$_C$ ⟷ N ⟷ I
              83   90   141

which has a gross overall molecular shape that is very similar to
that of paired cyano test ester, was attempted. An example of this
miscibility diagram of state is shown in Fig. 5 for the n-octyloxy
homologue. Again the smectic A thermal stability is elevated so
that the N-S$_A$ transition temperatures of some binary mixtures are
higher than for either of the two components. The smectic C phase
stability falls away quickly as one component is added to the
other. Thus, even this standard miscibility material, selected
because of its structural similarity to the paired system, has the
effect of splitting the pairs and depressing C formation. This

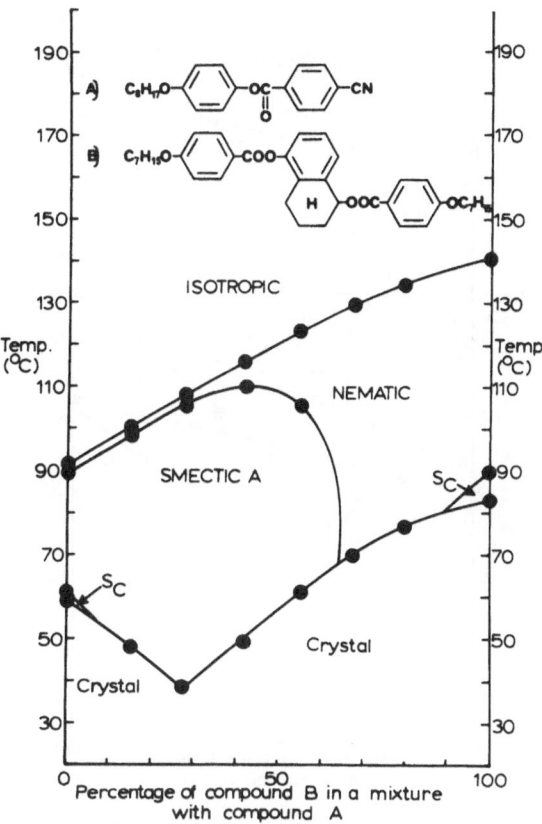

Fig. 5.   Diagram of state for mixtures (wt.%) of 4-n-octyloxy-
           phenyl 4'-cyanobenzoate (A) and bis-4-n-heptyloxyben-
           zoyloxy-1,5-tetrahydronaphthalene.

indicates that the C phase formed by these compounds has a very
specific structure and related properties, and is hence unlikely to
be co-miscible with other smectic C phases of compounds which don't
have similar molecular structures, i.e., the structure must have a
terminal cyano or nitro group.

     These observations are born out by miscibility studies between
members of the same homologous series which exhibit both A and C
phases, for example, Fig. 6 shows the co-miscibility between the
corresponding A phases and C phases of the n-octyloxy and the
n-dodecyloxy members of series (I).

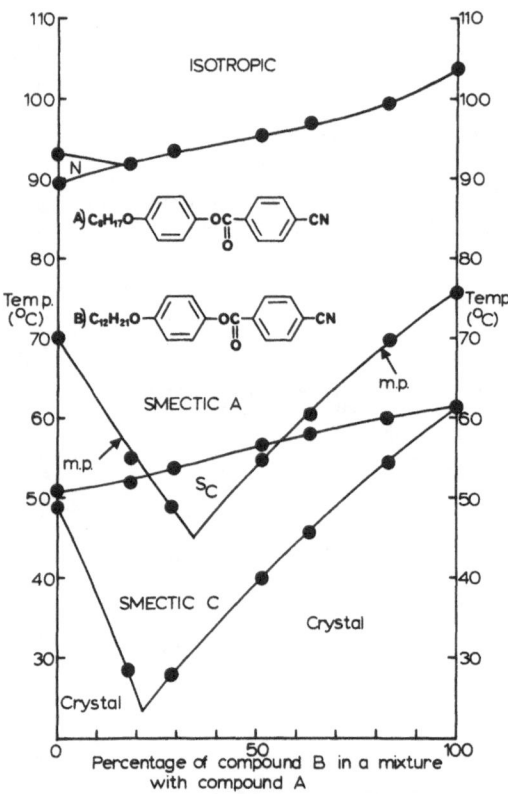

Fig. 6.   Diagram of state for mixtures (wt.%) of 4-n-octyloxyphenyl
          4'-cyanobenzoate (A) and 4-n-dodecyloxyphenyl 4'-cyano-
          benzoate (B).

     A third type of miscibility investigation was carried out be-
tween members of the two series that did not exhibit smectic properties and those that did; here a number of mixed systems were shown
to exhibit reentrant phenomenon. Miscibility studies involving 4-n-
hexyloxyphenyl 4'-cyanobenzoate (60PCB) (N phase) and the higher members of the same homologous series are of particular interest. The
miscibility diagram of state for mixtures of the n-hexyloxy and the
n-octyloxy materials is shown in Fig. 7. Binary mixtures of between
30 and 55% of the n-hexyloxy compound exhibit reentrant nematic phenomenon, but here the reentrant nematic is followed by a smectic C

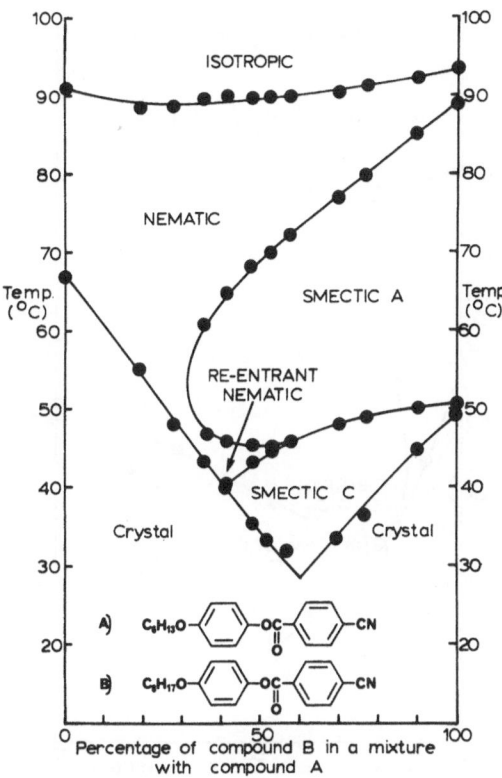

Fig. 7.   Diagram of state for mixtures (wt.%) of 4-n-hexyloxyphenyl
          4'-cyanobenzoate (A) and 4-n-octyloxyphenyl 4'-cyanoben-
          zoate (B).

phase on cooling.   Hence the mixtures have the following phase
sequence:

$$I - N - S_A - N - S_C$$

     Similar behavior is observed for binary mixtures of the n-hexy-
loxy and n-dodecyloxy (60PCB, 120PCB) compounds, as shown in Fig. 8.
This diagram shows that the reentrant phase temperature and composition ranges have been reduced by the increased stability of the A
phase in the n-dodecyloxy member.   However it is still surprising
that two materials of very different terminal alkoxy chain lengths
can still produce reentrant behavior.

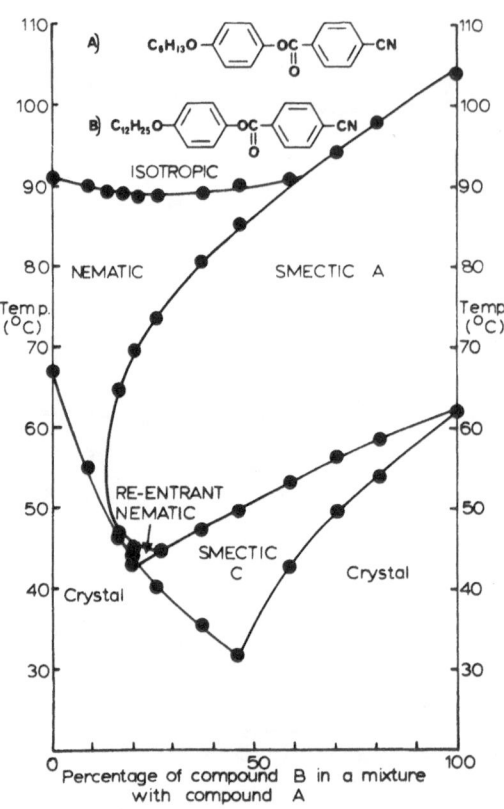

Fig. 8.   Diagram of state for mixtures (wt.%) of 4-n-hexyloxyphenyl
           4'-cyanobenzoate (A) and 4-n-dodecyloxyphenyl 4'-cyanoben-
           zoate (B).

Miscibility studies involving members of series (I) and one of
series (II) also show reentrant properties, but none show a N,A,N,C
phase sequence as the C phase usually falls away sharply on the addition of the non-smectic C material to the binary mixture, as shown
in Fig. 9 for binary mixtures of 4-n-dodecyloxyphenyl 4'-cyanobenzo-
ate (120PCB) and 4-cyanophenyl 4'-n-octyloxybenzoate (CP8OB).

Similarly, binary mixtures involving the more rigid materials
shown in Table 3 also exhibited reentrant phenomenon. Particularly
a miscibility study involving 4-(1"-butenoxy)phenyl 4'-cyanobenzoate
and 4-n-octyloxyphenyl 4'-cyanobenzoate provided some mixtures which
exhibited a N,A,N,C phase sequence, see Fig. 10, a triple point be-

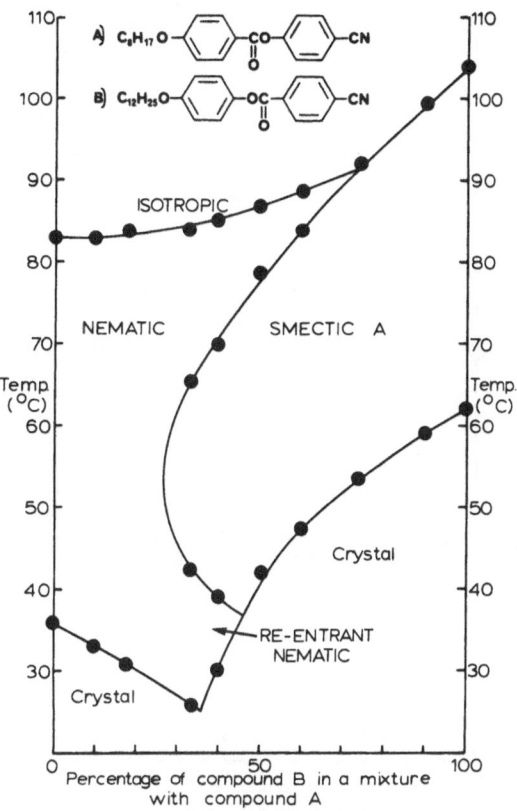

Fig. 9.   Diagram of state for mixtures (wt.%) of 4-cyanophenyl
          4'-n-octyloxybenzoate (A) and 4-n-dodecyloxyphenyl
          4'-cyanobenzoate (B).

tween the A,C, and nematic phases occurs at approximately 42° for
77% by weight of the n-octyloxy compound.

DISCUSSION

The smectic C phase had not been observed in materials which
contained a terminal cyano group directly attached, and therefore
conjugated to the central core structure, until a recent report by
Weissflog et al.[11] that 4(4-n-decyloxybenzoyloxy)-benzylidene-4'-
cyanoaniline

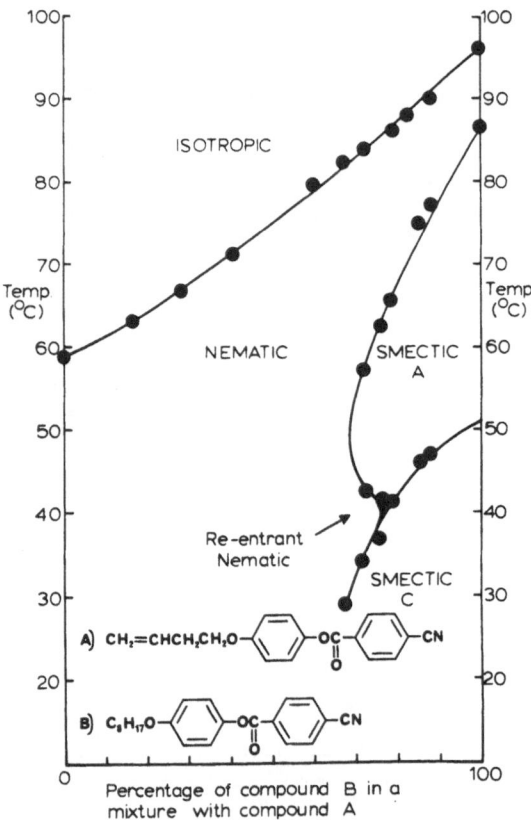

Fig. 10. Diagram of state for mixtures (wt.%) of 4-(1"-butenoxy)
          phenyl 4'-cyanobenzoate (A) and 4-n-octyloxyphenyl
          4'-cyanobenzoate (B).

exhibited a N,A,C,N phase sequence. In this present study we have
examined the properties of these two closely related systems

$$RO-\!\!\langle\bigcirc\rangle\!\!-OOC-\!\!\langle\bigcirc\rangle\!\!-CN \qquad\qquad (I)$$

and

$$RO-\!\!\langle\bigcirc\rangle\!\!-COO-\!\!\langle\bigcirc\rangle\!\!-CN \qquad\qquad (II)$$

which have almost identical molecular structures and hence steric

shapes. These two systems differ considerably however in their
dipolar distribution within each molecule. In series (I) the cyano
group is linked directly to the ketonic oxy atom of the ester function by mesomeric relay through the π cloud of the second ring as
shown below:

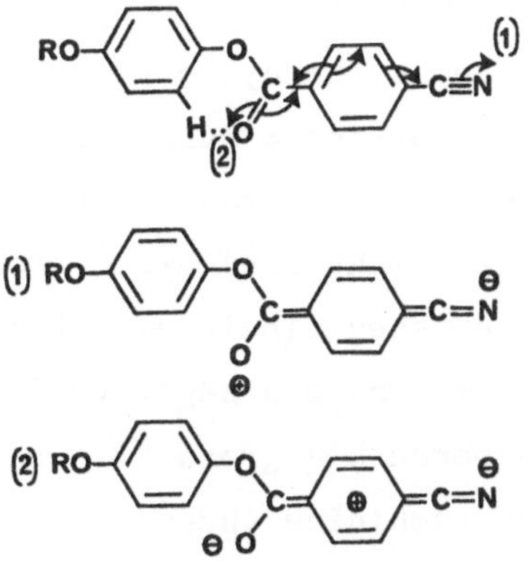

In the second system (II) the cyano group is only linked to the
ester oxy atom and not to the keto oxygen, therefore two separate
mesomorphic relays can be employed as shown.

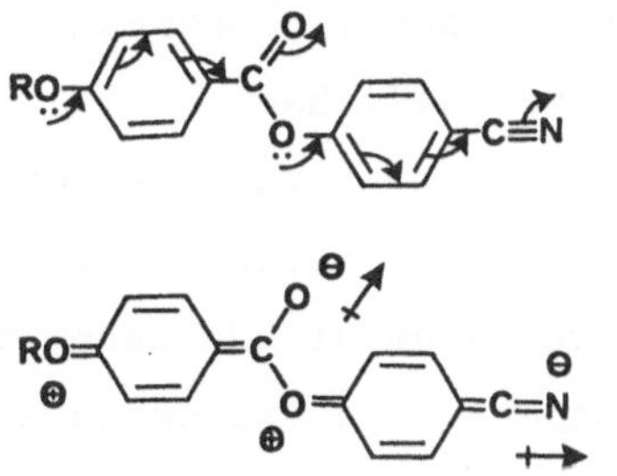

In the first system both of th electronegative groups (C=O and
CN) are trying to attract electrons from each other through the π
cloud of the second ring. Thus, the strength of the overall dipole
of each group is reduced in comparison to the second system (II) in
which these groups are separated and the longitudinal dipole is
smaller in case (I) than in case (II). The paired molecules are
overlapped such that the cyano group overlays the oxygen atom of
the alkoxy group as shown below:

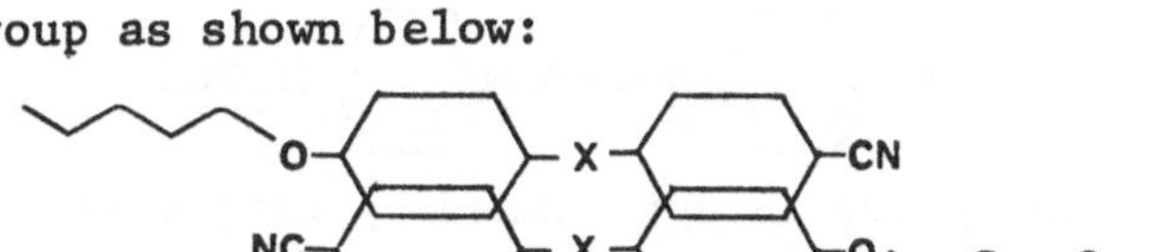

Thus, for the 4-n-alkoxyphenyl 4'-cyanobenzoates (I) we have a system almost acting as though there were compound formation. The
oxygen atom of the alkoxy chain plays two roles in the pairing;
because it is not totally conjugated into the first ring (due to
the lack of an electron withdrawing group in the para position) it
is more easily polarised by the cyano group of the neighboring molecule. Consequently, it carries a stronger crossed dipole due to
the lack of conjugation and the cyano polarisation. In essence,
the polarisation by the neighboring cyano group holds the pair together. The stronger crossed dipole aids the formation of tilted
phases, i.e., the pair may have a negative dielectric constant. As
the pair of molecules possess a gross "zig-zag" molecular shape; it
is possible to see how combining these three factors that the system
can exhibit smectic C properties:

Pairing   1)   reduces the longitudinal component of the monomer
               dipoles

          2)   enhances the lateral component

          3)   provides "zig-zag" shape.

     In the second series however this is not the case, the alkoxy
oxygen atom is linked through mesomeric relay in the first phenyl
ring to the ketonic oxygen atom of the ester function. Therefore
the electrons of this atom are polarised towards the ring system,
hence its crossed dipole strength is reduced and the "bonding" with
the cyano group of its nearest neighbor is weakened. Consequently,
the second series of materials does not exhibit tilted smectic
phases, even though the strength of the central ester dipole is
greater than in the first case.

     Similar results have been obtained by Pelzl et al.[13] for com-
pounds of the type

in which the cyano group is not linked by conjugation to the
polarisable π cloud electrons of the central core. Hence, this

group acts as an isolated strong longitudinal dipole at one end of
the molecule. Although, the alkoxy oxygen atom is intrinsically
linked to the ketonic group of the central ester function, it is
separated from it by the double bonded bridge of the cinnamic moiety
thus diminishing its overall effect. Hence, in this case we have a
similar dipolar situation to that of the 4-n-alkoxyphenyl 4'-cyano-
benzoates, and therefore these compounds also exhibit nematic,
smectic A and smectic C phases. Binary mixtures of these compounds
were also shown to exhibit reentrant behavior with similar phase
sequences to those exhibited by series (I), i.e., N,A,N,C.

These arguments provide an over-simplified theory for the
formation of the C phase by compounds (I) and not by (II), however,
if we try and extend these ideas to reentrant behavior the situation
becomes more complex.

Firstly, it is important to note that reentrant phenomenon for
two phenyl ring species is only observed in binary mixtures when
one of the materials exhibits only a nematic phase and the other
exhibits a smectic A phase, as shown in Table 4.

Reentrant behavior seems to be unaffected by small changes in
steric and dipolar properties. For example, some binary mixtures
of 60PCB and 120PCB exhibit reentrant nematic phases, these two
molecules have very different alkoxy chain lengths, thus producing
pairs of considerably different size. Binary mixtures of members

TABLE 4

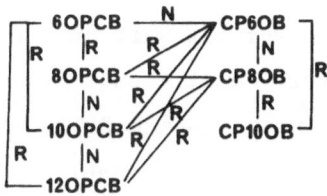

N - Normal
R - Reentrant

of series (I) and (II) also exhibit reentrant behavior, hence changes
in the dipolar character do not greatly affect the stability of the
phase. However, the phase has been shown to be dependent on the
degree of pairing in the nematic phase and somewhat on the distribution of the dipoles within the pair. In the first series of compounds (I) this pairing is very strong, but the lateral dipoles are
also large enough to produce a tilted phase and not directly reentrant behavior. The mixing of two homologues of the series may
disrupt the steric factors influencing the formation of the C phase
thus injecting reentrant properties.

Current theories of the formation of the smectic C phase tend
to be based primarily on dipolar factors with some steric influences
on the packing requirements of the molecules.[14,15] Most of these
theories rely on the coupling of lateral dipoles in order to produce
the necessary conditions for tilting of the molecules, a fact that
has been born out experimentally.[16] If the C phase does require
this strong lateral dipole it is probable that the bulk material
has an overall negative dielectric constant. It is possible for
compounds of series (I) to have a positive dielectric constant in
the higher temperature nematic phase where the pairing is less, and
for this constant to fall on passing through to the more ordered
smectic phase where the pairing is greater. Thus, the dielectric
constant may vary with temperature depending on the amount of
pairing which in turn depends on the attractive forces between the
molecules.

When two materials of group (I) are mixed to produce a reentrant phase one of the materials must only exhibit a nematic phase.
If this higher temperature nematic phase ($+\Delta\varepsilon$) is then mixed with a
system where the dielectric constant falls with temperature we
might obtain a mixture where a reentrant phase is formed in which
the nematic phase has a low value for its dielectric constant.
This may be the case for mixtures between members of series (I)

which exhibit a N,A,N,C phase sequence. The C phase might be expected to have a negative dielectric constant and the reentrant
phase above it may be weakly positive, with the higher temperature
nematic phase being more positive.

The results for the dielectric constants of these materials
and certain mixtures is a complex and lengthy study, initial
results seem to be partially in agreement with our speculations
concerning these homologous series, and a full report will be
published elsewhere.

EXPERIMENTAL

The materials were prepared from the corresponding carboxylic
acids and phenols by standard methods. They were purified by column
chromatography over silica gel (80-200 mesh 30 cm. x 2.5 cm) using
hexane/dichloromethane (1:1) as eluant and recrystallised from light
petrol (40-60°). The structures of the final products were checked
by infra-red spectroscopy and mass spectrometry. The transition
temperatures of the pure products and of binary mixtures for miscibility studies were determined by optical microscopy using a Zeiss
Universal polarising microscope in conjunction with a Mettler FP52
hot-stage and control unit.

CONCLUSION

We have produced a number of materials which when mixed show
novel reentrant behavior. The materials chosen provide us with
some complex relationships between molecular structure and reentrant
behavior.

Further physical studies made on these materials are given in
a separate paper. The quantity and diversity of some of the results
obtained provide a number of conflicting views to the true nature
of reentrant behavior. The correct interpretation of these observa-

tions can only follow further detailed investigations by dielectric
and mixture studies.

REFERENCES

1. P. E. Cladis, Phys. Rev. Lett. <u>35</u>, 48, 1975.

2. P. E. Cladis, R. K. Bogardus and D. Aadsen, Phys. Rev. <u>A18</u>,
        2292, (1978).

3. D. Guillon, P. E. Cladis, D. Aadsen and W. B. Daniles, Phys.
        Rev. Lett. <u>A21</u>, 658, (1980).

4. F. Hardouin, G. Sigaud, M. F. Achanel and H. Gasparoux, Phys.
        Lett. <u>71A</u>, 347, (1979).

5. P. S. Pershan and J. Prost, J. Phys. <u>40</u>, L27, (1979).

6. J. Prost, <u>Liquid Crystals of One- and Two- Dimensional Order</u>,
        eds. W. Helfrich and G. Heppke, Springer-Verlag, Berlin,
        Heidelberg, New York, 1980, p. 125.

7. P. E. Cladis, D. Guillon, F. Bouchet and P. L. Finn, Phys.
        Rev. <u>23A</u>, 2594, (1981).

8. P. E. Cladis, Mol. Cryst. Liq. Cryst. <u>67</u>, 177, (1981).

9. W. Weissflog, N. K. Sharma, G. Pelzl and D. Demus, Krist. und
        Tech., <u>15</u>, 35, (1980).

10. A. Boller, H. Scherrer, M. Schadt and P. Wild, Proc. IEEE 160,
        1002, (1972).

11. W. Weissflog, G. Pelzl, A. Wiegeleben and D. Demus, Mol.
        Cryst. Liq. Cryst. Lett. <u>56</u>, 295, (1980).

12. J. W. Goodby, unpublished results.

13. G. Pelzl, S. Diele, A. Wiegeleben and D. Demus, Mol. Cryst.
        Liq. Cryst. Lett. <u>64</u>, 163, (1981).

14. W. L. McMillan, Phys. Rev. <u>A8</u>, 1921, 1973.

15. A. Wulf, Phys. Rev., <u>A11</u>, 365, (1975).

16. J. W. Goodby, G. W. Gray and D. G. McDonnel, Mol. Cryst. Liq.
        Crystl. Lett. <u>41</u>, 145, (1978).

EFFECTS OF MOLECULAR LENGTH ON NEMATIC MIXTURES – IV: STRUCTURE EFFECTS ON VISCOSITY OF ESTER MIXTURES

J. David Margerum, Siu-May Wong, John E. Jensen, and
Camille van Ast

Hughes Research Laboratories, 3011 Malibu Canyon Road
Malibu, California 90265

ABSTRACT

The flow viscosity (η) of nematic ester mixtures is studied
as a function of average molecular length ($\bar{L}$), chemical structure,
and temperature. When $\bar{L}$ is increased by use of longer alkyl end
groups, the η of less polar mixtures increases while the η of more
polar mixtures changes slightly or even decreases. However, when
cybotactic nematic characteristics occur with increased $\bar{L}$, then
η increases sharply. Studies of η are made for eighteen different
classes of ester structures used as additives in 4-alkoxyphenyl
4-alkylbenzoate (RO-R') mixtures at fixed values of $\bar{L}$ for both
additive components and mixtures. Many interesting effects of
structure on η are observed, and approximate class viscosities
(η_{class}) at 25°C are assigned to each of the 18 classes. The
$\eta_{25°}$ of other ester mixtures is estimated by summing η_{class} times
the mole fraction of that class present in the mixture. These
η_{calc} values are generally within 10% of the actual $\eta_{25°}$ for
multicomponent ester mixtures containing some RO-R' components.
Temperature variations often result in a non-linear plot for
$\log \eta$ vs T^{-1} of ester mixtures. The apparent activation energy

between 25° and 40°C generally increases strongly with the η of the mixture, ranging from 6.7 to 12.0 kcal/mole between $\eta_{25°}$ values of 16 and 188 cP.

I.   INTRODUCTION

The flow viscosity (η) of a nematic liquid crystal (LC) is an anisotropic physical parameter which is relatively simple to measure and is often used to help characterize LC mixtures. It is often assumed that the electrooptical response time of a LC is faster for mixtures with lower η values. We have been studying the properties of ester LC mixtures for electrooptical applications, particularly in regard to dynamic scattering (DS) effects.[1-6] In each of three classes of ester LC mixtures, we found that their η values at a given temperature increased with the average molecular length (L) of the mixtures.[1-3] Mixtures within two of these classes of esters, 4-alkoxyphenyl-4-alkylbenzoates (RO-R') and 4-alkoxyphenyl 4-alkylcyclohexanecarboxylates (RO-[C]R'), showed DS. Their DS decay times increased with η in surface-perpendicular cells, although not in surface-parallel cells.[1-2] However, in six different multiclass ester mixtures, the DS on-times and decay times in surface-parallel cells were generally longer for the LCs with the higher η values.[4] When new mixtures are being considered for the optimization of various LC properties, it is important to understand the molecular structure effects of the components on η. There is also a need to predict the approximate η of mixtures, such as newly calculated eutectic LC mixtures formulated from several classes of nematic esters.

In this study, our main approach is to evaluate the viscosity contributions of various classes of LC ester structures by measuring the capillary flow viscosity of RO-R' mixtures which contain these other esters as added components. This permits us to study many different components in room-temperature nematics, to compare results at similar temperatures, to vary systematically the $\bar{L}$ of several series, and to compare structural effects on viscosity by measurements at fixed values of $\bar{L}$. Although these results are specifically related to the behavior of components in RO-R' mixtures, we find that very useful structural correlations are observed and that the η of new ester mixtures can be estimated with fairly good accuracy.

II. EXPERIMENTAL

The flow viscosity is measured in calibrated Cannon-Manning type viscometer tubes held in a temperature controlled water bath. The calibrated ranges are 3 to 15, 7 to 35, and 20 to 100 CS, respectively, in three different diameter tubes. In several cases, overlapping results from different size tubes are found to be in good agreement (i.e. there is no tube diameter effect on η). The density is measured in calibrated pycnometer tubes, at the same temperatures as the viscosity.

The class structures and the class code abbreviations for the LCs used in this study are shown in Figure 1. The structures are listed in decreasing class viscosities, as shown in the third column. These assigned η_{class} values are derived from the present studies, as described below in the section on results. The compounds used as additives are listed in Table I, where the numbers in the compound code refers to the number of carbons in each straight chain alkyl end group. For example, 7-6 is the abbreviation for 4-n-heptylphenyl 4-n-hexylbenzoate. The molecular lengths are measured using CPK molecular models in a fully extended configuration, as previously described.[1-3] The melting points listed are mostly crystalline to nematic transitions, except for a few crystalline to smectic transitions, and the clearpoints are nematic to isotropic transitions. The lower end of the transition temperatures are given, as determined from differential scanning calorimetry (DSC) analysis[1] or from a hot stage attachment (Mettler FP5) on a polarizing microscope. Most of the LC compounds are synthesized at the Hughes Research Laboratories (HRL), and the preparation of some of them has been reported previously.[1-9] The commercial LC samples are used as received, except for EK-11650 which is recrystallized prior to use. Many of the other LCs in Table I are prepared by standard methods from the corresponding commercially available benzoyl chlorides and phenols (or thiophenols), e.g. the RO-OR', RSOR' and R-R' compounds. The acid intermediates for the R-ϕOR' compounds are prepared by hydrolysis of the commercially available

STRUCTURE CLASS	CLASS CODE	APPROX CLASS VISCOSITY (η_{CLASS})
R—⬡—O—C(=O)—⬡—⬡—OR' (with CN)	R(CN)—ϕOR'	310
R—⬡—O—C(=O)—⬡—⬡—R' (with CN)	R(CN)—ϕR'	200
R—⬡—O—C(=O)—⬡—⬡—OR'	R—ϕOR'	160
R—⬡—O—C(=O)—⬡—O—C(=O)—⬡—R' (with Cl)	R—(Cl)OOCϕR'	153
R—⬡—O—C(=O)—⬡—O—C(=O)—⬡—R'	R—OOCϕR'	130
RO—⬡—S—C(=O)—⬡—OR'	RO-S-OR'	92
RO—⬡—O—C(=O)—⬡—OR'	RO—OR'	82
NC—⬡—O—C(=O)—⬡—R	NC—R	82
R—⬡—O—C(=O)—⬡—⬡—R'	R—ϕR'	77
RO—⬡—O—C(=O)—⬡—O—C(=O)—R'	RO—OOCR'	74

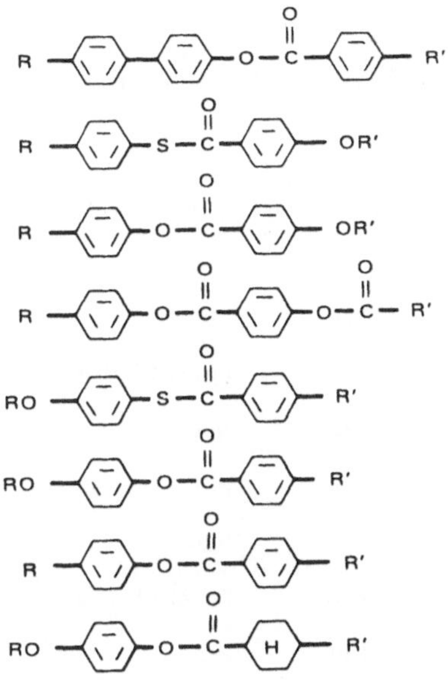

Rϕ–R'	65
R-S-OR'	58
R–OR'	55
R–OOCR'	43
RO-S-R'	37
RO–R'	36
R–R'	19
RO–[C]R'	16

Fig. 1 Structure, class code, and class viscosities of LC ester components. (The η_{class} at 25°C apply when 10 to 25% of these components are used in RO-R' mixtures with $\bar{L}$ ~ 22 Å.)

(BDH) 4-alkoxy-4'-cyanobiphenyls according to the procedure of Byron et al.[10] The phenol intermediates for the R-OOCϕR' compounds are prepared by refluxing the 4-alkylphenol with 4-hydroxybenzoic acid in toluene with sulfuric and boric acid added.[11] The resulting phenol is isolated and reacted with a benzoyl acid chloride to give the corresponding diester. The 4-butoxythiophenol intermediate for 40S05 and 40S3 is synthesized by the alkaline hydrolysis of the corresponding xanthate esters as described by Tarbell and Fukushima.[12] Thin-layer chromatography and liquid chromatography analysis are used to evaluate the purity of the recrystallized ester LCs. (Note that 7-4 is distilled, b.p. 220° at 0.05 mm.) The impurity content of the LC compounds is less than 0.5%, although the R(CN)-ϕOR' compounds and some of the commercial samples may be slightly less pure. The LC eutectic mixtures used in these studies are listed in Table II. The

Table I. Liquid Crystal Compounds Used as Additives in
          Viscosity Studies

Class Code	Compound Code	Molecular Length, Å	Melting Point, °C	Clearpoint, °C	Source
R(CN)-φOR'	4(CN)-φ03	25.82	103	137	HRL[c]
	4(CN)-φ05	28.14	80	126	HRL[c]
R(CN)-φR'	7(CN)-φ5	31.50	44	103	Merck(S1014)
R-φOR'	4-φ03	25.82	118	196	HRL
	4-φ05	28.14	92	165	HRL
R-(Cl)OOCφR'	2-(Cl)OOCφ2	23.79	60	126	HRL
	5-(Cl)OOCφ5	31.03	40	122	Kodak(11650)
R-OOCφR'	2-OOCφ2	23.79	143	190	HRL
	4-OOCφ4	28.31	88	186	HRL[d]
ROSOR'	10S04	21.20	74	107	HRL[e]
	10S06	23.90	65	100	HRL[e]
	40S05	26.40	84	104	HRL
RO-OR'	10-01	17.52	124	–	HRL
	10-04	21.09	92	79	HRL
	10-06	23.54	93	76	HRL
	20-01	18.72	97	93	HRL
	40-01	21.13	79	80	HRL
	40-05	26.03	71	84	HRL[e]
	60-01	23.60	55	77	HRL[e]
	60-05	28.50	55	84	HRL
NC-R	NC-4	20.18	68	39	HRL[d]
	NC-7	23.57	43	55	HRL[d]
R-φR'	5-φ5	28.98	96	176	Merck(S1011)

Class Code	Compound Code	Molecular Length, Å	Melting Point, °C	Clearpoint, °C	Source
RO-OOCR'	40-OOC4	25.40	67	84	HRL[e]
	60-OOC5	28.92	51	86	HRL[e]
Rφ-R'	3φ-4	25.55	93	182	Merck(S1013)
	5φ-4	27.95	97	172	Merck(S1012)
RSOR'	1S06	22.51	66	78	HRL
R-OR'	1-06	22.25	62	51	HRL[f]
	3-07	25.87	65	57	HRL[f]
	5-06	27.51	50	62	HRL[f]
R-OOCR'	3-OOC5	24.81	48[a]	55[a]	Merck(S1008)
	7-OOC5	28.88	42[b]	61[b]	Merck(S1008)
ROSR'	40S3	22.50	51	82	HRL
RO-R'	40-3	21.78	71	61	HRL[g]
	40-6	25.81	40	49	HRL[g]
	60-5	27.03	41	59	HRL[g]
R-R'	4-1	18.72	36	–	HRL[h]
	5-5	24.16	33	–	EMC[i]
	7-4	25.68	9[b]	15[b]	HRL
	7-6	27.92	32	23	HRL
RO-[C]R'	40-[C]4	23.84	40	70	HRL[j]
	60-[C]5	27.41	32	80	HRL[j]

[a]Lit. value,

[b]Lit. value, R. Steinsträsser, Z. Naturforsch. 27b, 774 (1972).

[c]See ref 7. [d]See ref 8. [e]See ref 4. [f]See ref 3. [g]See ref 1. [h]See ref 9.

[i]Electronic Materials Corp. [j]See ref 2.

Table II.   Liquid Crystal Eutectic Mixtures Used in
            Viscosity Studies

Mixture No.	Components	Length $\bar{L}$, Å	Melting Point, °C	Clearpoint °C
HRL-2N25[a]	20-3, 60-4, 60-01	24.48	0	54
HRL-2N40[b]	10-1, 20-3, 20-5, 40-1, 40-6, 60-01	22.68	0	58
HRL-2N42[c]	10-1, 20-3, 20-5, 40-1, 40-3	20.39	5	58
HRL-2N44[c]	20-5, 40-3, 60-4	24.31	-8	51
HRL-2N46[c]	40-3, 40-6, 60-5	25.92	16	55
HRL-2N48[c]	60-3, 60-5, 80-3, 80-6	27.14	18	56
HRL-2N54[d]	10-01, 10-04, 10-06, 20-01, 40-01, 40-05, 60-01	23.23	(25)[g]	(78)[g]
HRL-2P37[e]	2-(Cl)OOCφ2, 5-(Cl)OOCφ5	28.05	(21)[g]	(124)[g]
Merck S1008[f]	3-OOC5, 7-OOC5	26.12	20	56

[a] Component mole fractions are 0.132, 0.640, and 0.228 respectively.

[b] See ref 6.

[c] See ref 1.

[d] Component mole fractions are 0.045, 0.051, 0.033, 0.078, 0.118, 0.291, and 0.385 respectively.

[e] Component mode fractions are 0.412 and 0.588 respectively.

[f] Components mole fractions are reported to be 0.586 and 0.414.

[g] Calculated values.

component mole fractions are given here for those mixtures not
previously reported in the literature. All of the HRL mixtures
are formulated by calculations with the Schroeder-Van Laar
equation, using heat of fusion and melting points data obtained
by DSC analysis of the components.

III. RESULTS AND DISCUSSION

A.   Effects of Molecular Length

In three series of single class ester mixtures, namely RO-R',
RO-[C]R', and R-OR', the η values in each series increases as the
$\bar{L}$ of the mixtures increase.[1-3] For comparison purposes, these
results are summarized in Figure 2. In each series, $\bar{L}$ was increased by an increase in the average length of just the R and R'
alkyl end groups. The relationship between η and $\bar{L}$ in the RO-R'
series indicates that the increase in molecular length has a
larger effect on η than the relative decrease in overall molecular

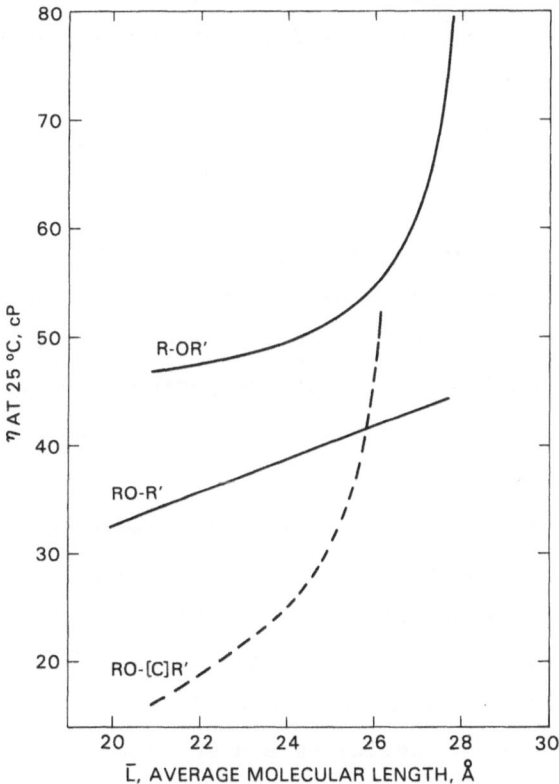

Fig. 2   Effect of $\bar{L}$ on the viscosity of three series of single
          class ester LC mixtures.

polarity caused by the longer alkane groups. Similar effects are
seen in isotropic liquids, such as n-alkyl acetates.[13] We believe
that the large (exponential) increases in the η of the RO-[C]R'
and R-OR' series at longer $\bar{L}$ is due to the strong cybotactic
nematic (short range smectic) characteristics identified in these
mixtures.[2,3] The flow viscosity of a nematic LC usually corresponds
approximately to the Helfrich viscosity η_2 (in which the director
is parallel to the flow direction) when the temperature is well
below the clearpoint.[14,15] The short range smectic ordering in
longer $\bar{L}$ RO-[C]R' and R-OR' mixtures impedes the flow parallel to
the director and increases η. In the shorter $\bar{L}$ (<23Å) of Figure 2
there are no cybotactic characteristics and the effects of
molecular structure can be compared between the three series. Use
of the cyclohexane ring in RO-[C]R', in place of the benzene ring
in RO-R', substantially decreases η. This is probably largely due
to the decreased polarity of the cyclohexane ester carbonyl as
compared to the aromatic ring and its ester carbonyl group. On the
other hand, when the alkoxy group is switched from the phenol side
in RO-R' to the acid side in R-OR' the polarity of the molecules
is increased and η increases. This is largely due to the electron-
donating and conjugation effects of the p-alkoxyl group, which
increases the polarity of the ester carbonyl in the R-OR' structure
as compared to the RO-R' structure.

Since it is difficult to prepare room temperature nematics of
most other single class ester mixtures, we studied the $\bar{L}$ effects
on the η of many ester structures as additives to the RO-R'
mixtures. The detailed data are given in Tables III, IV, V and VI;
graphical summaries are given in Figures 3, 4 and 5. The $\eta_{25°}$
data with 25 mole percent additives are shown in Figure 3, where
the effect of $\bar{L}$ is found to vary with the structure of the
additive ester. In the mixture with R-R' additives, η varies with
$\bar{L}$ as in the 100% RO-R' case. Mixtures with RO-OR' esters show
slightly less dependence of η on $\bar{L}$. The shorter $\bar{L}$ mixtures with

Table III. Effect of $\bar{L}$ on Viscosity: RO-R' With 25% R-R'

R-R' Component	RO-R' Components	$\bar{L}(\text{Å})$	Mixture With 75% RO-R'		
			Clearpoint °C	Viscosity (cP) $\eta 25°$	$\eta 40°$
4-1	HRL-2N42	19.97	45	29.1	15.1
5-5	HRL-2N42	21.33	45	28.5	14.6
7-4	HRL-2N42	21.71	42	31.1	17.3
7-6	HRL-2N42	22.27	42	32.1	16.1
4-1	HRL-2N48	25.03	47	37.7	-
7-6	HRL-2N48	27.34	48	39.9	21.0

Table IV. Effect of $\bar{L}$ on Viscosity: RO-R' With 25% RO-OR'

RO-OR' Components				Mixture With 75% RO-R'		
Compound	% In Mixture	RO-R' Components	$\bar{L}$ (Å)	Clearpoint °C	Viscosity (cP) $\eta 25°$	$\eta 40°$
⎧10-04	10.61⎫	HRL-2N42	20.92	62	46.4	-
⎩10-06	14.39⎭					
60-01	25.00	HRL-2N42	21.19	62	44.5	19.8
⎧40-05	15.00⎫	HRL-2N42	22.05	65	49.3	21.9
⎩60-05	10.00⎭					
60-01	(24.50)[a]	(HRL-2N40)[a]	22.68	58	46.4	20.8
⎧20-01	10.00⎫	HRL-2N44	23.27	57	48.4	22.5
⎩40-01	15.00⎭					
60-01	25.00	HRL-2N44	24.13	57	47.4	22.6
40-05	25.00	HRL-2N44	24.74	60	50.2	25.7
60-05	25.00	HRL-2N44	25.36	61	51.2	26.1
60-01	25.00	HRL-2N48	26.25	60	52.0	23.5

[a]HRL-2N40 contains 24.5 mole percent 60-01; see ref 6.

Table V.   Effect of $\bar{L}$ on Viscosity: RO-R' With 25% NC-R

NC-R Components[a] Additive $\bar{L}$ (Å)	RO-R' Components	$\bar{L}$ (Å)	Mixture With 75% RO-R'		
			Density At 25°C	Viscosity (cP) $\eta_{25°}$	
20.39	HRL-2N42	20.39	1.087	45.6	
21.86	HRL-2N42	20.76	1.070	45.6	
21.86	HRL-2N44	23.70	1.048	49.3	
21.86	HRL-2N46	24.90	1.044	55.3	
21.86	HRL-2N48	25.82	1.038	64.8	

[a]Only NC-4 and NC-7 are used as additives: The 21.86 Å length corresponds to this binary eutectic of 31.8 and 68.2 mole percent, respectively.

Table VI.   Effect of Mole Percent and $\bar{L}$ on Viscosity of Ester Mixtures With R-(Cl)OOCφR'

R-(Cl)OOCφR' Components[a]		RO-R' Components[b]		Total Mixture		Viscosity of Components[c]	
Mole %	$\bar{L}$ (Å)	Mole %	$\bar{L}$ (Å)	$\bar{L}$ (Å)	$\eta_{25°}$ (cP)	RO-R' η_{est}	R-(Cl)OOCR' η_{calc}
10	27.03	90	20.39	21.05	46.2	34.1	155
10	28.05	90	27.14	27.23	51.9	43.6	127
25	27.03	75	20.39	22.05	66.1	35.6	158
25	29.03	75	23.76	25.08	64.1	40.3	136
25	28.05	75	27.14	27.37	65.2	43.8	129
25	31.03	75	27.14	28.11	65.2	45.0	126
50	28.05	50	27.14	27.60	94.8	44.2	145
75	28.05	25	27.14	27.82	155.8	44.5	193
100	28.05	0	–	28.05	249.3	–	249

[a]Components are 2-(Cl)OOCφ2 and 5-(Cl)OOCφ5.

[b]Components are HRL-2N42 and/or HRL-2N48 mixtures.

[c]The RO-R' viscosities are estimated at the $\bar{L}$ of the total mixture from the plot their η vs $\bar{L}$.

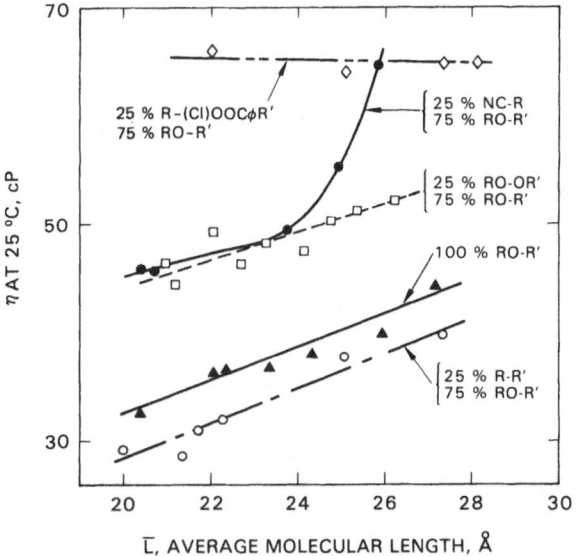

Fig. 3   Effect of $\bar{L}$ on $\eta_{25°}$ of RO-R' mixtures and of RO-R' mixtures with 25% of other LC components.

NC-R additives behave similarly, but η increases sharply at longer $\bar{L}$ values. The η of the mixtures with R-(Cl)OOCϕR' additive decrease slightly as $\bar{L}$ increases. These effects can be understood better by examining Figure 4, in which the $\bar{L}$ effects are shown on the calculated viscosity (η_{calc}) contribution attributed to the added ester in the RO-R' mixture. If x is the mole fraction of the additive, then we determine η_{calc} from equation (1) at the $\bar{L}$ of the overall mixture:

$$\eta_{obs} = \eta_{calc}(x) + \eta_{RO-R'}(1-x) \tag{1}$$

Thus, η_{calc} is the apparent viscosity of the additive in the mixture. (We have found that the most consistent results are obtained when η_{calc} and $\eta_{RO-R'}$ are both taken at the $\bar{L}$ of the overall mixture rather than at the $\bar{L}$ of each class prior to mixing.) The η_{calc} values in Figure 4 are for mixtures in which x is only

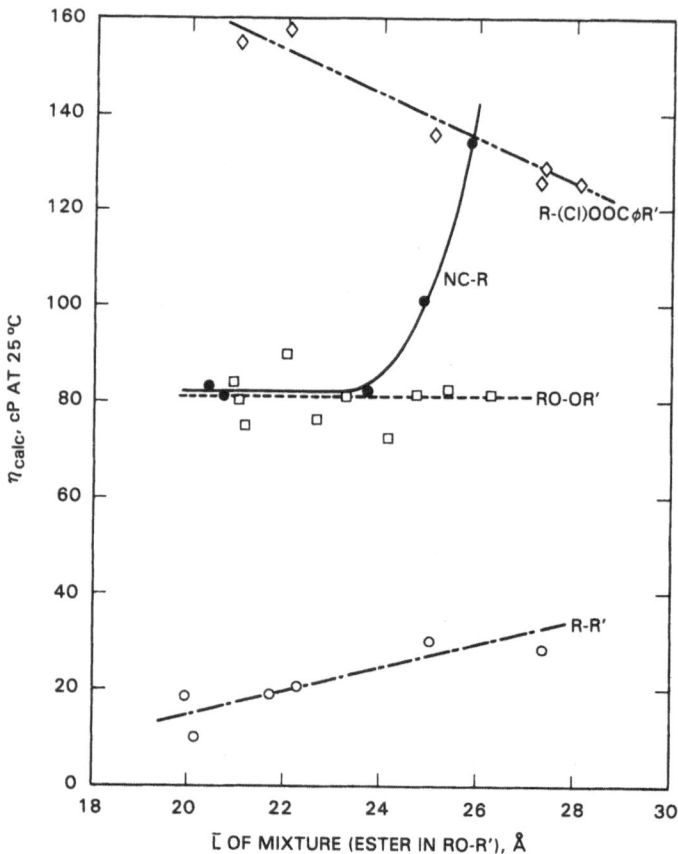

Fig. 4   Effect of $\bar{L}$ on η_{calc} of the additive components in RO-R'
          mixtures.

0.25 or 0.10, so that the additive is in the presence of a large
excess of RO-R' molecules. The results are strongly dependent on
the polarity of the additive ester structure. The η_{calc} of R-R'
esters increases with $\bar{L}$ slightly more steeply than does the η of
the 100% RO-R' mixtures. The R-R' compounds are not very polar:
the dominant effect on η_{calc} is the increased length of the end
groups. With the more polar RO-OR' additives, η_{calc} is approximately constant: the increase in $\bar{L}$ of the mixture is apparently
offset by the dilution of intermolecular dipole interactions as

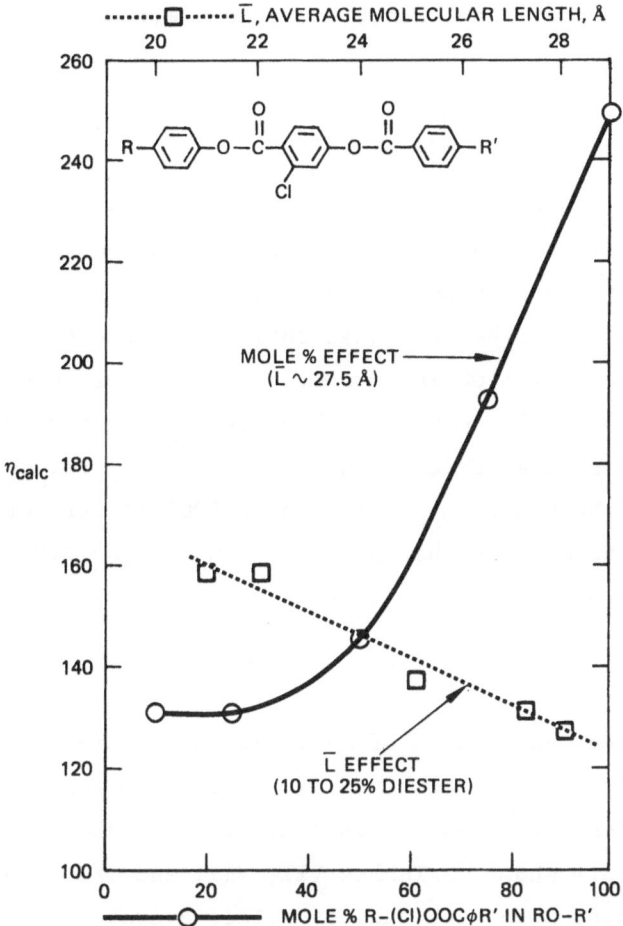

Fig. 5   Effect of $\bar{L}$ and concentration of a diester on its η_{calc} from $\eta_{25°}$ in RO–R' mixtures.

the non-polar end groups are lengthened. The NC–R additives are strongly polar and their η_{calc} is independent of $\bar{L}$ in the shorter mixtures; however, it increases sharply at longer $\bar{L}$ where cybotactic nematic structures are formed. This is a behavior related to the induced smectic phases observed by others[16-18] when strongly polar LCs are mixed with relatively non-polar LCs. In the mixtures with R–(Cl)OOCϕR' additives, the η_{calc} decreases as $\bar{L}$ increases: presumably, decreased dipole interactions occur as the alkyl end groups are lengthened and this overrides the effects of longer $\bar{L}$.

IV.   EFFECT OF ADDITIVE CONCENTRATION

The η_{calc} value of R-(Cl)OOCϕR' at a constant $\bar{L}$ strongly increases as its concentration in the mixture increases when more than 25 mole percent is used. This is shown by the data in Table VI and Figure 5. We believe that this is because the polar inter- actions between R-(Cl)OOCϕR' molecules is greater than those between R-(Cl)OOCϕR' and RO-R' molecules. As indicated below, we have observed similar effects in comparing η_{calc} for RO-OR' in 75% RO-R' versus the higher η_{obs} value of a 100% mixture of RO-OR' components, and also in the increase of η_{calc} for R(CN)-ϕOR' when the RO-R' components are 75% instead of 90%.

V.   EFFECTS OF STRUCTURE

We studied the effects of molecular structure by measuring the η of various ester additives (10 or 25 mole percent) with fixed $\bar{L}$ in RO-R' mixtures of fixed $\bar{L}$. Three such series of measurements are shown in Tables VII, VIII and IX. Table VII is the main series, with each additive class averaging 27.03$\overset{o}{A}$ in length and 25% added to HRL-2N42. Some of the components are not soluble enough for 25% mixtures, so the Table VIII series uses 10% of such length additives (27.03$\overset{o}{A}$) added to HRL-2N42. There is some overlap of classes in Table VII and VIII in order to compare concentration and $\bar{L}$ effects. The series in Table IX is measured to compare esters with the corresponding thioesters (of which we have available only relatively short length components) with 25% additives of 22.50$\overset{o}{A}$ added to HRL-2N42. In each table the additives are listed in decreasing η. The overall results are compared as η_{calc} values in Table X, and our assigned class viscosities are shown in Figure 1. These assigned η_{class} values are specifically

Table VII.   Structure Effects on Viscosity: Mixtures With 75% RO-R' and 25% Additives

25% Additive Components[a]			Mixture With 75% HRL-2N42[b]			
Class	Components	Mixture Mole %	Clearpoint °C	Density at 25°C	Viscosity, cP $\eta_{25°}$	$\eta_{40°}$
R(CN)-φOR'	4(CN)-φ03	12.28 }	76	1.103	118.0	44.7
	4(CN)-φ05	12.72				
R-(Cl)OOCφR'	2-(Cl)OOCφ2	13.37 }	73	1.120	66.1	28.3
	5-(Cl)OOCφ5	11.63				
R-OOCφR'	2-OOCφ2	6.72 }	88	1.105	60.6	26.0
	4-OOCφ4	18.28				
RO-OR'	40-05	15.00 }	65	1.083	49.3	21.9
	60-05	10.00				
RO-OOCR'	40-OOC4	13.64 }	65	1.087	45.6	20.7
	60-OOC5	11.36				
Rφ-R'	3φ-4	5.00 }				
	5φ-4	7.20 }	92	1.079	45.4	20.6
R-φR'	5-φ5	12.80 }				
R-OR'	5-06	25.00	58	1.065	39.8	18.8
R-OOCR'	3-OOC5	14.65 }[c]	56	1.080	38.0	18.1
	7-OOC5	10.35				
RO-R'	60-5	25.00	58	1.052	36.3	17.4
R-R'	7-4	15.07 }	42	1.058	31.7	16.6
	7-6	9.93				
RO-[C]R'	40-[C]4	2.82 }	62	1.056	31.1	15.5
	60-[C]5	22.18				

[a]Additives have $\bar{L}$ = 27.03 Å,

[b]Mixtures with additives have $\bar{L}$ = 22.05 Å; HRL-2N42 alone has $\bar{L}$ = 20.39 Å and $\eta_{25°}$ = 32.6,

[c]The R-OOCR' additives have $\bar{L}$ = 26.12 Å, and the mixture $\bar{L}$ = 21.82 Å.

for about 10 to 25% of the class structure in RO-R' mixtures with an overall average $\bar{L}$ of 22Å.

Many interesting correlations of η_{class} with structural features are indicated in Figure 1, including some effects already known from other studies. These semi-quantitative results clearly

Table VIII. Structure Effects on Viscosity: Mixtures with 90% RO–R' and 10% Additives

10% Additive Components[a]			Mixture With 90% HRL-2N42[b]
Class	Components	Mole % In Mixture	Viscosity $\eta_{25°}$, cP
R(CN)–ϕOR'	4(CN)–ϕ03	4.91 ⎫	57.0
	4(CN)–ϕ05	5.09 ⎭	
R–ϕOR'	4–ϕ03	4.91 ⎫	47.0
	4–ϕ05	5.09 ⎭	
R–(C1)OOCϕR'	2–(C1)OOCϕ2	5.35 ⎫	46.2
	5–(C1)OOCϕ5	4.65 ⎭	
R–OOCϕR'	2–OOCϕ2	2.69 ⎫	43.6
	4–OOCϕ4	7.31 ⎭	
ROSOR'	40S05[c]	10.00	41.3[c]
RO–OR'	40–05	6.00 ⎫	39.0
	60–05	4.00 ⎭	
Rϕ–R'	3ϕ–4	2.00 ⎫	
	5ϕ–4	2.88 ⎬	37.9
R–ϕR'	5–ϕ5	5.12 ⎭	
Rϕ–R'	3ϕ–4	3.70 ⎫	37.4
	5ϕ–4	6.30 ⎭	
RO–R'	60–5	10.00	34.8

[a]Additives have $\bar{L}$ = 27.03 Å

[b]Mixtures with additives have $\bar{L}$ = 21.05 Å

[c]Mixture contains 89.1% 2N42 and 0.9% 60-5 to give $\bar{L}$ = 21.05 Å

show the following relationships for the η_{class} of the same length ester structures used in relatively short length RO–R' ester mixtures:

Factor Decreasing η_{class}	Examples (Relative Size of η_{class})
• Replace benzoate with cyclohexane-carboxylate	RO-R' > RO-[C]R'
• Replace p-alkoxy with alkyl	RO-R' and R-OR' > R-R' RO-OOCR' > R-OOCR' RO-OR' > RO-R' and R-OR' ROSOR' > ROSR' and RSOR' R-φOR' > R-φR R(CN)-φOR' > R(CN)-φR'
• Move p-alkoxy from acid to phenol side	R-OR' > RO-R'
• Replace aromatic with aliphatic groups	Rφ-R' and R-φR' > R-R' R-φOR' > R-OR' R-OOCφR' > R-OOCR'
• Replace p-CN by alkoxy or alkyl	NC-R > RO-R' and R-R'
• Replace o-CN by hydrogen	R(CN)-φOR' > R-φOR' R(CN)-φR' > R-φR'
• Replace o-Cl by hydrogen	R-(Cl)OOCφR' > R-OOCφR'
• Replace thiobenzoate by benzoate	ROSOR' > RO-OR' RSOR' > R-OR' ROSR' > RO-R'
• Replace benzoyloxy with phenyl	R-OOCφR' > R-φOR'
• Replace phenyl with acyloxy	R-φR' > R-OOCR'

These results are partially consistent with the general rule that η is increased by more polar structures. Thus large increases in η are observed from strong dipolar groups (e.g. cyano), and by groups which increase the molecular dipole and polarizability (e.g. alkoxy para to the ester carbonyl). On the other hand, different factors must be considered in other cases. For example, the p-alkyphenyl substituent in R-φR' gives a larger η than does the more polar p-acyloxy group in R-OOCR'. The relatively large effect on η by phenyl groups (i.e. biphenyl esters) may be due to a combination of effects such as a larger molecular volume, more of a bent-rod molecular shape, and increased intermolecular association as compared to acyloxy and alkoxy groups. The high η_{class} values of the biphenyl esters is seen on either end of the ester substrate, i.e.,: R-φR' > R-OR' > R-OOCR' > R-R' and also

Table IX.   Effects of Thiobenzoates Versus Benzoates on Viscosity
            of Mixtures with 75% RO-R'

| | 25% Additive Components[a] | | Mixture With 75% HRL-2N42[b] | |
Class	Cmpds	Mole % In Mixture	Viscosity $\eta_{25°}$, cP	Density at 25°C
ROSOR'	10S04	12.96 ⎫	49.0	1.106
	10S06	12.04 ⎭		
RO-OR'	10-04	10.61 ⎫	46.4	1.098
	10-06	14.39 ⎭		
RSOR'	1S06	25.00	42.5	1.088
R-OR'	1-06	23.27 ⎫	42.0	1.081
	3-07	1.73 ⎭		
ROSR'	40S3	25.00	34.9	1.079
RO-R'	40-3	20.53 ⎫	34.7	1.073
	40-6	4.47 ⎭		

[a]Additives have $\bar{L}$ = 22.50 $\overset{\circ}{A}$.

[b]Mixtures with additives have $\bar{L}$ = 20.92 $\overset{\circ}{A}$.

Rϕ-R' > RO-R' > R-R'.

VI.   PREDICTING η OF NEW MIXTURES

We find that the $\eta_{25°}$ of short length ester LC mixtures can be
estimated with fairly good accuracy by using equation (2), where
x_i is the total mole fraction of each class.

$$\eta_{mix} = \Sigma (\eta_{class})_i \cdot x_i \tag{2}$$

The calculations from equation (2) are often more accurate than
estimates made by taking the average of the actual ηs of the pure
components.   For example, in mixing 25% of the R-(Cl)OOCϕR'
LC HRL-2P37 with 75% of the RO-R' LC HRL-2N48 the η_{25} is 65.2 cP

Table X. Calculated Viscosity of Additives When Used in RO-R' Mixture

	25%	25%	25%	10%	10 to 25%
Amount of Additives[a]	25%	25%	25%	10%	10 to 25%
$\bar{L}$ of Additives, Å	22.50	27.03	Various	27.03	Various
$\bar{L}$ of Mixture, Å	20.92	22.05	23.00	21.05	~22
η_{calc} at 25°C					η_{class}[c]
R(CN)-φOR'		363		257	310
R(CN)-φR'			(209)[b]	(192)[b]	200
R-φOR'				157	160
R-(Cl)OOCφR'		156	152	149	153
R-OOCφR'		134		~123	130
ROSOR'	92				92
RO-OR'	82	88	80	~77	82
NC-R			82		82
RO-OOCR'		74			74
Rφ-R';R-φR'		73		~66	71
RSOR'	66				66
Rφ-R'				~61	65
R-OR'	64	50			55
R-OOCR'		43			43
ROSR'	36				37
RO-R'	[35]	[36]	[37]	[35]	36
R-R'		18	22		19
RO-[C]R'		16			16

[a] In HRL-2N42.

[b] Addition of 7(CN)-φ5, whose L = 31.50 Å, gives a 25% mixture with 79.9 cP and 23.17 Å and a 10% mixture with 50.2 cP and 21.50 Å.

[c] Assigned class viscosity at 25°C in RO-R' mixtures.

and the calculated η_{mix} is 65.2 cP, whereas the η_{av} is 95.0 cP. On the other hand, differences are noted between our assigned η_{class} values and the η of those single class mixtures with more polar structures. This is shown in Table XI-A, where the $\eta_{25°}$, of 100% R-(Cl)OOCφR' and of 100% RO-OR' is much larger than the η_{class}

Table XI.   Comparison of Calculated vs Observed Viscosities of
            Ester Eutectic Mixtures

A.   Single Class Mixtures

Class	Mixture	$\overline{L}$ (Å)	$\eta_{25°}$(cP)	Assigned Viscosity[a]	
				η_{class}	Variation
R-(Cl)OOCϕR'	HRL-2P37[b]	28.05	249	153	-39%
RO-OR'	HRL-2N54[b]	23.23	106	84	-21%
R-OR'	HRL-2P15[c]	21.93	50.3	55	+9%
R-OR'	HRL-2P17[c]	22.15	46.6	55	+18%
R-OOCR'	Merck S1008[b]	26.12	41.0	43	+5%
RO-R'	HRL-2N43	22.37	36.5	36	-1%
RO-[C]R'	HRL-6N7[e]	21.20	16.3	16	-2%
RO-[C]R'	HRL-6N5[e]	22.71	20.5	16	-22%

B.   Multiple Class Mixtures

Mixture	No. Classes	% RO-R'	$\overline{L}$ (Å)	$\eta_{25°}$(cP)	Calc. Viscosity[a]	
					η_{calc}	Variation
HRL-2N25[b]	2	77.2	24.48	48.5	47	-3%
HRL-2N40[f]	2	75.5	22.68	46.4	48	+3%
HRL-2N52[g]	3	38.4	24.90	64.2	63	-2%
HRL-25N4[g]	4	42.0	23.92	59.2	58	-2%
HRL-26N3[g]	4	26.6	26.41	47.8	49	+2%
HRL-26N4[g]	4	22.8	23.29	49.3	54	+9%
HRL-246N1[g]	4	18.1	26.28	66.6	80	+20%
HRL-256N5[g]	6	14.9	23.32	58.2	59	+1%

[a]Using class viscosities calc. from η in RO-R' mixtures.   [b]This paper.
[c]See ref. 3.   [d]See ref. 1.   [e]See ref. 2.   [f]See ref. 6.   [g]See ref. 4.

value determined from 10 to 25% of these structures in RO-R'
mixtures. As noted above, the high η values of the 100% polar LCs
is attributed to increased intermolecular effects as compared to
their interaction when mixed with the less polar RO-R' esters. The
less polar single class mixtures, such as R-OR', R-OOCR', and
RO-[C]R' show better agreement with the η_{class} values.

    Multiple class mixtures containing some RO-R' components have

$\eta_{25°}$ values that are within 20% of the value predicted by our η_{mix} calculation. As illustrated in Table XI-B, the η_{mix} values are often quite accurate. We find that when the calculated and observed values are not very close, as in the HRL-26N4 and HRL-246N1 mixtures, the η_{mix} is generally higher than the η_{obs}. Thus, we use the η_{mix} as an approximate upper limit for estimating the $\eta_{25°}$ of new ester mixtures -- particularly those mixtures designed to have low viscosity and negative dielectric anisotropy.

VII. TEMPERATURE EFFECTS

Although 100% mixtures of RO-R' esters showed a linear plot of log η versus T^{-1}, other ester mixtures studied previously[2-4] as well as the ones in this study do not give a linear equation (3) plot. Instead,

$$\log \eta = \log A + E_\eta/RT \qquad (3)$$

the η values above room temperature are larger than expected from the variations near 25°C. This is the case even well below the clearpoint. (The η increases sharply at the clearpoint, in a manner similar to changes observed for η_2 measurements.[15]) For comparison purposes, we have calculated the apparent E_η values from the slope of equation (3) curves between the fixed temperatures of 25° and 40°C. These data are summarized in Figure 6 which includes results from prior papers.[1-3] The E_η values in the 100% RO-R' series (various $\bar{L}$s) does not vary appreciably. However, in the 100% RO-[C]R' series (various $\bar{L}$) and in the Table VII mixtures (constant $\bar{L}$, with 25% additive ester and 75% RO-R') the apparent E_η increases linearily with log $\eta_{25°}$. In fact, all of the data points for all of the short length mixtures in Figure 6 (the dark points) are roughly correlated by this relationship, with apparent E_η ranging from 6.7 to 12.0 kcal/mole. This indicates that in the 25° to 40°C range the change of η with temperature is greatest for the most viscous ester mixtures and smallest for the least viscous

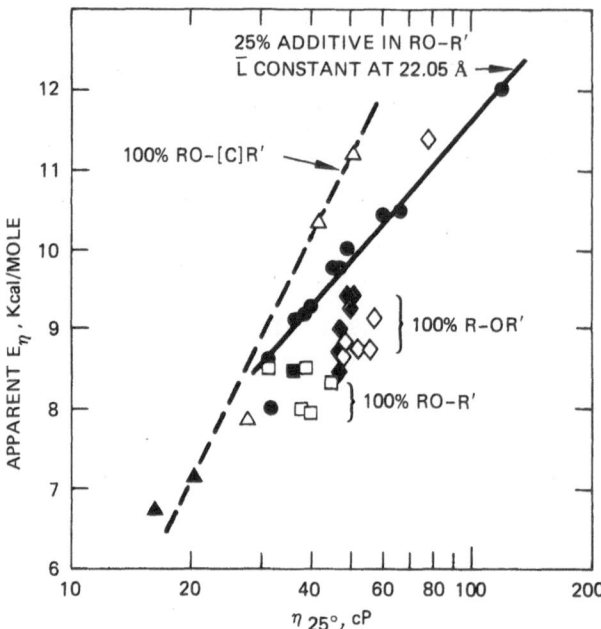

Fig. 6   Dependence of apparent activation energy of viscosity
          between 25 and 40°C on $\eta_{25°}$.

esters. We have also calculated an apparent E_η' for the mixtures
between fixed η values of 30 and 15 cP and the corresponding
(variable) temperature. These E_η' values range between 7.7 and
9.6 kcal/mole, with an average value of 8.5 kcal/mole. We have
found no obvious relationship between these apparent E_η' values and
the component structures, nor the $\bar{L}$ of the mixtures.

VIII. CONCLUSIONS

These studies give semi-quantitative values to the viscosity
characteristics of 18 different classes of LC ester structures.
When new LC mixtures are made from among these classes of esters,
the $\eta_{25°}$, can be predicted with fairly good accuracy. The
calculated value generally provides an upper limit for η, and is
often within 10% of the actual value of the mixture. Our

correlations of η_{class} with various structural features also provide a good basis for rough estimates of the η contributions of additional classes of ester LCs. The formulation of LC mixtures for electro-optical applications, generally involves trade-off of properties e.g. for a mixture with wide nematic range, low η, negative dielectric anisotropy ($\Delta\epsilon$) and high conductivity anisotropy for DS devices. In general the lowest η values are obtained with short $\bar{L}$ mixtures of the less polar esters. Short $\bar{L}$ mixtures also give less cybotactic nematic effects. The cyclo-hexanecarboxylates, RO-[C]R', are one of the few structures with low η, negative $\Delta\epsilon$, and fairly high clearpoints. The acyloxy esters (R-OOCR' and RO-OOCR') have relatively low ηs for polar structures of negative $\Delta\epsilon$. In contrast, the o-cyanophenyl biphenylcarboxylates such as R(CN)-ϕR' have strongly negative $\Delta\epsilon$ and high clearpoints, but they also have such high η_{class} values that even in small amounts they greatly increase the η of most mixtures. Relatively low η ester mixtures for DS are made by including components from classes such as RO-[C]R', RO-R', and R-OOCR'. The lower η mixtures also show less temperature dependence on η than do the higher η mixtures.

ACKNOWLEDGEMENTS

We are indebted to the Directorate of Chemical Sciences, Air Force Office of Scientific Research, Contract F49620-77-C-0017, for partial financial support of this research; to F. G. Yamagishi for assistance in preparing o-cyano components; and to W. H. Smith, Jr., for assistance with some DSC measurements.

REFERENCES

1. J. D. Margerum, J. E. Jensen, and A. M. Lackner, Mol. Cryst. Liq. Cryst. <u>68</u>, 137 (1981). (Paper I of this series.)

2.   J. D. Margerum, S. -M. Wong, A. M. Lackner, and J. E. Jensen,
      Mol. Cryst. Liq. Cryst. 68, 157 (1981). (Paper II of
      this series.)

3.   J. D. Margerum, S. -M. Wong, A. M. Lackner, J. E. Jensen, and
      S. A. Verzwyvelt, Mol. Cryst. Liq. Cryst., 84, 79
      (1982). (Paper III of this series.)

4.   J. D. Margerum and A. M. Lackner, Mol. Cryst. Liq. Cryst. 76,
      211 (1981).

5.   H. S. Lim, J. D. Margerum, and A. Graube, J. Electrochem. Soc.
      124, 1389 (1977).

6.   H. S. Lim and M. J. Little, U. S. Pat. 4129312 (Dec. 5, 1978).

7.   F. G. Yamagishi, L. J. Miller, J. E. Jensen, and J. D. Margerum,
      U. S. Pat. #4,225,454 (Sept. 30, 1980).

8.   Y. S. Lee, P. Y. Hsieh, and J. E. Jensen, U. S. Pat. #4,000,084
      (Dec. 28, 1976).

9.   S. -Y. Wong, U. S. Pat. #3,826,757 (July 30, 1974).

10.   D. J. Bryon, G. W. Gray, and R. C. Wilson, J. Chem. Soc., C,
      840 (1968).

11.   W. W. Lawrence, Tetrahedron Lett. 3464 (1971).

12.   D. S. Tarbell and D. K. Fukushima, Org. Synthesis, Col. Vol.
      III, John Wiley, New York, 1955, p. 809.

13.   Handbook of Chemistry and Physics, R. D. Weast, Ed., CRC
      Press, Boca Raton, FL, 1979, 60th Ed., pp. F52-57.

14.   A. E. White, B. E. Cladis, and S. Torza, Mol. Cryst. Liq.
      Cryst. 43, 13 (1977).

15.   W. H. deJeu, Physical Properties of Liquid Crystalline
      Materials, Gordon and Breach, New York, 1980, p. 121.

16.   C. S. Oh, Mol. Cryst. Liq. Cryst. 42, 1 (1977).

17.   G. Heppke and E. -J. Richter, Z. Naturforsch. 33a, 185 (1978).

18.   M. Bock, G. Heppke, E. -J. Richter, and F. Schneider, Mol.
      Cryst. Liq. Cryst. 45, 221 (1978).

ORIENTATIONAL DISORDER IN SMECTIC LIQUID CRYSTALS: AN ASPECT OF STRUCTURE THAT HAS TO BE INCLUDED[*]

Adriaan de Vries

Liquid Crystal Institute
Kent State University
Kent, Ohio 44242

ABSTRACT

A few years ago, we have shown that the generally observed difference between layer thickness and molecular length in monolayer smectic A phases can be adequately explained by orientational disorder. We want to emphasize here that this new explanation does not compete on an equal level with the older explanations involving special tilted structures, interdigitation, or kinking of the chains. Rather, the orientational-disorder effect takes precedence over all other explanations: Since it is certain that orientational disorder exists, and since it has been shown that the influence of this disorder on the observed layer thickness is quite significant, <u>the effects of orientational disorder have to be taken into account first.</u> This is true not only for smectic A phases, but also for all other liquid crystals with smectic order, and a simple calculation of the tilt angle in smectic C phases or skewed cybotactic nematic phases from layer thickness and molecular length, without considering orientational disorder, is incorrect. Moreover, the influence of orientational disorder goes beyond its effect on layer

[*] Research supported by NSF under grant No. DMR-78-26495.

137

thickness, and in these other areas, too, the effects of orientational disorder should be taken into account first, before further
theories are advanced or calculations are made. For instance,
the disorder will influence the intensities of the x-ray reflections, it will make the smectic layers appear less well-defined,
and it will facilitate the mixing of molecules of different lengths.

I.   INTRODUCTION

When smectic liquid crystal phases were first discovered and
described, it was thought that in all smectic liquid crystals the
molecules were approximately perpendicular to the smectic layers.[1,2]
From this it was logical to conclude that the layer thickness would
be approximately equal to the length of the molecule, and early experimental data appeared to confirm this for most compounds (Ref. 3,
pp. 81,82). Later and more accurate experiments have shown, however, that for monolayer smectic $\underline{A}$ phases--the most common kind of
smectic phases--the layer thickness $\underline{d}$ is significantly less than
the length $\underline{\ell}$ of the molecule in its most extended conformation.[4-7]
Several hypotheses were advanced to explain this difference between
$\underline{d}$ and $\underline{\ell}$: interdigitation,[4] kinking,[6,8] or tilting.[6,9,10] More recently, however, Leadbetter and co-workers[11-13] pointed out that
the mere existence of orientational disorder of the long axis in
smectic $\underline{A}$ phases will already cause $\underline{d}$ to be less than $\underline{\ell}$. De Vries,
Ekachai, and Spielberg[14] gave a detailed model for the incorporation
of the orientational order parameter $\underline{S}$ in the calculation of $\underline{d}$
from $\underline{\ell}$, and showed that for the smectic $\underline{A}$ phases of four quite
different compounds the differences between $\underline{d}$ and $\underline{\ell}$ could be explained completely by the effect of $\underline{S}$. Their model was subsequently expanded to deal with smectic $\underline{C}$ phases[15] and skewed cybotactic
nematic phases.[16,17] In the present paper, we will develop the
thesis, first advanced by de Vries et al.,[14] that in discussions
of observed $\underline{d}$ spacings one should always first take into account
the effect of $\underline{S}$, and that conclusions drawn from $\underline{d}$ data should

always be based on data corrected for the influence of orientational disorder. This will mean that much previous and current work will have to be re-evaluated.

II. SMECTIC A PHASES

For ordinary monolayer smectic A phases, it appears that the observed differences between d and ℓ can be completely explained by the effect of orientational disorder.[11-14,18] Thus, the d data no longer support any hypotheses for interdigitation, kinking, or special tilted structures.[14] This does not mean, of course, that these effects could not be present, but it does mean that the d data do not provide any evidence for them, contrary to earlier reports.[4-10]

For the recently extensively studied bilayer smectic A phases (those formed by molecules with a CN or NO_2 group at one end), inclusion of the effect of S is important, too. It means, e.g., that the statement of Litster et al.,[19] that for three compounds the layer thickness is "slightly less than twice the molecular length", might lose its perceived significance. Also, as pointed out by Leadbetter et al.,[20] reported values for the ratio d/ℓ should be treated with caution. They should not be interpreted as representing $d_{bilayer}/d_{monolayer}$, and using them as a basis for the calculation of the extent of molecular overlap within the layer[21] leads to errors. For instance, with d = 35.0 Å and ℓ = 25.0 Å, d/ℓ = 1.40. The layer thickness to be expected for a monolayer, however, would be 22.1 Å rather than 25.0 Å (if we take S = 0.70), so that d_{bi}/d_{mono} = 35.0/22.1 = 1.58 rather than 1.40. This is a quite significant difference.

III. SKEWED CYBOTACTIC NEMATIC PHASES

In his first paper on the structure of the skewed cybotactic nematic (N_{sc}) phase, de Vries[22] reported that his data on layer

thickness and tilt angle could be explained with a molecular length
of 38 Å, and that this length corresponded with the molecular
length that would be expected if the alkyl chains were in the cog-
wheel conformation proposed by Gray (Ref. 3, p. 214). The molecular
length appeared to be constant over a temperature range of 95°C.
Later, these results were essentially confirmed by the more accurate
data obtained by de Vries and Qadri.[23] From these data it can be
calculated that the tilt angle $\tau_{calc} = \cos^{-1} d/\ell$ agrees with the
observed tilt angle τ_{obs} to within 0.06° over a temperature range
of 76°C, with $\underline{\ell}$ = 39.20 Å (Table I). This extremely close agreement between τ_{obs} and τ_{calc} over such a long temperature range is
quite remarkable, and might be considered strong evidence for the
correctness of the model used for the determination of τ_{calc}. A
similar conclusion could be drawn from a new set of data obtained
recently by Sethna et al.[16,24] on the same compound as studied by
de Vries and Qadri. From the data given by Sethna et al., one ob-

TABLE I

Layer thickness and tilt angle data for an N_{SC} phase, based on the
work of de Vries and Qadri[23]

$t(°C)$ [a]	$d(Å)$ [a]	$\tau_{obs}(°)$ [a]	$\tau_{calc}(°)$ [b]	$\Delta\tau(°)$ [c]
60	27.04	46.43	46.38	+0.05
79	27.52	45.40	45.40	0.00
98	28.00	44.38	44.41	−0.03
117	28.75	42.76	42.82	−0.06
136	29.87	40.39	40.35	+0.04
155	31.36	36.06	36.86	−0.80
174	33.27	23.10	31.92	−8.82

[a] Data from Ref. 23.

[b] Calculated with ℓ = 39.195 Å.

[c] $\Delta\tau = \tau_{obs} - \tau_{calc}$.

tains differences of about 0.1-0.2° between τ_{obs} and τ_{calc} over a temperature range of 64°C (Table II). This means "perfect agreement", in view of the fact that the error in the τ_{obs} data is also 0.1-0.2°.[16] The data yield $\ell = 39.29$ Å, in close agreement with the value of 39.20 Å referred to above.

TABLE II

Layer thickness and tilt angle data for an N_{SC} phase, based on the work of Sethna et al.[16,24]

$T(K)$ [a]	d_{obs} (Å) [a]	τ_{obs} (°) [b]	τ_{calc} (°) [c]	$\Delta\tau$(°) [d]	R [e]
335.5	- - -	- - -	- - -	- - -	- - -
336.2	26.91	46.88	46.77	+0.11	1.002
343.0	27.06	46.48	46.47	+0.01	1.000
350.8	27.26	46.19	46.07	+0.12	1.003
359.0	27.48	45.71	45.62	+0.09	1.002
368.3	27.70	45.16	45.17	-0.01	1.000
378.2	28.03	44.47	44.49	-0.02	1.000
388.6	28.39	43.62	43.73	-0.11	0.997
399.8	29.02	42.20	42.39	-0.19	0.996
411.7	29.50	42.07	41.34	+0.73	1.018
424.2	30.41	38.62	39.29	-0.67	0.983
437.5	31.92	35.39	35.67	-0.28	0.992
453.7	- - -	- - -	- - -	- - -	- - -

[a] Data from Ref. 16, p. 20.

[b] Data correspond to Ref. 16, p. 20, but contain one more figure after the decimal point; also, a typographical error in the τ value in Ref. 16 for $T = 399.8$K has been corrected.

[c] Calculated with $\ell = 39.29$ Å.

[d] $\Delta\tau = \tau_{obs} - \tau_{calc}$.

[e] $R = \tau_{obs}/\tau_{calc}$.

Nevertheless, considering that all the above calculations completely ignore the effects of orientational disorder, these beautiful agreements between τ_{obs} and τ_{calc} must be regarded as merely
fortuitous, and the values of 39.20 Å and 39.29 Å for ℓ as rather
meaningless. Let us take, for example, the data from Sethna et
al.[16,24] at the low temperature end of the nematic range,[†] and
examine the effect of S on the conclusions that can be drawn from

---

[†] We chose the lowest temperature because at this temperature the
N_{SC} phase will most closely resemble the smectic C phase, and,
therefore, the orientational-disorder models, which were first
developed for smectic phases, will be most likely to be applicable.
Also, we want to establish whether or not the influence of orientational disorder is significant. At the lowest temperature, the
amount of disorder will be least, and the effect of including S in
the calculations will be smallest. Therefore, if the effect of S
is significant at the lowest temperature, it may be expected to be
even more so at the other temperatures.

TABLE II

An example of the effect of S on data for a skewed
cybotactic nematic phase

T[a]	d_{obs}[a]	τ_{obs}[a]	$d_{calc}(S=1)$[b]	$d_{calc}(S=0.826)$[c]	$\Delta d(S=1)$[d]	$\Delta d(S=0.826)$[d]
336.2K	26.91 Å	46.88°	28.56 Å	26.89 Å	-1.65 Å	0.02 Å

[a] Data from Table

[b] This d_{calc} is equal to $\ell\cos\tau_{obs}$, with ℓ=41.79 Å = the length of the molecule
in its most extended conformation.[16]

[c] This d_{calc} is obtained from ℓ and τ_{obs} with a simple orientational-disorder
model[16,25] using the S calculated from the Maier-Saupe theory[26]
($\Delta T = T_{NI} - T = 117.5K$; $S_{NI} = 0.437$).

[d] $\Delta d = d_{obs} - d_{calc}$.

these data (Table III). We note two things: (1) Even at this
relatively low temperature the influence of $\underline{S}$ is quite large: the
change in d_{calc} is 1.67 Å. (2) At this temperature, a simple orientational-disorder model[16,25] appears to give a very adequate explanation* of the N_{sc} phase: d_{calc} for this model is practically
equal to d_{obs}. These results confirm the proposition made above,
viz., that the agreement between τ_{obs} and τ_{calc} obtained with a
simplistic molecular-tilt model (i.e., ignoring the effect of orientational disorder, or, in other words, assuming S=1) is fortuitous
and does not mean that this simplistic model is valid. Rather, this
model leads to a too low value for $\underline{\ell}$ (ℓ = 39.3 Å), whereas the
calculations including the effect of orientational disorder show

----

*A somewhat more complicated orientational-disorder model[16,17]
appears to be necessary to explain the data at the higher temperatures, but this model is quite reasonable in light of the structure of the N_{sc} phase.[17]

TABLE IV

Examples of the effect of $\underline{S}$ on data for smectic $\underline{C}$ phases

Compound[a]	τ_{obs}[a]	d_{obs}[a]	ℓ_{calc}(S=1)[b]	ℓ_{model}[a]	ℓ_{calc}(S=0.70)[c]
HOAB	32°±2°	24.0±0.4 Å	28.3 Å	31.6 Å	32.7
OOAB	30°±2°	27.7±0.4 Å	32.0 Å	34.0 Å	36.9

[a]Data from Ref. 13; ℓ_{model} is the molecular length in the most extended
conformation.

[b]This ℓ_{calc} is equal to $d_{obs}/\cos\tau_{obs}$.

[c]This ℓ_{calc} is obtained from d_{obs} and τ_{obs} with an orientational-disorder
model[15] using S=0.70 (the $\underline{S}$ value given in Ref. 13 for OOAB).

that the data agree with an effective molecular length equal to the
length of the molecule in its most extended conformation (ℓ =
41.8 $\overset{\text{o}}{\text{A}}$).

It might be argued that a few kinks in the alkyl chains should
be expected, and that, therefore, the molecular length should be
less than the length ℓ of the fully extended molecule (i.e., less
than 41.8 $\overset{\text{o}}{\text{A}}$). However, as pointed out above, the length of 38 $\overset{\text{o}}{\text{A}}$
corresponds to a cog-wheel conformation of the alkyl chains, and
this would involve far too many kinks. Therefore, the length
should be greater than 38 $\overset{\text{o}}{\text{A}}$.   Also, in all calculations done on
smectic A phases,[11-14,18] it is found that the assumption of a fully
extended molecule leads to very reasonable S values; significantly
shorter molecular lengths would lead to S values that would be far
too high. Further, in our x-ray work on the isotropic phases of
mesogens, we have found that, as the temperature approaches the
clearing point (i.e., as one approaches the liquid crystal phase),
the molecular length calculated from the x-ray data approaches that
of the fully extended molecule.[7] Thus, all these x-ray data indicate that in the liquid crystal phase the effective molecular
length (i.e., that number that indicates how much space the molecule occupies in the direction of its long axis) is very adequately
given by the length of the molecule in its most extended conformation (see also our discussion of the smectic C data from Leadbetter
and Norris in Section IV).

IV. SMECTIC C PHASES

Since the smectic C phase is almost as disordered as the
smectic A phase, corrections for orientational disorder should be
quite significant in this phase, too. That it is possible to
make these corrections has been demonstrated by the use of an
orientational-disorder model for the calculation of the changing
layer spacing d through the full temperature range of a smectic C

phase and across the corresponding $\underline{C}$-$\underline{A}$ phase transition.[15] In most other studies on smectic $\underline{C}$ phases, however, such corrections have not been made, and, consequently, many of the data obtained and many of the conclusions drawn should be questioned. Three examples follow.

Bartolino et al.[27] report that their study of four smectic $\underline{C}$ phases (in compounds which also have smectic $\underline{A}$ phases) shows that the optical tilt angle τ_{obs} is significantly different from the tilt angle τ_{calc} calculated from layer thickness measurements [$\tau_{calc} = \cos^{-1}(d_C/d_A)$, where d_C and d_A are the layer thicknesses obtained from x-ray measurements in the smectic $\underline{C}$ phase and the smectic $\underline{A}$ phase, respectively]. On the basis of this difference between τ_{obs} and τ_{calc}, the authors propose a rather detailed molecular model for the smectic $\underline{C}$ phase. This very interesting model should now be seriously questioned, however, because the τ_{calc} data on which it is based were obtained from a formula which completely ignores the existence of orientational disorder.

Leadbetter and Norris[13] give for two compounds the molecular lengths of the molecules in their most extended conformation (ℓ_{model} in Table IV) and also the experimentally obtained values for $\underline{d}$ and τ in the smectic $\underline{C}$ phase, which allow the calculation of an "effective molecular length" as $d_{obs}/\cos\tau_{obs}$ [ℓ_{calc} (S=1) in Table IV]. From these data, Leadbetter and Norris conclude that "the effective molecular lengths are shorter than the most extended conformation by about 2-3 $\overset{o}{A}$" [see data in Table IV on ℓ_{calc} (S=1) and ℓ_{model}], and that "the above results imply very considerable disorder in the alkyl tails". Calculating $\underline{\ell}$ as $d_{obs}/\cos\tau_{obs}$, however, implies that one ignores the existence of orientational disorder. If one takes the existence of orientational disorder into account, e.g., with the model developed by us for the smectic $\underline{C}$ phase,[15] the situation turns out to be actually the opposite of what Leadbetter and Norris perceived it to be. The effective

molecular length [ℓ_{calc} (S=0.70) in Table IV] is now found to be
larger than the length of the most extended conformation,[†] and,
therefore, the data of Leadbetter and Norris actually imply the
absence of disorder in the alkyl chains, in agreement with the conclusion reached above for other liquid crystal phases (see last
paragraph of Section III).

Safinya et al.[28] find for the compound $\bar{8}$S5 that the ratio R =
τ_{obs}/τ_{calc} [$\tau_{calc} = \cos^{-1}(d_C/d_A)$; both τ_{obs} and τ_{calc} are obtained
from x-ray measurements] "is constant (1.2 $\pm$ 0.1) through the C
phase, supporting a simple molecular-tilt model[*] for the transition."
This statement warrants a few comments. (1) The smectic C phase of
$\bar{8}$S5 has a 25°C temperature range,[29] but the data of Safinya et al.
cover only the upper 2°C of this range. Thus, the statement that R
is "constant through the C phase" should be read as "constant through
the upper 2°C range of the C phase." (2) The fact that R = 1.2 is
actually in disagreement with the "simple molecular-tilt model."
If this model were correct, R should have been equal to unity, as
the authors themselves acknowledge.[28] (3) Even though the R of

---

[†]The fact that in this case the ℓ_{calc} values are considerably larger
than the ℓ_{model} values, rather than equal to them, can be the result
of a number of different factors: (1) The values given for ℓ_{model}
could be in error. For the compound HOAB, e.g., the value given is
31.6 Å, but we ourselves have obtained for this compound a value of
32.9 Å [approximately equal to the ℓ_{calc} (S=0.70) in Table IV].
(2) The S value of 0.70 might be too low. A higher S would result
in a lower value for ℓ_{calc}. (3) The observed d and τ values have
fairly large probable errors, as indicated in Table IV. As a consequence, the values obtained for ℓ_{calc} have fairly large errors,
too.

---

[*]This means, a model ignoring the existence of orientational disorder.

Safinya et al. is apparently constant within the accuracy of their measurements (R = 1.2 ± 0.1), it is far less constant than the R given in Table II, obtained from Sethna's data[16] for a skewed cybotactic nematic phase: these R data yield R = 1.000 + 0.003 over the first 64°C. Thus, the data of Sethna et al.[16,24] yield a far more constant R over a far greater temperature range. Nevertheless we concluded above that this beautiful apparent consistency of the data of Sethna et al. "must be regarded as merely fortuitous," since the calculations ignored the effect of orientational disorder. The same conclusion must be drawn with regard to the data of Safinya et al. Moreover, Ekachai[18] has shown that τ_{calc} in the C phase of 8S5 can be fitted very well, and over a quite large temperature range (15°C), with a model that includes the effect of orientational disorder. Thus, we find that the data of Safinya et al. do not provide any support for "a simple molecular-tilt model," contrary to the authors' statement.

V. THE SMECTIC A-C PHASE TRANSITION

Smectic A and smectic C phases have quite commonly been described as phases in which the molecular long axes are perpendicular and tilted, respectively, with respect to the planes of the smectic layers (see, e.g., Ref. 30). The smectic A-to-C phase transition is, then, the onset of molecular tilting.[28] In another molecular model of the phase transition, proposed by Wulf,[31] the molecules are tilted in both phases, with a constant tilt angle, and the phase transition is a change in the distribution of the molecular tilt directions, from one with infinite rotational symmetry in the A phase to one with only a mirror plane of symmetry in the C phase. The inclusion of orientational disorder in the structures of both phases, and the molecular models based on this, suggest that the smectic A-C phase transition has actually some aspects of both models simultaneously.[15] In the smectic A phase, there is indeed a distribution of molecular tilt directions, with infinite rotational

symmetry, as in Wulf's model, but there is also a broad distribution
of the molecular tilt angles.[14]   At the A-to-C phase transition,
the distribution of the tilt directions loses its rotational symmetry, as in Wulf's model, but there is also a sharp shift of the
distribution of the tilt angles toward larger values; this increase
in tilt angles corresponds to the molecular-tilting aspect of the
first model mentioned above for the A-C transition. Thus, both
models appear to have described one aspect of the more complete
picture now presented by the orientational-disorder model of the
phase transition.[15]

    Another aspect of the A-C phase transition for which the con-
cept of orientational disorder might have significant consequences
is the following. In many discussions of the A-C transition, the
tilt angle of the director, τ, is taken to be the important parameter describing the transition, and it is argued that the temperature dependence of τ can be predicted by theoretical arguments (see,
e.g., Ref. 32, p. 324). The orientational-disorder models, however, suggest that τ is only a "secondary" parameter, the temperature dependence of which is determined by those of a number of
other, more fundamental parameters. Thus, it appears that it might
well be more appropriate to use one of these more basic parameters
(e.g., θ_m, the preferred tilt angle of the molecular long axis[15])
as the main parameter in theoretical descriptions of the A-C phase
transition.

VI. FURTHER CONSEQUENCES

    We would like to suggest here that all the examples of the
effect of orientational disorder presented above, and further
similar cases, represent only "the tip of the iceberg", since all
these examples involve only the time-and-space-averaged effect of
the orientational disorder. The main impact of the existence of
orientational disorder will be, we believe, the implications of

this disorder for the actual molecular arrangement in a given area of space at a given time.

This impact starts already with the way in which the smectic phases are represented in schematic drawings of their structures. Generally, a smectic $\underline{A}$ monodomain is pictured as a set of equidistant planes with the molecules (represented by lines) perpendicular[30] or approximately perpendicular (Ref. 32, p. 13) to these planes (Fig. 1). A much better picture, however, appears to be the schematic representation first given by Kosterin[33] and later by Chistyakov,[34] reproduced in Fig. 2. This representation is, in our opinion, much more realistic than the ones usually given, for the following reasons: (1) It gives a better indication of the degree of orientational disorder. (2) It shows that the layer thickness is not constant. (3) It shows that the boundary planes of the layers are not flat. Still, even this representation probably does not yet go far enough: (1) The orientational disorder associated with S = 0.7-0.8 is probably significantly greater than that suggested by Fig. 2. (2) The planes through the centers of the molecules in each layer are probably significantly less flat than is apparent from Fig. 2.

In an attempt to arrive at a better representation of the actual molecular arrangement in a smectic liquid crystal, Jolly A. Rahman of our Institute has started calculations involving the effects of various kinds of intermolecular interactions. When such a better, detailed, and more quantitative representation has

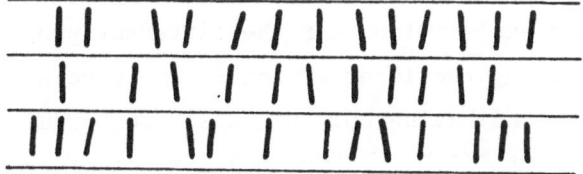

Fig. 1 Schematic drawing of the arrangement of the molecules in a monolayer smectic $\underline{A}$ phase according to De Gennes (from Ref. 32, p. 13).

Fig. 2   Schematic drawing of the arrangement of the molecules
         in a monolayer smectic $\underline{A}$ phase according to Chistyakov
         (from ref. 34).

been achieved, we expect it to lead to further developments in a
number of different areas. (1) It will allow the calculation of
the relative intensities of successive orders of reflections against
the smectic layers, and this will show, we expect, why the intensities of the higher-order reflections are always so weak. (2) Using
molecules of different lengths, it might give a better picture of
what happens in a mixture of the smectic $\underline{A}$ phases of two compounds,
and why the diffraction maxima of the mixtures appear to be as
sharp as those of the separate components.[35] (3) The effects of
changes in the various intermolecular interactions could be studied.
(4) The effects of boundary conditions (e.g., flat surfaces, free
smectic/air interfaces) could be studied. (5) New light might be
shed upon the mechanisms for undulation modes (Ref. 32, p. 307;
Ref. 36) and peristaltic modes[37] in smectic $\underline{A}$ phases. (6) The
model could be expanded to bimolecular smectic $\underline{A}$ phases, and
possibly offer some explanations for the different properties of
these phases, e.g., why the layer structure in these phases generally appears to behave as having only a single sinusoidal electron-
density wave component.[20,38] (7) The model might offer additional
insight into the so-called ripple phase.[39]

VII. SUMMARY

In this paper we have shown that neglecting the effect of orientational disorder on the thickness of smectic layers can lead to serious errors in the evaluation of the x-ray diffraction data of liquid crystals with smectic order. Examples have been given for monolayer and bilayer smectic $\underline{A}$ phases, for skewed cybotactic nematic phases, and for smectic $\underline{C}$ phases. These errors can be avoided by the use of appropriate models incorporating orientational disorder. Thus, <u>the effects of orientational disorder should be taken into account first,</u> before drawing conclusions from the data.

We have also pointed out that the influence of orientational disorder goes far beyond its effect on layer thickness, and that here, too, the consequences may be expected to be quite significant.

REFERENCES

1. G. Friedel, Ann. Phys. (Paris), <u>18</u>, 273 (1922).

2. G. H. Brown and W. G. Shaw, Chem. Rev., <u>57</u>, 1049 (1957).

3. G. W. Gray, <u>"Molecular Structure and the Properties of Liquid Crystals"</u>, Academic Press, London, 1962.

4. A. de Vries, Mol. Cryst. Liq. Cryst., <u>11</u>, 361 (1970).

5. D. Demus, S. Diele, M. Klapperstück, V. Link, and H. Zaschke, Mol. Cryst. Liq. Cryst., <u>15</u>, 161 (1971).

6. S. Diele, P. Brand, and H. Sackmann, Mol. Cryst. Liq. Cryst., <u>16</u>, 105 (1972).

7. A. de Vries, Mol. Cryst. Liq. Cryst., <u>20</u>, 119 (1973).

8. J. Doucet and A. M. Levelut, J. Phys. (Paris), <u>38</u>, 1163 (1977).

9. A. de Vries, Mol. Cryst. Liq. Cryst. Lett., <u>41</u>, 27 (1977).

10. W. H. de Jeu and J. A. de Poorter, Phys. Lett., <u>61A</u>, 114 (1977).

11. A. J. Leadbetter and R. M. Richardson, Mol. Phys., <u>35</u>, 1191 (1978).

12. A. J. Leadbetter in "The Molecular Physics of Liquid Crystals",
     G. R. Luckhurst and G. W. Gray, Eds., Academic Press,
     London, 1979, Chapter 13.

13. A. J. Leadbetter and E. K. Norris, Mol. Phys., 38, 669 (1979).

14. A. de Vries, A. Ekachai, and N. Spielberg, Mol. Cryst. Liq.
     Cryst. Lett., 49, 143 (1979).

15. A. de Vries, J. Chem. Phys., 71, 25 (1979).

16. V. M. Sethna, Ph.D. Thesis, Kent State University, Kent, Ohio,
     1980.

17. A. de Vries, V. M. Sethna, and N. Spielberg, to be published.

18. A. Ekachai, Ph.D. Thesis, Kent State University, Kent, Ohio,
     1980.

19. J. D. Litster, R. J. Birgeneau, M. Kaplan, C. R. Safinya, and
     J. Als-Nielsen in "Ordering in Strongly-Fluctuating Con-
     densed Matter Systems", T. Riste, Ed., Plenum Press,
     New York, 1980, pp. 357-382.

20. A. J. Leadbetter, J. C. Frost, J. P. Gaughan, G. W. Gray, and
     A. Mosley, J. Phys. (Paris), 40, 375 (1979).

21. P. E. Cladis, D. Guillon, W. B. Daniels, and A. C. Griffin,
     Mol. Cryst. Liq. Cryst. Lett., 56, 89 (1979).

22. A. de Vries, Mol. Cryst. Liq. Cryst., 10, 219 (1970).

23. A. de Vries and S. B. Qadri in "Liquid Crystals", S. Chandra-
     sekhar, Ed., Heyden, London, 1980, pp. 179-184.

24. V. M. Sethna, A. de Vries, and N. Spielberg, Mol. Cryst. Liq.
     Cryst., 62, 141 (1980).

25. A. de Vries in "Advances in Liquid Crystal Research and
     Applications", L. Bata, Ed., Akadémiai Kiadó, Budapest,
     1981, Vol. 1, pp. 71-80.

26. W. Maier and A. Saupe, Z. Naturforsch., A, 14, 882 (1959).

27. R. Bartolino, J. Doucet, and G. Durand, Ann. Phys. (Paris),
     3, 389 (1978).

28. C. R. Safinya, M. Kaplan, J. Als-Nielsen, R. J. Birgeneau,
    D. Davidov, J. D. Litster, D. L. Johnson, and M. E.
    Neubert, Phys. Rev., B21, 4149 (1980).

29. M. E. Neubert, R. E. Cline, M. J. Zawaski, P. J. Wildman, and
    A. Ekachai, Mol. Cryst. Liq. Cryst., 76, 43 (1981).

30. H. Sackmann and D. Demus, Mol. Cryst. Liq. Cryst., 21, 239
    (1973).

31. A. Wulf, Phys. Rev., A17, 2077 (1978).

32. P. G. de Gennes, "The Physics of Liquid Crystals", Oxford
    University Press, London, 1975.

33. E. A. Kosterin, Kristallografiya, 17, 639 (1972); Soviet Phys.-
    Cryst., 17, 549 (1972).

34. I. Chistyakov, Adv. Liq. Cryst., 1, 143 (1975).

35. J. E. Lydon and C. J. Coakley, J. Phys. (Paris), Suppl., 36,
    C1-45 (1975).

36. W. Helfrich, J. Phys. (Paris) Suppl., 40, C3-105 (1979).

37. F. Hardouin and A. M. Levelut, J. Phys. (Paris), 41, 41 (1980).

38. J. Als-Nielsen, R. J. Birgeneau, M. Kaplan, J. D. Litster,
    and C. R. Safinya, Phys. Rev. Lett., 39, 1668 (1977).

39. M. J. Janiak, D. M. Small, and G. G. Shipley, Biochemistry,
    15, 4575 (1976).

EFFECT OF MOLECULAR STRUCTURE OF RIGID - FLEXIBLE POLYESTERS ON THEIR THERMOTROPIC LIQUID CRYSTALLINE PROPERTIES

Leszek Makaruk and Hanna Polanska

Institute of Organic Chemistry
and Technology
Technical University (Politechnika)
Koszykowa 75
Warsaw, Poland

INTRODUCTION

Liquid crystalline polymers are of considerable current interest not only because of their unique properties for practical
purposes but also from the theoretical point of view. As early as
in 1941, Kargin and Slonimski (1) predicted that the backbone
polymers should exhibit the liquid crystalline properties. Taking
into account the fact that almost all of the low molecular weight
liquid crystals known so far consisted of rigid polarizable
molecules of prolate shape, Keller and coworkers (2,3) assumed that
long polymer molecules should behave the same way. That is, macromolecules in molten state or dissolved in the appropriate solvent
should form a kind of aggregates built up from colinearly arranged
segments. Most of long-chain polymers, however, possess quite
flexible chains which can easily coil due to effects of interactions and thermal fluctuations and therefore their paralleli-
zation is eliminated, so as the liquid crystalline properties. In
the mid fifties the required conditions which allow the polymer to
form a mesophase has been theoretically studied (4-6). In his

original papers Flory (7,8) concluded that spontaneous ordering
should occur in liquid state only for polymers of relatively stiff
macromolecules. This result has been justified for synthetic poly-
peptide (9,10) (poly-L-γ-benzyl glutamate), which tends to change
conformation from statistical coil to α-helix (stabilized by hydrogen bonds) when dissolved in a certain concentration range in some
solvents. This was the case of the typical lyotropic liquid crystal. There is, however, another group of polymers exhibiting mesophasic behavior in a molten state within a specific temperature
range (thermotropic liquid crystals). Vorlander (11) describes
oligomer of p-hydroxybenzoic acid which melts at 142°C to yield a
nematic liquid crystalline phase persisting to 282°C, when an isotropic phase is obtained.

As it is known, there are two types of thermotropic polymeric
liquid crystals, which have been widely investigated so far. The
first one are polymers with mesogenic side groups attached directly
to (or through spacers to polymer backbone) so called comb-like
polymers). The next are polymers with mesogenic units alternating
with flexible spacers in the main chain. The properties of the
comb-like liquid crystals can be easily modified by incorporating
the side groups of varying structure. The influence of the chemical
nature and the length of the substituents on the stability and the
temperature range of the liquid crystalline phase in the thermotropic polymers with mesogenic groups in the side chains have been
investigated by Shibaev and Platé (12) and Finkelmann and co-
workers (13). According to their studies, the stability and the
broadness of the mesophase increases with the length of the substituent. The same effect is observed when alkyl chain is exchanged by alkoxy chain. These results for the liquid crystalline
polymers of the general formula (13):

$$+CH_2-\underset{\underset{COO+CH_2\frac{1}{n}-O-\bigcirc-COO-\bigcirc-R}{|}}{\overset{\overset{CH_3}{|}}{C}}\frac{}{}x$$

$$R = +CH_2\frac{1}{m}H$$
$$-O+CH_2\frac{1}{m}H$$

are consistent with the behavior of low molecular weight liquid
crystalline benzoic acid phenylesters (14).

The theoretical considerations on the synthesis and properties of thermotropic liquid crystalline polymers with mesogenic
groups in the main chain were performed by de Gennes (15). His results stimulated a number of papers dealing with the influence of
the polymer chemical structure on the transition temperatures (16-
21). The following conclusions can be drawn from these works. In
aromatic-aliphatic polyesters temperatures of the transition from
solid to mesophase (T_m) and mesophase to isotropic liquid (T_i) depend upon the number of aromatic rings in the rigid part of the repeating unit. Transition temperatures increase with the increasing
length and stiffness of this part. The opposite phenomenon is
observed with the increasing length of the corresponding flexible
part in the macromolecule (for example with the increasing number
of methylene groups connecting p-hydroxyphenoxy radicals of the
polyester (16,20). The lowering of the transition temperatures
and the narrowing of the thermal stability of the mesophase with
an increase of alkyl chain length was also observed for a homologous series of mesophasic polymers prepared from 4,4'-dihydroxy-α-
methylstilbene and diols have various number of carbon atoms in the
molecule (22,23). For a homologous series of polyesters (21,24)
and poly(Schiff base ethers) (25) the transition temperatures T_m
and T_i both alter according to even-odd numbers of methylene
groups in the aliphatic segment of the macromolecule. As for the

low molecular weight thermotropic mesogens T_m and in most cases T_i
are always higher for the even members of the homologous series
than for the odd ones.

The transition temperatures are also sensitive to the chemical
stucture of substituents attached to the central aromatic ring of
the mesogenic part of the macromolecules (20). The polymers obtained from 1,10 - bis (p-carboxyphenoxy) decane and mono substituted
or asymetrically disubstituted hydroquinones reveal much lower T_m
than those of unsubstituted or symetrically substituted hydroquinones. The size (bulkiness) of the substituents as well as
their polarity and polarizability are also important factors controlling the liquid crystalline characteristics of the thermotropic
polymers. It was shown experimentally, that the introduction of the
substituents to the central aromatic ring of the mesogenic unit
of the macromolecule usually decreases T_i. Hence, it can be concluded that among many factors controlling the temperature range
of the mesophase of liquid crystalline polymers the most important
seem to be geometric and steric effects. It concerns also the substituents introduced between two phenolic rings in nonlinear bisphenols, playing the role of flexible spacers in copolyesters derived from terephthalic acid and either methylhydroquinone or chlorohydroquinone (19).

This work has been carried out with the aim to investigate the
relationship between the polymer chemical structure and the liquid
crystalline properties of backbone polyesters. We have earlier
prepared liquid crystalline copolyesters by modification of poly-
(ethylene terephthalate) and poly(ethylene sebacate) with p-acetoxybenzoic acid (26) according to the method described by Jackson and
Kuhfuss (27,28). Due to the quite complicated structure of so obtained polymers, we were not able to derive any regularities between the liquid crystal behavior and the chemical nature and the
length of their mesogenic moieties and flexible spacers. Therefore,

we have attempted another procedure to correlate the lengths of
the rigid and flexible parts of the backbone polymers with their
thermotropic liquid crystalline properties. We have synthesized
the low molecular weight compounds of desired structure and then
polymerized them in order to obtain the polymer. Relatively
stiff 4,4'-dihydroxybenzophenone and 1,4-di(p-hydroxybenzoylo)ben-
zene were chosen as a model of rigid segments, whereas aliphatic
dicarboxylic acids derivatives (sebacyl and suberyl chlorides)
were used as flexible spacers. The liquid crystalline behavior
of polymers obtained from the above mentioned compounds was investigated and compared with the properties of their low molecular
weight analogues.

EXPERIMENTAL

Preparation of Compounds

4,4'-Dihydroxybenzophenone (DHBP)

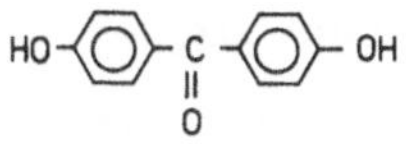

HO—⟨O⟩—C—⟨O⟩—OH
         ‖
         O

The compound, synthesized using anisoyl chloride and anisole
according to the Barclay's method described in "Condensation
Monomers" by Stille and Campbell (29), had melting point of 214°C
and its structure has been confirmed by combustion analysis and IR
spectrum.

1,4-Di (p-hydroxybenzoylo) benzene (DHBB)

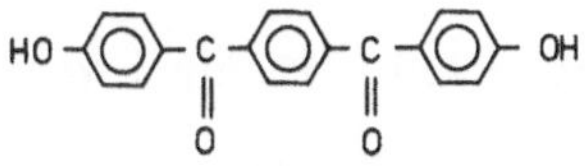

HO—⟨O⟩—C—⟨O⟩—C—⟨O⟩—OH
         ‖   ‖
         O   O

This three-aromatic-ring bisphenol was obtained from terephthaloyl dichloride and anisole via its methoxy derivative modifying the
method used for the synthesis of methoxynaphthalene described in (30).

In a 250 ml four-necked flask fitted with thermometer, dropping
funnel, stirrer and a reflux condenser with calcium chloride guard
tube attached to its top, a solution of 33.3 g (0.25 mole) of anhydrous $AlCl_3$ in 92 ml of dry nitrobenzene was cooled down to 0-2°C
during intensive stirring. Then keeping the mixture at 0°C (or
slightly above) the solution of 28 ml (0.26 mole) of anisole in
10 ml of dry benzene was slowly added. The rate of addition was
controlled, so that the temperature was maintained at 0±1°C. Then
keeping the same conditions, the solution of 20 g (0.1 mole) of terephthaloyl dichloride in 27 ml of dry nitrobenzene was added. Afterwards, stirrer, dropping funnel and the reflux condenser were removed and the flask, stopped with the calcium chloride guard tube,
was placed in the freezer and allowed to stand there for 4 days.
Then the contents of the flask was poured into a mixture of 600 g
of crushed ice and 66 ml of concentrated hydrochloric acid. After
the ice melted, the mixture was placed in 1 ℓ flask and nitrobenzene was distilled off with the steam. The residue (yellow-
greenish precipitate) was then filtered at the water pump using
Buchner funnel and washed with small amount of diethyl ether and
methanol. The resulting white-pale greenish crystals were dried
in a vacuum oven at room temperature for 4 hours. The yield was
25.3 g (i.e. 73% of theoretical one). The structure of the product, that is 1,4-di(p-methoxybenzoylo) benzene was confirmed by
IR, NMR and combustion analysis and its melting point was 241-
242°C. It dissolves only when heated in DMSO, o-dichlorobenzene,
DMF and tetrachloroethane and is not soluble in such common solvents
as alcohols (methyl, isopropyl), diethyl ether, benzene, CCl_4, acetone, ethyl acetate and acetic acid.

Reduction of 1,4-Di (p-Methoxybenzoylo)benzene (DMBB) to bisphenol

13.85 g (0.04 mole) of DMBB and 30 g (0.26 mole) of pyridine hydrochloride was placed in 100 ml flask and heated on a silicone bath at 214-216°C for 5 hours. The hot mixture was then diluted with 370 ml of 2.8% hydrochloric acid and allowed to stand at room temperature for 24 hours and in the refrigerator for the next 72 hours. The precipitated product was then filtered, washed with 300 ml of water and dried in a vacuum oven at 60°C for 15 hours. The yield of crude bisphenol, m.p. 308-316°C, was 12.7 g (100% of theoretical). After the recrystallization from the mixture of ethanol and water (1:5 by volume), its melting point was 310-311°C. Its structure was checked by IR and combustion analysis. 1,4-di(p-hydroxybenzoylo) benzene dissolves in acetone, DMSO and DMF, can hardly dissolve (after heating) in ethanol and isopropanol and is insoluble in tetrachloroethane, methylethylketone, diethyl ether, CH_2Cl_2, benzene and 30% acetic acid.

Dicarboxylic acids dichlorides

$$Cl - \overset{\overset{O}{\|}}{C} - (CH_2)_n - \overset{\overset{O}{\|}}{C} - Cl$$

n = 6 suberyl chloride
n = 8 sebacyl chloride

Dichlorides were prepared in reaction of tionyl chloride with dicarboxylic acids according to the procedure described in (31).

Polymer Synthesis

Two polyesters of different number of aromatic rings in the stiff part of the macromolecules were obtained from DHBB and DHBP with sebacyl chloride. Two series of copolyesters incorporating

various percentages of two-ring and three-ring moieties, as well
as different alkyl chain lengths, were also synthesized from sebacyl
chloride and the mixture of bisphenols (I Series), and DHBP and the
mixture of dichlorides (II Series), respectively. The polymers were
prepared by interfacial condensation according to the procedure described below. In 250 ml reaction flask fitted with double surface
reflux condenser, thermometer, dropping funnel and an efficient
mechanical stirrer 0.01 mole of bisphenol dissolved in 100 ml of
1 wt% aqueous solution of sodium hydroxide and 0.2 g of benzyltri-
ethylammonium chloride (TEBA) were placed. To a vigorously stirred
mixture the solution of proper amount of dicarboxylic acid dichloride in 100 ml of methylene chloride was slowly added from the
dropping funnel. The addition took about 30 minutes. The change
of color from yellow to orange has been observed. After additional
two hours of stirring the organic layer was separated and precipitated while pouring slowly to vigorously stirred 800 ml of acetone.
The white powdered precipitate was filtered, washed with 250 ml of
water and 100 ml of acetone and finally dried in vacuum at 50°C
for 8 hours.

Synthesis of Low-molecular-weight Polymer Analogues

    Low-molecular-weight analogues, having the same number of ali-
phatic carbon atoms and aromatic rings as the repeating units of
polysebacate prepared from DHBP or DHBB, are the esters of valeric
acid (VA) with the corresponding bisphenols. Their chemical
formulas are the following:

$$CH_3{+}CH_2{\tfrac{}{3}}\overset{\overset{O}{\|}}{C}-O-\!\!\bigcirc\!\!-\overset{}{\underset{\underset{O}{\|}}{C}}-\!\!\bigcirc\!\!-O-\overset{\overset{O}{\|}}{C}{+}CH_2{\tfrac{}{3}}CH_3$$

DHBP - VA

$$CH_3{+}CH_2{\tfrac{}{3}}\overset{\overset{O}{\|}}{C}-O-\!\!\bigcirc\!\!-\overset{}{\underset{\underset{O}{\|}}{C}}-\!\!\bigcirc\!\!-\overset{}{\underset{\underset{O}{\|}}{C}}-\!\!\bigcirc\!\!-O-\overset{\overset{O}{\|}}{C}{+}CH_2{\tfrac{}{3}}CH_3$$

DHBB - VA

The esters were prepared according to the following procedure, described in (32), using valeric chloride and the corresponding bisphenol. In 250 ml reaction flask provided with a mechanical stirrer, reflux condenser and dropping funnel 0.01 mole of bis-phenol (2.14 g of DHBP or 3.18 g of DHBB) was dissolved in 100 ml of 1 wt% aqueous solution of sodium hydroxide. During vigorous stirring of the mixture 0.2 g TEBA was added and 0.04 mole (4.82 g or 4.85 ml) of valeric chloride in 100 ml of methylene chloride was carefully dropped. After additional 4 hours of stirring the con-tents of the flask was poured to a separating funnel and on the next day the methylene chloride was distilled off from the organic layer. The crude precipitate was recrystallized from 25 ml of methanol. Yield was about 70% of theoretical. The structure of both esters was confirmed by elemental analyses and IR spectra.

Characterization

Infrared spectra of the new compounds and polymers were ob-tained on a Perkin-Elmer Model 577 spectrometer. Elemental analy-ses on all new compounds were performed on Perkin-Elmer Model 240 microanalyzer. Inherent viscosities of polymers were measured with an Übbelhode type viscometer at 25°C. Melting points for all mono-mers were determined using a Boëtius hot-stage microscope (PHMK VEB Analytik Dresden). Small-angle light scattering (H_V) measure-ments were carried out in room temperature with a He-Ne (λ = 6328 Å) laser fitted with a red filter on an optical bench with crossed polarizers. A pinhole 1 mm in diameter was placed behind the sample. For small-angle light scattering studies a small a-mount of polymer was placed between two microscope cover glasses on a Boëtius apparatus, which has been preheated to a temperature of 20°C above T_m. After the sample melted, it was quenched rapidly in cold water bath. The optical texture, the phase changes, as well as the transition temperatures from the crystalline (or semi-crystalline) to a mesomorphic (liquid crystalline state) (T_m)

and from mesomorphic to an isotropic melt (T_i) were determined by
means of visual examination on hot-stage microscope equipped with
a pair of crossed polarizers. T_m was taken as the point at which
the polymer changed upon heating into an intensively birefringent
liquid exhibiting the Schlieren texture characteristic of the liquid-
crystalline state. At this temperature the top cover glass was
moved forth and back with a microspatula in order to observe the
occurrence of stirr opalescence. T_i was the point at which the last
of the birefringent or irridescent entities melted giving an isotropic melt that allowed no light to pass through the polarizer-
sample-analyzer set. Two heating cycles on each sample were conducted. In the first cycle the heating rate was relatively fast,
being about 10°C/min up to the temperature about 20°C below the
melting point (T_i), then gradually changed to about 3°C/min. In
this run the sample was pressed between two microscope cover glasses
and allowed to cool to room temperature with the hot-stage after
turning off the heater. The second run has been performed with the
heating rate relatively slow near the transition points. For each
polymer sample at least five recordings were performed. The results of hot-stage microscopy were compared with DSC for the selected number of samples. DSC measurements were carried out on Du
Pont Model 990 apparatus under a nitrogen atmosphere with a heating
or cooling rate of 10°C/min. The maximum temperature of the endotherm was taken as the transition temperature. The reproducible
transition temperatures were obtained in both of two runs consisted of heating the sample approximately 30°C above its isotropic
transition. The temperature axis of the DSC was calibrated with
reference standard of high purity indium.

RESULTS AND DISCUSSION

In tables 1 and 2 data concerning the polymer preparation
(e.g. substrate compositions and the yields of the polycondensation reaction) along with the transition temperatures (T_m, T_i)

Table 1. The Properties of Polymers Obtained From the Mixture of Bisphenols and Sebacyl Dichloride

Bisphenol Composition		Polymer Yield	η_{inh} *	Transition Temperatures by			
DHBP	DHBB			Hot Stage Microscopy**		DSC	
mol%	mol%	%		T_m,°C	T_i,°C	T_m,°C	T_i,°C
100	-	74	0.191	105	140	97	145
80	20	62	0.245	97	147	-	-
60	40	85	0.320	126	190	125	211
50	50	81	0.358	133	188	-	-
40	60	81	0.259	135	204	132	216
20	80	78	0.198	180	231	-	-
-	100	68	0.380	205	241	201	236

* Measured on 0.5 % (w/v) solutions in o-chlorophenol at 25 °C.
** Mean values of at least five measurements.

Table 2. The Properties of Polymers Obtained from the Mixture of Dicarboxylic Acid Dichlorides and 4,4'-Dihydroxy-benzophenone

Composition of Dichlorides		Polymer Yield	η_{inh} *	Transition Temperatures by			
Sebacyl	Suberyl			Hot Stage Microscopy**		DSC	
mol %	mol %	%		T_m, °C	T_i, °C	T_m, °C	T_i, °C
100	-	74	0.191	105	140	~7	145
80	20	50	0.126	106	127	84	128
70	30	58	0.159	104	123	-	-
60	40	66	0.124	98	112	88	114
50	50	40	0.175	96	123	97	130
30	70	47	0.206	119	150	125	154
-	100	71	0.193	152	173	136	173

* Measured on 0.5 % (w/v) solutions in o-chlorophenol at 25 °C.
** Mean values of at least five measurements.

obtained by means of hot-stage microscopy and DSC are listed. The
yields of the polycondensation reaction range from 70 to 80% for
copolyesters with various amounts of two- and three-ring bisphenol
and from 40 to 60% for those with mixed aliphatic fragment in the
macromolecule. Taking into account the solution viscosities (as
inherent viscosities), their molecular weights are relatively low.
Hence, their low solubilities in various solvents are rather sur-
prizing. At room temperature the polymers of this work can be dissolved only in o-chlorophenol and phenol/1,1,2,2-tetrachloroethane
(1:1) mixtures, while they are practically insoluble in chlorinated
hydrocarbons, DMSO, DMF, nitrobenzene, chlorobenzene or o-dichlorobenzene.

   All aromatic-aliphatic polyesters and copolyesters containing
mesogenic units and flexible spacers in the main chain show thermotropic behavior as revealed by microscopic observations and DSC.
According to these results, it seems that stiff block of polymer
has to contain at least two aromatic rings in order to make the
polymer capable of attaining a thermotropic liquid crystalline
state, since quite similar to our polyesters, as for the chemical
nature is concerned, poly(decamethylene terephthalate), having a
single aromatic ring connected by ester bonds, did not show mesogenic properties (33). Therefore de Gennes' predictions that
linear polymers with rigid blocks interconnected by flexible spacers
in the main chain would behave as liquid crystals was confirmed
for the series of our polyesters. We also noticed, that the sub-
stitition of two-aromatic-ring blocks by three-aromatic-rings ones,
shifts the transition temperatures T_m and T_i towards higher values
(table 1). With the increasing percentage of DHBB moieties in
polyester based on DHBP and sebacyl chloride, T_i and T_m were raising up (Fig. 1), though the occurrence of the maximum is found
in the case of mesophase range broadness (Fig. 2). The maximum
of mesophase range was observed for equimolar content of DHBP and
DHBB fragments in macromolecule. The liquid-crystalline range

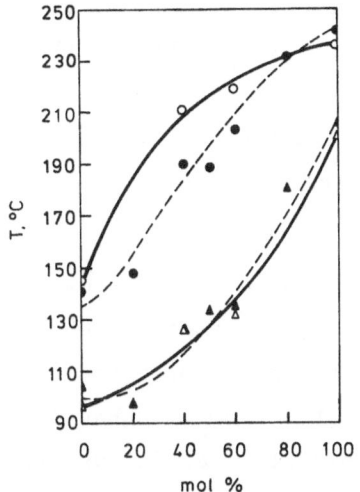

Fig. 1   Dependence of solid-to-liquid crystalline (△,▲)
and liquid crystalline-to-isotropic (O,●)
transition temperatures on the molar concentration
of DHBB in copolyesters based on the mixture of
bisphenols (DHBP + DHBB) and sebacyl chloride.
△,O - measured by DSC;   ▲,● - measured by
hot-stage microscopy.

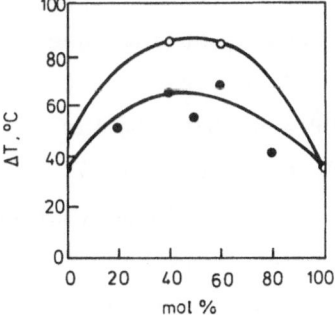

Fig. 2   Relationship between the temperature range of liquid
crystallinity and the molar concentration of DHBB in
copolyesters of Series I.
O - for DSC measurements;   ● - for hot-stage microscopy measurements.

(T_i-T_m) is almost doubled with respect to pure DHBP-sebacyl chloride polyester and tripled as far as DHBB-sebacyl chloride is concerned (Fig. 2). The similar behavior, that is the increase of the mesogenic transition temperatures and the broadening of T_i-T_m range with the length of the stiff block in macromolecule, has been observed in aromatic-aliphatic polyesters prepared by Liebert and Strzelecki (17).

We thought, that it could be also possible to broaden the mesophasic range of polyester by changing the structural regularity of its flexible part. Therefore we prepared the second series of copolyesters with different ratios of sebacyl and suberyl acid radicals in the macromolecule. As it was expected (16,20), polyester of DHBP and suberyl chloride showed higher T_m and T_i temperatures (approximately of about 36°C) than its sebacyl homologue of longer aliphatic chain (cf. table 2). There was, however surprising the fact, that the incorporation of the mixture of aliphatic chloride residues into the polyester macromolecule did not increase the solid-liquid crystalline and liquid crystalline- isotropic liquid transitions temperatures, as it should occur, according to the literature. We observed a decrease of both transition temperatures for all copolyesters containing less than 40 mol% of suberyl chloride (Table 2), followed by an increase at high concentrations. The lowest T_i and T_m transition temperatures were obtained for the copolyester based on DHBP and 40/60 (molar ratio) suberyl chloride/sebacyl chloride mixture (Fig. 3). The further increase of suberyl/sebacyl chloride ratio in polymer composition caused a steady increase in transition temperatures, though it did not influence very much the range of mesogenic phase (Fig. 4). So, it is proved that the incorporation of the shorter aliphatic chain flexible moieties to our copolyesters changes the mesogenic transition temperatures. The effect consists in shortening and shifting of the temperature range of liquid crystallinity.

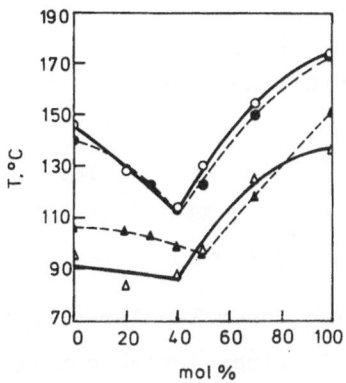

Fig. 3 Dependence of T_m ($\triangle$, $\blacktriangle$) and T_i ($\bigcirc$, $\bullet$)
on the molar concentration of suberyl chloride in
copolyesters obtained from the mixture of dicarboxylic
acid chlorides and DHBP
        $\triangle$,$\bigcirc$- measured by DSC
        $\blacktriangle$,$\bullet$- measured by hot-stage microscopy.

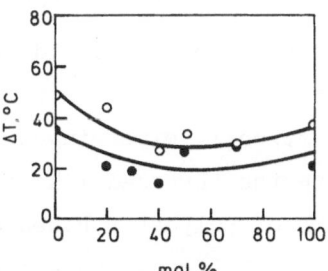

Fig. 4 Temperature range of liquid crystallinity versus
molar concentration of suberyl chloride in copolyesters of Series II.
        $\bigcirc$- DSC data
        $\bullet$- hot-stage microscopy data.

An attempt to compare the liquid crystalline properties of
polyesters above described with their low-molecular-weight analogues
gave only little information. Valerate of DHBP (which resembles the
monomeric unit of polyester based on DHBP and sebacyl chloride) exhibited the liquid crystalline properties from room temperature to
about 58°C, when it melted into isotropic liquid. This temperature was then about 80°C lower than T_i for macromolecular analogue.
On the contrary, valerate of DHBB did not show thermotropic behavior,
melting very sharply at 170–172°C.

We are also studying low-molecular-weight compounds with other
linkages between aromatic rings, which can be introduced as the
stiff moieties into polymer chain thus giving a chance to collect
more data concerning the relationship between the structure and
length of flexible and ridig blocks of polymers with their thermotropic behavior. It seems interesting to compare properties of copolyesters with physical mixtures of their homopolymers. In this
case the only difference would be that two- or three-aromatic-ring
blocks as well as flexible blocks of various methylene chain lengths
are not chemically bonded but incorporated to the separate macromolecular chains. The properties of such systems are studied.

Polyesters of both series formed turbid melts in mesophase
temperature range, that is up to the isotropic phase transition
temperature T_i, and were strongly birefringent, giving on a hot-
stage polarized microscope the threaded like texture. As an example, the photomicrograph of polyester based on DHBP and suberyl
chloride at the temperature of its mesophasic behavior (165°C) is
shown in Fig. 5a.

The small-angle light scattering carried out on polymer samples
gave us the idea about the shape of the scattering elements in melt.
A ± 45° crossed four-leaf-clover type pattern was obtained for all
polymers in their liquid crystalline states. The pattern intensity, however, as well as its size varied for the particular polymer

composition. The scattering pattern for DHBP-suberyl chloride polyester, at the temperature corresponding to that at which the photomicrograph was taken, is presented in Fig. 5b.

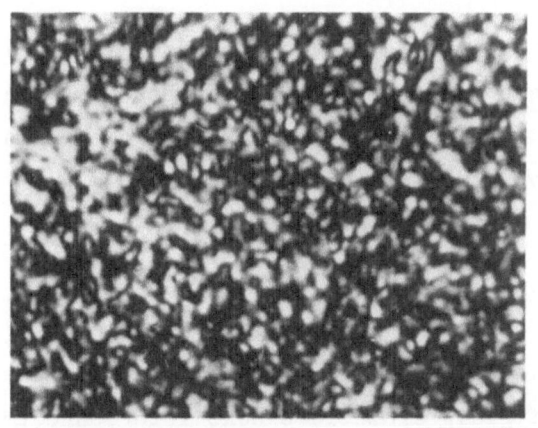

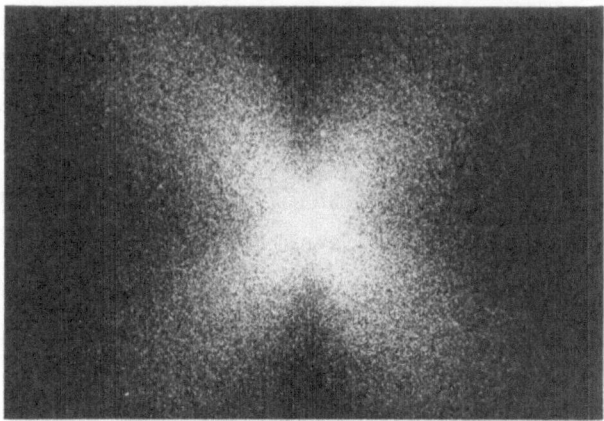

Fig. 5 a. Polarizing microphotograph of DHBP - suberyl chloride
           polyester taken at 165°C (magnification 480X).
        b. Small-angle light scattering (H_v) pattern of the
           same sample after rapid quenching in cold water.

As it is known, the small-angle light scattering characteristics depend on the shape and orientation of the ordered assemblies in the system. Four-leaf-clover, H_v patterns are reported for rod, disk or spherulite type of scattering elements (34). The results suggest that in our polymer melts we are dealing with rod-like elements, since the intensity of the scattered light decreases with increasing the distance from the center of the pattern (see Fig. 5b). There are informations in the literature that such a behavior is typical for nematics (20).

It has to be pointed out that no more scattering characteristic for mesophase was observed for samples heated above T_i and then quenched, thus confirming the disappearance of nematic orientation in polymer above its isotropic transition.

CONCLUSIONS

1. All the linear polymers described in the present study, consisting of ridig aromatic blocks interconnected by aliphatic flexible spacers were liquid crystalline, thus confirming the theroetical predictions of de Gennes.

2. Rigid moieties consisting of two aromatic rings connected by carbonyl groups are sufficient in preparations of the polyesters exhibiting the liquid crystalline phase. Moreover, this is the smallest length of mesogenic unit, which incorporated to macromolecule is capable of attaining the thermotropic liquid crystalline properties.

3. It is shown that selecting an appropriate molecular structure of the copolymer one can control mesophase range and temperature of its appearance.

4. The solid-mesophase and liquid-crystalline-isotropic transitions were for all samples reversible in full, and the liquid crystalline phase was found to be nematic.

5.   It is not always necessary to incorporate low-molecular-weight
     liquid crystal (mesogen) as the rigid segment of macromolecule
     in order to obtain a polymer exhibiting liquid crystalline
     properties.

REFERENCES

1.   W. A. Kargin, G. L. Slonimski, Zh. Phys. Khim., 15, 1022 (1941).

2.   J. Sanderman, A. Keller, J. Polym. Sci., 19, 40 (1956).

3.   A. Keller, A. Maradurin, J. Chem. Phys. Solids, 2, 301 (1957).

4.   L. Onsager, Ann. N.Y. Acad. Sci. 51, 627 (1949).

5.   A. Ishihara, J. Chem. Phys. 19, 1142 (1951).

6.   P. J. Flory, Proc. R. Soc. London, Ser. A, 234, 73 (1956).

7.   P. J. Flory, Proc. R. Soc. London, Ser. A. 234, 60 (1956).

8.   P. J. Flory, Ber. Bunsenges. Phys. Chem. 81, 885 (1977).

9.   A. E. Elliott, E. J. Ambrose, Discuss. Faraday Soc., 9, 246
     (1950).

10. C. Robinson, Trans. Faraday Soc. 52, 571 (1956).

11. D. Vorlander, Physiol. Chem. 105, 211 (1923).

12. V. P. Shibaev, N. A. Platé, Makromol. Chem. 181, 1393, 1381
     (1980).

13. H. Finkelmann, M. Portugall, H. Ringsdorf, Polym. Prepr., Am.
     Chem. Soc., Div. Polym. Chem. 19(2), 183 (1978).

14. R. Steinstrásser, L. Pohl, Angew. Chem. 85, 706 (1973).

15. P. D. de Gennes, C. R. Acad. Sci. Paris, Sèrie B, 281, 101
     (1975).

16. L. Strzelecki, D. van Luyen, Eur. Polym. J. 16, 299 (1980).

17. D. van Luyen, L. Strzelecki, Eur. Polym. J. 16, 303 (1980).

18. L. Liebert, L. Strzelecki, D. van Luyen, M. Levelut, Eur.
     Polym. J. 17, 71 (1981).

19. R. W. Lenz, J. I. Jin, Macromol. 14, 1405 (1981).

20. S. Antoun, R. W. Lenz, J. I. Jin, J. Polym. Sci., Polym. Chem.
     Ed., 19, 1901 (1981).

21. A. C. Griffin, S. J. Havens, J. Polym. Sci., Polym. Phys. Ed., 19, 951 (1981).

22. A. Roviello, A. Sirigu, Makromol. Chem., 180, 2543 (1979).

23. B. Millaud, A. Thierry, C. Strazielle, A. Skulios, Mol. Cryst. Liq. Cryst., 49, 299 (1979).

24. A. C. Griffin, S. J. Havens, Mol. Cryst. Liq. Cryst., 49, 239 (1979).

25. R. W. Wiercinski, Ph.D. Thesis, University of Connecticut, Storrs, 1978.

26. L. Makaruk, unpublished data.

27. W. J. Jackson, H. F. Kuhfuss, J. Polym. Sci., Polym. Chem. Ed., 14, 2043 (1976).

28. F. E. MacFarlane, V. A. Nicely, T. G. David, Cont. Topics in Polym. Sci., 2, 109 (1977).

29. J. K. Stille, T. W. Campbell, Eds.,"Condensation Monomers", J. Wiley, New York 1972.

30. W. Polaczkowa, Ed., "A Textbook of Practical Organic Chemistry", PWT, Warsaw 1954, p. 377.

31. A. P. Grigoriev, O. Ya. Fiedotova, Eds., "Laboratoryj Prakti-kum Po Technologii Plasticheskich Mass", Part II, Moscow 1977, p. 135.

32. G. Vogel, Ed., "A Textbook of Practical Organic Chemistry", 3rd Edition, Longmans 1959, p. 274.

33. I. Daniewska, unpublished observations.

34. R. S. Stein, "Optical Studies of the Morphology of Polymer Films"in"Structure and Properties of Polymer Films", R. W. Lenz and R. S. Stein, Eds., Plenum, New York 1973, p. 1.

RELATIONSHIPS BETWEEN MOLECULAR STRUCTURE AND THE INCIDENCE OF CRYSTAL B AND HEXATIC B PHASES

J. W. Goodby

Bell Laboratories
Murray Hill, New Jersey   07974

ABSTRACT

The liquid crystal phase – smectic B – has been shown to be subdivided into two different structural and miscibility groups. In one group the molecules have short-range in-plane positional ordering, and long-range bond orientational order, both within and between the layers (hexatic B). The other phase, however, shows three dimensional crystalline order (crystal B). The structure adopted by the B phase (either hexatic or crystal) is dependent on the molecular structure of the constituent molecules, particularly for central linkages or other functional groups. In this study a number of systems are examined which have been shown to exhibit B phases of one type or the other. The materials are categorized by B type and by functional group type. From the limited number of materials available for study some tentative conclusions can be drawn, for example, compounds containing -CH=N- or -COS- linkages invariably exhibit crystal B phases whilst -COO- containing materials usually exhibit hexatic B phases. The dipolar and steric properties of these functional groups are used in an attempt to rationalise this situation.

INTRODUCTION

The smectic B phase has been the subject of numerous structural investigations in the last few years. Only recently, however, has it been conclusively proved that two B phases with separate identities exist – the hexatic B and crystal B phases.[1] In the hexatic B phase the molecules have short-range in-plane hexagonal ordering with very little positional correlation between the layers, but yet the phase has long-range bond-orientational order in three dimensions. The second B phase has three dimensional crystalline order with an in-plane correlation length of over 1000 Å and lamellar correlation extending over 1000 layers.[2]

The first report of a two dimensional B phase was made on ester (–COO–) materials[3] with the definitive proof of the existence of a three dimensional hexatic B phase being made on the ester n-hexyl 4'-n-pentyloxybiphenyl-4-carboxylate (650BC).[4] Similarly, most characterisations of crystal B materials have been performed on Schiff's bases (–CH=N–).

Coupling results obtained from various structural studies by X-ray diffraction with results obtained on new materials permits some tentative relationships between the molecular structure (particularly for the central or primary functional group) of the mesogen and the structure of the phase to be drawn.

RESULTS

Affects of the Central Linkage

Esters. The first observation of an uncorrelated smectic B was made by Leadbetter, Frost and Mazid[3] for the 4-n-alkoxyphenyl 4'-n-octyloxybiphenyl-4-carboxylates:

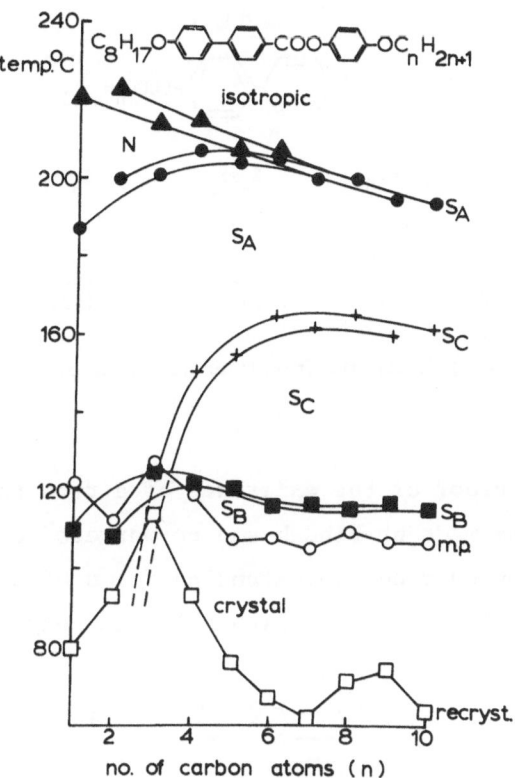

Fig. 1.   Plot of the transition temperatures against the number of
           carbon atoms (n) in the n-alkoxy chain of the 4-n-alkoxy-
           phenyl 4'-n-octyloxybiphenyl-4-carboxylates. Key: ▲,
           N-I; ●, S_A-N or I; +, S_C-S_A; ■ , S_B-S_C or S_A; ○, crystal-
           mesophase; □ , mesophase-crystal.

The variation of the transition temperatures with increasing terminal chain length (n) is shown in Fig. 1 for these materials.[5]
Leadbetter et al. demonstrated by X-ray diffraction of well aligned
samples that the n-pentyloxy, n-hexyloxy and n-octyloxy esters possessed two dimensional, uncorrelated smectic B phases. Similarly,
they showed that the closely related compound[6]

4-n-octylphenyl 4'-n-octylbiphenyl-4-carboxylate also exhibited an
uncorrelated B phase.

TABLE 1

$$C_5H_{11}O-\!\!\!\langle\bigcirc\rangle\!\!-\!\!\langle\bigcirc\rangle\!\!-COOC_nH_{2n+1}$$

n	$I \rightarrow S_A$	$S_A \rightarrow S_B$	$S_B - S_E$	mp
4	93.5	75.5	-	75.2
5	86.7	-	-	77.0
6	84.7	66.0	(60.0)	65.0
7	81.0	(65.0)	(59.0)	69.0

( ) Monotropic Transition Temperature.

Definitive proof of the existence of a true three dimensional hexatic phase was made by Pindak and co-workers[4] employing X-ray diffraction techniques on free-standing films of the smectic B phase of n-hexyl 4'-n-pentyloxybiphenyl-4-carboxylate (650BC):

$$C_5H_{11}O-\!\!\!\langle\bigcirc\rangle\!\!-\!\!\langle\bigcirc\rangle\!\!-COOC_6H_{13}.$$

Similar results were obtained for other members of the n-alkyl 4'-n-pentyloxybiphenyl-4-carboxylates, the transition temperatures for the n-butyl to n-heptyl homologues are given in Table 1. Typically, all the homologous series of the group of compounds – n-alkyl 4'-n-alkoxybiphenyl-4-carboxylates (nmOBC) – exhibit hexatic B phases at some point in their series.

$$C_mH_{2m+1}O-\!\!\!\langle\bigcirc\rangle\!\!-\!\!\langle\bigcirc\rangle\!\!-\underset{\underset{O}{\|}}{C}OC_nH_{2n+1}$$

When m reaches a value of seven however, only the early members of the series exhibit hexatic phases, namely the methyl, ethyl and n-propyl esters. The last series to exhibit the B phase through the first ten (and probably more) members is the n-alkyl 4'-n-hexyl-oxybiphenyl-4-carboxylates,[7] the transition temperatures for which

TABLE 2

$$C_6H_{13}O\text{-}\langle\text{biphenyl}\rangle\text{-}COOC_nH_{2n+1}$$

n	$I\text{-}S_A$	$S_A\text{-}S_B$	$S_B\text{-}S_E$	mp
1	139[+]	-	132	124
2	119	97	92	81
3	107	(74)	(67)	80
4	92	64	-	58
5	90	(58)*	-	83
6	86	(57.5)*	-	79
7	84	(57)*	-	76
8	82	(56)	-	74
9	80	(55)	-	71
10	78	(54.5)	-	59

( ) Monotropic Transition Temperature

* "Virtual" Transition Temperature

+ $I\text{-}S_{AB}$ Transition

are shown in Table 2, and as a function of increasing terminal ester chain length in Fig. 2.

The smectic B phases of all of these materials and other closely related materials, for example;

$$RO\text{-}\langle\text{biphenyl}\rangle\text{-}COO\text{-}\langle\text{phenyl}\rangle$$

$$RO\text{-}\langle\text{biphenyl}\rangle\text{-}COO\text{-}\langle\text{phenyl}\rangle\text{-}Hal$$

exhibit hexatic B and not crystal B phases. It is important to note that all of these materials are derivatives of biphenyl and possess only an ester linkage as the primary functional group. They all have varying terminal substituents; alkyl, alkoxy, and halogen groups.

Schiff's Bases. The situation with materials that incorporate a Schiff's base central linkage in their molecular structures is quite the reverse of that for esters. The first, and most famous, examples of crystal B phases were found in the N-(4-n-alkoxybenzylidene)-4-n-alkylanilines (nOms)

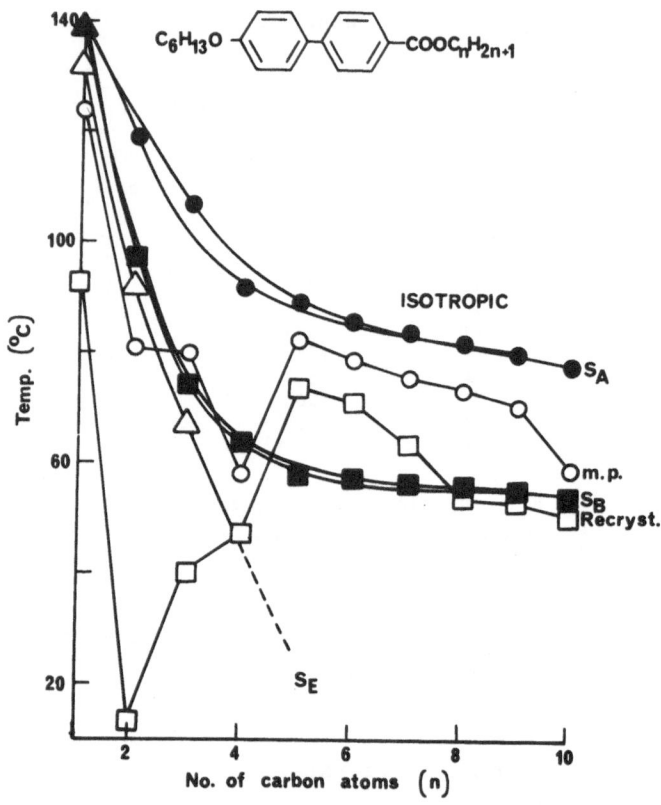

Fig. 2.   Plot of the transition temperature against the number of
           carbon atoms (n) in the ester chain of the n-alkyl
           4'-n-hexyloxybiphenyl-4-carboxylates. Key: ●, S_A-I;
           ■ , S_B-S_A; △, S_e-S_B; ▲, S_{AB}-I; ○, crystal-mesophase;
           □ , mesophase-crystal.

which have been the topic of many detailed structural investigations.[8,9]   Similarly, numerous other Schiff's base materials were
found to exhibit crystal B phases, for example, compounds of the
following series.[10,11,3]

RO—⟨benzene⟩—N=CH—⟨benzene⟩—⟨benzene⟩

RO—⟨benzene⟩—CH=N—⟨benzene⟩—⟨benzene⟩

RO—⟨benzene⟩—⟨benzene⟩—CH=N—⟨benzene⟩—Hal

RO—⟨benzene⟩—⟨benzene⟩—CH=N—⟨benzene⟩—R'

⟨benzene⟩—⟨benzene⟩—CH=N—⟨benzene⟩—CH=CHCOOR

have been shown to exhibit crystal B phases by X-ray diffraction or microscopic textural investigations.

Azo Compounds. One material however is suspected to exhibit both hexatic B and crystal B phases; 4-propionyl-4'-n-heptanoyloxy-azobenzene[12]

C_2H_5CO—⟨benzene⟩—N=N—⟨benzene⟩—$OCOC_6H_{13}$

is thought to exhibit a N,A,B,B phase sequence, demonstrating that the two phases have separate identifies.[13]

Thio Esters. In this present study, the investigation of smectic B properties has been extended to include thio-esters for a direct comparison with the properties of the n-alkyl 4'-n-alkyl (and alkoxy) biphenyl -4-carboxylates (the analogous Schiff's bases are unstable). Earlier studies of thio esters, particularly of the compound

$C_{14}H_{29}O$—⟨benzene⟩—COS—⟨benzene⟩—C_5H_9 $\overline{14}$S5

seem to indicate that this type of linkage may prefer crystalline properties.[14]

The directly analogous material to 650BC, n-hexyl 4'-n-pentyl-oxybiphenyl-4-thiocarboxylate 650SBC

$C_5H_{11}O$—⟨benzene⟩—⟨benzene⟩—$COSC_6H_{13}$

TABLE 3

$$C_5H_{11} \text{—} \langle\text{—}\rangle \text{—} \langle\text{—}\rangle \text{—} COSC_nH_{2n+1}$$

n	$I-S_A$	S_A-S_B	S_B-S_E
2	111.4	113.0	103.5
3	118.5	110.3	90.0
4	120.5	109.0	75.0
5	120.0	104.5	59.8
6	118.0	102.0	50.0
7	116.7	100.2	40.1
8	116.3	99.8	33.0
9	113.8	95.4	25.0
10	113.2	94.0	15.0

which has the following phase sequence

$$\text{Crystal} \longrightarrow S_B \longleftrightarrow S_A \longleftrightarrow I$$
$$\qquad\qquad 91 \qquad 121 \qquad 149.5$$

exhibits a crystal B phase. Deletion of the oxy atom of the alkoxy
terminal chain also provides further examples of smectic B phases
as shown in Table 3 for the n-alkyl 4'-n-pentyloxybiphenyl-4-thio-
carboxylates (n5SBC).[+]

The variation of the transition temperatures as a function of
increasing terminal ester chain length is shown in Fig. 3. Note,
that the melting points are not given because of the uncertainty of
observing a crystal to smectic E transition by optical methods for
some of these compounds. From Fig. 3, it can be seen that the smectic E properties of the thio esters fall considerably with increasing ester chain length. The I to A and A to B transition temperatures fall slowly with increasing terminal chain length, but the
temperature range of the A phase widens on passing up the series.
The alternation of the transition temperatures for the A to I
transitions is in the correct sense (i.e. even homologues higher
than odd) however the A to B temperatures alternate the same way
and therefore opposite to the situation that is usually observed

+ Note: This classification has since been confirmed by X-ray
        diffraction studies.

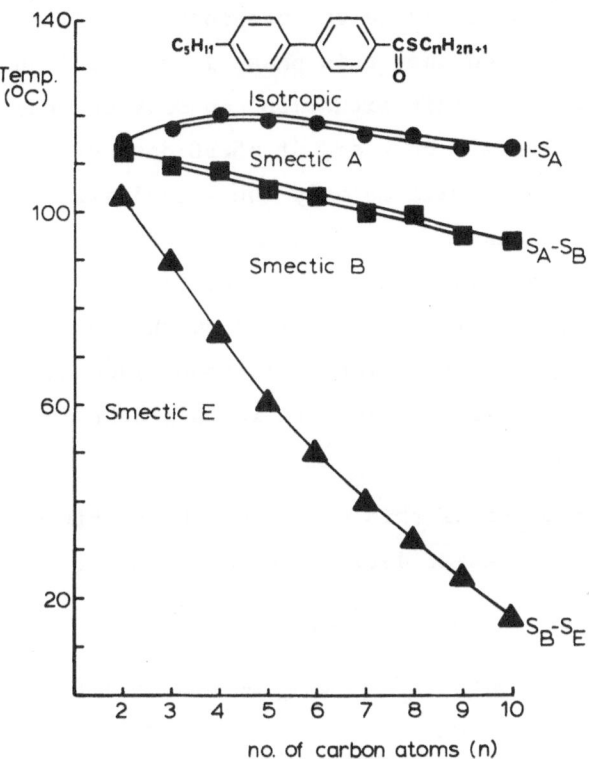

Fig. 3.   Plot of the transition temperatures abainst the number of
          carbon atoms (n) in the ester chain of the n-alkyl
          4'-n-pentylbiphenyl-4-thiocarboxylates. Key: ●,
          S_A-I; ■ , S_B-S_A; ▲, S_E-S_B.

The B to E transition temperatures fall so rapidly that the sense
of alternation cannot really be detected. A direct comparison of
Figs. 2 and 3 demonstrates the vast differences created by the
introduction of a sulphur atom into the molecular structure. However, it is important to note that all three of the more ordered
orthogonal phases (B hexatic, B crystal and E) are favored by
shorter ester chain lengths.

    Detailed head capacity studies performed on n-hexyl 4'-n-pentyl-
biphenyl-4-thiocarboxylate (65SBC)[16] show that the transition from
the A phase to the B phase takes place via a short-lived phase which
lasts only one degree or less. The microscopic textures observed

at transition do not indicate the presence of a tilted phase, thus
it can only be assumed that this phase is a hexatic B precursor to
crystal B formation. This situation can only be resolved by detailed x-ray analysis. However, if 65SBC does exhibit a short
hexatic phase then it will be only the second example of a material
exhibiting both B phases, thus hardening the view that the two
phases have separate identities and therefore should not be comis-
cible. It would also indicate, that even though the thio esters
strongly favor crystal B formation in comparison to the analogous
esters, they are still intermediate in this behavior with respect
to Schiff's bases.

Further examples of thio esters which exhibit smectic B phases
are the phenyl thio ester derivatives of biphenyl e.g.

$$C_5H_{11}O-\bigcirc-\bigcirc-COS-\bigcirc$$

$$cryst: \rightarrow (S_B) \longleftrightarrow S_A \longleftrightarrow N \longleftrightarrow I$$
$$\quad\quad\quad\ 152 \quad\quad 160 \quad\quad\ 182 \quad\quad 189$$

$$C_{10}H_{21}O-\bigcirc-\bigcirc-COS-\bigcirc$$

$$I \longleftrightarrow S_A \longleftrightarrow S_B \longleftrightarrow S_E \quad mp$$
$$\ 181 \quad\quad 145.5 \quad\quad 93 \quad\quad\quad 99$$

Directionality of the Central Linkage

The effect of the positioning of the central or coupling linkage with respect to the core sections of the molecule (e.g. COO or
OOC, CH=N or N=CH) can also complicate matters further. In systems
where the linkage is fully conjugated into the core sections; small
but not significant changes can occur. For example the two homologous series

$$\bigcirc-\bigcirc-CH=N-\bigcirc-OR$$

and

$$\bigcirc-\bigcirc-N=CH-\bigcirc-OR.$$

exhibit much the same trends in transition temperatures and phase changes.[10] However, if the reversal of the central or coupling linkage disrupts the conjugation of that linkage with the core then more prominent effects on the phase types are observed. This can be demonstrated by the reversal of the ester linkage in the compound 650BC to produce a compound of the type:

$$C_5H_{11}O-\langle\text{ring}\rangle-\langle\text{ring}\rangle-OOCC_5H_{11}$$

$$I \xleftrightarrow{106} B \xleftrightarrow{97.7} E$$

These materials show greatly enhanced smectic B properties over those of the 650BC homologous series. No smectic A phase is observed and the transition takes place directly from the isotropic liquid to the B phase. It is interesting to note that this phase segregates from the isotropic liquid in dendritic growth patterns which ultimately form a mosaic texture making it impossible to classify the phase as crystal or hexatic in type by microscopy.

The reversal of the ester function prevents the ketonic oxygen atom from conjugating to the central core, thus reducing the strength of its lateral dipole.

The removal of the oxygen atom of the alkoxy terminal chain has a similar effect, only this time the transition temperatures drop. Even so, the formation of the B phase is enhanced with respect to the A and E phases. For example the compounds

$$C_8H_{17}-\langle\text{ring}\rangle-\langle\text{ring}\rangle-COOC_2H_5$$

$$\text{cryst.} \xrightarrow{} I \xleftrightarrow{} S_{AB} \xrightarrow{} \text{cryst.}$$

$$\quad\quad 64 \quad\quad 61.4 \quad\quad\quad 49.5$$

$$C_5H_{11}-\langle\text{ring}\rangle-\langle\text{ring}\rangle-COOC_8H_{17}$$

$$\text{cryst.} \xrightarrow{} I \xleftrightarrow{} S_B$$

$$\quad\quad 29 \quad\quad 25$$

do not possess an alkoxy oxygen atom in the terminal chain, both
have lower transition temperatures than the analogous alkoxy compounds, but both have enhanced smectic B properties. In this case
the removal of the alkoxy oxygen atom reduces the overall strength
of the lateral dipole of molecule. Again this enhancement of B
properties makes it difficult to identify which B modification is
present when it segregates directly from the isotropic liquid.

However, it can be seen that the weaker the crossed dipole
becomes the more it favors B (particularly crystal B) and other
more ordered phase types.

Microscopic Textures

The two smectic B phases can be readily identified by careful
observation of their paramorphotic microscopic focal-conic fan
textures.[1] This requires that the preceding phase on cooling must
also exhibit a focal-conic texture. If the B phase segregates
directly from the isotropic liquid it usually does so in the form
of mosaic platelets. In this case it is almost impossible to identify one phase from the other. For example, Fig. 4 shows the B
phase of n-octyl 4'-n-pentylbiphenyl-4-carboxylate (85BC) separating
from the isotropic liquid in the form of mosaic platelets and lan-
cets. From previous correlations with molecular structure it is
expected that this phase will be of the hexatic type. However,
this distinction is difficult by microscopic observations, making
X-ray analysis the prime method for classification. If the B
phase of this compound is indeed hexatic then this is the first
example of a direct isotropic liquid to hexatic B phase change,
with the phase exhibiting its natural microscopic texture.

The two B modifications are best distinguished by observation
of their focal-conic textures in very thin preparations. The
sequence of Figs. 5-8 demonstrate a typical cooling sequence for
n-propyl 4'-n-pentylbiphenyl-4-thiocarboxylate (35SBC). Fig. 5
shows the focal-conic fan texture of the A phase formed on cooling

Fig. 4.   The smectic B phase of n-octyl 4'-n-pentylbiphenyl-4-
carboxylate separating from the isotropic liquid on
cooling (X100).

Fig. 5.   The focal-conic texture of the smectic A phase of
n-propyl 4'-n-pentylbiphenyl-4-thiocarboxylate (X100).

Fig. 6.   Transition bars at the A to B phase change for n-propyl
           4'-n-pentylbiphenyl-4-thiocarboxylate (X100).

Fig. 7.   The truncated focal-conic fan texture of the crystal B
           phase of n-propyl 4'-n-pentylbiphenyl-4-thiocarboxylate
           (X100).

Fig. 8.   The striated focal-conic texture of the smectic E phase
           of n-propyl 4'-n-pentylbiphenyl-4-thiocarboxylate (X100).

the isotropic liquid. Further cooling produces a transition to the
B phase, the transition is marked by transition bars, as shown in
Fig. 6. The transition bars are usually observed at a smectic A to
crystal B phase change. The bars cross the fans at the start of
the transition, widen, meet and disappear. This suggests that they
are caused by a two phase region coupled with a build-up of layer
correlation. Therefore, they would not usually be observed at a
smectic A to hexatic B phase transition when there is no layer
coupling and a lesser change in the in-plane correlations.

The B phase formed at this transition shows an almost truncated
focal-conic texture. The fans become angular and stepped in parts
with their back becoming less lined and flat in appearance, suggesting the layers are less curved in this texture than in the preceding
A phase (Fig. 6). In the corresponding situation for the hexatic B
phase the fans would almost be identical to those of the preceding
A phase.[1]

Subsequent cooling of this phase produces the striated fan
texture of the E phase. Again an indication of the true nature of
the preceding phase can be obtained from this texture. The arcs
are not particularly curved and tend to be angled around the fan
back giving the impression that the fan has a flattened surface.

In a given homologous series of crystal B materials as the
series is ascended so the crystal B phase textures become more
difficult to distinguish from those of the hexatic B phase, particularly in thicker preparations (compare Fig. 7 with Fig. 9 for
65SBC).

A direct comparison of the texture of the smectic B phase of
35SBC can be made with that of 650BC (Fig. 10). A full description
of the microscopic textures of the phase sequences of this compound
are given elsewhere.[1]

Fig. 9.   The focal-conic texture of the crystal B phase of n-hexyl
           4'-n-pentylbiphenyl-4-thiocarboxylate (X100).

Fig. 10.   The focal-conic texture of the hexatic B phase of n-hexyl
           4'-n-pentyloxybiphenyl-4-carboxylate (X100).

        Further examples of the focal-conic texture of the crystal B
phase are shown in Figs. 11 and 12 for N-(4-n-butyloxybenzylidene)-
4'-n-heptylaniline (40.7) and 4-phenylbenzylidene 4'-n-octyloxyani-
line respectively.

Miscibility Studies

        Miscibility studies involving materials of differing B types
usually prove inconclusive and the two phases appear co-miscible.
This is because either the transition between the two phases is not
readily observed or the immiscibility regions for the two phases
are so small that it is very easily missed.   Fig. 13 shows the
miscibility of n-hexyl 4'-n-pentylbiphenyl-4-thiocarboxylate
(65SBC) with N-(4-n-butyloxybenzylidene)-4'-n-octylaniline (40.8)
(N,A, and crystal B phases, left-hand side) and with n-hexyl
4'-n-pentyloxybiphenyl-4-carboxylate (650BC) (A hexatic B and S$_E$

Fig. 11. The focal-conic texture of the crystal B phase of
          N-(4-n-butyloxybenzylidene)-4'-n-heptylaniline (X100).

Fig. 12.   The focal-conic texture of the crystal B phase of
           4-phenylbenzylidene 4'-n-octyloxyaniline (X100).

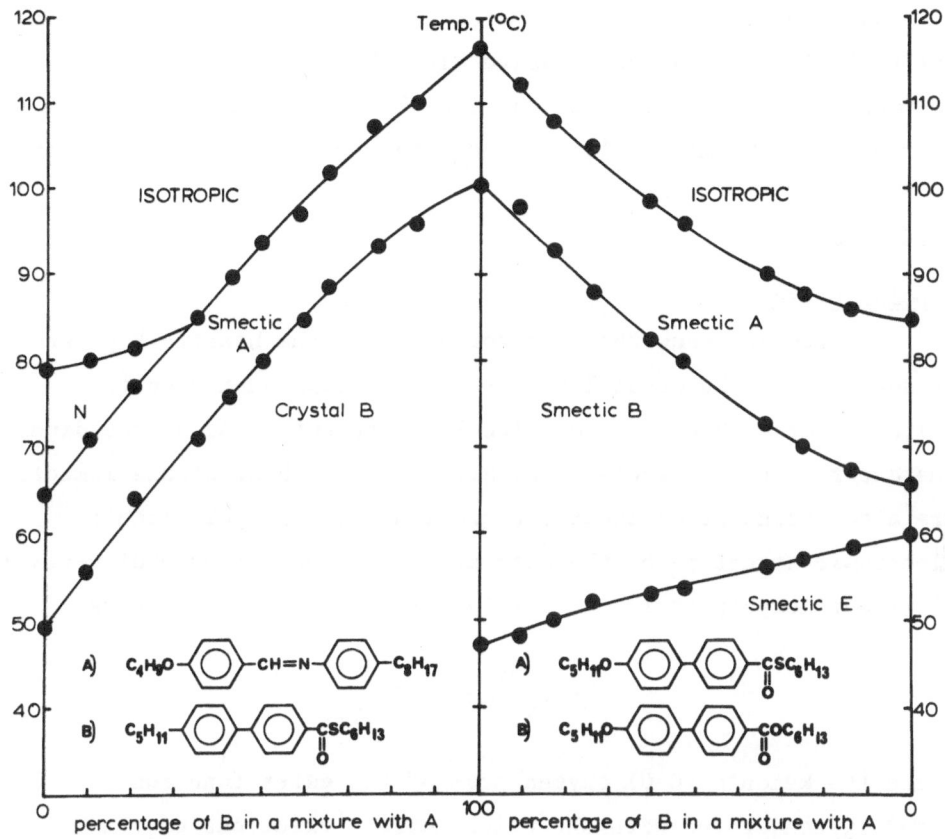

Fig. 13. Miscibility diagram of state of 65SBC with 40.8 (wt.%)
          left figure and 65SBC with 650BC (wt%) right figure.

phases, right-hand side). 65SBC is suspected of exhibiting an
A,B,B,E phase sequence, thus it appears co-miscible with both
crystal and hexatic B phases, yet no B-B transitions were observed
in any of the binary mixtures. This gives the impression that the
two phases are apparently co-miscible.

     In a separate study involving 4-propionyl-4'-n-heptanoyloxyazo-
benzene as the test material and 40.8 and 650BC as the reference
compounds, B-B transitions are observed in binary mixtures indicating
the phases have separate identities. This is possible to observe
because the hexatic B phase has a wider temperature range in this
test material than it does in 65SBC.

Subsequent miscibility studies involving 650SBC as the test material also proved inconclusive with the B phases appearing co-miscible, indicating that the two phases cannot be readily distinguished by miscibility methods involving optical microscopy.

DISCUSSION

Dipolar Effects

The results show that the formation of the hexatic B phase versus that of the crystal B phase is not dependent upon the overall steric shape of the molecular structure, but on the dipolar character of the molecule. The strengths of these dipole moments are also dependent on their integration into the polarisable electronic structure of the material. The central core of n-hexyl 4'-n-pentyloxybiphenyl-4-carboxylate (650BC) has a structure

where the ketonic (C=O) oxygen atom of the ester function is directly conjugated with the polarizable π electrons of the core. The strongly electronegative ketone group withdraws electrons from this electron rich source by mesomeric relay, and consequently the alkoxy oxygen atom pushes electrons back into the π system to produce the canonical structure

which is relatively stable. Thus, there is a polarisation of the electrons towards the ester function increasing its overall lateral dipole strength.

If the ester group is reversed with respect to the core, the ketonic oxygen atom is no longer conjugated with the polarisable core electrons

and therefore the strength of the ester dipole is greatly dimin-
ished, with the likely result that the B phase exhibited will be
crystalline in nature.

In Schiff's base compounds the situation is considerably
different because the linkage does not contain a strong electoro-
negative group. Hence, any charge separation produced by mesomeric
relay is relatively unstable

and the linkage remains relatively unpolarised. The dipolar forces
associated with this functional group are weaker than for the
ester. The situation for thio esters is not as clear cut as those
discussed previously. The keto function still pulls electrons
through the π cloud as before, but the second oxygen atom of the
ester is replaced by a more polarisable sulphur atom. The keto
group can therefore polarise its d electrons as well as those of
the core, and because sulphur is already less electronegative than
oxygen the overall strength of the crossed dipole of the –COS–
system drops even though its polarisability increases.

In essence the crystal B phase is favored by materials which
have a central linkage that carries a weak lateral dipole, whereas
the hexatic B phase is favored by linkages with stronger lateral
dipoles.

Further speculations for the preference of one B phase over the other can be made if we examine the closeness of approach of two neighboring molecules. This distance is probably governed by a Leonard-Jones potential energy function. Therefore, as the strength of the crossed dipole becomes larger so will the closest distance of approach because of increased repulsive forces. In this situation the molecules will be less tightly packed and the possibility of the formation of a hexatic phase occurs. The extent of the in-plane correlations is reduced by the stronger repulsive forces of the central linkage in esters.

EXPERIMENTAL

All the materials were prepared by standard synthetic methods and purified by column chromatography over silica-gel (60-200 mesh, 3x30 cm) using hexane/dichloromethane (1:1) as eluant, and then recrystallised from light-petrol (40-60°). Their identities were confirmed by infra-red spectroscopy, mass-spectrometry and elemental analysis.

Transition temperatures of the pure compounds and of binary mixtures for miscibility studies were determined by thermal optical microscopy using a Zeiss Universal polarising microscope in conjunction with a Mettler FP 52 hot-stage and FP5 control unit.

CONCLUSIONS

The type of B phase formed by a given smectic B phase material is dependent primarily on the strength of the lateral dipole. The stronger the lateral dipole is the more likely the material will exhibit hexatic properties.

REFERENCES

1. J. W. Goodby and R. Pindak, Mol. Cryst. Liq. Cryst. 75, 233, (1981).

2. D. E. Moncton and R. Pindak, Phys. Rev. Lett. 43, 701, (1979).

3. A. J. Leadbetter, J. C. Frost, and M. A. Mazid, J. de Physique Lett. 40, L-325, (1979).

4. R. Pindak, D. E. Moncton, S. C. Davey and J. W. Goodby, Phys. Rev. Lett. 46, 1135, (1981).

5. G. W. Gray and J. W. Goodby, Mol. Cryst. Liq. Cryst. 37, 157, (1976).

6. J. W. Goodby and G. W. Gray, J. de Physique 40, 363, (1979).

7. J. W. Goodby and G. W. Gray, J. de Physique 37, 17, (1976).

8. A. J. Leadbetter, M. A. Mazid, B. A. Kelly, J. W. Goodby and G. W. Gray, Phys. Rev. Lett. 43, 630, (1979).

9. D. E. Moncton and R. Pindak, Ordering in Two Dimensions, ed., S. K. Sinha, North Holland, New York, (1980).

10. D. J. Byron, D. A. Keating, M. T. O'Neill, R. C. Wilson, J. W. Goodby and G. W. Gray, Mol. Cryst. Liq. Cryst. 58, 179, (1980).

11. A. D. Clemson, Hull University Report, 1974.

12. G. Poeti, E. Fanelli and M. Braghetti, Mol. Cryst. Liq. Cryst. 61, 163, (1980).

13. J. W. Goodby, Mol. Cryst. Liq. Cryst. Lett. 72, 95, 1981.

14. R. Pindak, private communication.

15. G. W. Gray; Molecular Structure and the Properties of Liquid Crystals, Academic Press, New York, (1962).

16. C. C. Huang, private communication.

CHARACTERIZATION OF THE PHASES OF TWO ISOMERIC CYANO-ESTER

HOMOLOGOUS SERIES – II

P. E. Cladis, P. L. Finn, J. W. Goodby

Bell Laboratories
Murray Hill, New Jersey 07974

ABSTRACT

One of the puzzles of the ordering of Liquid Crystal Phases has been the observation that a less-ordered phase, the nematic phase, can occur at both a higher and a lower temperature than a more ordered phase, the smectic A phase. Although the physical properties of the lower temperature nematic are identical to the higher temperature phase, it is called the re-entrant nematic in recognition of the fact that similar phenomena have been noted in other ordered systems of Condensed Matter Physics.

In this paper, we have chosen to study two model systems based on cyano-ester isomers: (4-n-alkoxyphenyl 4'cyanobenzoate [nOPCB] where n = 6,7,8,9,10 and 12 and 4-cyanophenyl 4'-n-alkoxybenzoate [CPnOB] where n = 6, 8 and 10). X-ray powder diffraction patterns were made in the solid, smectic A and smectic C phases of these materials in order to understand the molecular features responsible for the occurrence of re-entrant nematic phases.

A striking feature revealed by x-rays is that lengths associated with monomers are not observed, neither in the solid phases,

nor the smectic phases. The data suggests that the same kind of
dimer formation we proposed for cyanobiphenyl and cyano-Schiff base
compounds is also present in the smectic phases of these new materials. Dimer formation is believed to be in part responsible for
the occurrence of re-entrant nematic phases because dimers do not
pack efficiently into smectic A layers. Decreasing the free volume around a dimer (by increasing pressure or decreasing temperature) easily destabilizes A-type smectic layering in favor of a
more dense nematic or smectic C type packing, depending, for example, upon the strength of the lateral dipoles. The occurrence of
a smectic C phase is totally unexpected in materials possessing
large longitudinal molecular dipoles. Its occurrence in the nOPCB
series, therefore, can be considered additional evidence favoring
our dimer model which predicts zero or a very small net longitudinal dipole.

The CPnOB series is found to behave like a shorter molecule
than its isomeric counterpart in the nOPCB series. For example:
1) CPnOB clearing temperature are about 10°C lower than those of
nOPCB. 2) CP8OB-100PCB mixtures are ideal. 3) Smectic phases
occur for $n \geq 7$ in nOPCB but only for $n \geq 10$ in CPnOB.

The sequence of solid transitions leading to melting in these
materials is very complicated and not always unique. We discuss
them in some detail.

INTRODUCTION

In order to predict the relative stability of the different
liquid crystal phases it is necessary to take into account many
different factors. The effective contribution of each one is difficult to assess because there are so many. The study of molecular
isomers affords one a rare opportunity to study the effect on
macroscopic physical properties effected by a change in just one
of the many molecular parameters. Here, we report the results of

our studies using x-ray diffraction, the optical microscope and differential scanning calorimeter for the two isomeric homologous series:

1)   4-n-alkoxyphenyl 4'-cyanobenzoate

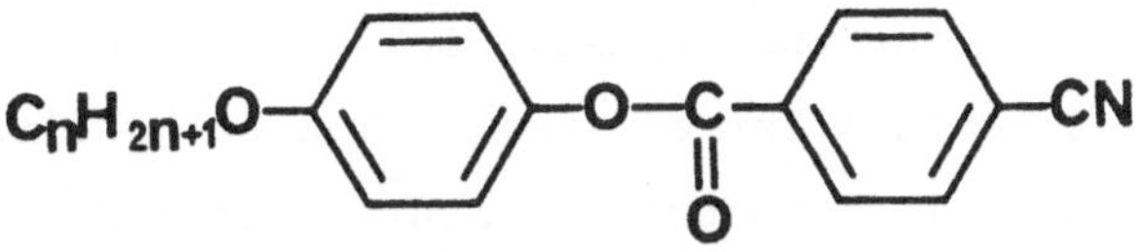

nOPCB

where n = 6,7,8,9,10 and 12

2)   4-cyanophenyl 4'-n-alkoxybenzoate

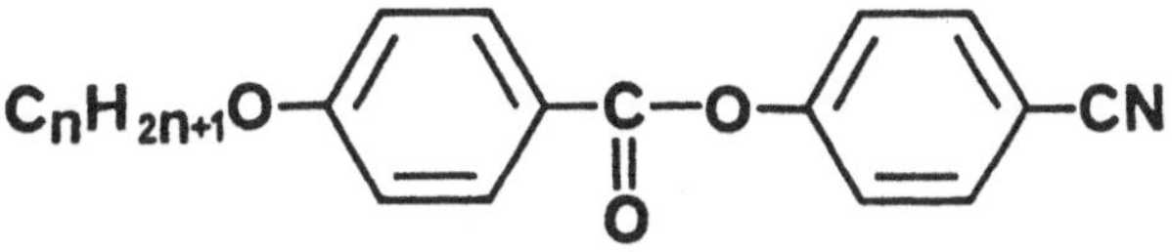

CPnOB

where n = 6,8 and 10. As we discussed in a previous paper[1], the two rings are not well conjugated and in the nOPCB series, the dipole associated with the ester function (-COO-) generally opposes that associated with the cyano function (-C = N) and reinforces it in the CPnOB series. The object was to study the effect this would have on the appearance of the re-entrant nematic phase and the structure of the smectic A phase which separates it from the usual higher temperature nematic phase.

The model which has been proposed to account for the layer spacing of the smectic A phase was originally based on evidence

found in the syano-biphenyl (nOCB) (and cyano-Schiff base) series
(top Fig. 1). In this compound, it was assumed that the electron
withdrawing power of the cyano function would extend the length of
the biphenyl system and be reinforced by the ether-oxygen donating
electrons to the benzene rings, resulting in a single large longitudinal dipole for the aromatic moiety. Dimers would tend to form
and neutralize this large dipole. The bulky dimer would not pack
so efficiently into smectic A layers. In order to pack more dense-
ly, the nematic phase would re-enter and/or even possibly a smectic
C phase could occur. At even lower temperatures, the occurrence of
another smectic A phase was also envisaged.

Scrutiny of Fig. 1 reveals that although the cyano-function is
present it is not clear that arguments used to justify the model
for the cyanobiphenyls apply to cyano-esters. For example, as discussed in the previous paper,[1] benzene rings of the nOPCB and
CPnOB series are not as conjugated as in the nOCB series.

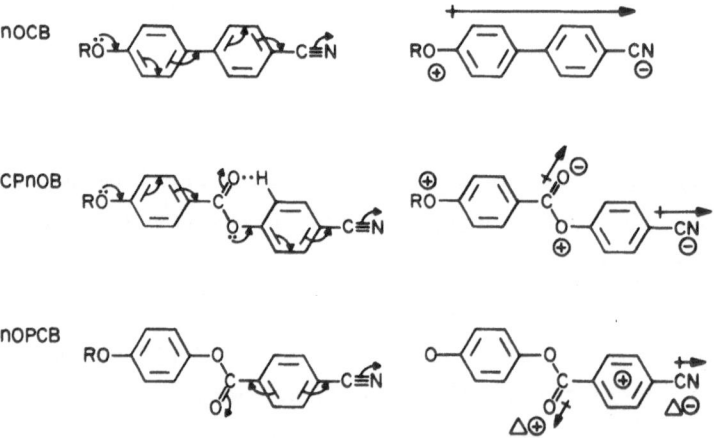

Fig. 1. A comparison of the electronic structure of the cyanobiphenyls with the n)PCB and CPnOB.

The dimer proposed for nOCB is with the benzene rings of two monomers forming an anti-parallel configuration with the cyano-end of one monomer close to or overlying the alkoxy end of the other. This leads to a configuration in which the longitudinal dipoles of the monomers cancel. Forming dimers of the sort discussed above with the two isomers nOPCB and CPnOB (Fig. 1) does not lead to any obvious cancellation of dipoles but rather the formation of two independent associated rings with opposing longitudinal dipoles (since the dipole associated with the cyano ring will be larger than that associated with the ester ring).

Arguments can be made (and we have made them in the previous paper[1]) how the combined effect of mesomeric relay (the cyano function pulls electrons through the σ framework as well as the labile π-cloud), and polarization (the cyanobenzoate moiety polarizes the electron cloud of a neighboring alkoxyphenol group so that nOCB type dimer formation is more akin to compound formation in nOPCB than in CPnOB) could lead to dimer formation. These arguments are not obvious from first principles, in fact, their credibility depends largely upon knowing what does happen.

After briefly describing our experimental arrangement in Section II, we present our results in Section III. They indicate that the same kind of dimers we proposed for cyanobiphenyl and cyano-Schiff base compounds are also present in the smectic A phases of these two new materials and re-entrant nematic phases abound. The lateral dipole of nOPCB introduces monotropic smectic C phases whose layer spacing we have only been able to measure in mixtures. Evidence for compound formation, rather than simple dimerization, is provided by the absence of monomer lengths in the x-ray pattern of the solid phases of nOPCB and CPnOB. As we recall, in the nOCB compounds, even the solid phases of mixtures exhibited the molecular lengths of the monomers.[4]

II. EXPERIMENTS

A. Materials

The chemical synthesis of nOPCB and CPnOB is described previously[1]. Both series of compounds were purified by column chromatography over silica gel (60-200 mesh), (30x2.5cm) using hexane
dichloromethane 1:1 as eluant. The samples were recrystallized
from light-petrol (40-60°) and their purity and structures elucidated by thin-layer chromatography, CHN elemental analysis, proton-
NMR and mass spectrometry.

B. Optical Observations

These were made using a Leitz polarizing microscope in conjunction with a Mettler FP52 hot-stage and FP5 control unit.

C. Differential Scanning Calorimetry

For these measurements we used a Perkin Elmer DSC1B operating
at 10°C/minute. The measurements were recorded on chart paper and
the area under each of the peaks evaluated with a planimeter.
Calibration was made using the melt of 1.172 mg indium.

D. X-ray Diffusion

Powder patterns were recorded using an X-ray diffraction system described as follows. X-rays were obtained from a rotating
anode generator operating at 4kW with a projected spot size of
0.2 mm x 0.2 mm. X-rays were monochromatized and focused by a
singly-curved asymmetrically cut quartz crystal using the $10\bar{1}1$
planes. The focal spot size at the detector was $\sim$0.5 mm x 0.1 mm.
Diffracted X-rays were recorded using a stable linear position-
sensitive detector. The anode is made of a nickel chromium alloy
and the detector is filled with 90% argon and 10% methane at
100 lb/in^2. Charge division position encoding was used. Typically
X-ray patterns were recorded in 15 min. The precision of spacing
measurements was on the order of 0.1 Å.

For the measurements, the samples were loaded into thin quartz capillaries. Their temperature was monitored by a thermocouple and controlled to ±0.02°C using the YSI controller.

III. RESULTS

The results of the differential scanning calorimetry are tabulated in Table I. The X-ray measurements have been gathered together in the Appendix. The transitions which are clearly observed in the microscope are shown in the previous paper.[1] Figs. 2 and 3 show the transitions found by DSC analysis for nOPCB and CPnOB, respectively. On the whole, the agreement is quite good. In these larger samples, however, crystallization occurred about 10°C, higher than in the thinner optical samples so that the C phase of pure nOPCB was not accessible for study by X-rays.

Table I. The heats of transition and transition temperature as determined from DSC measurements. The largest heats of melting have been underlined.

	Solid-Solid		Solid-Solid		Solid-Solid		$S_A \leftrightarrow I$		$S_A \leftrightarrow N$		$N \leftrightarrow I$	
	T(°C)	cal/mol	T(°C)	cal/mol	T(°C)	cal/mol	T(°C)	cal/mol	T(°C)	cal/mol	T(°C)	cal/mol
12OPCB	74	746					103.5	83				
10OPCB	61	182	65	460	74	107	97.	234				
9OPCB[1]	68	753					95	203				
8OPCB[2]	68	658							88	1.8	93	21
7OPCB	61	253	70	406					79		92	13
6OPCB	65	28	82.5	424							90	14
	65.5	18.1	68.5	35	79	757						
CP10OB	59	790	65	28.5					79		85	19
CP8OB	65	27	73	760							81	3.9
CP6OB	69.5	570									80	6.1

1. Two crystal forms appeared very close together in this melt.

2. Three crystal forms appeared very close together in this melt.

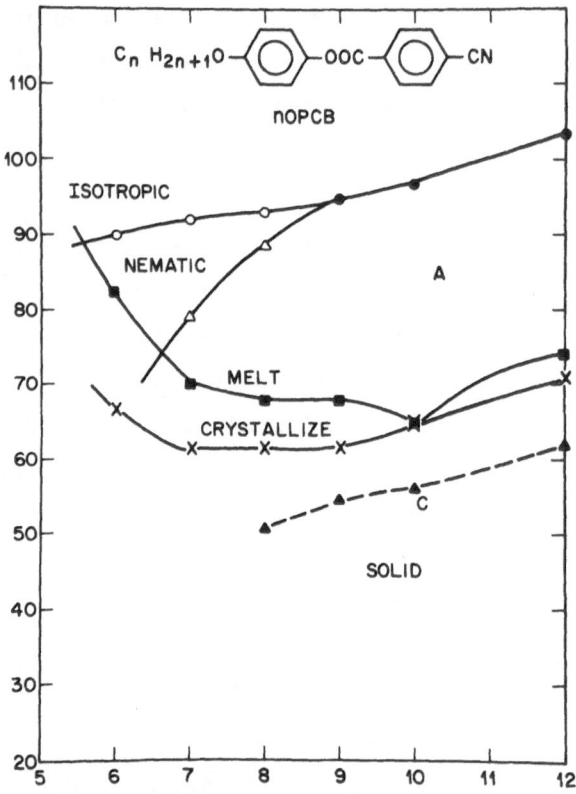

Fig. 2 Transition temperatures versus alkyl chain length (n)
        for nOPCB as determined by DSC.

A.   Solid Phases

      Usually, where there is multiplicity of solid phases, the true
melting point is taken to be the one with the maximum heat of
transition. In these compounds with several solid phases, the DSC
evidence is that the relative peak sizes depends upon how fast the
temperature is scanned as well as the thermal history of the sample.
For example, in the case of CP100B, on the initial run with increasing temperature, small peaks were observed at 65°C and 69°C
followed by a large peak at 79°C. After increasing the temperature
to the isotropic phase then cooling back to 41°C where a large
peak signalled recrystallization, the temperature was lowered to

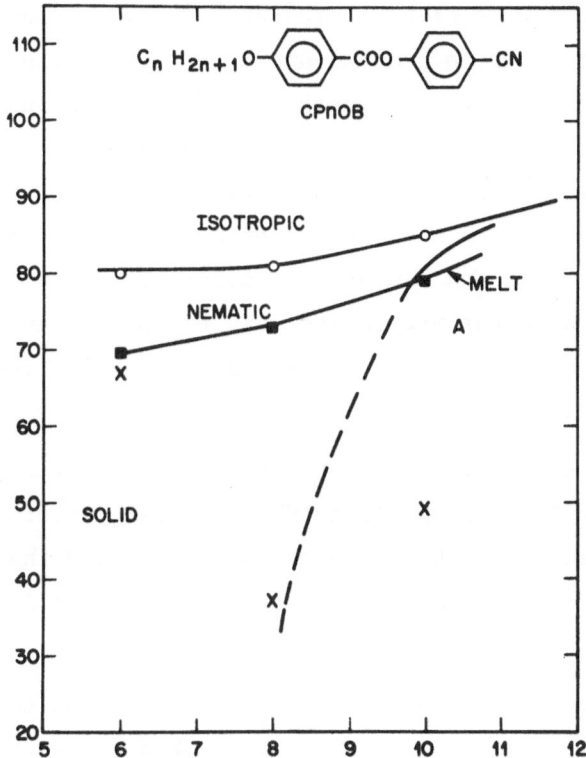

Fig. 3 Transition temperature versus alkyl chain length (n)
for CPnOB by DSC.

room temperature, then increased again. This time a large peak
was observed at 59°C and a smaller one at 65°C. In the first run,
the combined heats of transition are 810 cal/mol and in the second
818.5 cal/mol. The similarity of these two numbers seems to indicate that the initial phase (solid) and final phase (smectic A)
are the same but melting may proceed by a set of non-unique intermediate steps. A glance at the X-ray pattern (see Appendix for
CP10OB) only concurs that 79°C is the initial melt.

Frequent observations in the optical microscopic have shown
that depending upon how fast cooling proceeds, several of the pure
CPnOB compounds exhibit co-existence of smectic A and K_2 (K_2 being

our designation of the higher temperature solid phase) transforming at lower temperatures to nematic and K_2. Heating the nematic-K_2 coexistence region results in both nematic domains and K_2 domains melting to smectic A in CP100B. Co-existence of two phases is not expected in pure compounds and suggests that the smectic A dimers are not compatible with whatever other molecular association the solid phase requires and can only transform very slowly to it. The X-ray patterns frequently revealed weak but persistent diffraction characteristic of the smectic A dimer mixed in with the strongest lines of the solid (see Appendix).

In the case of 100PCB we note that the DSC determined melting transition at 65°C corresponds to the optically determined one but that another smaller peak was observed at 74°C and this correlates with what appears to be melting of a K_2 and smectic A co-existence region (as shown by Fig. A.5 in the Appendix).

Fig. 4 shows a weak odd-even effect in the sum total of the heats of melting versus position in the nOPCB series. The melting temperatures also show weak odd-even effects consistent with the idea of dimerization in the solid phase as well as the smectic A phase. Certainly X-ray diffraction supports this notion. In Fig. 5, the lattice parameters observed just before melting are shown as a function of n. The line, L_s, represents the length of each monomer as measured from a scaled model. With fully extended alkyl chain: $L_s = 15.71 + 1.32 n$. The line L_{p_1} represents the length when only the cyano-moiety is overlapped[1] in an anti-parallel arrangement: $L_{p_1} = 21.31 + 2.72n$ and can be seen to fit quite well the longer lengths observed in homologues of nOPCB (but not CPnOB). The shorter lengths then could be interpreted as second order diffraction corresponding to these longer lengths, L_{p_1}.

The longest lengths measured in the solid phase of CPnOB are also displayed and found to be much shorter - at least 10Å shorter than their isomeric counterpart. This suggests that the CPnOB

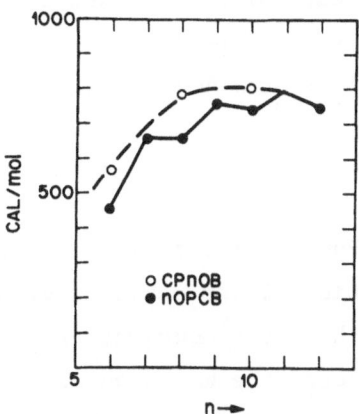

Fig. 4 Total heats of melting versus n for CPnOP (○) and nOPCB
(●)

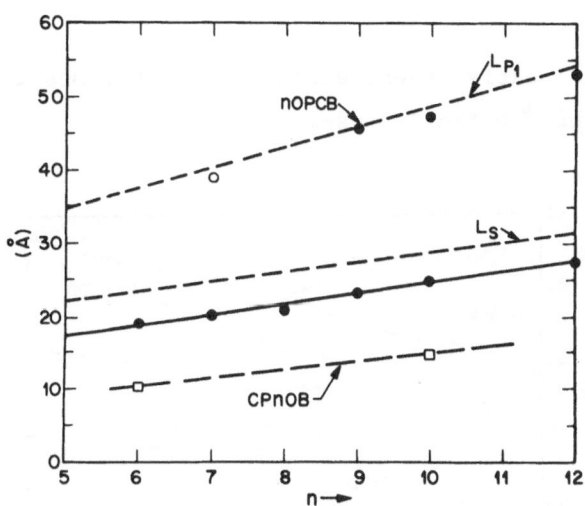

Fig. 5 Lengths observed in the solid just before melting nOPCB
(■) and CPnOB (□). The nOPCB lengths are much longer
then the CPnOB ones. L_s is the monomer length and L_{p_1} is
the length of a dimer formed by anti-parallel association
but only one benzene ring is overlapped.

solid is more dense than the nOPCB solid which in turn is consistent with its larger heats of melting and lower clearing temperatures (Figs. 2, 3 and 4). If we interpret these as second order
lines, the fundamental is somewhat shorter but close to the fully
extended monomer.

B.   Smectic A Phases

Evidently, the CPnOB series is less smectogenic than nOPCB.
Fig. 6 shows the smectic A layer spacings versus n along with L_{P_2} =
16. + 2.58n, the length of the model dimer originally proposed
for nOCB. As can be seen, the measured lengths of nOPCB smectic A
phases is consistent with this length whereas again the CP1OOB
length is shorter. In fact, it's the same as the A phase in 8OPCB.
The shorter length of CPnOB is consistent with its lack of smectic
phases. The temperature dependence of the smectic layering, d, was
fitted to a simple linear expression $d = a - bT(°C)$ and the results
are shown in Fig. 7. Several interesting results emerge:

1)   a vs. n is nearly linear except when n becomes too small to
      support the smectic A phase.

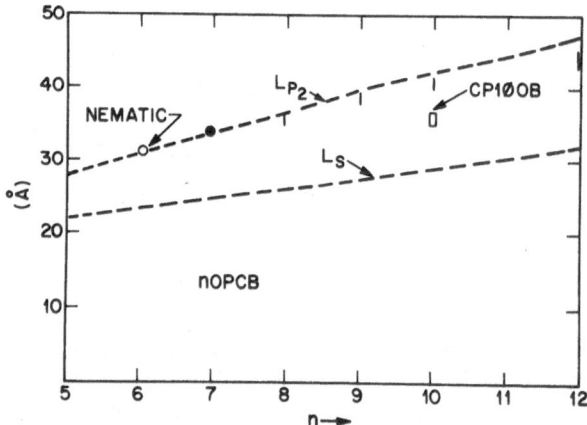

Fig. 6   Smectic A layer spacing versus alkyl chain length observed
         in nOPCB and CP1OOB. Lp_2 is the length of an antiparallel
         association with both benzene rings overlapped. L_S is the
         monomer length.

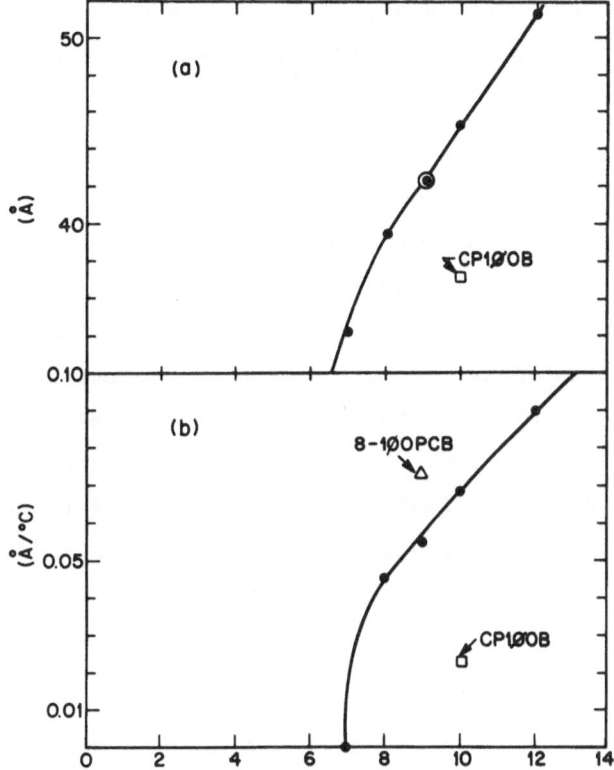

Fig. 7 The temperature dependence of the smectic A layer spacing
        is summarized in the form d = a - bT(°C) where a and b
        are shown here as a function of n for nOPCB. Note that
        both the length and temperature dependence of CP100B
        correlates with a shorter homologue in the nOPCB series.
        The open circle in (a) is for a 50% 8-10OCPB mixture.

2)   The length of a 50 wt. % mixture of the 8-10 homologues of the
      nOPCB series is the same as the nonyl-homologue but the
      temperature dependence of the mixture is even larger than
      100PCB.

3)   The temperature dependence of the layer spacing falls off as
      the chains get shorter being zero for the 70PCB and -0.09Å/°C
      for 120PCB.

4)   Both the temperature dependence, b, and the size of the 0°C
      layer spacing, a, for CP100B are comparable to shorter members
      of the nOPCB series about n = 7.5, in fact.

      Evidence that L_{P_2} type pairing occurs in the smectic A phase
of these compounds is shown in Fig. 8. Here we plot the lengths
observed in mixtures of 8OPCB and 8OCB (4-cyano 4'-n-octyloxybi-
phenyl) and the pair-breaker[4] 408 (N(-4-n-butyloxybenzilidene)-4'-
n-octylaniline). The linear variation of d in the former case is
evidence that the pairs are the same in both compounds. The decrease in layer spacing with increasing 408 is also similar to that
which is observed in 8OCB and has been interpreted[4] as being due to
the inhibition of the dimerization process owing to the presence of
the 408 molecules. Extrapolating the 408 rich side of these
measurements results in the molecular length of the single monomers
of 8OPCB as 25.8 Å and CP8OB as 24.7 Å. Even the CP8OB monomer is
shorter than the 8OPCB monomer!

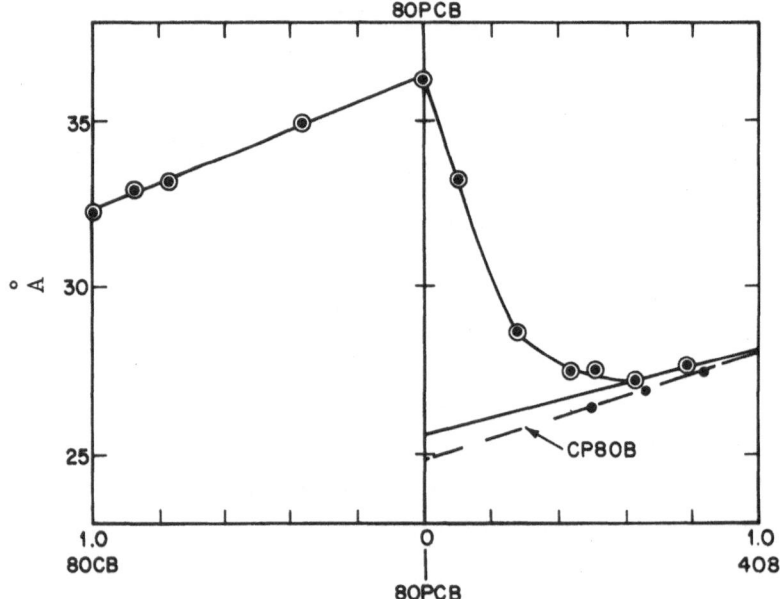

Fig. 8   Smectic A layer spacing if 8OPCB as a function of concen-
         tration (wt.%) in mixtures with 8OCB and 408. The extrapo-
         lated monomer length is 25.8Å for 8OPCB and 24.7Å for CP8OB.

Now, the fact that CP100B exhibits a smaller layer spacing
than 100PCB can be interpreted to mean that it is less paired than
the 100PCB. This interpretation makes it difficult to account for
the re-entrant nematic phase. It can also be interpreted to mean
that both 100PCB and CP100B are equally well paired but CP100B just
simply is shorter, because for example, its overlap is greater,
therefore its dimer is shorter. To support this notion, mixtures
of 80PCB and CP80B exhibit the re-entrant nematic phase (Fig. 9)
as do mixtures of CP80B with its higher homologue CP100B (Fig. 10).
The CP100B-CP80B mixture qualitatively resembles the 80CB-60CB
mixture.[5]

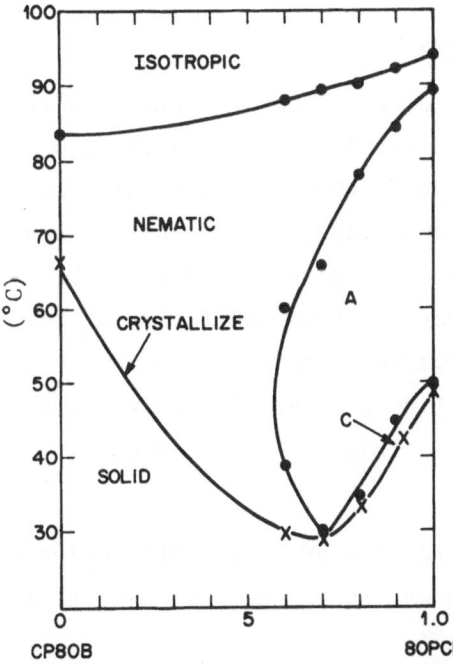

Fig. 9 The re-entrant nematic phase formed when CP80B is mixed
with 80PCB. The smectic C phase is suppressed.

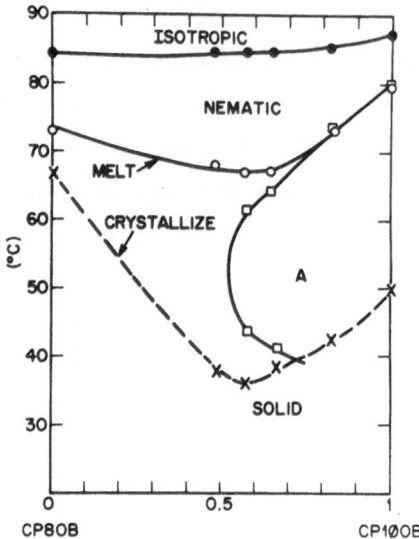

Fig. 10 The re-entrant nematic phase is formed when CP8OB is
         mixed with CP1OOB in a manner exactly analogous to mix-
         ture of 8OCB and 6OCB.

C.   Smectic C Phases

      The appearance of the smectic C phase in the nOPCB series is
remarkable. Since crystallization prevented our being able to
measure the evolution of the layer spacing through the A and C
phases in the pure compounds, we did this in a 50 wt. % mixture
of 8OPCB and 1OOPCB. We chose this mixture because it seemed[6]
that C phases are most stable when formed of compounds with alkyl
chains extending from each end of the aromatic moiety and which
differ by two methylene groups. In addition, the melting point of
mixtures is depressed from the pure compounds. In this case, it
went down to 58°C.

      Fig. 11 shows results of X-ray diffraction and in Fig. 12
optical micrographs are shown corresponding to the temperatures
marked a, b, c, and d.

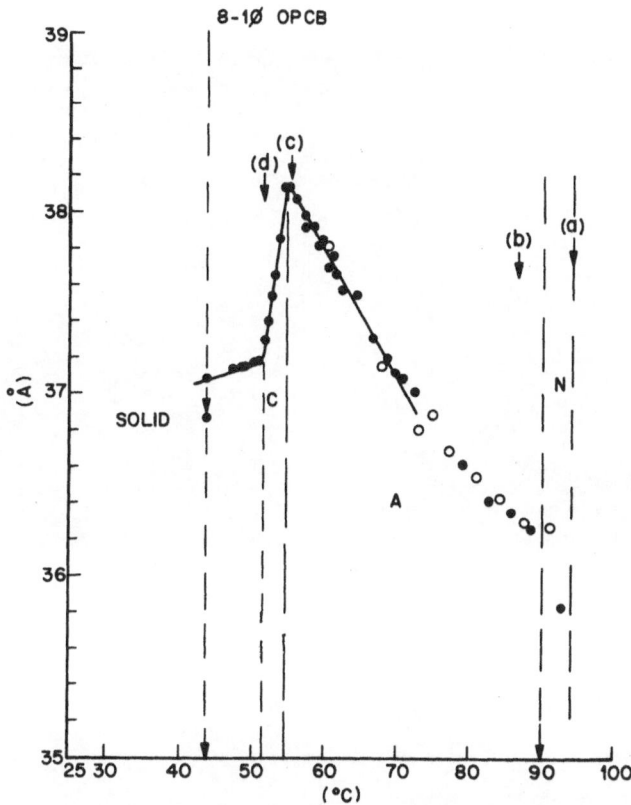

Fig. 11 Details of the change in layer spacing observed in the
50 wt. % mixture of 8OPCB and 10OPCB. Although the change
in layering in the C phase is quite sharp in the vicinity
of the A phase, the transition appears continuous.
Textures corresponding to the temperatures marked a, b,
c, and d are shown in Fig. 12.

The A-C transition temperature was found to be very reproducible upon cycling the temperature and from its behavior shown in
Fig. 11, it seems that the A-C transition is a continuous one.
Furthermore, it is interesting that the change in slope observed
in the middle of the C phase correlates with a texture change observed in the optical microscope due to the nucleation of a solid
phase.

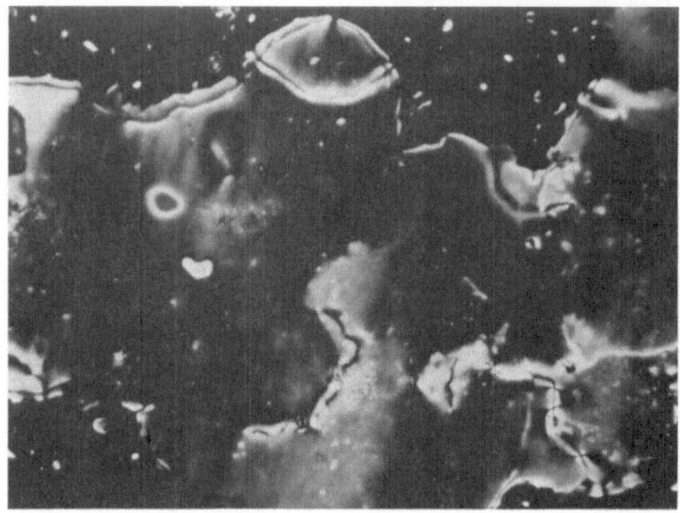

a

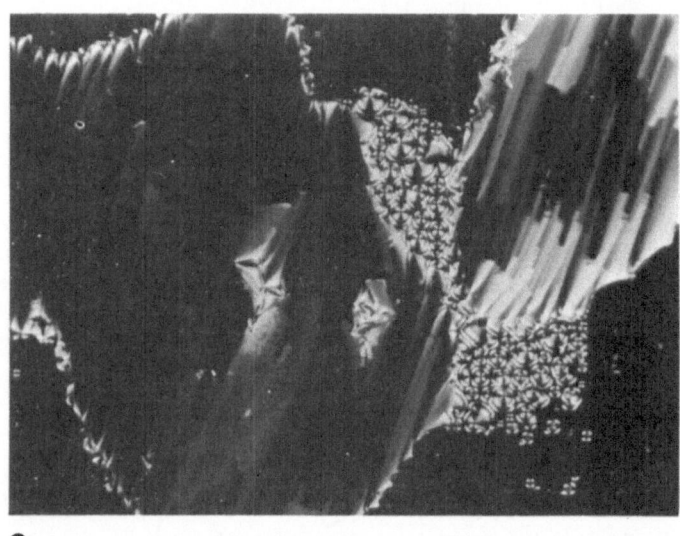

C

Fig. 12   Textures observed at the temperatures shown in Fig. 11.
          (a) nematic – isotropic, (b) smectic A, (c) smectic C –
          smectic A, (d) smectic C and solid.

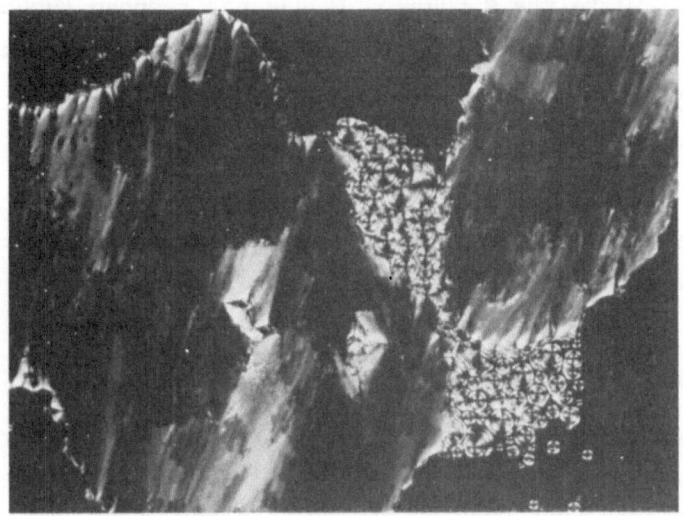

b

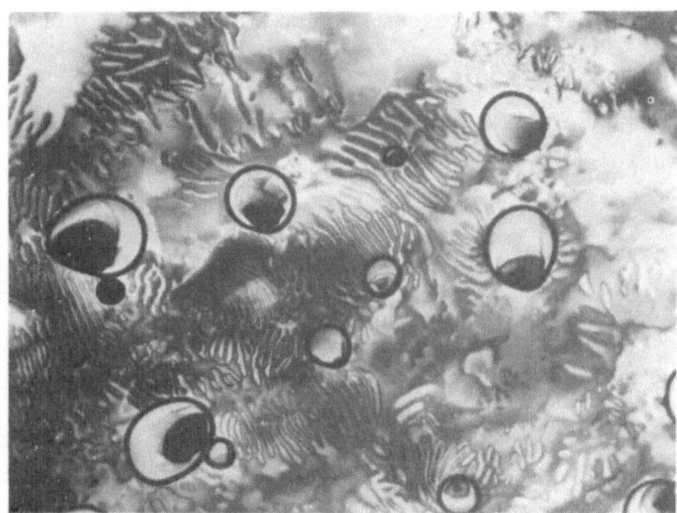

d

Fig. 12   Textures observed at the temperatures shown in Fig. 11.
          (a) nematic – isotropic, (b) smectic A, (c) smectic C –
          smectic A, (d) smectic C and solid.

One final observation we made was to apply an electric field
$[3 \times 10^3$ volts/cm] to the C phase. It did not respond suggesting
that the dielectric anisotropy of this phase is much weaker than
one would expect for a cyano-compound.

CONCLUSION

Simply altering the relative orientation of the ester link
has a profound effect on the stability of the various liquid crystal phases.

In order to characterize the phases of the two isomers nOPCB
and CPnOB, we have studied them using X-ray diffraction, the optical
microscope and differential scanning calorimetry. Our results
seem to indicate that even in the solid phase, these materials
form associations which behave like foreign material when mixed
with the different associations formed in the liquid crystal planes.
The melting transitions are complex and not unique and the monomer
lengths are conspicuous by their absence.

Dimerization of the sort previously proposed for the smectic A
phase of the cyanobiphenyls occurs in these compounds as well,
although the A phase of CP10OB appears much shorter than its isomeric counterpart 10OPCB. We conclude that CP10OB is as associated
as 10OPCB (since it also exhibits re-entrant nematic phases when
mixed with its shorter homologue CP8OB which has no smectic phase
as does 8OCB when mixed with 6OCB) but its dimer is simply shorter.

Another way of expressing this is in terms of the strength of
short-range anti-parallel correlations induced by longitudinal
molecular dipoles.[7] When the dipoles are strong, these correlations are strong and result in an increase in the orientational
order parameter. The reduction in the longitudinal dipole of a
dimer compared to a monomer implies a net decrease in anti-parallel
correlation of dimers (compared to monomers) with a resulting net
loss in the strength of the orientational order parameters, S.

Cyanophenyl monomers have large dipoles. They are thus expected to induce large anti-parallel correlations with correspondingly larger orientational order parameters (and more stable smectic A phases) than cyanophenyl dimers which interact via a reduced dipole compared to the monomer. For a given position, n, in a homologous series, CPnOB dimers cancel the longitudinal dipoles more effectively than nOPCB dimers resulting in a weaker orientational ordering in CPnOB dimers compared to nOPCB dimers at fixed n and dimer concentration. For example, CP8OB does not have a large enough S to stabilize smectic A phases whereas 80 PCB does. This kind of description is also compatible with the enhancement of the smectic A phases with the break-up of dimers.[4]

The appearance of smectic C phases in nOPCB is unexpected. The microscopic mechanism responsible in these materials for which the ring systems are not as conjugated is thought to be the polarization mechanism proposed previously,[1] i.e. polarization of the oxy-ether by the cyano function during dimerization.

REFERENCES

1. J. W. Goodby, T. M. Leslie, P. E. Cladis and P. L. Finn, (previous communication).

2. P. E. Cladis, P. K. Bogardus and D. Aadsen, Phys. Rev. A18 2296 (1978).

3. P. E. Cladis, R. K. Bogardus, W. B. Daniels and G. N. Taylor, Phys. Rev. Letters 39, 720 (1977).

4. P. E. Cladis, Mol. Cryst. Liq. Cryst. 67, 177 (1981).

5. P. E. Cladis, D. Guillon, F. R. Bouchet and P. L. Finn, Phys. Rev. A23, 2594 (1981); D. Guillon, P. E. Cladis and J. Stamatoff, Phys. Rev. Letters 41, 1598 (1989).

6. G. W. Gray and J. W. Goodby, Mol. Cryst. Liq. Cryst. 37, 157 (1976).

7. N. V. Madhusudana and S. Chandrasekhar, Pramana, Suppl. No. 1, 57 (1975).

APPENDIX

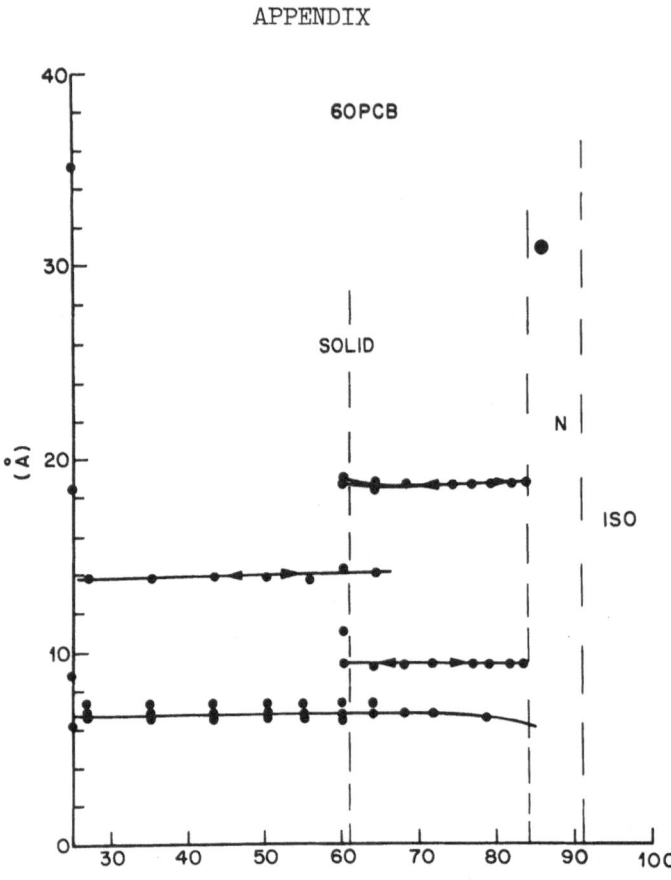

A1. 6OPCB. Starting at room temperature, one often observes
     a small peak at ∿35 Å. Upon heating, this disappears as
     supercooled or vitrified sample crystallizes to a solid
     with characteristic spacing at 14 Å with second harmonic
     and two other wide angle spacings of 6.8 and 6.6 Å. These
     wide angle peaks are very intense. About 60°C, the 14 Å
     solid melts to another solid-like phase with characteristic
     lengths at 18.6 Å (and second harmonic at 9.3 Å) and with
     small wide angle scattering at ∿6.8 Å, 6.4 and perhaps
     5.5 Å. About 84°, this solid-like phase melts to a nema-
     tic with broad peak centered about 31 Å. Cooling to 82°C,
     18.6, 9.3 and 6.8 Å lines re-appear.

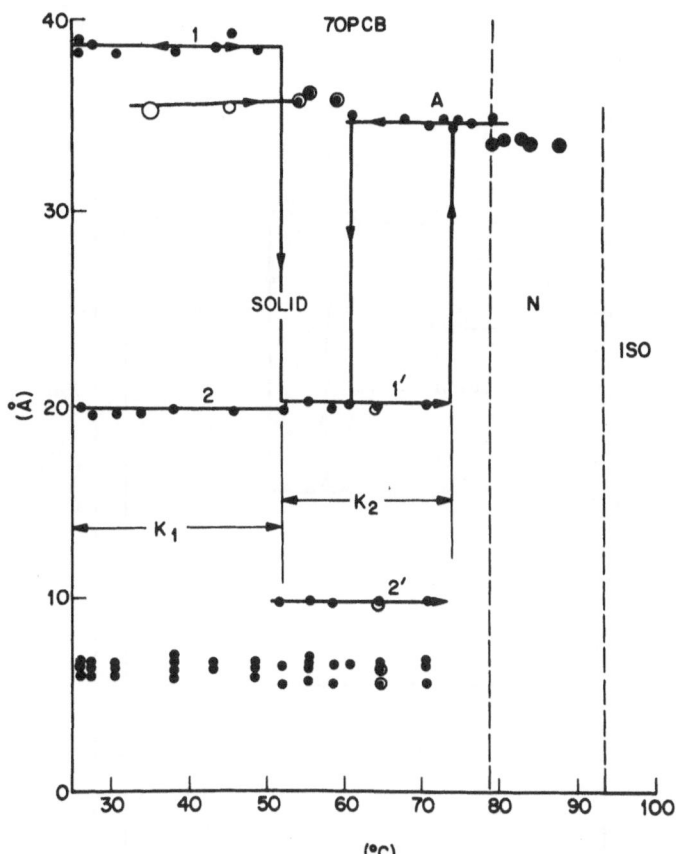

A2. 7OPCB. Heating from room temperature, this material showed
     solid-like lines at 39.0 Å with second harmonic (S.H.) or
     2H at 19.7 Å and several lines at wide angles 6.7, 6.4 and
     6 Å. About 50°C, the 39 Å lines melted to 20 Å (S.H. 9.9
     Å) with all the wide angle lines intact. About 74.1 °C,
     this melted to a smectic A type pattern with characteris-
     tic spacing ∿34.2 Å which in turn went over to the nematic
     phase ∿79°C (broad peak centered at 33.1 Å). Upon cooling,
     the smectic A crystallized to the 20 Å solid around 61°C.
     Large hysteresis at this transition suggests this to be
     solid. Frequently, weak lines at 35.6 Å in solid phases
     suggest smectic A phase co-exists with both solids. In
     S_A phase

$$d(\overset{\circ}{A}) = 34.1 + .006T \ (°C)$$

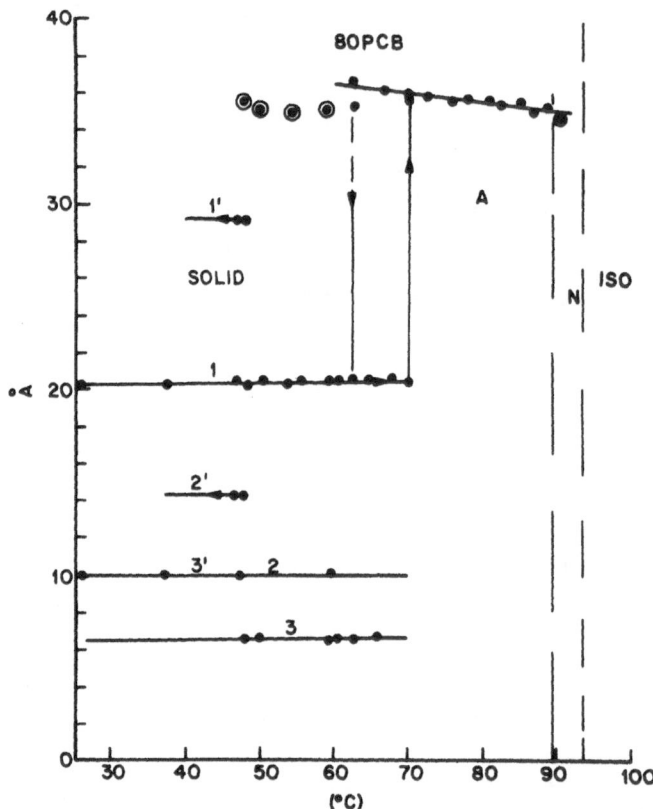

A3. 80PCB. Upon heating from room temperature, the 20.4 Å solid
     (2H = 10.0 Å, 3H ≈ 6.6 Å) persisted to 70.3°C. When heated
     through the S_A phase to isotropic (∿95°C) then cooled back
     through S_A phase, the sample crystallized back to the 20.4 Å
     solid with very small peaks ∿35.4 Å at 63.3°C. At 47.5°C
     another solid with spacing ∿29.3 Å nucleated, co-existed with
     20.4 Å solid. S_A layer spacing follows rule

$$d(\overset{\circ}{A}) = 39.3 - .045T(°C).$$

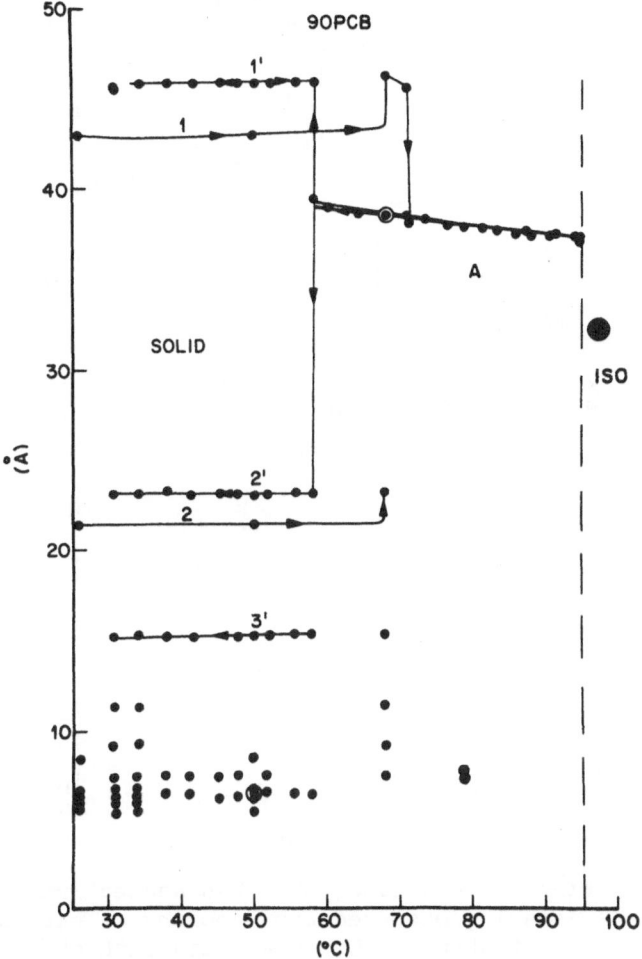

A4.   90PCB. At $69.2°C$, K_1 melts to K_2 which in turn melts to
      smectic A around $71°C$. Weak scattering observed in iso-
      tropic phase $(T=97.5°C$. Extremely broad peak centered
      around $32Å)$. S_A layer spacing changes quite a bit with
      temperature. For layer spacing

$$d = 42.1-0.055T(°C)$$

The layer spacing grows from $37 Å$ at the S_A isotropic vicinity to $39 Å$ just before crystallizing to K_2 at $59°C$. K_2
supercooled to room-temperature. Again large hysteresis
at K_2-S_A transition suggests that this is solid. K_2 layer
spacing is $\sim45.9 Å$ with 2nd, 3rd and 4 harmonics sometimes
observed. Lots of action in the $6.4-7 Å$ vicinity.

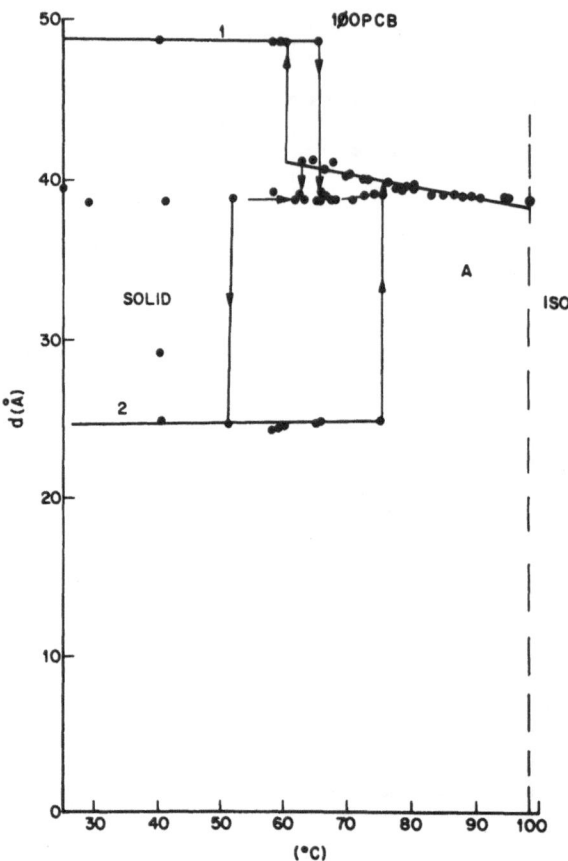

A5. 100PCB. First time around, this was heated rapidly to 80°C,
     then layer spacing observed upon cooling. This sample
     crystallized at about 60°C, to a solid with characteristic
     spacing ~48.9 Å (2.H. 24.5 Å). This solid persisted upon
     cooling to 40°C co-existing with another solid ~24.8 Å be-
     low 51°C. It melted about 65°C to what may be another
     solid with spacing ~38.7 Å. Rerunning this sample later,
     we observed the 39 Å "solid" at 26°C. Heating this, small
     peak at ~25 Å persisted to 78°C. Upon cooling, abrupt re-
     turn to this solid from S_A phase at about 62.5°C. The ori-
     ginal melt from the 48 Å solid could agree fairly well with
     initial melt. Second crystal melts about 78°C and re-
     crystalize 62.5°C which is above nominal S_A-S_C transition
     (55.8°C). Reappearance of 24.8 Å solid at 52°C agrees
     rather well with S_C-crystal except there is no S_C. S_A
     layer spacing follows rule:

$$d(\overset{\circ}{A}) = 45.2 - 0.069T(°C).$$

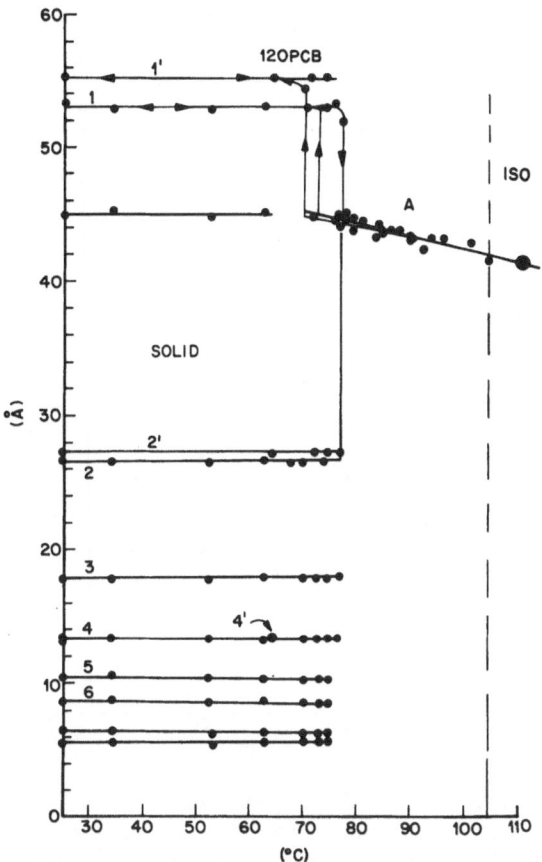

A6. 120PCB. Heating from 25°C, there are many, many lines 53.2, 26.7, 17.8, 13.2, 10.5, 8.7Å, which may all be various harmonics associated with the 53.2 Å fundamental. Perhaps co-existing smectic peak at 44.8 Å and two more sharp peaks at 6.5 Å and 5.7 Å. Upon heating, these lines persisted to 76.9°C. Heating through smectic A phase to 110.7°C then cooling back through S_A phase, sample crystallized at 64.5°C to another solid which showed lines at 54.5, and 13.1. This solid persisted to 26°C and no smaller spacings were observed. Upon heating, a very small 44.6 Å typical of S_A phase first appeared at 71.2°C but all other lines did not melt away until 76.9°C. Heating through the S_A phase to 110°C then leaving overnight in crystalline form and again, heating next day to 100°C and cooling back through S_A phase, the sample crystallized at 72.7°C to 53.7 Å solid, (with harmonics 1-6) and two more lines at 6.6 and 5.8 Å. S_A phase has relatively large temperature dependence of layer spacing: $d(Å) = 51.6-.090T(°C)$, nearly .1 Å/°C.

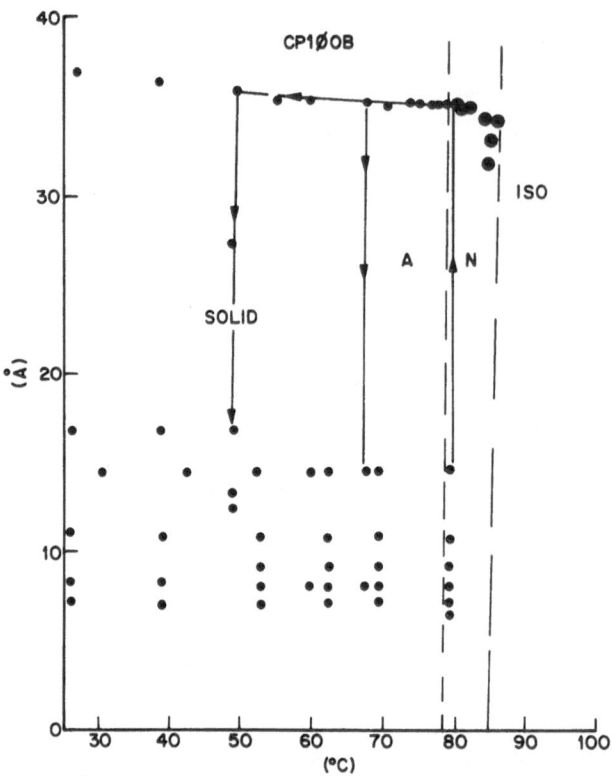

A7. C P1ØOB. Heat to 117.5°C to get rid of 14.7 Å solid! This is a
very mysterious compound! First sample would not melt until
117.5°C. Cooling this sample from 117.5°C through S_A phase,
it recrystallized at 49°C to a solid of 27.2 Å, 13.5 Å, 16.4 Å,
12.4 Å and nematic-like peak at about 35.8 Å. After leaving
overnight, this compound first: lost 16.9 Å at ∿40°C. There
were still many other peaks: 14.7 Å, 10.8, 9.3, 8.1, 7.2 which
persisted all the way to 80°C. Sample then melted to nematic
phase as previously advertised – then broke. Loading a new
sample which had been stuffed into the capillary with no prior
history of melting, again many lines 14.5, 10.7, 9.2, 8.0, 7.1
Å which persisted to 80°C. At 78.4°C, seemed like a new peak
was starting at 34.9 Å. At 83.1°C, there was only a nematic
like lump vaguely centered at 33.1 Å. Cooling, about 73.5°C,
an 8 Å peak started to grow and became hugh by by 63.3°C. A
sample which had been previously melted showed in addition to
previous peaks (14.5, 10.7, 9.2, 8.0 Å) the 16.9 Å peak which
melts away around 40°C. This sample melted ∿80°C. Upon heating
sample to 96°C, it recrystallized at ∿68°C to 14.7 Å, 8.04 Å
solid co-existing with 34.7 Å small peak. Cooling to room-temperature, 14.7 Å peak takes over and 8 Å became very weak. S_A
layer spacing follows- d(Å) = 37.1 - 0.024T(°C).

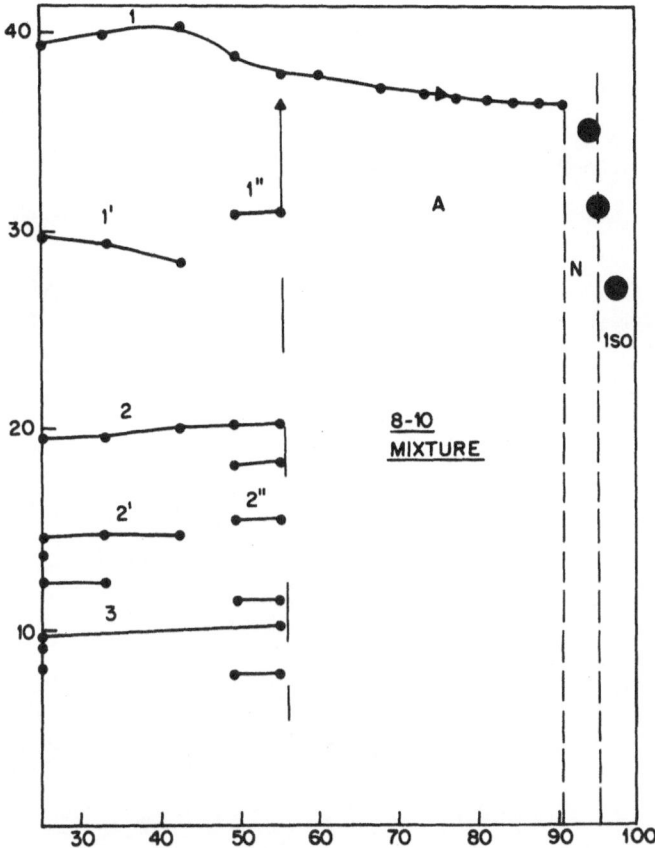

A8. 8-100B. The layer spacing in A phase follows

$$d(\overset{\circ}{A}) = 42.2 - 0.074T(^\circ C).$$

Intercept is same as 90PCB but temperature dependence stronger than 100PCB alone.

UNUSUAL GLASS TRANSITION OF SMECTIC LIQUID CRYSTAL IN p-n-
HEXYLOXYBENZYLIDENE-p'-BUTYLANILINE*

Michio Sorai, Hideki Yoshioka and Hiroshi Suga

Chemical Thermodynamics Laboratory and
Department of Chemistry, Faculty of Science
Osaka University, Toyonaka, Osaka 560, Japan

ABSTRACT

The heat capacities of the title compound with a purity of
99.93 mole per cent have been measured with an adiabatic calorimeter in the range from 13 to 393 K. The enthalpy and entropy of
the phase transitions (crystal $\rightarrow$ S_G, $S_G \rightarrow S_B$, $S_B \rightarrow S_A$, $S_A \rightarrow$ N,
N $\rightarrow$ isotropic liquid) were determined. A glassy state of the S_G
liquid crystalline phase was realized by rapid cooling. The heat
capacity of the glassy state gave rise to two stepped anomalies
due to the enthalpy relaxations around 207 and 222 K; a quite
unusual phenomenon in comparison with usual glass transition
phenomena. In response to the two relaxational heat capacity
anomalies, two kinds of relaxation time characterizing a rate at
which the molecular mode frozen-in at the glass transition temperature approaches to its equilibrium state were observed.
Possible frozen-in modes have been discussed. At zero Kelvin
the molar enthalpy of the glassy S_G state was by (9.27 ± 0.16)kJ
mol^{-1} higher than that of the crystalline state and the residual
entropy of the glassy state was (7.51 ± 0.63)J K^{-1} mol^{-1}.

*Contribution No. 34 from Chemical Thermodynamics Laboratory.

INTRODUCTION

The glassy state as a thermodynamically nonequilibrium state has been nowadays a widely accepted concept and has been realized not only for liquids but also orientationally disordered crystals and liquid crystals[1,2]. Based on thermodynamic studies of single-component liquid crystals, our group found the first example of a glassy cholesteric liquid crystal for cholesteryl hydrogen phthalate[3], a glassy nematic state for o-hydroxy-p-methoxy-benzylidene-p'-butylaniline (OHMBBA)[4,5], and glassy smectic states for a homologous series of benzylideneaniline[6], while Sackmann et al.[7] had reported a glassy state for a liquid crystal "mixture" of cholesteryl chloride and cholesteryl myristate, and Mahler and Panar[8] also found a glassy cholesteric state for binary and ternary "mixtures" of cholesterol derivatives. Proper studies aimed at elucidation of characteristic aspects inherent in the glassy liquid crystal have, since then, been gradually reported. The experimental methods cover thermal analyses (DTA and DSC)[3,4,6,9,10], heat capacity measurement[5], X-ray diffraction[7,11], Mössbauer spectroscopy[12-17], optical microscopy[18], infrared dichroism[19] and visible light absorption spectroscopy[8,20].

A solitary example for which a comprehensive thermodynamic study of the glassy liquid crystalline state has been made is the nematogenic OHMBBA[5]. One of the interesting features was that the residual entropy at 0 K of the glassy state of this compound was intermediate between those of ordinary glassy liquids and glassy crystals; indicating that the degrees of freedom frozen-in at the glass transition temperature, T_g, are less for the glassy liquid crystal than for the former but greater than for the latter.

On the other hand, the thermal analyses[6] for the smectogens of a homologous series of aromatic Schiff's bases of the form (abbreviated conventionally as "mO·n").

$$C_m H_{2m+1} O \underset{}{-\bigcirc-} CH=N \underset{}{-\bigcirc-} C_n H_{2n+1}$$

revealed quite different behavior of the glass transition in comparison with those hitherto known. The glass transition phenomena for the smectic glasses occurred over an extremely wide temperature interval of 30–60 K, while those for the nematic glasses and the isotropic liquid glasses appeared in a narrow interval of 10–15 K. This fact seemed to suggest that in the glassy smectic state the relaxation time characterizing a rate at which the molecular mode frozen-in at T_g approaches to its equilibrium state cannot be described by a single value[6].

The purpose of the present paper is to throw more light on this multi-relaxation phenomenon based on heat capacity measurements of p-n-hexyloxybenzylidene-p'-butylaniline (60·4 or denoted as HEXOBUTA according to a literature[21]). This compound exhibits three smectic and one nematic mesomorphic polymorphism[6,21]. Although the two high-temperature smectic phases were easily assigned to S_A and S_B[22,23], identification of the lowest temperature smectic phase settled at last down to S_G[24–28] after an itineracy[21–23,29]; the confusion was whether the smectic phase should be S_H or S_G. This S_G phase is easily undercooled below the melting point by rapid cooling and is finally transformed into a thermodynamically nonequilibrium glassy state via a glass transition phenomenon[6].

The present paper deals with the heat capacity measurements over a wide temperature range from 13 to 393 K, in which all the stable crystalline, mesomorphic and isotropic-liquid phases, the metastable undercooled S_G and the nonequilibrium glassy S_G phases are included. As the present paper will be mainly concerned with the glassy smectic state so the detailed discussion about the polymorphic transitions will be presented elsewhere[30].

EXPERIMENTAL

Material. The sample was synthesized by azeotropic dehydra-
tion of p-n-hexyloxybenzaldehyde and p-n-butylaniline (Tokyo Kasei
Kogyo Co., Ltd.) in a benzene solution by removing the water produced. The crude material was recrystallized five times from
absolute ethanol to give lath-like crystals. The material was
finally purified by molecular distillation under a vacuum. Anal.
Calcd. for $C_{23}H_{31}NO$: C, 81.85%; H, 9.26%; N, 4.15%. Found: C,
81.82%; H, 9.27%; N, 4.11%.

Heat Capacity Measurements. The heat capacities were measured
with an adiabatic-type calorimeter[31]. A calorimeter cell made of
gold and platinum[32] contained 18.0110 g ($\hat{=}$ 0.0533653 mol) of the
specimen and a small amount of helium gas to aid the heat transfer
inside the cell. A platinum resistance thermometer (Leeds &
Northrup Co., Ltd.) used in this experiment has been calibrated
based on the IPTS-68 temperature scale.

RESULTS

The results of the heat capacity measurements for HEXOBUTA
(60.4) in the whole temperature region investigated here are
plotted in Figure 1 for all the stable phases by open circles and
for the metastable undercooled and the nonequilibrium glassy S_G
phases by solid circles. The crystal melted into the S_G state at
306.597 K (the so-called melting point). Purity of the specimen
was determined by a fractional fusion method. Plot of the reciprocal of the fraction melted against the melting temperature gave
an approximately straight line, indicating nonexistence of solid-
soluble impurities, and the slope yielded the sample purity of
99.93 mole per cent. The triple point of pure material was estimated to be 306.622 K.

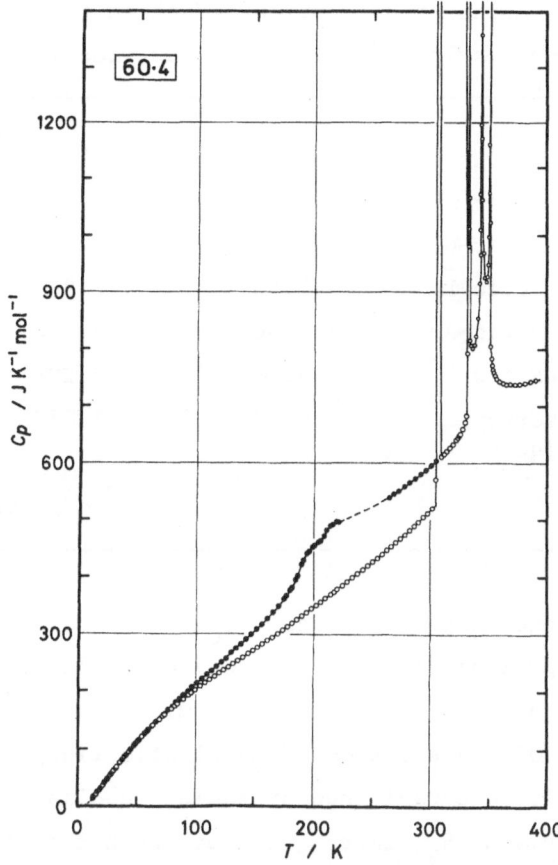

Fig. 1.   Molar heat capacities of HEXOBUTA (60.4). $\circ$:   The stable
           phases; the sequence of the phases being crystal, S_G, S_B,
           S_A, nematic and isotropic liquid. $\bullet$:   The glassy liquid
           crystal and the undercooled liquid crystal of the S_G
           phase.

The S_G phase was transformed into S_B at 331.56 K, immediately
followed by a transition to S_A at 332.86 K; the temperature interval of the S_B phase being as narrow as 1.30 K. The transition
from S_A to the nematic state took place at 343.24 K and finally
the nematic phase was transformed into the isotropic liquid at
350.92 K (the clearing point). The enthalpy and entropy of the
trnasitions were determined by combining the heat capacity measurements and the independent enthalpy measurements for each phase
transition. Table I summarizes these data.

TABLE I   Enthalpy and entropy of phase transitions in
p-n-hexyloxybenzylidene-p'-butylaniline

Transition	T_C/K	ΔH/kJ mol^{-1}	ΔS/J K^{-1}mol^{-1}
crystal → smectic-G	306.597	23.293	75.981
smectic-G → smectic-B	331.56	0.835	2.527
smectic-B → smectic-A	332.86	3.374	10.135
smectic-A → nematic	343.24	3.204	9.373
nematic → isotropic liquid	350.92	1.887	5.373
		Total	103.389

On the other hand, the glassy S_G state was realized by cooling the specimen in the S_A state down to liquid nitrogen temperatures at an average rate of 7.2 K min^{-1} (the first series experiment). Of course, from an idealistic viewpoint, the glassy S_G state should be established by rapidly cooling the specimen in one and the same S_G state. However, we had confirmed based on a preliminary DTA experiment that the glassy S_G state was obtained irrespective of the starting temperature of cooling and that the thermal behavior concerning the glass transition was identical among them. Since the cooling rate attained by the present calorimeter was slow compared with ca. 12 K min^{-1} of the DTA apparatus[33], the cooling from the S_A state in which no memory of the crystalline lattice persists seemed to be favorable to prevent the nucleation during rather slwo cooling process.

The heat capacity, C_p, of the glassy S_G state thus obtained progressively deviated from that of the crystal above ca. 60 K and its magnitude became larger with increasing temperature. The heat capacity of the glassy state finally gave rise to two stepped anomalies around 200 K, at which the glassy state was transformed into the undercooled S_G state (see Figure 1).

As shown in Figure 2, temperature drifts arising from the enthalpy ($\underline{H}$) relaxation characteristic of the glass trnasition phenomenon were detectable from as low as $\underline{ca.}$ 120 K. Interestingly, however, the temperature dependence of $d\underline{H}/d\underline{t}$ behaved quite

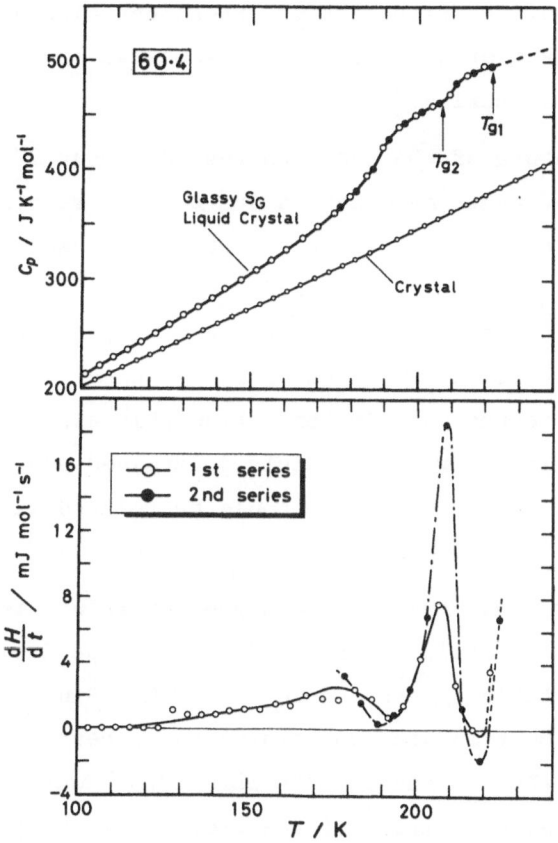

Fig. 2.   Temperature dependence of the rate of enthalpy relaxation, $d\underline{H}/d\underline{t}$, around $\underline{T}_g$ (bottom) and the heat capacities of HEXOBUTA in an enlarged scale for the same temperature range (top). The glass transition temperatures, $\underline{T}_{g1}$ and $\underline{T}_{g2}$, are taken to be those at which the relaxation time becomes 1 ks.

differently from those observed for usual glass transition pheno-
mena for which the $d\underline{H}/d\underline{t}$ versus $\underline{T}$ curve shows a pair of a maximum
and a minimum around $\underline{T}_g$. In this regard the present curve seems
as if a pair of glass transitions were superimposed. This fact
just corresponds to the existence of two stepped heat-capacity
anomalies. To confirm the unusual feature of the present glass
transition phenomenon, similar heat capacity measurements were
performed (the second series experiment). The results were identical with those of the first series experiment although the
average cooling rate of 6.7 K min^{-1} was somewhat lower than 7.2 K
min^{-1} for the first series.

When temperature of the specimen reached about 225 K, a
large evolution of heat was observed due to an irreversible transition from the undercooled S_G to the crystalline phase. On account
of this effect, $\underline{C}_p$ of the undercooled S_G state could not be further
measured in this series of experiment. However, the temperature
rise due to the crystallization could be followed by keeping adiabatic conditions in the calorimeter. From this experiment, the
heat of crystallization was accurately determined and thus the
molar enthalpy of the glassy state could be related with the
crystal enthalpy.

On the other hand, the $\underline{C}_p$ in a high-temperature region of the
undercooled S_G phase could be measured for the specimen slowly
cooled from the stable S_G state down to 270 K. Since further
cooling brought about spontaneous evolution of heat due to the
crystallization, the $\underline{C}_p$ measurements below 270 K were impossible.
Therefore, the heat capacities of the undercooled S_G phase in the
range from 225 to 270 K were estimated in reference to the enthalpy diagram so that the area under the assumed $\underline{C}_p$ curve in this
temperature interval may coincide with the enthalpy difference
between these two temperatures. The heat capacities thus estimated are illustrated in Figure 1 by a broken line.

The residual entropy of the glassy S_G state at 0 K was determined to be (7.51 ± 0.63)J K^{-1} mol^{-1} by assuming that the crystalline state at 0 K obeys the third law of thermodynamics. The
rather large error-bound is mainly originated from the uncertainties involved in estimations of both the interpolation of the heat
capacity in the range from 225 to 270 K and the extrapolation from
13 to 0 K. Of interest is the fact that the residual entropy of
7.51 J K^{-1} mol^{-1} for the present smectic glass is much smaller
than 12.69 J K^{-1} mol^{-1} for the nematic glass of OHMBBA[5]. This is
caused by the situation that the molecular order is much higher in
a smectic state than in a nematic one.

The enthalpy difference between the glassy S_G state and the
crystalline state at 0 K was (9.27 ± 0.16)kJ mol^{-1}. As the interpolation procedure mentioned above is not included in this case
so the error-bound becomes small.

DISCUSSION

The remarkable features inherent in the present smectic glass
are the existence of at least two dominant glass transitions and
a widely spread heat capacity anomaly arising from the enthalpy
relaxation. This situation is obvious if we compare the present
results (Figure 2) with those of the nematic glass of OHMBBA[5] as
shown in Figure 3.

The multi-glass-transition reflects the fact that there exist
multi-relaxation processes in a given system. Although this fact
itself is not so uncommon, the marvel is that the present smectic
liquid crystal gives rise to two distinct glass transition phenomena in the heat capacity curve. For instance, amorphous polymers
usually show two kinds of dielectric relaxation process; the high-
temperature process called the α-relaxation (primary relaxation)
has been attributed to the large scale conformational rearrange-
ments of the main chains while the low-temperature one called the

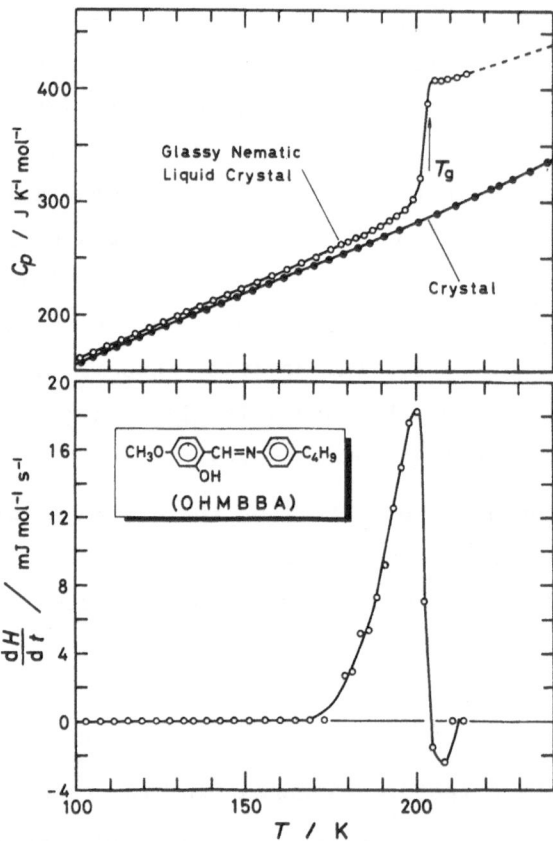

Fig. 3. Temperature dependence of the rate of enthalpy relaxation, dH/dt, around T_g (bottom) and the heat capacities of OHMBBA[5].

β-relaxation (secondary relaxation) results from the motion of side groups and/or the local relaxation mode of the main chain[34]. Such a secondary dielectric relaxation has also been observed for many glassy states of simple molecules[35-39]. Moreover, even a distribution of relaxation times has been proposed for inorganic glasses such as borosilicate crown glass[40] and GeO_2 glass[41]. However, since the secondary dielectric relaxation generally does not accompany any detectable enthalpy change so it might be next

to impossible to point out an existence of the β-relaxation solely
from heat capacity measurements. Although a few examples for
which the onset of the β-relaxation process has been calorimetri-
cally detected are known for the glassy crystals of cyclohexanol[42]
and $CFCl_2-CFCl_2$[43], the increment of the heat capacity anomaly is
not so sharp as in the case of the main glass transition. In
fact, until the dielectric measurement was made for cyclohexanol[38],
it had been difficult to attribute the diffuse heat capacity
anomaly unambiguously to the onset of the molecular motion of the
β-relaxation process.

Taking these situations into account, we can easily discard
the possibility that the present two stepped heat-capacity anomalies might be caused by the α- and β-relaxations, respectively.
The present smectic glass should be characterized by two kinds of
the α-relaxation. One of the powerful methods to corroborate this
prediction may be dielectric measurements about T_g although some
technical difficulties are involved because the undercooled S_G
state is easily transformed into the stable crystalline state,
especially in the range from 225 to 270 K.

It should be remarked here that there exist really the systems which thermally exhibit the phenomena of vitreous polymorphism accompanied by two discernable glass transitions; one
example is the glassy liquid of a binary mixture of glycerol and
water[44] and the other is the glassy crystal of pinacol hexadeuterate[45]. The occurrence of double glass transitions for the
former has been interpreted in terms of coexistence of two
amorphous phases resulted from phase separation; formation of
sufficiently large clusters with a rough composition of
$2C_3H_8O_3 \cdot 3H_2O$ in water-rich systems[44]. The problem of immiscibility or phase separation is often encountered in inorganic
glasses[46,47] and aqueous solutions of inorganic salt at low temperatures[48]. In the case of pinacol hexadeuterate, two kinds of
the glass transition have been attributed to the freezing pro-

cesses of the reorientational motions of the diol and the water molecules, respectively[2,45]. Therefore, from a viewpoint of vitreious polymorphism, the pinacol hexadeuterate crystal may be regarded as consisting of a mixture of 'the pinacol crystal' and 'the heavy water crystal'.

On the other hand, as the present material consists of a single component so the double glass transitions cannot be accounted for in terms of a phase separation in the glassy state. Of course one may well wonder whether traces of the S_A and/or S_B phases might persist in the glassy S_G phase because the present glassy state was realized by cooling the specimen in the S_A state down to liquid nitrogen temperatures. However, there is not the slighest shadow of doubt about the implausibility of the admixture from the following three reasons. (1) The intermesophase transition generally takes place rapidly and shows no noticeable undercooling effect. (2) As described in the previous section, we confirmed based on thermal analyses that the glassy S_G state was similarly realized irrespective of the starting temperature of cooling and that the thermal behavior concerning the glass transition was identical among them. (3) Final but decisive evidence is given by the fact that two distinct glass transitions have been observed for the present homologous series, PENTOBUTA (50.4; p-n-pentyloxy-benzylidene-p'-butylaniline), for which the glassy S_G state was realized by cooling the specimen in one and the same S_G state[49]. Thus we can conclude that the double glass transition phenomenon is intrinsic and characteristic of the glassy state consisting of smectic liquid crystals, at least the smectic-G state.

Another marvellous point is a question why the smectic glass, unlike the nematic glass[4-6], shows such a phenomenon. Possible molecular motions frozen-in at the glass transition region are generally believed to be collective motion of molecular orientations, translational self-diffusion and/or internal rotation of the constituent groups. However, these motional modes are common to both smectics and nematics, and even to isotropic liquids. The most characteristic feature of the smectic phases differentiated from other liquid crystalline phases is that all the smectic phases consist of a layer structure. One of the motional modes proper to such a layer structure is the undulation mode[50].

To examine whether the undulation modes are concerned with the glass transition phenomenon, we shall analyze the enthalpy relaxation in terms of the relaxation times as follows. Figure 4 represents a schematic diagram of the enthalpy against temperature around the glass transition region. Here, the configurational enthalpy of the glassy S_G state is assumed to be relaxed according to two relaxation times, $\underline{\tau}_1$ and $\underline{\tau}_2$, at a given temperature. Furthermore, the rate of each enthalpy relaxation, $d\underline{H}/d\underline{t}$, is

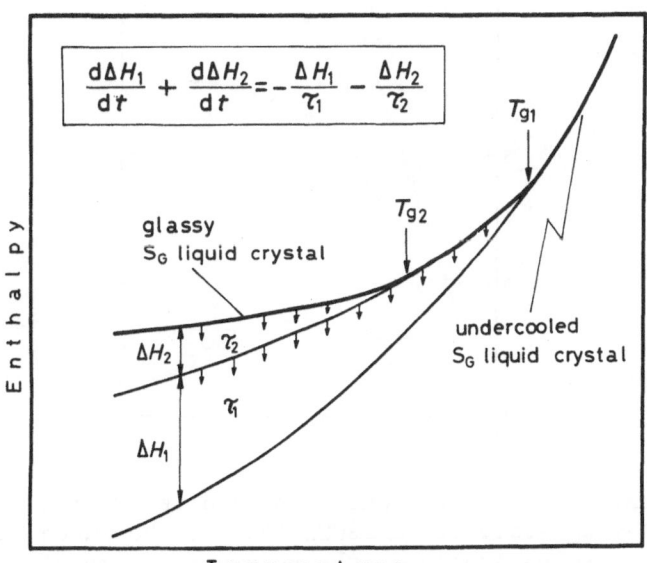

$$\frac{d\Delta H_1}{dt} + \frac{d\Delta H_2}{dt} = -\frac{\Delta H_1}{\tau_1} - \frac{\Delta H_2}{\tau_2}$$

Fig. 4.   Schematic diagram of the enthalpy against temperature in the glass transition region.

assumed to be proportional to $-\Delta H/\tau$ [51,52]. As a result, the relaxation rate can be expressed in the form of

$$\frac{d\Delta H_1}{dt} + \frac{d\Delta H_2}{dt} = - \frac{\Delta H_1}{\tau_1} - \frac{\Delta H_2}{\tau_2} \tag{1}$$

The quantities, ΔH_1 and ΔH_2, can be estimated by extrapolating the configurational heat capacities below T_{g1} and T_{g2}, respectively. In the temperature range of $T_{g2} < T$, the temperature drift during the heat capacity measurement is controlled only by the first relaxation process characterized by τ_1;

$$\frac{d\Delta H_1}{dt} = - \frac{\Delta H_1}{\tau_1} \tag{2}$$

The temperature dependence of τ_1 thus obtained is plotted in Figure 5 by open circles. Interestingly, the Arrhenius plots of

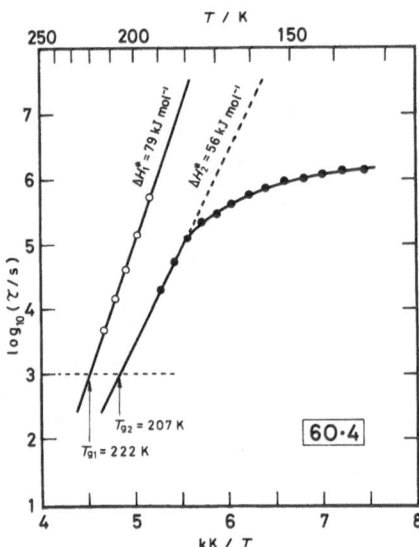

Fig. 5. The Arrhenius plots concerning the relaxation time for the glassy S_G state of HEXOBUTA (60.4). The curve through open circles indicates the relaxation time of τ_1 and the curve through solid circles shows the relaxation time of τ_2. The glass transition temperatures, T_{g1} and T_{g2}, are taken to be those at which the relaxation time becomes 1 ks.

this temperature region are linear;

$$\tau_1 = \tau_{01} \exp(\Delta H_1^*/RT) \tag{3}$$

$$\Delta H_1^* = 79 \text{ kJ mol}^{-1}$$

$$\tau_{01} = 1.8 \times 10^{-16} \text{ s}$$

where ΔH_1^* represents the activation enthalpy, τ_0 is a pre-exponential factor and R is the gas constant. To estimate τ_2 below T_{g2}, this temperature dependence of τ_1 was assumed to hold below this temperature and the contribution of this process to the temperature drift was subtracted from the observed one. As shown in Figure 5 by solid circles, the Arrhenius plots of this temperature region are nonlinear, especially at low temperatures. This fact strongly suggests that the glass transition at the low-temperature side, T_{g2}, cannot be described by a single relaxation time but by a distribution of relaxation time. If we force a few points just below T_{g2} to obey the linear Arrhenius relation (see Figure 5), τ_2 is expressed by

$$\tau_2 = \tau_{02} \exp(\Delta H_2^*/RT) \tag{4}$$

$$\Delta H_2^* = 56 \text{ kJ mol}^{-1}$$

$$\tau_{02} = 6.8 \times 10^{-12} \text{ s}$$

As shown in Figure 5, if we take 1 ks as the demarcation between dynamic and static disorder[2], the Arrhenius plots given by Equations 3 and 4 indicate that the glass transition will occur at T_{g1} = 222 K and T_{g2} = 207 K, respectively. Location of these temperatures are shown by arrows in Figure 2.

Ribotta et al.[53] have really succeeded in detecting the thermally excited undulations of layers in the smectic-A phase of p-cyanobenzilidene-p-octyloxyaniline (CBOOA) based on light scattering experiments. These modes are highly damped and follow an exponentially decaying correlation function characteristic of a dynamic undulation. The relaxation time (or the damping rate) of layer undulations was of the order of 10^{-3} s at 75°C at which their experiment was done. If we extrapolate the linear Arrhenius relations (Equations 3 and 4) to this temperature, the relaxation times $\underline{\tau}_1$ and $\underline{\tau}_2$ become 0.15×10^{-3} s and 1.8×10^{-3} s, respectively. These values are very close to the damping rate found for the undulation of smectic-A layers. Therefore, if we assume that such undulation modes also exist in the present smectic-G phase and that the undulation modes for both phases are damped in a similar time scale, it is quite likely that freezing of the undulation mode is responsible for one of the present double glass transitions. If this is the case, the undulation mode is a candidate for the glass transition at $\underline{T}_{g2}$ because a distribution of the relaxation time is known for the damping of undulations. And the candidates for the glass transition at $\underline{T}_{g1}$ may be collective motion of molecular orientations and/or translational self-diffusion as expected for ordinary glass transitions.

Alternative interpretation for the double glass transitions is the freezing of anisotropic translational self-diffusions parallel and perpendicular to a director of the S_G liquid crystal. LaPrice and Uhrich[16] reported the ^{57}Fe Mössbauer effect of 1,1'-diacetylferrocene doped in the glassy S_B phase of BBOA (or 40.8) and in the glassy S_G phase of HBPA (or 60.3: the authors have used the notation of S_H but the recent assignment[26,27] is S_G). According to their results, the Mössbauer-Debye temperatures parallel and perpendicular to a smectic layer are quite anisotropic; that is, $\underline{\Theta}_{//}$ is 64 K and $\underline{\Theta}_{\perp}$ is 49 K for the glassy S_B phase. Unfortunately, they failed in detection of the anisotropic nature for the S_G phase only because of the technical reason; the molecular tilt angle of about 43° was approximately midway between their experimental angles of 0° and 90°, and thus they observed similar values for $\underline{\Theta}_{//}$ and $\underline{\Theta}_{\perp}$ ($\sim$46 K). At any rate, their Mössbauer study clearly demonstrates that the molecular motion, probably a group diffusion, is easier in the direction perpendicular to the director than in the parallel direction. In other words, these molecular motions along the different directions are characterized by different time scales. Accordingly, if these motional modes are the candidates for the present glass transitions, they may be expected to be frozen-in at different temperatures, say, T_{g1} and T_{g2}.

We discussed above two possible origins responsible for the double glass transition phenomenon found for the smectic-G state. At the present stage, however, it is difficult to decide which origin is true. In this respect, it seems helpful to examine whether a similar double glass transition phenomenon will be observed for other smectic states such as S_A, S_B, S_C, etc.

REFERENCES

1. H. Suga, S. Seki, J. Non-Cryst. Solids 16, 171 (1974).

2. H. Suga, S. Seki, Faraday Discussions 69, 221 (1980).

3. K. Tsuji, M. Sorai, S. Seki, Bull. Chem. Soc. Japan 44, 1452 (1971).

4. M. Sorai, S. Seki, Bull. Chem. Soc. Japan 44, 2887 (1971).

5. M. Sorai, S. Seki, Mol. Cryst. Liq. Cryst. 23, 299 (1973).

6. M. Sorai, T. Nakamura, S. Seki, Pramana, Suppl. 1, 503 (1975).

7. E. Sackmann, S. Meiboom, L. C. Snyder, A. E. Meixner, R. E. Dietz, J. Am. Chem. Soc. 90, 3567 (1968).

8. W. Mahler, M. Panar, J. Am. Chem. Soc. 94, 7195 (1972).

9. S. E. B. Petrie, H. K. Bucher, R. T. Klingbiel, P. I. Rose, Eastman Org. Chem. Bull. 45, No. 2, 1 (1973).

10. J. Cognard, C. Ganquillet, Mol. Cryst. Liq. Cryst. 49, L-33 (1978).

11. J. E. Lydon, J. O. Kessler, J. Phys. (Paris) 36, C1-153 (1975).

12. R. E. Detjen, D. L. Uhrich, C. F. Sheley, Phys. Lett. 42A, 522 (1973).

13. R. E. Detjen, D. L. Uhrich, Mössbauer Effect Methodology,
       ed. by I. J. Gruverman, C. W. Seidel and D. K. Dieterly,
       Plenum Press, New York and London (1974), Vol. 9, p. 113.

14. V. O. Aimiuwu, D. L. Uhrich, Mol. Cryst. Liq. Cryst. 43, 295
       (1977).

15. W. J. LaPrice, D. L. Uhrich, J. Chem. Phys. 71, 1498 (1979).

16. W. J. LaPrice, D. L. Uhrich, J. Chem. Phys. 72, 678 (1980).

17. M. C. Kandpal, V. G. Bhide, Proceedings of the International
       Conference on Liquid Crystals, Bangalore (1979), ed. by
       S. Chandrasekhar, Heyden, London (1980), p. 421.

18. J. O. Kessler, E. P. Raynes, Phys. Lett. 50A, 335 (1974).

19. N. Kirov, M. Fontana, Mol. Cryst. Liq. Cryst. 56, L-195
       (1980).

20. E. Sackmann, J. Am. Chem. Soc. 90, 3569 (1968).

21. J. B. Flannery, Jr., W. Haas, J. Phys. Chem. 74, 3611 (1970).

22. G. W. Smith, Z. G. Gardlund, R. J. Curtis, Mol. Cryst. Liq.
       Cryst. 19, 327 (1973).

23. G. W. Smith, Z. G. Gardlund, J. Chem. Phys. 59, 3214
       (1973).

24. J. W. Goodby, G. W. Gray, Mol. Cryst. Liq. Cryst. 49, L-217
       (1979).

25. D. Demus, J. W. Goodby, G. W. Gray, H. Sackmann, Mol.
       Cryst. Liq. Cryst. 56, L-311 (1980).

26. A. Wiegeleben, L. Richter, J. Deresch, D. Demus, Mol.
       Cryst. Liq. Cryst. 59, 329 (1980).

27. J. W. Goodby, G. W. Gray, A. J. Leadbetter, M. A. Mazid,
       Liquid Crystals of One- and Two-Dimensional Order, ed.
       by W. Helfrich and G. Heppke, Springer Verlag, Berlin
       (1980), p. 3.

28. G. W. Gray, Mol. Cryst. Liq. Cryst. 63, 3 (1981).

29. A. J. Leadbetter, M. A. Mazid, B. A. Kelly, J. W. Goodby,
       G. W. Gray, Phys. Rev. Lett. 43, 630 (1979).

30. H. Yoshioka, M. Sorai, H. Suga, to be published in Mol. Cryst. Liq. Cryst.

31. M. Yoshikawa, M. Sorai, H. Suga, S. Seki, to be published in J. Phys. Chem. Solids.

32. K. Tsuji, M. Sorai, H. Suga, S. Seki, Mol. Cryst. Liq. Cryst. 55, 71 (1979).

33. H. Suga, H. Chihara, S. Seki, Nippon Kagaku Zasshi 82, 24 (1961).

34. Y. Ishida, J. Polymer Sci. A2 7, 1835 (1969).

35. M. Goldstein, J. Chem. Phys. 51, 3728 (1969).

36. G. P. Johari, M. Goldstein, J. Chem. Phys. 53, 2372 (1970).

37. G. P. Johari, M. Goldstein, J. Chem. Phys. 55, 4245 (1971).

38. K. Adachi, H. Suga, S. Seki, S. Kubota, S. Yamaguchi, O. Yano, Y. Wada, Mol. Cryst. Liq. Cryst. 18, 345 (1972).

39. G. K. Gupta, V. P. Arora, V. K. Agarwal, A. Mansingh, Mol. Cryst. Liq. Cryst. 71, 77 (1981).

40. P. B. Macedo, A. Napolitano, J. Res. Natl. Bur. Stand. 71A, 231 (1967).

41. A. Napolitano, P. B. Macedo, J. Res. Natl. Bur. Stand. 72A, 425 (1968).

42. K. Adachi, H. Suga, S. Seki, Bull. Chem. Soc. Japan 41, 1073 (1968).

43. K. Kishimoto, H. Suga, S. Seki, Bull. Chem. Soc. Japan 51, 1691 (1978).

44. R. L. Bohon, W. T. Conway, Thermochimica Acta 4, 321 (1972).

45. M. Oguni, T. Matsuo, H. Suga, S. Seki, Bull. Chem. Soc. Japan 53, 1493 (1980).

46. J. Zarzycki, Discuss. Faraday Soc. No. 50, 122 (1970).

47. J. P. de Nerfville, D. Turnbull, Discus. Faraday Soc. No. 50, 182 (1970).

48. C. A. Angell, E. J. Sare, J. Chem. Phys. 49, 4713 (1968); 52, 1058 (1970).

49. M. Sorai, K. Tani, H. Suga, to be published.

50. P. G. de Gennes, J. Phys. (Paris) $\underline{30}$, C4-65 (1969).

51. T. Matsuo, M. Oguni, H. Suga, S. Seki, Proc. Japan Acad. $\underline{48}$, 237 (1972).

52. T. Matsuo, M. Oguni, H. Suga, S. Seki, J. F. Nagle, Bull. Chem. Soc. Japan $\underline{47}$, 57 (1974).

53. R. Ribotta, D. Salin, G. Durand, Phys. Rev. Lett. $\underline{32}$, 6 (1974).

SOME HIGH ELECTRIC FIELD EFFECTS IN NEMATIC LIQUID CRYSTALS

E. F. Carr, R. W. H. Kozlowski and Mojtaba Shamsai

Physics Department, University of Maine
Orono, Maine 04469

ABSTRACT

A model to explain molecular alignment and material flow due to electric fields (conduction regime) in nematic liquid crystals has been discussed previously. This model involves the formation of walls (defects) perpendicular to the electrodes and shear flow alignment. In much of the earlier work fields of a few hundred V/cm were applied to samples that were initially well ordered, but observations are reported here that clearly indicate walls for continuously applied fields up to 12,000 V/cm in a sample with an electrode separation of 0.5 cm. Observation at the free surface and the electrode-liquid crystal interface indicate that these walls are separated by approximately 0.05 cm and extend vertically an appreciable distance into the sample. For continuously applied fields in bulk samples, walls are more easily observed at several thousand V/cm then at lower fields. This work on bulk samples is related to dynamic scattering (primary and secondary) in thin samples.

INTRODUCTION

A model proposed earlier[1] to explain molecular alignment and material flow due to electric fields is illustrated in Fig. 1. One of the objectives of the work discussed here is to present additional evidence which supports this model and also indicates that the behavior of nematics in an electric field of 12 kV/cm is more easily understood than at much lower fields. This involves nematic materials in the conduction regime with a positive conductivity anisotropy (ionic conduction a maximum parallel to the director). These materials must also exhibit flow alignment. The presence of walls (defects) are a very important aspect of this model. Previous work has shown how these walls can form perpendicular to the electrodes if the sample was initially well ordered with the director parallel to the electrodes before applying the electric field. After the walls have formed, charge will accumulate at the walls because of the conductivity anisotropy. Forces due to the interaction of the electric field with the space charge at the walls tend to shear the sample. Because of shear flow, the director associated with the sample between the walls is turned toward the electric field giving rise to the "flow-alignment angle" θ. Although the walls should appear to be stationary, the material making up the walls is constantly changing and moving toward the electrodes.

In samples with a depth of approximately 1 cm and an electrode separation of 0.5 cm, walls tend to extend to the bottom of the samples if they were initially well ordered before applying the field. In samples where the depth is much greater than the electrode separation, we expect the walls to be perpendicular to only the electrodes, and not necessarily the free surface. Although the model applies to walls that are perpendicular to the electrodes, it has been shown[2] that walls which are formed at angles less than 90° with respect to the electrodes tend to align perpendicular when the field is applied. Also walls perpendicular to the electrodes

tend to remain perpendicular except near the surface of the electrodes. This indicates that walls at almost any angle can be
involved in producing molecular alignment.

Although much of the earlier work concerning this model involved making observations shortly after an electric field was
applied to an initially well ordered sample, a recent article[3]
presented evidence that walls are present when high electric fields
(5000 V/cm) are applied continuously. These observations were made
at the free surface for an electrode separation of 1.2 cm. The
most interesting aspect was not the fact that walls existed, but
rather that they were so evenly spaced. The movement of dust
particles indicated that material making up the wall moved as indicated in Fig. 1. Walls were being created and destroyed and they
were fluctuating some, but they could be clearly observed. If the
model in Fig. 1 applied, there should have been lines of maximum
flow velocity and lines of zero flow velocity on the surface.

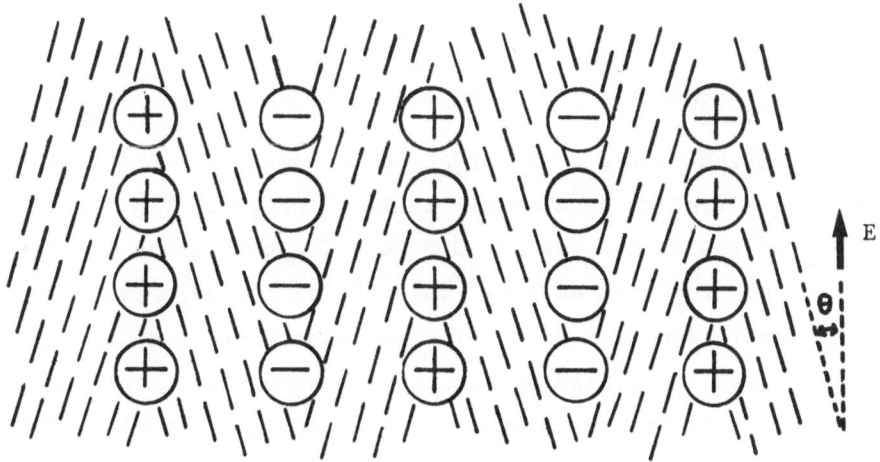

Fig. 1   Model for molecular alignment and material flow due to
          electric fields.

According to the Bernoulli effect, the height of the liquid crystal
could vary by a few microns making the free surface look like a
corrugated surface. If this was the case, the reflection of a laser
beam from the surface could lead to a diffraction pattern which was
observed.[3] This pattern fluctuated as expected but there were
moments when very clear patterns were observed.

Although evidence of walls in the presence of continuously
applied fields was presented, information concerning the depth of
these walls was not obtained. One of the objectives of the work
presented here is to show that these walls do extend an appreciable
distance vertically into the sample.

EXPERIMENTAL

The sample cell was constructed from two transparent conductive
coated glass electrodes separated by a teflon spacer of thickness
0.5 cm. The length and depth of the cell were 1.7 cm and 1.0 cm
respectively. One or more laser beams were directed into the cell
and the observations were made using the light that was internally
scattered. The sample was MBBA with a resistivity of approximately
10^9 ohm-cm.

DISCUSSION OF RESULTS

A photograph of the free surface of a sample of MBBA in the
presence of a 12 kV/cm dc electric field is shown in Fig. 2a. The
separation of the electrodes was 0.5 cm. The photograph is very
similar to that reported earlier[3] in a field of 5 kV/cm with an
electrode separation of 1.2 cm. The pattern fluctuated more at the
higher field and there also appeared to be some structural changes
in the sample that will be discussed later in this article. Defects,
which we believe were walls, can be seen extending out from the
electrodes to more than one-half the distance to the opposite electrode. The movement of dust particles showed that the fluid

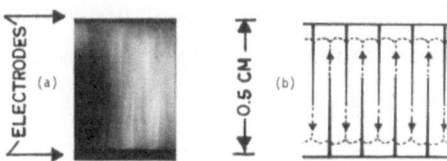

Fig. 2   (a) Walls due to a continuously applied 12 kV/cm dc
            electric field at the free surface of MBBA.   (b) Schematic
            diagram for the material flow.

velocity was a maximum at the defects.   The flow pattern for Fig.
2a is illustrated in Fig. 2b and it is consistent with the model in
Fig. 1.   These walls, which could be observed with the naked eye,
were associated with the electric field, because when the field was
removed the walls disappeared.   They appeared to fade out rather
than disappear instantaneously which was to have been expected if
they were walls.

In earlier work[1,2,3] the extension of walls vertically into
the sample was not shown for continuously applied fields.   The
reasons they were not observed was probably due to the presence of
many other defects and some turbulent motion near the electrodes.
It has been very difficult to make observations which indicate that
the walls extend an appreciable distance vertically into the sample,
but we have made these observations and they are shown in Fig. 3.
In order to obtain this evidence, it was necessary to direct a
laser beam vertically into the center of the sample.   The observations were made at the electrode-to-liquid crystal interface and
at an angle of approximately 45 degrees (in a horizontal plane)
with the electrode surface.   The patterns at the electrode-to-liquid
crystal interface appeared when the field (12 kV/cm) was turned on
and faded as the field was removed as did the defects at the free
surface (Fig. 2a).   The separation of the defects at the electrode-
to-liquid crystal interface and those at the free surface appeared

to be the same. The time periods associated with the fluctuations
of the defects in both cases also appeared to be the same. There
doesn't seem to be much doubt that the defects at the surface were
associated with the defects at the electrode-to-liquid interface,
therefore we believe that we were observing the top and the edge of
walls in the presence of continuously applied high electric fields.
Many of the walls appeared to extend to more than one-half the
distance to the bottom of the cell.

Although we have made observations at the electrode-to-liquid
crystal interface showing evidence of walls at 12 kV/cm, it is
not clear why we have not been able to obtain clear evidence of
walls at this interface in continuously applied fields of 5 kV/cm
and below. It is also not clear why we have been unable to make
good observations at the free surface for continuously applied fields
below a few thousand V/cm. These results suggest that the behavior
of nematic materials in bulk samples may be more easily understood
in high rather than low fields.

In order to gain some insight as to why the behavior of a liquid
crystal may be better understood at the highest field strengths, we
considered some observations in the dynamic scattering mode.
Sussman[4] and Nehring and Petty[5] have observed that above a voltage
of 20-30V, the mechanism responsible for dynamic scattering produces
changes that can be observed optically. This effect was also studied
independently by Chang[6] and was more recently investigated by
Krupkowski and Ruszkiewicz.[7] The transition to secondary dynamic
scattering appeared to depend only on the voltage and was independent
of the sample thickness. It was reported that for voltages above
20-30V secondary dynamic scattering does not appear at the instant
the field is applied, but rather appears as spots of increased
scattering. The area of these spots increases with time until the
entire sample exhibits secondary dynamic scattering. The rate of
increase in area increases with the applied voltage. It was also

reported that when the field was removed, the relaxation (decay time) to the original state (director parallel to the electrode and no defects) was longer for secondary dynamic scattering. We found this to be the case except at voltages much above the threshold for secondary dynamic scattering.

In a 135 micron sample of MBBA we found the decay time to be longer at 50 volts and shorter at 150 volts than the decay time for primary dynamic scattering. In 25 micron samples we also observed much shorter decay times for voltages much above the threshold for secondary dynamic scattering. In the 135 micron samples secondary dynamic scattering was difficult to distinguish from ordinary dynamic scattering in the neighborhood of 50 volts. It was necessary to use a sequence of steps involving a high voltage (150 volts and higher) to identify the spots exhibiting secondary dynamic scattering. It is interesting that many threads and other defects remain after removal of voltages in the neighborhood of 50 volts, but at much higher voltages there appears to be fewer defects than in the ordinary dynamic scattering mode.

A voltage of 150 volts applied to a 135 micron sample gives a field comparable to that used in obtaining the results (0.5 cm electrode separation) shown in Figures 2 and 3. This may imply that fewer threads and other defects are created in both bulk and 135 micron samples at the higher fields. If the model in Fig. 1 applies to the 135 micron samples, then walls would be present, but the results on bulk samples indicated that walls relax rather quickly. The presence of only walls could explain the short decay time in the higher fields. The absence of threads and other defects for the 12 kV/cm field in the bulk samples may be the reason that we could find evidence of walls at the electrode-to-liquid crystal interface, but did not see them at much lower fields.

Although the absence of threads in thin samples and the ability to observe walls in bulk samples indicates a consistency, the

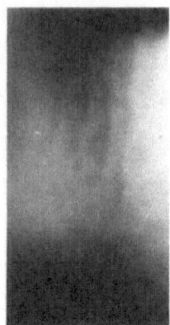

Fig. 3   Walls due to a continuously applied 12 kV/cm dc electric
         field at the electrode-to-liquid crystal interface.

reason for this behavior is not clear. One possibility that can be
considered is that the structure of the walls may vary with the
electric field strength. A change in wall structure with the
increased velocity of the fluid may lessen the tendency for other
defects to form. Although we do not believe that the walls can be
classified as a particular type of alignment-inversion wall as
discussed by Helfrich,[8] the ideas presented may be very helpful in
understanding these walls. The use of a polarizer indicated that
walls, formed at low fields in a sample that was initially well
ordered, appeared to involve considerable twist as in twist walls.
At 5 kV/cm in a continuously applied field, they appeared to have
less twist. At 12 kV/cm we were unable to find evidence of an
appreciable amount of twist. Therefore, at 12 kV/cm they probably
behaved more like splay-bend walls. It might be possible that the
amount of twist in a wall is a factor in producing threads and
other defects.

Although it has been pointed out[5] that threads play an important role in secondary dynamic scattering, the change from primary
to secondary dynamic scattering probably involves more than just
the formation of threads. The reason is because at higher voltages,

secondary dynamic scattering does not appear to exhibit a greater density of threads than ordinary dynamic scattering.

It has been suggested[4] that the flow in ordinary dynamic scattering may be nearly circular as in Williams domains, but as the sample changes to secondary dynamic scattering the flow shows up as elongated vortices. This suggestion may be correct under some circumstances, but it does not explain primary dynamic scattering shortly after a high field has been applied. Before a sample had completely changed to secondary dynamic scattering in the presence of a high field, the appearance of primary dynamic scattering did not appear to have vortices comparable to dynamic scattering below 20 volts. It looked more like secondary dynamic scattering.

Earlier work[9] involving flow cell widths as a function of the electric field intensity in a 135 micron sample did not indicate any differences between primary and secondary dynamic scattering. Although it was assumed in this work that the patterns observed in a continuously applied field represented flow cells bounded by walls, walls were also observed shortly after applying an electric field to an initially well ordered sample. There is the possibility that the depth of the walls changed in passing from primary to secondary dynamic scattering. In primary dynamic scattering the depth of the walls was probably much greater than the separation of the electrodes but in secondary dynamic scattering it could have been less than the electrode separation.

CONCLUSION

At an electric field of 12 kV/cm in a bulk sample we have observed defects at the free surface and electrode-to-liquid crystal interface which we believe represents walls (defects). The movement of the fluid implies that for regions not close to the electrodes, the model in Fig. 1 is applicable. One of the most interesting results of this work is that observations of walls are more easily made at very high fields than at the lower fields. It appears that fewer defects are created at the higher fields which enable an observer to make clearer observations. In order to better understand the effect of the field strength on the ability to make observations, we probably need to know a lot more about the structure and behavior of walls in electrohydrodynamic instabilities. Walls may play a much greater role in electrohydrodynamic instabilities than originally thought. The inability of an electric field to create walls in very thin samples (less than 10 microns) may help explain the transition from the dynamic scattering mode to the variable grating mode.[10]

Although we have been able to provide information about walls in high electric fields, we were not able to measure a flow-alignment angle θ using optical observations. However a flow-alignment angle θ (see Fig. 1) has been measured in MBBA in the presence of 6 kV/cm electric field using NMR techniques.[11] There is some disagreement among investigators concerning the value of this angle. We believe that the disagreement is probably due to impurities in the different samples. Impurities may affect the flow alignment to a much greater extent than previously realized.

REFERENCES

1.   E. F. Carr, Liquid Crystals and Ordered Fluids III, eds.
          J. F. Johnson and R. S. Porter, Plenum Press, 1978, 165.

2.   E. F. Carr and R. W. H. Kozlowski, Liquid Crystals, ed.
          S. Chandrasekhar, Heden, London, 1980, pp. 287-295.

3.   R. W. H. Kozlowski and E. F. Carr, Mol. Cryst. and Liq. Cryst.
          (Lett.) 64, 299 (1981).

4.   Alan Sussman, Appl. Phys. Lett. 21, 269 (1972).

5.   J. Nehring and M. S. Petty, Phys. Lett. 40A, 307 (1972).

6.   R. Chang, J. Appl. Phys. 44, 1885 (1973).

7. Teodor Krupkowski and Wieslaw Ruszkiewicz, Mol. Cryst. and Liq. Cryst. (letters) 49, 47 (1978).

8. W. Helfrich, Phys. Rev. Lett. 21, 1518 (1968).

9. E. F. Carr, P. H. Ackroyd and J. K. Newell, Mol. Cryst. and Liq. Cryst. 43, 93 (1977).

10. W. Greubel and U. Wolff, Appl. Phys. Lett. 19, 213 (1971).

11. C. E. Tarr and E. F. Carr, Solid State Commun. 33, 459 (1980).

APPLICATION OF GENERALIZED VAN DER WAALS THEORY OF HOMOLOGOUS

NEMATOGENS – PART 1: TRANS-4-ETHOXY-4'-N-ALKANOYLOXYAZOBENZENES

Andrew C. Pineda, Todd J. Jones and
Gerald R. Van Hecke

Department of Chemistry
Harvey Mudd College
Claremont, California 91711

ABSTRACT

In the generalized van der Waals theory (GVDW) of Cotter, nematogens are viewed as rigid spherocylinders moving in a mean field potential given by $V(\theta) = -\lambda_o v_o \rho - \lambda_2 v_o \rho \eta P_2(\cos\theta)$ where λ_o and λ_2 are potential parameters independent of temperature. Employing the equation of state developed by Cotter, values of λ_o, λ_2, and the order parameter are calculated for a homologous series of trans-4-ethoxy-4'-alkanoyloxyazobenezenes from observed volume and temperature data. Calculated λ_o values were found to be 3 to 5 times those calculated from heat of vaporization estimates. In calculating the temperature dependence of the order parameter, the λ_2 values were found to vary markedly with temperature. In addition, calculated values of the order parameter were generally about one-half to one-third of the experimentally measured values.

I.   INTRODUCTION

Nematic liquid crystals are fluids characterized by long-range orientational order. Such compounds are composed of long, rigid, rod-like molecules whose ordering is described by their alignment

with a space-fixed axis known as the director. Current debate over
how the behavior of these nematogens may be modeled centers around
the question of the relative magnitudes of the contributions to the
observed ordering from short-range repulsions and longer-range anisotropic intermolecular attractions. Theories which emphasize each
of these views have been developed.[1]

There exist two versions of the generalized van der Waals
theory, one developed by Cotter[2] and the other by Gelbart and co-
workers.[3,4] Each incorporate both short-range repulsions and long-
range anisotropic attractions. The two theories differ in the
form of the pseudopotential used, and furthermore Gelbart's version
allows for coupling between isotropic attractions and angle dependent hard core repulsions. Both versions model nematic molecules
as hard rigid spherocylinders moving in a mean field potential
that has both isotropic and anisotropic components.

Calculations employing these GVDW theories have been performed
on several compounds in order to predict their thermodynamic properties. Both Baron and Gelbart[5] and Cotter[2] have performed calculations on p-azoxyanisole (PAA) which yielded only qualitative
agreement with experiment. Dunmur and Miller have reported calculations on the homologous series of alkylcyanobiphenyl liquid
crystals $C_nH_{2n+1}\emptyset-\emptyset CN$ for n = 5 to 9. Using Cotter's version of the
theory, Dunmur and Miller obtained reasonable predictions for the
order parameter.[6]

We employ Cotter's version of the theory in our calculations
on the homologous series of trans-4-ethoxy-4'-n-alkanoyloxyazo-
benzenes,

$$CH_3CH_2O-\emptyset-N:N-\emptyset-O(C=O)(CH_2)_nCH_3 \quad \text{for } n = 3 \text{ to } 12.$$

The expressions that were used are presented in Section II. The
method of calculation is described in Section III and the results
presented in Section IV. Finally, the significance of these re-

sults are discussed in Section V.

II. THEORY

In the model system employed by Cotter[2], nematogens are viewed
as collections of rod-like molecules with spherocylindrical hard
cores. Each molecule may interact with its neighbors through hard
core repulsions and an attractive mean field produced by all of its
neighbors. This mean field is of the form

$$V(\theta) = -\lambda_o v_o \rho - \lambda_2 v_o \rho \eta P_2(\cos \theta) \qquad (1)$$

where ρ is the number density of molecules, v_o is the molecular
volume, θ is the angle that the molecule's long axis makes with
the director, η is the order parameter and λ_o and λ_2 are potential
parameters which are assumed independent of temperature. The first
term in the above expression may be interpreted as the energy change
involved in bringing one molecule into an isotropic solution from
an infinite separation. Thus it is related to the work required
to vaporize the isotropic liquid while the second term provides a
potential that serves to align the molecules in a preferred
direction.

Applying this expression to a collection of sphere cylinders
of length to width ratio, x, Cotter obtained the following equation
of state[2]:

$$\frac{Pv_o}{kT} = \Pi - \frac{1}{2} (\lambda_o + \lambda_2 \eta^2) \frac{v_o^2 \rho^2}{kT} \qquad (2a)$$

where

$$\Pi = \frac{v_o \rho}{(1-v_o\rho)^3} \left[1 + v_o\rho + \frac{2(3x^2-1)}{(3x-1)^2} v_o^2 \rho^2 \right.$$
$$\left. + \frac{3(x-1)^2}{(3x-1)} [1 + \frac{(x+1)}{(3x-1)} v_o\rho](1 - \frac{5}{8}\eta^2)v_o\rho \right] \qquad (2b)$$

and P is the pressure, k is Boltzmann's constant, η is the order
parameter. The self-consistency condition[3], is

$$\eta = \frac{\int_0^\pi P_2(\cos\theta)\,\exp[\phi\eta P_2(\cos\theta)]\sin\theta\,d\theta}{\int_0^\pi \exp[\phi\eta P_2(\cos\theta)]\,\sin\theta\,d\theta} \tag{3a}$$

where

$$\phi = \frac{15(x-1)^2 v_o \rho}{4(3x-1)(1-v_o\rho)^2}\left[1 - \frac{(x-1)v_o\rho}{3x-1}\right] + \lambda_2\,\frac{v_o\rho}{kT} \tag{3b}$$

Still other expressions were derived for the configurational chemical potential and Helmholtz free energy, but were not used in our
calculation.

III. METHOD OF CALCULATION

In performing our calculations, we followed the method used by
Dunmur and Miller. They fixed the transitional densities and
temperatures and varied the potential parameters to reproduce the
experimental results.

Equations 2 and 3 above relate eight variables: p, P, T, v_o,
x, η, λ_o, λ_2; we must specify six of them. Length to width ratios
and molecular volumes were estimated from van der Waals radii and
bond lengths. Experimental data allowed the pressure, temperature, and nematic density at the transition temperature to be
fixed. The boundary condition that the order parameter be zero
in the isotropic phase then allowed us to calculate the isotropic
potential parameter, λ_o, by the use of the isotropic density at
the transition temperature (Eqn 2 with $\lambda = 0$). Thus with the
above parameters set, the equations are then solved for the aniso-

tropic potential parameter, λ_2, and the order parameter. The
temperature dependence of η and λ_2 was calculated in an identical
fashion using the experimental nematic densities for temperatures
below the transition temperature. As mentioned earlier, expressions are available for the Helmholtz free energy and from there
the chemical potential. These equations at the nematic-isotropic
transition would have provided another relationship between the
eight variables through equating the chemical potentials of the
nematic and isotropic phase. Since we were interested in calculating η as a function of temperature in the nematic phase we chose
not to use the chemical potential restriction to make the calculations in the nematic phase and at the nematic to isotropic transition as comparable and consistent as possible. Inclusion of the
chemical potential restriction at the phase transition might
change those results but cannot affect the calculations of order
parameter in the nematic phase.

IV. RESULTS

The data used in our calculations at the transition temperature are given in Table 1. Values of the nematic densities were
obtained by fitting the volume data to the equation suggested by
Van Hecke and Stecki.[7]

$$V = a + bT + c\left|T-T_{sp}\right|^{1/2} \tag{6}$$

The isotropic densities were obtained by a linear fit of the
isotropic volume data. Results of the calculations at the transition temperature are presented in Tables 2 and 3. Several features are apparent in Table 2. First of all, generally smaller
values of the length to width ratio than those predicted from
bond lengths and bond angles had to be used in order to produce
consistent results. The criterion used to judge the consistency
of the results was that the anisotropic potential parameter
be non-negative for all temperatures when the parameter was calcu-

Table 1

Data used in the calculation of the order and potential parameters at T_{NI}.

Carbons is alkane chain	T_{NI}/K [a]	$v_o/\mathring{A}^3$ [b]	$\rho(T_{NI})/(\# /10^3\mathring{A}^3)$ [c]	
			nematic	isotropic
5	400.2	273	1.9029	1.8974
6	401.6	291	1.8047	1.8018
7	392.3	309	1.7204	1.7150
8	392.8	327	1.6399	1.6348
9	387.7	345	1.5681	1.5652
10	386.1	363	1.5000	1.4968
11	383.1	381	1.4403	1.4377
12	382.5	399	1.3843	1.3813
13	379.7	417	1.3339	1.3315
14	379.4	435	1.2867	1.2843

(a) Data from Ref. 8.   (b) Estimated from van der Waals radii and bond lengths.   (c) Data from ref. 9. Nematic density data was fitted to $V = a + bT + c|T-T_{sp}|^{1/2}$.

lated as a function of temperature. As can be seen in the table, we were able to use reasonable length to width ratios for some of the smaller members of the series. Secondly, we note the large values for the estimated heats of vaporization. Typical molecules of comparable size have heats of vaporization on the order of 10 to 15 kcal/mole, thus the calculated values are three to five times too large. Thirdly, we observe that the calculated transitional order parameters are fairly low.

The results in Table 2 show that our criterion for choosing length to width ratios has produced an unrealistic trend, i.e., length to width ratios decreasing with increasing chain length. To remedy this defect, further calculations were performed in which

length to width ratios were calculated relative to the smallest
allowed value using the calculated lengths of the molecules. Table
3 shows the results of this calculation. At least two features are
worth noting. First, the isotropic potential parameter, λ_o, increases regularly with chain length while the anisotropic potential
parameter, λ_2, decreases. This would seem to indicate that hard
core repulsions become more dominant with increasing chain length.
Intuition might suggest that as a molecule becomes longer it should
become more flexible because of the increasing length of the alkane
chains. The result that the anisotropic potential becomes less
important while the hard rod repulsions become more so seems
counter-intuitive. Yet in this model which allows for no coupling
between the isotropic and anisotropic potentials or for molecular
flexibility, experiment is best fit by describing the longer
molecules as increasingly hard rod-like. Second, the order parameter is found to demonstrate an almost regular odd-even effect.
This alternation closely follows the alternation in the observed
density changes at the nematic to isotropic transition.

Figures 1-3 show plots of the order parameter and the potential parameters as functions of molecular volume and length to
with ratio for one member of our series: heptanoate. These plots
illustrate several general features. If we look at molecular
volumes fixed at reasonable values, we see that larger order
parameters and smaller isotropic potential parameters can be
obtained by decreasing the length to width ratio. This seems
to be a counter-intuitive finding, but we reserve comment on it
since it may only be an artifact of the way in which the calculation was done. For constant length to width ratio, increasing
the molecular volume results in increased order parameters and
isotropic potential parameters. Thus, by decreasing the length
to width ratio, it is possible to obtain more reasonable values
for the order parameters, and the heats of vaporization.

Table 2

Results calculated using the criterion that λ_2 be non-negative when temperature dependence of η was calculated.

| Carbons in | X_{est} (a) | X_{max} (b) | Calculated for X_{max}, v_o at T_{NI} | | | ΔH_{vap} (estimated)(c) kcal/mole |
			λ_o/eV	λ_2/eV	η	
5	3.07–4.89	3.25	3.39728	0.0227818	0.186249	40.6
6	3.21–5 11	3.10	3.43109	0.142486	0.142486	41.5
7	3.36–5.34	2.94	3.34989	0.0478979	0.206070	41.1
8	3.50–5.57	3.00	3.49415	0.0316052	0.206670	43.1
9	3.64–5.80	2.90	3.48858	0.0428480	0.162237	43.6
10	3.78–6.01	2.94	3.57772	0.0298941	0.176108	45.0
11	3.91–6.23	2.91	3.62556	0.0290793	0.162239	45.9
12	4.06–6.47	2.88	3.67342	0.0263958	0.184761	46.6
13	4.20–6.68	2.85	3.71807	0.0263622	0.169175	47.7
14	4.34–6.91	2.82	3.77271	0.0239744	0.190868	48.7

(a) Length to width ratio estimated from bond lengths and bond angles. Two numbers are presented: one representing the use of the aromatic rings as a basis calculating the width of the molecule, the other representing the use of the alkane chain.

(b) This was the largest value of the length to width ratio that did not violate our criterion that the anisotropic potential parameter be non-negative for all temperatures when the temperature dependence of η was calculated.

(c) Assuming the enthalpy of vaporization approximately equals the energy of vaporization.

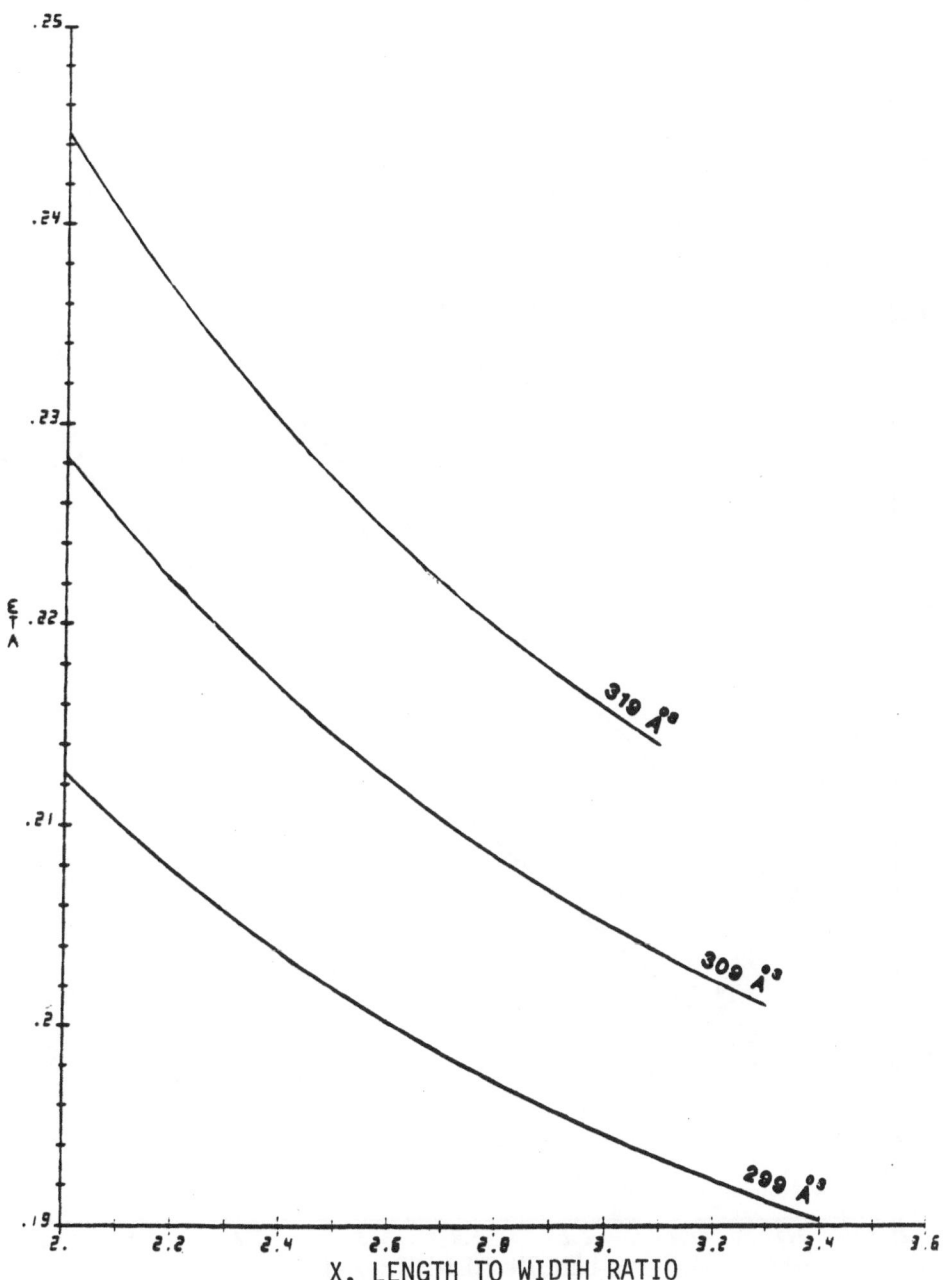

Fig. 1 Order parameter as a function of length to width ratio, X, and molecular volume, v_o, for heptanoate.

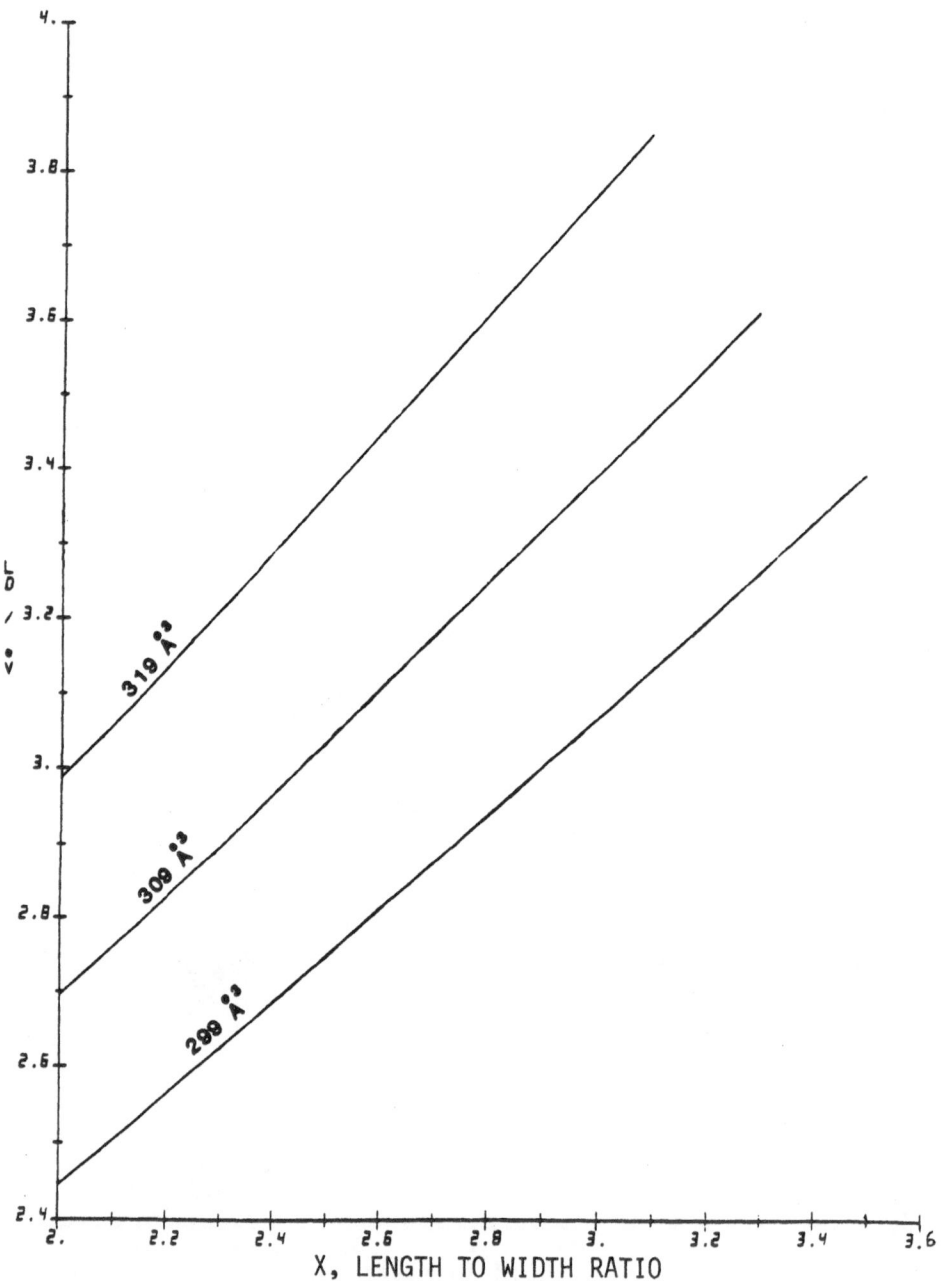

Fig. 2   Isotropic potential parameter as a function of length to
width ratio, X, and molecular volume, v_o, for heptanoate.

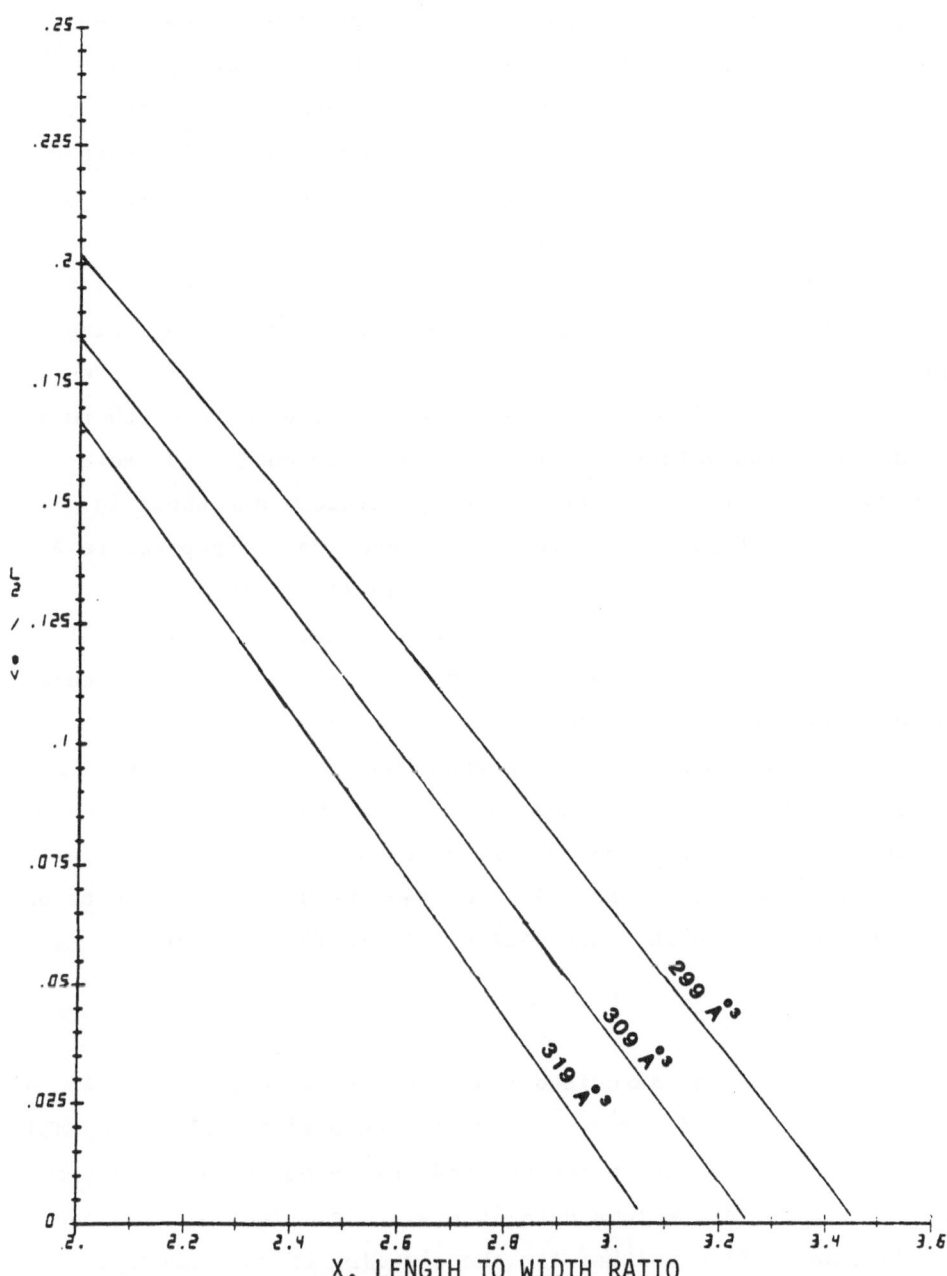

Fig. 3 Anistropic potential parameter as a function of length to
width ratio, X, and molecular volume, v_o, for heptanoate.

Figures 4-7 show the calculated order parameters and anisotropic potential parameters plotted as a function of the reduced
temperature, $T-T_{NI}$, for fixed x (=x_{max}) and v_o. Order parameters
are compared in Figures 4-7 with order parameters calculated by
Van Hecke, et al. from the Vuks and Neugebauer models.[10] In each
of these plots, one observes that our calculated results are consistently low; however, they do follow the trends in the order parameter fairly closely. As noted from Figures 1-3, it is in fact
possible to obtain even better agreement with the data by using a
smaller length to width ratio, but we felt that it was more realistic to use as reasonable a value for the length to width ratio
as possible. The values of the anisotropic potential parameter
corresponding to the calculated order parameters are shown in
Figure 7. This Figure shows that as a function of temperature λ_2
ranges roughly over one order of magnitude (λ_o is being held
constant). This is a somewhat disturbing result because λ_2 is
in principle supposed to be a constant independent of temperature.
However, since it varies over only a relatively small range
of values, it is difficult to ascertain whether this effect is a
consequence of the theory, uncertainties in the data, or the method
of calculation. (The method of calculation fixes all variables
except for η and λ_2, if there is some temperature dependence to one
of these other variables, then its effect is forced into λ_2.)

V. CONCLUSIONS

The calculations described above show that Cotter's version of
the generalized van der Waals theory can be used to make reasonable
predictions of the order parameter and its temperature dependence
when the appropriate volume data is available. The calculations we
have done here allow neither for the coupling of the isotropic and
anisotropic potentials, nor for any molecular flexibility. However,
as Dunmur and Miller[6] noted, the coupling effects included by Gelbart

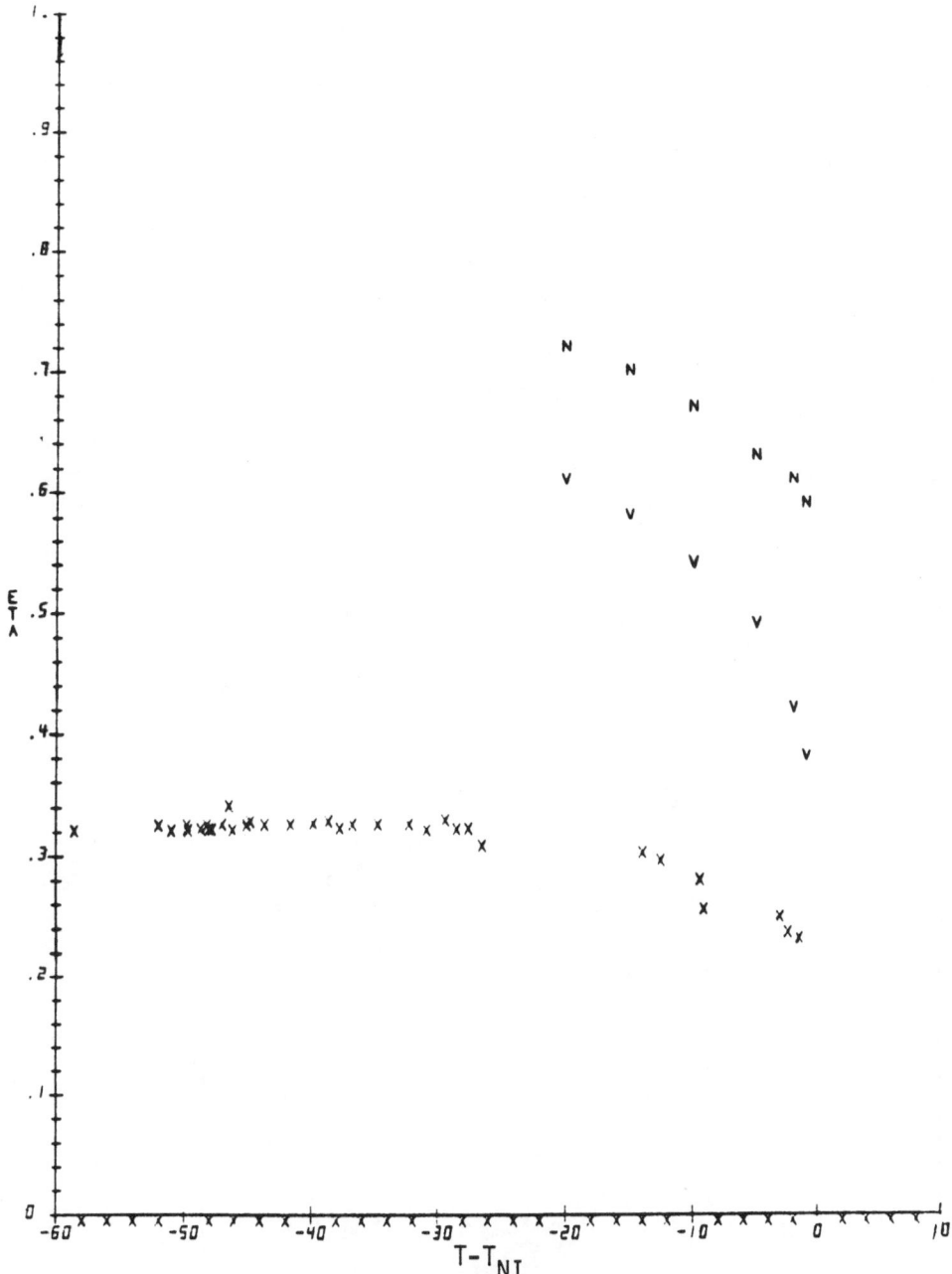

Fig. 4 Order parameter versus reduced temperature, $T-T_{NI}$, for heptanoate V = calculated from Vuks model, N = calculated from Neugebauer model, see Ref. 10.)

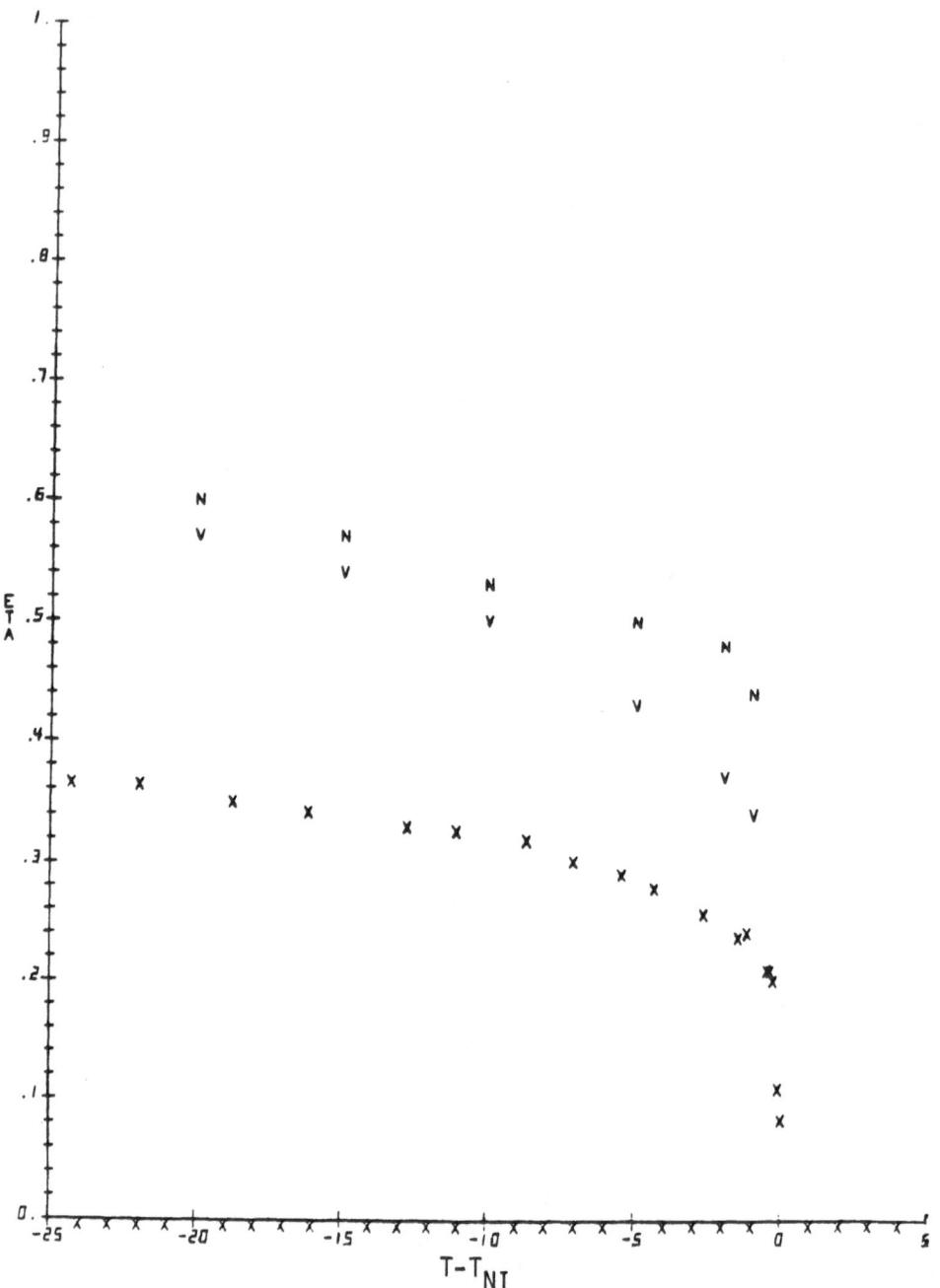

Fig. 5   Order parameter versus reduced temperature, T–T$_{NI}$, for non-
         anoate (V = calculated from Vuks model, N = calculated from
         Neugebauer model, see Ref. 10.)

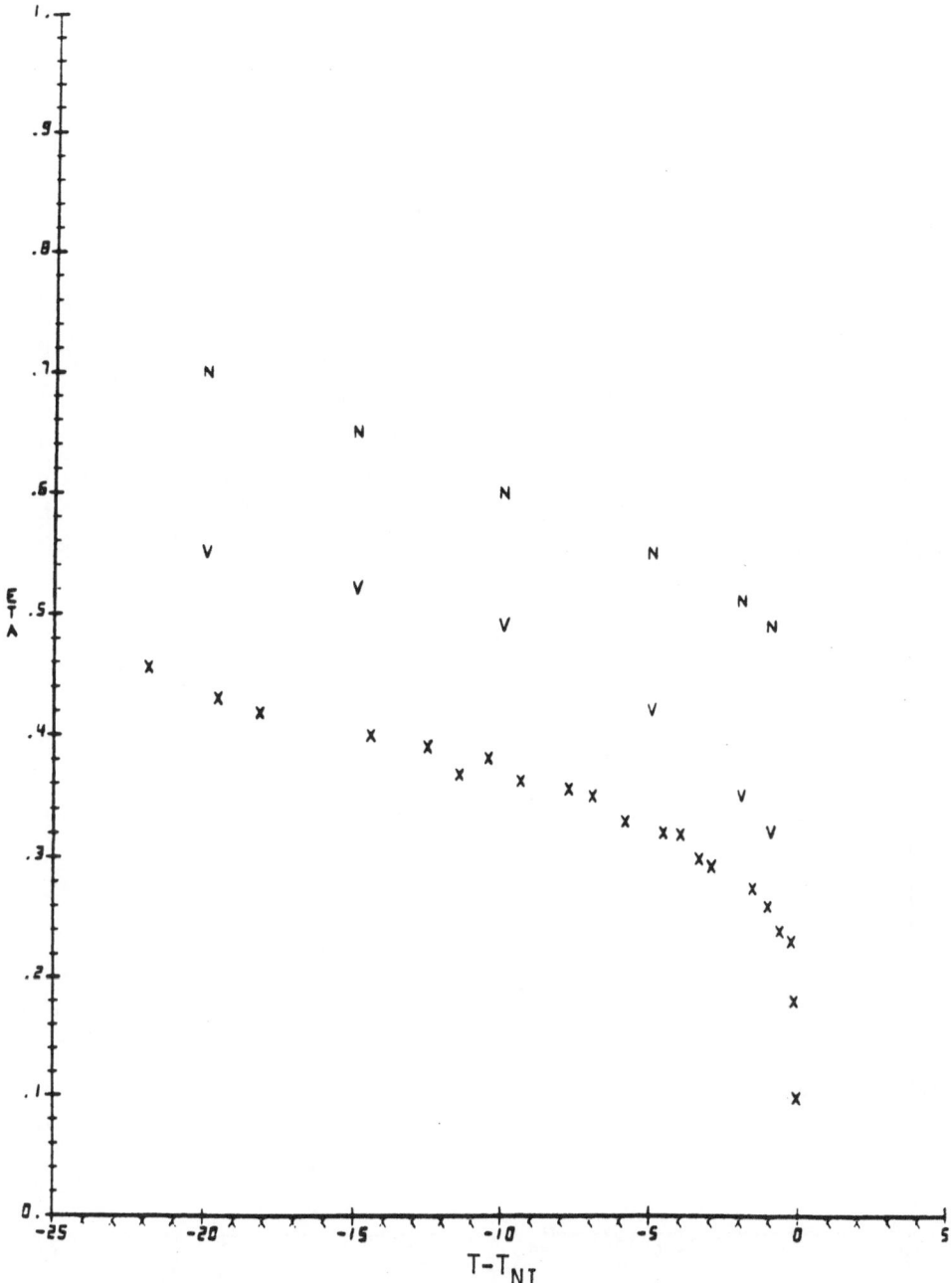

Fig. 6 Order parameter versus reduced temperature, $T-T_{NI}$, for tetra-
decanoate (V = calculated from Vuks model, N = calculated
from Neugebauer model, see Ref. 10.)

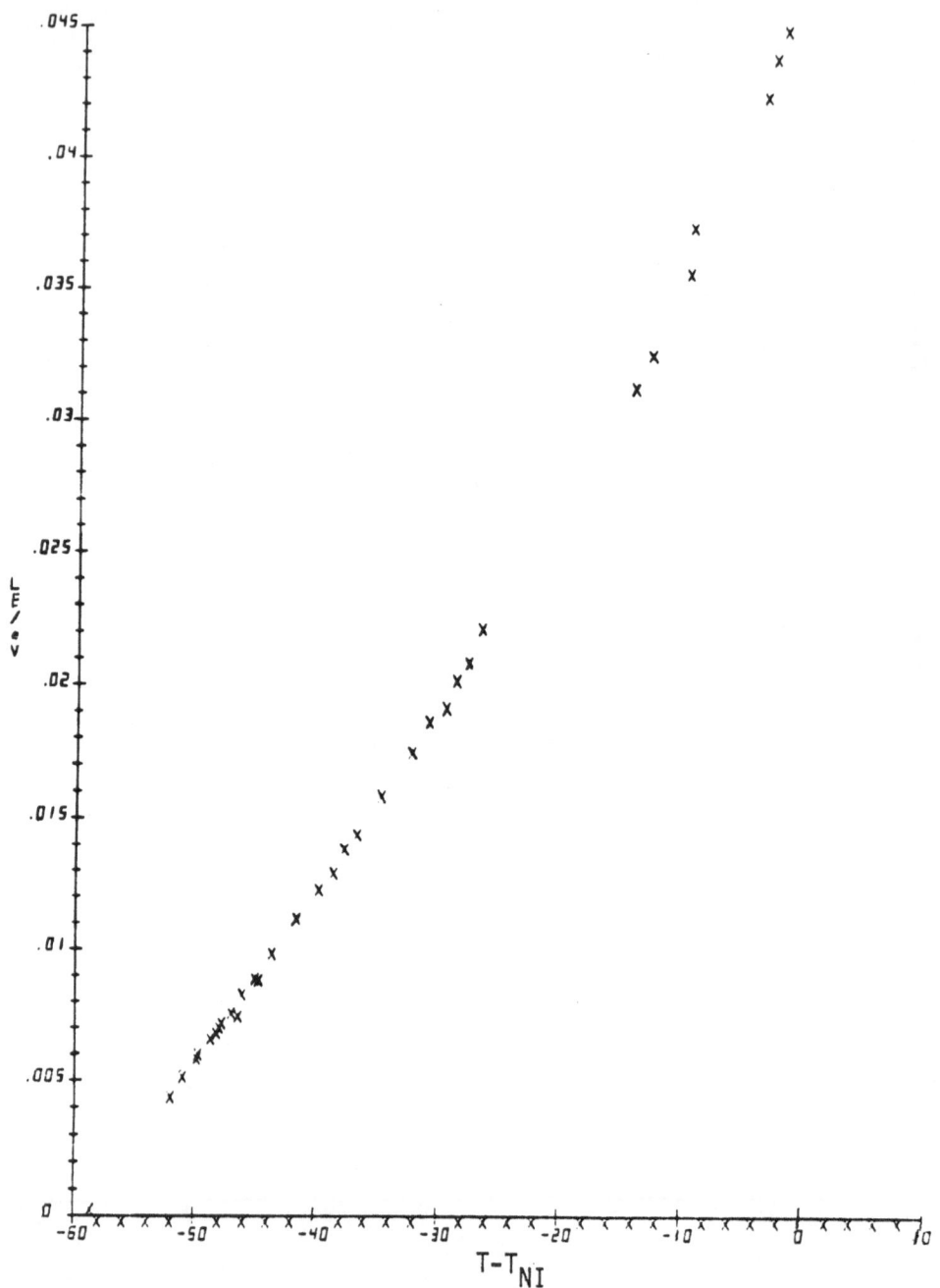

Fig. 7   Anisotropic potential parameter versus reduced temperature, $T-T_{NI}$ for heptanoate.

Table 3

Results calculated so that length to width ratios increase along the series in proportion to molecule length.

Carbons in alkane chain	estimated length/Å	relative length[a] to width ratio, X_{rel}	Calculated for X_{rel} and v_o at T_{NI}		
			λ_o/eV	λ_2/eV	η
5	21.5	1.99	2.56254	0.20287	0.208059
6	22.5	2.09	2.7307	0.188397	0.156413
7	23.5	2.18	2.81596	0.159499	0.222614
8	24.5	2.27	2.96105	0.141528	0.222413
9	25.5	2.37	3.09186	0.1231	0.171349
10	26.45	2.45	3.20234	0.105144	0.18514
11	27.4	2.54	3.33491	0.0865448	0.168527
12	28.45	2.64	3.48024	0.064264	0.189403
13	29.4	2.73	3.61921	0.0454894	0.171293
14	30.4	2.82	3.77271	0.0239744	0.190868

(a) Calculated relative to C_{14}.

are probably important since their inclusion would probably allow
larger, more reasonable order parameters to be calculated with
smaller potential parameters and larger length to width ratios. Our
data also seems to support Cotter's[2] idea that the values of the
length to width ratio we found required by the calculations may be
attributed to the fact that molecular flexibility is not taken into
account in the theory. Further work would be most productive, then,
including coupling and/or molecular flexibility.

REFERENCES

1. See for example:

    A. Wulf and A. G. DeRocco, J. Chem. Phys., $\underline{55}$, 12 (1971).

    S. Marcelja, J. Chem. Phys., $\underline{60}$, 3599 (1974).

    R. Alben, Mol. Cryst. Liq. Cryst., $\underline{13}$, 193 (1971).

    W. Maier and A. Saupe, Z. Naturforsch, $\underline{A14}$ 882 (1959).

2. M. A. Cotter, J. Chem. Phys., $\underline{66}$, 1098 (1977).

3. W. A. Gelbart and B. A. Baron, J. Chem. Phys., $\underline{66}$, 207 (1977).

4. W. A. Gelbart and Barboy, Acc. Chem. Res., $\underline{13}$, 290 (1980).

5. B. A. Baron and W. A. Gelbart, J. Chem. Phys., $\underline{67}$(12), 5796
    (1977).

6. D. A. Dunmur and W. H. Miller, J. de Physique, $\underline{C3}$, $\underline{40}$, C3-141
    (1979).

7. G. R. Van Hecke and J. Stecki, Phys. Rev. A., $\underline{25}$, 1123 (1982).

8. C. L. Hillemann, G. R. Van Hecke, S. R. Peak, J. B. Winther,
    M. A. Rudat, D. A. Kalman, and M. L. White, J. Phys. Chem.,
    $\underline{79}$, 1566 (1975).

9. B. D. Santarsiero and L. J. Theodore, unpublished results.

10. G. R. Van Hecke, B. D. Santarsiero, and L. J. Theodore, Mol.
    Cryst. Liq. Cryst., $\underline{45}$, 1 (1978).

APPLICATION OF REGULAR SOLUTION THEORY TO DISCOTIC MESOPHASES: CALCULATION OF PHASE DIAGRAMS EXHIBITING MINIMA

Ralph A. Wheeler and Gerald R. Van Hecke

Department of Chemistry
Harvey Mudd College
Claremont, CA 91711

ABSTRACT

Binary phase diagrams of the discotic nematogenic benzene-hexa-n-alkanoates show minimum azeotrope-like behavior. Regular solution theory is successfully applied to calculate phase diagrams for binary mixtures of four members of this homologous series. Furthermore, an empirical correlation between the regular solution parameters and molar volume ratio of the components is proposed.

INTRODUCTION

Near the turn of the century, Schroeder and van Laar provided a thermodynamic description of ideal binary mixtures in terms of transition temperatures and enthalpies for the pure materials.[1,2] More recently, Reisman, and Domon and Billard have discussed the wide variety of ideal two phase spindles that can exist depending upon the magnitudes of enthalpies and temperatures.[3,4] Ideal solution theory does not, however, admit the occurrence of phase diagrams exhibiting maxima or minima. The azeotrope-like behavior shown by some binary liquid crystal mixtures can nevertheless be described using regular solution theory to account for non-ideality.

283

For binary mixtures of calamitic mesogens, both maximum and minimum deviations from ideality have been interpreted using regular solution theory and the non-ideality of the minima forming systems correlated with molar volume ratio of the components.[5,6] The recent discovery of discotic mesophases[7] and the demonstration that their structures differ from those of calamitic mesophases[8] questions whether or not the non-ideality shown by some discotics also correlates with component molar volume ratio. We present here the results of applying the regular solution approximation to describe phase diagrams of the disc-like molecules, benzene-hexa-n-alkanoates. In addition, we demonstrate an empirical correlation between non-ideality and molar volume ratio of the components. Although reasons for this relationship remain unclear, we show its utility for estimating regular solution parameters. These parameters then allow preliminary prediction of the liquid two phase region shown by these systems.

THEORY

Phase diagrams exhibiting non-idealities, indicated by maximum or minimum azeotrope-like behavior, can be calculated using regular solution theory.[5,6] The chemical potential for component 1 in phase α is written as

$$\mu_{1\alpha} = \mu_{1\alpha}^{o} + RT \ln x_{1\alpha} + A_{\alpha} x_{2\alpha}^{2} \qquad (1)$$

where the last term in equation 1 accounts for non-idealities. A_{α}, the regular solution parameter for phase α, is assumed independent of temperature and composition. This treatment of non-ideal behavior leads to two equations describing the coexistence lines, or temperature-composition lines, for two phases α and β in equilibrium.

$$\ln (x_{1\beta}/x_{1\alpha}) + (A_{\beta}x_{2\beta}^{2} - A_{\alpha}x_{2\alpha}^{2})/RT = H_{1} \qquad (2a)$$

$$\ln \ (x_{2\beta}/x_{2\alpha}) \ + \ (A_\beta x_{1\beta}^2 - A_\alpha x_{1\alpha}^2)/RT = H_2 \tag{2b}$$

$$H_i = \Delta H_{i\alpha\beta}^o (T-T_i)/RTT_i \tag{3}$$

To arrive at this result, transition enthalpies are assumed independent of temperature and the difference of heat capacities for phases
α and β is assumed to be negligible. Iterative solution of these
equations for $x_{1\alpha}$ and $x_{1\beta}$ is possible given T, properties of the
pure components, and some estimate for A_α and A_β. As pointed out
earlier and reviewed below, the presence of a maximum or a minimum
in the phase diagram allows estimation of these parameters.

Regular solution parameters can be estimated by considering
the "azeotrope" point, where $T = T_m$, $x_{1\alpha} = x_{1\beta} = x_{1m}$, and $x_{2\alpha} =$
$x_{2\beta} = x_{2m}$. Using these relations simplifies equations 2a and 2b to
give

$$(A_\beta - A_\alpha) \ x_{2m}^2 = RT_m H_1 \tag{4a}$$

$$(A_\beta - A_\alpha) \ x_{1m}^2 = RT_m H_2 \tag{4b}$$

Solution of these equations yields

$$x_{1m} = (1 + H_1/H_2)^{1/2})^{-1} \tag{5}$$

$$(A_\beta - A_\alpha) = RT_m H_2 \ (1 + (H_1/H_2)^{1/2})^2 \tag{6}$$

Therefore, only T_m and the pure component transition temperatures
and enthalpies are needed to calculate the composition of the
azeotrope point and the difference of the regular solution parameters.

If one of the phases can be treated as ideal, for example the
equilibrium between a liquid and an ideal solid, one of the parameters is zero and equations 2a and 2b become

$$\ln x_{1\beta} + A_\beta(1 - x_{1\beta})^2/RT = H_1 \tag{7}$$

where the α phase has been taken to be ideal pure solid for which

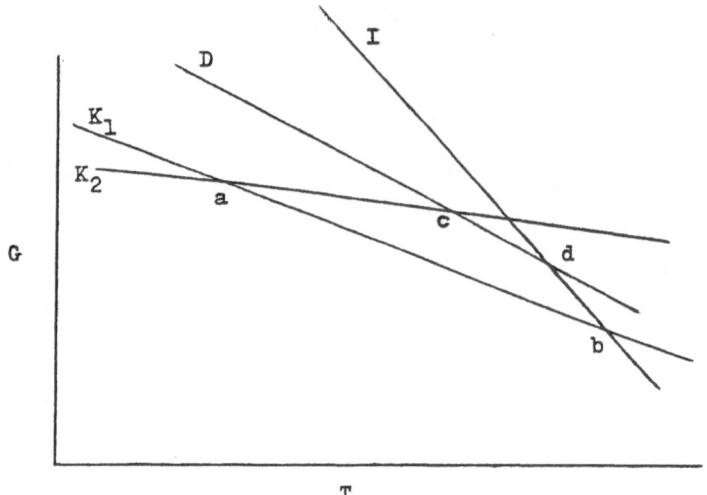

Fig. 1.   Idealized, qualitative free energy vs. temperature
           diagram for benzene-hexa-n-hexanoate.

$A_\alpha = 0$, $x_{1\alpha} = 1$, and $x_{2\alpha} = 0$. Hence, if one liquid phase is in
equilibrium with an ideal solid phase, and one point on the liquid-
solid coexistence line is known, properties of the pure components
will allow calculation of A_β. A_α can then be calculated from
equation 6.

     For experimentally unobservable "virtual transitions", pure
component transition enthalpies can be calculated using a method
previously described.[4] In the idealized chemical potential vs. temperature diagram for benzene-hexa-n-hexanoate presented in Fig. 1,
points a and b refer to experimentally observed transitions. Point
a represents the crystal K_2 to crystal K_1 transition and point b
represents the crystal K_1 to isotropic liquid I transition. To
calculate the transition enthalpy for the virtual transition of
discotic liquid D to isotropic liquid (point d in Fig. 1) requires
solution of the equations

$$\Delta H_{K_2 K_1} + \Delta H_{K_1 I} = \Delta H_{K_2 D} + \Delta H_{DI} \qquad (8)$$

$$\Delta S_{K_2 K_1} + \Delta S_{K_1 I} = \Delta S_{K_2 D} + \Delta S_{DI} \qquad (9)$$

where $\Delta H_{\alpha\beta}$ and $\Delta S_{\alpha\beta}$ are assumed independent of temperature. These equations can be solved for ΔH_{DI} by using the approximation that $\Delta S_{\alpha\beta} = \Delta H_{\alpha\beta}/T_{\alpha\beta}$. Thus, ΔH_{DI} can be calculated from experimentally determined transition enthalpies and entropies for the K_2 to K_1 and the K_1 to I transitions, along with $T_{K_2}D$ and T_{DI}, obtained by extrapolating the appropriate coexistence lines to pure hexanoate on the experimentally determined phase diagrams.

APPLICATIONS AND RESULTS

To illustrate the application of regular solution theory to discotic mesogens, we will consider binary phase diagrams of benzene-hexa-n-alkanoate mixtures reported by Billard and Sadashiva.[9] These authors present phase diagrams, obtained by the contact method, for four of the possible six binary mixtures of benzene-hexa-n-alkanoate homologues from the hexanoate to the nonanoate. Transition temperatures for all four pure compounds and transition enthalpies for the heptanoate, octanoate, and nonanoate are available.[9,10] The discotic to isotropic liquid transition enthalpy for the hexanoate was calculated by the method described earlier.

TABLE I. Transition temperatures and enthalpies for the discotic liquid to isotropic liquid transition of benzene-hexa-n-alkanoates.

Compound	T_{DI}(K)	ΔH_{DI}(kcals/mole)
hexanoate[a]	362.3	4.3
heptanoate[b]	359.3	5.15
octanoate[c]	356.9	4.5
nonanoate[c]	351.9	3.4

[a] Transition enthalpy was calculated from T_{K2D} = 361.4K, obtained by extrapolation, and thermodynamic data provided in references 9 and 11.

[b] Data taken from reference 10.

[c] Data taken from reference 9.

These thermodynamic data were used to estimate A_D and A_I, the regular solution parameters characterizing the discotic and isotropic liquid phases, and an iterative computer program based on the dual Newton's method was written to solve equation 2. A_D was estimated by assuming the discotic liquid to be in equilibrium with an ideal solid phase. Equation 7 was thus solved for A_D at the eutectic temperature and composition, taken from the experimental phase diagram. Solution of equation 6 at the azeotrope temperature then allows estimation of A_I. As seen in Table II, the magnitude of A_I is less than A_D for every mixture considered. Since the magnitude of these regular solution parameters indicates the extent of deviation from ideality ($A_\alpha = 0$ implies ideality of the α phase), relative magnitudes of A_I and A_D indicate that ideality is partially recovered as orientational order is destroyed. Thus, the isotropic liquid phase is more nearly ideal than the discotic phase.

Calculated values of the A parameters allow iterative solution of equation 2 for the four phase diagrams of benzene-hexa-n-alkanoate mixtures previously reported (Figs. 2-5). These T-x diagrams were plotted using a Megatek computer graphics terminal and compared with both experiment and ideal solution calculations. Although ideal solution theory closely approximates the experimental phase diagrams, regular solution theory provides a more realistic description of the discotic-isotropic liquid coexistence lines.

TABLE II. Eutectic temperatures, eutectic compositions, azeotrope temperatures, and regular solution parameters for mixtures of benzene-hexa-n-alkanoates.

components	T_e(K)	x_{2e}	T_m(K)	A_I/R	A_D/R	$(A_I-A_D)/R$
octanoate, nonanoate	343.9	0.522	351.5	370.8	422.0	-51.19
heptanoate, octanoate	333.2	0.552	355.5	82.7	150.4	-67.73
heptanoate, nonanoate	338.2	0.416	348.7	-78.3	82.4	-160.69
hexanoate, heptanoate	339.5	0.693	357.4	388.7	471.0	-82.27
hexanoate, octanoate	-----	-----	-----	92.2	276.6	-184.4
hexanoate, nonanoate	-----	-----	-----	133.8	401.3	-267.55

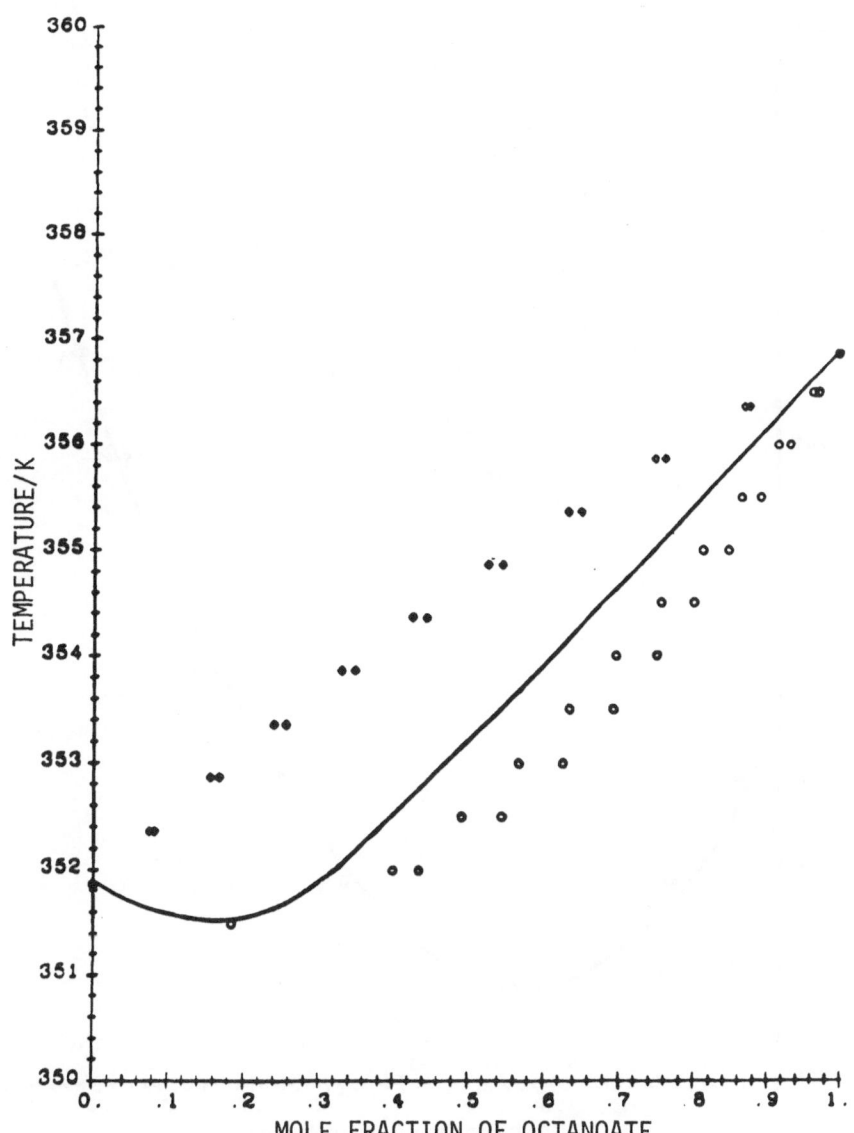

Fig. 2.   Isobaric phase diagram for benzene-Hexa-n-octanoate and nonanoate.

—— Experimental Phase Diagram
◇ Ideal Solution Theory
○ Regular Solution Theory

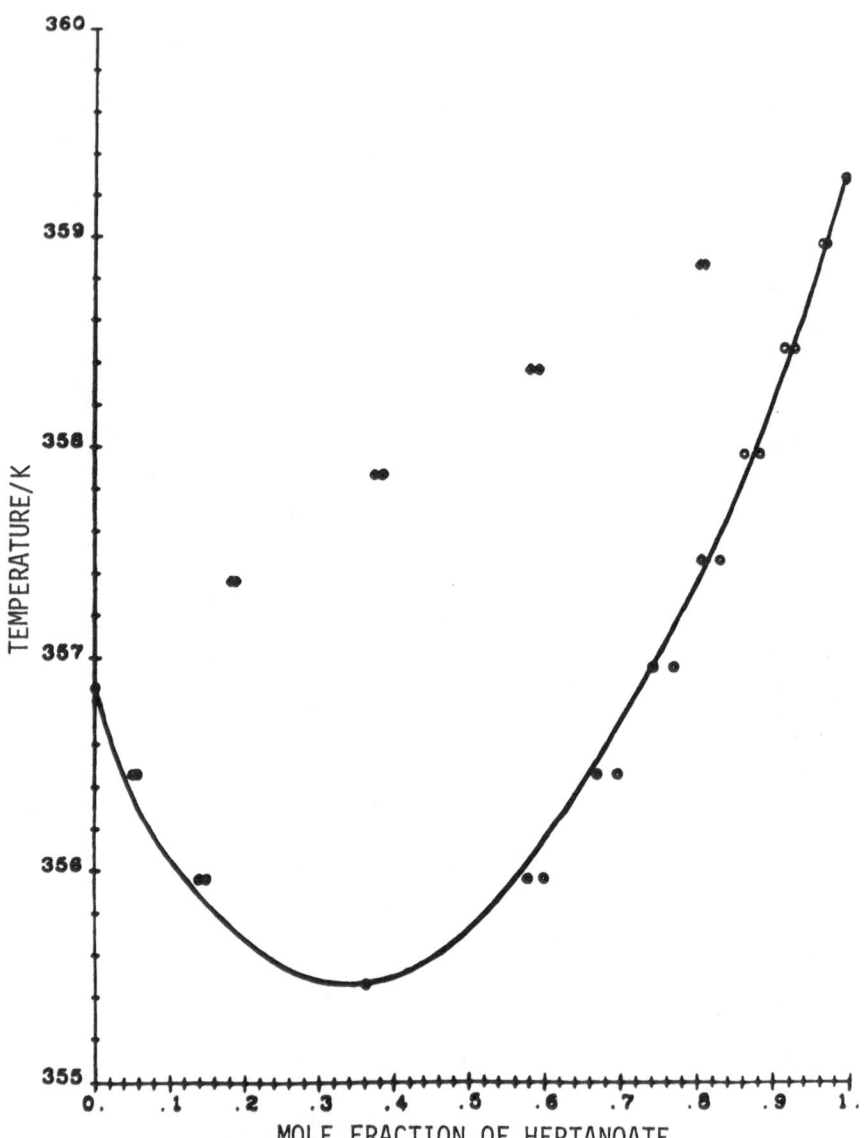

Fig. 3.   Isobaric phase diagram for benzene-hexa-n-heptanoate and
           octanoate.
                    ── Experimental Phase Diagram
                    ◇ Ideal Solution Theory
                    ○ Regular Solution Theory

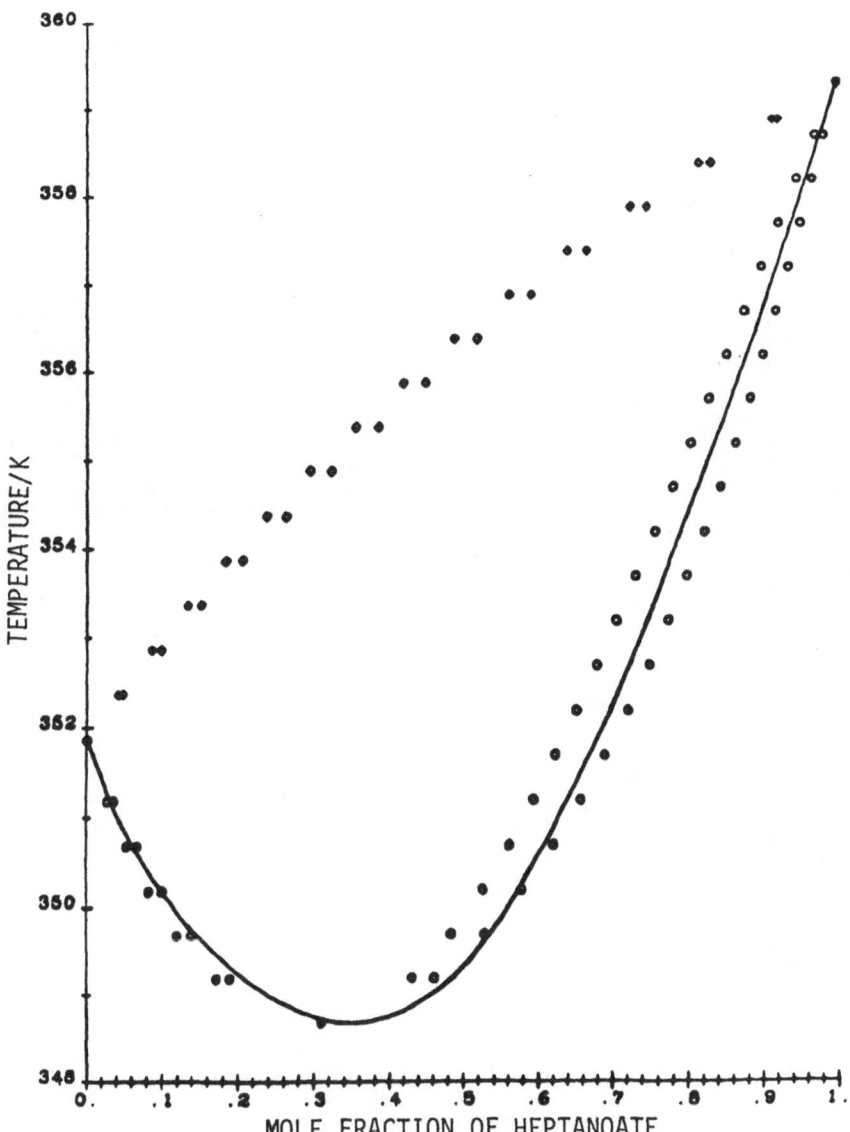

Fig. 4.   Isobaric phase diagram for benzene-hexa-n-heptanoate and
          nonanoate.
                    —— Experimental Phase Diagram
                    ◇ Ideal Solution Theory
                    ○ Regular Solution Theory

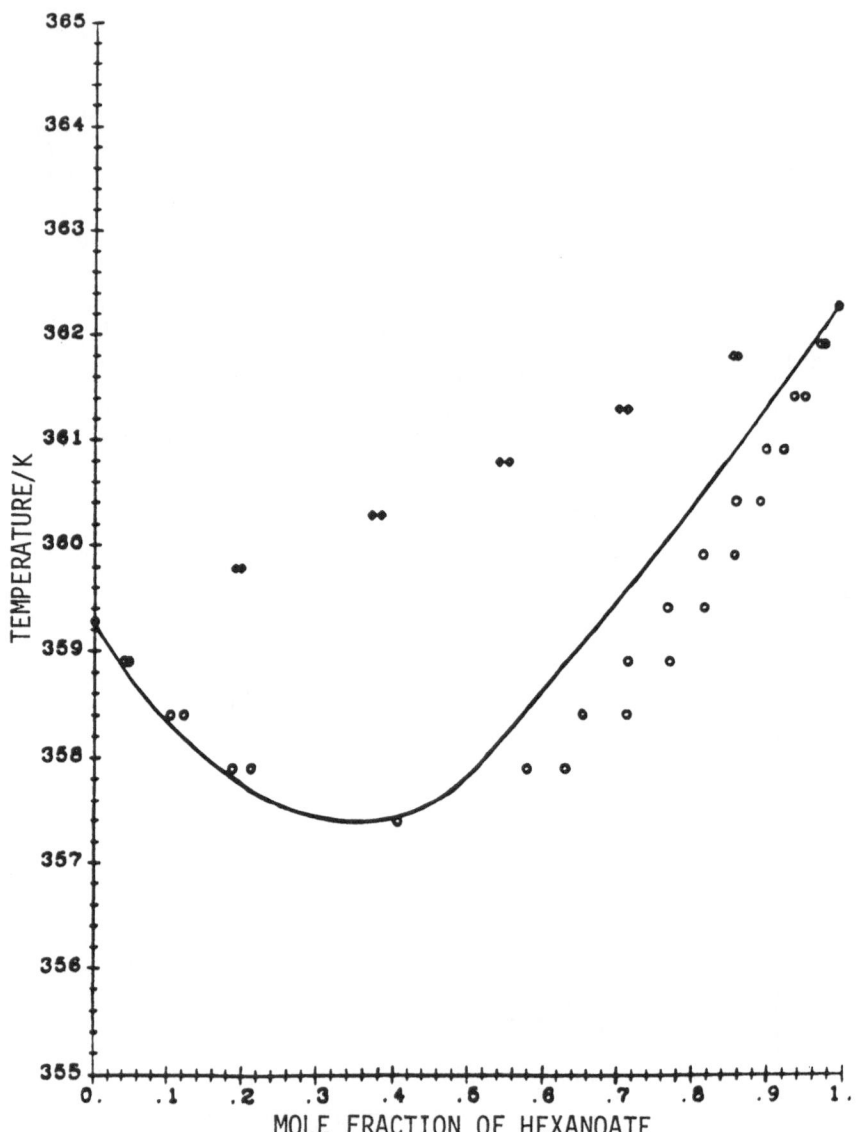

Fig. 5.   Isobaric phase diagram for benzene-hexa-n-hexanoate and
           heptanoate.
           —— Experimental Phase Diagram
           ◇ Ideal Solution Theory
           ◯ Regular Solution Theory

TABLE III. Estimated molar volumes for benzene-hexa-n-alkanoates at the discotic to isotropic liquid transition temperature.

Compound	Molar Volume (cm^3 $mole^{-1}$)
hexanoate	767.37
heptanoate	871.36
octanoate	975.35
nonanoate	1079.34

Next, an empirical correlation between the difference of regular solution parameters, $A_I - A_D$, and the molar volume ratio of the components is noted. Molar volumes at the discotic to isotropic transition temperature were estimated from the relationship $V = 871.36 + 6(n-6)(17.33)$, where 871.36 cm^3 $mole^{-1}$ is the molar volume for benzene-hexa-n-heptanoate and 17.33 cm^3 $mole^{-1}$ is the estimated molar volume per additional methylene group in the alkanoate chain.[12] The linear plot of $(A_I - A_D)/R$ vs. molar volume ratio provides a basis for estimating the difference of regular solution parameters without the detailed knowledge provided by the phase diagram. These parameters can be estimated from component molar volume ratio if they are chosen arbitrarily to be equally spaced about the difference, that is

$$A_D = -1.5(A_I - A_D) \qquad A_I = -0.5(A_I - A_D) \qquad (10)$$

Other choices of A_D and A_I make little difference in calculations as long as their difference is kept at the specified value. This approximation allows the preliminary prediction of the phase diagrams for binary mixtures of benzene-hexa-n-hexanoate with the octanoate and with the nonanoate (Figs. 7-8). Transition temperatures also show an approximately linear relationship with molecular weight,[9] but unfortunately transition enthalpies do not demonstrate such a simple relationship. Therefore, calculation of phase diagrams for other members of the homologous series of benzene-hexa-n-alkanoates requires further experimental data.

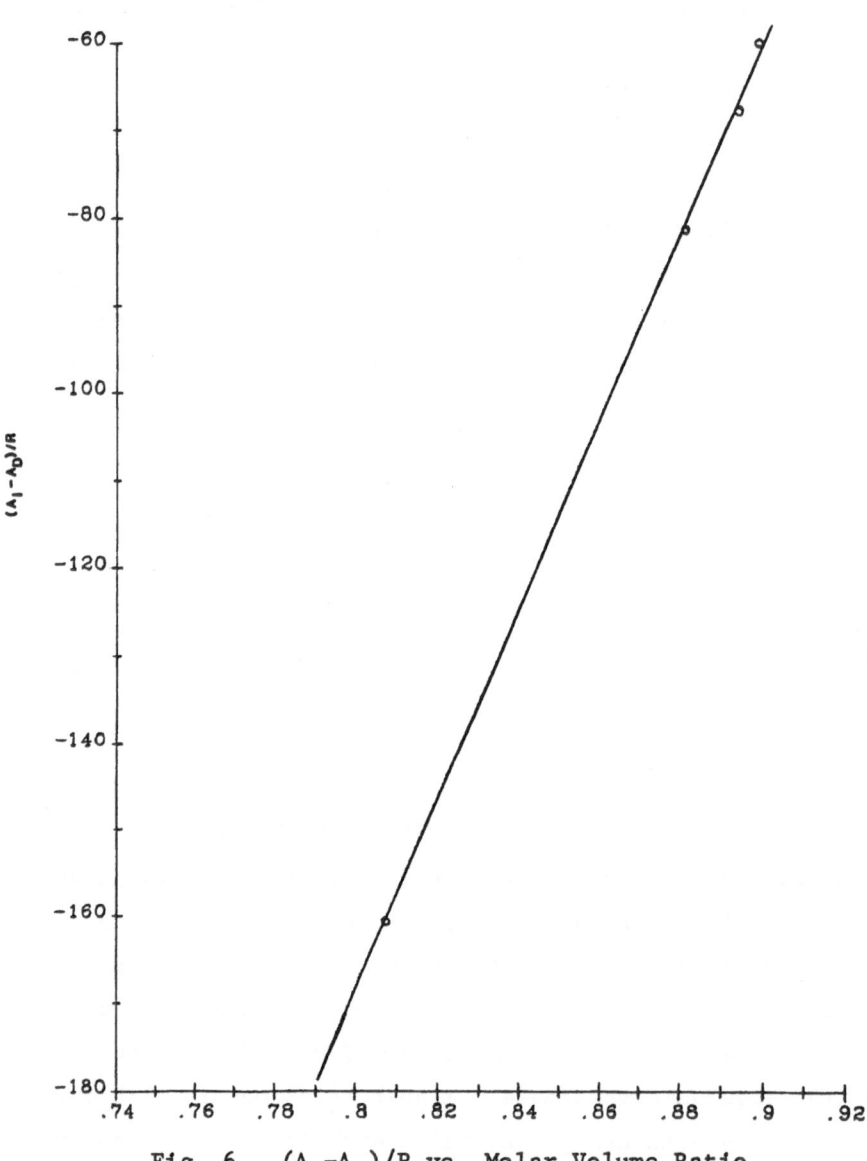

Fig. 6. $(A_1-A_D)/R$ vs. Molar Volume Ratio

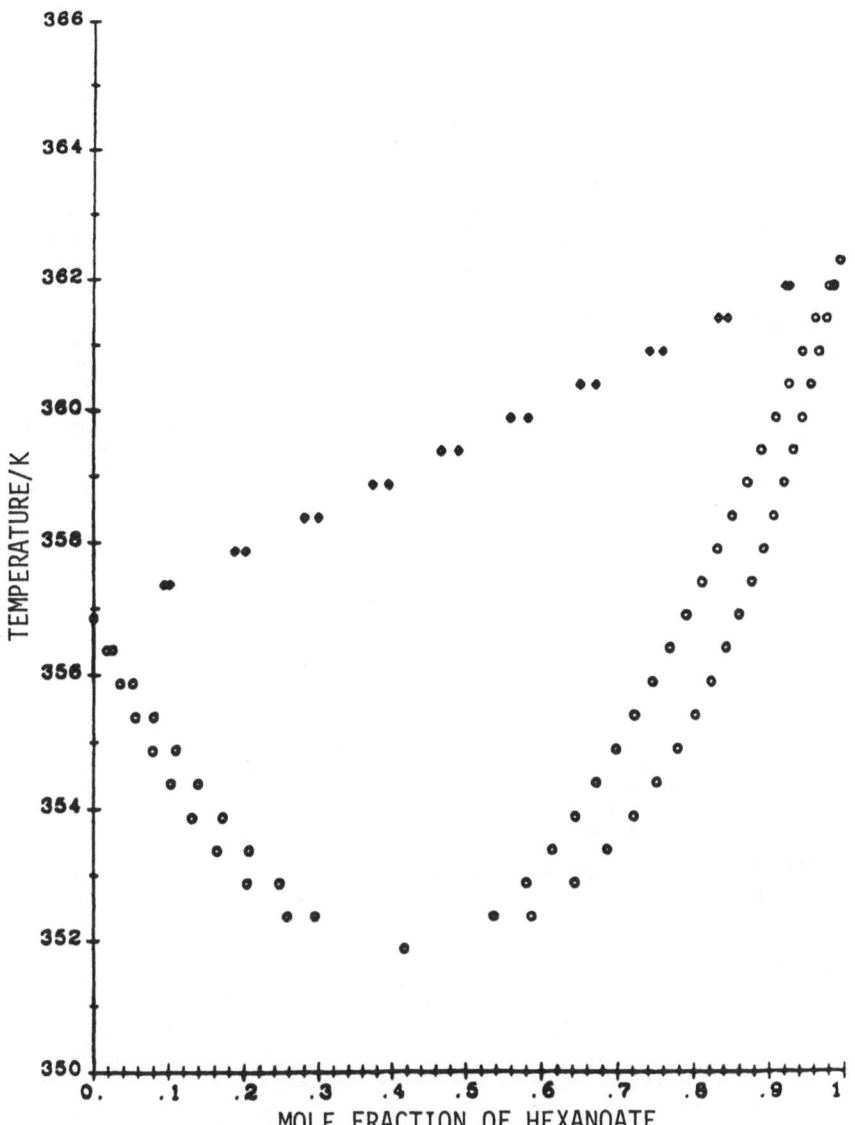

Fig. 7. Calculated isobaric phase diagram for benzene-hexa-n-
hexanoate and octanoate.
◇ Ideal Solution Theory
○ Regular Solution Theory

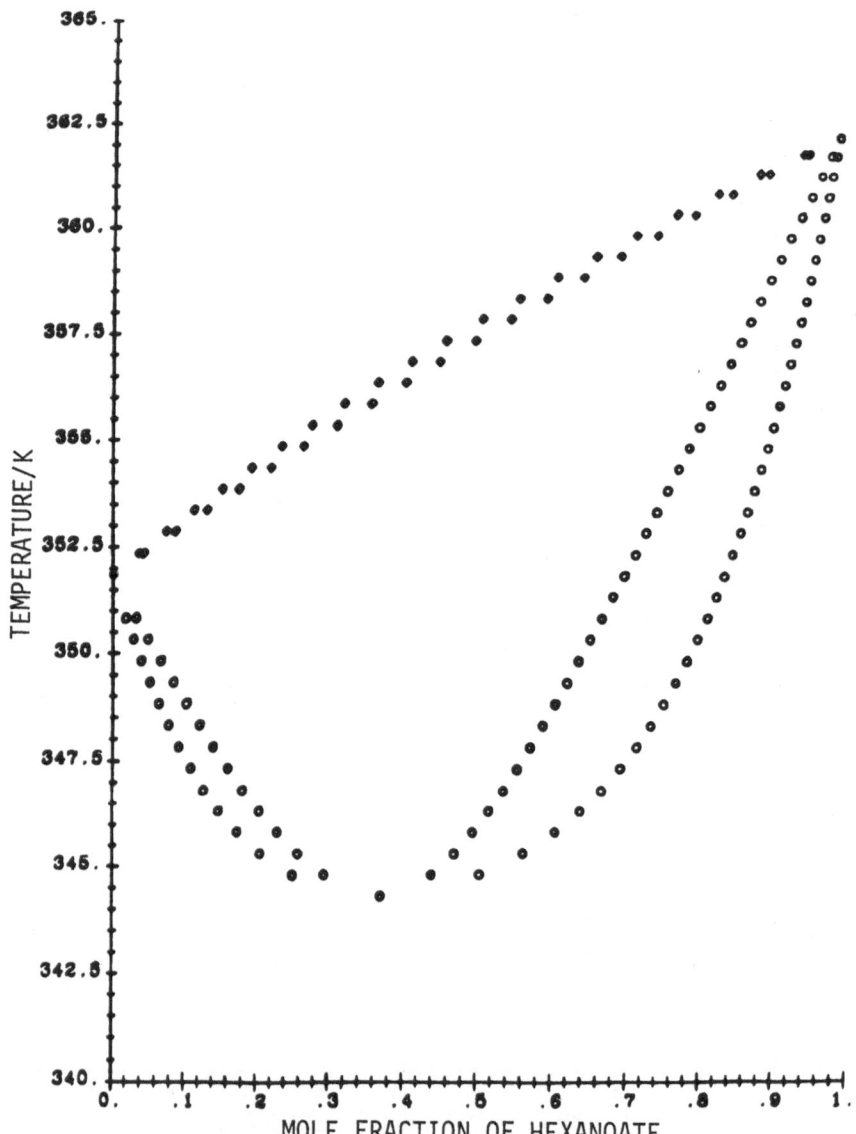

Fig. 8.   Calculated Isobaric Phase Diagram for Benzene-hexa-n-
           hexanoate and Nonanoate.
                ◇ Ideal Solution Theory
                ○ Regular Solution Theory

SUMMARY

The liquid two phase region in binary discotic liquid crystal mixtures that show minimum azeotrope-like behavior can be quantified using simple regular solution theory. Furthermore, the relative magnitudes of the regular solution parameters A_I and A_D show that ideality is partially recovered in the discotic to isotropic transition, as the orientational order of the discotic phase is destroyed. This correlation of non-ideality with orientational order, together with the linear relationship between the difference of regular solution parameters and molar volume ratios, suggests that any molecular description of non-ideality should include dependence on molar volumes.

REFERENCES

1. I. Z. Schroeder, Z. Phys. Chem. <u>11</u>, 449 (1893).

2. J. J. van Laar, Z. Phys. Chem. <u>63</u>, 216 (1908).

3. A. Reisman, <u>Phase Equilibria</u>, Academic Press Inc., New York, N.Y., 1970.

4. M. Domon and J. Billard, Pramana, Suppl. No. 1, 131 (1975).

5. G. R. Van Hecke, J. Phys. Chem. <u>83</u>, 2344 (1979).

6. G. R. Van Hecke, T. S. Cantu, M. Domon, and J. Billard, J. Phys. Chem. <u>84</u>, 263 (1980).

7. S. Chandrasekhar, B. K. Sadashiva, K. A. Suresh, N. V. Madhusudana, S. Kumar, R. Shashidhar, G. Venkatesh, J. Phys. <u>40C3</u>, 120 (1979).

8. S. Chandrasekhar, B. K. Sadashiva, K. A. Suresh, Pramana, <u>9</u>, 471 (1977).

9. J. Billard, B. K. Sadashiva, Pramana, <u>13</u>, No. 3, 309 (1979).

10. M. Sorai, H. Suga, Mol. Cryst. Liq. Cryst., <u>73</u>, 47 (1981).

11. M. Sorai, K. Tsuji, H. Suga, S. Seki, Mol. Cryst. Liq. Cryst., <u>80</u>, 33 (1980).

12. T. H. Smith, G. R. Van Hecke, Mol. Cryst. Liq. Cryst., <u>68(1-4)</u>, 23 (1981).

EXTENSION OF McMILLAN'S MODEL TO LIQUID CRYSTALS OF DISC-LIKE
MOLECULES

S. Chandrasekhar, K. L. Savithramma and
N. V. Madhusudana

Raman Research Institute
Bangalore 560 080, India

ABSTRACT

McMillan's mean field model of smectic A is extended so that
the translational order parameter is now periodic in two dimensions.
The theory is developed for a face-centered rectangular lattice
composed of liquid-like columns, taking the molecular cores to be
circular discs normal to the columnar axes but with an asymmetrical
disposition of the chains. When the axial ratio $b/a = \sqrt{3}$, the
rectangular lattice reduces to a hexagonal one and the theory
becomes essentially identical to that proposed very recently by
Feldkamp, Handschy and Clark.

When $b/a = \sqrt{3}$ or departs from it only slightly, it turns out
that the transition from the columnar to the isotropic phase may
take place either directly or *via* a nematic phase, depending on the
model potential parameters (α). Interpreting α to be a measure of
the chain lengths as in McMillan's theory, the phase diagram is in
broad agreement with the trends exhibited by the hexa-n-alkoxybenzoates of triphenylene. For higher values of b/a the theory predicts
a columnar-smectic A transition as well. The new smectic A phase

is biaxial. The possibility of such a phase occurring in real
systems is discussed briefly.

INTRODUCTION

A number of disc-shaped molecules are now known to exhibit
thermotropic mesomorphism[1]. The mesophases so far discovered fall
into two distinct categories, the columnar and the nematic. In the
columnar (D) type, the discs are stacked one on top of the other,
the different columnar stacks forming a two-dimensionally periodic
array. Several variants of this structure have been identified -
hexagonal, rectangular, tilted, etc. (Figs. la-e). The nematic
(N_D) is an orientationally ordered arrangement of the discs with no
long range translational order (Fig. lf). Transitions between
columnar and nematic phases have been observed in a few cases[2,3].
An an example we present in Table I the data for the hexa-n-alkoxybenzoates of triphenylene (HABT).

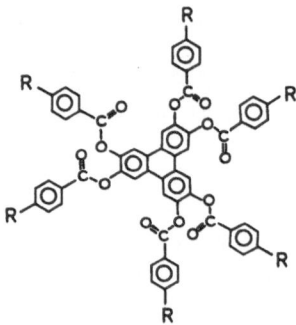

Table I: Hexa-n-alkoxybenzoates of triphenylene:
Transition temperatures in °C [2]

$R = C_n H_{2n+1} O$	K		D_t		D_r		N_D		I
n=4	.	257	-		-		.	>300	.
5	.	224	-		-		.	298	.
6	.	186	.	193	-		.	274	.
7	.	168	-		-		.	253	.
8	.	152	-		.	168	.	244	.
9	.	154	-		.	183	.	227	.
10	.	142	-		.	191	.	212	.
11	.	145	-		.	179	.	185	.
12	.	146	-		.	174	-		.

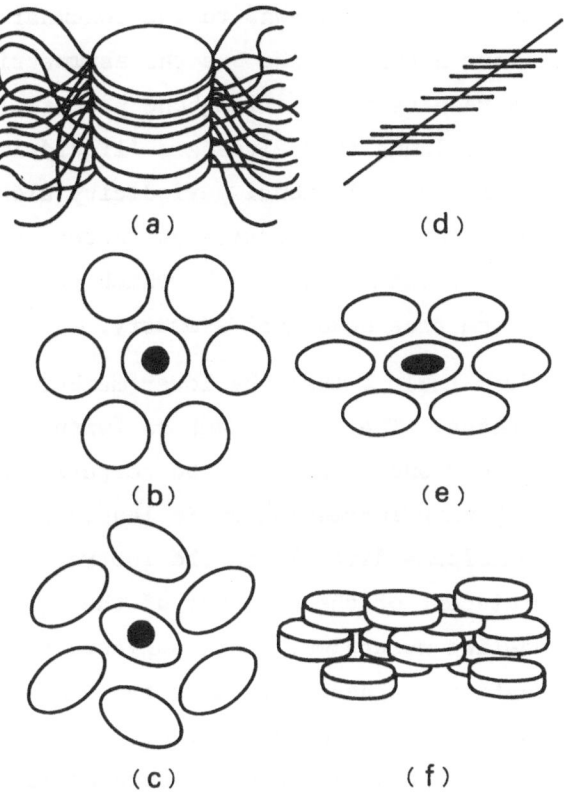

Fig. 1.   (a) Schematic representation of a 'liquid-like' stack of
           disc-shaped molecules (from Levelut [4]). (b) Hexagonal
           modification D_h of an upright columnar structure. The
           filled circle depicts the molecular core which is normal
           to the columnar axis. (c) Rectangular modification D_r of
           an upright columnar structure. The circular cores are
           normal to the columnar axes and form a face-centered
           rectangular lattice, but the asymmetric disposition of
           the chains results in a herringbone pattern of ellipses,
           whose ellipticity is highly exaggerated here for the sake
           of clarity [2,4]. (d) Tilted columnar structure D_t.
           (e) Face-centered rectangular lattice of D_t. [2] The cores
           are tilted with respect to the columnar axes [2]. (f) Nematic
           phase N_D of disc-shaped molecules [2].

Here D_t signifies tilted columns and a face-centered rectangular

(FCR) lattice (Figs. 1d and e), and D_r upright columns and a

rectangular arrangement as in Fig. 1c. Levelut [4] has shown that in

the latter case the discs are *normal* to the columnar axes: thus the
'circular' cores form a FCR lattice but the asymmetrical disposition of the chains results in a herringbone pattern (Fig. 1c). In
both D_r and D_t, the columns themselves are 'liquid-like', i.e.,
there is no long range translational periodicity along the columnar
axes. For the n=11 homolog the lattice parameters have been determined to be a = 32.6 Å and b = 51.8 Å [2], which represents only a
slight departure from true hexagonal symmetry.

It is seen from Table I that the lower members, n = 4 and 5,
show only the N_D phase. The higher members (with the exception of
n = 7) show both the D and N_D phases, the temperature range of the
N_D phase decreasing with increasing chain length till at n = 12 the
columnar phase transforms directly to the isotropic phase. Broadly,
the trend is reminiscent of the behavior of the S_A-N-I transitions
in systems of rod-like molecules. This suggests that one may be
able to give a qualitative description of the D-N_D-I transitions by
extending McMillan's mean field model of S_A [5] so that the density
wave is now periodic in two dimensions. Such an idea had in fact
been considered very briefly by Katz[6], but no calculations were presented by him. We undertook a detailed theoretical study of the
problem for the hexagonal and FCR columnar structures. When our
work was nearing completion there appeared a paper by Feldkamp et
al.[7] on essentially the same theory for the hexagonal case and we
were gratified to find that their conclusions are in exact agreement
with ours. However, as our formulation differs slightly from theirs
in some details we shall first outline the theory briefly.

THEORY

We consider a two dimensional FCR lattice composed of liquid-
like columns, the molecular cores being circular discs normal to
the columnar axes. (Therefore, the departure from hexagonal symmetry is supposed to be solely due to the asymmetric disposition of

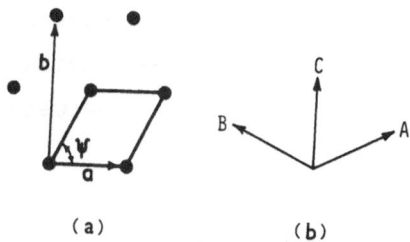

(a)   (b)

Fig. 2.   Two-dimensional face-centered rectangular lattice showing
         (a) the lattice parameters a and b and the primitive cell
         of the direct lattice, (b) the reciprocal lattice vectors
         $\vec{A}$, $\vec{B}$ and $\vec{C}$.

the chains as in the D_r structure (Fig. 1c) except that for simplicity we ignore the herringbone pattern.) The FCR lattice can be
described by a superposition of three density waves with wave
vectors (Fig. 2).

$$\vec{A} = 2\pi\left(\frac{\vec{i}}{a} + \frac{\vec{i}}{b}\right) \quad ;$$

$$\vec{B} = 2\pi\left(-\frac{\vec{i}}{a} + \frac{\vec{i}}{b}\right) \quad ; \quad \text{and}$$

$$\vec{C} = \frac{4\pi\vec{i}}{b} = (\vec{A} + \vec{B})$$

where $\vec{i}$ and $\vec{j}$ are unit vectors along the x and y axes respectively.

     Since the orientational ordering of the molecules is obviously
necessary for the existence of the columnar structure, we need to
define order parameters which couple the two types of order. In
the smectic A phase of rod-like molecules, the orientational order
is normally defined with respect to the director, which also
represents the direction of the wave vector of the density wave.
On the other hand, in the present case the director is parallel to
the columnar axis (z-axis say) whereas the wave vectors $\vec{A}$, $\vec{B}$, $\vec{C}$ are
in the orthogonal xy plane. Therefore, we couple each density wave
to the appropriate component of the orientational order parameter

along $\vec{A}$, $\vec{B}$ or $\vec{C}$, rather than to the component along z. Retaining only the leading terms in the Fourier expansion of the density waves, the single particle potential in the mean field approximation is assumed to be of the form

$$V_1(\vec{r},\theta,\phi) = -V_0[-\eta_1\, P_2(\sin\,\theta\,\cos\,\phi)\, -\eta_2\, P_2(\sin\,\theta\,\sin\,\phi)$$

$$-2\alpha_1\sigma_1\, P_2(\sin\,\theta\,\cos\,\phi)\,\cos\,\vec{C}\,\cdot\,\vec{r}$$

$$-\alpha_2\sigma_2\{P_2(\sin\,\theta\,\cos(\psi+\phi))\,\cos\,\vec{A}\cdot\vec{r}$$

$$+P_2(\sin\,\theta\,\cos(\psi-\phi))\cos\,\vec{B}\cdot\vec{r}\}]$$

where V_0 determines T_{NI}, α_2 is an interaction strength related to density waves along $\vec{A}$ and $\vec{B}$ and α_1 that along $\vec{C}$, $\psi = \tan^{-1}(b/a)$, θ and ϕ are the polar angles, $\vec{r}$ is the position vector, and P_2 is the Legendre polynomial of order two. This form of the potential ensures that the energy of the molecule is minimum when the disc is centered in the column with its plane normal to the z axis.

Since the molecular core is assumed to be circularly symmetric, we get

$$\alpha_1 = 2\exp\{-(2\pi r_0/b)^2\}$$

$$\alpha_2 = 2\exp[-\{\pi r_0(a^2+b^2)^{1/2}/ab\}^2]$$

where r_0 is the range of the intermolecular attractive potential.

In general there are four order parameters, η_2 and η_1 the orientational order parameters measured along the x and y axes, and σ_2 and σ_1 the order parameters coupling the orientational and translational orders along the $\vec{A}$ (or $\vec{B}$) and $\vec{C}$ directions respectively. We have chosen to define the order parameters in such a way that they vary from 0 in the disordered system to +1 in the perfectly ordered system. We then have

$$\eta_1 = <-2P_2(\sin\theta\cos\phi)>$$

$$\eta_2 = <-2P_2(\sin\theta\cos\phi)>$$

$$\sigma_1 = <-2P_2(\sin\theta\cos\phi)\cos\vec{C}\cdot\vec{r}> \qquad (1)$$

$$\sigma_2 = <-\{P_2(\sin\theta\cos(\psi+\phi))\cos\vec{A}\cdot\vec{r}$$
$$+ P_2(\sin\theta\cos(\psi-\phi))\cos\vec{B}\cdot\vec{r}\}>,$$

where the angular brackets represent a statistical average, and the normalised single partition distribution function

$$f_1(\vec{r},\theta,\phi) = \frac{\exp\{-V_1(\vec{r},\theta,\phi)/kT\}}{\int d\vec{r}\int_0^1 d(\cos\theta)\int_0^{2\pi} d\phi \exp\{-V_1(\vec{r},\theta,\phi)/kT\}}$$

where $\int d\vec{r}$ is over the primitive cell.

The molar internal energy of the oriented system can now be written as

$$\frac{\Delta U}{NkT} = -\frac{V_0}{2kT}\{\frac{1}{2}\eta_1^2 + \frac{1}{2}\eta_2^2 + \alpha_1\sigma_1^2 + 2\alpha_2\sigma_2^2\}$$

where N is the Avogadro number, the entropy as

$$\frac{\Delta S}{Nk} = -\int_0^1\int_0^{2\pi}\int f(\vec{r},\theta,\phi)\ln f(\vec{r},\theta,\phi)d(\cos\theta)d\phi\, d\vec{r}$$

$$= -\frac{V_0}{kT}(\frac{\eta_1^2}{2} + \frac{\eta_2^2}{2} + \alpha_1\sigma_1^2 + 2\alpha_2\sigma_2^2)$$

$$+ \ln\frac{1}{\pi ab}\int_0^1\int_0^{2\pi}\int\exp\{-V_1(\vec{r},\theta,\phi)/kT\}d(\cos\theta)d\phi\, d\vec{r}$$

and hence the molar free energy as

$$\frac{\Delta F}{NkT} = \frac{\Delta U - T\Delta S}{NkT} = \frac{V_0}{2kT}(\frac{\eta_1^2}{2} + \frac{\eta_2^2}{2} + \alpha_1\sigma_1^2 + 2\alpha_2\sigma_2^2)$$

$$-\ln\frac{1}{\pi ab}\int_0^1\int_0^{2\pi}\int\exp\{-V_1(\vec{r},\theta,\phi)/kT\}d(\cos\theta)d\phi\, d\vec{r}$$

There are four possible solutions to the equations:

1) $\eta_1 \neq \eta_2 \neq 0$, $\sigma_1 \neq \sigma_2 \neq 0$ (biaxial rectangular
                                                           columnar phase)

2) $\eta_1 \neq \eta_2 \neq 0$, $\sigma_1 \neq 0$, $\sigma_2 = 0$ (biaxial smectic phase)

3) $\eta_1 = \eta_2 \neq 0$, $\sigma_1 = \sigma_2 = 0$ (uniaxial nematic phase)

4) $\eta_1 = \eta_2 = \sigma_1 = \sigma_2 = 0$ (isotropic phase)

The free energy corresponding to the different solutions can be evaluated to determine the phase diagram as a function of the α coefficients for a given value of the axial ratio b/a.

When $b/a = \sqrt{3}$, $\psi = 60°$, and we have a hexagonal lattice. It then follows that $\alpha_1 = \alpha_2$ and the solutions take the simpler form

1) $\eta_1 = \eta_2 \neq 0$, $\sigma_1 = \sigma_2 \neq 0$ uniaxial hexagonal
                                                          columnar phase

2) $\eta_1 = \eta_2 \neq 0$, $\sigma_1 = \sigma_2 = 0$ uniaxial nematic phase

3) $\eta_1 = \eta_2 = \sigma_1 = \sigma_2 = 0$ isotropic phase.

CALCULATIONS AND RESULTS

The consistency conditions (1) were solved by using PDP-11 and DEC-10 computers. The numerical integrations were performed in double precision by using the Gaussian quadrature technique with 12 (or sometimes 16) quadrature points.

The phase diagram for the hexagonal case is shown in Fig. 3. This is exactly identical to the one presented by Feldkamp et al.[7]. As pointed out by these authors, the hexagonal-nematic transition is always first order, unlike the S_A-N transition which is predicted to become second order for $\alpha < 0.70$.

For the biaxial rectangular case, the nature of the phase diagram depends on the value of b/a. In the present paper, we confine our attention to $b/a > \sqrt{3}$. If b/a is only slightly greater than $\sqrt{3}$, the phase diagram is rather similar to that for the hexagonal lat-

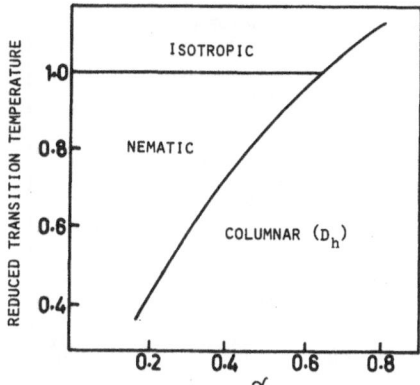

Fig. 3.   Theoretical diagram showing the hexagonal, nematic and
           isotropic phase boundaries. All the transitions are of
           first order. This diagram is identical to that presented
           by Feldkamp et al.[7]

tice (Fig. 3). It is seen that the temperature range of the nematic
phase decreases with increasing α, and for higher values of α ($\alpha >$
0.64 for the hexagonal case) the columnar phase transforms directly
to the isotropic phase. If, as in McMillan's theory, α is interpreted to be a measure of the chain length these results are in
qualitative accord with the observed trends for HABT (Table I).

     As the asymmetry of lattice is increased we get solutions cor-
responding to a smectic phase. Fig. 4 presents the phase diagram
for b/a = 1.95 (ψ = 62°51'). In this case, for values of $\alpha_1 < 0.52$,
the rectangular columnar phase transforms to a smectic A phase (with
the layer normal along $\vec{C}$) which in turn undergoes a second order
transition to the nematic phase at a higher temperature. Both the
columnar and smectic phases are weakly biaxial. For example, for
α_1 = 0.4, η_1 = 0.85 and η_2 = 0.84 in the columnar phase while η_1 =
0.812 and η_2 = 0.806 in the smectic phase at the columnar-smectic
transition point. The columnar phase goes over to the nematic
phase for 0.52 $<\alpha_1<$0.70, and to the isotropic phase for $\alpha_1 > 0.70$.

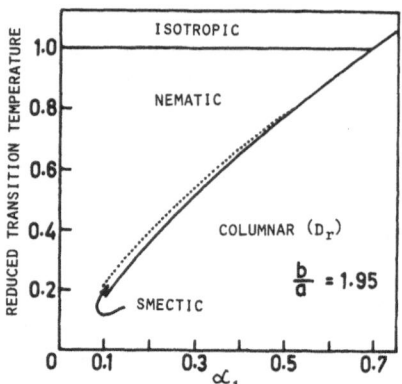

Fig. 4.   Theoretical diagram showing the rectangular, smectic,
           nematic and isotropic phase boundaries. The calculations
           are for an axial ratio b/a = 1.95 or ψ = 62°51'. Full
           lines represent first order transitions and the broken
           line a second order transition.

        Rectangular phases have been observed with disc-like molecules
having essentially circular cores, but in all these cases b/a does
not depart significantly from $\sqrt{3}$. According to our theory the
smectic A phase can intervene between the rectangular and nematic
phases only if the lattice is sufficiently anisotropic and moreover
the α-values (and hence the chain lengths) have to be small. It is
doubtful if an asymmetric distribution of relatively short chains
can give rise to the necessary values of b/a. On the other hand,
rectangular lattices of high anisotropy may be expected if the
molecular core itself is elliptical as in the case of rufigallol[8]
(or in effect elliptical as in the tilted columnar structure which
of course would give rise to a smectic C phase). The extension of
the model to allow for the ellipticity of the cores and also to
explicitly take into account the herringbone arrangement will
result in a very large number of order parameters (as well as free
parameters) and the calculations will no doubt be extremely tedious.
Nevertheless, the simple model discussed above serves to illustrate
two points, namely that (a) the origin of the two-dimensional trans-

lational order in the columnar phase is similar to that of one-dimensional order in smectic A in so far as the large mutual attraction between the aromatic cores and the role of the end chains are concerned, and (b) the possibility exists of observing smectic phases in certain discotic systems.

REFERENCES

1. For an up-to-date review see, S. Chandrasekhar, Advances in Liquid Crystals, Vol. 5 (in press).

2. N. H. Tinh, H. Gasparoux and C. Destrade, Mol. Cryst. Liq. Cryst. 68, 101 (1981).
   C. Destrade, N. H. Tinh, H. Gasparoux, J. Malthete and A. M. Levelut, Mol. Cryst. Liq. Cryst. 71, 111 (1981).

3. C. Destrade, H. Gasparoux, A. Babeau, N. H. Tinh and J. Malthete, Mol. Cryst. Liq. Cryst. 67, 37 (1981).

4. A. M. Levelut, Proc. Int. Liquid Crystals Conference, Bangalore, December 1979, Ed. S. Chandrasekhar, Heyden, London (1980) p. 21.

5. W. L. McMillan, Phys. Rev. A4, 1238 (1971).

6. E. I. Katz, Sov. Phys. JETP, 48, 916 (1978).

7. G. E. Feldkamp, M. A. Handschy and N. A. Clark, Phys. Lett. 85A, 359 (1981).

8. A. Queguiner, A. Zann, J. C. Dubois and J. Billard, Proc. Int. Liquid Crystals Conf., Bangalore, December 1979, Ed. S. Chandrasekhar, Heyden, London (1980) p. 35.

THERMOTROPIC LIQUID-CRYSTALLINE POLYMERS WITH MESOGENIC GROUPS AND FLEXIBLE SPACERS IN THE MAIN CHAIN

A. Blumstein, J. Asrar and R. B. Blumstein

Polymer Science Program
Department of Chemistry
University of Lowell
Lowell, MA 01854

INTRODUCTION

Polymers displaying liquid crystalline behavior have been known since 1933 (1,2) and the theoretical foundation for solutions of long rods was established by Onsager (3) and Flory (4) over 25 years ago. Thermotropic mesomorphism in polymers, however, was largely ignored until the recent development of ultra-high strength, ultra-high modulus fibers spun from nematic solutions of rigid aromatic polyamides. It was then soon realized that synthetic polymers have a strong tendency to promote the formation of liquid-crystalline phases in solution or in melt, provided that mesogenic moieties be incorporated either into side groups or into the backbone itself (5). The field of liquid-crystalline polymers, originally confined mainly to helical rodlike macromolecules, has recently expanded into the realm of semi-flexible and flexible synthetic macromolecules and is now at the forefront of industrial and academic research.

Rigid macromolecules, such as poly(p-benzamide) or poly(p-phenyleneterephthalamide), decompose before melting and are soluble only in solvents such as concentrated sulfuric acid. Considerable syn-

thetic effort has recently been devoted to the development of more
easily processed and more soluble thermotropic structures which
would nevertheless retain the remarkable mechanical properties of
the rigid rod-like polymers.

Thermotropic polymers can be prepared by depressing the melting
points of rigid backbone polymers through the following approaches,
often used in combination: random copolymerization, introduction of
kinks or bends into the chain, asymmetric substitution, inclusion
of randomly spaced flexible aliphatic chain fragments (flexible
spacers). A review of such structures can be found in reference
(6). Typically, one deals with random aromatic copolyesters, such
as the poly(p-oxybenzoate-co-ethyleneterephtalates) described by
Jackson and Kuhfuss (7). An added element of structural complexity
can occur as a result of reorganization of the initial structure
into multi-block arrangements (8).

In this review we shall deal exclusively with polymers in
which rigid mesogenic elements _regularly_ alternate with flexible
spacer groups. In addition to homopolymers, random copolymers can
be prepared in this fashion, for example by condensing a mesogenic
biphenol with a combination of two different aliphatic diacids or a
mixture of mesogenic biphenols with an aliphatic diacid. Examples
of such structures are provided in the Tables[*]. Other types of
mesomorphic polymers are reviewed in references (5,9,10,11).

Since the pioneering work of Roviello and Sirigu (12) performed on polymeric Schiff base derivatives of p,p'-dihydroxy-2,2'-
dimethylbenzalazine [1,14] all three basic types of mesophases have
been discovered in rapid succession: nematic (12-17), smectic
(12,13,18,19,27), and cholesteric (20,21,47).

The synthesis of polymeric liquid crystals with moderate or
depressed mesophase transition temperatures and increased solubility

---

*Compounds are referenced to the tables by means of brackets

provides the opportunity of comparison with low molecular weight liquid crystals. Such comparison allows one also to test recent theoretical developments and to promote new applications. Flexible backbone polymers have a unique application potential. Their position is central inasmuch as they are suitable simultaneously for the development of fibers, coatings and films.

As synthesis of new systems progresses in various laboratories structure-property relationships are established, so that the influence of structure of mesogen and spacer on mesophase formation is reasonably well documented. The first part of this review will be concerned with a discussion of such structure-property relations.

A detailed knowledge of the long range order, conformation, and chain dynamics in both the solid and mesomorphic state of flexible backbone polymers, is still missing. Rheological investigations and studies of thermotropic systems are just beginning (75). Synthesis of model compounds, influence of molecular weight and molecular weight distribution on properties need to be addressed. Such data as are available to date will be presented in the second part of this review.

DISCUSSION

Role of the mesogen

The role of the mesogen in the formation of thermotropic mesophases is critical. The mesogen influences the transition temperatures, the nature and stability of the mesophase and the solubility characteristics of the polymers. Because the mesogenic moieties are connected to each other, the rules which govern the structure-property relations in free, low molecular weight mesogens (22), though extremely useful, may not be indiscriminately applied in the case of polymers. In the case of backbone polymers the linear nature of the macromolecule tends to promote one dimensional order (23),

Table I. LIQUID CRYSTALLINE POLYESTERS FROM MESOGENIC DIOLS AND FLEXIBLE CHAIN DIACIDS

No.	Diols	$HOO-(CH_2)_n-COOH$	Transition Temperatures	References
1a	HO—⬡—⬡—OH	5	K245N255I	76
b	"	6	K245N325I	76
c	"	7	K231N246I	78
d	"	8	K210N275I	76
e	"	8	K245S295I	13,14
f	"	8	K208L.C.261I	78
g	"	10	K205L.C.245I	78
h	"	12	K205L.C.245I	78
i	"	12	K195N255I	76
2a	$HO-(CH_2)_2-O-$⬡$-O(-CH_2-)_2-OH$	2	K191N*196I	57
b	"	8	K155N*162I	57
3	HO—⬡—CH=CH—OH	8	K203S	13,14
4	$HO-(CH_2)_2-O-$⬡$-CH=CH-$⬡$-O-(CH_2)_2-OH$	8	K114L.C.129I	57
5a	HO—⬡—C(CH₃)₂—⬡—OH	4	K176N220N+I>300I	47
b	"	MAA*	K110Ch166Ch+I240I	47
c	"	6	L.C.289I	79

K-crystal , S-smectic, N-nematic, N*-potentially nematic, L.C. - Liquid Crystal (not identified),

Ch - Cholesteric, I - isotropic.

No.	Diols	HOOC-(CH$_2$)$_n$-COOH	Transition Temperatures	References
d	HO-⬡-C(CH$_3$)=CH-⬡-OH	8	L.C.241I	79
e	"	10	K186N218I	50
f	"	12	I177L.C.	80
6a	HO-⬡-COO-⬡-OH	3	K150-17GN320-350I	37
b	"	4	K140-160N320-345I	37
c	"	5	K130-150N290-345I	37,76
d	"	6	K150-170N300-360I	37,76
e	"	7	K130-150N280-345I	37
f	"	8	K140-170N300-340I	37,76
g	"	9	K130-160N265-320I	37
h	"	10	K145-170N255-300I	37
i	"	11	K150-170N245-300I	37
j	"	12	K150-180N250-300I	37,76
k	"	14	K120-140N230-250I	37
l	"	20	K120-140N220-250I	37
7	HO-⬡-CH=N-⬡-OH	8	K152S220I	13,14
8a	HO-⬡-N=N-⬡-OH	5	K240N262I	45
b	"	6	K245N300I	45
c	"	7	K206N255I	45
d	"	7	K211N250I	46

(continued)

Table I. LIQUID CRYSTALLINE POLYESTERS FROM MESOGENIC DIOLS AND FLEXIBLE CHAIN DIACIDS (Continued)

No.	Diols	$HOOC-(CH_2)_n-COOH$	Transition Temperatures	References
e	HO-⬡-N=N-⬡-OH	8	K215N278I	45
f	"	9	K195N235I	45
g	"	10	K222N241I	45
9a	HO-⬡(CH₃)-N=N-⬡-OH	5	K160N209I	46
b	"	6	K175N249I	46
c	"	7	K146N191I	46
d	"	8	K120N205I	46
e	"	10	K158N193I	46
10a	HO-⬡-N=N(O)-⬡-OH	5	K235N260I	45
b	"	6	K200N270I	45
c	"	7	K190N255I	45
d	"	7	K222N271I	46
e	"	8	K190N267I	45
f	"	9	K190N219I	45
g	"	10	K222N242I	45
h	"	10	K216N265I	20,29
i	"	MAA*	K221Ch294.5I	29
11a	HO-CH₂-CH₂O-⬡-N=N-⬡-OCH₂-CH₂-OH	6	K185L.C.191I	29
b	"	7	K140I	46

No.	Diols	HOOC-(CH$_2$)$_n$-COOH	Transition Temperatures	References
12a	(structure)	7	K135N151I	46
b	"	10	K142N165I	29
13a	(structure)	2	K243.8I	36
b	"	3	K185.8I(167.8N)134.8K	36
c	"	4	K198.2N255I	36
d	"	5	K135.8N173.8I	36
e	"	6	K160.3N230.3I	36
f	"	7	K98.3N164.6I	36
g	"	8	K136.8N188.3I	36
h	"	9	K121N144I	36
i	"	10	K118.2N163.5I	28,29,36
j	"	11	K100.8N132.8I	36
k	"	12	K110.8N143.8I	36
l	"	14	K121N134I	29
m	"	MAA*	K151N210I	28,29
4a	(structure)	6	K238N295I	12
b	"	8	K203N256I	12
c	"	10	K203N241I	12

(continued)

Table I. LIQUID CRYSTALLINE POLYESTERS FROM MESOGENIC DIOLS AND FLEXIBLE CHAIN DIACIDS (Continued)

No.	Diols	$HOOC-(CH_2)_n-COOH$ $Cl-C(=O)-O-(CH_2)_n-O-C(=O)-Cl$	Transition Temperatures	References
15	$HO-\bigcirc-\underset{CH_3}{\overset{CH_3}{C=N-N=C}}-\bigcirc-OH$			
a	"	8	K208N257I	49
b	"	10	K180N235I	49
c	"	12	K175N213I	49
d	"	14	K167N194I	49

whereas in systems with mesogenic moieties in the side-groups two-
dimensional order is strongly promoted (24,26). This does not
preclude formation of nematic "side-group" (See for example references 24,25) or smectic "backbone" polymers (14,19,27, 51). Smectic
backbone polymers have been prepared using strongly smectogenic
mesogens such as the biphenyl (14,27,51,57) or triphenyl (51)
moieties.

The transition temperature increases very rapidly as a function of the axial ratio in low molecular weight liquid crystals.
This also seems to be the case in high molecular weight flexible
thermotropic polyesters in which the rigid core was formed form a
sequence of 2 to 4 p-oxyphenylcarboxy units (76).

In low molecular weight nematics structure broadening by substitution in the aromatic ring drastically reduces the stability of
the nematic phase (22). This does not seem to be the case to the
same extent in backbone polymers; indeed, structure broadening is
widely used in the preparation of thermotropic nematic polyesters
with rigid and semi-rigid backbones (6). In the case of flexible
backbone polymers, condensation of the acid chloride of dodecane-
dioic acid with 4,4$^{\underline{\,}}$hydroxyazoxybenzene, (polymer DDA-8),[1,10h]
and 4,4$^{\underline{\,}}$hydroxy-2,2$^{\underline{\,}}$methylazoxybenzene, (polymer DDA-9),[I,13ℓ] ,
respectively, illustrates the effects of structure broadening
(20,28,29). Both polymers have a nematic range of some 50°C, but
the melting point of DDA-9 is some 100°C lower than that of DDA-8.
In contrast, introduction of methyl groups in the 2 and 2' positions
of 4,4$^{\underline{\,}}$methoxyazoxybenzene (PAA) completely destroys the nematic
phase (30).

In low molecular weight compounds strongly dipolar substituents
in the aromatic ring tend to enhance the stability of the smectic
mesophase (31). In contrast data obtained on laterally substituted
polyesters prepared from 1,10-bis(p-carboxyphenoxy) decane and hydroquinones laterally substituted with chlorine (32) seem to point

towards a nematogenic behavior of these polymers [III,8,9,10].
Lateral substitution using phenyl or fused phenyl rings also lowers
the transition temperature and gives a narrower mesomorphic interval [III,11,12,13] (83).

The value of the axial ratio x of most mesogens listed in
Table I is ∿3-4, clearly below the critical value x_c=6.4 (no solvent present) predicted by theories describing the macromolecules
as a series of independent rigid rod-like elements connected by
highly flexible spacers (33,34). Orientation dependent interactions
(82) lower the axial ratio required for a stable nematic phase, but
presently available theories do not yet quantitatively account for
the behavior of such systems. In a series of strictly alternating
rigid-flexible copolyamides (35) lyotropic liquid crystallinity was
found to exist in solutions of 25 to 40% of polymer for axial ratios
of the rigid element between 3 and 3.5. For values of x between 4
and 4.5 the anisotropic phase appeared at a concentration range of
15-25%.

Influence of the spacer

It is clear from what precedes that connection of mesogenic
groups through the spacer imparts to the thermotropic flexible
backbone polymers some unique features of behavior. The following
examples will illustrate this point.

In terms of a model consisting of incompatible flexible aliphatic chains and independent rigid anisotropic mesogenic moieties,
one can anticipate in the light of theories developed for lyotropic
solutions of rods a rather rapid collapse of liquid crystallinity
with increasing length of spacer. The experimental data, however,
show that, far from callapsing, the mesomorphic interval often
widens at first with increasing number n of methylene groups and
then narrows slowly. The maximum stability of mesomorphic interval
occurs for 6<n<10. For example in the case of poly(4,4^{1}oxy-2,2^1
methylazoxybenzenealkanediyls) [I,13] the nematic interval ΔT_N

reaches 70°C for n=6, then slowly narrows to $\sim13^{\circ}$C for n=14 (36).
For the series of polyesters derived from 4,4$^{\perp}$oxyphenylbenzoate and
alkanedioic acids [I,6] the nematic interval of some 180–150°C narrows extremely slowly and is reported to be 100°C for n=20 (37).
If one keeps in mind that n=14 and n=20 correspond to a weight fraction of mesogen w$\sim$0.4 and w$\sim$0.3, respectively, for an axial ratio
of the mesogen of 2.5<x<3.5, it becomes obvious that a model of the
nematic mesophase based on a dispersion of short anisotropic, hard
cores surrounded by flexible coiled polymethylene sequences cannot
be sustained for these systems. These examples also show that in
contrast with low molecular weight mesogens, an increase in the
length of the flexible sequence of methylene units does not promote
the formation of smectic mesophases. In the case of derivatives
containing the 4,4$^{\perp}$oxycarboxydiphenyl moiety described in (37)
[I,6], no smectic phase was detected and all polymers displayed a
wide range of nematic mesomorphism for values of n between 3 and
20. In the realm of low molecular weights, analogues such as for
example 1,4-bis(4n-alkanoyloxybenzoyloxy)benzenes display a smectic
C phase for n$\geqslant$7 (38). In 4,4$^{\perp}$alkyl and 4,4$^{\perp}$alkoxyazoxybenzenes a
smectic mesophase appears for n$\geqslant$7 (44) while in the related polymer
[I,10h] only a nematic phase is reported (20). Similarly in the
substituted poly(4,4$^{\perp}$oxy-2,2$^{\perp}$dimethylazoxybenzenealkanediyls) [I,13]
only nematic mesophases were reported when n was varied from 2 to
14 (36). Similar conclusions can also be reached from the results
obtained with derivatives of 4,4$^{\perp}$dihydroxy-α,ω-diphenoxyalkanes (6)
[III,1] as well as for derivatives of 4,4$^{\perp}$dicarboxy-α,ω- diphenoxy-
alkanes [III,2-4] (39). Nevertheless, in some polyesters obtained
from 4,4$^{\perp}$dicarboxy-α,ω-diphenoxyalkanes and p-hydroquinone [III,6]
a change from a nematic to a smectic phase was reported for n$\geqslant$6
(19), in conflict with data on similar polymers (39) in which no
smectic phase was reported for n=10.

    In homologous series of low molecular weight mesogens the well
known even-odd alternation of T_{IN} and ΔS_{IN} as well as the increase

of ΔS_{IN} with increasing terminal chain length have been interpreted in terms of ordered chain ends (40,41). Chains linked to a meso-genic unit are constrained to conform to an orientation imposed by the mean field potential of the mesophase. Deuteron magnetic resonance spectra of nematics with short perdeuterated alkyl chains indicate that the order parameter along the alkyl chain decreases from the linkage point to the mesogen reaching rapidly a "free chain" value (42,43). This is reflected in the rapid damping of the odd-even effect which is practically erased in 4,4$^\perp$alkoxyazoxy-benzenes for n$\geqslant$6 (44).

In the case of polymers an odd-even alternation of K$\longrightarrow$LC and L.C.$\longrightarrow$I transition temperatures was reported (19,27,32,36,37,39, 45,46,78). The good reproducibility of ΔH_{NI} and T_{NI} values is a common feature of thermal analysis performed on flexible backbone polymers (36,39), contrasting with less certain values for ΔH_{KN} and T_{KN}. The N$\longrightarrow$I transition temperatures are generally higher for even values of n, though Jin et al (6) have reported higher tempera-tures for odd values of n. In some instances the odd-even effect in polymers is remarkable by its persistence and intensity. For the high molecular weight polyester series based on substituted azoxybenzene (36) there was practically no damping of odd-even oscillations of T_{NI} and ΔS_{NI} for 3$\leqslant$n$\leqslant$14 [I,13]. The increment per methylene group was $\Delta(\Delta H_{NI})$=0.46KJ and $\Delta(\Delta S_{NI})$=1.52$\dfrac{J}{deg}$. A persistent alternation of T_{NI} was found for polymers containing the 4,4$^\perp$oxyphenylbenzoate mesogen (39) but no regular alternation of ΔH_{NI} or ΔS_{NI} was reported. In some instances a regular increase in ΔH_{NI} with n was reported (32).

In many cases the results of thermal analysis, although point-ing towards an alternation, are not as clear. One has to bear in mind that most of data were collected on unfractionated polymers of rather low molecular weight containing sizable amounts of oligomers, as well as on copolymers with irregular sequence distributions.

Polymers with high isotropisation temperature could also contain low molecular weight degradation products. Since presence of oligomers appears to widen the stability of a biphasic region (nematic and isotropic) which, though clearly visible in the polarizing micro- scope, is not detected by DSC (30,47) more data obtained for well fractionated, low melting systems are needed to confirm the remark- able persistence of the odd-even effect in polymers at the N$\longrightarrow$I transition. Values of ΔH_{NI} between 5-15$\frac{KJ}{m.r.u.}$, substantially higher than the values of ΔH_{NI} for low molecular analogues have been reported for a variety of polymers (6,32,36,39). For those few poly- mers for which the degree of crystallinity was independently mea- sured $\Delta S_{NI}/\Delta S_{KI}$ was found to be 0.1-0.3 (32,36). This is over one order of magnitude above the $\Delta S_{NI}/\Delta S_{KI}$ for low molecular weight nematogens (48). This high isotropisation entropy and its persis- tence over a wide range of values of n points toward an ordering of the flexible spacer and suggests a rather extended conformation with high population of trans conformers in the nematic state (28,29,36, 39). In contrast data reported by Roviello and Sirigu for poly(3,3$^{\perp}$ oxy-2,2$^{\perp}$dimethylbenzalazinealkanediyls) [I,15] indicate an approxi- mately constant value of ΔS_{NI} and ΔH_{NI} with n (49,50). This result is at variance with most data obtained for homologous series of nematogens.

From the limited amount of data available it appears that the geometry of the spacer influences the mesomorphic behavior of ther- motropic flexible polyesters. Thus, for example, comparing adipic acid and methyl adipic acid as spacers for the (4,4$^{\perp}$oxy-2,2$^{\perp}$methyl- azoxybenzene) mesogenic moiety [I,13c,m] (28,36) one observes a strong decrease of the transition temperatures. Similar trends have been observed for the series of stilbenes [1,5ab] (78) and terphenyls and [II,3] (51). In particular the replacement of -CH$_2$- by a -C(R)$_2$- segment in the spacer substantially reduces the clearing point and the effect is much more pronounced as the size of the substituent increases. In the series of terphenyls for example, the replacement of 2,2'methylpropyl sequence by a 2,2'ethylpropyl sequence produces a drop in the isotropisation temperature from 348°C to 187°C. The substitution in the spacer of the hydrogen by a methyl group broadening the molecule and depressing the transition temperature can be more harmful to a nematic than to a smectic phase. Thus, terphenyl derivatives, with a sequence $-CH_2-C(Me)_2-CH_2-$ of flexible bonds show transitions K $\longrightarrow$ S $\longrightarrow$ N $\longrightarrow$ (decomposition); substitution with ethyl groups eliminates the nematic phase altogether while maintaining the smectic mesophase (51). This is not astonishing in the light of the behavior of low molecular weight L.C. systems in which moderate branching of the terminal alkyl or alkoxy groups depresses the nematic transition more than the smectic and may even stabilize the smectic phases through lateral contacts between side groups (52). The replacement of a hydrocarbon sequence by an ethyleneoxide sequence of an equivalent length produces no substantial differences in the mesophase in the case of terphenyl derivatives (51). The oxygen in the (CH_2-CH_2-O) sequence is in this case stereochemically equivalent to a methylene unit. This appears not to be the case for the $-CH_2-$ group adjacent to a carbonyl in the spacer. According to Roviello and Sirigu (49) the replacement of $(-O-\overset{O}{\overset{\|}{C}}-CH_2-)$ by a $(-O-\overset{O}{\overset{\|}{C}}-O-)$ linkage leads to lower melting points and wider L.C. stability ranges [I,15]. However, it is reported that replacement of alkanediols by oligoethyleneglycols in the polyurethane derived from 3,3'Me,4,4'biphenyldiyldiisocyanate [II,6] results in the collapse of the mesophase (53). In other cases of replacement reported by the same authors (45,54) the situation is not so clear. The replacement of $-CH_2-$ groups in the spacer by dimethylsiloxane $-Si(Me)_2O-$ units [III,14] decreases considerably phase and glass transition temperatures of the mesomorphic polymers (55,56).

It was found that the position of the $-\overset{\|}{\underset{O}{C}}-O-$ group in the spacer is of importance to the stability and nature of the mesophase. In

the case of 2,2' substituted azoxybenzene derivatives the most
stable nematic mesophase is formed when the carbonyl is in the β
position relative to the aromatic ring (29). Inversion to the α
position or removal to the ε position produce a drastic narrowing
and collapse of the mesophase [I,10h] and [II,5g]. Similar effects
were observed for azobenzene derivatives [I,8,II,4] (45,53) and biphenyl derivatives [II,1g;I,1b] (27,76). These results are explained
by geometric considerations: when the carbonyls are in the β position
to the aromatic ring, two consecutive mesogenic units can be roughly
collinear without significant distortion of the spacer from an extended conformation (29). If this explanation is correct it indi-
rectly supports the view of the extended nature of the flexible
spacer because the deviations from collinearity become apparent only
for extended conformations.

The copolymerization of various spacers (or mesogens) leads to
an additional lowering of the transition temperatures (See for ex.
20,28,29,39) and generally leads to nematic (or cholesteric) systems
(20,21,28,47) with an increase in mesophase stability. Phase diagrams for copolymers of various spacer composition have been reported in particular for copolymers of spacers containing asymmetric
carbon atoms such as (+3)methyladipic acid (29,58). The pitch of
the cholesteric helix depends on the proportion of chiral centers
(19,20, 28,47) and polymers with various irridescent colors have
been prepared.

Miscibility of liquid crystalline "backbone" polymers.

Miscibility of liquid-crystalline thermotropic polymers with
other low molecular weight liquid-crystals was studied by several
authors (15,47,58,59,60,61). The applicability of Sackman and Demus'
miscibility rules for high molecular weight compounds was confirmed
for several nematogenic polymers (15,58,60,61). Smectic derivatives
of terphenyls have also been identified by the contact method (61).
Attempts to calculate the eutectic temperature and composition for

mixtures of low molecular weight and high molecular weight nematogens using the Van Laar equations were less successful (15), the discrepancy being explained as arising from nonideality of nematic mixtures due to limited compatibility of polymeric chains with small molecules. Nematic polymers were not compatible with smectic phases of model compounds (62). Thus, the limited experimental evidence available to date suggests the applicability of L.C. miscibility rules to high molecular weight compound although the high viscosity displayed by liquid crystalline polymers can complicate the identification of polymeric mesophases by miscibility.

Influence of molecular weight and molecular weight distribution

The influence of molecular weight and molecular weight distribution on the properties of L.C. flexible backbone polymers has not been investigated in a systematic fashion. From the limited information available it appears particularly strong in the realm of low molecular weights. Flexible center "Siamese-Twin" liquid crystalline diesters structurally related to 4-alkoxyphenyl-4^{l}alkoxybenzoates have been synthesized and compared to the corresponding high polymers [III,2-4] and also to the corresponding 4-alkoxyphenyl-4^{l}alkoxybenzoate (62) as a function of the number n of $-CH_2-$ groups in the spacer. Here the previously mentioned tendency for "backbone" polymers to favor nematic over smectic mesophases is vividly illustrated: whereas increasing n in low molecular weight analogues leads to the appearance of smectic mesophases, this tendency is lost in the "Siamese Twins" prepolymers and in the high polymers themselves. As pointed out previously, the odd-even effect persists for polymers while vanishing for $n \geq 6$ in the low molecular compounds and assuming intermediate values for "Siamese Twins" prepolymers. In the case of polymer DDA-9 [I,131] the phase transition temperatures and stability of nematic phase increase rapidly with increasing molecular weight and reach a plateau of $\bar{M}_n \sim 6,000-8,000$ (29). Model compounds and prepolymers are not mesomorphic (30) and the appearance of a

Table 2. LIQUID CRYSTALLINE POLYMERS FROM MESOGENIC DIACIDS AND FLEXIBLE CHAIN DIOLS

No.	Dicarboxylic Acid	HO-(CH$_2$)$_n$-OH	Transition Temperature	References
1a	HOOC-⟨O⟩-⟨O⟩-COOH	2	K336;340I	27
b	"	2	K314S350I	51
c	"	3	K253S261I	27
d	"	4	K279S310I	27
e	"	4	K274S293I	51
f	"	5	K205;212I	27
g	"	6	K204S223I	27
h	"	6	K209S229I	**51**
i	"	10	K154;160I	27
2a	HOOC-⟨O⟩-CH=CH-⟨O⟩-COOH	5	K185L.C.248I	51
b	"	6	K238L.C.256I	51
c	"	10	K197L.C.200I	51
d	"	HO-CH$_2$-C(Me)$_2$-CH$_2$-OH	K205I	51
3a	HOOC-⟨O⟩-⟨O⟩-⟨O⟩-COOH	2	K322L.C.393I	51
b	"	4	K340L.C.407I	51
c	"	5	K226S366I	51
d	"	6	K235S355I	51
e	"	10	K256S311I	51

(continued)

Table 2.   LIQUID CRYSTALLINE POLYMERS FROM MESOGENIC DIACIDS AND FLEXIBLE CHAIN DIOLS (Continued)

No.	Dicarboxylic Acid	HO$-$(CH$_2$)$_n$$-$OH	Transition Temperature	References
	HOOC$-$◯$-$◯$-$◯$-$COOH	HO$-$(CH$_2$$-CH_2$O)$_n$$-CH_2$$-CH_2$$-$OH		
f	"	1	K213S389I	51
g	"	2	K187S337I	51
h	"	3	K125S253I	51
i	"	9	K70L.C.117I	51
j	"	HO$-$CH$_2$$-$C(Me)$_2$$-CH_2$$-$OH	K256L.C.348I	51
k	"	HO$-$CH$_2$$-$C(Et)$_2$$-CH_2$$-$OH	K117L.C.187I	51
4a	HOOC$-$◯$-$N=N$-$◯$-$COOH	6	K268I	45
b	"	8	K232I	45
c	"	10	K212I	45
d	"	12	K192I	45
		HO$-$(CH$_2$$-CH_2$$-O-$)$_n$$-$H		
e	"	2	K137N188I	81
f	"	3	K126M146I	81
g	"	4	K87M97I	81
5	HOOC$-$◯$-$N=N$\rightarrow$O$-$◯$-$COOH	HO$-$(CH$_2$)$_n$$-$OH		
a	"	5	K235I(I222S188K)	45
b	"	6	K235I	45
c	"	7	K205I(201S160K)	45

No.	Dicarboxylic Acid	HO-(CH$_2$)$_n$-OH	Transition Temperature	References
5	HOOC-⬡-N=N-⬡-COOH	HO-(CH$_2$)$_n$-OH		
d		8	K218I	45
e	"	9	K181I	45
f	"	10	K200I	45
g	"	10	K198; 201I	29
h	"	HO-(CH$_2$)$_{12}$-OH	K188I	45
		HO-(CH$_2$-CH$_2$-O)$_n$-H		
i	"	2	K85L.C.220I	81
j	"	3	K94L.C.160I	81
k	"	4	K87L.C.100I	81
6	Isocyanate O=C=N-⬡(CH$_3$)-⬡(CH$_3$)-N=C=O	HO-(CH$_2$)$_n$-OH		
a	"	5	K248L.C.257I	53
b	"	6	K220L.C.234I	53
c	"	8	K231L.C.237I	53
d	"	10	K202L.C.221I	53
e	"	12	K174L.C.186I	53

Table 3. MISCELLANEOUS STRUCTURES

No.	Diols	Dicarboxylic Acid	Transition Temperature	Ref.
1	$HO-\bigcirc-O-(CH_2)_n-O-\bigcirc-OH$	$HOOC-\bigcirc-COOH$		
a	5	"	K242N355I	6,32
b	6	"	K267N330I	6,32
c	7	"	K237N322I	6,32
d	8	"	K245N280I	6,32
e	9	"	K229N297I	6,32
f	10	"	K237N268I	6,32
2	$HO-\bigcirc-O-(CH_2)_6-O-\bigcirc-OH$	$HOOC-\bigcirc-O-(CH_2)_n-O-\bigcirc-COOH$		
a	"	2	K298.5N333.5I	39
b	"	3	K273N283I	39
c	"	4	K271N306.5I	39
d	"	5	K237N270I	39
e	"	6	K256.5N282I	39
f	"	7	K230.5N259I	39
g	"	8	K233.5N261.5I	39
h	"	9	K205N244.5I	39
i	"	10	K214.5N238.5I	39
j	"	12	K201N213.5I	39
3a	$HO-\bigcirc-O-(CH_2)_8-O-\bigcirc-OH$	2	K261N295.5I	39
b	"	3	K228N248I	39
c	"	4	K250.5N289.5I	39

No.	Diols	Dicarboxylic Acid	Transition Temperature	Ref.
3d	HO–⬡–O–$(CH_2)_8$–O–⬡–OH	5	K220.5N248.5I	39
e	"	6	K233N258.5I	39
f	"	7	K203N237.5I	39
g	"	8	K208N236I	39
h	"	9	K181.5N218.5I	39
i	"	10	K188.5N221I	39
j	"	12	K182N205I	39
4	HO–⬡–O–$(CH_2)_{10}$–O–⬡–OH	HOOC–⬡–O–$(CH_2)_n$–O–⬡–COOH		
a	"	2	K233N292I	39
b	"	3	K211.5N232.5I	39
c	"	4	K222.5N265.5I	39
d	"	5	K195.5N228I	39
e	"	6	K221N251.5I	39
f	"	7	K181N212I	39
g	"	8	K205.5N227I	39
h	"	9	K179.5N202.5I	39
i	"	10	K185N212I	39
j	"	12	K175N193I	39
5	HO–⬡–O–$(CH_2)_n$–O–⬡–OH	HOOC–⬡–O–$(CH_2)_n$–O–⬡–COOH		
a	4	4	K210N300I	76
b	5	5	K165N230I	76

(continued)

Table 3.   MISCELLANEOUS STRUCTURES (Continued)

No.	Diols	Dicarboxylic Acid	Transition Temperature	Ref.
5	HO–◯–O–$(CH_2)_n$–O–◯–OH	HOOC–◯–O–$(CH_2)_n$–O–◯–COOH		
c	6	6	K220N245I	76
d	8	8	K175N230I	76
e	9	9	K160N200I	76
6	HO–◯–OH	HOOC–◯–O$(CH_2)_n$–O–◯–COOH		
a	"	2	K365N390I	19
b	"	3	K350N380I	19
c	"	4	K305N380I	19
d	"	5	K275N315I	19
e	"	6	K220S_{II}280N380I	19
f	"	8	K240S_{II}300N350I	19
g	"	9	K220S_{II}250N305I	19
h	"	10	K105S_I215S_{II}225N270I	19
i	"	10	K236N294I	6,32
j	"	11	K125S_I190S_{II}205N250I	19
7	HO–◯(CH_3)–OH	10	K162N274I	6,32
8	HO–◯(Cl)–OH	10	K157N279I	6,32
9	HO–◯(Br)–OH	10	K146N270I	6,32
10	HO–◯(Cl)(Cl)–OH	10	K200N255I	6,32

No.	Diol	Dicarboxylic Acid	Transition Temperature	Reference
11	HO⟨◯⟩–OH, C$_6$H$_5$	10	K151N168I	83
12	HO⟨◯⟩⟨◯⟩⟨◯⟩–OH	10	K258N374I	83
13	HO⟨◯⟩⟨◯⟩–OH	10	K224N248I	83
14	CH$_2$=CH–CH$_2$–O–⟨◯⟩–COO–⟨◯⟩–OOC–⟨◯⟩–OCH$_2$–CH=CH$_2$	H–(Si–O)$_n$–Si–H (CH$_3$, CH$_3$, CH$_3$, CH$_3$)		
a	"	2	K35L.C.103I	55
b	"	3	5L.C.114I	55
c	"	5	g–13L.C.94I	55
	CH$_2$=CH–CH$_2$–O–⟨◯⟩–COO–⟨◯⟩–OOC–⟨◯⟩–O–CH$_2$–CH=CH$_2$			
d	"	2	g–14L.C.166I	55
e	"	3	g–S151I	55
f	"	4	g–47L.C.127I	55
g	"	5	g–57L.C.121I	55
15	H$_2$N–⟨◯⟩–O–(CH$_2$)$_{12}$–O–⟨◯⟩–NH$_2$	OHC–⟨◯⟩–CHO	K240S290N	16

Table 4. COPOLYMERS

$$\left[\overset{O}{\underset{O}{C}}-R_1-\overset{O}{\underset{O}{C}}-O-R-O\right]_n \left[\overset{O}{\underset{O}{C}}-R_2-\overset{O}{\underset{O}{C}}-O-R-O\right]_n$$

1	R	R_1	R_2	Transition Temperatures	Reference
1		$-(CH_2)_4-$ .50	$-CH_2-\overset{*}{CH}-(CH_2)_2-$ $\quad\quad CH_3$ $(+)\quad\quad$ .50	K199Ch282>300I	47
2a		$-(CH_2)_6-$ .75	$-(CH_2)_{10}-$ .25	K201M287I	17
b	"	.50	.50	K127M274I	17
c	"	.75	.25	K173M254I	17
3		$-(CH_2)_{10}-$ 5	$-CH_2-\overset{*}{CH}-(CH_2)_2-$ $\quad\quad CH_3$		
a	"	.20	.80	K184.8Ch291.2I	59
b	"	.30	.70	K175.0Ch292.9I	59
c	"	.40	.60	K170.4Ch289.6I	59
d	"	.50	.50	K162.0Ch278.5I	20,59,6
e	"	.60	.40	K165.0Ch280.3I	59
f	"	.70	.30	K160.0Ch277I	59
g	"	.75	.25	K158.0Ch272I	59
h	"	.90	.10	K216N265I	59

1	R	R$_1$	R$_2$	Transition Temperatures	Reference
4	(azo structure, N=N, CH$_3$)	$-(CH_2)_{10}-$	$-CH_2-\overset{*}{C}H-(CH_2)_2-$ CH_3		29
a	"	.25	.75	K132Ch193I	29
b	"	.50	.50	K76Ch178I	28,29
c	"	.75	.25	K96Ch169I	29
5	(biphenyl)	$-(OH_2)_8-$	$-CH_2-\overset{*}{C}H-(CH_2)_2-$ CH_3	K220Ch275I	78
		.50	.50		
6	$-(CH_2)_2-$				27
a	"	.70	.30	K288I	27
b	"	.50	.50	K203I	27
7	$-(CH_2)_6-$	.80	.20	K193L.C.202I	27

nematic phase in this system seems to be associated with the development of a polymeric structure indicating cooperativity between
repeating units. The value of ΔS_{NI} and the order parameter in the
nematic state increase with increasing molecular weight to reach a
plateau at $\overline{DP} \sim 8$ (30).

Conformation of mesomorphic flexible backbone polymers

The study of conformation and organization of thermotropic L.C.
"backbone" polymers in the oriented state presents a particular interest in view of the potential applications of such systems. Already in 1975 de Gennes had predicted the possibility of mechanically induced transitions and unusual stress-strain behavior (23).
It can be expected that flexible backbone polymers are randomly
coiled in solution and in the melt above their isotropisation temperature. This has indeed been shown for solutions (16,29,63) and
also for the melt above the N→I transition (63,64). The study of
the magnetic birefringence (Cotton-Mouton effect) in fields of 10-12
Tesla in the case of isotropic melts of thermotropic backbone polyesters reveals strong orientational effects in the pretransitional
state. Polymers displaying a stable nematic phase when cooled
towards the isotropic nematic transition temperature T_c show a pre-
transitional orientational ordering similar to that observed in
classical nematics. However, the extrapolated second order transition temperature T^* (obtained by extrapolation of $\Delta n^{-1}(T)$ to 0) is
unusually low: $T_c - T^* \sim 27°C$ for polymer DDA-9, as opposed to $3.4°C$
for p-azoxyanisole (64). These results suggest that N→I transition is much more first-order in nature than that of ordinary nematics such as p-azoxyanisole.

Polymers with very narrow nematic intervals (sometimes difficult to detect) also show an increase of Δn as one approaches T_c
but Δn decreases precipitously in the immediate vicinity of the
transition (59,65). This phenomenon is attributed to the nucleation of a crystalline phase by the aligned pretransitional nematic

domains so that formation of a nematic and a polycrystalline phase
occurs almost simultaneously. Such polymers designated as "potentially nematic" have been described (57,59,65).

Flexible nematic backbone polymers have been found to align in
magnetic fields (29,65,66,67) and crystalline fiber x-ray diagrams
have been reported for derivatives of the 4,4$^{\perp}$oxyazoxybenzene mesogene (29,59,66) as well as for polymers containing the 4,4$^{\perp}$oxycar-
boxybenzene moiety [I,6c] (66). In both cases a rather strong influence of the molecular weight on the orientation was reported.
Low molecular weight (up to a few thousand) polymers oriented well
even in low intensity magnetic fields of 0.3T. In contrast, high
molecular weights (exceeding 10,000) were rather difficult to
orient (66,68). Cholesteric polymers could not be oriented even
with fields exceeding 12T, presumably due to the very high critical
fields necessary to unwind the cholesteric helix (59,65). It was
also found that the perfection of the fiber obtained by cooling a
nematic melt in a strong magnetic field was dependent on the mechanism of crystallization; some polymers such as derivatives of
unsubstituted 4,4$^{\perp}$oxyazoxybenzene moiety gave fibers displaying a
high degree of crystallite alignment, others such as the 4,4$^{\perp}$oxy-
2,2$^{\perp}$methylazoxybenzene derivatives crystallized in a more random
fashion (29). Random crystallization occurs sometimes in the magnetic field and is often responsible for lack of orientation in the
crystalline sample. For example, (DDA-9) crystallizes from the
nematic melt according to a spherulitic morphology after athermal
nucleation (77). In contrast to magnetic field orientation, melt
extrudates of thermotropic nematic polyesters of high molecular
weights ($\bar{M}_n \sim 20,000$) gave highly oriented fibers (29,50). The dode-
cylcarboxy derivatives of the 1,4-phenylene-2-methylvinylene-1,4-
phenylene moiety [I,5e] displayed fiber period spacings corresponding to an extended repeat unit of the polymer (50). A fiber period
corresponding to an extended repeat monomer unit was also observed
for DDA-8 [I,10h] (29). In both cases very high orientation para-

meters (S∿.9) were observed. In contrast, the 2,2'dimethyl substituted polymer DDA-9 [I,131] displayed low orientation functions for
the crystallized fiber obtained by cooling the melt in magnetic
fields of 12T (S∿.6) and a very high orientation function (S∿.9) for
melt extruded fibers. In this last case the low angle spacing was
different from the corresponding spacing obtained in the magnetic
field, indicating a fiber not only with a different degree of order
but a different crystal lattice (29). The x-ray picture of an
oriented nematic phase of 4,4'oxycarboxybenzeneheptyldiyl [I,6c] indicates a low angle spacing slightly below the extended repeat monomer unit (19.7Å vs 21Å). This period is preserved during crystallization (66). A value of the order parameter in the neamtic melt
at 200°C was deduced from x-ray data (S∿0.64)(66). In contrast, the
x-ray picture of a quenched oriented nematic phase of DDA-9 showed
a high degree of alignment and an x-ray diffraction picture very
similar to the oriented low molecular weight n-alkoxyazoxybenzenes
(69) with two diffuse equatorial peaks and four sharp inner peaks
at 50°C from the meridien corresponding to the fiber period (29).
The phase geometry of polymer [I,6c] is interpreted in terms of a
model corresponding to a simple nematic (66). In the case of DDA-9,
however, the analogy with the low molecular azoxybenzene derivatives
suggests a cybotactic nematic model (29). In both cases, x-ray results indicate a significant extension of the spacer, in agreement
with other data. Even though the amount of x-ray work performed on
mesogenic "backbone" polymers is very limited it points unambiguously
toward the extention of the flexible spacer in the nematic state and
thereby reinforces the evidence from less direct thermal analysis
and structure-property correlations.

    Very little work was devoted to the study of the electric field
effects on flexible backbone polyesters. Such experiments are ham-
pered by high purity requirements not often achieved in polymers as
well as high viscosity of the polymeric nematic phases. Nevertheless, some orientation effects were reported for poly(ethylenetere-

phthalate-co-1,4-oxybenzoate) and compared with orientation of p-azoxyanisole. It was suggested that the method could be used for identification of polymeric nematic phases (78).

The orientation of polymeric nematic phases in electric and magnetic fields is dependent on melt viscosity and is in contrast to the orientation achieved in hydrodynamic fields; the former orients more efficiently low molecular weight polymers while the latter induces high orientation in nematic melts of high molecular weight polymers. The orienting effect of the magnetic field on nematogenic copolymers derivatives of hydroquinone and terephtalic acid was used for the determination of anisotropy of magnetic sus-ceptibility of the mesogenic core. The alignment of the polymer takes place at H=.5-1 Tesla (72).

McFarlane, Nicely and Davis (73) have used broad line 1H NMR spectra to identify the nematic phase of some poly(p-oxybenzoate-co-ethyleneterephtalates), but order parameters were not measured. In the case of DDA-9 of low molecular weight ($\overline{M}_n \sim 4,000$), the sample aligns macroscopically in a field of 1.17 Tesla at 100°C (74). The nematic order parameter associated with the mesogenic group was found to vary from S=0.84 at T_{KN} to S=0.69 at T_{NI} (68,74). In the vicinity of T_{NI} there was an excellent agreement between the value of S measured by NMR and that obtained from magnetic birefringence using the Landau-deGennes theory (64). This very high degree of order was required for the spacer as well as for the mesogen for a satisfactory agreement between the experimental and the simulated spectrum of homogeneously aligned DDA-9 (68). An equally high order parameter was found for a poly(4,4$^\perp$oxy-azoxybenzenedodecanediyl-co-4,4$^\perp$oxy-azoxybenzene-3-methyladipoyl)[IV,3]. These results tend to indicate that the entire repeating unit is oriented by the nematic potential. It thus becomes increasingly clear in the light of these and other data presented here, that the spacer is an integral part of the mesophase and actively participates in the molecular ordering process within the nematic fluid. Consequently such systems cannot be considered solely as a dispersion of rod-like mesogenic elements in an isotropic diluent. Orientation dependent interactions (82) will need to be introduced to account for the behavior of mesomorphic flexible backbone polymers.

ACKNOWLEDGEMENT

The authors acknowledge the support of the National Science Foundation's Polymer Program under Grant No. DMR-7925059.

REFERENCES

1. D. Vorlander, Trans. Faraday Soc. 29, 907 (1933).

2. C. Robinson, Trans. Faraday Soc. 52, 571 (1956).

3. L. Onsager, Ann. N. Y. Acad. Sci. 51, 627 (1947).

4. P. J. Flory, Proc. Roy. Soc., Ser. A, 234, 73 (1956).

5. See for ex. Liquid Crystalline Order in Polymers, A. Blumstein, Editor, Academic Press, 1978.

6. J. I. Jin, S. Antoun, C. Ober and R. W. Lenz, Brit. Polymer J. 12, (4), 132 (1980).

7. W. J. Jackson, Jr. and H. F. Kuhfuss, J. Polymer Sci. (Polymer Chem. Ed.) 14, 2043 (1976).

8. R. W. Lenz and K. A. Feichtinger, Polymer Preprints, 20(1), 114 (1979).

9. Y. B. Amerik, Itogi Nauki, Tekhniki i Tekhnologii Vysokomolekularnykh Soedinenii 12, 177 (1978).

10a. A. Blumstein, Contemporary Topics Polym. Sci., Vol. 3, 79 (1979), Plenum Press.

  b. A. Blumstein, Polymer News 5, 254 (1979).

11. E. T. Samulski, D. B. DuPre, Adv. Liq. Cryst. 4, 121 (1979).

12. A. Roviello and A. Sirigu, J. Polymer Sci. (Polymer Letters Ed.) 13, 455 (1975).

13. K. N. Sivaramakrishnan, A. Blumstein, S. B. Clough and R. B. Blumstein, Polymer Preprints (ACS) 19, 2, 190 (1978).

14. A. Blumstein, K. N. Sivaramakrishnan, S. S. Clough and R. B. Blumstein, Mol. Cryst. Liq. Cryst. (Letters) 49, 255 (1979).

15. A. C. Griffin and S. J. Havens, J. Polymer Sci. (Polymer Letters Ed.) 18, 259 (1980).

16. B. Millaud, A. Thierry, C. Strazielle and A. Skoulious, Mol. Cryst. Liq. Cryst. (Letters) 49, 299 (1979).

17. A. Roviello and A. Sirigu, Europ. Polymer J. 15, 61 (1979).

18. D. Guillon and A. Skoulious, Mol. Cryst. Liq. Cryst. (Letters) 49, 119 (1978).

19. L. Strzelecki and D. Van Luyen, Europ. Polymer J. 16, 299 (1980).

20. S. Vilasagar and A. Blumstein, Mol. Cryst. Liq. Cryst. (Letters) 56, 263 (1980).

21. D. Van Luyen, L. Liebert and L. Strzelecki, Europ. Polymer J. 16, 307 (1980).

22. "Molecular Structure and Properties of Liquid Crystals", G. W. Gray, Academic Press (1962).

23. P. G. deGennes, C. R. Acad. Sci. (Ser. B) Paris, 281, 101 (1975).

24. A. Blumstein, Macromolecules 10, 872 (1977).

25. H. Finkelmann, M. Happ, Portugal and H. Ringsdorf Makromol. Chemie 179, 2541 (1978).

26. V. P. Shibaev, V. M. Moiseenko, Ya. S. Freidzon and N. A. Plate, Europ. Polymer J. 16, 277 (1980).

27. W. R. Krigbaum, J. Asrar, H. Toriumi, A. Ciferri and J. Preston, J. Polymer Sci. (Letters) 20, 2, 109 (1982).

28. A. Blumstein and S. Vilasagar, Mol. Cryst. Liq. Cryst. (Letters) 72, 1 (1981).

29. A. Blumstein, S. Vilasagar, S. Ponrathnam, S. B. Clough, R. B.
    Blumstein, J. Polymer Sci. (Polymer Phys. Ed.) 20, 877
    (1982); Polymer Preprints 23 (1), 309 (1982).

30. R. B. Blumstein and E. Stickles, Proceedings of the Macro-
    IUPAC 1982, Amherst, Mass., p. 799.

31. Liquid Crystals and Plastic Crystals, G. W. Gray, Academic
    Press, 1974, v. 1, p. 103.

32. S. Antoun, R. W. Lenz and J. I. Jin, J. Polymer Sci. (Polymer
    Chem. Ed.) 19, 190 (1980).

33. P. J. Flory, Macromolecules 11, 1141 (1978).

34. R. R. Matheson Jr., and P. J. Flory, Macromolecules 14, 954
    (1981).

35. S. M. Aharoni, J. Polymer Sci. (Polymer Phys. Ed.) 19, 281
    (1981).

36. A. Blumstein and O. Thomas, Macromolecules (to be published).

37. L. Strzelecki and L. Liebert, Europ. Polymer J. 17, 1271
    (1981).

38. Flussige Kristalle in Tabellen, D. Demus, H. Demus and H.
    Zaschke, Ve. Grundstoff Industrie, Leipzig, 1976, p. 165.

39. A. C. Griffin and S. J. Havens, J. Polymer Sci. 19, 951
    (1981).

40. Handbook of Liquid Crystals, H. Kelker and R. Hatz, Vg Chemie,
    1980 (Section 84.2 and references therein).

41. S. Marcelja, J. Chem. Phys. 60, 3599 (1974).

42. P. J. Boss, J. Pirs, P. Ukleja and J. W. Doane, Mol. Cryst.
    Liq. Cryst. 40, 59 (1977).

43. B. Deloche and J. Charvolin, J. Phys. (Paris) 37, 1497 (1976).

44. H. Arnold, Z. Phys. Chem. 226, 146 (1964).

45. K. Iimura, N. Koide and R. Ohta, Reports Progr. Polymer Phys.
    Japan 24, 231 (1981).

46. J. Asrar, O. Thomas, Qi-Xiang Zhou and A. Blumstein, Proceed.
    Macro-IUPAC 1982, Amherst, Mass., p. 797.

47. W. R. Krigbaum, A. Ciferri, J. Asrar, J. Preston and H. Toriumi, Mol. Cryst. Liq. Cryst. 76, 79 (1981).

48. Liquid Crystals and Plastic Crystals, E. M. Barrall II and J. F. Johnson, G. W. Gray and P. A. Windsor Eds., 1974, Hallsted Ellis Harwood Publishers, vol. 2, p. 254.

49. A. Roviello and A. Sirigu, Europ. Polymer J. 15, 423 (1979).

50. A. Roviello and A. Sirigu, Makromol. Chem. 181, 1799 (1980).

51. P. Meurisse, C. Noel, L. Monnerie and F. Fayolle, Brit. Polymer J. 13, 55 (1981).

52. Polymeric Liquid Crystals, Science and Technology, G. W. Gray, Proceedings Conference Santa Margarita, Academic Press Publishers, May 1981.

53. K. Iimura, N. Koide, H. Tanabe, M. Takeda, Makromol. Chem., 182, 2569 (1981).

54. K. Iimura, N. Koide and R. Ohta, Reports Progr. Polymers Phys. Japan 24, 233 (1981).

55. H. Ringsdorff and A. Schneller, Brit. Polymer J. 13, 43 (1981).

56. C. Aguilera, Thesis, University of Mainz, Mainz, 1981.

57. A. Blumstein, K. N. Sivaramakrishnan, R. B. Blumstein and S. B. Clough, Polymer 23, 47 (1982).

58. J. Billard, A. Blumstein and S. Vilasagar, Mol. Cryst. Liq. Cryst. (Letters) 72, 163 (1982).

59. G. Maret, A. Blumstein and S. Vilasagar, Polymer Preprints, 22, 1, 246 (1981), Div. Polymer Chem. A.C.S.

60. C. Noel, J. Billard, Mol. Cryst. Liq. Cryst. (Letters) 41, 269 (1978).

61. B. Fayolle, C. Noel and J. Billard, J. Physique Collogne C3, suppl. no. 4, 40, 485 (1979).

62. A. C. Griffin and T. R. Britt, J. Am. Chem. Soc. 103, 4957 (1981).

63. A. Blumstein, G. Maret and S. Vilasagar, Macromolecules 14, 1534 (1981).

64. G. Maret, F. Volino, R. B. Blumstein, A. F. Martins and A.
     Blumstein, Proc. 27th Inst. Symp. on Macromolecules,
     Strasbourg, vol. II, 973 (1981).

65. G. Maret and A. Blumstein, Mol. Cryst. Liq. Cryst. 88, 295
     (1982).

66. L. Liebert, L. Strzelecki, D. Van Luyen and A. M. Levelut,
     Europ. Polymer J. 17, 71 (1981).

67. F. Volino, A. Farinha Martins, R. B. Blumstein and A. Blumstein,
     C. r. Acad. Sc. Paris, 292, II, 829 (1981).

68. A. F. Martins, B. Ferreira, F. Volino, A. Blumstein and R. B.
     Blumstein, Macromolecules 15, 000 (1982).

69. I. G. Chistyakov and W. M. Chaikovski, Mol. Cryst. Liq. Cryst.
     7, 269 (1969).

70. A. deVries, Mol. Cryst. Liq. Cryst. 10, 219 (1970).

71. W. R. Krigbaum, H. J. Lader and A. Ciferri, Makromol. 13, 554
     (1980).

72. C. Noel, L. Monnerie, M. F. Achard, F. Hardoin, G. Sigaud and
     H. Gasparoux, Polymer 2, 578 (1981).

73.   F. E. McFarlane, V. A. Nicely and T. G. Davis, Contemporary
     Topics in Polymer Science, E. Pearce Ed., Plenum Press,
     vol. 2, 1977, p. 109.

74. F. Volino, A. F. Martins, R. B. Blumstein and A. Blumstein, J.
     Physique (Letters) 42, L-305 (1981).

75. K. F. Wissbrun and A. C. Griffin, J. Polymer Sci. (Polymer
     Phys. Ed.) 20, 1835 (1982).

76. D. Van Luyen and L. Strzelecki, Europ. Polymer J. 16, 303
     (1980).

77. J. Grebowicz and B. Wunderlich, J. Polymer Sci. (Polymer Phys.
     Ed.) (submitted for publication).

78. J. Asrar, H. Toriumi, J. Watanabe, J. Preston, W. R. Krigbaum
     and A. Ciferri, J. Polymer Sci. (in print).

79. A. Roviello and A. Sirigu, Proc. 26th Instn. Symp. on Macro-
     molecules, Florence, 03, 290 (1980).

80. A. Roviello and A. Sirigu, Makromol. Chemie, 180, 2543 (1980).

81. K. Iimura, N. Koide, R. Ohta and M. Takeda, Makromol. Chem.
        182, 2563 (1981).

82. P. J. Flory and G. Ronca, Mol. Cryst. Liq. Cryst. 54, 311
        (1979).

83. B. W. Jo and R. W. Lenz, Makromol. Chem. Rapid Commun. 3, 23
        (1982).

EFFECT OF MESOGENIC UNIT AND SPACER STRUCTURES ON THE THERMOTROPIC PROPERTIES OF MAIN CHAIN LIQUID CRYSTAL POLYESTERS

Robert W. Lenz

Chemical Engineering Department, University of
Massachusetts, Amherst, MA 01003

Jung-Il Jin

Chemistry Department, College of Sciences, Korea
University, 1-Anam Dong, Seoul 132, Korea

INTRODUCTION

The thermotropic liquid crystal behavior of polymers is of strong current interest among polymer scientists both because of their possible development into new materials with unique properties and to expand our knowledge on structured fluid phases of polymers.[1-6] The structures of polymers which are capable of forming thermotropic liquid crystal phases are of great variety. However, essentially all of them can be assigned to one of two categories: either to those which have mesogenic units attached, directly or through spacers, to a polymer backbone ("side chain liquid crystal polymers") or those having mesogenic units and rigid or flexible spacers in the polymer backbone ("main chain liquid crystal polymers"). We have been particularly interested in delineating relationships between the structure and liquid crystal properties of main chain polyesters,[7,8,10-13] and this report critically analyzes the effect of structures of both the mesogenic units and the non-mesogenic spacers on the thermotropic

liquid crystal properties of such polymers prepared in this
laboratory.

RESULTS AND DISCUSSION

Mesogenic Unit Structure Dependence

The dependence of the thermotropic, liquid crystal properties
on the structure of the mesogenic units in main chain polyesters
has been studied for the effect of three different aspects of their
chemical structure: (1) stereochemistry, (2) length, and (3) the
presence of lateral substituents. It has been found in this and
other laboratories that a slight change in the chemical structure
of the mesogenic groups can result in a significant alteration in
the thermal properties of the mesophase. Such an effect can be
demonstrated by comparing the liquid crystal-to-isotropic phase
transition temperature, the clearing temperature, T_i, of the first
three polymers in Table 1.[7,10,11] These polymers contained three
different aromatic ester triad mesogens connected by a single
flexible spacer, the decamethylene group. The major structural
differences between those polymers was in the central aromatic
ester of the mesogenic units. Polymer A had a central hydroquinone
moiety while Polymers B and C had a central terephthaloyl residue.

As seen in Table 1, T_i for Polymer A was significantly higher
than those of either Polymers B and C, indicating a greater thermal
stability of the liquid crystal phase of the former. This observation can be explained on the basis of an expected coplanar and
colinear geometry of the mesogenic units of Polymer A in contrast
with the non-linear conformation resulting from the middle
terephthaloyl moiety in the other two, as shown in Figure 1. A
coplanar molecular geometry would favor a more effective molecular
packing and alignment between the polymer molecules in the liquid
crystal phase, which in turn would stabilize the mesophase.

The slightly higher thermal stability of the mesophase of

Table 1. Thermal Behavior of Liquid Crystal Polyesters with Different Mesogenic Units

Designation	Polymer Repeat Unit	T_m, °C	T_i, °C	ΔT, °C
A	$-O-\text{(O)}-\overset{O}{\overset{\|}{C}}-O-\text{(O)}-O-\overset{O}{\overset{\|}{C}}-\text{(O)}-O(CH_2)_{10}$	236	297	61
B	$-O-\text{(O)}-O-\overset{O}{\overset{\|}{C}}-\text{(O)}-\overset{O}{\overset{\|}{C}}-O-\text{(O)}-O(CH_2)_{10}$	237	265	28
C	$-\overset{O}{\overset{\|}{C}}-\text{(O)}-O-\overset{O}{\overset{\|}{C}}-\text{(O)}-\overset{O}{\overset{\|}{C}}-O-\text{(O)}-\overset{O}{\overset{\|}{C}}-O(CH_2)_{10}$	220	267	47
D	$-O-\text{(O)}-\overset{O}{\overset{\|}{C}}-O-\text{(O)}-\text{(O)}-O-\overset{O}{\overset{\|}{C}}-\text{(O)}-O(CH_2)_{10}$	258	374	116
E	$-\overset{O}{\overset{\|}{C}}-\text{(O)}-O-\overset{O}{\overset{\|}{C}}-\text{(O)}-\overset{O}{\overset{\|}{C}}-O(CH_2)_6-O-\overset{O}{\overset{\|}{C}}-\text{(O)}-\overset{O}{\overset{\|}{C}}-O-\text{(O)}-\overset{O}{\overset{\|}{C}}-O(CH_2)_6-O-$	170	190	20

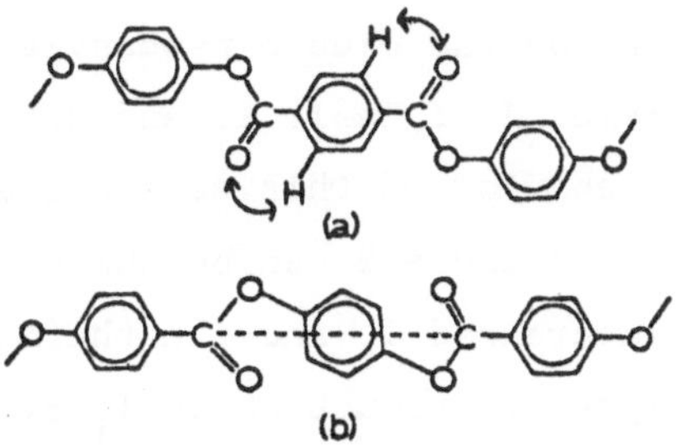

Fig. 1   Repeating unit conformation of Polymers A, B, and C (Table 1).

Polymer C over that of Polymer B is understandable in light of the
fact that the mesogenic unit in Polymer C is further extended
through the terminal carbonyl groups, which should be in resonance
interaction with the neighboring phenyl rings. Even though this
effect would seem to be minor, the presence of the two terminal
carbonyl groups enabled the formation of smectic layers of the
mesogens. Polymer C formed a smectic phase while the other two
formed nematic phases, as will be further discussed later.

Increasing the length of the linear rigid mesogenic unit also
enhanced the thermal stability of the mesophase as shown by a
comparison of the properties of Polymers A and D in Table 1.[13]
The replacement of the middle p-phenylene unit in Polymer A with
the biphenylene unit was accompanied by an increase in T_i of about
80°C and a much greater thermal stability of the mesophase.

In contrast, on shortening the length of the mesogenic unit,
as expected, the thermal stability of the liquid crystal phase was
decreased as seen in the data for Polymers A and E in Table 1[11]
for triad vs. dyad units, respectively.

In addition to the factors discussed above, the presence of
lateral substituents in the mesogenic units plays a very important
role in controlling the thermal behavior of the mesophase of the
main chain polymers. In this study, we have systematically varied
the lateral substituent on the middle p-phenylene ring of the
mesogenic unit of Polymer A of Table 1, and the results from a
study of the thermal behavior of this series are tabulated in
Table 2.[7,10] The data in Table 2 can be summarized as follows:
(1) monosubstitution decreased T_i and the thermal stability of the
mesophase, (2) the degree of reduction in T_i by a substituent
approximately paralleled its size, (3) two substituents (Polymer
A-5 in Table 2) lowered T_i approximately twice as much as one
indicating the possible existence of additivity in substituent
effect, and (4) polymers based on a monosubstituted hydroquinone

Table 2.   Effect of Lateral Substituents on the Liquid Crystal
Properties of Main Chain Polyesters

$$-O-\bigcirc-\overset{\overset{O}{\|}}{C}-O-\underset{\underset{Y}{}}{\overset{\overset{X}{}}{\bigcirc}}-O-\overset{\overset{O}{\|}}{C}-\bigcirc-O\text{-}(CH_2)_{10}$$

Designation	Polymer Repeat Unit		T_m,°C	T_i,°C	ΔT,°C	ΔH_i Kcal/mol	ΔS_i e.u.
	$\underline{X}$	$\underline{Y}$					
A	H	H	236	297	61	0.97	1.7
A-1	H	Cl	157	279	122	2.5	4.6
A-2	H	CH_3	162	274	112	7.6	2.9
A-3	H	Br	146	270	124	2.8	5.2
A-4	H	C_6H_5	151	168	117	1.6	3.6
A-5	Cl	Cl	200	255	55	0.94	1.8

unit (Polymers A-1 through A-4 in Table 2) exhibited higher values
of ΔS_i than those with an unsubstituted (Polymer A) or a
symmetrically disubstituted unit (Polymer A-5 in Table 2).

All of these observations can be rationalized on the basis
of either of two effects: (1) steric hindrance by the substituents
causing an increased separation of the mesogenic units in adjacent
polymer chains or (2) interlocking by the substituent on adjacent
chains decreasing molecular mobility,[14] in the liquid crystal
phase.   The former would decrease T_i while the latter would increase ΔS_i, as observed in Table 2.   Polar effect does not seem to
exert a predominant role on the thermal stability of the mesophase
of the polymers being discussed.

With the exception of Polymer C, all of the polymers in Tables
1 and 2 exhibited only a nematic liquid crystal phase.   Figure 2
shows the optical texture observed for Polymer A which is typical

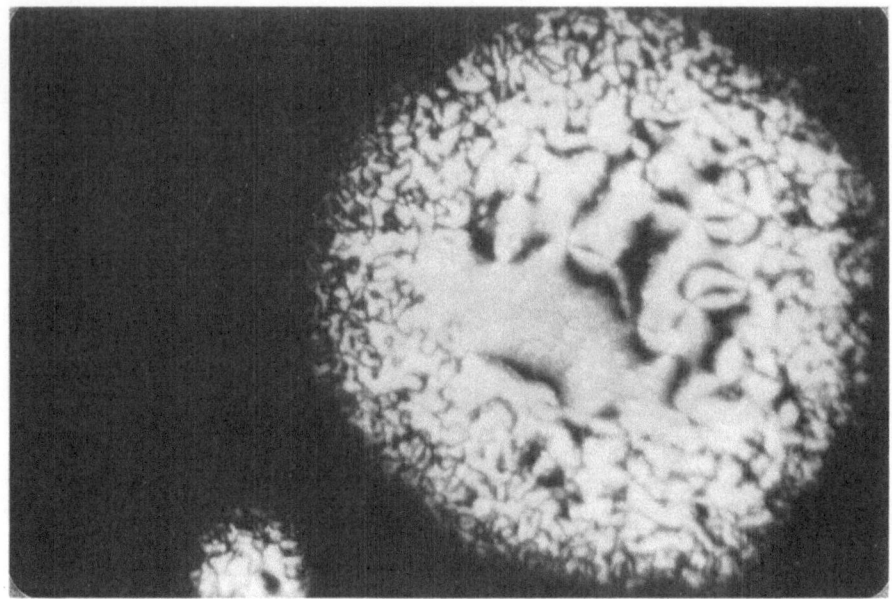

Fig. 2   Photomicrograph of Polymer A taken at 255°C
           (magnification X320).

of the nematic phase. A wide-angle X-ray diffraction photograph
of the quenched melt of Polymer A-5 in Table 2 is shown in Figure
3. The diffraction pattern is also typical of a nematic liquid
crystal. The formation of the liquid crystal phase by the polymers
was reversible, meaning that they are enantiotropic. In general,
super-cooling during reformation of the liquid crystal phase from
the isotropic phase was much less pronounced than that for recrystallization.

Spacer Unit Structure Dependence

     Spacer units can be classified into two different groups,
either rigid or flexible spacers. We have studied the influence
of the structure of both rigid and flexible spacers on the
properties and nature of the mesophase of main chain liquid
thermotropic behavior.[7,8]

W A X S

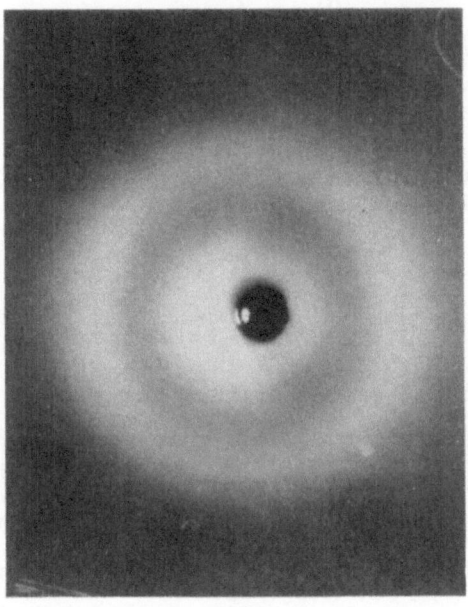

X=Y=Cl

Fig. 3   Wide-angle X-ray diffraction pattern of a sample quenched
         from the melt for the Polymer A-5.

polymers, as discussed in the following sections.

Rigid Spacers. The types of rigid spacers studied in this
work were those based on bisphenols containing different central
substituents between the two phenolic rings: HO —$\langle$O$\rangle$— X —$\langle$O$\rangle$— OH,
where X was $C(CH_3)_2$ (BPA), CH_2 (BPM), O (ODP), S (TDP), SO_2 (SDP)
and none (BP). Resorcinol (RES) was also included in this series.
A series of copolyesters having o-chloro-p-phenylene terephthalate
and bisphenol terephthalate units was prepared covering a wide
range of comonomer compositions, as shown in Figure 4. These
polymers were characterized by thermal analysis and optical
examination to find the effect of the structure of the rigid
spacers and the composition of the copolyesters on the

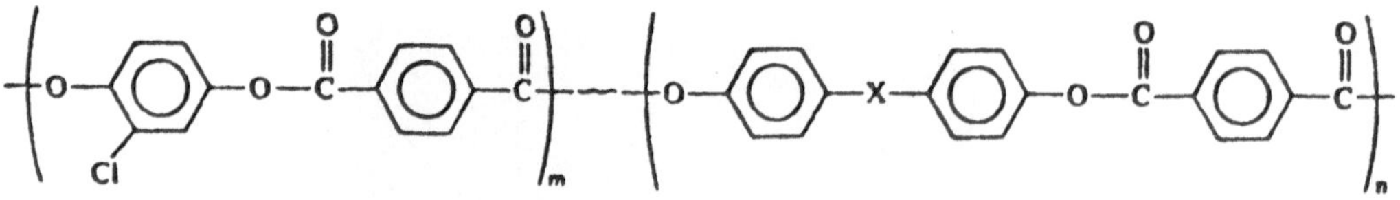

Fig. 4   Structure of liquid crystal aromatic polyesters with
rigid bisphenol spacers.

More specifically, we determined the maximum or threshold
amount of each bisphenol comonomer which could be incorporated into
the copolymer without complete destruction of the liquid crystal
nature of the resulting copolymers, with the results shown in
Table 3. All of the liquid crystal copolymers formed nematic
phases as judged by microscopy (Figure 5) and by X-ray diffraction
analysis (Figure 6) of the quenched melts.

The results collected in Table 3 clearly demonstrate that the
greater the bulkiness of the central substituent, X, in the rigid
bisphenol spacer unit, the lower the threshold comonomer amount
which could be accommodated in the copolymer without completely
losing the liquid crystal characteristics. The differences in
nonlinearity of the non-mesogenic units caused by the presence of
the middle substituents of the bisphenols were all within about
a 5° angle, which indicates that the degree of molecular bending
caused by X was approximately the same.

Certainly the larger groups, for example the $C(CH_3)_2$ and
SO_2 groups, must have caused an increased separation of the
adjacent polymer chains to destabilize the mesophase, so the
stereogeometry or space-filling characteristics of the polymer
units and the bulkiness of X are considered to be the two prime,
important factors in controlling the formation of the liquid

Table 3. Maximum Amount of Each Bisphenol Which Could Be
Copolymerized without Complete Destruction of the Liquid
Crystallinity of the CHQ Copolyester[a]

Bisphenol	Max. Amount, mol %	Bisphenol	Max. Amount, mol %
BPA	40	RES	60
SDP	50	ODP	70
BPM	60	BP	100
TDP	60		

[a]See text and Figure 4 for polymer unit and bisphenol structures.

Fig. 5   Photomicrograph of the quenched melt of 70 CHQ/30 ODP
copolyester (magnification X320).

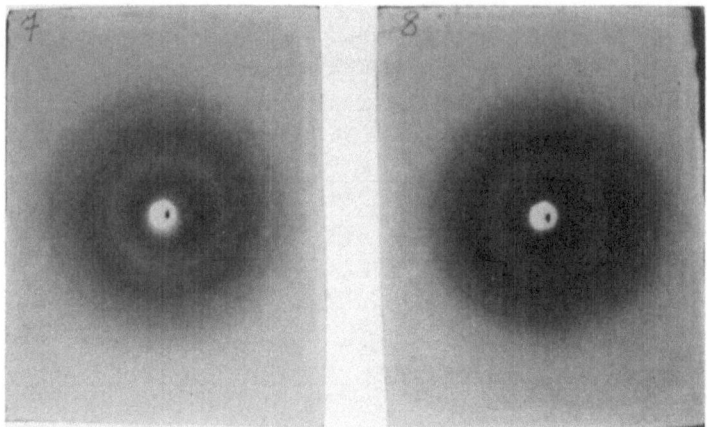

Fig. 6   Wide-angle X-ray diffraction patterns of 70/30 CHQ/ODP
           copolyester: (A) diffraction by the original sample;
           (B) diffraction by the quenched sample.

crystal phase of these polyesters. On the other hand, the
electronic or polar effects of the substituents on the liquid
crystal properties were not as evident as the steric effects and
can be relatively minor in comparison. The copolymers of
chlorohydroquinone (CHQ) and RES were unique in that the resorcinol
unit did not have a central substituent, X, as in the other bisphenols. This unit was also able to maintain the rigidity of the
polymer chain, but it induced a bend angle of 120° along the
polymer backbone, destroying the linearity and reducing the
parallel association of polymer chains in the nematic state, which
in turn should decrease the stability of mesophase. McFarlane and
co-workers earlier arrived at the same conclusion from their study
on the liquid crystal properties of polyesters based on p-oxy-
benzoate modified poly(ethylene terephthalate).[9] Unfortunately,
in these studies any information on the clearing transitions of
the liquid crystal phases could not be obtained, because all of
their polymers underwent thermal degradation before reaching the
T_i transition.

Flexible Spacers. Several series of polyesters containing mesogenic units interconnected through flexible spacers in the main chain were prepared. The structure and length of the spacers were systematically varied, and the thermotropic liquid crystal properties of the polymers were examined with the results shown in Tables 4 and 5.[7,10-13]

For polymers of Series C in Table 4, those with spacers up to n = 8 formed a nematic phase, while those with n = 9 and 10 formed a smectic phase.[11] Figure 7 shows the focal-conic optical texture observed for the liquid crystal phase of Polymer C-10 in Table 4, which is characteristic of smectics. The clearing temperatures showed a regular trend with a zig-zag decrease in T_i for polymers up to n = 9, but then T_i increased for the 9 and 10 polymers as seen in the data in Table 4 and Figure 8. The transition temperatures for the polymers with an even n were generally higher than those with an odd n. Griffin and Havens,[5] and Strzelecki and Luyen[6] also observed the same trend for polymers with similar structural features.

From these results it appears that longer flexible spacers render higher degrees of freedom to the mesogenic units for alignment to form smectic layers. Another possible explanation for the change from nematic to smectic order may be that a conformational change of the polyalkylene spacer from the fully extended trans conformation to the one with a central gauche unit can occur in the longer spacers.[15] Such a conformational change would represent a higher energy state but it would stabilize the mesophase order.

The liquid crystal behavior of the polymers of Series D in Table 4 showed a very strong dependence on the length of the polyethyleneoxy spacer, n.[11] The polymers with medium values of n formed smectic as well as nematic phases, while the liquid crystal polymer with the longest spacer (n = 9 for Polymer D-9 in

Table 4. Effects of Structure and Length of Flexible Spacers on
          the Liquid Crystal Properties of Main Chain Polyesters

Designation	Polymer Repeat Unit	T_m, °C	T_i, °C

$$-\left[\overset{O}{\underset{\|}{C}}-\langle O\rangle-O-\overset{O}{\underset{\|}{C}}-\langle O\rangle-\overset{O}{\underset{\|}{C}}-O-\langle O\rangle-\overset{O}{\underset{\|}{C}}-O(CH_2)_n-O\right]-$$

C-2	n = 2	340	N365I
C-3	3	240	N315I
C-4	4	285	N345I
C-5	5	175	N267I
C-6	6	227	N290I
C-7	7	176	N253I
C-8	8	165	N220I
C-9	9	174	S233I
C-10	10	220	S267I

$$-\left[\overset{O}{\underset{\|}{C}}-\langle O\rangle-O-\overset{O}{\underset{\|}{C}}-\langle O\rangle-\overset{O}{\underset{\|}{C}}-O-\langle O\rangle-\overset{O}{\underset{\|}{C}}-O(CH_2CH_2O)_n-\right]-$$

D-1	n = 1	342	N365I
D-2	2	185	S222N288I
D-3	3	180	S203N257I
D-4	4	121	S211N245I
D-9	8.7	102	N242I
D-13	13.2	91	no l.c.

Table 4) formed only a nematic phase. When the length of the
spacer was further increased (Polymer D-13 in Table 4), the
polymer did not form a liquid crystal phase. These observations
emphasize the fact that the thermal stability and the nature of
the mesophase, as well, strongly depend on a combination of both
chemical structure and spacer length. The fact that a relatively
short polyethyleneoxy spacer (for example, that with n = 2 as in
Polymer D-2 in Table 4) promotes the formation of a smectic phase
indicates that the presence of an oxygen atom in the spacer may
have exerted a polar effect, which strengthened the lateral intermolecular attraction between adjacent polymer chains, thereby
helping the formation of smectic layers of the mesogens.

     The E series of polymers in Table 5 contained a bis(dimethyl-

Fig. 7   Focal conic optical texture of Polymer C in Table 1
          (magnification X320).

silyl)ether spacer which considerably reduced both T_m and T_i of
the polymers.[12] The bulkiness and flexibility of this spacer
probably caused this phenomenon. The copolymer with mixed
flexible spacers showed intermediate values of T_m and T_i (Polymer
E-3 in Table 5), which were very close to the average of the
transition temperatures of the corresponding homopolymers,
Polymer A-3 in Table 2 and Polymer E-2 in Table 5. Hence, a
combination of large lateral substituents in the mesogenic unit
and bulky and flexible spacers can significantly reduce T_m of the
polymer while maintaining a fairly high T_i as seen for Polymer E-2
in Table 5. All of the polymers with silylether spacers formed
only nematic mesophases. The mesophase of Polymer E-2 had a much
broader nematic range than that of E-1 because T_m was decreased to
a much greater extent than T_i by the bromine substituent.

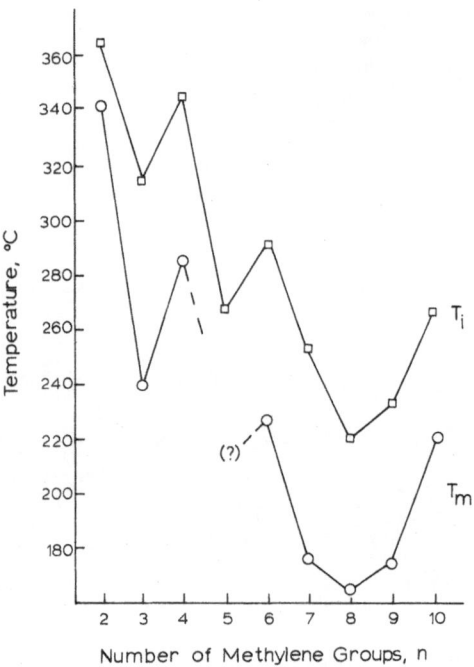

Fig. 8   Dependence of T_m and T_i on the length of the flexible
          spacer for the polymers of the C series in Table 4.

Table 5.   Thermal Properties of Liquid Crystal Polyesters with
           Bis(dimethyl silyl) ether Spacers.

Designation	Mesogen	Spacer	T_m,°C	T_i,°C
E-1	-O-⬡-O-C(=O)-⬡-C(=O)-O-⬡-O-	$-CH_2Si(CH_3)_2OSi(CH_3)_2CH_2-$	199	N208I
E-2	-O-⬡-C(=O)-O-⬡(Br)-O-C(=O)-⬡-O	$-CH_2Si(CH_3)_2OSi(CH_3)_2CH_2-$	43	N177I
E-3	1:1 copolymer of siloxane spacer and $+CH_2+_{10}$ with E-2 mesogen		101	N176I

EXPERIMENTAL

The synthetic procedures used for the preparation of the polymers discussed in this report were all described in our earlier studies.[7,8,11-13] The thermal behavior of all polymers was examined by differential scanning calorimetry (Perkin-Elmer DSC-1B) and on a hot-stage (Mettler FP-2) of a polarizing microscope (Leitz, Ortholux). X-ray diffraction analysis was made on a pelletized sample obtained from quenched melts that were made by pouring liquid nitrogen on the melt. When the T_i was too high to be measured by DSC, it was determined by observing the disappearance of stir-opalescence of the melt placed between two microscope cover slides on a Fisher-Johns melting point apparatus.

ACKNOWLEDGEMENT

The authors are grateful to the Office of Naval Research and the NSF-sponsored Materials Research Laboratory of the University of Massachusetts for the support of this work. The many and important contributions of our coworkers made this report possible, and we are most pleased to acknowledge the efforts of S. Antoun, C. Ober and B. -W. Jo in the investigations described.

REFERENCES

1. A. Roviello and A. Sirigu, J. Polymer Sci., Polymer Lett. 13, 455 (1975).

2. P. G. de Gennes, C. R. Acad. Sci. Paris, B281, 101 (1975).

3. W. J. Jackson and H. F. Kuhfuss, J. Polymer Sci., Polymer Chem. Ed., 14, 2043 (1976).

4. A. Blumstein, K. N. Sivaramakrishnan, S. B. Clough, and R. B. Blumstein, Mol. Cryst. Liq. Cryst. (Letters), 49, 255 (1979).

5. A. C. Griffin and S. J. Havens, J. Polymer Sci., Polymer Phys. Ed., 19, 951 (1981).

6.   L. Strzelecki and D. van Luyen, Europ. Polymer J., <u>16</u>, 299
     (1980).

7.   J. -I. Jin, S. Antoun, C. Ober, and R. W. Lenz, Brit. Polymer
     J., <u>12(3)</u>, 132 (1980).

8.   R. W. Lenz and J. -I. Jin, Macromolecules, <u>14</u>, 1405 (1981).

9.   W. J. Jackson, Jr., British Polymer J., <u>12</u>, 154 (1980).

10. S. Antoun, R. W. Lenz, and J. -I. Jin, J. Polymer Sci.,
     Polymer Chem. Ed., <u>19</u>, 1901 (1981).

11. C. Ober, J. -I. Jin, and R. W. Lenz, Polymer J. (Japan), <u>14</u>,
     9 (1982).

12. B. -W. Jo, J. -I. Jin, and R. W. Lenz, Europ. Polymer J., in
     press.

13. B. -W. Jo, J. -I. Jin, and R. W. Lenz, Makromol. Chem. -
     Rapid Comm., <u>3</u>, 23 (1982).

14. G. R. Luckhurst and G. W. Gray, Eds., <u>The Molecular Physics
     of Liquid Crystals</u>, Academic Press, New York, 1979,
     p. 15 and references cited therein.

15. G. W. Gray and A. Mosley, J. Chem. Soc., Perkin II, 97 (1976).

THE ROLE OF SEQUENCE DISTRIBUTION ON LIQUID CRYSTALLINE PROPERTIES OF AROMATIC COPOLYESTERS

J. Tsay, W. Volksen, and J. Economy

IBM San Jose Research Center
San Jose, California 95193

INTRODUCTION

A number of aromatic copolyesters have been studied to develop processable polymers which would retain strength properties at temperatures of up to 300°C.[1,2,3,4] The copolymers of p-hydroxybenzoic acid (PHBA) with biphenol and terephthalic acid (BPT) within the compositional range of 2/1 to 1/2 of PHBA/BPT are melt processible and display outstanding mechanical properties. Furthermore, the melt state of these copolymers has been conclusively shown to be liquid crystalline and display the typical birefringence and shear sensitivity.[5] The melt flow characteristics of the copolyesters are not only dependent on the shear stress but also on the molecular weight and the sequence distribution of the PHBA/BPT units. However, neither the molecular weight nor the sequence distribution can be determined by solution techniques since the copolymers of PHBA/BPT are insoluble in any known solvents. Since the polymerization reaction of the mixture of three monomers is always carried out under heterogeneous conditions, there is good reason to believe that the copolymer structure may be blocky. This is further substantiated from cross polarization solid state C-13 NMR spectra which indicate that in the as prepared copolymer, 50 percent of the PHBA

is present in more ordered regions as compared to the BPT and remaining PHBA.[5] The present study was undertaken to better characterize the effect of sequence distribution on the melt properties.
Copolyesters of PHBA/BPT (2/1) ranging from a perfectly alternating
structure to a highly blocky structure were prepared and the melt
flow properties characterized.

EXPERIMENTAL

Monomer Synthesis

The p-acetoxybenzoic acid was prepared under Schotten-Baumann
conditions by dissolving 1 equivalent p-hydroxybenzoic acid in 2 eq.
NaOH solution in a separatory funnel, cooling it internally with ice
and shaking vigorously with 10 percent excess of acetic anhydride for
10 minutes. The product was then acidified by dilute HCl solution,
collected by suction filtration, vacuum dried and recrystallized from
chloroform.

The p,p'-diacetoxybiphenyl was synthesized by refluxing the p,p'-
biphenol with a 10 mole percent excess of acetic anhydride at 140°C
for 12 hours. The excess acetic anhydride was then distilled off
under reduced pressure, and the residue crystallized three times
from toluene.

Polymerization Reaction

The starting materials for the syntheses of the conventional
2/1 PHBA/BPT copolyester, long block and short block copolyesters
were p-acetoxybenzoic acid, p,p'-diacetoxybiphenyl and terephthalic
acid (Fig. 1). The conventional copolymer[1] was prepared by starting
the polymerization with a heterogeneous mixture of the three monomers.
The short and long blocks copolymers were prepared by reacting the
p-acetoxybenzoic acid for a period of time before adding the other
two monomers. The alternating copolymer was prepared by synthesizing the HBA-BP-HBA unit first and then reacting with terephthalic

PHBA/BPT Copolyesters

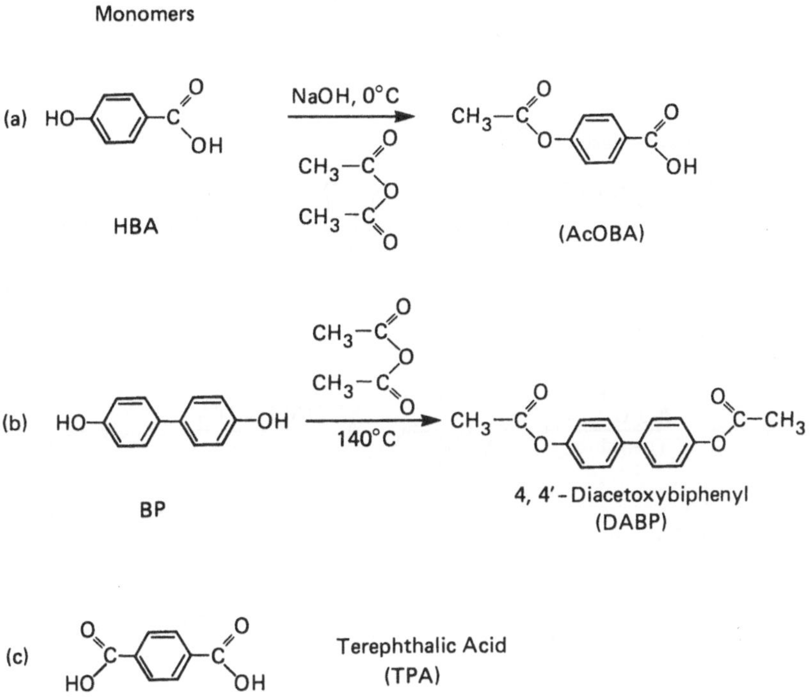

Monomers

Fig. 1 The preparation of three monomers, p-acetoxybenzoic acid, p,p'-diacetoxybiphenyl, and terephthalic acid.

acid to form a highly ordered copolymer consisting of HBA-BP-HBA-TPA as a repeating unit (Fig. 2).

(a) Conventional 2/1 PHBA/BPT copolyester. Two equivalents of p-acetoxybenzoic were mixed with 1 equivalent of p,p'-diacetoxy-biphenyl and 1 equivalent of terephthalic acid in Therminol 66. The polymerization proceeded by heating the mixture from 200°C to 280°C for 1 hour and then kept at 320-350°C for 12 hours with rapid stirring and under a constant stream of dry nitrogen. After completion of the reaction the polymer was washed with acetone, ground in a micro-

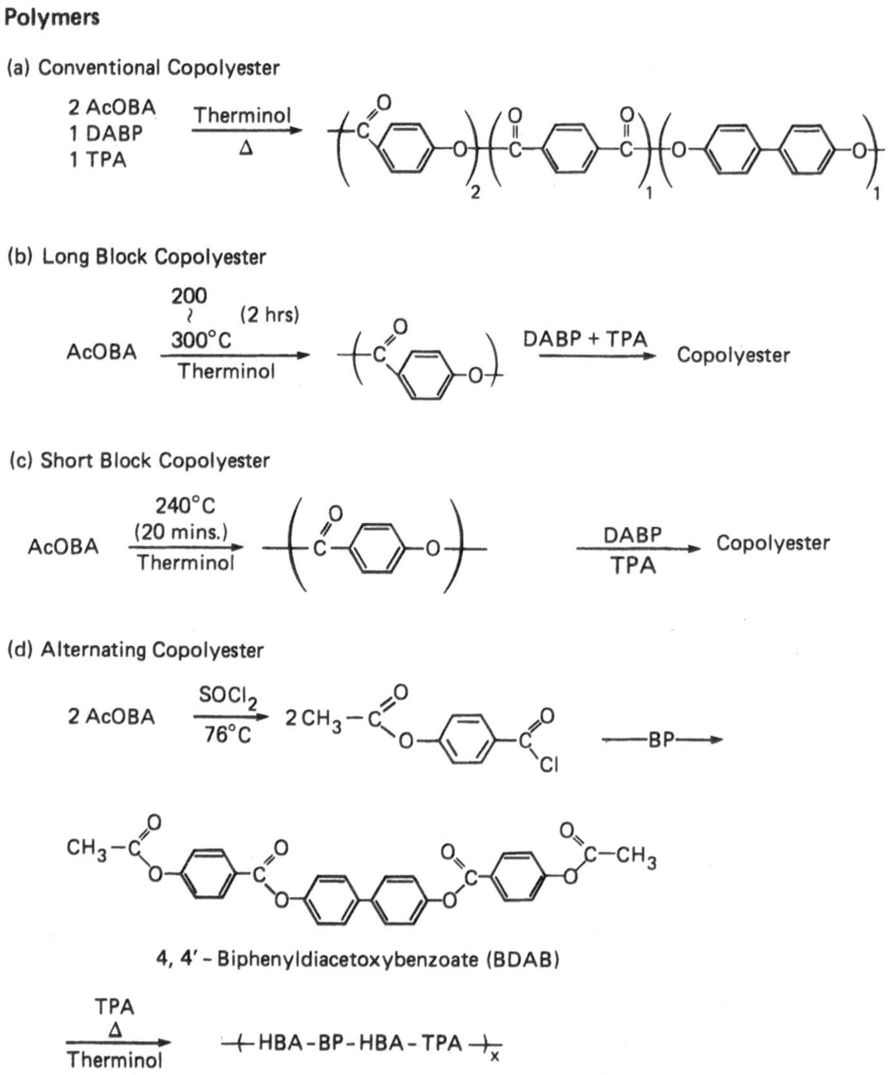

Polymers

(a) Conventional Copolyester

(b) Long Block Copolyester

(c) Short Block Copolyester

(d) Alternating Copolyester

4, 4' – Biphenyldiacetoxybenzoate (BDAB)

Fig. 2   The synthetic scheme for the conventional PHBA/BPT (2/1) copolyester and model compounds with long block, short block, and alternating sequences.

mill, extracted in a Soxhlet extractor with boiling acetone, and then dried in a vacuum oven at 110°C overnight.

(b) Long block copolyester. The p-acetoxybenzoic acid was polymerized in Therminol 66 from 200 to 300°C for two hours after which approximately 80 percent of the acetic acid evolved. At this point some homopolymer had reached a molecular weight at which precipitation occurs (15>DP>8 measured by end group analysis). Then 0.5 equivalent of 4,4'-diacetoxybiphenyl and 0.5 equivalent of terephthalic acid were added to the reaction mixture with continued heating at 300-340°C for another 12 hours. The work-up procedures were the same as in the preceeding reaction.

(c) Short block copolyester. The preparation method for the short block copolyester was similar to that of the long block copolymer except that p-acetoxybenzoic acid was heated at 240°C for 20 minutes to the first stage of homopolymerization. Although 30 percent of the acetic acid had distilled there was no evidence of precipitation of the homopolymer, indicating that the maximum repeating unit of PHBA must be less than 8.

(d) Alternating copolyester. The p-acetoxybenzoic acid was refluxed with excess thionyl chloride and a catalytic amount of pyridine at 76°C overnight. After distillation of the unreacted thionyl chloride, p-acetoxybenzoyl chloride was collected by vacuum distillation and recrystallized from hexane. Ten mole percent excess of the p-acetoxybenzoyl chloride was dissolved in pyridine then added dropwise to the biphenol and pyridine solution with stirring for 3 hours. The product was poured into dilute HCl solution to remove pyridine. The solid product, 4,4'-biphenyldiacetoxybenzoate (BDAB), was collected by filtration and recrystallized three times from dioxane. BDAB has a melting transition at 220°C converting into a nematic mesomorphic state. Equivalent amounts of BDAB and terephthalic acid were mixed in Therminol 66 and polymerized from 180 to 300°C at a heating rate of 10°C/hr. and then kept at 320°C for 12 hours. A white powder was collected after filtration, extraction and drying.

Polymer Characterization

The degree of polymerization and the number average molecular weight of PHBA homopolymers were measured by end group analysis. The PHBA was completely hydrolyzed first by concentrated potassium hydroxide solution and then neutralized by phosphoric acid. The concentration of the acetic acid in the hydrolyzed solution were measured by gas chromatography technique using a column packed with Porapak Q coated with 3 percent phosphoric acid.[6] Propionic acid was used as an internal standard in this quantitative analysis. The detailed results of the molecular weight measurement of PHBA homopolymers will be reported elsewhere in the near future.

Thermal analysis was carried out on a DuPont 990 Thermal analyzer coupled with a Differential Scanning Calorimetry module and a DuPont 951 Thermogravimetric Analyzer. All thermograms were obtained under a constant flow of dry nitrogen (30 ml/min) with a heating rate of 20°C/min.

Wide angle X-ray powder diffraction was carried out on a vertical scanning Norelco diffractometer using copper Kα radiation.

RESULTS AND DISCUSSION

In earlier work, an extremely rapid polymerization rate was observed for the homopolymerization of p-acetoxybenzoic acid (PHBA) at temperature of 270-300°C.[7] However, the polymerization of diacetoxybiphenyl and terephthalic acid (BPT) at the same temperature proceeded at a much slower rate as evidenced by very slow evolution of acetic acid, and the reaction took 2-3 hours to reach 50 percent conversion (DP=2). Since the polymerization reaction of the mixture of p-acetoxybenzoic acid, diacetoxybiphenyl and terephthalic acid is always carried out under heterogeneous conditions, and a sharp difference exists in polymerization rate of PHBA and BPT, there are good reasons to believe that the conventional copolymer may have a blocky structure.

In the magic-angle solid state C-13 NMR cross-polarization (CP) dynamics study,[8] a series of spectra of the conventional copolyester were taken at various CP contact times. Since the static dipolar interaction is greater in the ordered regions as compared to the less ordered region, the carbon resonance from the more ordered regions will cross-polarize more rapidly. It was found that the PHBA resonance had already grown in at a contact time of 0.13 milliseconds, while the BPT peak began to grow at longer times in parallel with further growth of the PHBA peak. This two stage growth for the PHBA suggests that about 50 percent of the PHBA exists in more ordered regions as compared to the BPT and the remaining PHBA. One can infer from these results that the conventional PHBA/BPT (2/1) copolyester has sequences of ordered PHBA.

The DSC thermogram of the PHBA/BPT (2/1) conventional copolyester shows an irreversible endotherm around 390 to 430°C which corresponds to the temperature at which the material flows (Fig. 3). This endothermic transition is very dependent on the molecular weight of the polymer. For example, if the polymerization was maintained at 300-320°C for several hours, the endotherm of this product was observed at 380°C. Solid state thermal aging of this low molecular weight sample (after grinding into powder form) under vacuum at 300°C for several hours shifts this transition from 380 to over 400°C. The polymer melts and flows under stress above this temperature. The birefringence and shear sensitive properties indicate a mesomorphic order in the melt state. The irreversible character of the transition suggests the formation of a supercooled liquid crystalline state after one heating-cooling cycle. Wide angle X-ray scattering showed that much of the crystallinity was lost in the thermal cycling, presumably due to supercooling of the liquid crystalline state

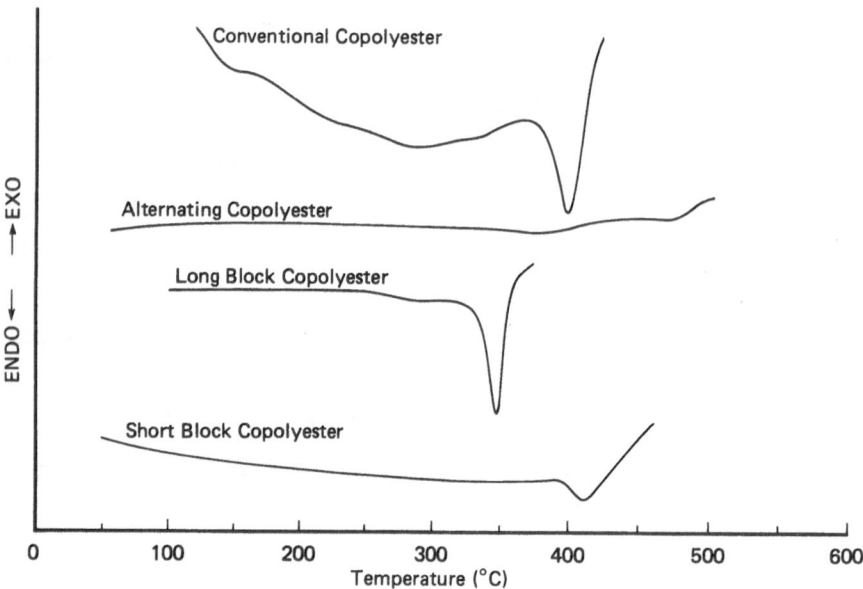

Fig. 3 The differential scanning calorimetry diagrams of the
        various copolyesters.

(Fig. 4). Interestingly, with lower molecular weight copolymer,
there was some memory of the endotherm during second heating.

The DSC thermogram of the alternating copolymer showed no melting peak before decomposition at 500°C. Thermal annealing of this
polymer at 360°C for 2 hours did not change the thermogram. WAXS
shows a high percentage of crystallinity with four sharp peaks at
19.8°, 22°, 27.5°, and 29.1° (2 θ), corresponding to 4.48°A, 4.04°A,
3.24°A and 3.07°A d spacings respectively. The intractability of
the alternating copolymer clearly is due to the high crystallinity
resulting from the highly ordered sequence distribution. The absence of any melt behavior for this structure resembles more closely
that of the BPT homopolymer as opposed to the PHBA homopolymer which
displays a reversible endotherm transition at 330–340°C. Above
this transition the PHBA homopolymer will flow but only under high
pressure.

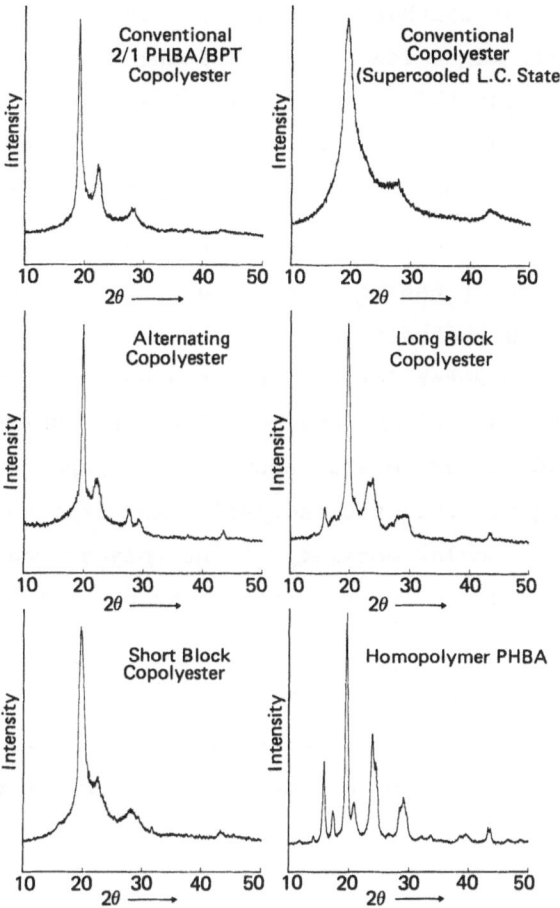

Fig. 4 The wide angle X-ray scattering diagrams of the various
        copolyesters.

The DSC of the long block copolymer showed a sharp reversible
340°C peak with 20°C supercooling. It should be noted that this
340°C reversible endotherm transition is very similar to that of
PHBA homopolymer, and raises the question as to whether this system
is a blend of the two homopolymers or a true block copolymer. The
multiple peaks on WAXS diagram of the long block copolyester also
resemble that of the PHBA homopolymer. Since the homopolymerization
of p-acetoxybenzoic acid does proceed very fast at 300°C, it seems

reasonable that the synthesis procedure used for preparing the long
block copolyester might lead to significant amount of the homopolymers of PHBA and BPT. In fact, the WAXS and DSC diagrams of
the prepared long block copolyester were almost identical to those
of a blended product of PHBA and BPT homopolymers (in 2 to 1 ratio).

The possibility that reorganization may occur was also tested
particularly since it has been reported that reorganization of random copolyester in the melt can occur at elevated temperatures in
the presence of an ester interchange catalyst.[9] We have annealed
the alternating, long block, and polyblend of PHBA/BPT (2/1) copolyesters at 400°C for several hours with no apparent reorganization. This suggests that the reorganization of a copolyester requires certain molecular mobility of the polymer chains (e.g. in
the melt state) and the presence of strong ester interchange catalyst.

The DSC and WAXS of the short block copolyester were similar to
the diagrams for the conventional copolyester. These results along
with the solid C-13 NMR experiments indicate that the conventional
copolyester made from three monomers mixture does have a short
block sequence.

ACKNOWLEDGMENTS

The authors appreciate very much the help of Mr. K. Erickson
in wide angle X-ray scattering experiments.

REFERENCES

1.   S. G. Cottis, J. Economy, and B. E. Nowak, US Patents 3,637,595
         and 3,962,314 (1972); 3,772,250 (1973).

2.   C. R. Payet, Ger Offen 27 51 653 (1978).

3.   H. F. Kuhfuss and W. J. Jackson, US Patents 3,778,410 (1973)
         and 3,804,805 (1974).

4.   G. W. Calundann, US Patents 4,161,470 (1979) and 4,184,996 (1980).

5. W. Volksen, B. L. Dawson, J. Economy, and J. R. Lyerla Jr.,
    Polymer Preprints, 20 (1), 86 (1979).

6. W. R. White and J. A. Leenheer, J. Chromatog. Sci., 13, 386
    (1975).

7. J. Tsay, W. Volksen, and J. Economy, Polymer Preprints, 23 (1),
    266 (1982).

8. J. Economy and W. Volksen, Book Chapter in "The Strength and
    Stiffness of Polymers", Marcel Dekker, in press.

9. R. W. Lenz and K. A. Feichtinger, Polymer Preprints, 20 (1),
    114 (1979).

DIFFUSION OF RIGID POLYAMIDES THROUGH SWOLLEN GEL OF SAME

Shaul M. Aharoni

Corporate Research and Development
Allied Corporation
Morristown, New Jersey   07960

ABSTRACT

Linear aromatic poly(p-benzanilide nitroterephthalamide)s of increasing molecular weight, were synthesized and used as the macromolecular diffusant through gels of their crosslinked analogs. Intrinsic viscosity measurements in concentrated H_2SO_4 yielded the relationship

$$[\eta] = 9.0 \times 10^{-5} M_n^{1.07} .$$

Crosspolarized light microscopy indicated that concentrated solutions of the linear polyamides in sulfuric acid or in N,N-dimethylacetamide (DMAc) were lyotropic in nature. Both observations indicate the polyamides to be rigid or semi-rigid.

The crosslinked network was synthesized such that the average axial ratio (L/d) of the stems between branch points was L/d = 12. The linear analogs had L/d = 10, 19, 31, 98 and 137. They were made to diffuse across the swollen network gel, and their respective diffusion constants, D, were determined from colorimetric measurements. The receiving solvent, solution solvent of the linear polyamide, and swelling solvent in the gel, were all pure DMAc. The relationship of diffusion constants and L/d showed two branches.

The long aromatic polyamide macromolecules showed a relationship of
D∝1/L, reflecting a one-dimensional mobility through the crosslinked
gel. The short macromolecules strongly deviated from this behavior,
indicating a three-dimensional mobility, parallel and perpendicular
to the axes of the rodlike molecules.

INTRODUCTION

In the last decade there arose a great interest in the way in
which macromolecules move from point a to point b in space. This
translational motion is reflected by the change in the spatial position of the center of gravity of the mobile polymer chain. For the
process describing the movement, or diffusion, of such a chain,
deGennes[1] coined the term reptation. As conceived originally[1] the
model dealt with the reptation of a single linear flexible polymer
chain through a three dimensional network such as a polymeric gel.
Subsequent work expanded the treatment to describe the translational
motion of polymeric chains by reptation in the melt and concentrated
solution[2,3]. The reptation theory, and the scant experimental work
aimed at proving it[4], were so far limited to flexible chains reptat-
ing through a medium comprised of crosslinked or entangled flexible
chains in the presence or absence of solvent. Less attention was
paid to the problem of rigid or rodlike molecules. Starting with a
treatment of the Brownian motion, Doi[5] and then Doi and Edwards[6]
developed a theory describing the translational diffusion of a
given rodlike molecule through concentrated solutions of such molecules. The treatment did not consider the case in which a rodlike
macromolecule diffuses through a crosslinked gel. A very recent
treatment by deGennes[7] finally addressed the problem of motions of
a stiff molecule in a polymer network, the latter being either a
gel or a melt. In the following, we would like to describe the
preparation of, and some diffusion results obtained from a system
in which the polymer network part comprises rodlike sections
between the branch points in the crosslinked gel.

The polymer we chose to use both as the diffusant and as the rigid segments between crosslinks in the network gel, was poly(p-benzanilide nitroterephthalamide)(PBNT)

prepared by the Yamazaki et al[8] procedure from 4,4'-diaminobenzanilide and nitroterephthalic acid. Past experience[9] with this particular polymer has taught us that the substitution made the polymer very soluble in the reaction mixture yielding substantially higher molecular weights than the less soluble unsubstituted counterpart, and very much higher molecular weights than the poorly soluble poly(p-phenylene-terephthalamide). Unlike the unsubstituted polyamides, the nitrated one is highly soluble in pure N,N-dimethylacetamide (DMAc). The solubility of PBNT in DMAc made it especially attractive for the purpose of the present work since it obviates the need to add LiCl to this solvent or to use a strong acid as a solvent, which may lead to complications due to possible polyelectrolyte effects. A polyelectrolyte behavior imposed on poly(m-phenyleneisophthalamide) due to the presence of LiCl in DMAc was recently demonstrated[10].

EXPERIMENTAL

Solvents, reagents and monomers, of the highest grade available were used without further purification. Rigid aromatic polyamides, and their crosslinked analog, were prepared by means of the Yamazaki et al[8] procedure. In all preparations the aromatic diamine was 4,4'-diaminobenzanilide (DABA) and the aromatic diacid was nitroterephthalic acid (NTPA). For the low molecular weight linear polyamides, amounts of the monofunctional m-nitrobenzoic acid (NBA) were added, calculated to limit the molecular weight, M, of the products to certain desired averages. The degree of polymerization

of the higher-M linear polyamides was controlled by the duration
the polymerization reaction was allowed to continue. To prepare
crosslinked networks, a predetermined amount of the trifunctional
1,3,5-benzenetricarboxylic acid (BTCA) was added to the reaction
mixture, while keeping the total amount of acid and amine groups
exactly equal. The quantities of materials introduced into the
reaction vessel were such that the amount of polymer in the final
reaction mixture, or the polymer network in the resulting gel, was
about 10%.

In a typical procedure for the preparation of linear polymer,
there were placed in a 500 ml round bottom flask 100 ml dimethylace-
tamide (DMAc), 5.0 gr LiCl and 50 ml pyridine. The mixture was
stirred by means of a magnetic stirrer and heated to about 100°C.
At this point there were introduced the monomers in Table I and the
hot mixture stirred until a slightly turbid solution was obtained.
Then, 30 ml triphenylphosphite were added. Within a few minutes
the turbid mixture became transparent and after about 20 minutes
its viscosity was noticeably increased, indicating the polymerization reaction to be well in progress. With increased viscosity the
reaction mixtures became first translucent and later opaque. After
the mixtures became opaque, the reaction was allowed to continue
for 2 hrs. before termination. The reaction product was poured
into a large excess of methanol, comminuted in a blender and washed
about 6 times in methanol. Tests of the wash solvents indicated
that a very small amount of residual monomers and a trace of dimers
were washed out of the polymeric product that was then dried under
vacuum at 130°C. Mass balance indicated the yield of the linear
polyamides to be very close to 100% of theory.

All crosslinked samples employed in the study were of the same
composition. In a typical preparation there were introduced into
the reaction mixture described about, 11.35 gr DABA, 7.91 gr NTPA
and 1.88 gr 1,3,5-benzenetricarboxylic acid, BTCA, such that the

total amount of amine and acid groups was equal. Once the reaction
mixture in the round bottom flask became transparent and started
increasing its viscosity, it was poured into a concentric mold
built from two glass vessels. Special care was taken to insure
that the distance between the inner and outer glass walls was the
same all around, producing a cup-shaped mold cavity with uniform
wall thickness. The mold was kept at $110\pm5^{\circ}$C throughout. The
reaction mixture was constantly stirred by a magnetic stirring bar,
until it was immobilized by the increasing viscosity of the mixture. From that point, the reaction was allowed to continue an
additional two to three hours before termination. The cooled assembly was carefully dismantled, often sacrificing either or both
glass vessels, to produce a heavy-bottomed gel cup of sufficient
mechanical strength to support its own weight and shape. The magnetic stirring bar remained in the lowest part of the cup, a section
that was later trimmed-off with a sharp scalpel. The trimmed cup
was placed in a large amount of pure DMAc, which was changed daily
for a period not shorter than two weeks or longer than one month.
The frequent replacement of DMAc gradually leached out all the
solvent mixture, unreacted monomers, and polymerized material not
covalently bounded or permanently entangled into the network. The
complete removal of these ingredients was determined by a colorimetric comparison of pure DMAc and the leaching solvent; the polymers imparting to it an intense yellow color, and the light-colored
monomers being leached out faster than the polymers. It should be
noted that all ingredients in the reaction mixture and all linear
polyamides prepared in this work, are highly soluble in pure DMAc
at room temperature. A sol-gel fractionation performed on the
crosslinked polymer indicated that over 95% of the monomers introduced into the reaction vessel ended up in the gel phase. The
fractionation was performed as follows: tests verified the insolu-
bility of DABA and BTCA in methanol and the sparse solubility of

NTPA in it. Samples of swollen gel, straight from the mold, were placed in large amounts of methanol, which was frequently replaced. The methanol with the leached DMAc, pyridine, triphenylphosphite and LiCl was water-clear. The samples collapsed and reached a new constant volume. At that point the samples were dried in a vacuum oven at 130°C to a constant weight. The dried samples were then swelled and repeatedly leached in DMAc. The DMAc-solubles imparted to the solvent a yellow color. The solvent was replaced until it became water-clear. At that point the samples were redried in the oven at 130°C. The sol-gel ratios were calculated from the difference in weight between the initial and final dry samples.

Parallel polymerization runs, in which various amounts of BTCA were first allowed to react completely with DABA (extent of reaction was determined by end group analysis by means of IR spectroscopy) followed by the addition of NTPA which was allowed to react for about two additional hours, similar to the procedure above, produced polymers that were within experimental error the same as when all monomers were added at once. This indicated that BTCA and NTPA interact with DABA at about the same rate. This similar reactivity results in a random placement of branches in the crosslinked network. The amount of BTCA used in the network preparation was calculated to produce one branch point per twelve aromatic rings. The expected random placement of the branch points along the chains, and the large gel fraction led us to conclude that the average L/d ratio between branch points in the network polymer is about 12.

The diffusion of the linear polyamide samples across the gelcup walls was effected in a setup described in Fig. 1, and was measured colorimetrically. Quantities of 10 ml each of the fluid in the receptor compartment in Fig. 1 were siphoned (and returned after measurement) at intervals and the % transmittance of light through these samples was measured in a Bausch & Lomb Spectronic 20

spectrophotometer set at 450nm wavelength. The % transmittance of
the pure DMAc was 100%, and once the intensely yellow linear polymer
diffused clear across the gel cup wall, the % transmittance of
light decreased. For each polymer sample in Table I there was
prepared a light transmittance calibration curve covering the
concentration interval of 2.00% to 0.06125% in DMAc. From the
volumes of the solution in the gel cup (100-150 ml) and in the
receptor compartment surrounding it (about four times the volume of
the solution in the cup), and from a comparison of % transmittance
of the solution in the receptor compartment with the calibration
curves, one could calculate the amount of polymer actually diffused
across the gel cup walls. At the end of the run, the solution in
the gel cup was siphoned out, the polymer precipitated in methanol,
retrieved and dried, and its intrinsic viscosity $[\eta]$ measured in
concentrated sulfuric acid. The intrinsic viscosity of the corresponding polymer in Table I was measured as well. At the end of
each diffusion run the gel cup was broken and its wall thickness
was carefully measured in order to determine the diffusion coefficient D.

Intrinsic viscosities were measured in concentrated H_2SO_4 at
$25^{\circ}C$ in Cannon-Ubbelohde internal dilution glass viscometers.
Infrared scans were obtained from KBr pellets of the pulverized
samples with a Perkin Elmer Model 283B spectrometer. Concentrated
solutions of the linear polyamides in DMAc were studied by means of
a Reichert cross-polarized light microscope operating at 100 x
magnification, in order to observe their mesomorphic behavior.

RESULTS AND DISCUSSION

In order to insure that the diffusant linear polyamides, and
the crosslinked network will be as compatible as possible, both
were synthesized from the same monomers, except for the end groups
in low molecular weight linear polyamides and the trifunctional

branch points in the crosslinked analog. By so doing, a possible
rejection of the diffusant polymers by the network polyamide was
eliminated or, at least, minimized.

Crosspolarized light microscopy performed on solutions of
linear PBNT in DMAc and in concentrated sulfuric acid, showed that
at concentrations in the order of 20-30% lyotropic liquid crystallinity at room temperature was manifest. This indicates that the
linear PBNT is endowed with backbone rigidity sufficient for the
chain to be labeled as rigid or, at least, semi-rigid.

The composition of the polymers prepared in the study, and
their intrinsic viscosity values, are recorded in Table I. It is
well known[11] that the axial ratio of length to diameter (L/d) of
each aromatic repeat unit was calculated to approximately be
L/d≈1.0. From this one may calculate the number of average axial
ratios of polymers I, II and III as L/d = 10, 19 and 31, respectively. This corresponds to number average molecular weights of
M_n = 1340, 2550 and 4150. When a log-log plot of $[\eta]$ against M_n
was prepared, the three points for polymers I, II and III fell on a
straight line

$$[\eta] = 9.0 \times 10^{-5} \times M_n^{1.07} \tag{1}$$

with $[\eta]$ being defined in units dl/g. The fact that no scatter was
observed among the three points indicated to us that eq (1) is
reasonably accurate and led us to use it in order to calculate M_n
for polymers IV and V. Accordingly, the calculated molecular
weight for polymer IV is $M_n \approx 13200$ and for polymer V, $M_n \approx 18300$.

A value of a=1.07 in the Mark-Houwink eq (1) above is usually
taken to indicate a considerable level of polymer backbone rigidity.
This value, however, should be regarded cautiously. It is far
lower than the values 1.76 obtained for poly(p-benzamide)[12] or 1.36
obtained for poly(p-phenyleneterephthalamide)[13], both of relatively
low molecular weight and both in concentrated H_2SO_4 (96.5±0.1%)

TABLE I

COMPOSITION OF POLYAMIDES PREPARED IN THIS STUDY

Polymer Code	DABA	NTPA in grams	NBA	BTCA	DABA	NTPA in Moles	NBA	BTCA	Intrinsic Viscosity in conc. H_2SO_4
I	11.35	7.04	5.57	0	0.050	0.0333	0.0333	0	0.20 dl/g
II	11.35	8.79	2.78	0	0.050	0.0416	0.0166	0	0.40 dl/g
III	11.35	9.50	1.67	0	0.050	0.0450	0.0100	0	0.67 dl/g
IV	11.35	10.50	0	0	0.050	0.0502	0	0	2.30 dl/g
V	11.35	10.55	0	0	0.050	0.0500	0	0	3.28 dl/g
GEL	11.35	7.91	0	1.88	0.050	0.0375	0	0.009	Crosslinked

similar to the sulfuric acid used in our intrinsic viscosity measurements. When compared with these results, our value of a=1.07 may
indicate the backbone of PBNT to be significantly less rigid than
either of its un-nitrated analogs. Poly(n-alkylisocyanates), however, yield exponent values in their Mark-Houwink equation also in
the range of a=1.05±0.07[14] even though their backbone chains are
known[14-18] to be highly rigid and extended. The smaller than expected a values in the case of poly(n-alkylisocyanates) are most
likely due to the strong interaction between the aliphatic part of
their repeat units and the, usually, non-polar or moderately polar
solvent. Therefore, it may be that the value of a=1.07 is due in
part to lowered chain rigidity and in part to strong interactions between the nitro groups and the solvent. Either of these variables
may be responsible for the rather high solubility that PBNT have
shown in both H_2SO_4 and DMAc. The relative weight of the two variables is not known as of now, but work on swelling and de-swelling
of crosslinked PBNT gels[19] supports a backbone not as rigid as
poly(p-benzamide) or poly(p-phenyleneterephthalamide).

The diffusion runs were conducted in setups as shown in Fig. 1.
At the end of each run, the solution in the receptor compartment
was colorimetrically compared with the corresponding calibration
curve. It was thus determined that in all runs between 30 and 45%
of the linear polymer had diffused through the walls of the gel
cup. Intrinsic viscosities were obtained from each starting polymer and from its analog retrieved from the gel cup at the end of
the diffusion run. The intrinsic viscosities for each such pair
were equal with an experimental error of ±0.01 dl/g. This indicates
that no segregation according to molecular weight took place during
the diffusion experiment. Had the reverse been true and the low
molecular weight fraction of each sample preferentially diffused
through the gel cup walls, the molecular weight and intrinsic viscosity of the polymer remaining in solution in the gel cup would
have been higher than those of the parent polymer.

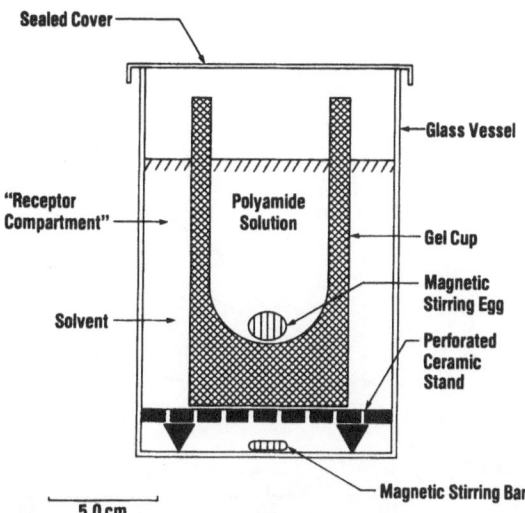

Figure 1.   A cross-section of the diffusion set-up including the
gel cup, through the walls of which the polyamide diffused from the solution to the solvent in the "receptor compartment".   Thickness of the gel cup walls
varied, for different cups, from 3.5 mm to 7.0 mm.

The results of the diffusion measurements are plotted in
Fig. 2.   In it the % transmittance of 450 nm wavelength light
through the solution retreived from the receptor compartment, was
plotted as a function of time from the start of each run.   In all
cases, after several hours in which the transmittance remained at
100%, the yellow polymer finally diffused through the gel cup walls
and the % transmittance of the solution in the receptor compartment
started dropping off.   The intercept of the % transmittance curve
of each linear polymer with the time axis, defines the point in
time at which the diffusing polymer first made it across the gel
cup wall, and is usually called the time-lag, t.   Knowledge of the
time lag and the thickness d of the gel cup walls allows us to
calculate the diffusion constant D, assuming it to be independent
of polymer concentration, C, in the gel cup[20]:

$$D = D^2/6t \qquad\qquad (2)$$

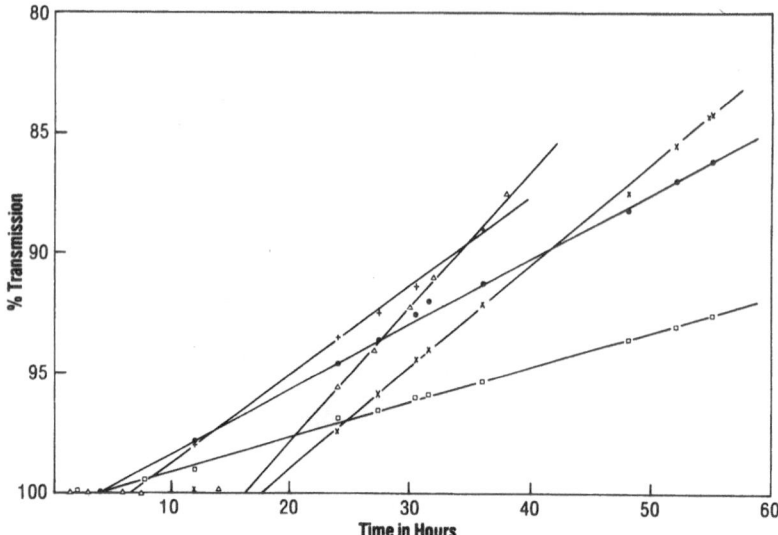

Figure 2.   Diffusion of linear polyamide as a function of time.
Plotted in terms of % transmission of light of 450nm
wavelength against time in hours. Polymer I: ●, Polymer
II: +, Polymer III: ☐, Polymer IV: Δ, Polymer V: X.

The linearity of the diffusion curves in Fig. 2 indicates that the
diffusion is indeed independent of C and the use of eq (2) is justified. The results are tabulated in Table II.

When the diffusion constant D is plotted against the number
average molecular weight, M_n, or L/d of the diffusing polyamides
then a curve is obtained as in Fig. 3, indicating a rather abrupt
break in the dependence of D on L/d centered at L/d of around 25.
This value is about twice the average L/d∽12 calculated for the
stems of the polymer network in the gel cups, between branch
points.

The translational motion of a rodlike macromolecule in a
dilute solution or through a crosslinked gel in which the average
mesh size is far larger than the rod length L, is governed by an
overall translational mobility μ. If the angle θ between the rod
and the direction of flow is randomly distributed, then the overall
mobility contains contributions from the mobility μ_{11} parallel to

TABLE II

DIFFUSION OF LINEAR AROMATIC POLYAMIDE THROUGH GEL-CUP WALLS

Polymer	$[\eta]$ dl/g	M_n	L/d	Cup #	d in cm	t in seconds	D in cm^2/sec
I	0.20	1340	10	1	0.65	14400	4.89×10^{-6}
II	0.40	2550	19	2	0.52	24300	1.85×10^{-6}
III	0.67	4150	31	3	0.35	14400	1.42×10^{-6}
IV	2.30	13200	98	4	0.70	58800	1.39×10^{-6}
V	3.28	18300	137	5	0.70	63700	1.28×10^{-6}

the rod axis and mobility μ_+ transverse to the rod[7]:

$$\mu = <\cos^2\theta>\mu_{11} + <\sin^2\theta>\mu_+ \tag{3}$$

Here μ_{11} and μ_+ are related as[6,7,21,22]:

$$\mu_+ = \tfrac{1}{2}\mu_{11} \tag{4}$$

When the rodlike macromolecule is immersed in a concentrated solution, in the melt or in a crosslinked gel in which the average mesh size is of the order of L or smaller, then the perpendicular mobility μ_+ is expected[7] to be enormously smaller than μ_{11} and approach zero. Accordingly, the overall mobility relates to the parallel mobility as[7]

$$\mu = <\cos^2\theta>\mu_{11} + <\sin^2\theta>\mu_+ \simeq <\cos^2\theta>\mu_{11} = 1/3\ \mu_{11} \tag{5}$$

Thus, in a system in which short rods diffuse through a crosslinked gel whose average mesh size is large relative to the rod length, a three dimensional overall mobility containing contributions from both the parallel and transverse mobilities (eq. 3), is expected. When the rod length increases such that the transverse mobility is suppressed, then essentially only a one dimensional parallel mobility is expected to contribute to the overall mobility (eq. 5). If anisotropy is imposed on the angular distribution of the diffusing rods, then the angular dependence will change the numerical values in eq. 5, but not its overall form.

The dilute solution situation was considered by Kirkwood and co-workers[21,22] while the concentrated solution, melt or crosslinked gel conditions were considered by Doi and Edwards[5,6] and deGennes[7]. According to both, in the concentrated systems the translational mobility is envisioned as a one dimensional Brownian motion of a rodlike molecule in its own "tube". This motion is described by a tube diffusion constant[7]

$$D_t = kT\mu_{11} \simeq kT/\eta_o L \tag{6}$$

in which kT retain their common meaning, η_o is the local viscosity,

and any other numerical factors have been omitted[7]. The one dimensional diffusion constant is, then, inversely dependent on the
length of the diffusing rod. Combining eqs. (5) and (6) one finds,
as expected, that in the concentrated solution, melt or crosslinked
gel with relatively small mesh size, the overall mobility μ and its
associated experimentally measured D, are inversely proportional to
L[7,23,24]:

$$D \stackrel{\sim}{=} D_t \stackrel{\sim}{=} D_{11} \propto 1/L \qquad (7)$$

and

$$D \propto \mu \stackrel{\sim}{=} \mu_{11} \propto 1/L \qquad (8)$$

and reflect a one dimensional mobility. In the dilute regime, or
when the rods are short relative to the mesh size of the gel, a
three dimensional translational mobility is reflected in the measured D. This mobility includes a substantial contribution from
the perpendicular μ_+ mobility.

Thus, we hypothesize that the break in the curve in Fig. 3
occurs at the point where the contribution of the perpendicular
diffusion to the overall diffusion vanishes: The measured diffusion coefficients of polymers I and II contained a substantial
contribution from the perpendicular diffusion, while no such contribution was observed in the diffusion coefficients of polymers
III, IV and V.

It should be pointed out that the break in the curve occurs at
a diffusant average axial ratio which is twice the average axial
ratio (mesh size) of the crosslinked network. Visualizing the gel
as a three dimensional sieve with average mesh size of L/d=12, it
becomes apparent that once the average diffusant axial ratio sur-
passes L/d=24, each of its molecules is constrained by at least two
"holes" in said sieve and its orientational freedom and perpendicular mobility are drastically reduced. Therefore, one may estimate that the point where the perpendicular diffusion ceases to

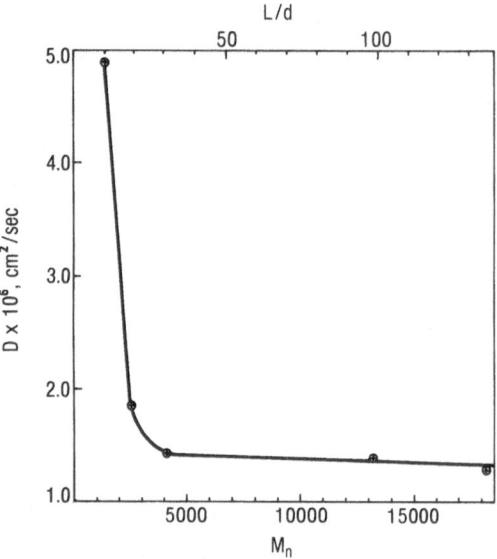

Figure 3.   Diffusion constant D plotted against M_n and L/d of the
linear polyamide diffusants.

contribute to the overall diffusion is when the axial ratio of the
thin rod is twice or more the average mesh size (measured in the
same L/d units) of the crosslinked polymer network.

The D values of polymers III, IV and V, show a straight line
with a negative slope when plotted in Fig. 3, against L/d.   When
replotted (not reproduced here) against d/L ($\propto 1/L$), the corresponding points fell, with a minimal scatter, on a straight line with a
positive slope.   It is obvious that both slopes, D vs. L/d and D
vs. d/L, indicate a linear dependence of D on d/L ($D \propto 1/L$).

The increased overall diffusion coefficient of polymer II and
especially polymer I, reflect the increasing contribution of the
transverse mobility which becomes very substantial (eq. 4) once the
average diffusant rodlength becomes shorter than the average mesh
size (e.g., polymer I).

It is thus concluded that the measured diffusion coefficient D
shows a dependence of $D \propto 1/L$ for polymers III, IV and V, where the

diffusion is in essence a one dimensional diffusion of the rodlike diffusant molecules in "tubes" constructed of the stems in the three dimensional crosslinked gel network. This is in agreement with deGennes[7] and Edwards[24]. In the case of polymer II and the shorter polymer I, the diffusion becomes three dimensional in nature, with contributions from both parallel and perpendicular translational mobility.

REFERENCES

1. P. G. deGennes, J. Chem. Phys. 55, 572 (1971).

2. M. Doi and S. F. Edwards, J. Chem. Soc., Faraday Trans. II 74, 1789 (1978).

3. J. Klein, Macromolecules 11, 852 (1978).

4. J. Klein, Nature 271, 143 (1978).

5. M. Doi, J. Physique 36, 607 (1975).

6. M. Doi and S. F. Edwards, J. Chem. Soc. Faraday Trans. II 74, 560 (1978).

7. P. G. deGennes, J. Physique 42, 473 (1981).

8. N. Yamazaki, M. Matsumoto and F. Higashi, J. Polymer Sci., Polym. Chem. Ed. 13, 1373 (1975).

9. S. M. Aharoni, unpublished observations.

10. D. D. Harwood and J. F. Fellers, Macromolecules 12, 693 (1979).

11. S. M. Aharoni, J. Polymer Sci. Polym. Phys. Ed. 19, 281 (1981).

12. S. Ali and M. M. B. El-Sabbah, Acta Polymerica 31, 638 (1980).

13. D. G. Baird and J. K. Smith, J. Polymer Sci. Polym. Chem. Ed. 16, 61 (1978).

14. M. N. Berger and B. M. Tidswell, J. Polymer Sci. Polym. Symp. 42, 1063 (1973).

15. M. N. Berger, J. Macromol. Sci., Rev. Macromol. Chem. 9, 269 (1973).

16. L. J. Fetters and H. Yu, Macromolecules 5, 385 (1971).

17.   A. J. Bur and L. J. Fetters, Chem. Rev. $\underline{76}$, 727 (1976).

18.   S. M. Aharoni, Macromolecules $\underline{12}$, 94 (1979).

19.   S. M. Aharoni and D. H. Wertz, in preparation.

20.   R. M. Barrer, Trans. Faraday Soc. $\underline{35}$, 628 (1939).

21.   J. G. Kirkwood and P. L. Auer, J. Chem. Phys. $\underline{19}$, 281 (1951).

22.   J. Riseman and J. G. Kirkwood, J. Chem. Phys. $\underline{18}$, 512 (1950).

23.   P. G. deGennes, Plenary Lecture, IUPAC 27th Symposium on
         Macromolecules, Strasbourg, France, July 5, 1981.

24.   S. F. Edwards, personal communication, July 31, 1981.

THE MOLECULAR WEIGHT DEPENDENCE OF THE SPLAY AND BEND ELASTIC CONSTANTS OF A POLYMER LIQUID CRYSTAL

Jacob R. Fernandes and Donald B. DuPré

Department of Chemistry
University of Louisville
Louisville, Kentucky 40292

ABSTRACT

The molecular weight dependence of the splay and bend elastic constants of a liquid crystal has been measured utilizing the Frederick's distortion of homeotropically aligned specimens in a slowly varying magnetic field. The splay modulus (K_{11}) is found to be much greater than the bend elastic constant (K_{33}) for three polymer molecular weights examined. Large K_{11} values for polymer liquid crystals are in accord with recent theoretical predictions. The molecular weight dependence of K_{11}/K_{33} observed in the polymer/solvent system of this study is, however, contrary to theory. The discrepancy may be due to substantial macromolecular associations which make the molecular length a difficult parameter to define.

INTRODUCTION

Research on the physical properties of lyotropic liquid crystals has gained increasing importance in recent years because of the occurance of these ordered fluids in biological structures. Lyotropic phases have also been found to form important molecularly ordered precursors in the spinning of high strength fibers from

solutions of elongated, rigid chain polymers. While lyotropic
liquid crystals are similar to their thermotropic, small molecule
counterparts in a number of ways, marked influence on some of their
physical properties may be expected because of the much larger
molecular dimensions involved. Other external variables such as
solvent disposition and concentration should also have a significant influence on the physical properties of this class of liquid
crystals.

A physically measurable manifestation of the molecular forces
that hold the liquid crystalline phase together and determine its
response to external stimulii is the set of three elastic constants
K_{ii} (i=1,2,3), corresponding to distortions of the director field
in splay, twist, and bend deformations. A number of studies have
been carried out to measure these elastic constants in thermotropic
liquid crystals, but very little is known about them in lyotropic
systems. While the elastic modulii have been found to be generally of the same order of magnitude and to obey the inequality
$(K_{33}>K_{11}>K_{22})$ in small molecule thermotropic liquid crystals,[1] a
high degree of variability and significant departures from the
above inequality have been predicted for liquid crystals formed by
macromolecules.[2-5]

Poly-γ-benzyl-L-glutamate (PBLG) is a rigid polymer with a
large intramolecular persistence length which forms a liquid crystalline phase in a wide range of solvents and is currently the best
characterized of the polymeric, lyotropic liquid crystals. While
many of the properties of PBLG have been extensively studied, there
are very few studies of the elastic constants of liquid crystal
phases of this polymer in the literature. Those studies[6-14] reported are primarily restricted to discussions of the twist modulus
(K_{22}). In an earlier publication[15] we described our attempts to
measure the splay and bend elastic constants (K_{11} and K_{33}, respectively) of PBLG liquid crystals using the magnetic field induced
Frederick's transition on homeotropically aligned specimens.

Reported here are the results of further measurements by this procedure of the molecular weight dependence of K_{11} and K_{33}.

EXPERIMENTAL SECTION

The bend and splay elastic constants were measured by the Frederick's transition procedure following the method of Saupe.[16] Homeotropically aligned samples were placed between the poles of an electromagnet with the direction of orientation being perpendicular to the field direction. A He–Ne laser was directed on to the sample along the uniaxial direction and an analyser and polarizer placed on either side of the sample crossed at an angle of 45° with respect to the field direction. The field was increased from 0 to 10 kG continuously at the rate of $\sim$12 G per minute and the transmitted intensity was measured with a photodiode and displayed on an X-Y plotter, the X-axis of the plotter being driven by a signal proportional to the magnetic field. Such a slow rate of sweep assures the observation of an equilibrium mode of distortion in this viscous liquid crystal. The critical field for the onset of distortion of the director in this geometry is related to K_{33} through $H_{c,3} = (\pi/d)\sqrt{K_{33}/\Delta\chi}$ where d is the sample thickness and $\Delta\chi (>0)$ is the diamagnetic susceptibility anisotropy. The intensity of light transmitted above the critical field is given by $I = I_o \sin^2(\delta/2)$ where

$$\delta = \frac{2\pi n_o d}{\lambda}\left\{\frac{2H_{c,3}}{\pi H}\int_0^{\pi/2}\left[\frac{1+\kappa\sin^2\theta_m\sin^2x}{(1-\sin^2\theta_m\sin^2x)(1+\nu\sin^2\theta_m\sin^2x)}\right]^{\frac{1}{2}}dx - 1\right\} \quad (1)$$

and λ is the wavelength; $\nu = (n_o^2 - n_e^2)/n_e^2$; n_o and n_e are the ordinary and extraordinary refractive indices; $\kappa = (K_{11} - K_{33})/K_{33}$; and $H(>H_{c,3})$, the magnetic field strength. The angle θ_m is the maximum deformation angle (obtained at the center of the sample) and is given by

$$\frac{Hd}{2}\left(\frac{\Delta\chi}{K_{33}}\right)^{\frac{1}{2}} = \int_0^{\pi/2}\left[\frac{1+\kappa\sin^2\theta_m\sin^2x}{1-\sin^2\theta_m\sin^2x}\right]^{\frac{1}{2}}dx \quad (2)$$

Special problems encountered in applying this method to this polymeric liquid crystal with low values of the birefringence and small accessible sample thicknesses were discussed in our earlier publication[15] along with our procedure to circumvent the difficulties. At this point, it is appropriate to note another limitation imposed by the large inequalities in the elastic constants. When the splay constant, K_{11}, is very much larger than the bend constant, K_{33}, the deviation of the director of the molecules above the critical field from its undistorted position is extremely small. This causes large uncertainties in the detection of the critical field, $H_{c,3}$. However, the upper limit of the critical field is clearly discernable, and in the range of $K_{11} >> K_{33}$, the intensity versus field curve above $H_{c,3}$ is quite sensitive to changes in K_{11}. The splay elastic constant determined by a computer curve fitting procedure[15] to Eqns. (1) and (2) is accurate to within 10%.

Samples of PBLG with three different molecular weights (MW = 50,000; 150,000; and 296,000) were prepared in dioxane with 2% trifluroacetic acid (TFA) at concentrations just above the critical concentration, $\emptyset \overset{\sim}{=} \emptyset_c$, for uniform molecular alignment. The degree of orientational order for all solutions studied should therefore be approximately the same. Solutions were allowed to mature for at least two weeks to assure solubilization of the polymer. Homeotropically aligned samples were obtained by sandwiching the solutions between glass plates which had been coated with SiO at oblique incidence. Such a coating has been found to produce homogeneous alignment in thermotropic liquid crystals, but provided homeotropically aligned samples with this polymer with thicknesses (d) much greater than those obtained in our earlier work.[15,17] Correspondingly, the thickness of the specimens was defined by spacers at either 23 μ or 50 μ.

TABLE I

Splay and Bend Elastic Constants for PBLG Liquid

Crystals in Dioxane/2% TFA

MW	50,000	150,000	296,000
$\phi \overset{\sim}{=} \phi_c$	0.209	0.166	0.144
K_{11} (dyne)	$1.0 \pm 0.1 \times 10^{-5}$	$3.0 \pm 0.3 \times 10^{-6}$	$6.0 \pm 0.6 \times 10^{-7}$
K_{33} (dyne)	1.0×10^{-8}	7.0×10^{-9}	1.2×10^{-9} (Upper bounds)

RESULTS AND DISCUSSION

The results of our measurements are summarized in Table I. As pointed out above, the values of the bend elastic constant can only be taken as upper bounds due to uncertainties in the measurement of the critical field, $H_{c,3}$. K_{11} is significantly greater than K_{33} which is in accord with recent theory[4,5] on the elasticity of elongated macromolecular liquid crystals. The strong decrease in K_{11} with increasing molecular weight however is contrary to these predictions. This departure could be because of the high degree of association between polymer molecules that is known to occur in dioxane. Such associations make it difficult to define the molecular length of the actual responding unit in these field induced deformations. We are currently studying the effects of increased TFA concentrations in dioxane and the use of polar molecules as the major solvent. These conditions should reduce macromolecular associations and perhaps bring experiment and theory into better accord.

ACKNOWLEDGEMENTS

Portions of this work were supported by the National Science Foundation under Grant DMR-7903760 and the National Institutes of Health under Grant HL24364. We also wish to thank Dr. Robert W. Filas of Bell Telephone Laboratories for supplying the coated glass plates.

REFERENCES

1. W. H. deJeu, Physical Properties of Liquid Crystalline Materials, (Gordon and Breach, New York, 1980), chapter 6.
2. J. P. Straley, Phys. Rev. A8, 2181 (1973).
3. R. G. Priest, Phys. Rev. A7, 720 (1973).
4. P. G. deGennes, Mol. Cryst. Liq. Cryst. (Letters) 34, 177 (1977).

5. R. B. Meyer in Polymer Liquid Crystals, A. Ciferri, W. R.
        Krigbaum and R. B. Meyer, eds., Academic Press, New York,
        in press.

6. R. W. Duke and D. B. DuPré, J. Chem. Phys. 50, 2759 (1974).

7. R. W. Duke and D. B. DuPré, Macromolecules 7, 374 (1974).

8. D. B. DuPré and R. W. Duke, J. Chem. Phys. 63; 143 (1975).

9. R. W. Duke, D. B. DuPré, W. A. Hines and E. T. Samulski, J.
        Amer. Chem. Soc., 98, 3094 (1976).

10. D. B. DuPré, R. W. Duke and E. T. Samulski, Mol. Cryst. Liq.
        Cryst. 40, 247 (1977).

11. D. L. Patel and D. B. DuPré, J. Polym. Sci., Polym. Lett. Ed.,
        17, 299 (1979).

12. D. L. Patel and D. B. DuPré, Mol. Cryst. Liq. Cryst. 53, 323
        (1979).

13. D. L. Patel and D. B. DuPré, J. Polym. Sci., Polym. Phys. 18,
        1599 (1980).

14. C. Guha-Sridhar, W. A. Hines and E. T. Samulski, J. Chem.
        Phys. 61, 947 (1974).

15. J. R. Fernandes and D. B. DuPré, Mol. Cryst. Liq. Cryst.
        (Letters) 72, 67 (1981).

16. A. Saupe, Z. Naturforsch 15a, 815 (1960).

17. D. B. DuPré in Polymer Liquid Crystals, A. Ciferri, W. R.
        Krigbaum and R. B. Meyer, eds., Academic Press, New York,
        in press.

DSC, MISCIBILITY AND X-RAY STUDIES OF THE THERMOTROPIC LIQUID
CRYSTALLINE POLYESTERS WITH AROMATIC MOIETIES AND FLEXIBLE SPACERS
IN THE MAIN CHAIN

L. Bosio, B. Fayolle*, C. Friedrich, F. Laupretre,
P. Meurisse, C. Nöel, J. Virlet**

Laboratoire de Physicochimie Structurale et
Macromoléculaire, et Laboratoire de Physique des Liquides
et Electrochimie, ESPCI, 10 rue Vauquelin, 75231 Paris
Cedex 05, France

*Rhône-Poulenc, Centre de Recherches des Carrières
 69190 Saint-Fons, France

**Département de Physicochimie, CEN-Saclay, B.P. n°2
  91191 Gif Sur Yvette

INTRODUCTION

The discovery and development of a new polymer system is
nearly always a stimulating event in macromolecular science.
Besides academic interest, sufficient interest is aroused on
account of its potential utility in various fields. In 1975,
Roviello and Sirigu[1] prepared new polyalkanoates from p,p'-di-
hydroxy-α,α'-di-methylbenzalazine and appropriate acyl chlorides.
All the examined polymers melted to give fluid anisotropic phases
whose textures and properties appeared quite similar to those
observed with conventional thermotropic liquid crystals, hence
the denomination of "thermotropic liquid crystalline polymers".
In the past few years, a search for newer and newer liquid
crystalline polymers with mesogenic moieties and flexible spacers
in the main chain has begun on a much wider scale. So far, most

401

of the work has focused on the synthesis of new polymers.
Relatively little work has been devoted to the study of their
structures as liquid crystalline phases. Indeed, high viscosity,
broad molecular weight distribution, coexistence of polycrystalline
and amorphous material make the identification of mesophases much
more difficult for mesomorphic polymers than for the conventional
low molecular weight liquid crystals. A careful identification
of thermotropic polymeric mesophase usually requires a combination
of different techniques.

In order to understand the relationship between liquid
crystallinity and chemical structure the homologous series of
polyesters of the following structural formula:

$$\left\{ OCO-\bigcirc\!\!-\!\!\bigcirc\!\!-\!\!\bigcirc-COO\_R \right\}_n$$

R = $\left\{ CH_2 \right\}_n$, $\left\{ CH_2 - CH_2 - O \right\}_n CH_2 - CH_2 -$, substituted alkyl
segments were synthesized. The polymorphism of these compounds
was investigated and the type of the mesophases was established
through a combination of DSC measurements, microscopic observations,
miscibility studies and X-Ray investigation. The results provide
definite proof for the nature of the liquid crystalline states.

EXPERIMENTAL

Polyesters were prepared, as described elsewhere,[2] via the
scheme given in Fig. 1. The compound labelling is given in Table
I.

Inherent viscosity [$\eta_{inh} = \ln(\eta_{rel})/c$] was determined at 25°C
and for a polymer concentration (c) of 0.6 g/100 ml of solution.
The traditional solvent was dichloroacetic acid. Alternatively
70/30 tetrachloroethane/antimony trichloride mixture can be used.

High resolution ^{13}C NMR spectra were measured on $CDCl_3$

Table I: Properties of Aromatic-Aliphatic Polyesters

Polyester	R	Solvent	Inherent Viscosity dl.g^{-1}	T$_m$°C	S$_2$	S$_1$	N	I
T2	$-(CH_2)_2-$	Insoluble	–	.(286)	.322	.381	393	.
T4	$-(CH_2)_4-$	Insoluble	–	. 318	.340	.358	407	.
T5	$-(CH_2)_5-$	CHCl$_2$COOH	0.25	.(210)	.226	.366		.
T6	$-(CH_2)_6-$	Insoluble	–	. 235		.355		.
T10	$-(CH_2)_{10}-$	Tetrachloroethane/SbCl$_3$ 70:30	1.37	. 256		.311		.
T2Me	$-CH(CH_3)-CH_2-$	CHCl$_2$COOH	0.21	. 250	.278	.307	372	.
T2Me2	$-CH(CH_3)-CH(CH_3)-$	CHCl$_2$COOH	0.28	. 196	.234	.248	290	.
T3Me2	$-CH_2-C(CH_3)_2-CH_2-$	CHCl$_2$COOH	0.15	.(226)	.256	.283	348	.
T3E t2	$-CH_2-C(CH_2-CH_3)_2-CH_2-$	CHCl$_2$COOH	0.14	. 117		.187		.
T5Me	$-(CH_2)_2-CH(CH_3)-(CH_2)_2-$	CHCl$_2$COOH	0.33	. 135		.320	352	.
T5Tol	$-(CH_2)_2-CH(CH_2-C_6H_5)-(CH_2)_2-$	CHCl$_2$COOH	0.29	. 158		.190		.
T05	$-(CH_2-CH_2-O)CH_2-CH_2-$	CHCl$_2$COOH	0.5	. 213		.389		.
T08	$-(CH_2-CH_2-O)_2CH_2-CH_2-$	CHCl$_2$COOH	0.6	. 187		.337		.
T011	$-(CH_2-CH_2-O)_3CH_2-CH_2-$	CHCl$_2$COOH	0.6	. 125		.253		.
T029	$-(CH_2-CH_2-O)_9CH_2-CH_2-$	CHCl$_2$COOH	0.69	. 70		.117		.

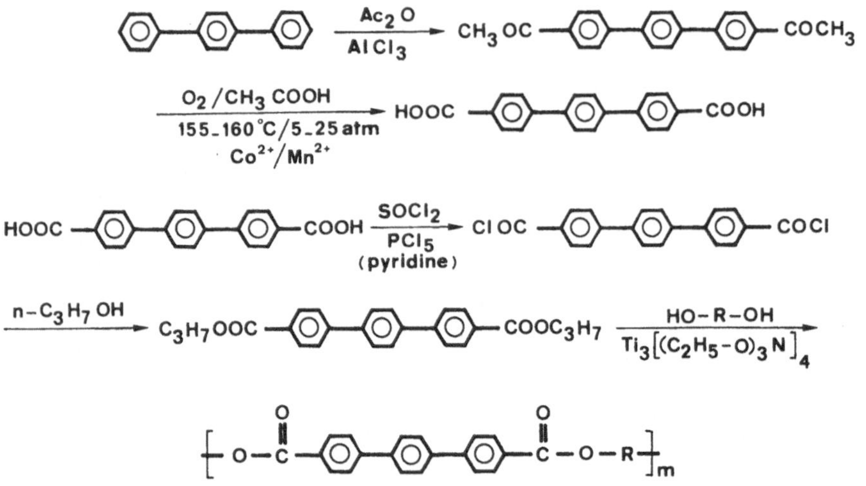

Fig. 1   Synthetic route for the preparation of polyesters with
various flexible spacers.

solutions of the polyesters at 62.9 MHz with a Brüker WP-250
spectrometer.

Cross Polarization, Magic Angle Spinning ^{13}C NMR spectra were
collected at 12.07 MHz on a home built spectrometer constructed
around a 12-in Varian electromagnet and employing ^{2}D field-
frequency stabilization, solid-state class-A transmitters and
double-tuned single coil probes.[3] Spectra were obtained at room
temperature using magic angle spinning in a Henriot-Huguenard
design[4] hollow rotor made out of Kel-F and spinning speed of 2 kHz.
The spectra were all obtained by cross-polarization from the spin-
locked protons[5] with the Hartmann-Hahn[6] matching condition met
followed by high-power proton decoupling. Amplitudes of the radio-
frequency fields expressed in frequency units, for both the carbons
and the protons, were 32 kHz. Spin-temperature inversion techniques were systematically employed to minimize base line noise
and roll.[7] Flip-back[8] allowed to shorten the delayed time between
two successive pulse sequences.

The DSC data were obtained with a Du Pont 990 differential thermal analyzer. All samples were under 10 mg and were heated at 10°C/min under a flow of dry nitrogen.

The transition characteristics were surveyed with a polarizing microscope (Olympus BHA-P) equipped with either a Mettler FP5 or a Reichert hotstage. All samples were viewed between crossed polarizers. In this way it was possible to determine all transition temperatures and to identify solid, smectic and nematic phases.

Binary mixtures were prepared by the contact procedure[9] and phase diagrams were constructed by polarized light microscopy of these mixtures using a hot stage. It should be noted that well-defined compositions were examined for the phase diagrams to insure a degree of accuracy for the composition coordinate. The reference compounds used were the terephthalidene-bis-(4-n-butylaniline) [K 113 S_H 144.5 S_C 172.5 S_A 199.6 N 236.5 I][10] and the diethyl-p-terphenyl-4,4" carboxylate [K_1 152 K_2 173 S_E 188.5 S_A 259 I].[11]

For the X-ray measurements, the samples were contained in 0.7 mm d. Lindemann glass tubes. Low-angle diffraction patterns were recorded on flat films with pinhole collimation or, for un-oriented samples, using a diffractometer with slit collimators. The sealed capillary was thus mounted in an electrically heated oven, the temperature of which was controlled with a precision of 0.2 K using a platin resistor as sensing element but absolute temperatures were only known to within about 1 K. The whole sample holder and the heater might be contained inside a tank filled with helium to reduce air scatter.

For experiments requiring wide-angle diffraction data, X-ray study was done at room temperature in a standard cylindrical camera.

In all cases, Ni-filtered CuK_α radiation was used.

RESULTS AND DISCUSSION

General Properties

The polyesters of these series showed surprisingly low
solubilities. Thus, the polyesters, T2, T4 and T6 prepared by
reaction of di-n-propyl-p-terphenyl-4,4" carboxylate and 1,2-
ethanediol, 1,4-butanediol and 1,6-hexanediol, respectively, were
insoluble in virtually all common organic solvents at room
temperature (Table I). For the polyester T10 solubility was
obtained only with a tetrachloroethane/antimony trichloride
mixture. The other polyesters were soluble in dichloroacetic acid,
although decomposition occurred. The polyesters TO11, TO29 and
T3Et2 were soluble in chloroform.

The inherent viscosities of the polyesters are shown in Table
I. The molecular weights of the polyesters prepared with branched
aliphatic diols were relatively low, as reflected by their solution
viscosity numbers. On the other hand, the value of η_{inh} =
1.37 dl.g^{-1} which has been determined for the polyester T10 strongly
supports high molecular weight for this sample. No other molecular
weight data were obtained owing to the limited solubility of the
polymers.

Conventional high resolution NMR cannot be utilized as a
characterization technique for these polymers which are essentially
insoluble unless subjected to extensive degradation so that the
structure of polyesters was verified by solid state ^{13}C NMR.
Representative MAS CP ^{13}C NMR spectra are shown on Figs. 2-4.
Solid state ^{13}C NMR spectrum of polyester TO11 consists of three
lines which were identified from left to right, in order of increasing magnetic field, as a line due to the carbonyl carbons, a broad
poorly-resolved line due to the aromatic carbons and a line due to
the methylene carbons. The chemical shifts compare nicely with
those deduced from the liquid state ^{13}C NMR spectrum and are in
good agreement with data reported for poly(tetramethyleneoxy)

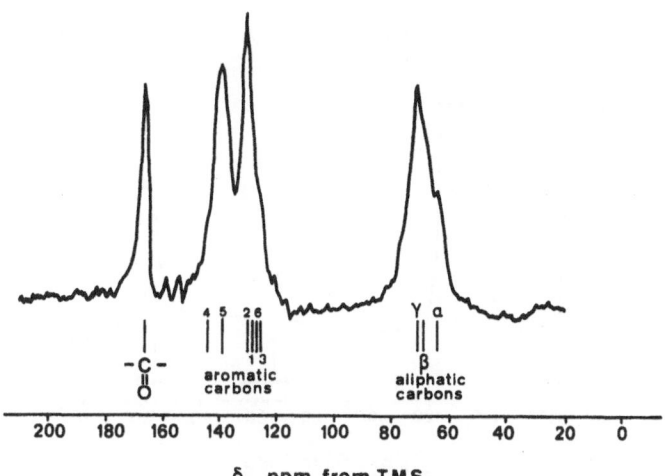

Fig. 2   MAS CP ^{13}C NMR spectrum of polyester T011 at 25°C.
Spectrum obtained from 10,000 accumulations with a CP
contact time of 1 ms and experiment repetition time of
0.5 s.

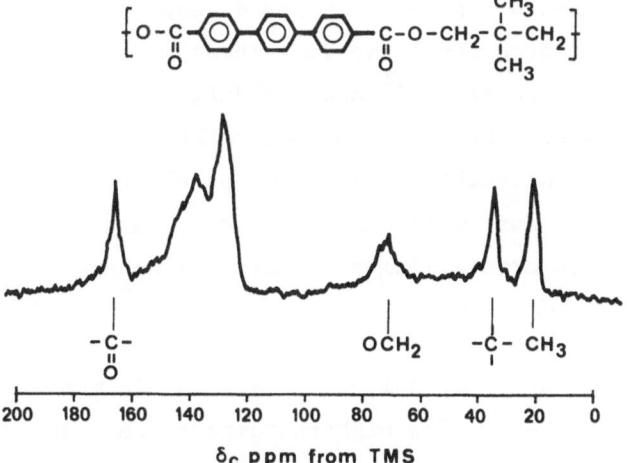

Fig. 3   MAS CP ^{13}C NMR spectrum of polyester T3Me2 at 25°C.
Spectrum obtained from 10,000 accumulations with a CP
contact time of 1 ms and experiment repetition time of
0.5 s.

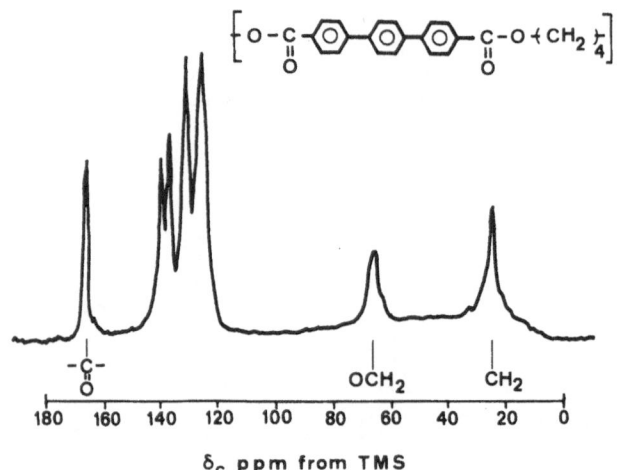

δ_c ppm from TMS

Fig. 4   MAS CP ^{13}C NMR spectrum of polyester T4 at 25°C.
          Spectrum obtained from 10,000 accumulations with a CP
          contact time of 1 ms and experiment repetition time of
          0.5 s.

terephthalate.[12] In the case of polyester T3Me2 a broad pattern
is again observed for the aromatic nuclei; the carbonyl, methylene,
quaternary and methyl lines are readily assigned. The spectrum of
polyester T4 presents a greatly improved resolution. The assignment of the non-protonated aromatic carbons is supported by the
delayed decoupled spectra[13,14] and the differences in cross
polarization times. The two methylene peaks are clearly separated
as expected from the deshielding effect of the oxygen atom on the
alpha carbon. The narrow resonances observed are likely to
reflect the higher crystallinity of this polyester.

Thermal Analysis

    All the DSC curves of polyesters show a minimum of two
endotherms (Figs. 5-9). The highest temperature transition in all
cases is broad and corresponds to the clearing point. At lower
temperatures, several of the polyesters exhibit distinct
endotherms. Such a behavior may indicate i/ true solid phase

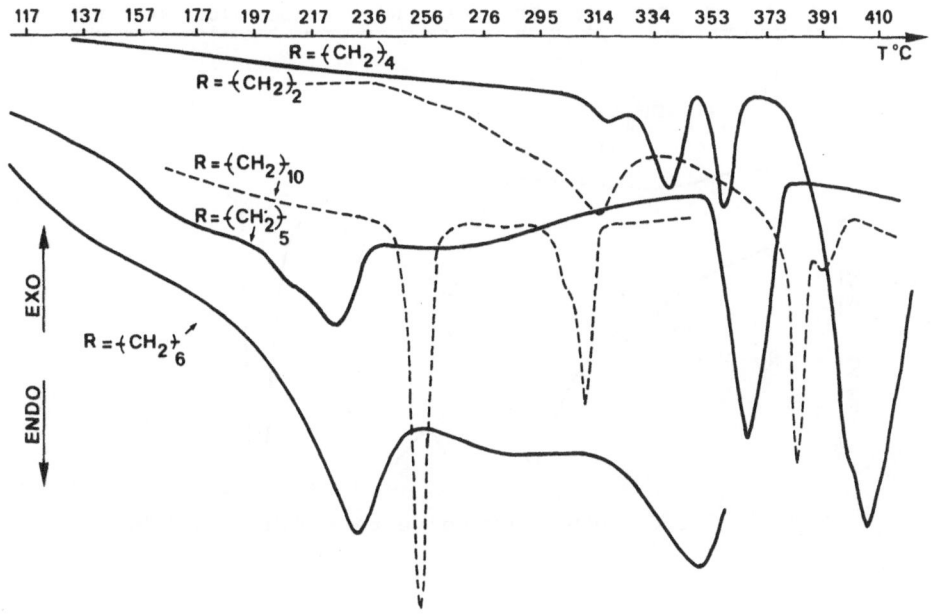

Fig. 5   DSC curves for polyesters T2, T4, T5, T6 and T10

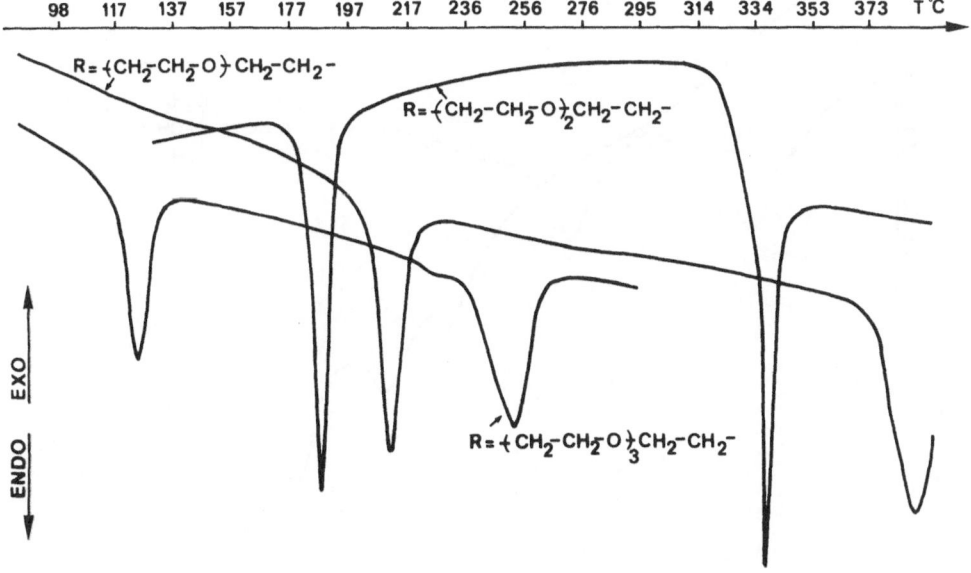

Fig. 6   DSC curves for polyesters T05, T08 and T011

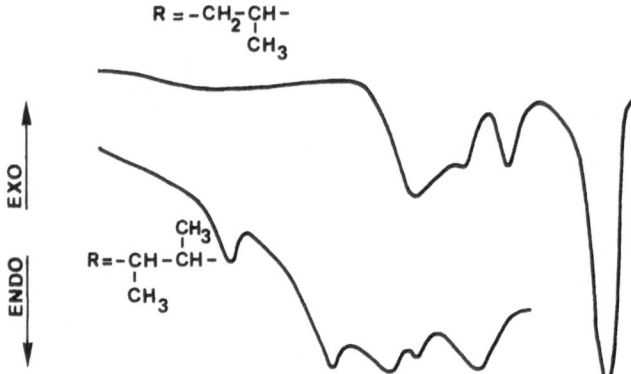

Fig. 7   DSC curves for polyesters T2Me and T2Me2

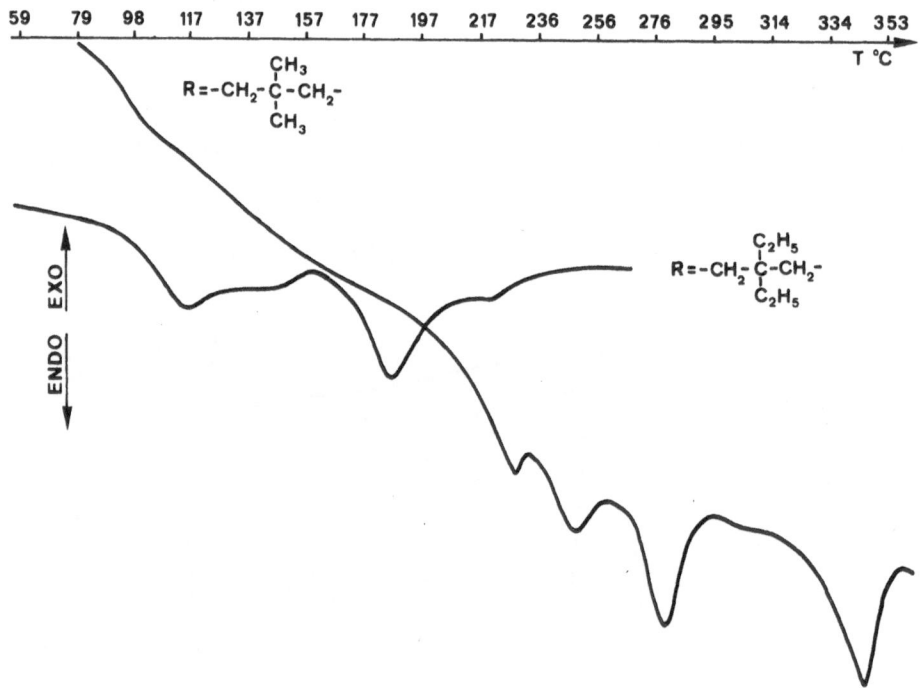

Fig. 8   DSC curves for polyesters T3Me2 and T3Et2

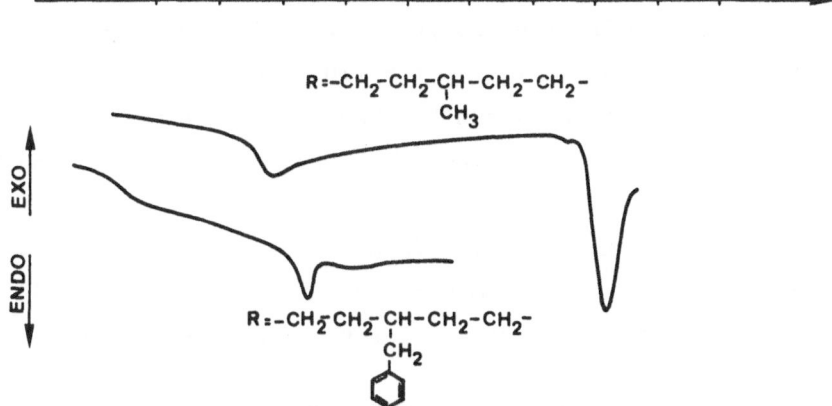

Fig. 9   DSC curves for polyesters T5Me and T5Tol

transitions ii/ the melting of different crystalline phases that are present together in the crystallized polymer iii/ solid-mesophase and mesophase-mesophase transitions. In aromatic polyesters and liquid crystal polymers with mesogenic moieties and flexible spacers in the main chain, multiple melting endotherms have been found to arise after various stages of annealing and to be dependent upon heating rate.[15-18] The melting phenomena for p-terphenyl derivatives are made complex by the occurrence of solid-solid transitions[19,20] but, several dialkyl-p-terphenyl-4,4" carboxylates, which can be considered as model compounds for polyesters under investigation, are also reported to exhibit smectic polymorphism involving S_E, S_C and S_A phases.[21,11] Having drawn attention to the problems involved in determining the type of transition which a given endotherm reflects, the crystal-mesophase transition temperature was taken as the temperature at which evidence of fluidity was found by hot stage microscopy. Other evidence was drawn from the cooling scans. Indeed, as noted by Griffin et al.[16] and Roviello and Sirigu,[1] on cooling from the isotropic state, the isotropic-mesophase and mesophase-mesophase

transitions are almost reversible while a rather marked supercooling is observed for the mesophase-crystal transition. This
effect is illustrated in Fig. 10 for T10. Upon heating two widely
separated endothermic peaks can be observed at 256°C and 311°C.
However, on the corresponding cooling curve two exothermic peaks
at 215°C and 304°C are obtained. Such a behavior is common to
many types of small molecule liquid crystals.[22] Consequently the
crystal-mesophase transition may be identified from DSC study
assuming that supercooling occurs for the mesophase-crystal

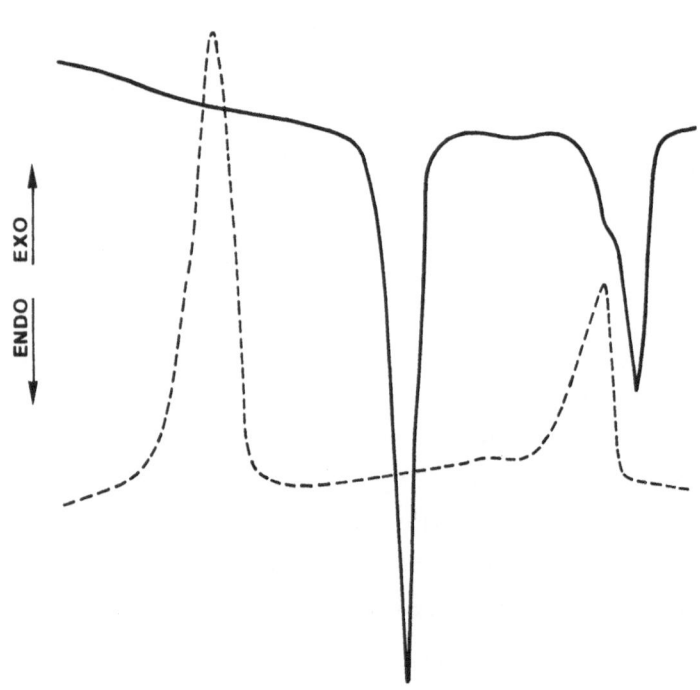

Fig. 10   DSC curves for polyester T10
           —————— first heating run
           --------- first cooling run

transition in cooling cycles and mesophase-mesophase or isotropic liquid-mesophase transitions characteristically exhibit little or no supercooling. The thermal properties of polyesters under investigation are summarized in Table I.

Compounds T2-T10 and T05-T029 show, by the decrease in clearing point, the marked influence of increasing the length of the flexible spacer. The same general trend has been reported for other polymeric liquid crystals.[24,23] This reflects the decreasing thermal stability of mesophases with decreasing polarity and molecular rigidity. If the length of the flexible spacer is excessively increased as in sample T029 the resultant polymer shows distinct phases of liquid crystal and isotropic regions (Fig. 11). The effect of $\{CH_2\}_{3 \, n}$ segments and $\{CH_2 - CH_2 - O\}_n$ segments on the mesophase-isotropic liquid transition temperature is nearly

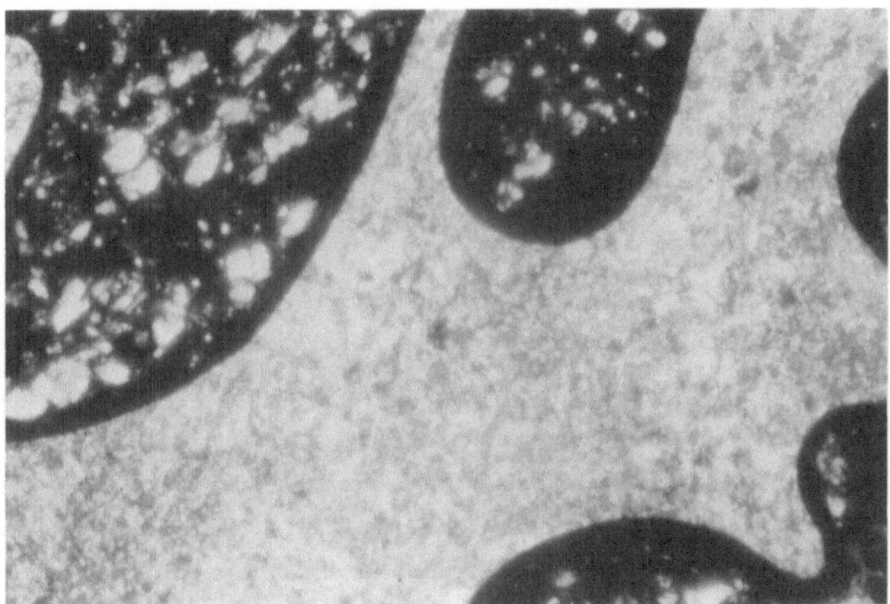

Fig. 11   Liquid crystalline and isotropic phases of sample T029
            at 90°C.

identical. Therefore, the oxygen in an ether effectively replaces a CH_2 group. We will now consider the effects of broadening a molecule by introducing a substituent in place of a hydrogen. With terphenyl derivatives, the tendency to form liquid crystals is so marked that substituents as large as $C_6H_5-CH_2-$ can be introduced in aliphatic segments without destroying the liquid crystal formation. However, this increases intermolecular separation and gives less thermally stable mesophases. The clearing points fall in proportion to the size of the substituent. Referring to Table I, the change H to Me involves a small decrease in the clearing point (~15–20°C) whereas the change H to $-CH_2Ph$ involves a considerable decrease in the isotropic-mesophase transition temperature (~170–180°C).

As observed for many types of small molecule liquid crystals,[25] the melting points do not show regular trends. The highest melting points are those of polyesters T2 (mp : 286°C) and T4 (mp : 318°C). Indeed, macromolecules incorporating rigid mesogenic groups (e.g. p-terphenyl) and short aliphatic chains (e.g. $(CH_2)_2$ or $(CH_2)_4$) tend to pack efficiently. The introduction of a branched alkyl group, such as -Me, -Et or $-CH_2Ph$ increases the separation of the long axes of neighbouring molecules and, as a consequence, strongly depresses the melting point as compared to the analogous n-alkylpolyesters.

Identification and Characterization of Mesophases

i/ Linear alkyl segments R = $(CH_2)_n$. For the lower homologues T2 and T4, the observed textures have suggested the assignment N and S_A for the mesophases which appear at high temperatures.[2] The nematic liquid crystals exhibit "threaded" or "schlieren" textures (Fig. 12). The transition bars appear at the transition to the S_A phases. The textures change on standing for some time into the stable focal conic or fan shaped textures (Fig. 13). However, there is some indication that these polyesters also form a more

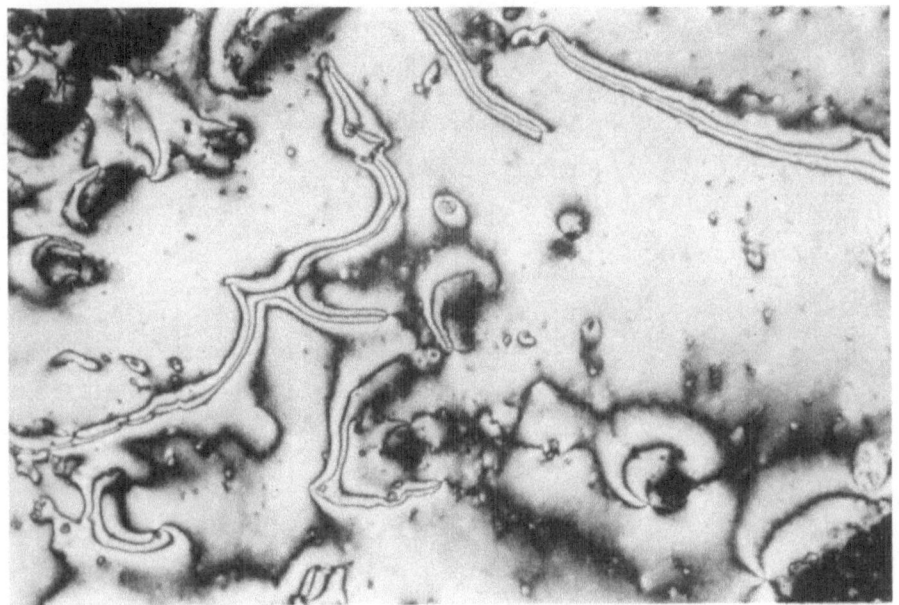

Fig. 12   Photomicrograph of quenched nematic mesophase from
           polyester T2.   Threaded-Schlieren texture.

ordered smectic phase at lower temperatures.   Indeed, in case of
fan shaped texture, concentric arcs appear at the transition to the
S_2 phase.   The resultant striated fan shaped texture suggests the
existence of S_E phase.   The X-ray study of the mesophases in these
polymers is difficult because of the high values of the mesomorphic
transition temperatures (Table I); during the time required to
record the scattered intensity, the degradation occurred.   So, the
X-ray investigation did not yield conclusive information except
the spacing in the crystalline phases (43.0, 20.0, 16.6, 4.56, 4.38,
3.79 Å and 18.4, 9.06, 4.84, 4.54, 4.30, 4.18, 3.92 Å for T2 and
T4 respectively).

Extending the alkyl segment length increases the smectic
tendencies relative to the nematic tendencies of the system but,
as observed for small molecule liquid crystals,[10] it also results

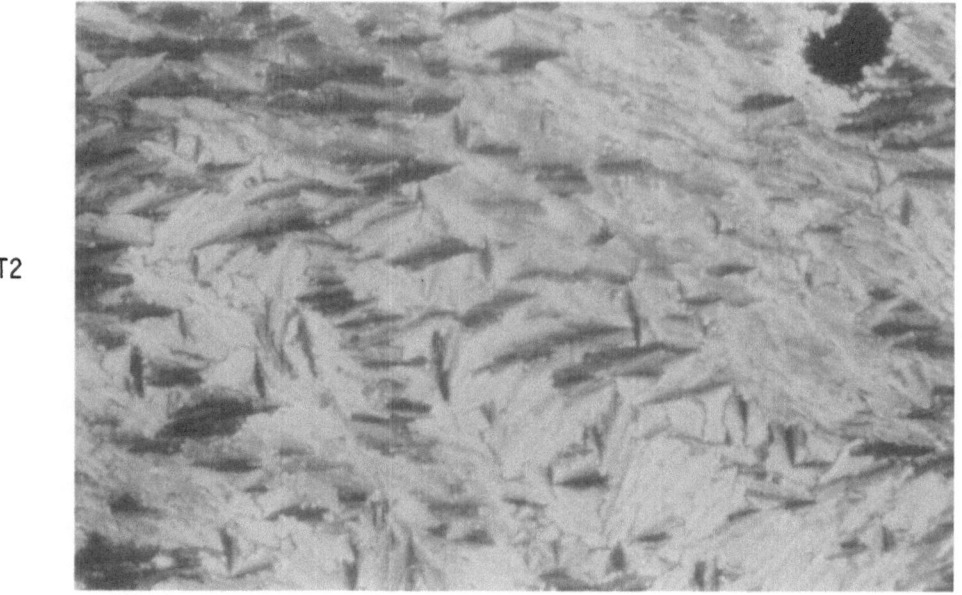

T2

T4

Fig. 13   "Focal-conic" and "Fan-shaped" textures observed by
           quenching polyesters T4 and T2 from the isotropic
           state.   (Crossed polarizers)

in the disappearance of the S_E phases so that a stage is reached
when nematic properties are extinguished (T5) and polyesters are
purely S_A (T6, T10). Such a behavior is partly confirmed by X-ray
data.

Compound T5 showed a poor crystallinity which remained unchanged even after annealing up to 185°C; the more apparent rings
corresponded to the lattice spacings: 19.5, 9.71, 4.7, 4.5 and
4.02 Å. A pattern recorded at 217°C, i.e. after the shoulder of
the first DSC endotherm (Fig. 5) showed some extra rings which
might be attributed to the transient formation of a 3-dimensional
order smectic phase. At 242°C, the diffraction photograph was
characterized by two sharp rings in the small-angle region, namely
a base reflection and a second order reflection, corresponding to
a lattice dimension of 19.3 Å; in the wide-angle region, a diffuse
halo lied in the wave-vector $q = 4 \pi \sin \Theta/\lambda$ range 1.3 to 1.45 Å^{-1}.
Such results are typical features of the diffraction patterns of
S_A and S_C phases. With unoriented samples it was impossible to
draw an inference on the S_A or S_C nature of this smectic liquid
crystal but the optic observations had removed this indetermination.

Heat treatment at about 180°C greatly increased the crystallinity of the compound T6; the first diffraction rings of the
crystalline phase at this temperature corresponded to the lattice
spacings: 21.7, 10.8, 4.52 and 4.09 Å. At a temperature just
above the first endothermic peak (Fig. 5), the degradation occurred
as indicated by a strong ring at 5.43 Å; however, the diffraction
pattern allowed to distinguish an inner ring related to a lattice
spacing of 21.7 Å in addition to a diffuse halo around $q = 1.35$ Å^{-1},
both features consistent with a S_A phase.

At room temperature, T10 showed a good crystallinity (the
observed lattice spacings are 23.4, 18.0, 11.6, 4.61, 4.35, 4.25
and 3.88 Å) and by selecting samples, slightly oriented samples
were found with the inner rings crescent-shaped parallel to the

capillary axis. At 272°C, the outer rings were replaced by a
diffuse halo in q range 1.20 to 1.45 $\overset{\circ}{A}^{-1}$ whereas a strong ring and
a weak second order reflection corresponding to a lattice dimension
of 26.5 $\overset{\circ}{A}$ was seen in the small-angle region. This distance,
related to the regular packing of smectic layers, is nearly identical
to the calculated length of the monomer unit in its most extended
conformation, the terphenyl group being oriented perpendicular
with respect to the lamellae.

To summarize the X-ray data of these polyesters, we have drawn
in Fig. 14 the variation of the lamellar thickness d with the number
of carbon atoms in the aliphatic chain, in both the crystalline
phase and the smectic modification. Solid lines illustrate the
linear variation of the spacings as calculated from the length of
the various monomer units (except T5) assumed entirely extended in
the *trans*-conformation. One can see that in the crystalline state,
the lamellar thickness of T2 is somewhat higher than a dimerized
molecule whereas, in the other homologous polyesters, the spacings
d are smaller than the length of the monomer unit. The assignment
S_A for the smectic phases from optical observations thus imply that
polyesters under investigation may not adopt a single elongated
conformation in both crystalline and mesomorphic modifications.
It is possible however that the parallel alignment of rigid core
parts of the molecule constrains movement within the alkyl chains
to some extent and that, although the chains may not adopt a
single or unique conformation, the conformations that are adopted
are more or less extended. The crystal structures of similar
polymers having the general formula:

$$\text{OCO} \underset{}{\bigcirc} \text{COO} \left(\text{CH}_2\right)_n$$

and abbreviated n-GT have been determined.[26-32] In the unstrained
form, the methylene section of the chain has the conformation
trans, gauche-trans-gauche (e.g. $\overline{G}\overline{G}TGG$), *trans-gauche-gauche-trans*

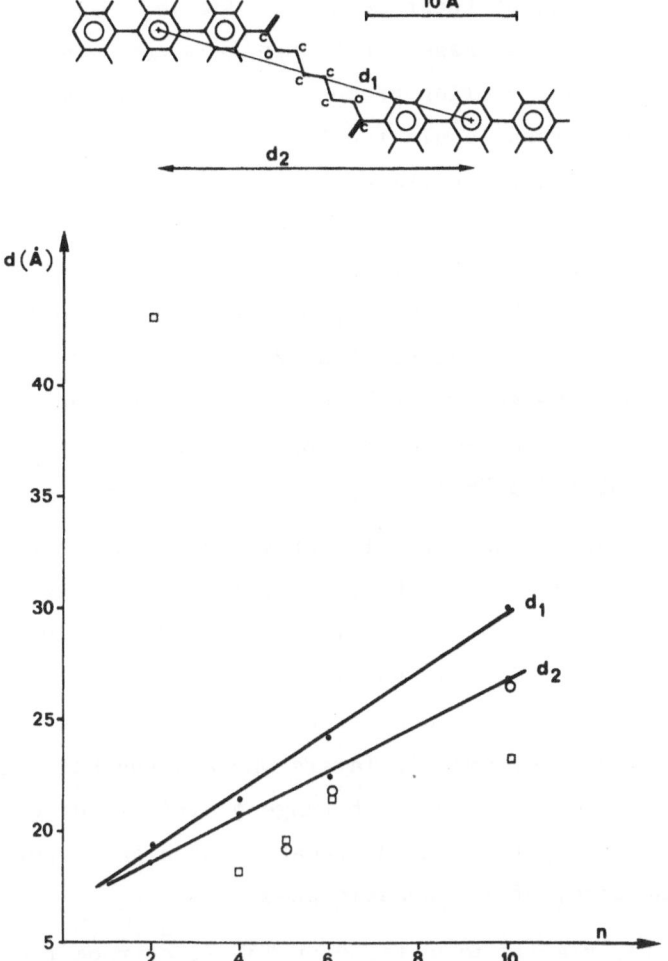

Fig. 14   The low-angle X-ray spacing d as a function of the
number of carbon atoms in the aliphatic chains (squares:
in the crystalline phase; circles: in the smectic
liquid crystal; solid lines: calculated values).

and *trans-trans-trans-trans-trans* for 2GT, 4GT, 5GT and 6GT,
respectively. For 4GT and 5GT it is noteworthy that, on straining,
twist of the *gauche* towards the *trans*-conformation occurs.

   <u>ii/ Branched alkyl segments</u>. Polymorphism is exhibited by
T2Me, T2Me2, T3Me2 and T5Me since both smectic and nematic

mesophases are shown by these members. Optical microscopy revealed that a typical nematic phase was formed from the isotropic liquid on cooling. Further cooling produced a smectic A phase which exhibited typical focal-conic fan texture. With the exception of T5Me, further reduction in temperature produced another smectic phase. The fans were lightly chequered and some arcs ran laterally across them, as in a smectic E phase. X-ray data yielded no further information about T2Me since this polyester was found to decompose at temperature higher than 270°C without any phase transition; only the spacing related to a 3-dimensional ordered phase (crystal or quenched smectic E) could be established: 18.0, 4.78, 4.57, 4.30 and 3.78 Å.

T2Me2 showed in the X-ray investigation no phase transition in temperature range 25 to 206°C, the lattice spacings observed at the higher temperature being equal to 19.4, 10.0, 8.39, 5.32, 4.91, 4.51 and 4.19 Å. At 240°C, the small-angle sharp diffraction corresponding to a lattice dimension of 21.7 Å was attributed to the regular packing of smectic layers whereas the halo at $q \sim 1.26$ Å^{-1} was due to the short range positional order of the chain within the layers. The degradation at high temperature did not allow the study of the nematic phase.

Some X-ray results were found on T3Me2; at room temperature, this compound exhibited diffraction rings at 22.5, 4.57, 3.92 Å and also ring giving proof of a degradation during the polymerization. Near 272°C the sharp diffraction rings related to the lattice dimension of 22.7 Å and the halo around $q = 1.4$ Å^{-1} were consistent with the formation of a smectic A phase.

DSC data and microscopic observations suggested that T3Et2 and T5Tol only exhibit a S_A phase. Actually, X-ray investigation on the former polymer showed metastability phenomena on both sides of the first endothermic peak, thus giving an explanation for the sluggish transition process (Fig. 8): the smectic phase,

characterized by a strong ring and a weak second order reflection
related to a layer spacing of 20.5 $\overset{\circ}{A}$ and a halo at q ~ 1.4 $\overset{\circ}{A}^{-1}$,
might be supercooled down to the room temperature. After annealing,
crystalline solid, together with some smectic phase remains, showed
reflection rings with the spacings 25.9, 20.5, 12.1 and 4.63 $\overset{\circ}{A}$.

T5Me and T5Tol exhibited clear behaviors; the crystalline
modifications, with lattice spacings respectively equal to 20.4,
15.7, 10.2, 4.42 $\overset{\circ}{A}$ and 31.4, 4.46, 3.90 $\overset{\circ}{A}$ at room temperature,
showed a smectic phase formation at higher temperature with a
lamellar thickness respectively equal to 20.4 and 30.0 $\overset{\circ}{A}$.

The assignment of N and S_A phases essentially founded on
texture observations was further confirmed by miscibility studies.
They were shown to be miscible with the corresponding phases of
the standard materials terephthalidene-bis-(4-n-butylaniline) and
diethyl-p-terphenyl-4,4" carboxylate. An example is given in Fig.
15. For the system T2Me2/diethyl-p-terphenyl-4,4" carboxylate the
melting curve exhibits a maximum. As a consequence, over a great
part of the composition range, the liquid crystalline state may
lie in the unstable region and it was not possible to prove that
the smectic low-temperature modification S_2 of the polyester T2Me2
is of the type E.

iii/ Ether segments R = $\{CH_2 - CH_2 - O\}_n$. Polarized light
photomicrographs depicting the appearance of the mesophases of
polyesters T05, T08 and T011 exhibited characteristic features of
smectic phases as illustrated in Fig. 16. The smectic C phases of
these compounds were identified by isomorphy with the smectic C
phase of terephthalidene-bis-(4-n-butylaniline).[33] Moreover the
S_C phases of T05, T08 and T011 were miscible with the left handed
twisted S_C phase of either 4'-(2-methyl hexyloxy) biphenyl-4-
carboxylic acid [K $\xleftrightarrow{171}$ S^*_{Cg} $\xleftrightarrow{215}$ N^*_g $\xleftrightarrow{229}$ I][34] or
terephthalylidene-bis-4-((+)-4'-methylhexyloxy)-aniline
[K $\xleftrightarrow{144}$ S^*_{Bg} $\xleftrightarrow{160.4}$ S^*_{Cg} $\xleftrightarrow{207.6}$ N^*_g $\xleftrightarrow{224.5}$ I].[35] Notice

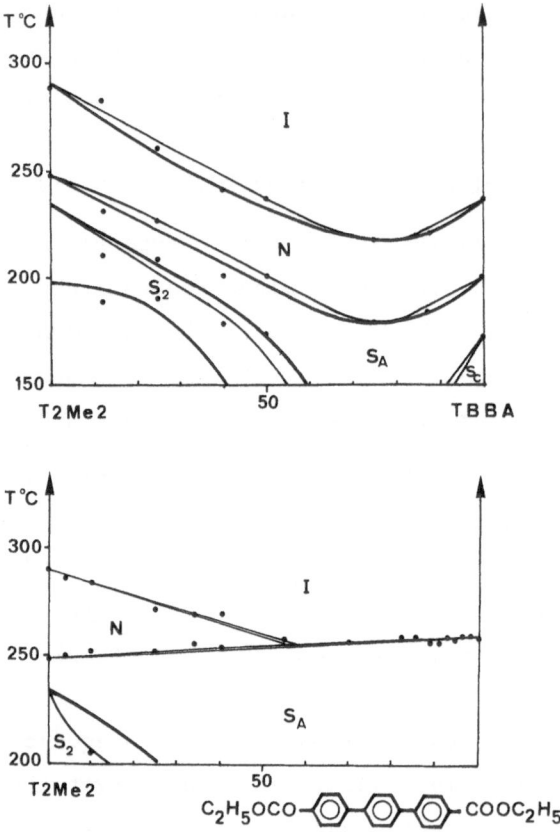

Fig. 15   Isobaric phase diagrams of T2Me2 and either
           terephthalidene-bis-(4-n-butylaniline) or diethyl-p-
           terphenyl-4,4" carboxylate.

particularly that these chiral compounds on addition to the
polyester T011 causes the formation of the typical twisted
smectic textures with equidistant lines.[33]   Such a situation can
be explained by the periodicity of layers with equal twist angle.

    The three polymers studied by X-ray scattering, namely T05,
T011 and T029, showed in their crystalline modifications some
structural analogies since the wide-angle diffractions occurred
approximatively at the same angles with identical relative
intensity.   Thus, their crystal lattice probably differs only in

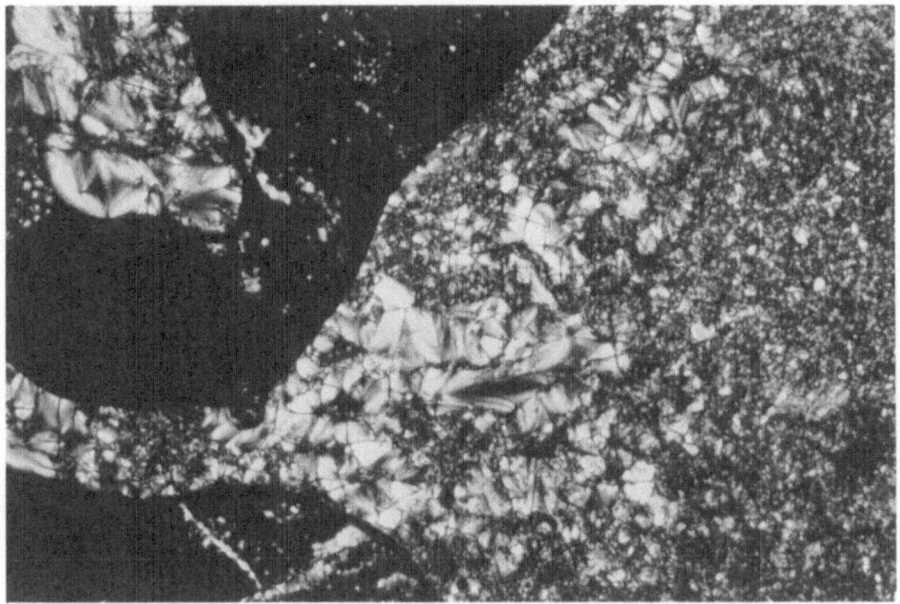

Fig. 16   Photomicrograph of smectic mesophase from polyester T011
           at 240°C. Focal-conic texture (crossed polarizers).

the length of the unit-cell parameter whose value could be deduced
from the low-angle rings, i.e. 20.8, 26.0 and 38.0, respectively;
these distances are in good agreement with the length of the
monomers in their most extended *trans*-conformation (Fig. 17).

On heating the behavior of the polymer T029 was not clear.
Whereas microscopic observations showed distinct phases of liquid
crystal and isotropic liquid in the temperature range 70-117°C,
X-ray measurements indicated that the small-angle ring disappeared
at 80°C and the wide-angle rings vanished in a halo at 1.4 $\overset{\circ}{A}^{-1}$
but remained apparent at temperatures higher than the clearing
point (117°C). These features probably depended on the poly-
dispersity of the commercial OH-terminated low molecular weight
(H 400) fraction of PEO used for preparing this polyester.

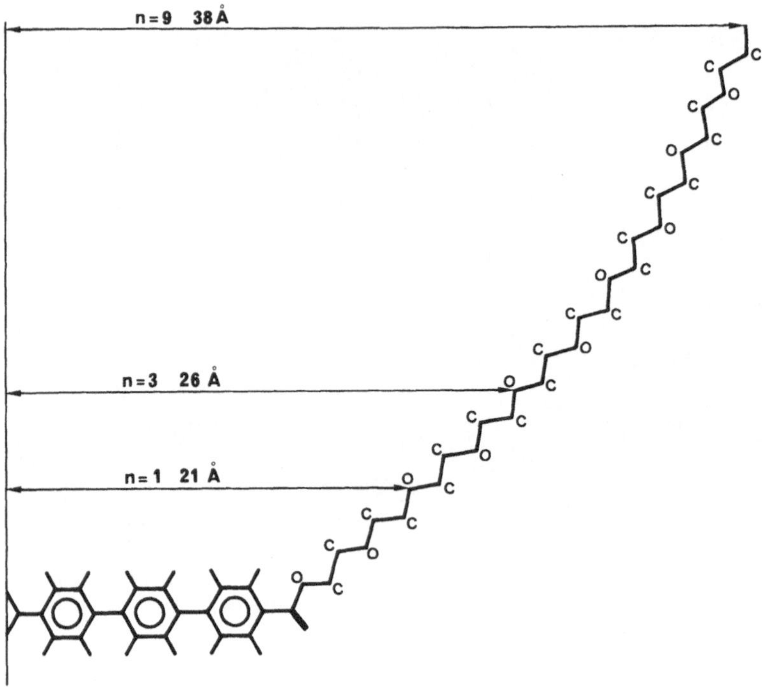

Fig. 17   Proposed model of packing within layers in crystalline

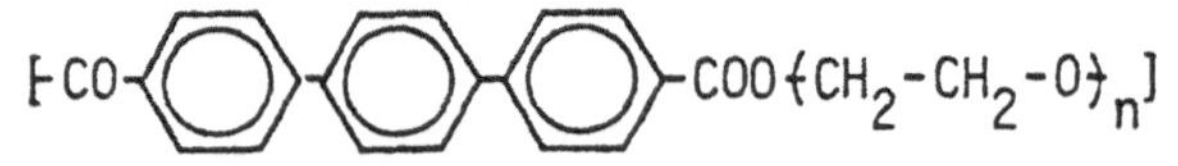

$$[CO-\!\!\bigcirc\!\!-\!\!\bigcirc\!\!-\!\!\bigcirc\!\!-COO(CH_2\!-\!CH_2\!-\!O)_n]\ \ polymers$$

For T05, the smectic phase was characterized by three inner rings corresponding to a layer spacing of 20.2 Å and a broad diffuse ring at $q \sim 1.35\ \text{Å}^{-1}$.

Usually, the synthesized polymer T011 gave unoriented specimens (Fig. 18, plate 1 and 2); but by selecting fibers, it was possible to find a well-aligned sample (plate 3). Above the crystal-mesophase temperature (plate 4), the two sharp small-angle reflections on meridian have to be attributed to regular smectic layer distance equal to 25.2 Å, whereas the diffuse arcs observed along the equator are due to the short-range positional order of the chains within the smectic layers. The relative position of

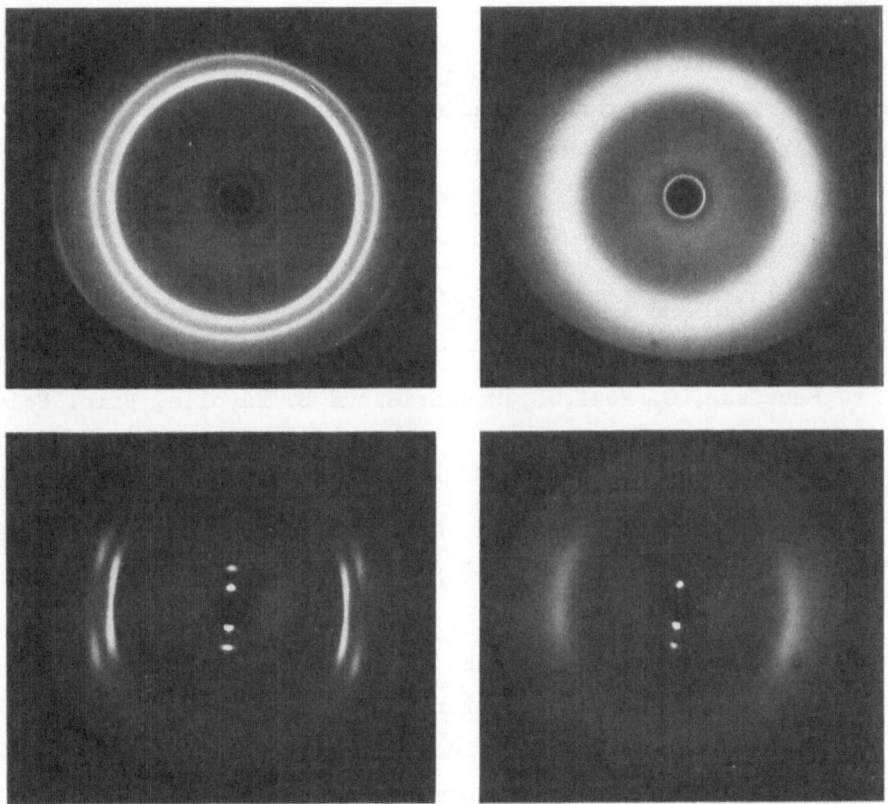

Fig. 18 X-ray diffraction photographs of crystalline solid (plates
1 and 3) at room temperature, and smectic modifications
at 145°C (plates 2 and 4) from powder and oriented samples
T011.

the small-angle reflections with respect to the diffuse crescents
yields X-ray patterns similar to those of the class S_A. On
account of the results quoted above and concerning the microscopic
observations in addition with the miscibility test, we have to
admit that the rigid mesogenic groups are azimuthally disordered
with a tilt angle relative to the layer normal.[36] This assumption
is supported by the fact that the smectic layer spacing (25.2 Å)
is smaller than the greatest distance in the crystalline solid
leading to a tilt angle of about 10°.

ACKNOWLEDGMENTS

   We would like to thank Dr. B. Jasse and Miss F. Bastard who
have kindly performed the synthesis and purification of
terephthalidene-bis-(4-n-butylaniline) and diethyl-p-terphenyl-
4,4"-carboxylate samples.

REFERENCES

1.   A. Roviello and A. Sirigu, J. Polym. Sci., Polym. Lett. Ed.
        13, 455 (1975).

2.   P. Meurisse, C. Noel, L. Monnerie and B. Fayolle, Brit. Polym.
        J. 13, 55 (1981).

3.   J. Virlet, to be published.

4.   E. Henriot and E. Huguenard, C. R. Acad. Sci. 180, 1389 (1925).

5.   A. Pines, M. G. Gibby and J. S. Waugh, J. Chem. Phys. 59, 569
        (1973).

6.   S. R. Hartmann and E. L. Hahn, Phys. Rev. 128, 2042 (1962).

7.   E. O. Stejskal and S. Schaefer, J. Magn. Reson. 18, 560 (1975).

8.   J. Tegenfeldt and U. Haeberlen, J. Magn. Reson. 36, 453 (1979).

9.   L. Kofler and A. Kofler, Thermomikromethoden, Verlag Chemie,
        Weinheim, 1954.

10. H. Kelker and R. Hatz, Handbook of Liquid Crystals, Verlag
        Chemie, Weinheim, Deerfield, 1980.

11. D. Demus, K. H. Kolz and H. Sackmann, Z. Phys. Chemie,
        Leipzig 252, 93 (1973).

12. L. W. Jelinski and J. J. Dumais, A. C. S. Polym. Prepr. 22(2),
        273 (1981).

13. M. Alla and E. Lippmaa, Chem. Phys. Lett. 37, 260 (1976).

14. S. J. Opella and M. H. Frey, J. Am. Chem. Soc. 101, 5854 (1979).

15. H. J. Lader and W. R. Krigbaum, J. Polym. Sci., Polym. Phys.
        Ed. 17, 1661 (1979).

16. A. C. Griffin and S. J. Havens, J. Polym. Sci., Polym. Phys.
        Ed. 19, 951 (1981).

17. A. Roviello and A. Sirigu, Makromol. Chem. 180, 2543 (1979).

18. A. Antoun, R. W. Lenz and J. I. Jin, J. Polym. Sci., Polym.
        Chem. Ed. 19, 1901 (1981).

19. P. Culling, G. W. Gray and D. Lewis, J. Chem. Soc. 2699 (1960).

20. M. E. Neubert and L. T. Carlino, Mol. Cryst. Liq. Cryst. 42, 353 (1977).

21. J. W. Goodby and G. W. Gray, Mol. Cryst. Liq. Cryst. Lett. 49, 165 (1979).

22. S. E. B. Petrie, Liquid Crystals, the Fourth State of Matter, F. D. Saeva, Ed., Marcel Dekker, New York and Basel, p. 170.

23. A. C. Griffin and S. J. Havens, Mol. Cryst. Liq. Cryst. Lett. 49, 239 (1979); J. Polym. Sci., Polym. Lett. Ed. 18, 259 (1980).

24. L. Strzelecki and D. Van Luyen, Europ. Polym. J. 16, 299, 303 (1980).
    D. Van Luyen, L. Liebert and L. Strzelecki, Europ. Polym. J. 16, 307 (1980).

25. G. W. Gray and P. A. Winsor, Liquid Crystals and Plastic Crystals, Halsted, Chichester, 1974, Vol. 1.

26. H. Klare and G. Reinisch, Polymer Science USSR 21, 2727 (1980).

27. A. M. Joly, G. Nemoz, A. Douillard and G. Vallet, Makromol. Chem. 176, 479 (1975).

28. I. H. Hall and B. A. Ibrahim, J. Polym. Science, Polym. Lett. Ed. 18, 183 (1980).

29. M. Yokouchi, Y. Sakakibara, Y. Chatani, H. Tadokoro, T. Tanaka, and K. Yoda, Macromolecules 9, 266 (1976).

30. I. H. Hall and M. G. Pass, Polymer 17, 807 (1976).

31. I. J. Desborough and I. H. Hall, Polymer 18, 825 (1977).

32. I. H. Hall and N. N. Rammo, J. Polymer Science, Polym. Phys. Ed. 16, 2179 (1978).

33. B. Fayolle, C. Noel and J. Billard, J. de Physique 40, C3-485 (1979).

34. M. Leclercq, J. Billard and J. Jacques, C. R. Acad. Sci. 266, 654 (1968).

35. D. Coates and G. W. Gray, Mol. Cryst. Liq. Cryst. Lett. 34, 1 (1976).

36. A. J. Leadbetter and E. K. Norris, Molec. Phys. 38, 669 (1979).

INFLUENCE OF THE LEPROSY DRUG, DAPSONE, ON THE PHASE TRANSITIONS OF THE LYOTROPIC, DIPALMITOYL PHOSPHALTIDYLCHOLINE (DPPC)

K. Usha Deniz[*], P. S. Parvathanathan[*], E. B. Mirza[+],

V. Amirthalingam[+] and S. Gurnani[**]

Bhabha Atomic Research Centre, Trombay, Bombay 400 085

INTRODUCTION

4,4'-Diaminol Diphenyl Sulfone (DDS, also called Dapsone),

$$H_2N - \bigcirc - \overset{\overset{O}{\|}}{\underset{\underset{O}{\|}}{S}} - \bigcirc - NH_2$$

is widely used for the treatment of malaria, leprosy and other infectious diseases. However, nothing is known about its mode of action at the molecular level. Studies of lipid-Dapsone systems are therefore important in understanding the effects of this drug on the structure and function(s) of biomembranes.

In this paper we describe differential scanning calorimetric studies carried out with the lyotropic liquid crystal (DPPC + water), one of the simpler model membranes, in which DDS is dissolved. The pure (DPPC + water) system with the range of water content used in our experiments is known[1] to exhibit the following phase transitions

---

[*] Nuclear Physics Division

[+] Reactor Chemistry Section

[**] Biochemistry and Food Technology Division

429

in the temperature range, -10°C to 65°C:

$$L_{\beta'} \xrightarrow{\longleftarrow} P_{\beta'} \xrightarrow{\longleftarrow} L_{\alpha}$$

The nomenclature for lyotropic phases is that of V. Luzzati.[2] $L_{\beta'}$ is an ordered lamellar phase in which the hydrocarbon chains of the lipid molecules are in an extended conformation and are inclined with respect to the lamellar normal. The chains have a hexagonal (pseudo-hexagonal) packing in the lamella, thus forming a 2-d ordered lattice. The $P_{\beta'}$ phase is similar to $L_{\beta'}$ but with the lamellae being distorted by a periodic ripple. The L_{α} phase is also a lamellar phase in which the chains are disordered and hence each lamella can be thought of as a 2-d liquid. Thus the $(P_{\beta'} \rightarrow L_{\alpha})$ transition is called a chain melting (CM) transition and therefore the $(L_{\alpha} \longrightarrow P_{\beta'})$ transition can be called a chain ordering (CO) transition. The order-disorder transitions of the lipid-hydrocarbon chains play an important role in biomembrane functions.

Our results clearly indicate that there is a phase separation in the (DDS + DPPC + water) system for large drug concentration. Our results are discussed in the light of theories of De Verteuil and Scott Jr.,[3,4] taking into account the possible interactions between the drug, DPPC and water molecules.

EXPERIMENTAL DETAILS

The compounds, DPPC and DDS were obtained from Sigma Chemicals and Burroughs Welcome, India Ltd. respectively. The stock solution of the former was prepared in analar chloroform and that of the latter in analar methanol. These solutions were stored at temperature, $T < 0°C$. Appropriate volumes of these were mixed so as to get the required mole fraction (R_m) of DDS in DPPC. R_m ranged from 0 to 0.6 in our experiments. The mixture was dried with a stream of nitrogen. Any remaining traces of solvent were removed by evacuation. The drug-lipid mixture was then hydrated by the addi-

tion of requisite quantity of triply distilled water to obtain
the desired weight fraction, χ, of water in DPPC. In our measurements, we have used samples containing 25% (χ = 0.25) and 50% (χ =
0.5) by weight of water in DPPC. A thorough dispersion of water
with the rest of the sample was obtained by alternately heating
in a water bath to 65°C and vorticising. 7-12 mg of the samples
were hermetically sealed in aluminum pans for use in the differential scanning calorimetric (DSC) experiments. These measurements
were carried out using, in general, a Perkin-Elmer DSC-2C instrument. A Perkin-Elmer DSC-1B model was also used in a few experiments. The scanning speeds used were 10°/min and 5°/min and phase
transitions were investigated in the temperature range, -60°C < T <
65°C. The samples used in the DSC scans were subsequently used for
X-ray diffraction studies, which were mainly carried out to investigate the nature of the room temperature phase in the system containing a large concentration of the drug. These diffraction experiments were done using CuKα radiation and a modified Laue camera
to obtain reflections at small angles.

RESULTS AND DISCUSSION

Our DSC scans show that for pure lipid-water systems, there are
three transitions in the temperature range scanned, the ice-water
transition, the pretransition ($L_{\beta'}$-$P_{\beta'}$) and the chain melting
($P_{\beta'}$-L_{α}) transition. All three of these are seen both while heating
and cooling, and are almost independent of the thermal history of
the sample. The ice to water peak (Fig. 1) is quite sharp for χ =
0.50 and appears around 3°C for the 10°/min scan. We believe this
peak to be associated with the 'free' water in the system. However,
the situation is quite different when the samples contain DDS
($R_m \neq 0$). The pretransition disappears even for small drug concentrations. The other two transitions become strongly dependent
on the thermal history of the sample.

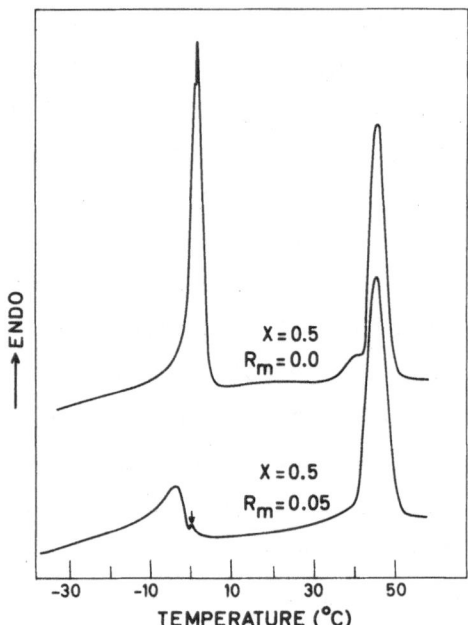

Fig. 1 Typical DSC scans for DDS–DPPC–water systems. The arrow
        points to the "free" water transition peak.

Ice–water transition, ($R_m \neq 0$): Even for small values of R_m,
the area of this transition peak decreases in comparison with that
for $R_m = 0$, this decrease being dramatic for $\chi = 0.5$ (Fig. 1). In
addition, this transition changes its shape with thermal cycling and
broad humps are present for $T < 0°C$ (Fig. 1). However the ice to
water transition around 3°C continues to remain sharp though considerably reduced in intensity for non-zero values of R_m. Its temperature hardly changes with thermal cycling.

Chain melting transition: In general the CM transition temperature, T_{CM}, is found to decrease slightly with thermal cycling between
–60°C and 65°C, for small values of R_m. The width of the transition
peak increases with increasing drug concentration, culminating in a
split peak for $R_m \geqslant 0.5$ (Fig. 2). The split becomes more pronounced
with repeated thermal cycling, as seen in Fig. 3, where (a) and (b)

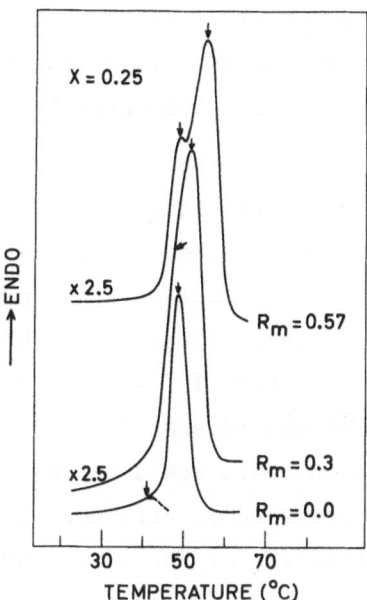

Fig. 2 DSC scans of the chain melting transition for χ = 0.25 and different R_m values. The dashed curve shows the pretransition for DDS free system. The curve for R_m = 0.3 shows a hump (indicated by the arrow) and for R_m = 0.57, there is a clear split peak.

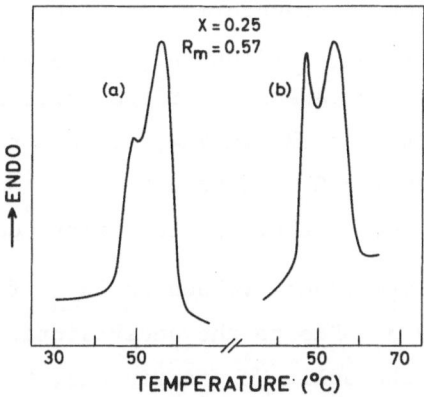

Fig. 3 Evolution of the R_m = 0.57 (Fig. 2) chain melting peak with thermal cycling. (a) first heating and (b) second heating.

refer to first and second heating of the sample. It is interesting
to note that the temperatures as well as the widths of these double
peaks lie above and below their respective values for $R_m = 0$.

The dependences on R_m, of the transition temperatures, T_{CM} and
T_{CO}, for $\chi = 0.25$ and 0.5, are shown in Fig. 4. The values shown are
those obtained for a scanning speed of $10°/min$. for first heating
and cooling. However, in the case of split peaks, the temperatures
were obtained from the last cycle in which the two peaks were well
resolved. It is seen that both T_{CM} and T_{CO} hardly change for
$0 \leqslant R_m \leqslant 0.3$, except for the xase of $\chi = 0.25$, where T_{CM} increases
slightly. This slight increase could be due to the CM peak consist-
ing actually of split peaks which were too close together to be
distinguished. The splitting of the CM and CO peaks for $R_m \geqslant 0.5$
leads to both an increase and a decrease of T_{CM} and T_{CO} with increasing R_m, as seen in Fig. 4.

The full width at half maximum (FWHM), of the CM transition is
given as a function of R_m in Fig. 5. It is found that for both $\chi =$
0.25 and $\chi = 0.5$, FWHM increases with increasing R_m. However, the
increase is much larger for the former. The result of split peaks
leading to both an increase and decrease of FWHM with respect to
its $R_m = 0$ value is clearly seen for the case of $\chi = 0.25$. The
ratio, A_R, of the heat of transformation, (ΔH_{iw}), of the ice to
water transition to that of the chain melting transition (ΔH_{CM}),
equal to the ratio of the areas under the corresponding peaks is
also given as a function of R_m in Fig. 5. As stated earlier, A_R,
decreases with increasing R_m for both $\chi = 0.25$ and 0.5 but the
initial decrease is much larger for the latter case.

Heats of transformation: Values of ΔH_{CM}, for $\chi = 0.5$ are given
for values of $R_m \leqslant 0.3$. Due to the uncertainties in the measurement of areas under the split CM transitions, the corresponding
ΔH_{CM} values have not been given for $R_m \geqslant 0.3$.

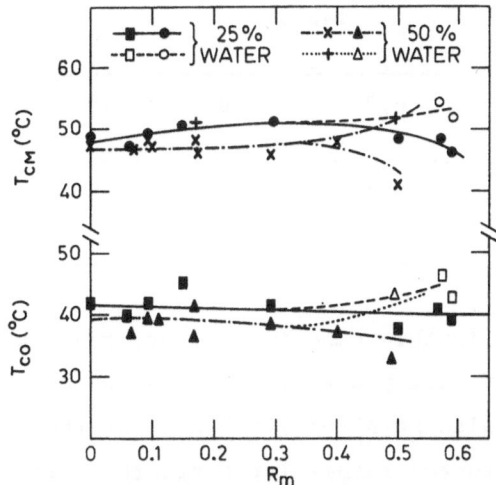

Fig. 4 Chain melting and chain ordering temperatures, T_{CM} and T_{CO}, as a function of R_m. The higher transition temperatures in the case of split peaks are given by crosses, open circles, open rectangles and open triangles. The lines are guides to the eye.

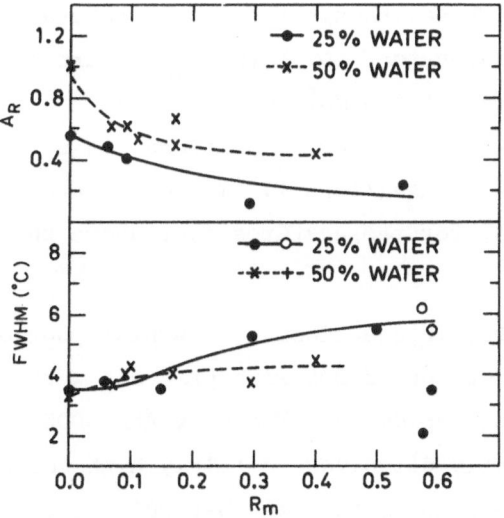

Fig. 5 Variation of A_R and FWHM as a function of R_m. For calculating A_R, in the case of split CM transition, the total areas of the split peaks is considered. The FWHM values of the high temperature CM peak, in the case of split peaks, is given by open circles. The lines are guides to the eye.

Table I

100 R_m	0.0	6.7	8.9	10.1	29.6
ΔH_{CM} (KCal/Mole)	7.15	8.17	8.27	8.17	7.50

It can be seen that (within experimental errors) the ΔH_{CM} values do not change significantly with increasing DDS concentration. The same trend was also seen for $\chi = 0.25$, but these values are not given here due to the rather large errors in the measurements of areas. Our value of ΔH_{CM} agrees fairly well with values quoted elsewhere.[5,6]

X-ray diffraction results: Our preliminary x-ray experiments have indicated that (1) even for large values of R_m the sample remains in the $L_{\beta'}$ phase, though the spotty nature of the diffraction photographs indicates that for large R_m, a part of the sample consists of aggregates of DDS molecules forming a 3-d lattice, and (2) for $R_m \geqslant 0.3$ there are two CM transitions, as evidenced by the presence of, both a diffuse $((4.6 \text{ Å})^{-1})$ and a sharp $((4.2 \text{ Å})^{-1})$ outer diffraction peaks, for temperatures just above the lower T_{CM}, indicated by DSC.

DDS-DPPC-Water Interactions: To explain our observations, the likely interactions of DDS with DPPC and water in the $L_{\beta'}$ phase, have to be considered. The fact that DDS is hardly soluble in water but readily dissolves in fairly high concentration in the lipid-water system, seems to show that DDS has preferential affinity for hydrophobic environment. The benzene rings of DDS which are hydrophobic, would prefer to interact with the hydrocarbon chains of DPPC, while the end amino groups and oxygens being hydrophilic would choose to interact with the polar groups of DPPC and with water. The amino groups of DDS can hydrogen bond with the oxygen

of the carbonyl and phosphate groups of DPPC and of water, whereas
the oxygens of DDS can only H-bond with water molecules. From considerations of the tetrahedral structure of DDS and the distance
between its end amino groups, we find that the following molecular
arrangements with respect to the lipid bilayer are possible (fig. 6):
(1) The end amino groups of DDS can H-bond with the carbonyl groups
of neighbouring DPPC molecules, while the oxygens of DDS (a) hydrogen
bond with water molecules, or (b) remain embedded within the hydrocarbon chain layer. (2) One of the amino groups of DDS can H-bond
with an acyl carbonyl group of DPPC while the other amino group H-
bonds with the phosphate group of the neighbouring lipid molecule.
(3) The DDS molecule can be completely embedded in the acyl chain
region of the bilayer without forming any hydrogen bonds (not shown
in fig. 6). Arrangement (3) favours only Van der Waals interactions
between DDS and the hydrophobic acyl chains of DPPC whereas the
other three arrangements favour to different extents, the above-
mentioned interaction as well as H-bonding.

Theoretical Considerations and Discussion: We will consider
here two theoretical models, those of De Verteuil et al.[3] and of
Scott,[4] since they are relevant to the systems we have examined.

The theory of De Verteuil et al.[3] assumes that each site of a
triangular lattice is occupied by a lipid chain. The guest (drug)
molecule which is considered to be small, occupies an interstitial
site. Model III of this theory assumes that the guest molecules
interact with the polar head groups (electrostatic forces) and with
the hydrocarbon chains (dispersion forces) of the lipid (DPPC) and
with each other. If the first two interactions are such that one
enhances and the other reduces the already existing attractive interactions between lipid molecules, the chain melting transition is
sharp. ΔH_{CM} varies little for small drug molecule concentration
(R_m). T_{CM} can increase or decrease with increasing R_m, depending
on whether the attractive lipid-lipid interactions are enhanced or

reduced by the presence of the drug molecule. Phase separation into
drug-rich and drug-poor phases can result, if the drug-drug interaction is strong. In our systems, different types of DDS-DPPC
interactions are possible, depending on the site that the DDS
molecule occupies. It is also possible that steric considerations
would prevent the DDS molecule from occupying an interstitial site.
In spite of these differences between our system and the theoretical
model, the latter admirably explains some of our results: (1) ΔH_{CM}
is independent of R_m for $R_m \leqslant 0.3$. (2) Double CM transition peaks
are seen for large R_m. The tendency of the two T_{CM}s (split peak
values), one to be larger and the other to be smaller with respect
to the $R_m = 0$ value could be due to the different sites that DDS
occupies in the bilayer, in the two phases. We feel that arrangements (1a) and (3) (see fig. 6) could cause an upward and down-
ward shifts of T_{CM} respectively.

The above model cannot explain the pretransition peak in (DPPC
+ water) system. For this we turn to Scott's model.[4] This is based
upon the 2-d packing of hard rods of varying lengths, the rods representing projections of lipid hydrocarbon chains and head groups on to
the bilayer plane. This model successfully predicts both a pretransition and a chain melting transition in the above system. In
this model the pretransition peak is due to the onset of orientational disorder of the hard rods in 2-d, which is in turn due to
the onset of rotational disorder in the acyl chains about their
long axis. In our systems, for $R_m \neq 0$, the H-bonding of the amino
group of DDS with the carbonyl group of acyl chains of neighbouring
molecules ((1a) and (1b) in fig. 6) could inhibit the onset of
rotational disorder for $T < T_{CM}$, thus destroying pretransition.

The reduction of ΔH_{iw} with increasing R_m could be due to the
H-bonding of the oxygens of DDS ((1a) of fig. 6) with the available
'free' water molecules. This would decrease the number of free H_2O
molecules taking part in the ice-water transition. The humps ob-

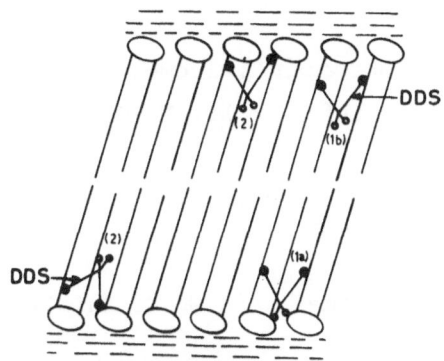

Fig. 6 Likely interactions of DDS molecule with (DDP+water)
        system in the $L_{\beta'}$ phases. In the DDS molecule, the filled
        circles represent the (NH_2) groups and the open circles,
        oxygen atoms.

served by us in DSC for T < 0°C could possibly be due to transitions
in water bound to different extents.

CONCLUSIONS

     From the above discussion a fairly consistent, though not com-
plete, picture emerges of the DDS interactions with the (DPPC +
water) system in the $L_{\beta'}$ phase. It would seem that for $R_m \leqslant 0.3$,
the DDS molecules predominantly occupy site (1a) (fig. 6), H-
bonding both with the acyl carbonyl groups and with the water molecules. Site (1b) might also be occupied to a certain extent but it
is rather unlikely that the DDS molecule would be at site (2). This
is because of the already existing H-bond between the water molecules and phosphate oxygen.[8] For larger drug concentrations, the
DDS molecules start occupying site (3). Due to the large interactions between DDS molecules, a phase separation occurs into
drug-poor and drug-rich phases, the former having DDS in sites
(1b) and (3) and the latter with DDS predominanting in the sites
(1a) and (1b). Our results imply that inclusion of DDS in biological membranes probably leads to alteration in membrane function.

ACKNOWLEDGEMENT

We wish to thank Burroughs Welcome, India Ltd. for the gift
of DDS. We are also grateful to Dr. M. D. Sastry for useful
discussions.

REFERENCES

1.   M. J. Janiak, D. M. Small and G. G. Shipley, Biochemistry, 15,
          4575 (1976).

2.   V. Luzzati, Biol. Membr. 1, 71 (1968).

3.   F. De Verteuil, D. A. Pink, E. B. Vadas and M. J. Zuckermann,
          Biochim. Biophys. Acta, 640, 207 (1981).

4.   H. L. Scott Jr., Biochim. Biophys. Acta 643, 161 (1981).

5.   K. Jacobson and D. Papahadjapoulos, Biochemistry 14, 152 (1975).

6.   D. J. Vaughan and K. M. Keough, FEBS Lett. 47, 158 (1974).

7.   C. Dickinson, J. McDonalt Steward, and H. L. Ammon, J. Chem.
          Soc. D 15, 920 (1970).

8.   H. Hauser and M. C. Philips, "Progress in Surface and Membrane
          Science" Eds. D. A. Cadenhead and J. F. Danielli, Academic
          Press, 13, 336 (1979).

NONAQUEOUS LYOTROPIC LIQUID CRYSTALS OF LECITHIN AND OLIGOMERS OF POLYETHYLENE GLYCOLS

M. El-Nokaly and S. E. Friberg

Chemistry Department
University of Missouri-Rolla
Rolla, Missouri 65401 USA

and

D. W. Larsen

Chemistry Department
University of Missouri-St. Louis
St. Louis, Missouri 63121 USA

ABSTRACT

Liquid crystals of lecithin plus mono-, di-, tri- and tetra-ethylene glycols were identified as lamellar with low angle X-ray diffraction determinations and from their optical patterns in polarized light.

The results showed enhanced perturbation of the lecithin molecules with concentration of the solvent and increased chain length. Replacing one terminal OH group with an ethoxide group enhanced this trend, as did substitution of the ethylene group with a propylene group.

INTRODUCTION

Lyotropic liquid crystals are formed by the action of a solvent on an amphiphilic substance[1]. The latter may be a solid, in

441

which case the structure transfer may be considered as a result of
a disordering influence of the solvent on the rigid network of the
polar groups of the amphiphile. The perturbation and disordering
of the polar group lattice reduces its ordering influence on the
hydrocarbon chains, causing a transition from a crystalline state
to the kind of order found in the liquid crystalline state[2,3].

Lyotropic liquid crystals may also be found from the influence
of a solvent on a liquid amphiphile. Examples of these systems are
water plus a nonionic surfactant of the polyethylene glycol alkyl
(aryl) ether type[3]. In this case, the formation of the liquid crystalline implies an enhanced order from the liquid to the liquid
crystalline state.

The research on these systems[1,4-8] has focused entirely on
water as the solvent. Water plus an amphiphile has been the constituent of the lyotropic liquid crystals, except for a single mentioning of one system[9]. In that case, a hydrocarbon was used to infer
sufficient disorder in the nonpolar part of the structure. Rigidity
of the polar layer was already moderate due to the non-metal cation
and the polar network obviously could not retain a crystalline order
when the hydrocarbon layer was penetrated by the added hydrocarbon.

Nonaqueous lyotropic liquid crystals have been found in a mul-
titude in the polymer/organic solvent combinations. These are
interesting systems[10,11]; the extensive investigations on polypep-
tide/hydrocarbon systems[12] deserve special mentioning.

For amphiphilic substance/solvent systems, a lamellar liquid
crystal containing lecithin and ethylene glycol was recently discovered and a comparison with the corresponding aqueous system was
made[13]. The lamellar phase could be formed also with higher homologues of diols; the series up to and including pentane diol has
recently been described[14]. A preliminary investigation of the dynamics, using quadrupolar splitting[15] showed only one ethylene
glycol molecule per lecithin molecule to have a measurable aniso-

tropic motion; a considerable difference from the conditions in the aqueous liquid crystals with lecithin, where several water molecules are found in sufficiently close vicinity to the lecithin to show a highly directed motion.

Since these liquid crystals obviously present a novelty in the field, we found a continued investigation into their properties to be of interest. The present publication describes liquid crystals formed from lecithin and a series of polyethylene glycols.

EXPERIMENTAL

Preparation of Materials

Lecithin. The lecithin, Epikuron 200 (Lucas Meyer, Hamburg, Germany) gave four spots on silica gel precoated plates and was purified according to the following procedure.

Epikuron 200 (17 gm) dissolved in a minimum amount of chloroform ($\sim$ 75 ml) was fractionated on a 4 x 87 cm alumina (625 gm) column[9]. The sample was washed in the column with chloroform (500 ml) and eluted with chloroform:methanol (9:1 by volume). The progress of the fractionation was followed by TLC and a qualitative test with Dragendorff's reagent. The first fractions (400 ml) were clear. The following 175 ml were turbid yellow and positive to Dragendorff. TLC showed this fraction to contain neutral fat and some lecithin. The next 300 ml of eluate were cloudy and contained only lecithin. The final 200 ml were clear and contained a decreasing amount of eluted material.

The fractions containing only lecithin were vacuum distilled (< 50°C) under nitrogen atmosphere to recover pure translucent and colorless lecithin giving one spot on TLC.

An anti-oxidant, Progallin P (NIPA Labs, Ltd., Pontypridd Glen, Wales), was added in the proportions of 2 mg/10 gm lecithin to the lecithin fractions before distillation.

<u>Solvents: Glycols</u>.

Ethylene glycol, 0.04% water (Fisher Certified)

Diethylene glycol, 0.04% water (BDH Analar)

Triethylene glycol, 0.05% water (Aldrich)

Tetraethylene glycol, 0.14% water (Aldrich)

Propylene glycol monomethyl ether, 0.07% water (Fisher Purified)

Ethylene glycol monoethyl ether (Cellosolve), 0.069% water (Fisher Purified)

The solvents had an original water content as high as $\sim$0.3%. They were dried to the water contents given and kept sealed under dry nitrogen in a dessicator over drierite[14].

<u>Preparation of Samples for Optical Microscopy and X-ray Diffraction</u>

The dried lecithin and the glycol solvents were weighed into small glass vials with screw tops and flushed with nitrogen after filling with the sample. The samples were mixed in a vortex vibro-mixer with intermittent heating (< 50°C) to obtain homogeneous samples and centrifuged to remove air bubbles. The final equilibration was made by slow cooling in a water bath from 50°C.

The sample was examined in a microscope in polarized light for proper mixing, absence of air, and to study its optical patterns.

A small amount of the equilibrated sample was drawn into a fine glass capillary, then sealed for interlayer spacing determination. X-ray diffraction was done by a Kiessig low-angle camera from Richard Seifert. Ni filtered Cu radiation was used and the reflection determined by a Tennelec position sensitive detection system (Model PSD-1100).

RESULTS

Low Angle X-ray Diffraction Measurements

The low angle X-ray diffraction patterns gave two reflections for most of the samples, which enabled the structure to be identified as lamellar[17]. The patterns from optical microscopy confirmed the lamellar structure.

The interlayer spacings were calculated[17] and plotted against the solvent weight ratio, giving a straight line for all the glycols involved (Fig. 1). The onset of the two-phase area for the

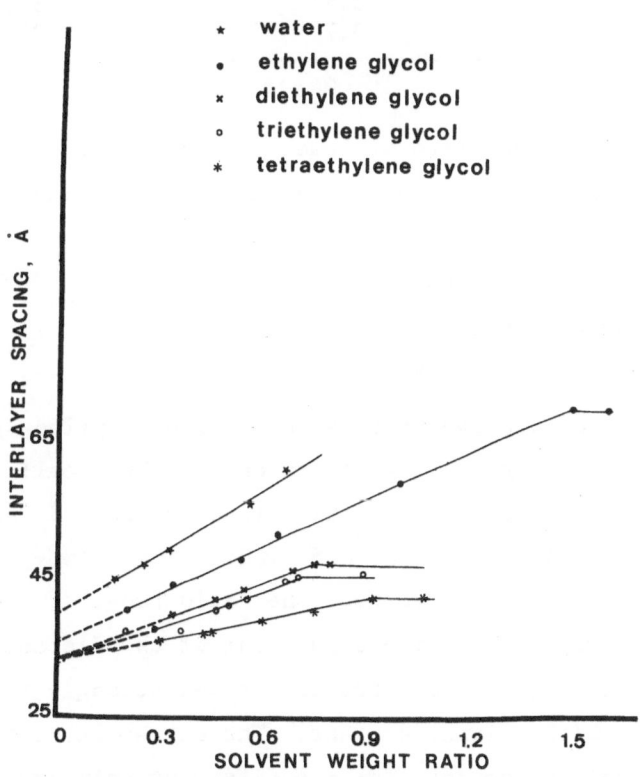

Figure 1. The interlayer spacing in lamellar liquid crystals of lecithin and a solvent.

TABLE I
Interlayer Spacing for Lamellar Liquid Crystals
of Lecithin and Solvents

Solvent	Solvent/lecithin weight ratio	Interlayer spacing, A
Ethylene glycol	0.25	33.2
monoethyl ether	0.30	32.4
	0.50	30.0
	0.75	28.6
	1.00	29.9
	1.25	28.0
Propylene glycol	0.2	32.9
monomethyl ether	0.3	34.5
	0.5	33.1
	0.7	32.3

di-hydroxy glycols occurred at lower solvent weight ratio with
increased chain length for compounds up to triethylene glycol

Ethylene glycol	1.5
Diethylene glycol	0.75
Triethylene glycol	0.7
Tetraethylene glycol	0.92

The onset of the two-phase area for the propylene glycol monomethyl ether and ethylene glycol monoethyl ether (cellosolve) could
not be determined exactly, but the values in Table I show constant
from a solvent weight ratio of 0.5, indicating a two-phase range at
solvent ratios above that value. The ratio range of the one-phase
area was too small for reliable judgment of the distance dependence
on the solvent weight ratio, but the values strongly indicate a constant spacing with solvent content. The extrapolated interlayer spacing at zero solvent content for the different solvents was as follows:

Ethylene glycol	36Å
Diethylene glycol	34Å
Triethylene glycol	34Å
Tetraethylene glycol	34Å

These values for the interlayer spacing extrapolated to zero solvent content were smaller than the value for dried lecithin in noncrystalline form (42.5Å).

Optical Microscopy

The optical patterns for liquid crystals with the different glycols containing one and two terminal hydroxyl groups showed the pattern of oily streaks and maltese crosses, as seen in Figure 2.

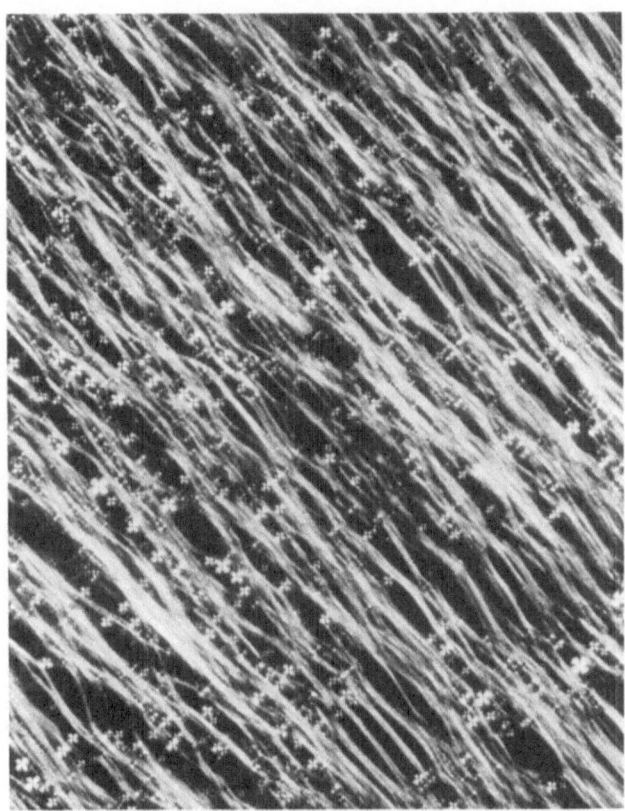

Figure 2.   The optical pattern was typical of a lamellar liquid crystal.

DISCUSSION

The results for the polyethylene glycols are similar to those obtained for the alkane diols[14]. The interlayer spacing extrapolated to zero solvent content gave a lower value than that for the water system, and increased molecular weight of the solvent gave smaller augments of the interlayer spacing with enhanced solvent content (Fig. 1). A formal calculation of percentage penetration between the molecules[14], assuming no change of amphiphile angle nor of chain disorder, gave the result shown in Table II.

As has earlier been emphasized[14], such calculations are purely an illustration of the combined effects of penetration of solvent molecules between the amphiphilic molecules, of changed disorder and tilt angle of the amphiphile, and should be viewed only as a parameter indicating the influence on the state of the amphiphile molecule by the enhanced concentration of the solvent molecules. With the same degree of caution, the difference between the different interlayer spacings extrapolated to zero solvent content may be viewed as an indication of the combined effect of the three phenomena for the amphiphilic molecules as such when in the liquid crystalline state.

The results make evident the fact that the changes in the extrapolated interlayer spacing are smaller than those at high concentrations (Fig. 1). This fact can be accommodated within a model with a distribution of solvent molecules between two sites: 1) between the amphiphilic layers, and 2) between the individual amphiphi-

TABLE II

"Percent Penetration" of a Solvent between Amphiphile Molecules for Polyethylene Glycols. EO = Ethylene Glycol Units.

Solvent	EO	$(EO)_2$	$(EO)_3$	$(EO)_4$
"Percent Penetration"	14	41.5	43.6	63.1

lic molecules. With increased chain length of the solvent molecule, it appears reasonable to conclude an enhanced perturbation of the amphiphile molecules by the presence of the solvent, presumably because of an increased degree of penetration.

This tentative conclusion is supported by the fact that the presence of polyethylene glycols with the highest molecular weight and presumably the highest penetration also caused destabilization of the lamellar structure at a lower fraction in the liquid crystalline structure.

It is further supported by the results from the mono alkyl ethers of the ethylene glycol and propylene glycols. These solvents gave no increase of the interlayer spacing with increased amount of solvent; in fact, the propylene glycol probably gave a reduction. Calculations of "percentage penetration" would now yield 100% and in excess of that value and the numbers are of little meaning. The obvious conclusion is a pronounced penetration between the lecithin molecules by the solvent molecules, causing excessive perturbation of the former ones.

This conclusion is supported by the fact that the liquid crystalline structure will tolerate even lower fractions of these solvents before phase separation takes place. It appears reasonable to conclude that the replacement of one terminal OH group with a short alkyl group has a significant influence on the location of the solvent molecule.

ACKNOWLEDGEMENT

The research was supported by NSF Grant INT 78 05868.

REFERENCES

1. P. Ekwall, in Advances in Liquid Crystals, G. H. Brown, ed., Academic Press, New York (1974), Vol. 1, p. 1.

2. S. Friberg, Mol. Cryst. Liq. Cryst. <u>40</u>, 49 (1977).

3. S. Friberg, Naturwiss. <u>64</u>, 612 (1977).

4. D. M. Small and M. Bourges, Mol. Cryst. Liq. Cryst. <u>1</u>, 541 (1966).

5. K. Larsson, in <u>Surface and Colloid Science</u>, E. Matijevic, ed., Wiley & Sons, New York (1973), Vol. 6, p. 261.

6. P. A. Winsor, in <u>Liquid Crystals and Plastic Crystals</u>, G. W. Gray and P. A. Winsor, eds., Ellis Harwood, Chichester (1974), Vol. 1, p. 199.

7. D. Chapman, in <u>Liquid Crystals and Plastic Crystals</u>, G. W. Gray and P. A. Winsor, eds., Ellis Harwood, Chichester (1974), Vol. 1, p. 288.

8. G. H. Brown and J. J. Wolken, <u>Liquid Crystals and Biological Structures</u>, Academic Press, New York (1979).

9. P. A. Winsor, Chem. Rev. <u>68</u>, 1 (1968).

10. <u>Liquid Crystalline Order Polymers</u>, A. Blumstein, ed., Academic Press, New York (1978).

11. A. Blumstein, R. B. Blumstein, S. B. Clough and E. C. Hsu, Macromolecules <u>8</u>, 73 (1975).

12. A. V. Tobolsky and E. T. Samulski, in <u>Liquid Crystals and Plastic Crystals</u>, G. W. Gray and P. A. Winsor, eds, Ellis Harwood, Chichester (1974), Vol. 1, p. 175.

13. N. Moucharafieh and S. E. Friberg, Mol. Cryst. Liq. Cryst. <u>49</u>, 231 (1979).

14. M. A. El-Nokaly, L. D. Ford, S. E. Friberg and D. W. Larsen, J. Colloid Interface Sci. <u>84</u>, 228 (1981).

15. D. W. Larsen, S. E. Friberg and H. Christenson, J. Am. Chem. Soc. <u>102</u>, 6565 (1980).

16. W. S. Singleton, M. S. Gray, M. L. Brown and J. L. White, J. Am. Oil Chem. Soc. <u>4</u>, 53 (1965).

17. K. Fontell, in <u>Liquid Crystals and Plastic Crystals</u>, G. W. Gray and P. A. Winsor, eds., Ellis Harwood, Chichester (1974), Vol. 2, p. 80.

THE DIAMAGNETIC ANISOTROPY OF LYOTROPIC NEMATIC MESOPHASES

Maria Elisa Marcondes Helene, Leonard W. Reeves* and
Carol Robinson

Chemistry Department
University of Waterloo
Waterloo, Ontario, Canada N2L 3G1

ABSTRACT

Successive replacement of aliphatic surfactant chains in type
II DM nematic phases by an aromatic amphiphile, potassium heplyoxy-
benzoate increases the susceptibility anisotropy of the mesophase,
which ultimately passes through zero and becomes positive at the
higher replacement concentrations. The magnetically isotropic condition has been named a type 0 DM nematic while subsequent behaviour
with $\Delta\chi > 0$ has been called a type I DM mesophase. The velocity constant for alignment of the type I DM mesophase is a linear function
of the concentration of the aromatic surfactant. A simple model
for the effect observed accounts in a qualitative manner for the
experimental results.

INTRODUCTION

The nature of a broad range of aqueous lyotropic nematic
phases[1] is now well established from studies of phase equilibria[2]
polarised microscopy[3-5], nuclear magnetic resonance[6-9] and low
angle X-ray diffraction[10-13]. There are two principal classes of

nematic systems, one based on finite disk-bilayer micelles[6,7] and the other on rod-like bilayer micelles[6,7]. The nematic phases align in a magnetic field with the director either parallel (Type I)[8] or perpendicular to the applied static field (Type II)[8,9]. For amphiphiles which are constituted from aliphatic hydrocarbon chains, the co-operative alignment leads to a disposition of the pseudo-extended chains, in micelles, perpendicular to the applied field[6,7]. The pseudo-extended chain[7] for the case of aliphatic chains there-fore, manifests a negative diamagnetic anisotropy[6,9]. When the bilayer micelles are of the disk-type for such a case, the director of the mesophase coincides with the mean direction of the extended chains and so it aligns perpendicular to the field. Such mesophases, to both indicate their diamagnetic anisotropy and micelle shape, have been named Type II DM[2]-the type II specifies the diamagnetic susceptibility anisotropy $\Delta\chi < 0$ for the phase and DM specifies disk micelles. In contrast the mesophases based on aliphatic hydrocarbon detergents which from finite rod micelles are called "Type I CM" for $\Delta\chi > 0$ and finite cylindrical micelles[2,7]. For type I CM meso-phases the director or symmetry axis of the mesophase is perpendicu-lar to the radially extended chains enclosed in the cylindrical micelle and so it becomes the equilibrium direction of the magnetic field[2].

Recently mesophases of the disk-type have been prepared in this laboratory, in which the diamagnetic anisotropy has been systemati-cally varied by incorporating mixtures of para-substituted aromatic detergents and aliphatic chain detergents together to form a single disk-micelle nematic phase[14]. The aromatic ring has the opposite sign of $\Delta\chi$ to the extended aliphatic chain and consequently several mesophases were taken through the transition Type II DM → Type 0 DM → Type I DM by adding the aromatic amphiphile. Boden et al[4] re-ported recently that fluorocarbon detergents behaved like Type I DM nematics, but no intermediate magnetically isotropic mesophase was prepared. These transitions in the sign of $\Delta\chi$ occur without a

change of phase because the disk-micelles are preserved in the controlled variation of $\Delta \chi^{14}$. The present study is a study of the
change in $\Delta \chi$ with systematic variation of detergents.

EXPERIMENTAL SECTION

p-heptyloxybenzoic acid (Frinton labs) was converted to the
potassium salt by neutralisation with potassium hydroxide in ethanol
and recrystallisation several times, after filtering, from methanol.
The product is abbreviated as KHB in the subsequent text. Decanoic,
undecanoic, and lauric acids from Aldrich were converted in a similar manner to the potassium salts and purified. The products are
abbreviated respectively as KC_{10}, KC_{11} and KC_{12}. Sodium lauroyl
sarcosinate (Gardol) was kindly donated by Colgate-Palmolive Co.
Potassium lauroyl alaninate (LAK) was used from a stock prepared
previously[15]. Decanol (DeOH) was redistilled from Eastmann and the
D_2O was 99.7% deuterium from Merck, Sharp, and Dohme. The type II
DM stock mesophases were prepared according to previously reported
procedures[8] with the following weight ratios.
A. KC_{10}/KCl/DeOH/D_2O = 586.7/145.4/122.0/1102.0;
B. KC_{11}/KCl/DeOH/D_2O = 597.1/96.5/93.0/1185.0;
C. KC_{12}/KCl/DeOH/D_2O = 654.3/50.6/125.0/1454.0;
D. Gardol/Na_2SO_4/DeOH/D_2O = 610.0/116.0/111.0/1102.0;
E. LAK/Na_2SO_4/DeOH/D_2O = 418.0/84.4/72.8/1109.0.

The samples were investigated in 10mm NMR tubes in a Bruker
W.P. 80 spectrometer with a deuterium Larmor frequency of 12.28 MHz
and an external deuterium lock. The ambient probe temperature was
generally 32.8°C and for the experiments involving the measurements
of velocity of alignments in a series of related samples, the temperature was controlled to ±0.1°C. The alignment behaviour was
characterised as type I or type II by observing the initial powder
pattern on immersion in the field and the final sharp doublet spectra for the aligned sample[8,9]. The alignment velocity of type I

mesophases was obtained in the following manner. The samples were
immersed in the magnetic field for at least 18 hours or until the
line width of the individual quadrupole splittings did not change
overnight in the magnet. A rotation of the sample was roughly
calibrated as $\sim25^{o}$ by using a pointer mounted on the sample tube
and a simple fixed plastic protractor attached to the magnet housing.
The actual angle rotated is obtained by using the quadrupole splitting as a goniometric tool[8,9]. The equation for the quadrupole
splitting, $\Delta\nu$, is:

$$\Delta\nu = \frac{3}{2} Q_c S_a \frac{1}{2} (3 \cos^2\Omega - 1) \qquad (1)$$

where, Q_c is the quadrupole coupling constant for the deuterium
nucleus observed, S_a is the degree of order of the principal axis
of the electric field gradient and Ω is the angle between the director of the mesophase and the magnetic field. For type I behaviour the equilibrium final quadrupole splitting with $\Omega = 0^{o}$ cali-
brates the value of "$\frac{3}{2}Q_c S_a$" as equal to $\Delta\nu$. The reduced splitting
on rotation with $\Omega \propto 25^{o}$ gives the angle of rotation rather precisely. The re-alignment of the type I system from $\Omega \approx 25^{o}$ to a
final equilibrium $\Omega = 0$ starts immediately and is followed by the
values of $\Delta\nu$ changing with time. The angle Ω between the director
and the magnetic field as a function of time is given by[9,16]:

$$\ln \tan \Omega = -kt + \ln \tan \Omega_o \qquad (2)$$

Ω_o is the initial angle of rotation of the director from the field
direction at time 't' = 0 and 'k' is the velocity constant of
alignment.

$$k = \Delta\chi H^2/\lambda_1 \qquad (3)$$

$\Delta\chi$ is the diamagnetic susceptibility anisotropy of the medium
(i.e. $\chi_{11} - \chi\perp$), H is the applied magnetic field strength and λ_1
is a twist viscosity coefficient defined earlier[17]. In this study
'H' remains constant at about 18.79 k Gauss while $\Delta\chi$ is varied

by substituting aliphatic amphiphile with aromatic amphiphile in
the mesophase and consequently λ_1 changes also. The values of 'k'
were measured experimentally for type I DM mesophases obtained after
the sign transition in $\Delta\chi$ from negative through 0 to positive.

RESULTS

The aromatic amphiphile KHB was substituted on a mole for mole
basis in successively larger amounts for the pure aliphatic chain
detergents with the composition mole % detergents, water, electrolyte and water always preserved at the values calculable from mesophases A-E listed in the experimental section. The resultant multicomponent phases always formed easily, except near the extremity of
solubility of the aromatic detergent, and they proved to be nematic
in nature. The position of the transition in the sign of $\Delta\chi$ could
be qualitatively determined by increasing the amount of KHB and observing the change from type II to type I behaviour bracketed between two trial concentrations[14]. The measurement of the velocity
constant 'k' of alignment was then accomplished in the region of
type I behaviour for a series of concentrations of KHB. A typical
plot for the determination of 'k' is given as figure 1.

A plot of the rate constant 'k' versus the mole fraction of
aromatic rings in the micelles of the mesophases A-E yields a
straight line in all cases, which extrapolates to the mole fraction
at k=0 where $\Delta\chi=0$, thus a precise determination of this point is
always possible in such experiments, if the mesophases exists. This
graph of rate constants versus mole fraction aromatic rings is shown
in figure 2 for all mesophases A-E.

In order to make the addition of the aromatic rings more quantitative, figure 2 is reproduced using a mole fraction of aromatic
rings, with a unit methylene group as the aliphatic component. For
this purpose end methyl groups and oxygen in the chains are approximated as methylenes in the calculation. For example, for a chain

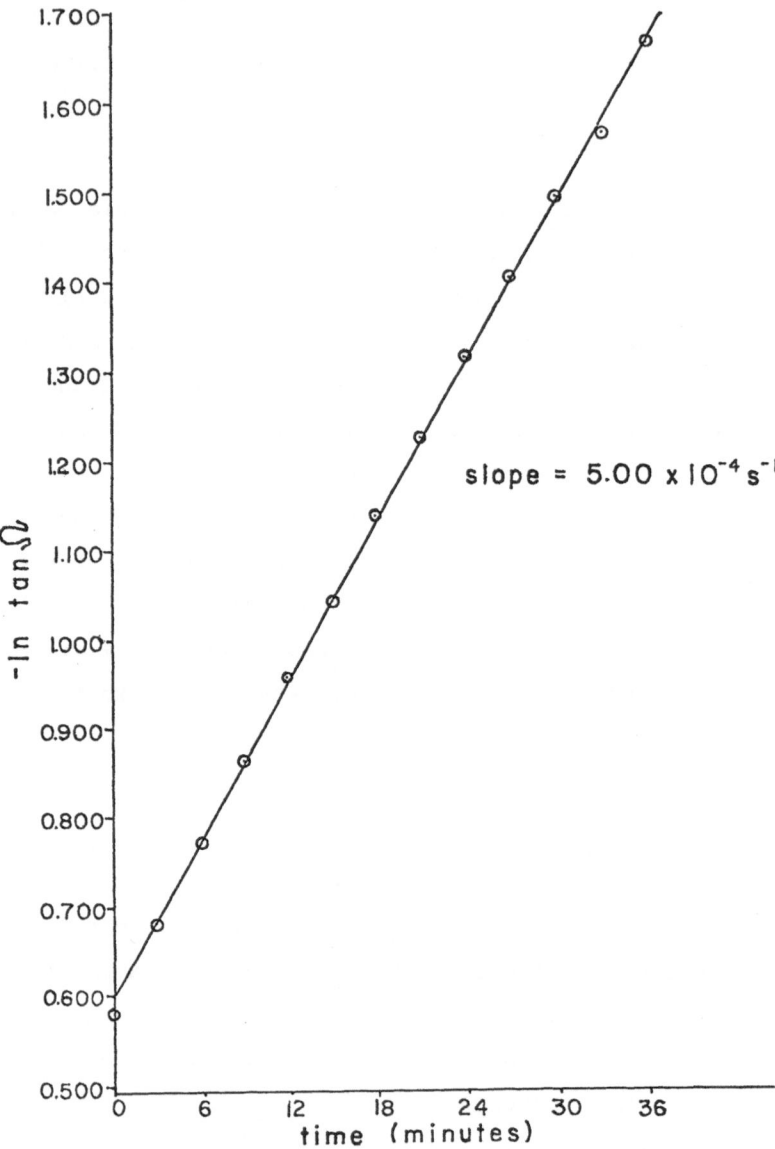

Figure 1. A plot of equation (2) for the system KC_{12}/KHB using
experimental data for Ω and k in the region of type I
behaviour. The rate constant 'k' in this case is 5 x
10^{-4} secs^{-1}.

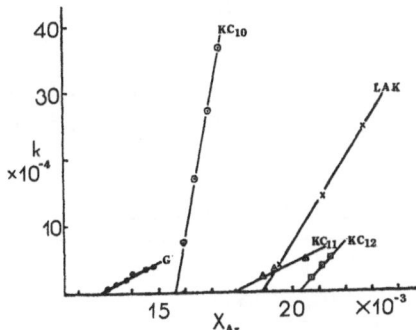

Figure 2.   Rate constants 'k' x 10^4 sec^{-1} plotted versus the mole
fraction of aromatic rings in the total detergents of
the disk micelles in the mesophases based on compositions A-E in experimental section for the type I DM
region.

of n carbons in the host detergent, (decanoic acid is considered to
have 9 methylenes, lauric, 11 etc.), 8 carbons in the heptyloxy
group (count the -O- as a methylene), and 10 methylenes in the decanol we have the following general mole fraction of aromatic rings
'x_{Ar}' in the micelle,

$$x_{Ar} = \left[\frac{n_{KHB}}{n\left(n_{KC_{10}}\right) + 8\left(n_{KHB}\right) + 10\left(n_{DeOH}\right)} \right] \tag{4}$$

$n_{subscript}$ = number of moles of subscripted amphiphile in the
mixture of surfactants.

The graph in figure 2 may be re-calculated simply by using the
mole fraction KHB in the micelle, but this tends to overemphasize
the mole fraction of aromatic rings because of the consequent neg-
ligence of the attached heptyloxy groups. The linear dependence is
still very good if this is done.

In order to confirm that no phase transition has taken place
the quadrupole splitting was measured from extrapolated concentrations at the $\Delta\chi=0$ point from both type II and type I limits. The

TABLE I

Compositions of the type 0 DM mesophases with $\Delta\chi=0$. x_{Ar} is the mole fraction of aromatic rings in the bilayer expressed as in equation (4) of the text. The temperature of the experiments was $32.8\pm0.1^{o}C$.

Mesophase	Aliphatic Components	x_{Ar} at $\Delta\chi-0$
A	KC_{10}/DeOH	1.561×10^{-2}
B	KC_{11}/DeOH	1.799×10^{-2}
C	KC_{12}/DeOH	2.024×10^{-2}
D	Gardol/DeOH	1.285×10^{-2}
E	LAK/DeOH	1.875×10^{-2}

ratio of splittings was always $|2|$ within a very small error $\pm0.1\%$ for all systems A-E, in accordance with equation (1) for a change in Ω of 90° to 0° without change in 'S_a'.

The velocity constants 'k' depend greatly on temperature and this is illustrated in figure 3 for the KHB/KC_{12} system where the linear dependence on mole fraction aromatic has a considerably different slope at $32.8^{o}C$ and $31.0^{o}C$. The mole fraction at which $\Delta\chi=0$ is not altered by the small temperature change. In Table I the composition of the mesophases that are exactly Type 0 DM, namely diamagnetically isotropic ($\Delta\chi=0$), are listed.

DISCUSSION

(a) Distinction between a sign change in $\Delta\chi$, with and without a phase change

A recent study of the genuine phase transition rods $\rightarrow$ disks[18] has shown that by adding a neutral amphiphile or by adding electrolyte to lyotropic nematic systems[8,19] the cylindrical micelles decrease in size, reach the point of the phase transition at a small

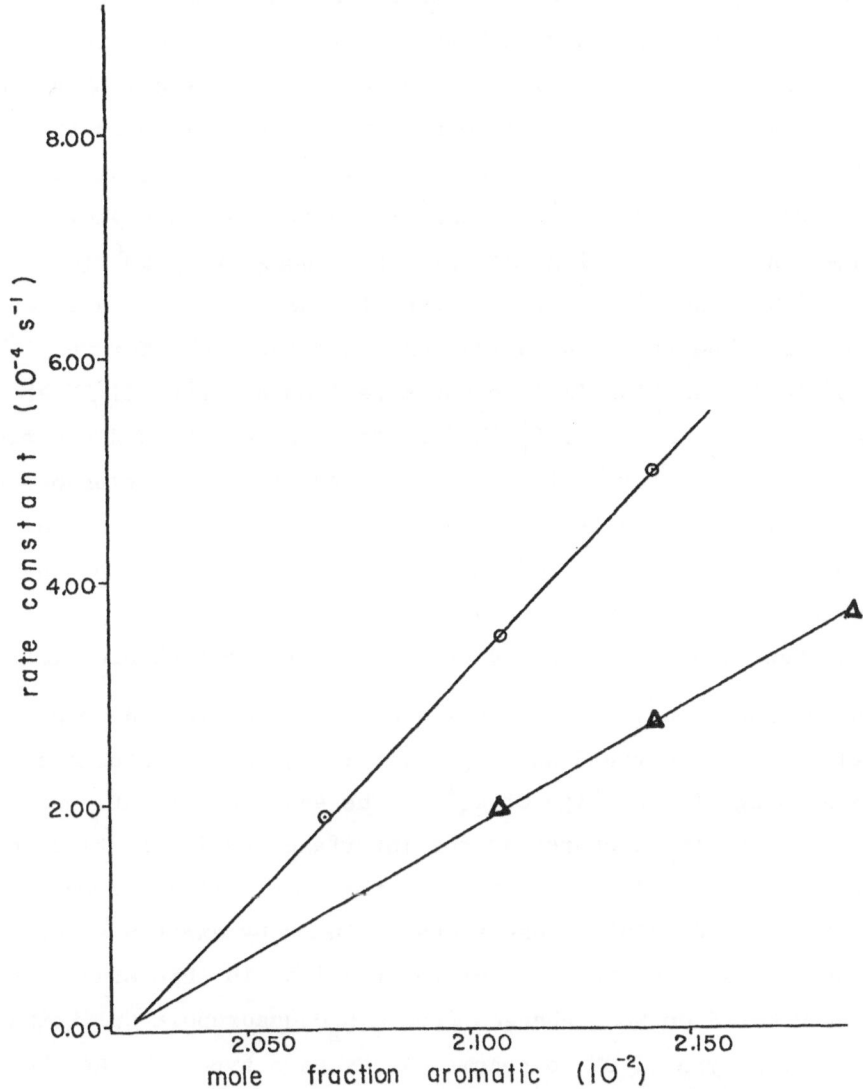

Figure 3. Rate constant 'k' versus mole fraction aromatic detergent in KC_{12}/KHB mixtures in mesophases at two different temperatures; (a) $31.0\pm0.1^{\circ}C$ and (b) $32.8\pm0.1^{\circ}C$.

size   estimated at $\simeq$ 500 Å in length and then with further chemical additions small disks form of about the same diameter (500 Å) with an accompanying change in the sign of the diamagnetic anisotropy similar to the changes observed here. There is no direct relationship between the micelle motion of the rods in the type I CM phase and the disks in the type II DM phase consequently although there is a sign change in $\Delta\chi$ the quadrupole splitting does not alter at the transition by a factor $|2|$. This is equivalent to saying that S_a changes in the transition, as well as Ω changing by 90° (see equation (1)). The 'S_a' is the degree of order along the axis of the director. The factor usually found at around $30^{\circ}C$ for the ratios of '$\Delta\nu$' (equation (1)) in the type I CM and type II DM at the phase transition is $\simeq |1.6| \rightarrow |1.7|$. This is well beyond experimental error for a factor of 2 and in any event the addition of decanol or electrolytes is unlikely to change the sign of $\Delta\chi$ for the mesophase. The origin of the sign change in $\Delta\chi$ is the change in micelle shape. Apart from a very recent report[14] therefore, all prior studies were genuine phase transitions rod $\rightarrow$ disk micelles.

The present study was chemically designed to avoid a phase transition, preserve the disk shaped micelles, but nevertheless achieve a change in the sign of $\Delta\chi$[14]. The addition of KHB, which has an aromatic ring anchored at the interface and is rotating about its para-axis perpendicular to the bilayer, must without a phase change at the $\Delta\chi=0$ point, cause a change in $\Delta\nu$ by exactly a factor $|2|$. The precision of this factor is excellent in this study. e.g., Mesophase C based on KC_{12} changes from a D_2O quadrupole splitting of 322 Hz as a type II DM to become 645 Hz as a type I DM at the $\Delta\chi$ transition. Mesophase D based on Gardol has corresponding splittings type II DM 428 Hz and type I DM, 856 Hz. Without further evidence, it is clear that the chemical synthesis of disk shaped micelle mesophases with variable $\Delta\chi$ has been achieved[14].

(b) Estimates of the relative diamagnetic anisotropies of methy-
     lene groups and aromatic rings.

The finite disk micelles are bilayers[2,7,20] where there is considerable deviation from all trans aliphatic chains[21]. It is reasonable to assume that the origin of the diamagnetic anisotropy of
the mesophase derives from the pseudo-extended hydrocarbon in the
bilayer, and the ordered interface of the micelle, which contains
some ions, water and the polar or ionic head groups of the surfactants. The contribution of the interstitial water to the overall
diamagnetic anisotropy can only occur for the adsorbed water on the
micelle. This interface is a relatively small fraction of the bilayer structure, so that to a first approximation the contribution
can be ignored. Assuming that the interface head groups etc. have
negligible effect, a simple calculation based on the hydrocarbon
chains can be made. At the point $\Delta\chi=0$ for the mesophase, we can
assume that:

$$\chi^{\perp} B \; x_{Ar} = x_{\|M} \; (1 - x_{Ar}) \tag{5}$$

where $\chi\perp_B$ is the mean magnetic susceptibility of the benzene ring,
measured in the direction $\perp$ to the plane of the ring. $\chi_{\|M}$ is the
mean or effective magnetic susceptibility of a unit methylene group
in an all trans rotating chain, measured in a direction parallel to
the chain. 'x_{Ar}' has been defined in equation (4). Now a similar
consideration for the susceptibility perpendicular to the all trans
chain gives us another relation:

$$\frac{1}{2} (\chi_{\|B} + \chi\perp B) x_{Ar} = \chi\perp_M (1 - x_{Ar}) \tag{6}$$

where $\chi_{\|B}$ is the susceptibility parallel to the axis of a benzene
ring and $\chi\perp_M$ is the mean susceptibility observed for a rotating all-
trans chain, of a unit methylene in the chain, and perpendicular to
that chain. Subtracting (5) from (6) and re-arranging gives:

$$\frac{x_{Ar}}{(1 - x_{Ar})} = \frac{- 2\Delta\chi_M}{\Delta\chi_B} \tag{7}$$

where $\Delta\chi_M = (\chi_{\parallel M} - \chi_{\perp M})$

and $\Delta\chi_B = (\chi_{\parallel B} - \chi_{\perp B})$

$\Delta\chi_M$ is a simplified diamagnetic anisotropy of a methylene group, which has three independent susceptibility components; except that two can be considered equal in a rapidly rotating extended hydrocarbon chain. The value of the ratio in equation (7) varies with the mesophase between 1.29×10^{-2} and 2.02×10^{-2} [table I, assuming $(1 - x_{Ar}) \approx 1$]. While these variations are remarkably constant considering the scale of the approximations and they do represent strong evidence for the dominance of the hydrophoic compartment in the susceptibility properties of the mesophase, they also manifest regularities that bear further comment. For the series of mesophases A, B, and C the only significant chemical change is the length of the aliphatic chain for a carboxylate amphiphile. A regular increase is observed (table I) from 1.56×10^{-2} to 2.02×10^{-2} as the chain lengthens from 10 to 12 (9 to 11 effective methylene groups) carbons. Any chain defects such as kinks, jogs and single gauche rotations[7] serve to decrease the mean susceptibility anisotropy per methylene in the aliphatic chain. The degree of order profiles of laurate and decanoate chains in disk micelles have been studied in some detail[22]. The degrees of order of decanoate chain methylene segments are considerably lower than those of laurate chains (compare figs. 1 and 4 of reference 22) and thus the mean contribution per methylene segment to a diamagnetic anisotropy of the mesophase is expected to be smaller. In order to be categoric about the trends in the ratio of equation (7) a knowledge of the rigidity of anchoring of the aromatic ring is essential as well as some other factors. The extrapolation of the ratios in table I to very long aliphatic chains does give a true relative measure of $\Delta\chi_M$ and $\Delta\chi_B$ in a rigid system and thus we can say the susceptibility ratio $\Delta\chi_M/\Delta\chi_B$ is certainly greater than 1.012×10^{-2} for such a case.

(c)   Variation of 'k' with aromatic detergent content.

The velocity constants 'k' for alignment of the type I mesophases are seen to be a linear function of x_{Ar} in figure 2. Based
on the simple all trans chain model in (b) the susceptibility
anisotropy '$\Delta\chi_{phase}$' of the mesophase becomes:

$$\Delta\chi_{phase} = \frac{\Delta\chi_B}{2} x_{Ar} - \Delta\chi_M \qquad (8)$$

neglecting $\Delta\chi_M$ in comparison to $\Delta\chi_B$ in their sum. The velocity
constant of alignment for the type I DM mesophases becomes:

$$k = - \frac{\left(\frac{\Delta\chi_B}{2} x_{Ar} + \Delta\chi_M\right)}{\lambda_1} H^2 \qquad (9)$$

with k=0 at

$$x_{Ar} = - \frac{2\chi_M}{\Delta\chi_B} \qquad (10)$$

The linear relation observed experimentally between k and x_{Ar} in
figure 2 at constant magnetic field H, implies that λ_1 over the
small concentration range near $\Delta\chi_{phase}$=0 is not a determining factor
in the linear relation. It can be assumed that its variation is
small compared to that in the term $\Delta\chi_B x_{Ar}/2$. The slopes of 'k'
versus 'x_{Ar}' in figure 2 are therefore approximate comparisons of
the inverse of the twist viscosity coefficient λ_1. The mesophases
have the following relative values of λ_1 near the $\Delta\chi$=0 transition
point $A(KC_{10}) < E(LAK) < C(KC_{12}) < D(KC_{11}) < D(Gardol)$.

In figure 3 two slopes of k versus x_{Ar} are available at a
temperature difference of $1.8^{o}C$. Based on the simple model above
it is apparent that λ_1 decreases with temperature for mesophase C
doped with aromatic surfactant and this is a large effect.

CONCLUSIONS

The results reported here confirm the hypothesis that the dominant contribution to the diamagnetic susceptibility anisotropy of lyotropic aqueous nematic phases resides in the hydrocarbon chains of the hydrophobic bilayer region. A mesophase with a $\Delta\chi = 0$ can be prepared, which has been called type O DM, by adding potassium heptyoxybenzoate as a replacement for aliphatic chain surfactants in the micelle bilayers. While the mesophase does behave as a magnetically isotropic medium at a critical concentration of aromatic surfactant the first derivative of this $\Delta\chi$ with respect to composition changes is not zero at $\Delta\chi = 0$ unlike the situation in an isotropic solution where this derivative is zero. This is the first reported case of a physical property, in a liquid crystal, which can be rendered isotropic.

ACKNOWLEDGEMENTS

This work was supported by operating funds awarded by the National Science and Engineering Research Council of Canada. The assistance of B. J. Forrest is gratefully acknowledged.

REFERENCES

1. K. D. Lawson and T. J. Flautt, J. Am. Chem. Soc. 89, 5489 (1967).

2. F. Y. Fujiwara and L. W. Reeves, J. Phys. Chem. 84, 653 (1980).

3. A. Saupe, J. Colloid and Interface Science 58, 549 (1977).

4. N. Boden, P. H. Jackson, M. C. Mullen and M. C. Holmes, Chem. Phys. Letters 65, 476 (1979).

5. M. Acimis and L. W. Reeves, Can. J. Chem. 58, 1533 (1980).

6. F. Fujiwara, L. W. Reeves, M. Suzuki and J. A. Vanin in "Solution Chemistry of Surfactants", 1, 63 (1979), ed. K. L. Mittal.

7. B. J. Forrest and L. W. Reeves, Chem. Revs. 81, 1 (1981).

8. K. Radley, L. W. Reeves and A. S. Tracy, J. Phys. Chem. 80, 174 (1976).

9. F. Y. Fujiwara and L. W. Reeves, Can. J. Chem. 56, 2178 (1978).

10. J. Charvolin, J. Phys. Lett. (Paris) 40, L587 (1980).

11. L. Q. do Amaral, C. A. Pimentel, M. R. Tavares and J. A. Vanin, J. Chem. Phys. 71, 2980 (1979).

12. L. Q. do Amaral and M. R. Tavares, Mol. Cryst. Liqu. Cryst. Lett. 56, 203 (1980).

13. A. M. Figuredo Neto, Ph.D. Dissertation, University of Sao Paulo, Brazil (1981).

14. B. J. Forrest, L. W. Reeves and C. J. Robinson, J. Phys. Chem. 85, 3244 (1981).

15. B. J. Forrest, L. W. Reeves, M. R. Vist, C. Rodger and M. E. Marcondes Helene, J. Am. Chem. Soc. 103, 690 (1981).

16. F. M. Leslie, G. R. Luckhurst and H. J. Smith, Chem. Phys. Lett. 13, 368 (1972).

17. P. G. Gennes, in 'The Physics of Liquid Crystals', Claredon Press, Oxford, England, 1974.

18. B. J. Forrest and L. W. Reeves, J. Am. Chem. Soc. 103, 1641 (1981).

19. D. M. Chen, F. Y. Fujiwara and L. W. Reeves, Can. J. Chem. 55, 2396 (1977).

20. B. J. Forrest and L. W. Reeves, Mol. Cryst. Liqu. Cryst. 58, 233 (1980).

21. B. J. Forrest, F. Y. Fujiwara and L. W. Reeves, J. Phys. Chem. 84, 622 (1980).

22. M. Acimis and L. W. Reeves, J. Phys. Chem. 84, 2279 (1980).

PHYSICAL MODELS OF LIPID MEMBRANES IN LATENT CANCER CELLS:

ORDERING EFFECTS OF PETROLEUM HYDROCARBONS

A. Wesley Horton[1,2] and George R. Penk[1]

(1) Section Chemical Biology and Oncology
    Department of Public Health and Preventive Medicine
(2) Department of Biochemistry
School of Medicine
Oregon Health Sciences University
Portland, Oregon 97201

Chemical carcinogenesis is known for the long quiescent or latent period between the time of first exposure to the carcinogen and the stage of relatively uncontrolled multiplication of tumor cells. In wax pressmen who operated the plate and frame presses to remove the first crude wax from chilled petroleum distillates this latent period before the onset of scrotal cancer lasted 24 years, on the average[1]. A very much longer delay (> 50 yrs. average) was noted among workmen, known as "mule spinners", who operated machinery for spinning cotton thread and were exposed to poorly refined, but dewaxed, lubricating oil sprayed from the unsealed bearing surfaces of the spindles[2]. This difference in latent periods in comparison with differences in composition between the paraffin distillates and the mainly naphthenic lubricants provided the important clue that ultimately led to an understanding of the carcinogenicity of petroleum fractions[3].

It turns out that, given a material as low in carcinogen content as petroleum, the controlling factors are non-carcinogenic,

promoting components. As simple a compound as n-dodecane may increase the effect of a low concentration of the carcinogen, benzo-
[a]pyrene, by three orders of magnitude[4]. Such promoting activity
has been shown by animal experiments to be characteristic of long
chain compounds up to at least $C_{20}H_{42}$, n-eicosane, including some
alkylcyclohexanes and alkylbenzenes[5,6]. Since skin and scrotal
cancer has not been a problem among workmen exposed to wood tars,
it is assumed that the terpenes of these tars do not have the promoting activity of long chain hydrocarbons.

The low concentrations of polycyclic aromatic carcinogens in
petroleum fractions are readily reduced to an insignificant level
by solvent extraction[3]. However neither this process nor catalytic
hydrogenation removes long chain promoters. Indeed the hydrocrack-
ing that may accompany the latter process tends to increase the
concentration of alkanes. The potential problem about these non-
carcinogenic but still promoting hydrocarbons is that repeated
contact with these materials may cause the progression of latent
cancer cells, initiated by earlier contact with carcinogens, to the
stage of uncontrolled multiplication characteristic of cancer[7].

Our research therefore has been focussed on the identification
of the molecular structures associated with promoting activity and
the cellular mechanisms by which they enhance the progression of latent cancer cells. The considerable cost and lengthy time requirements of full animal tests has mandated the development of rapid
short-term assays for promoting activity.

Consideration of likely cellular mechanisms of action of such
hydrocarbons focussed attention on the phospholipid-rich membranes
of the epithelial cells involved. This is the region in the cell
where such hydrocarbon oils would be localized after absorption
into the tissue. Simplified physical models of the cell membranes
have been developed utilizing naturally occurring lipid components
of membranes, including cholesterol.

In our previous research the lipids were dispersed as inverted micelles in a cycloparaffinic solvent. Changes in the interfacial properties of an emulsion between this organic phase and buffered aqueous KCl, resulting from incorporation of various petroleum hydrocarbons, were monitored by changes in the kinetics of transfer of anionic dyestuff probe from the aqueous phase into the lipid miscelles[8]. The extent of retardation of the kinetics by certain alkanes correlated well with their promoting activity in skin carcinogenesis in mouse tests. Pristane (2, 6, 10, 14-tetramethylpentadecane), which had not been tested biologically, produced a greater retardation of the transfer of the probe than did any n-alkane. Subsequent biological tests confirmed the correlation[9]. Pristane proved to be the most active promoter among the alkanes tested to date.

The question accompanying these experimental correlations was what important physical property of the model system and of the membranes of latent cancer cells is critically altered by promoting hydrocarbons. The most likely effect was hypothesized to be a change in micellar or membrane fluidity or molecular order in the hydrophobic regions. In this paper we present the results of experiments to estimate the effects of various petroleum hydrocarbons on molecular order in lyposomal models of cell membranes. The relevance of these systems to the membranes of latent cancer cells will be discussed in terms of their specific sensitivity in one direction, reduced order, to those hydrocarbons that are capable of promoting the progression of these cells to cancer. The converse, increased order induced by hydrocarbons that counteract such progression, would add further significance to this approach.

MATERIALS AND METHODS

Fluorescence Polarization Technique

Materials. Egg phosphatidylcholine (egg lecithin) solutions obtained from either Grand Island Biological Company (Gibco, Grand

Is., NY) or Calbiochem – Behring Corporation (LaJolla, Calif.) are
evaporated under prepurified N_2 and dried under vacuum for more than
3 days. Each shows only a trace of lysophosphatidylcholine (lysolecithin) by TLC on silica gel when spotted with 300 µg of the lecithin and developed with 65:25:4 chloroform: methanol: water. Both
are used without further purification. Egg lysolecithin (> 98%)
and oleic acid (> 99%) are obtained from Sigma Chemical Co. (Saint
Louis, MO.) and used as received. USP cholesterol (Merck, Rahway,
New Jersey) is purified by triple recrystallization from methanol.
Pristane (2,6,10,14-tetramethylpentadecane) (Aldrich Chemical Co.,
Milwaukee, Wisc.) and the normal paraffins, C_{14}, C_{16}, C_{18}, C_{20}, C_{22},
C_{24}, (Humphrey Chemical Co., North Haven, Conn.) are purified by
chromatography on activated silica gel to remove any ultraviolet ab-
sorbers. 1,6-diphenyl-1,3,5-hexatriene (DPH) (98%, Aldrich Chemical
Co.) is used as received. All solvents are from Burdick & Jackson,
"distilled in glass" (Muskegon, Michigan), and are used without pre-
servatives, taking precautions to avoid degradation from light or
oxygen.

Preparation of Liposomes. Solutions of 1.00 mM egg lecithin,
egg lysolecithin, cholesterol, and oleic acid are prepared in 1:1
chloroform: methanol. The lipids are pipetted into 4 dram vials
(3.6 ml of the lecithin solution and the others in the desired
ratios). The solvents are evaporated under prepurified nitrogen
and dried at 60-70° under vacuum for 2 hours.

1.0 mg of the desired hydrocarbon is added directly onto the
co-precipitate (as either a solid or a liquid), the atmosphere
flushed with prepurified nitrogen, and the vials heated at 60° for
1½ hours. Six ml of 0.13M KCl buffered with phosphate to pH 6.8 is
added to each vial. The mixtures are then sonicated for 6 minutes
under an atmosphere of dry nitrogen using a Branson sonicator, Model
185 (Branson Sonic Power Co., Danbury, Conn.), fitted with a ½" tip,
and operating at 60% full power output. The initial 1 min., or less,

of sonication is performed without cooling to allow the mixture to reach ∿65°. During this first minute, 4.0 μl of 1.00 x 10^{-3} $\underline{M}$ DPH in tetrahydrofuran (THF) is added, or in the case of blanks without probe, 4.0 μl of THF alone. The samples are then kept cooled in an ice/water bath for the remaining time under sonication.

The resulting liposome preparations are centrifuged at 9,400 x g for 30 min. at 27° and the supernatent transferred to test tubes equilibrating in a water bath at 27°. Samples remain in the dark until ready to use and are kept at least 1 minute in the thermostatically controlled spectrophotofluorometer cell holder (in the dark) before being exposed to the excitation beam.

Fluorescence Measurements. Fluorescence measurements are made on an Aminco-Bowman spectrophotofluorometer fitted with interchangeable plastic laminated polarizing filters (Edmond Scientific, Barrington, N.J.) in both the excitation and emission beam paths. Our instrument also utilizes a thermostatically controlled cell holder. For all measurements in this study the temperature is controlled within ± 0.1°. A 1.0 cm. cell (path length) is used for all measurements.

The monochromatic excitation light (355 nm) is polarized vertically, and the emission intesities I_{vv} and I_{vh} (at 434 nm) observed after being passed through either vertically or horizontally polarized filters, respectively. Due to the scattering of light, the observed emission intensities I_{vv} and I_{vh} must be corrected by subtracting the emission intensities of appropriate blanks (identical lipid-hydrocarbon dispersions which do not contain DPH). The fluorescence anisotropy, r, is calculated from the equation

$$r = \frac{I_{vv}G - I_{vh}}{I_{vv}G + 2I_{vh}}$$

where the correction factor $G = I_{hh}/I_{hv}$, I_{hh} and I_{hv} being the fluorescence intensities observed in the horizontally and verti-

cally polarized orientations after being excited with horizontally
polarized light. Because of the diffraction gratings employed a
selective reduction of the intensity of one polarization orientation (the vertical component) occurs and therefore the correction
factor must be applied to the term I_{vv}[10].

RESULTS AND DISCUSSION

It is probable that only a small proportion of the cells in
the epidermis of a mouse that has received a single threshold application of 7,12-dimethyl-benz[a]anthracene (DMBA) are "initiated"
and thus have a latent potential for cancer. Analysis of the membranes of epidermal cells following initiation by DMBA and after a
few weeks for the initial inflammatory reaction to subside would
give little information about the critical changes that have occurred
in the relatively few latent cancer cells. Indeed, it might well be
argued that, anyway, the critical changes probably occurred in the
cell nuclei and might or might not be reflected in the lipid composition of the cell membranes.

Hence a pragmatic approach had to be taken in the development
of an appropriate composition of phospholipids and cholesterol to
use as a model of the cell membrane of a latent cancer cell. The
only proof of the suitability of the composition chosen would be that
a measurable property of the system is altered in one direction by
those hydrocarbons known to be promoters of carcinogenesis in mice
(and by epidemiologic correlation, in man) and unaltered or altered
in the opposite direction by non-promoters. The ground rules of the
game therefore are based upon the previously established cocarcinogenic or promoting activities of the hydrocarbons listed in Table 1,
as derived from previous work[6,9].

In the next to last column of Table 1 is shown a comparison of
the changes in fluorescence anisotropy induced in egg phosphatidylcholine liposomes resulting from inclusion of various C_{14}-C_{24} al-

TABLE 1

Effects of Alkanes on the Fluorescence Anisotropy
of DPH in Lipid Liposomes.

| Alkane | | Relative Cocarcinogenic Activity | Anisotropy, r, as % of Control | |
Chain Length	Molar Ratio Alkane/Lecithin		Egg Lecithin Alone	Mixed Lipid Liposomes[2]
$n\text{-}C_{14}H_{30}$	1.40	1.5	91	92
$n\text{-}C_{16}H_{34}$	1.23	1.7	76	87
$n\text{-}C_{18}H_{38}$	1.09	1.9	72	87
$iso\text{-}C_{19}H_{40}$ (pristane)	1.03	2.2	82	89
$n\text{-}C_{20}H_{42}$	0.98	2.0	95	87
$n\text{-}C_{22}H_{46}$	0.89	----	105	117
$n\text{-}C_{24}H_{50}$	0.82	1.1	116	118
Control	0.00	1.0	100[1]	100[3]

1. r value for control = 0.095 ± .004.
2. Egg Phosphatidyl Choline, Egg Lysophosphatidyl Choline, Cholesterol, Oleic Acid. Molar ratios, 2:1:1:0.5.
3. r value without added alkanes = 0.160 ± .006.

kanes versus their relative cocarcinogenic activities. It is seen that at 27° this relatively unsaturated phospholipid used alone does not serve as a suitable model, the maximum decrease in molecular order experienced by the probe DPH being produced by the $n\text{-}C_{16}$ and C_{18} alkanes. These chain lengths correspond to the most common ester chain lengths in egg lecithin. The very active promoter n-eicosane, had no significant effect. The gradual decrease in r values to C_{18} followed by a gradual increase to C_{24} is not characteristic of the biological activity in this range. One might speculate that, if the molecular order in the lipid domains of the membranes of latent

cancer cells were as low as that in the liposomes of egg phosphatidylcholine, then n-octadecane might be the most active of the alkane
promoters.

It may be noted that in our earlier research using inverted
phospholipid micelles as membrane models[8] the egg lecithin available
in 1975 also failed to give a satisfactory correlation with promoting activity until the micellar properties were modified by inclusion of other classes of membrane lipids, particularly cholesterol
and lysophosphatidyl choline. Recent studies using fluorescence polarization of DPH[11] and ^{31}P-NMR[12] indicate a considerable degree of
heterogeneity in such mixed phase vesicles and give evidence that
such lipid domains are typical of plasma membranes of lymphocytes.
The coexistence of bilayer and micellar regions were observed at
molar ratios of phosphatidyl choline to lysophosphatidyl choline of
2:1. Apparently, such heterogeneity may be essential to provide the
model we are seeking that will distinguish promoters from non-
promoters.

Accordingly, we investigated the changes in fluorescence anisotropy, r, of the egg lecithin liposomes resulting from inclusion of
cholesterol, egg lysophosphatidyl choline, and oleic acid (the latter had proved essential in our studies using inverted micelles to
show a reversed physical effect of n-$C_{24}H_{50}$, correlating with its
lack of promoting activity)[6]. Both cholesterol and, to a lesser extent, lysolecithin increased the r value at 27°.

In Table 2 are shown the effects of a series of concentrations
of oleic acid on liposomes containing egg phosphatidyl choline, egg
lysophosphatidyl choline, and cholesterol in a 2:2:1 molar ratio.
As in the previous work using mixed inverted micelles, it is seen
that, without oleic acid, the r value for tetracosane would correlate with that of a weak promoter (lower than that for the 3-lipid
control). In the presence of oleic acid, however, the C_{24} alkane
tends to counteract its disordering effects to some extent, increas-

TABLE 2

Effect of Oleic Acid on the Fluorescence Anisotropy
of DPH in Mixed Lipid Liposomes*

Conc. of Oleic acid (Molar ratio)*	Polarization Anisotropy, r		
	Lipids Alone	Lipids + Pristane	Lipids + $n-C_{24}H_{50}$
0	0.206	0.156 (76%)	0.188 (91%)
0.84	0.165	0.140 (85%)	0.177 (107%)
1.0	0.161	0.141 (88%)	0.173 (107%)
1.16	0.158	0.137 (87%)	0.172 (109%)
1.33	0.158	0.146 (92%)	0.174 (110%)

* Egg Phosphatidyl Choline, Egg Lysophosphatidyl Choline, and
  Cholesterol, 2:2:1 molar ratio, respectively.

ing the anisotropy compared to each 4-lipid control. The spread between the r values for $n-C_{24}H_{50}$ and the C_{19} promoter, pristane, remains comparatively constant.

In the last column of Table 1 are shown the effects of a series
of alkanes from C_{14} to C_{24}, all n-paraffins except for the C_{19} pristane. The significant disordering effects of the $C_{16}-C_{20}$ alkanes
correlate with their high cocarcinogenic activity in contrast with
the effects of the C_{24} compound. N-docosane has not been tested biologically but the indication is that it would be completely non-
promoting.

The sharp decrease in cocarcinogenic activity from $n-C_{20}$ to
$n-C_{24}$ in contrast to the comparatively gradual increase from $n-C_{12}$
to $n-C_{20}$ is notable. This sharp break corresponds to a qualitative
shift in the effects on the mixed lipid liposomes, from a reduction
in molecular order by the hydrocarbons up to C_{20} to a sudden increase
in order by C_{22}.

Thus the lipid composition (2:1 lecithin: lysolecithin) that provided the most effective inverted micelles for predicting the cocarcinogenic activity of hydrocarbons[6] is again proving useful in our current fluorescence polarization study. As before, the promoting hydrocarbons are proving to be the ones that tend to counteract the ordering effect of cholesterol. In this regard, it is notable that Shinitzky and Inbar[13] found the fluorescence anisotropy of DPH significantly lower in malignant lymphoma cells than in the corresponding normal lymphocytes, and correlated the difference with the lower cholesterol content in the membranes of the malignant cells.

In subsequent reports[14], it was shown that the higher the r value for DPH in plasma membranes, the higher the rotational mobility of protein receptors for lectins that in some situations have shown cancer-inhibiting activity. It was postulated that vertical displacement of the receptors resulted from the increase in molecular order (or microviscosity) among the lipids of the membrane, permitting greater mobility of the proteins at the membrane/water interface. Such ability to change conformation or to migrate laterally along the membrane/cytoplasm interface may be a critical requirement for the regulatory activity ascribed to certain integral membrane proteins.

ACKNOWLEDGEMENTS

This research was supported in part by Chesebrough-Pond's, Inc.

REFERENCES

1.   J. G. Lione and J. S. Denholm, AMA Arch. Ind. Health 19, 530
           (1959).

2.   S. A. Henry, Brit. Med. Bull. 4, 389 (1947).

3.   A. W. Horton, M. J. Burton, R. Tye, and E. L. Bingham, Sympos-
        ium on Composition and Analysis of Petroleum. Petroleum
        Chemistry Division, Amer. Chem. Soc. 8, C-59 (1963).

4.   E. Bingham and A. W. Horton, Advances in Biology of the Skin,
        Wm. Montgna (ed.), Pergamon Press Ltd., Oxford, England,
        7, 183 (1966).

5.   A. W. Horton, D. T. Denman and R. P. Trosset, Cancer Res. 17,
        758 (1957).

6.   A. W. Horton, W. H. Perman, D. N. Eshleman and A. R. Schuff,
        J. Nat. Cancer Inst. 56, 387 (1976).

7.   I. Berenblum, A Re-evaluation of the Concept of Cocarcinogene-
        sis. Prog. Exp. Tumor Res. 11, 21 (1969).

8.   A. W. Horton, C. K. Butts and A. R. Schuff, Colloid Interface
        Sci. 5, 159 (1976).

9.   A. W. Horton, L. C. Bolewicz, A. W. Barstad and C. K. Butts,
        Biochim. Biophys. Acta 648, 107 (1981).

10. R. F. Chen and R. L. Bowman, Science 147, 729 (1965).

11. R. D. Klausner, A. M. Kleinfeld, R. L. Hoover and M. J.
        Karnovsky, J. Biol. Chem. 255, 1286 (1980).

12. C. J. A. Van Echteld, B. DeKruijff, J. G. Mandersloot and J.
        DeGier, Biochim. Biophys. Acta 649, 211 (1981).

13. M. Shinitzky and M. Inbar, J. Mol. Biol. 85, 603 (1974).

14. M. Shinitzky and M. Inbar, Biochim. Biophys. Acta 433, 133
        (1976).

ADIABATIC COMPRESSION: A NEW METHOD TO MEASURE LATENT HEATS IN
PHOSPHOLIPID BILAYERS*

Ned D. Russell and Peter J. Collings

Department of Physics, Kenyon College, Gambier, Ohio 43022

The standard method of measuring latent heats of phase transitions which occur in phospholipid bilayers is DSC or DTA at or
near atmospheric pressure. A recurring problem with this method
is that changes in the specific heat near the phase transition
cannot easily be separated from the latent heat itself. Adiabatic
compression offers an alternative method of performing a latent heat
measurement, and because the sample is under continuous pressure,
pre-transitional effects depend on a different combination of thermodynamic variables. As a result, adiabatic compression measurements
should not be subject to the same errors as DSC or DTA experiments.
Adiabatic compression techniques are described, and measurements on
the phospholipid dipalmitoylphosphatidylcholine (DPPC) are presented.
The results agree with prior measurements, but reveal changes in
the latent heat at high pressure never observed before.

INTRODUCTION

Latent heats of transition in condensed matter phases are

*Research supported in part by the Research Corporation under
Grant No. C-903.

typically measured using differential scanning calorimetry (DSC) or
differential thermal analysis (DTA). Significant improvements in
these techniques over the years have produced more accurate data,
as well as allowed investigation of a wider variety of phase transitions. One problem which occurs with this technique, however, is
the difficult separation of pre-transitional changes in the specific
heat from the actual latent heat of the transition.

It would therefore be useful to compare latent heat measurements
from DSC or DTA with the results of another technique which is not
affected by pre-transitional effects in the same way as DSC and DTA
measurements. Adiabatic compression techniques not only serve this
function well, but also permit the first direct measurements of the
latent heat of transition under pressure.

In this paper, we first discuss the theory of adiabatic compression through a phase transition, followed by a description of the
apparatus used to perform these measurements. In order to compare
adiabatic compression measurements with DSC results, we present
adiabatic compression data up to 1 Kbar for the phospholipid
dipalmitoylphosphatidylcholine (DPPC). Results using this new
technique agree with prior measurements in DPPC, as well as indicate
that the latent heat of transition in DPPC changes at high pressure.

THEORY

A change from one phase to another can result from adiabatic
compression near a coexistence curve. This is shown schematically
in Figure 1, where the thick line represents adiabatic compression
starting in the higher temperature liquid crystal phase, intersecting
and following the coexistence curve as the sample changes into the
lower temperature gel phase, and finally leaving the coexistence line
and ending in the gel phase. Since the entire line is an adiabat,
the specific entropy change in going from point a (in the liquid
crystal phase) to point d (in the gel phase) must be zero. This

PHOSPHOLIPID BILAYER PHASE DIAGRAM

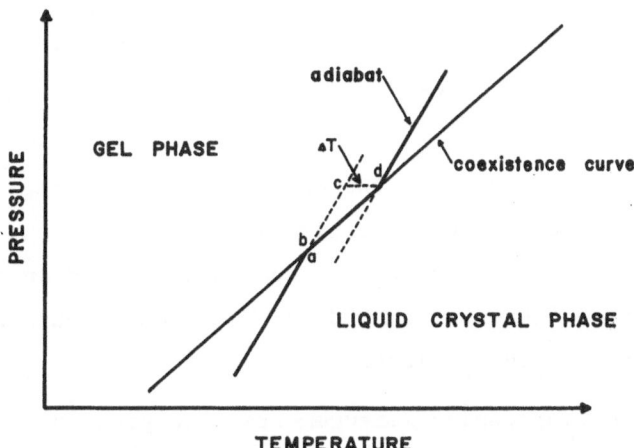

Fig. 1   Schematic diagram of adiabatic compression (heavy line)
           through a phase transition.

same specific entropy change can also be determined by following the
dashed path abcd. The contribution to the specific entropy change
from point a to point b is just $\frac{L}{T}$ where L is the specific latent heat
of transition and T is the temperature. Assuming the line bc is an
adiabat (it is drawn parallel to the gel phase adiabat), there is no
change in the specific entropy from point b to point c. The specific
entropy change in going from point c to point d is simply $c_p \Delta T/T$,
where c_p is the specific heat and $\Delta T \ll T$. Equating the two expressions
for the specific entropy change results in a simple expression for
the specific latent heat:

$$L = - c_p \Delta T.$$

In any experiment, of course, the adiabats will be rounded at
the coexistence line, so ΔT must be measured from extrapolation of
the two pure phase adiabats. Any pre-transitional effects which
change the slope of the adiabat near the phase transition will
therefore affect the latent heat measurement. The slope of a pure

phase adiabat is given by

$$\left(\frac{\partial P}{\partial T}\right)_S = \frac{c_p \rho}{T \beta},$$

where ρ is the density, S is the entropy, and β is the coefficient of volume expansion. Pre-transitional effects will cause errors in the latent heat measurement due to combined changes in c_p and β, which is quite different from DSC and DTA measurements where errors are due to changes in c_p alone.

Adiabatic compression measurements also produce accurate data on the coexistence curve itself. By combining the latent heat results with measurements of the slope of the coexistence curve, specific volume changes can be determined using the Clausius-Clapeyron equation:

$$\left(\frac{\partial P}{\partial T}\right)_{coex} = \frac{L}{T\Delta V}$$

EXPERIMENTAL METHOD

The sample was contained in a stainless steel bellows, approximately 3 cm in diameter and 4 cm long. As shown in Figure 2, the bellows was located inside a pressure vessel/double oven arrangement. Inside the bellows were two coaxial cylindrical baffles to prevent heat transfer via convection. The baffles were constructed from very thin copper sheet. In the middle of the sample vessel was a single Chromel/Alumel thermocouple, connected via a high pressure feed-thru to equipment outside the pressure vessel. The pressure was generated by a hydraulic pump and intensifier system.

The thermocouple was placed in series with a reference thermocouple at 0°C, and the total voltage was amplified and connected to the X axis of an XY recorder. Active and reference manganin resistors

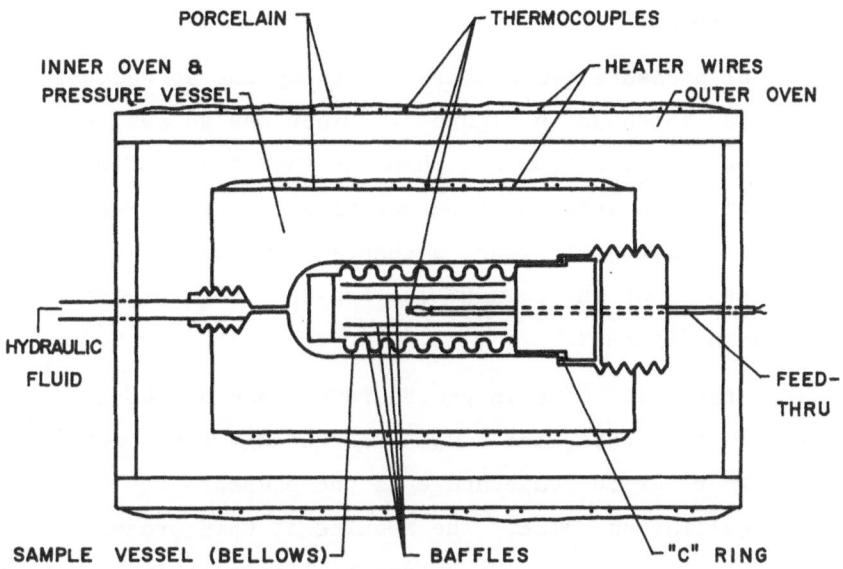

Fig. 2   Adiabatic compression apparatus used in this experiment.

were used in an AC bridge arrangement with a lock-in amplifier to
generate a DC voltage proportional to the pressure.   This was
connected to the Y-axis of an XY recorder.   Both axes were calibrated
beforehand for linearity and also during each adiabatic compression
run by direct thermocouple voltage measurement and the use of either
a 0-10,000 psi or 0-100,000 psi Heise bourdon-tube pressure gauge.

    The DPPC used in this experiment was obtained from Calbiochem
and was not further purified.   A 4.26% (by weight) mixture of
phospholipid in water was made which was then stirred rapidly for
about two minutes at a temperature above the gel to liquid crystal
phase transition.

    To obtain adiabatic compression data, equilibrium was first
established at a temperature and pressure on the coexistence curve.
The pressure was then quickly reduced by approximately 150 bars and
then increased at a rate of 3 bars/second until the pressure was
about 150 bars above the equilibrium pressure.

EXPERIMENTAL RESULTS

A typical pressure-temperature recording for an adiabatic
compression run is shown in Figure 3. Since the adiabat always
appeared to be parallel from one phase to the other, ΔT was measured
by extrapolating the straight line portions of the pure phase adiabats
to the coexistence region, where a constant pressure line was drawn
through the center of the coexistence region. This procedure is
demonstrated in Figure 3.

An important part of this procedure depends on the process being
adiabatic. To test this, the sample was quickly taken off equilibrium
and allowed to relax in temperature as the pressure was kept constant
at the off-equilibrium value. The results of this procedure are
shown in Figure 4, along with a theoretical calculation for heat
conduction in a cylindrical sample of water with a radius and length
equal to the bellows diameter.[1] The similarity of the experimental

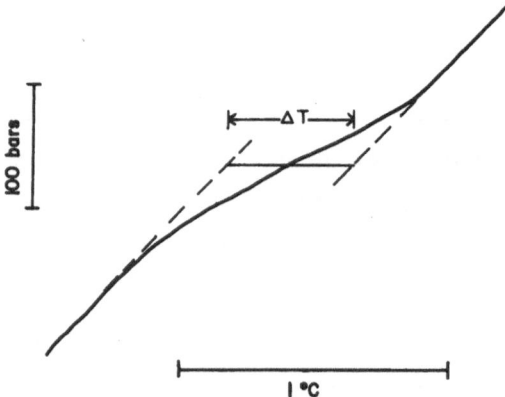

Fig. 3   Typical adiabatic compression data in DPPC. Dashed lines
          are extrapolations of the straight-line portion of the two
          pure phase adiabats. The horizontal line has been drawn
          through the middle of the transition region.

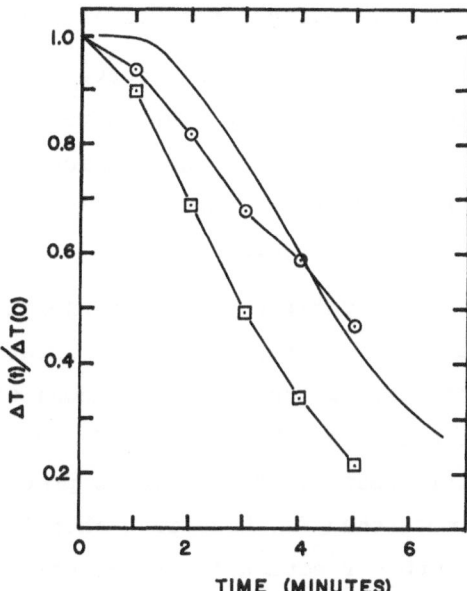

Fig. 4   Thermal relaxation for the apparatus used in this experi-
          ment when an off-equilibrium condition is caused by a
          sudden change in pressure.   ΔT is the temperature
          difference from the equilibrium value. Data points are
          the actual temperature values for two different trials,
          while the solid line is the result of a theoretical cal-
          culation involving heat conduction in a cylindrical sample
          (see text).

and theoretical curves (especially since the theory assumes an
instantaneous temperature shift while experimentally it took at least
10 seconds) confirms that heat transfer in this technique is solely
via conduction. These results also point out that little heat is
lost or gained if the sample is off equilibrium for less than a
minute at a time. From these results we estimate the error due to
the non-adiabatic character of the process to be less than 1%, mainly
because the heat transfer causes changes in the slope of the pure
phase adiabats which shift the ΔT interval on the temperature axis,
rather than change its length. Since the ΔT values are reproducable

only to ±2.5%, the non-adiabatic effect is of negligible significance.

The gel/liquid crystal coexistence curve as determined from these results is shown in Figure 5. The curve represents pressure-temperature points in the middle of the coexistence region; the points turn out to follow a fairly straight line, but possess definite curvature upward at higher pressures.

The latent heat and specific volume change measurements are contained in Figure 6. The specific heat of water was used to determine the latent heat, which in turn allowed the specific volume change to be calculated using the Clausius-Clapeyron equation. Actually, the specific heat of the sample is probably lower than 1 cal/(g-°C) by no more than 2%, but since this is within the range of the random error, no correction has been applied.

DISCUSSION

Previous measurements on DPPC using DSC have produced latent heat values of 8.7 Kcal/mole[2] and 8.5 Kcal/mole[3] at atmospheric

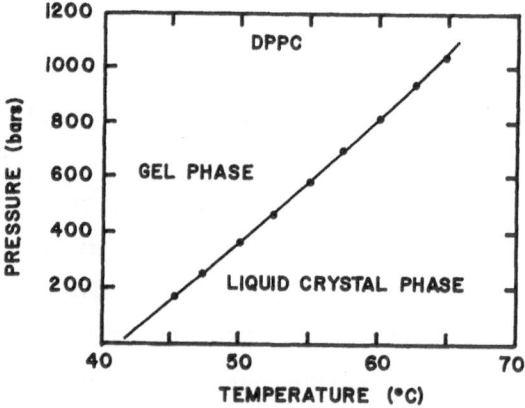

Fig. 5   Phase diagram for DPPC as determined in this experiment. The solid line has been added to aid the eye.

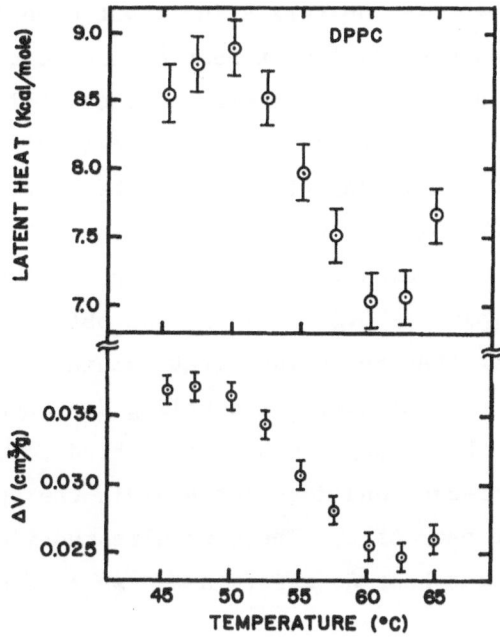

Fig. 6   Specific latent heat and specific volume change at the gel
          to liquid crystal transition in DPPC as a function of
          temperature. Some of the decrease is due to settling of
          the phospholipid vesicles over time (see text).

pressure. Our low pressure results fall within this range as
demonstrated by Figure 6. Previous measurements of the specific
volume change range from 0.032 cm^3/g to 0.037 cm^3/g at atmospheric
pressure.[3-6] Again, our low pressure values fall within these limits.

Previous work under pressure has determined the coexistence curve
to be straight with a slope of roughly 44 bars/°C and a one atmosphere
transition temperature of 41.4°C.[5-7] Our results of $\left(\frac{dp}{dT}\right)_{coex}$ range
from 42 bars/°C to 44 bars/°C depending on what part of the co-
existence curve is examined; the extrapolated transition temperature
at one atmosphere is 41.4 $\pm$ 0.1°C.

The latent heat of transition in DPPC up to about 300 bars has
been measured before,[5,6] with no change from the atmospheric value.

Our results show exactly the same effect, but also demonstrate
that at higher pressure the latent heat does seem to change. Figure
6 reveals that first a decrease seems to occur, followed by a slight
increase. Some of the decrease could be due to a phospholipid
concentration decrease in the vicinity of the thermocouple due to
slow settling of the bilayer vesicles. To test this, another
measurement at 52.5°C was performed after the entire experiment was
completed, and a value of 8.0 Kcal/mole was obtained. This is
significantly lower than the value first obtained, 8.6 Kcal/mole,
and does indicate that significant settling did occur over the 12 hour
period. However, this change is only one-third to one-half of the
total observed decrease, and does not explain the apparent increase
in the latent heat near 65°C. These results therefore indicate that
changes in the latent heat do exist in the zero to one kilobar range
for DPPC.

CONCLUSION

     Adiabatic compression does afford a useful method to measure
latent heats and volume changes under pressure. Good agreement
between these adiabatic compression results and prior DSC measurements indicate that the accuracy of the technique is within the 2.5%
estimated error limit. In addition, these measurements extend the
investigation on DPPC to almost three times the pressure ever investigated before, and seem to demonstrate that changes in the latent
heat and volume change at the gel/liquid crystal transition occur in
this region.

     Additional work with other phospholipid systems as well as
investigation of the effect of settling are in progress.

ACKNOWLEDGEMENT

     The idea of performing adiabatic compression measurements near
a phase transition originated with Richard Alben and James R. McColl

over ten years ago. McColl actually performed an adiabatic compression experiment, but the results seemed to indicate a problem
existed in the technique. A brief reference to McColl's work was
made by Srinivasan, Kay, and Nagle[5] in 1974.

The authors would like to acknowledge the valuable contribution made by V. A. Post during the early portion of this work.

REFERENCES

1.   M. Jakob, Heat Transfer, Wiley, New York, 1949, p. 265.

2.   S. Mabrey and J. M. Sturtevant, Proc. Natl. Acad. Sci. USA 73,
        3862 (1976).

3.   D. A. Wilkinson and J. F. Nagle, Biochemistry 20, 187 (1981).

4.   J. F. Nagle, Proc. Natl. Acad. Sci. USA 70, 3443 (1973).

5.   K. R. Srinivasan, R. L. Kay and J. F. Nagle, Biochemistry 13,
        3494 (1974).

6.   N. I. Liu and R. L. Kay, Biochemistry 16, 3483 (1977).

7.   J. R. Trudell, et al., Biochem. Biophys. Acta 373, 436 (1974).

ANISOTROPY OF THE ELECTRIC CONDUCTIVITY IN AMPHIPHILIC LIQUID CRYSTALS

P. Photinos and A. Saupe

Liquid Crystal Institute
Kent State University
Kent, Ohio 44242

ABSTRACT

The anisotropy of the ionic conductivity in amphiphilic liquid crystals with disc-like aggregates is studied theoretically. It is assumed that only the interstitial water phase is conducting (no ion-transport through the aggregates) and that the conductivity of the water phase is uniform and isotropic. Under these assumptions, the anisotropy depends on the size, the shape, the concentration and the distribution of the aggregates. Comparison with experimental results on a system consisting of decylammonium chloride/NH_4Cl/H_2O, shows that the aggregates are disc-like, with a diameter to thickness ratio of about 4. The diameter to thickness ratio decreases by about 30% as we move from the lamellar-nematic to the nematic-isotropic transition.

INTRODUCTION

Combinations of polar solvents and amphiphilic compounds give a variety of smectic liquid crystals, and also, under suitable conditions, nematic mesophases. For recent reviews see Refs. 1-4.

Depending on the shape of the aggregates, we distinguish two types of uniaxial nematics, namely the N_C phase, consisting of elongated cylindrical aggregates, and the N_L phase, consisting of discoid aggregates.[5] The amphiphilic molecules forming the aggregates are arranged with their polar heads on the external surface, while the interior of the aggregates is occupied by the hydrocarbon tails. The arrangement of the symmetry axes of the aggregates parallel to a preferred direction results in a nematic order.

Approximate expressions for the conductivities and numerical calculations on model systems approximating the liquid crystalline phases have been presented by several authors, and date back to Maxwell. Examples of such work can be found in Ref. 6. Francois[7] derived expressions for the principal values of the conductivity of the hexagonal phase at high amphiphile concentrations. The expressions are in good correlation with the measured conductivity of unoriented samples, while the predicted anisotropy is incorrect, as numerical calculations show.[8] In the present work we focus on the electric conductivity of an N_L phase. We use specific spatial arrangements of the aggregates in the N_L phase, and calculate the electrical conductivity along the primary axes, for different concentrations and differently shaped discoid aggregates. We derive plots of the anisotropy versus the diameter-to-thickness ratio, at constant concentrations of the mixture. The plots are then used to evaluate earlier conductivity measurements,[9] carried out on a mixture consisting of 45 wt% decyl ammonium chloride (DACl), 5 wt% ammonium chloride, and 50 wt% water. The mixture forms an N_L phase between 41°C and 61°C. Below 41°C, it forms a lamellar phase and above 61°C an isotropic solution. The discoid aggregates of the N_L phase are surrounded by an aqueous solution of chlorine and ammonium ions. The interpretation of the experimental findings, permits estimates on the diameter-to-thickness ratio, within the framework of the model.

THE MODEL

The aggregates in the N_L phase have a random arrangement, and a degree of orientational order $S<1$. For the calculation we assume that the aggregates are perfectly aligned, and that they are arranged in an orderly layered structure. The random arrangement is approximated by averaging over different stacking patterns. The following assumptions are made:

a. The aggregates are cylindrical in shape, rigid and immobile.

b. The aggregates are impermeable to the charge carriers.

c. The conductance of the mesophase is due to the neutral fluid surrounding the aggregates. In what follows, the conductance of this fluid is assumed isotropic, and is set equal to k_o.

d. The aggregates are arranged in layers, with their symmetry axes normal to the layers. The aggregates in each layer are arranged on a hexagonal lattice. The primitive cell is described by the vectors $\vec{a}$ and $\vec{b}$ ($a=b$). The layers are equidistantly stacked, along the preferred direction (the z-axis). We differentiate between A, B, and C layers, which are defined as follows: (A) the centers of the aggregates have the same xy-coordinates as the reference layer; (B) the coordinates differ by the vector $(2\vec{a}+\vec{b})/3$; and, (C) the coordinates differ by the vector $(\vec{a}+2\vec{b})/3$.

The xy-coordinates of the centers for the A, B, and C positions are shown in Figure 1. We will examine three particular stacking patterns, namely the AAAAAA..., ABABAB..., and ABCBABCBA... At this point, it is convenient to introduce several symbols relating to the geometry described so far. We call the diameter of the (cylindrical) discs D, and their thickness h. The distance of closest approach between discs of adjacent layers is h_2, and the distance of closest approach between discs of the same layer is h_1. The fraction of the volume occupied by the aggregates, i.e., the concentration by

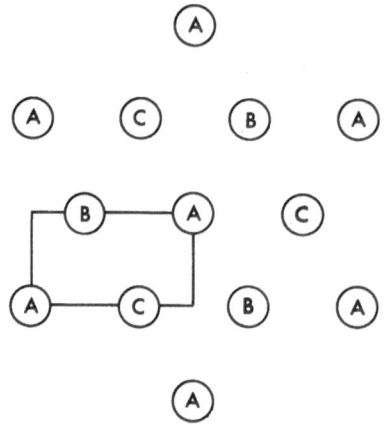

Fig. 1 Arrangement of the disc-centers.

volume, is v. Apart from an overall scaling factor, the situation
can be described completely by three quantities. Here we choose v,
the ratio D/h, and the ratio h_1/h_2.

The concentration of DACl in the mixtures used for the conductivity measurements of Ref. 9 is 45 wt%. Since these are the only
measurements available to us at this time, the value of v in our
analysis will be in the range of 40 to 50% (by volume). The ratios
D/h and h_1/h_2 are not known, and will be allowed to vary. In determining a reasonable range of values for the ratio h_1/h_2, one has to
note that, in the range of concentrations discussed here, both h_1
and h_2 are primarily determined by the repulsive forces. The repulsive forces we are concerned with, vary strongly with distance. In
addition, it is reasonable to assume that there is no strong dis-
parity between the forces acting on the aggregate from different
directions. Thus, h_1 and h_2 are of the same order of magnitude. In
view of the above argument, a value of the ratio h_1/h_2 a little less
than unity, seems justifiable. Thus, in our calculations we choose
values in the range 0.5 and 1.0. From the nature of the numerical

results, one could retrospectively assess the importance of the
choice of the h_1/h_2 value. As it will turn out, in the range of
concentrations studied here, the maximum change in the conductivity
is under 15% as h_1/h_2 changes from 0.5 to 1. To calculate the con-
ductivity parallel, $k_{||}$, and normal, $k_\perp$, to the director, we solve
Laplace's equation with the appropriate boundary conditions by
numerical integration. This is done for currents flowing in each
of the two principal directions, and for several combinations of the
three parameters, v, D/h and h_1/h_2. The principal conductivities are
evaluated by appropriate surface integration of the electric field
vector. The numerical procedures are discussed briefly in the
Appendix.

RESULTS AND DISCUSSION

    The principal conductivities of the three different stackings
versus D/h, are shown in Figs. 2-4, for $h_1/h_2=1$. For each stacking we considered three values of v, viz., (a) 0.4, (b) 0.45 and (c)
0.5. As expected, the conductivities for a given stacking increase
as the concentration decreases. Also, for given v, $k_{||}$ decreases
with increasing D/h, while $k_\perp$ increases slightly, achieving a value
near 1-v. These features can be readily understood in terms of the
geometry of the arrangement. We also note that the stackings ABCBA
and ABAB do not differ substantially in the $k_{||}$ values, while $k_\perp$
is relatively insensitive to the stacking pattern.

    A quantitative measure of the difference between the two
principal values of the conductivity, is given by the anisotropy:

$$\alpha = (k_{||} - k_\perp)/k_{||} + k_\perp).$$ (1)

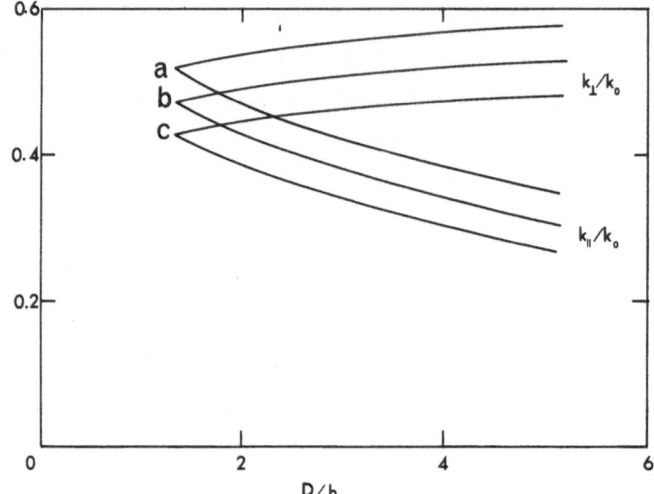

Fig. 2   Reduced conductivities versus D/h, for AAA
         stacking. (a) v=0.4, (b) v=0.45, (c) v=0.5.
         h_1/h_2=1.

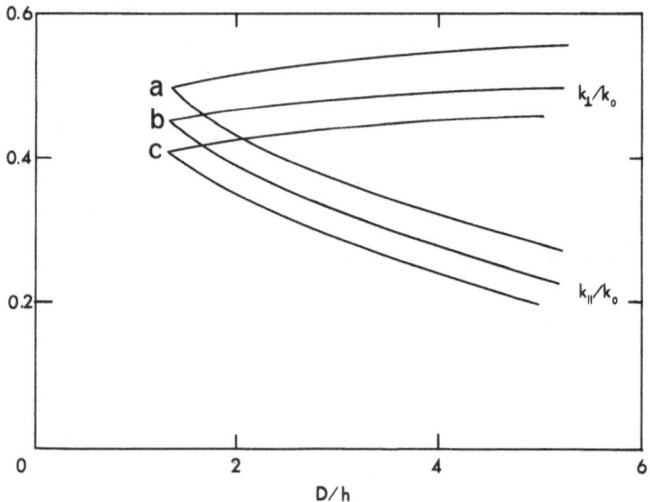

Fig. 3   Reduced conductivities versus D/h, for ABAB
         stacking. (a) v=0.4, (b) v=0.45, (c) = 0.5.
         h_1/h_2=1.

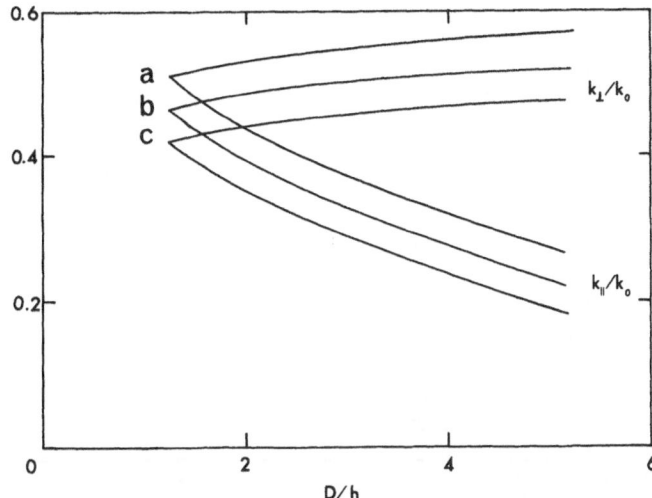

Fig. 4   Reduced conductivities versus D/h, for ABCBA
stacking. (a) v=0.4, (b) v=0.45, (c) v=0.5.
h_1/h_2=1.

The anisotropy versus D/h is plotted in Figure 5.   The ABA stacking
is omitted, since the resulting anisotropy is very close to the values
of the ABCBA stacking.   We note that the model becomes isotropic in
the vicinity of D/h=1, and that the anisotropy becomes positive for
D/h<1, i.e., for cylindrical aggregates.   It should be mentioned that
our earlier calculations[8] on hexagonal arrangements of infinitely
long cylinders also give a positive anisotropy.

To evaluate the dependence of our results on the value of the
ratio h_1/h_2 we repeat the calculations for h_1/h_2=0.5.   The resulting
conductivity values are plotted in Figures 6-8.   We encounter the
same features pointed out earlier.   In addition, we note that $k_{||}$
decreases and $k_\perp$ increases with decreasing h_1/h_2 as one should expect, in view of the geometric significance of the ratio h_1/h_2.
Consequently, the absolute value of the anisotropy increases with
decreasing h_1/h_2.   The anisotropy versus D/h is plotted in Figure 9.
Comparing with Figure 5, we note that the change is most marked for
the AAAAA stacking, reaching about 10% in absolute value, at large D/h.

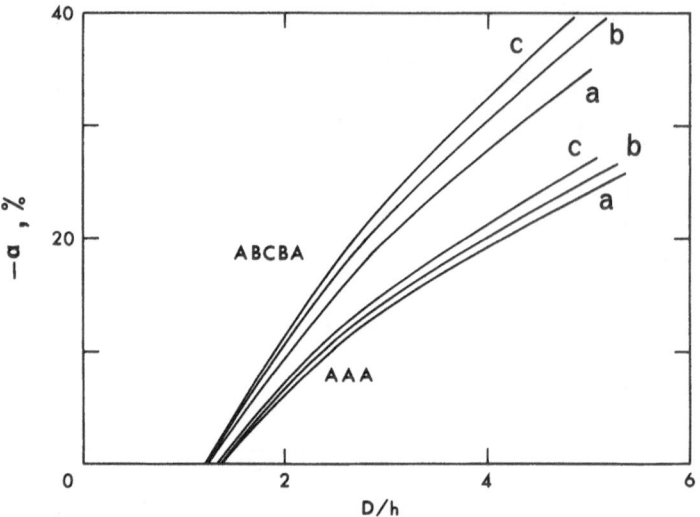

Fig. 5 Anisotropy versus D/h for AAA and ABCBA stackings.
(a) v=0.4, (b) v=0.45, (c) v=0.5. h_1/h_2=1.

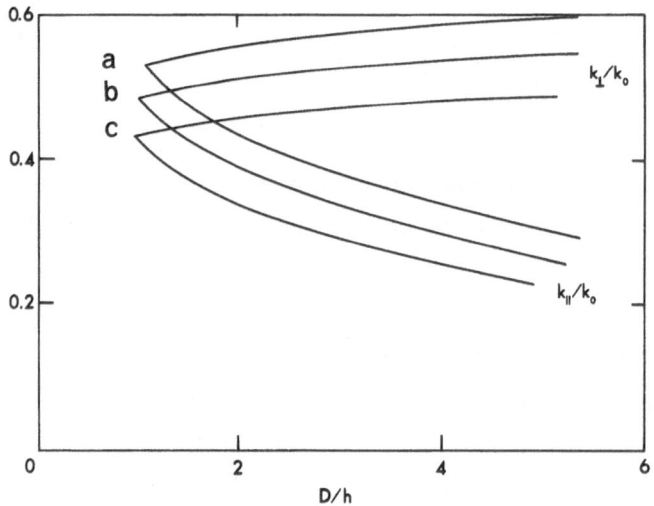

Fig. 6 Reduced conductivities versus D/h, for AAA
stacking. (a) v=0.4, (b) v=0.45, (c) v=0.5.
h_1/h_2=0.5.

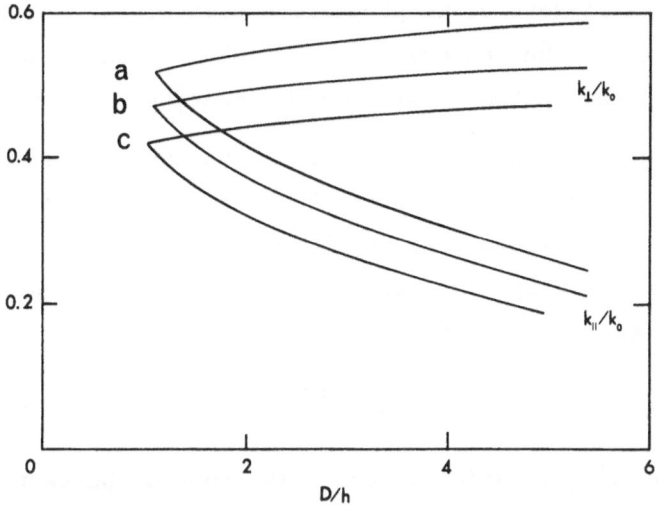

Fig. 7 Reduced conductivities versus D/h for ABAB stacking.
(a) v=0.4, (b) v=0.45, (c) v=0.5. h_1/h_2 = 0.5.

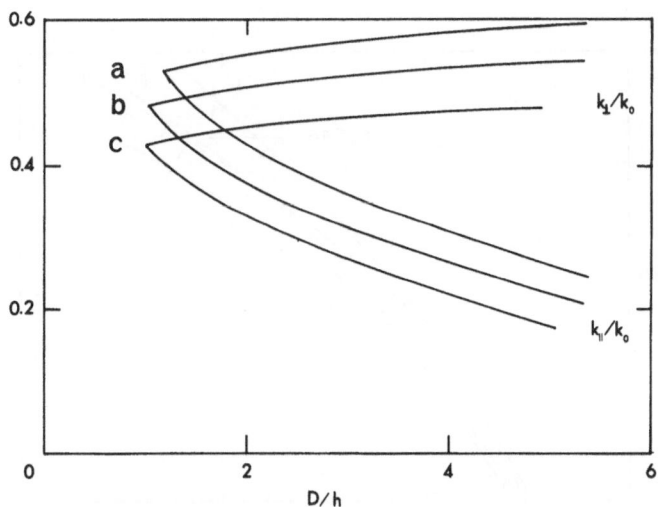

Fig. 8 Reduced conductivities versus D/h for ABCBA
stacking. (a) v=0.4, (b) v=0.45,(c) v=0.5.
h_1/h_2=0.5.

From data of Ref. 9, we prepared a plot of the measured anisotropy, α_e, versus the temperature. The plot is shown in Figure 10,
and refers to the system $DACl/NH_4Cl/H_2O$ mentioned in the introduction.

Since our calculations assume that the aggregates are perfectly
ordered, the anisotropy values should be duly corrected. As a first
approximation, we assume that in the nematic range the experimental
value of the anisotropy is a linear function of the degree of orientational order, and we write

$$\alpha_e = \alpha S \qquad\qquad (2)$$

which allows the evaluation of the anisotropy at perfect order, α,
from the experimental data, once S is known. The variation of S with
temperature can be obtained from the refractivity data.[10] Assuming
that S=0.8 at the N_L to lamellar phase transition, we plotted in
Figure 10 the ratio α_e/S in the nematic range. We note that α_e/S
reaches –30% at the N_L to lamellar transition.

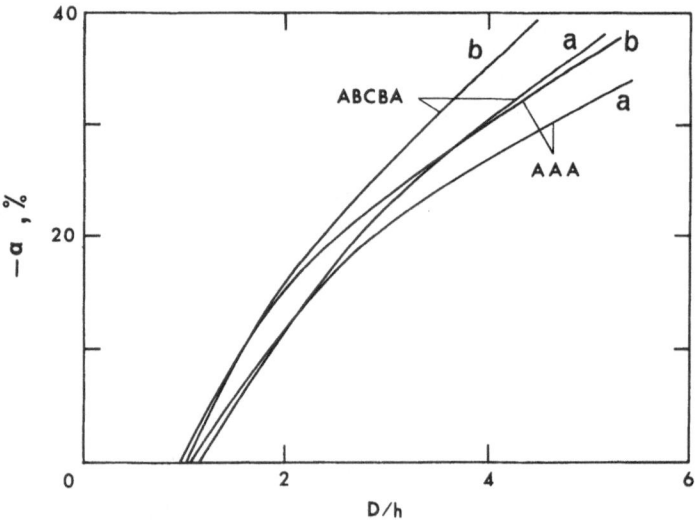

Fig. 9   Anisotropy versus D/h for AAA and ABCBA
stackings.   (a) v=0.4, (b) v=0.5. h_1/h_2=0.5.

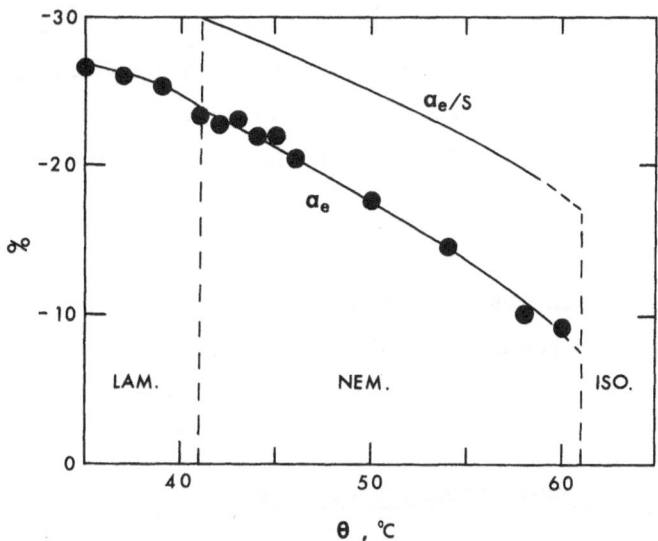

Fig. 10 Conductivity anisotropy versus temperature,
for a mixture of DACl/NH$_4$Cl/H$_2$O. Values
from Ref. 9.

Given the experimental values of the anisotropy, one can use
the theoretical anisotropy diagrams to evaluate the ratio D/h. In
particular, for DACl, the molecular structure suggests a value of
h=20 Å, hence one can evaluate the diameter of the aggregates in
the N$_L$ phase. For our comparisons we will use an average of the
stackings ABCBA and AAA, which we effect by taking a weight 2 for
the former and weight 1 for the latter. By this procedure we find
from Figure 5 that α=-30% corresponds to D/h=4.1, and from Figure 9,
D/h=3.7, or in terms of diameters D=75 to 82 Å.

From Figure 10 we see that the anisotropy increases as the
temperature rises, reaching a value of -17% at the N$_L$ to isotropic
transition. Repeating the above procedure we find that the diameter of the aggregates decreases to about 50 Å at the N$_L$ to
isotropic transition.

The dimensions of the discoid aggregates in a mixture of
sodium decyl sulfate/water/decanol/sodium sulfate, were the subject

of x-ray diffraction experiments by Amaral et al.[11] and by Charvolin
et al.[1,12]. The reported results are in obvious disagreement. More
specifically, Amaral et al. find D>500 Å, while Charvolin et al. find
65 Å. Our results seem to support the latter value.

ACKNOWLEDGEMENT

    This research was supported in part by a grant from the
National Science Foundation, DMR-79-07789.

APPENDIX

Outline of the Computational Procedure

    By symmetry considerations, it is sufficient to solve the
boundary value problem in a volume with xy cross-section given by
the rectangle of Figure 1. This is so because the long sides of
the rectangle are along the xy projections of symmetry planes,
while the short sides of the rectangle represent periodic boundaries.
The remaining boundaries of the volume are an xy-plane through the
disc-centers of an A layer, and its nearest parallel symmetry plane
in the structure.

    In order to apply the finite difference method, a three-dimen-
sional network is constructed, using 20x34 points in the xy plane
(thus approximating tan 30° to 0.3%) and 65 points in the z direction. Since the discs are treated as insulators, the component of
the electric field normal to the surface, must vanish at the surface.
This boundary condition is introduced by methods similar to the
one outlined in Ref. 8. The values of the solutions are stable
to 5 parts per 10^5.

REFERENCES

1. Y. Hendrikx and J. Charvolin, J. Physique 42, 1427 (1981).

2. W. Helfrich, <u>Physics of Defects</u>, Les Houches, Session XXXV,
        1980. North Holland Publishing Co. (1981), Course 11,
        <u>Amphiphilic Mesophases Made of Defects</u>.

3. B. J. Forrest and L. W. Reeves, Chem. Rev. <u>81</u>, 1 (1981).

4. G. J. Tiddy, Phys. Rep. <u>57</u>, 1 (1980).

5. L. J. Yu and A. Saupe, J. Am. Chem. Soc. <u>102</u>, 4879 (1980).

6. R. M. Barrer, J. A. Barrie and M. G. Rogers, J. Polym. Sci.
        Part A, <u>1</u>, 2565 (1963); M. Bishop and E. A. DiMarzio, Mol.
        Cryst. Liq. Cryst. <u>28</u>, 311 (1974) and references therein.

7. J. Francois, Kolloid Z. Z. Polym. <u>246</u>, 606 (1971).

8. P. Photinos and A. Saupe, J. Chem. Phys. <u>75</u>, 1313 (1981).

9. P. J. Photinos, L. J. Yu and A. Saupe, Mol. Cryst. Liq. Cryst.
        <u>67</u>, 277 (1981).

10. T. Haven, K. Radley and A. Saupe, Mol. Cryst. Liq. Cryst. <u>75</u>,
        87 (1981).

11. L. Q. Amaral, C. A. Pimentel, M. R. Tavares and J. A. Vanin,
        J. Chem. Phys. <u>71</u>, 2940 (1979).

12. J. Charvolin, A. M. Levelut and E. T. Samulski, J. Phys. Lett.
        <u>40</u>, L-587 (1979).

PHASE TRANSITION PROPERTIES OF BILAYERS OF MIXED LIPIDS AND THEIR
COVALENT ANALOGS

Mahendra Kumar Jain,[1] Joe Rogers,[1] and Fausto Ramirez[2]

[1]Department of Chemistry, University of Delaware, Newark
Delaware 19711

[2]Department of Chemistry, State University of New York
at Stony Brook, Stony Brook, NY 11794

The elucidation of the role played by mixtures of lipids in
biomembranes presupposes information on the modes of organization
that prevail among different phospholipid species in bilayers.[1]
One of the manifestations of the organization of phospholipids in
bilayers is the thermotropic phase transition from gel to liquid
crystalline phase. In the gel phase, almost all the acyl chains are
in an all-trans conformation, and they are predominantly close-
packed except at the defect sites such as grain boundaries and line
defects. In the liquid crystalline phase the presence of 1 to 3
gauche conformers per acyl chain on the average increases the
disorder in lateral packing. This description of molecular events
is consistent with a variety of experimental observations.[2-4]
Additional conclusions reached from such studies can be summarized
as follows: (a) the phase transition characteristics depend on all
the steric factors that determine the packing of the acyl chains in
bilayers, such as length, unsaturation and branching of acyl chains;
(b) additives incorporated into bilayers modify the transition
profile; (c) bilayers of phospholipid mixtures exhibit either an

ideal mixing or a phase separation of the molecular species in the
bilayer depending on the structure of the chains and the head group
in the molecules.

Several hypotheses have been proposed to interpret the phase
transition characteristics of phospholipid bilayers.[5] However,
the various molecular parameters that govern the phase properties
of bilayers remain uncharacterized and unexplained. At the present
state of our knowledge it is safe to assume that both, the intermolecular interactions between lipid components, and the intramolecular interactions operating in each lipid, will determine the
phase properties and the motional characteristics of lipid molecules
in bilayers. Since lipid molecules are close-packed in gel phase,
the gel to liquid crystalline phase transition is a sensitive
measure of intermolecular interaction. To establish the contributions of such factors, we have compared the phase properties of
aqueous codispersions of two lipid components with the properties
of dispersions of the corresponding covalent analog, i.e. the
compound in which two components of the codispersion are covalently
bonded. In such models one can systematically vary the motional and
geometrical constraints of packing of phospholipids in bilayers.
For example, in a previous investigation,[6] we have interpreted the
properties of 1,2-diacylphosphatidylcholesterol dispersions by
considering this compound to be a "conformationally and orientationally restricted analog" (CORA) of diacylphospholipid +
cholesterol (1:1) in a codispersion. Analogous considerations have
been applied to codispersions of 1-acyl lysophosphatidylcholine +
fatty acid in equimolar proportions in relation to their covalent
analog, 1,2-diacylphosphatidylcholine.[7] The properties of codispersions and those of corresponding CORA proved to be quite
similar. For example, the aqueous dispersions of lysophosphatidylcholine or fatty acid alone form micelles, however, their codispersions form bilayers as demonstrated by a variety of techniques
including freeze-fracture electron microscopy, NMR for the various

nuclei, birefringence, and x-ray studies. Furthermore, the phase
transition characteristics of the lysophosphatidylcholine + fatty
acid codispersions are very similar to those of diacylphosphatidylcholine of the same chain length. Thus, not only the transition
temperature (T_m), enthalpy of transition, and the cooperativity of
transition are comparable for codispersions and the CORA dispersions,
but they exhibit similar dependence on the acyl chain length, and
the phase transition is not observed if cholesterol is also incorporated into codispersions. Such similarities in the phase transition properties of bilayers of codispersions suggest that the
motional freedom of the components and certain changes in the head
group region have little effect on the phase transition characteristics. It should, however, be emphasized that several subtle
differences are observed in the properties of codispersions compared
to the properties of the dispersions of the corresponding CORA.[6,7]

Two classes of naturally occurring phospholipids possess four
acyl chains per molecule: diphosphatidylglycerol (cardiolipin), and
phosphatidyldiacylglycerol ("bisphosphatidic acid"). The present
paper extends the concept of CORA by comparing the thermotropic
phase transition characteristics of diacylphospholipid + diacylglycerol codispersions with their covalent analogs 1,2-diacyl-
phosphatidyl-1,2-diacylglycerol dispersions. In this manner, we
hope to obtain information on the extent of intermolecular interaction between the two diacylglycerol moieties in these systems.

RESULTS AND DISCUSSION

The premise that a given 1,2-diacylphosphatidyldiacylglycerol
(4R-bis-PA) in bilayers is a conformationally and orientationally
restricted analog (CORA) of an equimolar mixture of the corresponding phosphatidic acid (DRPA) + 1,2-diacylglycerol (DR) is supported
by the differential scanning calorimetric data shown in Table I.
For both the palmitoyl- and myristoyl-series (the first set of data

Table I.

Phase Transition Temperatures (°K) of Lipid Dispersions[a]

Components[b]		Codispersions	Covalent Analog bis-PA
I	II	I + II (1:1)	
DPPA	1,2-DP	346°	348° (4P-bis-PA.K$^+$)
			344° (4P-bis-PA.Na$^+$)
DMPA	1,2-DM	330°	332° (4M-bis-PA.K$^+$)
DPPA	1,3-DP	342°	
DPMe	1,2-DP	342°	
DPMe	1,3-DP	337°	
1,2-DM	--	314°	
1,2-DP	--	325°	
1,2-DS	--	336°	
1,3-DP	--	347°	
DPPA	--	343°	
DMPA	--	326°	
DPMe	--	320°	
DSPA	--	343°	

[a]All measurements were made in 200 mM Tris + 100 mM KCl at pH 8.0 on Perkin Elmer DSC 1B at scanning rate 5°K/min. For experimental details see ref. 3, 7, and 8. The half-height widths for all these samples are about 1-3°K. Some of these samples also exhibit considerable hysteresis.

[b]R = acyl chain; P = palmitoyl, M = myristoyl, S = stearoyl. DRPA = diacylphosphatidic acid. DR = diacylglycerol. DRMe = phosphatidyl-methanol. 4R-bis-PA = phosphatidyldiacylglycerol.

in Table I) the T_m for the codispersions is only 2° lower than the T_m for the corresponding bis-PA dispersions. Such a small difference can be attributed to entropic factors, since $T_m = \Delta H/\Delta S$. We do not yet understand the precise nature of such thermodynamic factors, it should, however, be noted that T_m for bis-PA depends somewhat on the nature of the monovalent cation bound to the head group. For example, T_m for the sodium and potassium salts of 4P-bis-PA is 344°K and 348°K respectively.[8] The T_m for the dispersions of the pure components (last set) of the codispersion are lower than T_m for the mixture. Similarly, T_m for the codispersions containing 1,3-dipalmitoylglycerol (1,3-DP) is lower than that for the codispersions containing 1,2-DP. These differences suggest that the exact geometry of the head group of the two components is important. This is further illustrated by a lower T_m for the codispersions containing phosphatidylmethanol (DPMe) + 1,2- or 1,3-DP. In this case both the charge and the geometrical factors influence T_m, presumably by influencing the acyl chain interactions.

In comparing bis-PA and the corresponding codispersions one observes a 16° change in T_m by changing the acyl chain length by two carbon atoms. Similar changes in T_m have been observed for a large variety of phospholipids.[2-4,9,10] Such observations suggest that the packing of the two 1,2-diacylglycerol moieties in codispersions, as well as in bis-PA dispersions, are identical in the gel phase. The acyl chains in the gel phase are in an all-trans conformation, therefore, they are probably stacked with notch-to-notch alignment of the zig-zag chains. In symmetrical 4R-bis-PA, such an arrangement may be achieved only on a time-averaged basis. As shown in Fig. 1, the theoretical analysis of ^{31}P-NMR signal of 4R-bis-PA dispersions suggests that the phosphodiester group is indeed slightly tilted.[11] Such a tilt would be stabilized in the codispersions where the two diacyl glycerol moieties are not structurally identical, and this difference would be further accentuated if the two diacylglycerol moieties are not identical. Indeed,

$45° < \beta < 58°$

$45° < \chi < 90°$

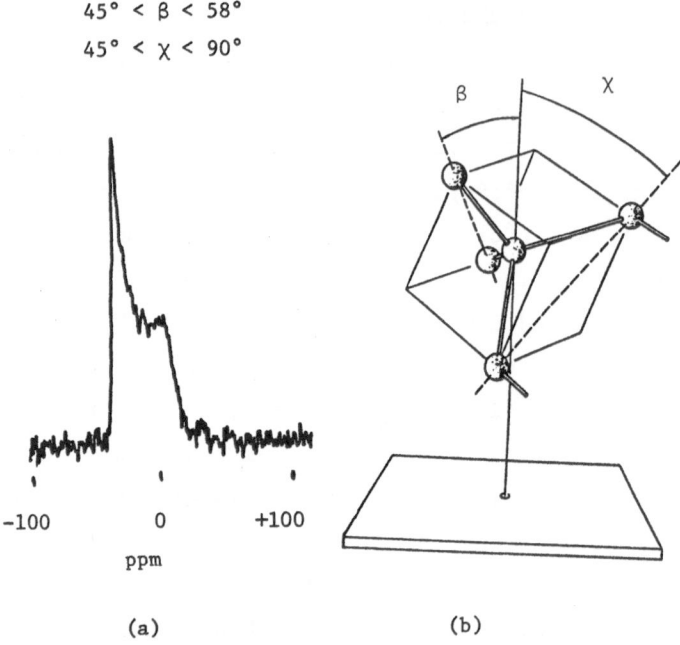

(a)   (b)

Fig. 1   (a) ^{31}P-NMR spectra of aqueous dispersions of Na-salt of
           4P-bis-PA above the phase transition temperature. This
           line shape predicts that the characteristic angles β and
           χ as shown in (b) describing orientation of phosphate
           tensors with respect to the bilayer surface normal are
           $45° < \beta < 58°$ and $45° < \chi < 90°$. Details are given in (11).

as shown in Table II, T_m for DMPA + DP codispersions is lower than
the T_m for DPPA + DM codispersions, 332° and 338°K, respectively.
If the two diacylglycerol moieties were symmetrically aligned, the
T_m for the asymmetric codispersions (DMPA + DP or DPPA + DM) would
be 338°K, that is, the averaged of the two symmetrical homologous
codispersions. The values for DPPA + DM and DPMe + DM are consistent with this prediction.

The difference between the T_m for "isomeric asymmetric" codispersions (DPPA + DM vs. DMPA + DP) can be rationalized if one assumes
that the overlap of acyl chains as manifested in T_m values is
determined by the shorter chain available in the close-packed

Table II

Phase Transition Temperatures of
Codispersions of DRPA + DR[a]

I/II	DM	DP	DS
DMPA	330°	332°	336°
DPPA	338°	346°	344°
DPMe	332°	342°	343°
DSPA	342°	345°	347

[a]Experiment conditions, abbreviations, and other details are given
as footnote to Table I.

bilayer. This concept has been used to account for the difference
in T_m of asymmetrical diacylphosphatidylcholine + fatty acid codispersions.[7] As shown in Table II, T_m for DPPA (or DMPA or DPMe)
+ DR codispersions increases as the chain length for the diacylglycerol increases. However, the increase in T_m with the increase
in chain length is always smaller for codispersions containing the
higher homolog of diacylglycerol. This implies that the two diacylglycerol moieties are not exactly aligned notch-to-notch,
although they are parallel. A lengthwise shift in the position of
one moiety with respect to the other in the codispersions should
reduce the net number of overlapping methylene groups, and hence
lower the T_m.

CONCLUSIONS

The results summarized here and elsewhere[6-8] demonstrate that
the aqueous dispersions of mixtures of certain lipids form bilayers
even though the dispersions of the individual components do not.
The bilayers of mixed lipid codispersions exhibit phase transition
characteristics which are very similar to those of the dispersions

of the corresponding covalent analog in which the two components are
conformationally and orientationally restricted (CORA) in the bilayer.
Such a behavior suggests that intermolecular association between the
components of a codispersion can lead to a packing of the acyl chains
that is very similar to that of their covalent analog.

The fact that T_m for codispersions and their CORA dispersions
are essentially identical raises several interesting questions.
Since the gel to liquid crystalline T_m depends on the acyl chain
conformation, it implies that the acyl chains in both the systems
have essentially identical conformation. Below T_m in the gel
phase, acyl chains have all-trans conformation in both the systems,
and the phase transition properties do not depend upon the motional
constraints introduced by the covalent attachment of the acyl
moieties. The geometrical constraints of packing acyl chains in a
bilayer, such as chain length, manifest in T_m. Such differences
are best accounted for by assuming that the notch-to-notch stacking
of zig-zag acyl chains occurs in the gel phase.

ACKNOWLEDGMENT

This work was supported by grants from Merck Research
Foundation (MKJ), NSF (INT-7925400, MKJ), and NIH (HL-23126, FR).

REFERENCES

1.   M. K. Jain and H. B. White, Adv. Lipid Res. 15, 1 (1977).

2.   M. C. Phillips, Progr. Surface Membrane Sci. 5, 139 (1972);
         D. Chapman, Q. Rev. Biophysics 8, 185 (1975); A. G. Lee,
         Biochim. Biophys. Acta 472, 237-281, 285-344 (1977).

3.   M. K. Jain, N. M. Wu, and L. V. Wray, Nature 255, 494 (1975);
         M. K. Jain and N. M. Wu, J. Membrane Biol. 34, 157 (1977).

4.   J. M. Boggs, Canad. J. Biochem. 58, 755 (1980).

5.   J. F. Nagle, Ann. Rev. Phys. Chem. 31, 157 (1980).

6.   M. K. Jain, F. Ramirez, T. M. McCaffrey, P. V. Ioannou,

J. F. Marecek, and J. Leunissen-Bijvelt, Biochim. Biophys. Acta 600, 678 (1980).

7. M. K. Jain, C. J. A. van Echteld, F. Ramirez, J. de Gier, G. H. deHaas, and L. L. M. van Deenen, Nature 284, 486 (1980); M. K. Jain and G. H. de Haas, Biochim. Biophys. Acta 642, 203 (1981).

8. S. Rainier, M. K. Jain, F. Ramirez, P. V. Ioannou, J. F. Marecek, and R. Wagner, Biochim. Biophys. Acta 558, 187 (1979).

9. J. R. Silvius, B. D. Read, and R. N. McElhaney, Biochim. Biophys. Acta 555, 175 (1979).

10. K. M. W. Keough and P. J. Davis, Biochemistry 18, 1453 (1979).

11. J. H. Noggle, J. F. Marecek, S. B. Mandal, C. van Echteld, R. van Vanetie, J. Rogers, M. K. Jain, and F. Ramirez, Biochim. Biophys. Acta, in press.

ON THE ORIENTATION OF LIQUID CRYSTALS BY MONOLAYERS OF AMPHIPHILIC MOLECULES

K. Hiltrop and H. Stegemeyer

Department of Physical Chemistry
University of Paderborn
4790 Paderborn
Warburger Straße 100, Gebäude J
Postfach 1621

ABSTRACT

The orientation of nematic liquid crystals (LC) by monolayers of suitable amphiphilic compounds on solid substrates is influenced mainly by the following parameters: 1. temperature; 2. physical properties of the substrate surface; 3. chemical properties of the substrate surface. Changes in LC orientation with film packing density and temperature are explained as phase transitions of the system "amphiphilic monolayer + embedded LC."

1. INTRODUCTION

Because of the existence of a long range orientational order in LC phases "liquid single crystals" may be produced by boundary layer action. Particularly the homeotropic orientation is obtained by covering the substrates with thin layers of amphiphilic compounds.

Perez et al.[1] and Porte[2] showed that anisotropic polar-, dispersion- and steric interactions could be of significant influence on LC orientation. Macroscopic models regarding only the surface tensions of substrate and LC are not generally applicable[3,4]. The influence of the substrate polarity was examined by Oghawara et al.[5] and by Naemura[6]. It is the aim of this paper i) to determine the dependence of LC homeotropization on experimental parameters and ii) to develop a molecular model of the interaction between LC and amphiphilic monolayers.

2. EXPERIMENTAL

The structures of the investigated amphiphilic compounds (lecithins, cephalins, fatty acids) as well as those of the LC are given in a preceding paper[7]. Monomolecular films of amphiphilic compounds were transferred to very hydrophilic substrates (polished and unpolished glass and quartz slides) by the method of Langmuir[8] and Blodgett[9]. Packing density and temperature of the film were controlled electronically. Further details are described elsewhere[10].

3. RESULTS

3.1. Temperature dependence of LC orientation

Cooling down a suitable nematic LC orientated homeotropically by action of an amphiphilic monolayer below a well defined temperature $T*$ the director field orientation continuously changes to a more and more tilted one. The transition temperature $T*$ characterizes an actual system monolayer/LC. In case of observing this temperature by polarizing microscopy it is denoted $T*_p$. The orientational transition homeotropic/tilted is reversed by heating up again and may be determined with an accuracy of about 1 K; obviously there is no hysteresis. For some nematics and substrates homeotropic orientation occurs only very nearby the clearing temperature T_c ($T_c - T \sim 0.5$ to 0.1 K).

The temperature dependence of the orientating action of a coated surface can also be proved by observing the helix unwinding of induced cholesteric phases[11]. For so-called "strong anchoring",

what we prefer to call "firm orientation", a calculation of Fischer[12]
results in a critical layer thickness d_o at which a helical texture
transforms into a detwisted homeotropic one:

$$d_o = p \cdot K_3 / (2 K_2) \qquad\qquad (*)$$

($p \equiv$ intrinsic pitch, K_2, $K_3 \equiv$ elastic constants for twist and bend
respectively). The calculated function $d_o/p = f(T/T_c)$ is plotted
in Figure 1 together with experimental results.

The measured values may be interpolated by two curves. The
high-temperature graph roughly matches with the calculated one.
However, below a defined reduced temperature T_H^*/T_c (index H for
helix unwinding) the interpolated curve strongly deviates from the
calculated one. We conclude the deviating values d_o/p at low
temperatures to result from a tilt of the boundary layer director

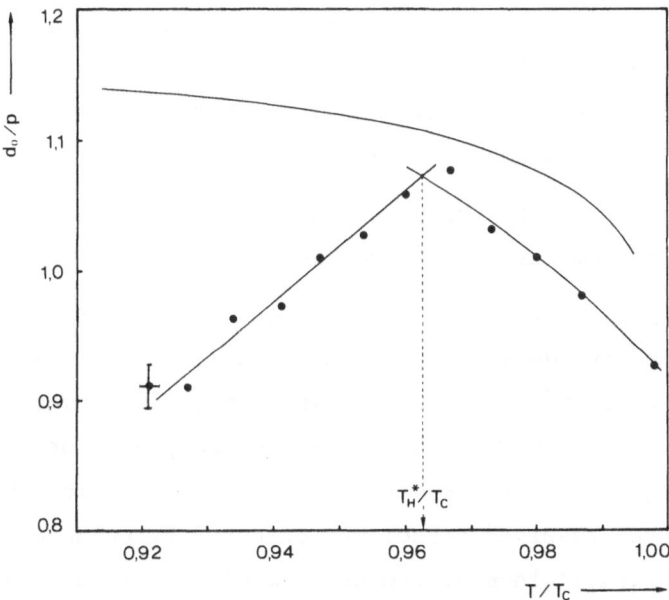

Figure 1:   Reduced maximum layer thickness d_o/p of helix unwinding
            vs. reduced temperature T/T_c. Upper curve: calculated
            after equ. (*), lower curves: interpolated experimental
            results (L-β,γ-dilauroyl-α-lecithin (DLL), A $\sim$63 · 10^{-2}
            nm^2/molecule, trans-4-n-penthyl-(4-cyanophenyl)-cyclo-
            hexane (PCH 5) + 1.41 mole % cholesteryl propionate).

field; this interpretation is consistent with the director field model of the fingerprint texture proposed by Press and Arrott[13].

To find out whether the striking change of orientation with temperature is accompanied by a change of interfacial tension we measured contact angles of some liquid crystals on glass plates coated with lecithin monolayers. In all cases we found only a weak temperature dependence. For example trans-4-n-heptyl-(4-cyanophenyl)-cyclohexane (PCH 7) on L-β,γ-dimyristoyl-α-lecithin (L-β,γ-DML) in the region of 20 to 60°C gave a mean temperature coefficient of -0.05 ± 0.02 degree/K, which is the same order of magnitude for isotropic fluids. Specially in the neighbourhood of the transition temperatures T* no anomalous alterations of contact angles could be observed.

3.2. <u>Physical properties of the substrates</u>

3.2.1. <u>Packing density of coating materials</u>. The packing density of the amphiphilic monolayers which obviously is preserved during the transfer to our hydrophilic substrates severely influences the LC orientation.

Measurements of the helix unwinding for different packing densities gave the results plotted in Figure 2.

The diagram shows the transition temperatures T_H^* to decrease monotonically with decreasing packing density. In order to look for transition temperatures at packing densities between zero and the lowest value occurring in Figure 2 we had to dilute the monolayers on the aqueous subphase with suitable amphiphilics; minimal disturbance of the observed system was expected by using an LC (PCH 7) as diluting agent. Indeed, the helix unwinding induced by diluted monolayers above a certain mean area per lecithin molecule gave a rise of the transition temperature T_H^*. Complete experimental results are plotted in Figure 3.

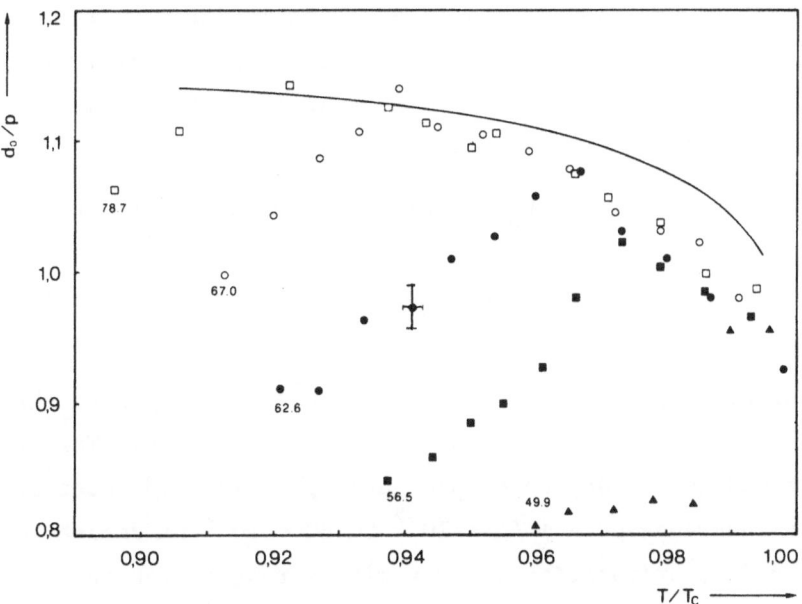

Figure 2: Reduced maximum layer thickness d_o/p of helix unwinding vs. reduced temperature T/T_c at different packing densities. Parameter: mean areas per lecithin molecule [10^{-2} nm^2/molecule].

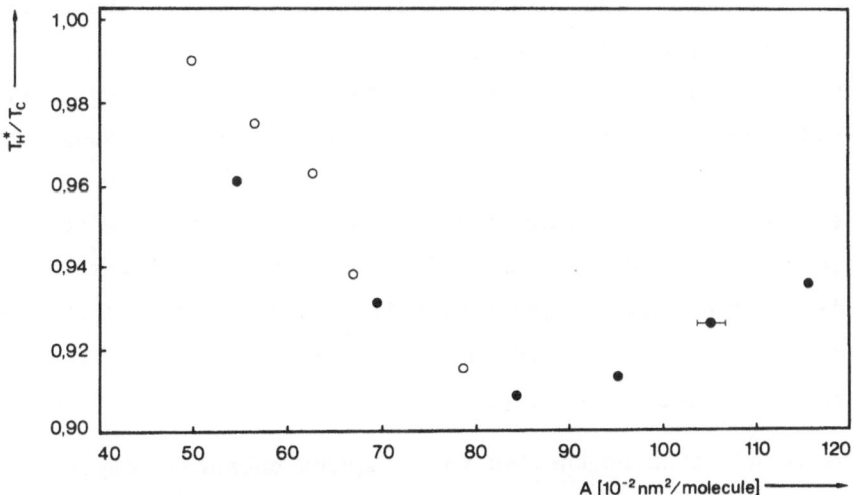

Figure 3: Reduced transition temperature T_H^*/T_c of helix unwinding vs. mean area per lecithin molecule A. Open circles: pure L-β,γ-DLL monolayers, closed circles: PCH 7 diluted L-β,γ-DLL monolayers.

T_H^*/T_C is minimal at a mean area per lecithin molecule of
$(88 \pm 3) \cdot 10^{-2}$ nm^2, which may be seen in comparison to the cross
section areas of $19 \cdot 10^{-2}$ nm^2 for a single alkyl chain and of
$44 \cdot 10^{-2}$ nm^2 for a dialkyl lecithin molecule.

Adhesion energies W_A calculated from contact angle measurements of PCH 7 on substrates coated with lecithin monolayers
increase monotonically with increasing mean area per lecithin
molecule.

After a method described by Zisman[14] we calculated the critical interfacial tensions of some coated substrates. The results
were found not to depend on mean area per lecithin molecule within
$\pm 5\%$ in the range of A $\sim$ $45 \cdot 10^{-2}$ to $80 \cdot 10^{-2}$ nm^2/molecule; yet
the boundary orientation of LC varies strongly with this parameter.
This fact does not fit with the global model of Creagh and Kmetz[3,4]
as shown already in a preceding paper[7].

3.2.2. Polymorphism of amphiphilic monolayers. Phospholipid
monolayers on aqueous subphases may exhibit several two-dimensional
phases. Sackmann[15] and more recently Firpo et al.[16] gave theoretically as well as experimentally based suggestions of four different
phases: isotropic liquid, liquid crystalline (S_C), tilted crystalline and non-tilted crystalline. The terms "liquid-expanded" and
"condensed" phase used in the early literature[17] correspond to the
smectic C and the tilted crystalline one. We observed a strong
difference in the orientating ability of films transferred in the
liquid-expanded phase compared with those transferred in the condensed phase, all other free parameters kept constant. Typical
results are listed in Table I.

For both lecithins the two investigated subphase temperatures
differed by 24 K. In contrast to dilauroyl-lecithin (DLL) dipalmitoyl-
lecithin (DPL) exhibited different phases on the water surface. Ob-

Table I

Influence of the phase type of the phospholipid
monolayer on the transition temperature T^*.

Lecithin	Subphase temperature T_S [°C]	Phase type of the monolayer	Transition temperature of LC layer T_P^* [°C]
L-β,γ-DPL	26	heterophase	53.5
	51	liquid-expanded	38
L-β,γ-DLL	15	liquid-expanded	31
	40	liquid-expanded	29

viously the great difference of the transition temperatures T* observed with DPL is caused by this phase difference. By the way,
films transferred in a heterophasic state can produce domain-like
structured LC orientations. It seems that micro-structures of the
monolayers are made visible by the LC, a useful application in monolayer and membrane research[19].

3.3. Chemical properties of the substrates

3.3.1. Influence of the uncoated substrate. The orientating
power of coated substrates was found to be independent on the nature
of the investigated substrate materials (different types of glasses
and quartz). A very loosely packed monomolecular coating of a suitable amphiphilic compound strongly changes boundary orientation of
LC compared to an uncoated substrate. These observations show the
amphiphilic monolayer to play the crucial role in orientating LC.

However, the cleaning procedure of the substrate influenced
the orientating ability of the subsequently transferred monolayers.
Treatment with a basic detergent followed by chromosulfuric acid is

a suitable preparation to obtain a perfect homeotropic orientation
by the coating. Without detergent treatment the glass plates were
less hydrophilic. We suppose the monolayers to be anchored the
stronger the more hydrophilic the substrate surface becomes. Nevertheless the LC orientation on uncoated plates did not depend on the
cleaning procedure (unregularly tilted or planar texture in any
case).

    3.3.2. <u>Chemical structure of the amphiphilics</u>. The chirality
of α-lecithins does not seem to influence the texture of the
orientated LC, as already discussed in a former paper[10].

    In order to compare the orientating power of different leci-
thins one has to consider comparable conditions for the films on
the substrates. After Sect. 3.2.2. the type of phase is essential.
Partly the films were parepared at constant temperature differences
to the so-called *Krafft*-temperature[18]. Alternatively we transferred different lecithins at same temperatures under same phase
conditions.

    The influence of the number of aliphatic chains per lecithin
molecules as well as of its position within the glycerine skeleton
was investigated with palmitoyl-lecithins. The results are shown
in Figure 4.

    At equal mean areas per molecule the LC was orientated homeo-
tropically by the monoalkyl lecithin palmitoyl-lyso-lecithin (PLL)
at much lower temperatures than the two corresponding dialkyl lecithins. This fact is made plain considering not the mean area per
molecule (A) but the mean area per alkyl chain (A*). Choosing A*
as abscissa unit, the T* values measured for the monoalkyl lecithin
roughly follow the trend of those for the dialkyl lecithins. This
result demonstrates the crucial role of packing density not of the
amphiphilic molecules at all but especially of the alkyl chains.
Further, it points to the role of holes in the monolayer.

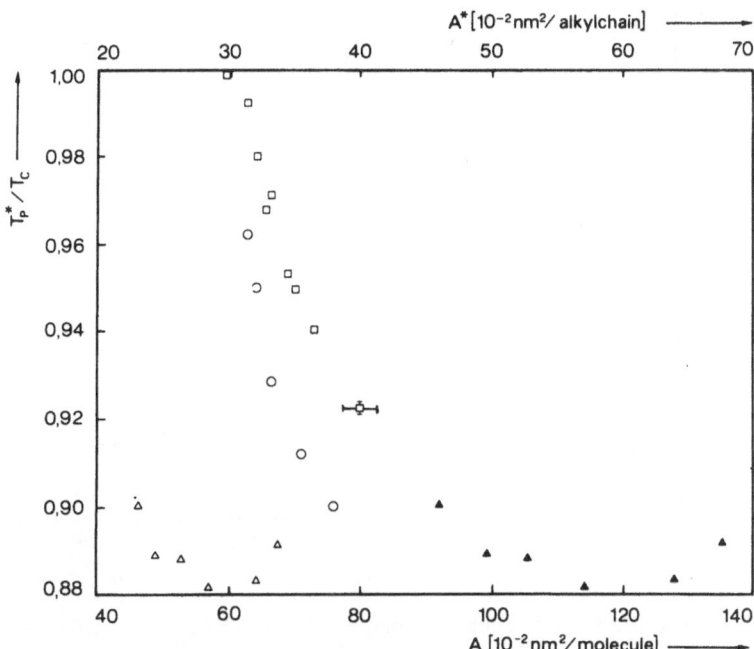

Figure 4: Reduced transition temperature T_p^*/T_c vs. mean area per *lecithin molecule* A (lower abscissa units, open symbols), and vs. mean area per *alkyl chain* A* (upper abscissa units, open circles and squares, closed triangles) (Palmitoyl-lyso-lecithin: triangles, dipalmitoyl-α-lecithin: squares, dipalmitoyl-β-lecithin: circles).

The transition temperatures T* increases with increasing chain-length of the investigated dialkyl lecithins as shown in Figure 5.

The energy of adhesion of LC on lecithin-coated substrates was found to be less for *long*-chained lecithins as well as for *dí*alkyl lecithins[19].

Comparing experiments were done with different linear and branched fatty acids as well as with L-β,γ-dipalmitoyl-α-cephalin. The investigated LC gave high values of T* ($T*/T_c \sim 0.98$ to 1 depending on packing density). Little ability to orientate LC homeotropically was found for fatty acids, too, although they easily could be transferred in the liquid-expanded phase and with low packing densities.

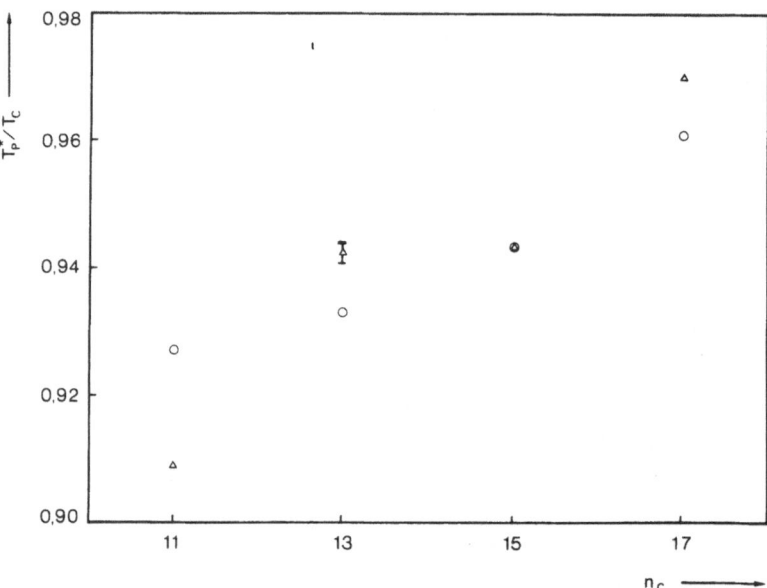

Figure 5 : Reduced transition temperature T_p^*/T_c vs. number of
carbon atoms n_c of the aliphatic chains of dialkyl
lecithins. Triangles: subphase temperature during film
transfer $T_s \sim 51^{\circ}C$, circles: $T_s \sim 10$ K above Krafft-
temperature.

4. DISCUSSION

4.1. Fundamental results

The following fundamental results must be taken into account
for any model describing the interaction of nematics with an amphiphilic monolayer:

Temperature:

a) Decreasing the temperature of an LC orientated homeotropically
by suitable amphiphilic monolayers may result in the loss of
homeotropy. The reversible transition to a tilted texture starts
at a temperature T*, which is characteristic for the actual system
substrate/LC; the tilt angle increases smoothly with decreasing
temperature (c.f. Sect. 3.1.).

b) The energy of adhesion of LC does not change anomalously with changing LC boundary orientation. Its temperature coefficient is of the same order of magnitude as for isotropic liquids (c.f. Sect. 3.1.).

Packing density of the film:

c) The transition temperature T* strongly varies with the packing density of the amphiphilics. At a defined packing density depending on the system substrate/LC values of T* are minimal (c.f. Sect. 3.2.1.).

d) The energy of adhesion decreases with increasing packing density.

Phase type of the monolayer:

e) Perfect homeotropic LC orientation is produced by lecithin monolayers transferred in the liquid-expanded phase. Those films transferred in the condensed phase usually orientate LC only at high temperatures of the LC sample (c.f. Sect. 3.2.2.).

Structure of the amphiphilics:

f) T* values of monoalkyl lecithins are less than those of dialkyl lecithins at equal mean areas per molecule. For short-chained dialkyl lecithins T* is less than for those with long chains. Fatty acids (C_nH_{2n+1} COOH, n = 13, 15, 17, 19) orientate LC only a few degrees below the clearing temperature, even when they were transferred in the liquid-expanded phase (c.f. Sect. 3.3.2.).

g) The energy of adhesion decreases with increasing chain length of amphiphilic molecules and is less for monoalkyl lecithins compared to dialkyl lecithins.

4.2. Interaction Model

A loosely packed film contains holes of molecular dimensions. LC molecules enter such holes thus attaining a preferred orientation, which is transferred into the LC bulk by elastic interaction. The available holes of the film are occupied (dynamically) by LC

molecules at any packing density and temperature. The amphiphilic
film and the LC always interact strongly. Consequently the boundary orientation of the LC is always the same as the mean orientation of the holes within the amphiphilic monolayer. Changes of the
LC orientation thus are caused by changes of the orientation of the
film molecules.

4.3. Test of the model by experimental results

The experimental results summarized in Sect. 4.1.are explained
as follows:

a) The temperature dependence of the director field orientation may
be seen in connection with the polymorphism of phospholipid monolayers. If one assumes a type of phase diagram as found by Sackmann[15] for such films on aqueous subphase (Figure 6) to be valid in
principle on solid substrates, too, decreasing temperature leads
from an isotropic liquid phase (I) to a tilted crystalline one (III).

One might argue[20] that one should look at a π/A or T/A diagram
inspite of the π/T (Figure 6), because the mean area per lecithin
molecule A is kept constant within our immobilized monolayers on
solid substrates. The system regarded here, however, is far away
from being a pure lecithin monolayer discussed by Sackmann[15]. Actually, the film must be regarded as strongly disturbed by the LC. As
the holes of the film are filled with LC molecules conditions of a
quasi-constant lateral pressure π seem to be plausible in the system
monolayer/LC. Thus the model proposes the homeotropic orientation
to be induced by phase I of a system "amphiphilic monolayer + entered
LC molecules". The transition to a tilted orientation is caused by
a phase transition I → II or I → III, respectively.

b) The weak temperature dependence of the energy of adhesion is
plausible because no change of the interaction mechanism is postulated, even with changing LC orientation.

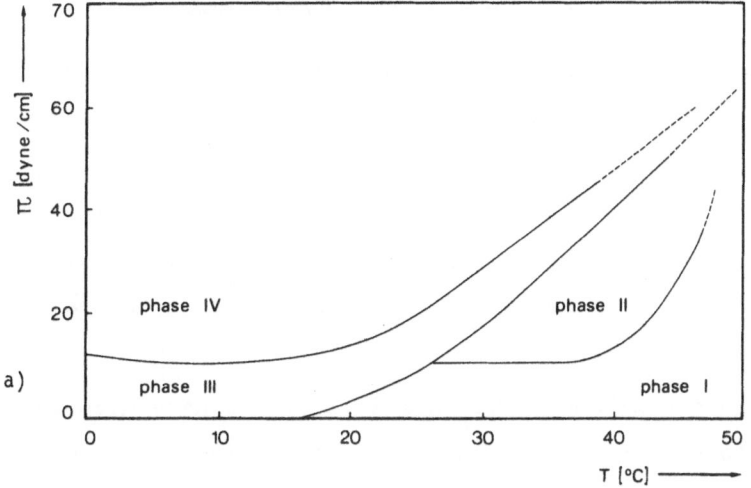

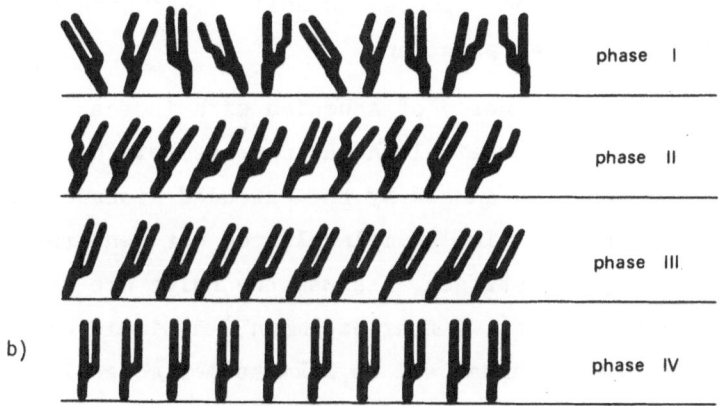

Figure 6:   a) Phase diagram of a dipalmitoyl-lecithin monolayer on
               an aqueous subphase after Sackmann[15]
               I: isotropic liquid, II: smectic C, III: tilted
               crystalline, IV: normal crystalline,
            b) sketch of the phases

c) The dependence of T^* on the mean area per molecule A becomes
plausible within the range of A $\underset{\sim}{} 50 \cdot 10^{-2}$ to $\underset{\sim}{} 90 \cdot 10^{-2}$ nm^2/mole-
cule by means of the phase diagram (Figure 6, A roughly varies in-
versely to the surface pressure π). The increase of T^* for larger
values of A is supposed to be due to the increased dilution of the
film: as the coherence of the film vanishes, the sterical interac-

tion between the monolayer and the LC decreases more and more.

d) The energy of adhesion decreases with increasing packing
density because the number of holes per unit area decreases.

e) At the condensation of a liquid-expanded monolayer on an aqueous
subphase obviously two-dimensional crystallites are formed whose
local packing density is much greater than the mean value. Thus
the transition temperatures T* must be higher at the same reasons
discussed under c).

f) On aqueous subphases monoalkyl lecithins have much lower phase
transition temperatures than dialkyl lecithins, the same holds for
short-chained lecithins compared with long-chained ones. An
analogous behaviour may be expected on solid substrates and indeed
matches with the observed phenomena.

g) The decrease of the energy of adhesion with increasing chain
length is explained by the simultaneous decrease of the polar part
of the energy of adhesion caused by the extended distance between
the interaction dipoles. The LC molecules do not penetrate film
holes down to the glass surface, because the holes are shortened by
kinks of the alkyl chains. (After Sackmann[21] a liquid-expanded
phospholipid monolayer contains 1 to 2 kinks per molecule.)

5. SUMMARY

    Nematic LC orientate definitely on solids coated with amphi-
philic monolayers. By means of several experimental methods, e.g.
helix unwinding of induced cholesteric LC, we have found the
following results. Below a well defined *temperature* T* a homeotropic director field transforms reversibly to a tilted one. T*
passes through a minimum value with the *packing density* of the
amphiphilic film changing. These phenomena lead us to a model
explaining the orientating action of amphiphilic films by a steric
effect: LC molecules penetrate molecular holes of the amphiphilic

monolayer, thus accepting a preferred orientation which is transferred into the LC bulk by elastic interaction. Experimentally
found changes of the orientation obviously are caused by *phase
transitions* of a system "amphiphilic film + embedded LC molecules."
which also occur in amphiphilic monolayers on aqueous subphases[15].

This work has been supported by the Fonds der Chemischen Indus-
trie and the Deutsche Forschungsgemeinschaft.

REFERENCES

1. E. Perez and J. E. Proust, J. Colloid Interf. Sci. <u>68</u>, 48
    (1979).

2. G. Porte, J. Physique <u>37</u>, 1245 (1976).

3. L. T. Creagh and A. R. Kmetz, Mol. Cryst. Liqu. Cryst. <u>24</u>, 59
    (1973).

4. T. Uchida, K. Ishikawa and M. Wada, Mol. Cryst. Liq. Cryst.
    <u>60</u>, 37 (1980).

5. M. Ohgawara and T. Uchida, Jpn. J. Appl. Phys. <u>20</u>, L 237
    (1981).

6. S. Naemura, J. Appl. Phys. <u>51</u>, 6149 (1980).

7. K. Hiltrop and H. Stegemeyer, Mol. Cryst. Liq. Cryst. <u>49</u>
    (Lett.), 61 (1978).

8. I. Langmuir, J. Amer. Chem. Soc. <u>39</u>, 1848 (1917).

9. K. B. Blodgett, J. Amer. Chem. Soc. <u>57</u>, 1007 (1935).

10. K. Hiltrop, H. Stegemeyer, Ber. Bunsenges. Phys. Chem. <u>82</u>, 884
    (1978).

11. M. Brehm, H. Finkelmann, H. Stegemeyer, Ber. Bunsenges. <u>78</u>,
    883 (1974).

12. F. Fischer, Z. Naturforsch. <u>31a</u>, 41 (1975).

13. M. J. Press, A. S. Arrott, J. Physique <u>37</u>, 387 (1976).

14. A. W. Adamson, <u>Physical Chemistry of Surfaces</u>, Interscience
    Publishers, New York, London, Sidney 1967, p. 363 f.

15. O. Albrecht, H. Gruler, E. Sackmann, J. Physique __39__, 301
    (1978).

16. J. L. Firpo, J. J. Dupin, G. Albinet, J. F. Baret, and A.
    Caillé, J. Chem. Phys. __74__, 2569 (1981).

17. G. L. Gaines, <u>Insoluble Monolayers at Liquid-Gas Interfaces</u>,
    Interscience Publishers, New York, 1966, p. 156 ff.

18. M. C. Phillips, D. Champman, Biochim. Biophys. Acta __163__, 301
    (1968).

19. K. Hiltrop, Dissertation, Paderborn, 1979.

20. H. A. van Sprang, private communication.

21. E. Sackmann, Ber. Bunsenges. Phys. Chem. __82__, 891 (1978).

REFRACTIVE INDEX MEASUREMENTS USING A DOUBLE ARM CONVERGING BEAM INTERFEROMETER

D. Balzarini and P. Palffy-Muhoray

The University of British Columbia
Vancouver, British Columbia, V6T 1W5

ABSTRACT

The temperature dependences of the ordinary and extraordinary refractive indices for the liquid crystal Butyl p-(p-Ethoxyphenoxy-carbonyl) Phenyl Carbonate have been measured. The measurements are made using a double arm, converging beam interferometer. The sample, sealed in a glass cell with parallel windows, is placed between the two converging lenses. The sample is aligned with a magnetic field between a polariser and an analyser. The interference of the sample beam and the reference beam produces a concentric fringe pattern. The temperature dependence is measured by monitoring the intensity at the centre of the interference pattern and the absolute values of the indices are obtained from the spacing of the fringes in the pattern.

Many organic materials exhibit a liquid phase displaying certain optical properties which depend on the ordered nature of the material. This paper reports research investigating the temperature dependence of the ordinary and extraordinary refractive indices for the liquid crystal Butyl p-(p-Ethoxyphenoxycarbonyl)

531

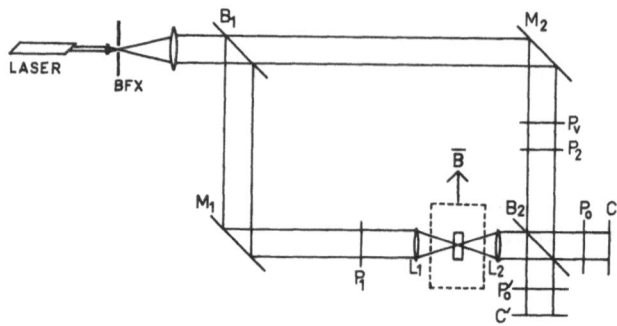

Fig. 1.   Diagram of the optical system.

Phenyl-Carbonate.   An optical system is described which allows precise measurements of the temperature dependence of the refractive
indices n and also provides measurements of the absolute values.
The temperature dependence is measured more accurately than the absolute values resulting in data for which the shape of n versus T
is very accurately determined, but for which the whole curve can be
shifted up or down within the uncertainty with which the absolute
values are measured.

The optical system is shown in Fig. 1. A laser beam $\lambda = 6328\mathring{A}$
is rendered uniphase and collimated by a beam expanding telescope.
This consists of a microscope objective, a 10 μm pinhole, and a
f2.8, 305 mm focal length lens and is labelled BFX in the diagram.
The use of the large lens allows for a reasonably uniform wave over
a large area. The expanded beam is split at B_1. The reference portion travels to mirror M_2, through attenuator P_v which allows for
variation of reference intensity, through P_2 which allows for selection of polarization, and to beam combiner B_2. The sample portion
of the beam travels to mirror M_1, through polarizer P_1 which allows
for selection of polarization, through lens L_1 which converges the
beam through the sample, through an identical lens L_2, and to beam
combiner B_2.

The position of lens L_2 relative to L_1 determines the diver-
gence of the beam exiting the sample vessel. The combined beams

travel through the polarisers P_0 and P_0' and interference is measured at C and/or C'.

If the focal point of L_2 coincides with the focal point of L_1,
the beam emerging from L_2 is again collimated and a uniform interference field is observed. If the foci of L_1 and L_2 are not coincident, but are collinear along the beam, the resulting interference pattern will be concentric circles if the sample is isotropic.
If the sample is not isotropic, the interference pattern depends on
the alignment of the sample material and the polarization selected.
The sample vessels used in this experiment are quartz spectrophotometer cells with graded seals to pyrex allowing for evacuation and
sealing. The inside dimension between windows is 2 mm. The sample
is aligned with a magnetic field $\bar{B}$ oriented as shown in Fig. 1. In
the following discussion $P_|$ and $P_{||}$ are used to denote polarization
perpendicular and parallel to the nematic alignment direction. The
symbols n_o and n_e are used to denote the perpendicular (ordinary)
and parallel (extraordinary) indices of refraction, n_i is the index
of the isotropic phase.

Each of the lenses L_1 and L_2 can be adjusted in three directions. The procedure usually followed is to adjust for a uniphase
interference pattern with the empty sample vessel in place, or in
some cases with the isotropic sample in place. If the collinear
lenses L_1 and L_2 are adjusted along the beam direction for a "flat"
interference pattern with the empty cell in place, the introduction
of the sample produces various interference patterns depending on
alignment of liquid crystal and selection of polarization.

Fig. 2 exhibits the interference pattern observed with the
sample in the isotropic state. The orientation of the polarization
is not important in this case. Fig. 3 exhibits the interference
pattern observed with the sample in the nematic state and the polarizers perpendicular to the nematic director. Fig. 4 exhibits
the interference pattern observed with the sample in the nematic

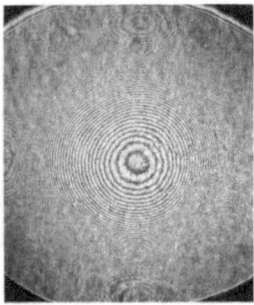

Fig. 2.   Interference pattern
          obtained with an iso-
          tropic sample.

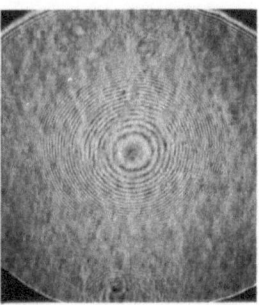

Fig. 3.   Interference pattern
          obtained with the po-
          larization perpendicu-
          lar to the nematic
          director.

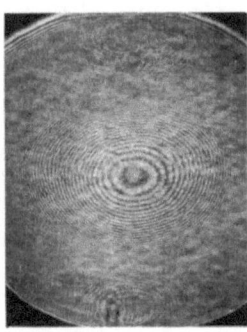

Fig. 4.   Interference pattern
          obtained with the polar-
          ization parallel to the
          nematic director.

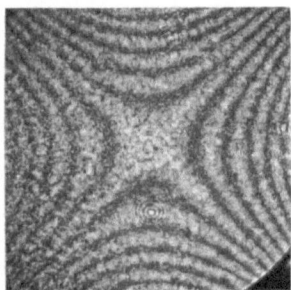

Fig. 5.   Interference pattern
          obtained with polarizers
          crossed and at angles
          of 45 degrees to the
          nematic director and
          with the reference beam
          blocked.

state and the polarizers parallel to the nematic director. Fig. 5
exhibits the interference pattern observed with the sample in the
nematic state, the polarizers aligned at 45° to the nematic director, and the reference beam blocked.

The temperature dependence of n_i, n_0, n_e, and $|n_e-n_0|$ is ob-
tained by monitoring the intensity at the centre of the patterns of

Figs. 2, 3, 4 and 5 respectively. The fringe number is related to
the wavelength of the laser, sample thickness, and the change in refractive index of the sample. As an index changes, fringes move
into the centre or out of the centre of the dupteau patterns of
Figs. 2, 3, and 4. For Fig. 5, the fringes move into the centre
from two opposite directions and out of the centre from the other
two directions. Data for the temperature dependence of the indices
can be obtained by simply plotting fringe number versus temperature.

The graphs of fringe number versus temperature produce a precise determination of the shapes of the functions for the indices
versus temperature, but they are only determined to within additive
constants. These additive constants can be obtained from analysis
of the positions of the fringes in the interference patterns. The
spacing of the fringes and their distances from the centre of the
pattern can be related to the wavelength, cell thickness, focal
length of L_2, distance to film plane, and refractive index. The
fringes are most easily analysed along a vertical direction in the
pattern, i.e., through the centre and perpendicular to the nematic
direction. A plot of the square of the vertical distance from the
centre of the fringe pattern versus fringe number yields a straight
line to first order. The value for the index obtained in this way
has a certain amount of uncertainty, i.e., the additive constant
needed for the temperature dependence data is determined to within
a certain error. This error can be reduced by repeating the fringe
spacing analysis at many temperatures. The errors are also reduced
by utilizing the additive constant obtained for the quantity
$|n_e - n_o|$ with those obtained for n_e and n_o separately.

The phase at the centre of the interference pattern is monitored with a photodetector. Photographs of the interference pattern are made at intervals as desired. An alternative option for
monitoring the fringes is to have a vertical slit through the pattern. A film strip can be made to travel slowly in a horizontal

direction behind this vertical slit. The movement of the fringes
towards (or away from) the centre of the pattern can be recorded on
the film strip. A small photodetector could also be used in conjunction with the strip film by mounting it at the centre. The vertical slit for the film strip covers the region of the interference
pattern which is used in the analysis for the additive constants.
Therefore this film strip can provide the same information. Caution must be exercised with this, however, since changes in the
alignment of the interferometer would not be as easily noticed in
the film strip as they would be in the full photographs of the interference patterns.

The use of two observation planes, C and C', at the same time
is possible but requires the use of good quality optical components.
Care must be exercised in the choice of polarizers; glass enclosed
polarizers are often wedged and are not suitable. This optical
system can also be used for studying dependence of the indices on
other parameters, such as applied electric and magnetic fields.

The sample vessel is contained in an inner copper block which
is insulated from an outer aluminum block with polyurethane foam.
Each block is wound with heating wire of resistance 0.067 Ω/m. The
temperature of the outer block is controlled by regulating the current input to the heating wire of the outer block. Precise thermo-
static control of the inner block is obtained by using a thermistor
embedded in the block as one arm of a d.c. bridge. The error signal
from this bridge is amplified and the output is conditioned by a
lead-lag network to insure stability. The conditioned bridge output drives the control input of a stabilized power supply which
provides power to the heating wire on the inner block. Temperature
can be maintained to within 0.0002 K.

Results for BEPC are shown in Fig. 6. The meaning of the
error bars is that a whole set of points can be moved up or down
within the error bar.

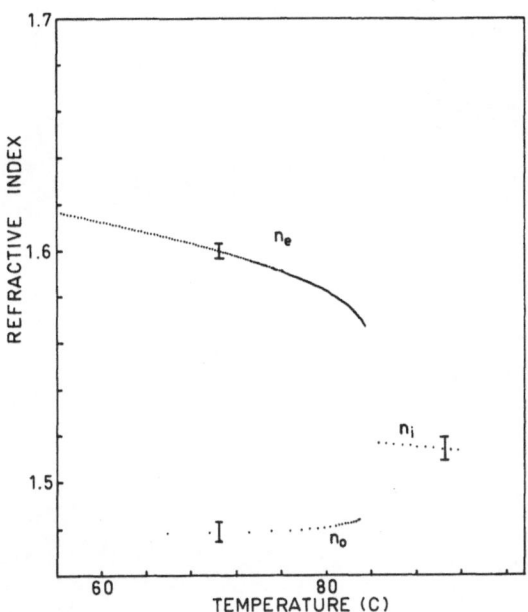

Fig. 6. Temperature dependence of the refractive indices for the liquid crystal Butyl p-(p-Ethoxyphenoxycarbonyl) Phenyl Carbonate.

A PULSED NMR STUDY OF TRANSIENT AND PERSISTENT MOLECULAR ORDER IN

NEMATIC LIQUID CRYSTALS*

P. A. Mitchel and C. E. Tarr

Department of Physics and Astronomy, University of Maine
at Orono, Orono, ME   04469

ABSTRACT

The proton absorption spectrum of a typical nematic liquid
crystal shows one or more well separated pairs of lines arising from
the large dipolar interaction of protons on the phenyl rings. Because these interactions have cylindrical symmetry with respect to
the nematic director, observation of the splittings gives a direct
measure of the orientation of the director with respect to the
magnetic field. Experiments are reported in which pulsed NMR
techniques were used to determine persistent molecular order and
orientation in response to applied electric fields. From these
experiments it was possible to determine the temperature dependence
of the flow alignment angle for PAA. These data show the first
evidence of flow alignment in the super-cooled nematic state. In
addition studies of CBEOA yielded information about the time evolution of molecular order and orientation in response to the sudden
application of an electric field.

---

*This work was partially supported by National Science Foundation
 Grant DMR-78-10313-02.

I.   INTRODUCTION

A variety of factors influence the orientation of the nematic director. Of particular interest are the effects of magnetic and electric fields. In general, the molecular axes are approximately parallel to the director which aligns parallel to an applied magnetic field. The alignment by electric fields is somewhat more complicated as the sign of the dielectric anisotropy may be either positive or negative. Further, many nematic materials exhibiting negative dielectric anisotropy and positive conductivity anisotropy show rather complicated alignment in electric fields. In particular, there may exist a cut-off frequency for electric fields above which the orientation of the director will be determined by the dielectric anisotropy and below which the orientation will be determined by the conductivity anisotropy.[1,2] For electric field frequencies near the cut-off frequency there may be competing effects due to a mixture of both. The cut-off frequency is strongly dependent upon sample purity and rises with increasing impurity concentration.[3]

Nuclear magnetic resonance provides a convenient way to monitor the molecular orientation of liquid crystals. A typical nematic liquid crystal molecule consists of two or more phenyl rings connected by short linkage groups and may have end groups or chains of various length. In most cases the para-axis of the molecule makes a small angle with respect to the molecular axis. Rapid molecular motions give rise to cylindrical symmetry of the para-axis about the molecular axis and the molecular axis about the director. The large dipolar interaction between adjacent protons on the phenyl rings produces large line splittings proportional to $P_2(\cos\alpha) = \frac{1}{2}(3\cos^2\alpha - 1)$, where α is the angle between the nematic director and the magnetic field H_o of the NMR experiment. Thus, the proton NMR spectrum of a nematic liquid crystal aligned with the director parallel to H_o will in general be rather simple. For example, the proton spectrum of partially deuterated cyano-benzylidene-ethoxy-aniline (CBEOA$_{d5}$) whose structure is given in Fig. 1 shows three broad lines, a center line resulting from the motional averaging of the single proton on the linkage group and a pair of lines arising from the dipolar inter-action of the phenyl protons. If the nematic director is rotated with respect to the magnetic field, then this splitting will change as $\Delta\omega = \Delta\omega_o \left|3\cos^2\alpha - 1\right|$, where $\Delta\omega_o$ is the splitting for $\alpha = 0$. This behaviour is illustrated in Fig. 2, where a high-intensity a.c. electric field with a frequency well above the cut-off frequency was used to vary α.

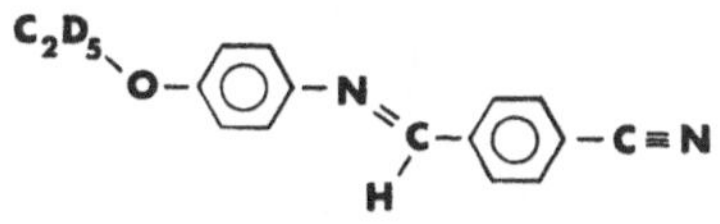

Fig. 1   Structure of partially deuterated cyano-benzylidene-ethoxy-aniline (CBEOA$_{d5}$).

II. DIELECTRIC ORIENTATION

If the nematic liquid crystal CBEOA, which has a positive di-electric anisotropy, is subjected to transient electric fields, relatively complicated alignment may result. If the high-intensity, high-frequency electric field above is switched on in a direction perpendicular to the magnetic field, the molecules will reorder and will approach uniform alignment with the director parallel to the direction of the electric field. The torque producing the reorien-tation of the nematic director is proportional to E^2, where E is the applied electric field. For an electric field of E = 1020 V/cm and

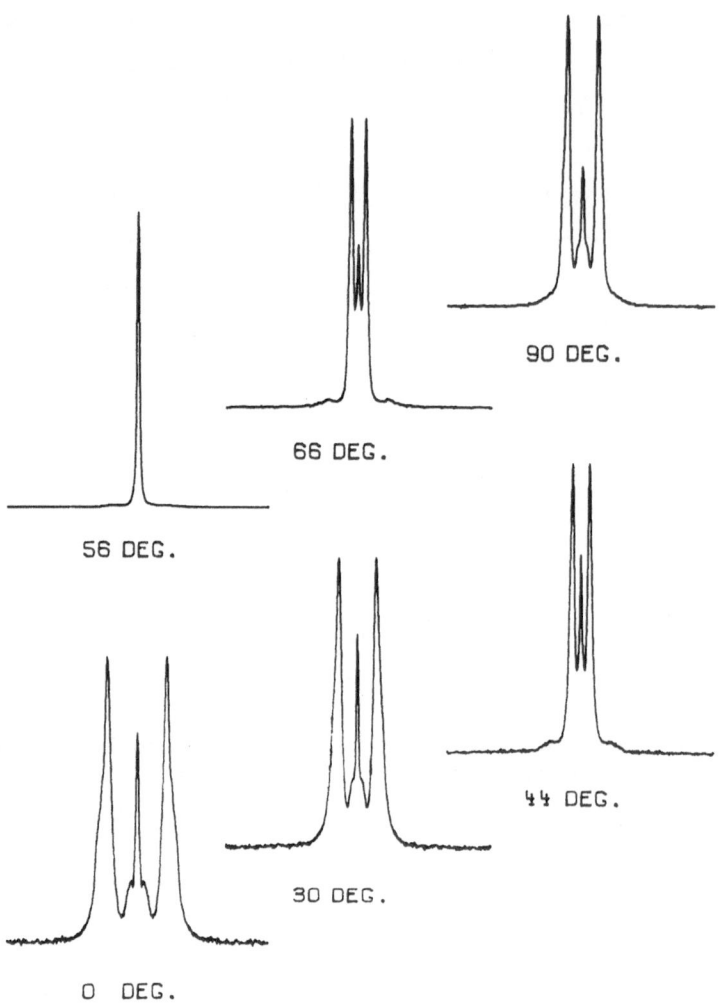

Fig. 2   Proton NMR spectra of CBEOA$_{d5}$ as a function of the angle
between the nematic director and H$_o$.

a magnetic field H$_o$ = 7 kG (30 MHz proton Larmor frequency) the
reorientation approaches equilibrium in approximately 600 milliseconds.

A.   Transient Electric Field Experiment

In order to determine the time evolution of this reordering
process, the proton NMR spectrum was monitored as a function of time

following the sudden application of the electric field. The rapid
reorientation of the sample required that the experiment be under
computer control. The pulse sequence used for the data collection,
r.f. pulse generation, and electric field control is shown in Fig. 3.
First a series of r.f. pulses was applied to bring the magnetization
to an equilibrium value in order to be able to use the full dynamic
range of the digitizer. Second, the electric field was turned on,
and following a fixed delay, an r.f. pulse was applied and the free
induction decay (F.I.D.) of the magnetization produced by this pulse
was digitized and stored in memory. Additional signals were acquired
with inter-pulse delays of 50 milliseconds. This 50 millisecond
delay was sufficiently long to prevent echo effects and undue
saturation of the magnetization. A total of five acquisitions were
recorded for each electric field cycle. Computer memory size limited
the number of observations to five per field cycle. In order to
improve the signal to noise ratio, the field cycle was repeated 100
times and signal averaged. The initial delay in the sequence was
then varied in increments of 5 milliseconds to produce data with
delays ranging from 11 to 256 milliseconds. The minimum possible

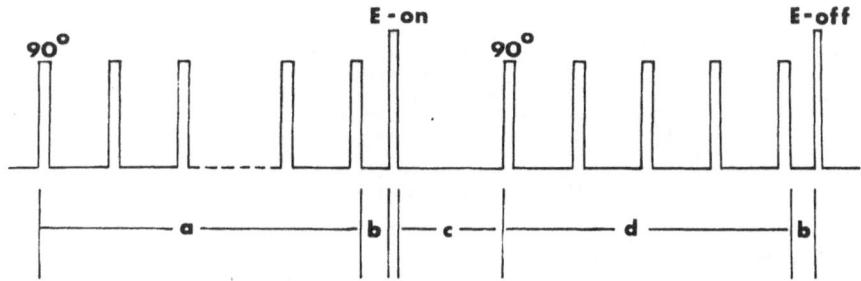

Fig. 3   Pulse sequence for dielectric reordering studies: a
          denotes comb of pulses to bring magnetization to
          equilibrium, b denotes reprogramming delay, c is the
          initial delay, and d the data acquisition sequence
          (see text).

delay was the pulse generator reprogramming time of 6 milliseconds.
Each signal-averaged F.I.D. was then Fourier transformed to produce
a proton NMR absorption spectrum corresponding to the molecular
orientation at each time. The time evolution of the spectrum is
shown in Fig. 4.

B.   Model for Reorientation

It had been observed earlier[4] that the equilibrium orientation
of a nematic in competing orthogonal magnetic and electric fields
could be well fit by assuming that the molecules were in one of

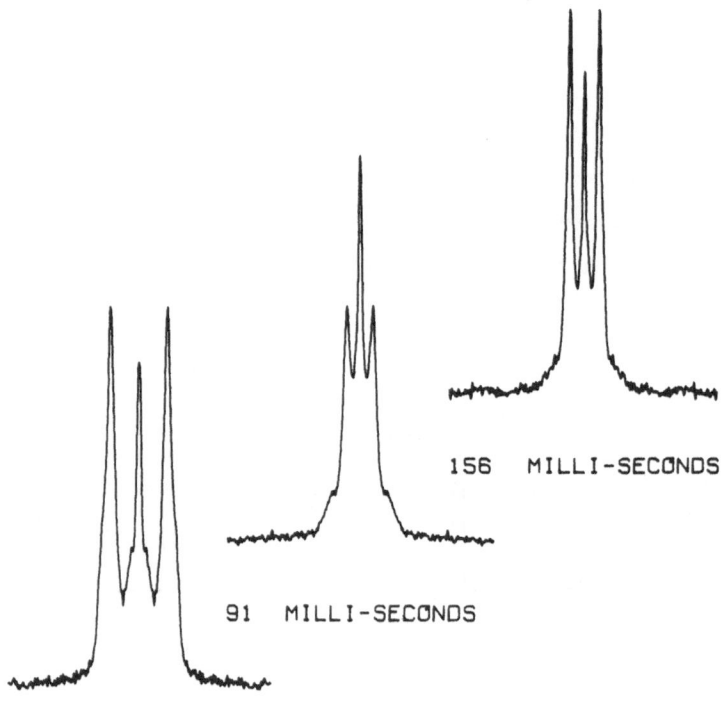

156   MILLI-SECONDS

91   MILLI-SECONDS

31   MILLI-SECONDS

Fig. 4   Partial time evolution of the proton NMR spectrum of
          CBEOA$_{d5}$ after application of electric field at time 0.

three groups: those aligned with the nematic director parallel to the magnetic field, those aligned with the director parallel to the electric field, and those with random orientation of the nematic director. It was determined that the time dependent data could also be well fit by such a model.

The time evolution data were least squares fit to three normalized spectra corresponding to the three orientation groups of the model. The spectrum corresponding to random orientation was calculated by adding normalized spectra taken with the nematic director at angles from 0° to 90° with respect to the magnetic field in 2° increments. The three basis spectra used in the fits are shown in Fig. 5. The least squares fit then yielded the percentage distribution of molecules associated with each of the groups as a function

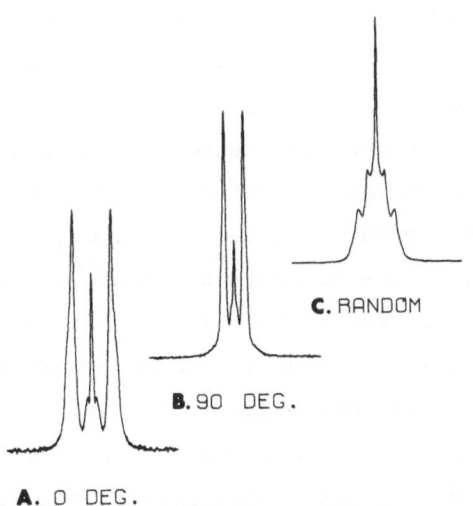

C. RANDOM

B. 90 DEG.

A. 0 DEG.

Fig. 5   Proton NMR spectra of CBEOA$_{d5}$ with the nematic director:
A.   parallel to H$_o$, B. perpendicular to H$_o$, and
C. oriented at random with respect to H$_o$.

of time following the application of the electric field. The fits
were in excellent agreement with the experimental data, with r.m.s.
errors in the area typically less than 5 per cent. A typical fit
is shown in Fig. 6. The time evolution of each of the three groups
of molecules is shown in Fig. 7. From these data we see that the
initial uniform alignment with the director parallel to the nematic
field is disrupted, and the molecules become randomly oriented. The
random orientation then more gradually decays and the molecules
approach uniform alignment with the director parallel to the electric
field. The reordering is seen to be complete in approximately 600
milliseconds.

III. NEMATIC FLOW ALIGNMENT

If a nematic liquid crystal with negative dielectric anisotropy
and positive conductivity anisotropy is subjected to a magnetic
field parallel to an electric field whose frequency is well below
the cut-off frequency, then macroscopic motion of the sample may
occur. This motion gives rise to shear that may produce a tilting
of the nematic director with respect to the direction of motion.
This tilting may be characterized by a nematic 'flow-alignment'
angle which is temperature dependent. If the electric field intensity
is such that the torque due to the dielectric anisotropy is approximately equal to the torque produced by the magnetic, then any motion
induced by the anisotropy in the ionic conductivity will not be
complicated by the effects of the fields. If the flow alignment so
produced is reasonably uniform, then NMR techniques may be used to
determine the flow alignment angle.[5] In particular, in the uniformly
aligned regions the director will be rotated away from its orientation
parallel to H_o with a resulting decrease in the side peaks associated
with the phenyl protons. This decrease in splitting may then be
related to the flow alignment angle, Θ, by: $\Delta\omega = \Delta\omega_o P_2(\cos\Theta)$,
where $\Delta\omega_o$ is the peak splitting with no electric field, and $\Delta\omega$ is
the splitting with the electric field applied. This is illustrated

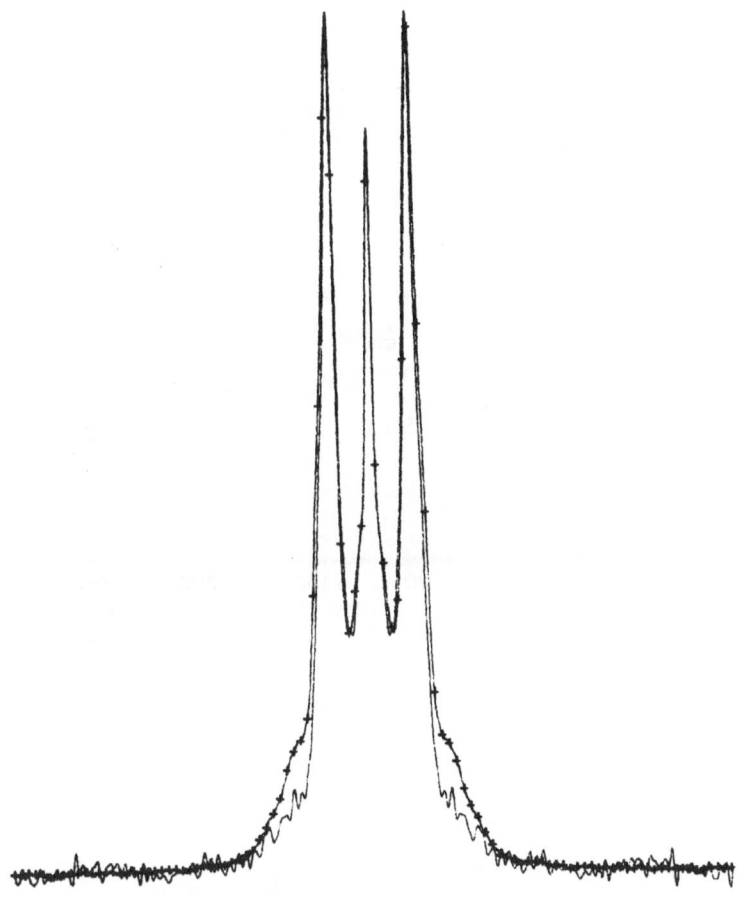

Fig. 6   Typical data. Fit to spectrum at t = 256 milliseconds.
The fit is denoted by + - + - + - +.

in Fig. 8. As $\Delta\omega_o$ depends upon the nematic order parameter, it is
temperature dependent, thus it is necessary to measure both $\Delta\omega$ and
$\Delta\omega_o$ to determine the temperature of Θ.

The temperature dependence of Θ has been measured in PAA in the
above manner and the results are shown in Fig. 9. The relatively
large error bars arise from a broadening of the peaks due to
imperfect alignment. As can be seen from these data, flow alignment
appears to persist below the melting point in PAA. Since the NMR
technique used here does not disturb the sample mechanically, as

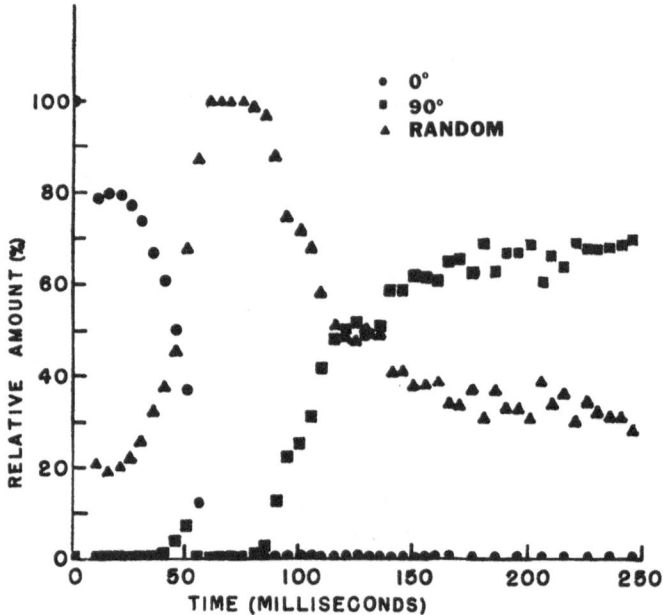

Fig. 7   Time evolution of the molecular orientation of CBEOA$_{d5}$
          following the application of electric field, see text.

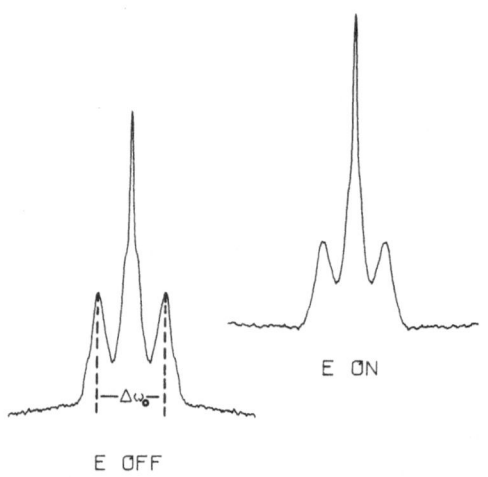

E OFF

E ON

Fig. 8   Proton NMR spectrum of PAA with no electric field and
          with a 2000 V/cm, 10 HZ electric field parallel to a
          magnetic field of 2.34 kG.

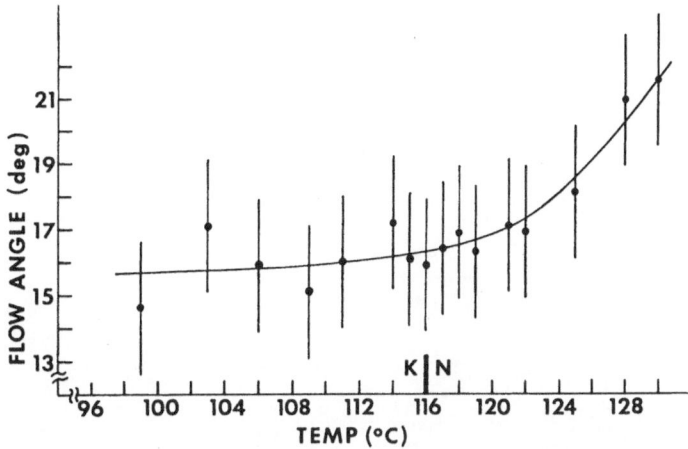

Fig. 9   Flow alignment angle for PAA as a function temperature.
The line is intended to aid the eye. Note persistent
alignment below solid-nematic phase transition.

would be the case of a viscometer measurement, it is possible to

observe this persistent motion.

REFERENCES

1.   E. F. Carr, Adv. in Chem. Ser. 63, 76 (1967).

2.   W. Helfrich, J. Chem. Phys. 51, 4092 (1969).

3.   D. P. McLemore, Ph.D. Thesis, Univ. of Maine, 1973
           (unpublished).

4.   T. E. Kubaska and C. E. Tarr, Mol. Cryst. Liq. Cryst. 29, 155
           (1974).

5.   C. E. Tarr and E. F. Carr, Solid State Commun. 33, 459 (1980).

THE EFFECT OF TWIST ON BIAXIAL ORDERING IN THE CHOLESTERIC PHASE*

Z. Yaniv, M. E. Neubert and J. W. Doane

Liquid Crystal Institute, Kent State University, Kent
Ohio 44242

ABSTRACT

Deuterium magnetic resonance and optics have been used to study chiral nematic materials composed of compounds selectively deuterated at specific sites. The deuterated materials are 4-methoxybenzylidene-4'-butylaniline (MBBA) selectively deuterated in the α position (MBBA-α-d_2) and on one aromatic ring (MBBA-2'-3'-5'-6'-d_4). To each of these materials was added 4-methoxybenzylidene-4'-[(+)-2-methyl-butyl]aniline (MBMBA) in various quantities to obtain samples of different pitch length. From deuterium spectral patterns, values were obtained for the asymmetry parameter η, from different sites of the molecule. These data show that the mechanism responsible for ^{2}H-NMR biaxiality is anisotropic fluctuations of the molecular long axis. Measurements of η in the vicinity of the blue phase suggest that this phase may be initiated by a characteristic value of η in these materials. New ^{2}H-NMR spectral patterns of the blue phase are shown.

---

*This work was supported by NSF under grant DMR79-04393 and
 facilities grant DMR78-09046.

Both NMR[1] and optical[2] observations show the nematic phase of thermotropic materials to be uniaxially ordered. When the nematic phase becomes twisted, however, by the addition of a chiral material, deuterium NMR observations show the resulting phase to be biaxially ordered.[3] This feature was in fact predicted by several theoretical investigators long before our ^{2}H-NMR experiment.[4-8] Biaxiality has not yet been observed optically in the cholesteric phase but it is apparently a weak optical effect in this phase as suggested by some experiments.[9,10]

The ^{2}H-NMR observation raises new questions for the cholesteric as well as the blue phases which appear between the cholesteric and the isotropic phase in materials which are twisted to sufficiently short pitch lengths. Some of these questions are: What is the mechanism responsible for the observed biaxiality? How does the biaxiality depend upon the pitch length? And, is there any connection between biaxial ordering and the creation of the blue phases?

In this paper, more detailed ^{2}H-NMR measurements are presented which provide some answers to these questions. Emphasis is placed on the question of the mechanism for ^{2}H-NMR biaxiality. Schiff base materials are selectively deuterated in two different positions from which it is possible to distinguish between two possible mechanisms: a) anisotropic fluctuations of the molecular long axis relative to the direction of the pitch axis, or b) "birotational freezeout" where the short axes of the molecule are partially aligned relative to the pitch axis direction. Data are shown which clearly demonstrate mechanism a) to be dominant in these materials. Furthermore, new data are shown in materials which have a short enough pitch length to exhibit the blue phase. Biaxiality measurements are reported for the first time up to onset of the blue phase and new spectral patterns for the blue phase are reported.

In the ^{2}H-NMR measurement, two physical parameters are normally measured. For the ith deuterated site of the molecule, these are the quadrupole coupling constant $\nu_Q = e^2 Q \, V_{zz}^i / h$ and the asymmetry parameter, $\eta^i = (V_{xx}^i - V_{yy}^i)/V_{zz}^i$ where V_{kk} are the electric field gradient tensors associated with the principal axis system $k = x,y,z$.[1] In the liquid crystal phases ν_Q^i and η^i are time averaged quantities and in the case of the twisted nematic the principal axis x,y,z frame is common to each deuterated site. The measured value of ν_Q^i at a particular site in the cholesteric phase is related to the degree of order of that site by the equation:[11,12]

$$\nu_Q^i = \nu_Q^s \, S_i = \nu_Q^s < \frac{3}{2}\cos^2\sigma_i - \frac{1}{2}> \tag{1}$$

where ν_Q^s is the value of the coupling constant measured in the solid state for that site and σ_i is the angle between z axis and the $(C-^2H)_i$ bond direction. The value of ν_Q^s is nearly the same for all sites of a deuterated molecule. Of more interest in this work is the measured value of η^i which can be expressed as:[11,12]

$$\eta^i = \frac{4}{3} \frac{\nu_Q^s}{\nu_Q^i} B_i = \frac{2}{3} \frac{\nu_Q^s}{\nu_Q^i} <\sin^2\sigma_i \, \cos 2\xi_i> \tag{2}$$

where σ_i and ξ_i are the polar coordinates of the $(C-^2H)_i$ bond direction in the x,y,z frame. With the convention $|V_{xx}| \leq |V_{yy}| < |V_{zz}|$, x has been determined to be the pitch axis[3] and z the long axis director which is twisted about x. In transforming to a molecular frame the biaxiality, B_i, can be written:[11,12]

$$B_i = \frac{9}{8}(r_i S_{2,0} + s_i S_{2,2}) \tag{3}$$

where $S_{2,0} = <\sin^2\theta \, \cos 2\phi>$ expresses anisotropic fluctutaions of the long molecular axis and $S_{2,2} = <\frac{1}{2}(1 + \cos^2\theta)\cos 2\phi \, \cos 2\psi -$

$\cos\theta \cos2\phi \sin2\psi> \approx <\cos2(\phi+\psi)>$ expresses partial ordering of the molecular short axis. The Euler angles ϕ,θ,ψ give the orientation of the molecular frame M_x, M_y, M_z in the x,y,z principal axis frame. In this paper we choose the long molecular axis, M_z, to be parallel to the <u>para</u> axis of the aromatic rings. The coefficients $r_i = (\frac{3}{2} \cos^2\beta_i - \frac{1}{2})$ and $s_i = (\sin^2\beta_i \cos2\alpha_i)$ are conformation parameters where β_i and α_i are the polar angles of the $(C-^2H)_i$ bond in the molecular frame.

The materials used in this study are the selectively deuterated materials:

MBBA-α-d$_2$, CH$_3$O-◯-CH=N-◯-C(D)(D)-CH$_2$CH$_2$CH$_3$

MBBA-2',3',5',6'-d$_4$, CH$_3$O-◯-CH=N-◯(D)(D)(D)(D)-CH$_2$CH$_2$CH$_2$CH$_3$

to which various percentages of 4-methoxybenzylidene-4'-[(+)-3-methylbutyl]aniline (MBMBA)

CH$_3$O-◯-CH=N-◯-CH$_2$C*H(CH$_3$) CH$_2$CH$_3$

were added.

For a sufficient amount of chiral additive cylindrical type spectral patterns like those reported earlier were obtained.[3,13,14] Values for the asymmetry parameter were extracted from the data in the usual manner.[3,13] The pitch length in the materials was determined optically by a simple light scattering technique.[15]

Binary mixtures of MBBA-α-d_2 with 11.3 wt% MBMBA and of MBBA-2',3',5',6'-d_4 with 20.0 wt% MBMBA were found optically[15] to both have the same pitch length of 1.4 μm. Measured values of the asymmetry parameter for these two mixtures are shown in Fig. 1 below. Using Eq. (2) the values of B_i are calculated and their ratio displayed in Fig. 2.

Since both deuteriums of the MBBA-α-d_2 are equivalent we can choose one of the molecular frames, M_y, to bisect the ^{2}H-C-^{2}H bonds.[16] Since, by definition, M_z is the most ordered axis, we chose the axis to be parallel to <u>para</u> axis of the aromatic ring. With this choice of the molecular frame:

$$B_\alpha = -0.37\, S_{2,0} + 0.84\, S_{2,2}. \tag{4}$$

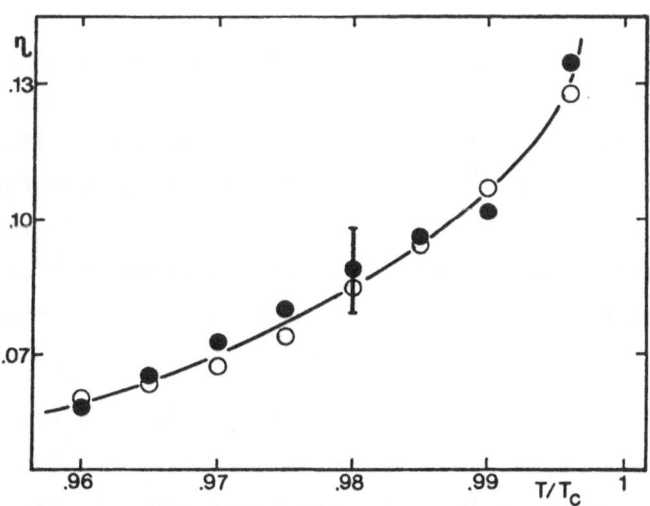

Fig. 1   Reduced temperature plot of η for binary mixtures of MBMBA with: MBBA-α-d_2, open circles; MBBA-2',3',5',6'-d_4, closed circles for a pitch length of 1.4 μm in each sample.

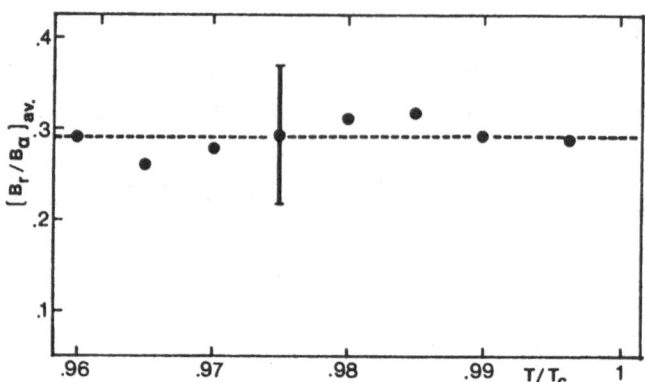

Fig. 2   Reduced temperature plot of the ratio B_{ring}/B_α taken from
          the values of η in Figure 1, where B_{ring} is the biaxiality
          measured of MBBA-2',3',5',6'-d_4 and B_α^{ring} is that of
          MBBA-α-d_2.

The aromatic ring is believed to be rotating about the para-axis
direction relative to the α position in which case we write:

$$B_{ring} = -0.12\ S_{2,0}.\tag{5}$$

From the data of Fig. 2 and Eqs. (4) and (5) it is seen that
the order parameter $S_{2,0}$ dominates at all temperatures for this
particular pitch length. It has also been shown that if $S_{2,0}$
dominates the value of η^1 is independent of the deuterated site[12]
which is the result of Fig. 1. From this, we conclude that there
is no measureable ordering of the short molecular axis and that the
^{2}H-NMR biaxiality is a result of anisotropic fluctuations of the
long molecular axis.

In an earlier publication[3] we reported that the biaxiality
increases with decreasing pitch length in long pitch length
materials. In order to determine how large biaxial ordering can
become we have measured values of η in materials with sufficiently
large concentrations of chiral MBMBA to obtain the blue phases at
high temperatures. Fig. 3 shows the temperature dependence of η

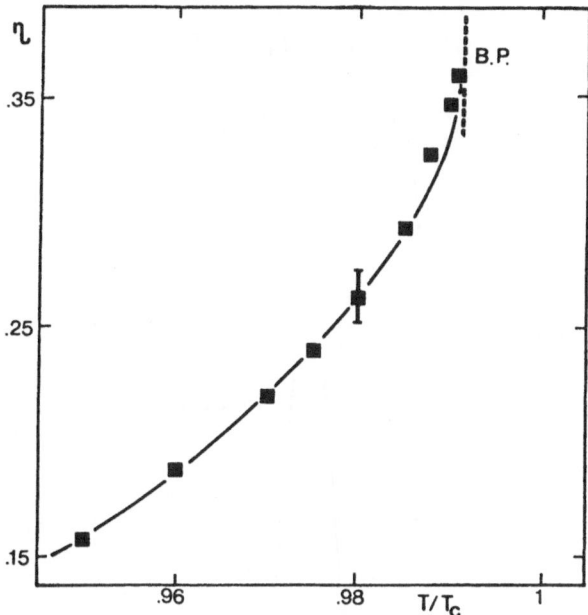

Fig. 3   Reduced temperature plot of η for MBBA-2',3',5',6'-d₄
            with 70 wt% MBMBA which exhibits the blue phase in the
            vicinity of the isotropic phase, transition, T_c = 295°K.

in MBBA-2',3',5',6'-d₄ with 70 wt% MBMBA. It is seen that the
maximum value of η before the onset of the blue phase is ∿0.35.
An interesting feature of this value is that our preliminary
measurements suggest that for higher concentrations of MBMBA and
shorter pitch lengths, the blue phase always appears at this value.
Further study on this point is currently in process.

    Upon the onset of the blue phase we observe a new spectral
pattern in these materials unlike those reported for cholesterol
containing materials.[17,18] This pattern is illustrated in Fig. 4
alongside of the spectral pattern of the twisted nematic phase.

    In conclusion, we report that the principal contribution to
NMR biaxiality in the twisted nematic phase is anisotropic fluctuation of the long molecular axis. We also report measurements of
biaxiality in the vicinity of the blue phase which suggest that this

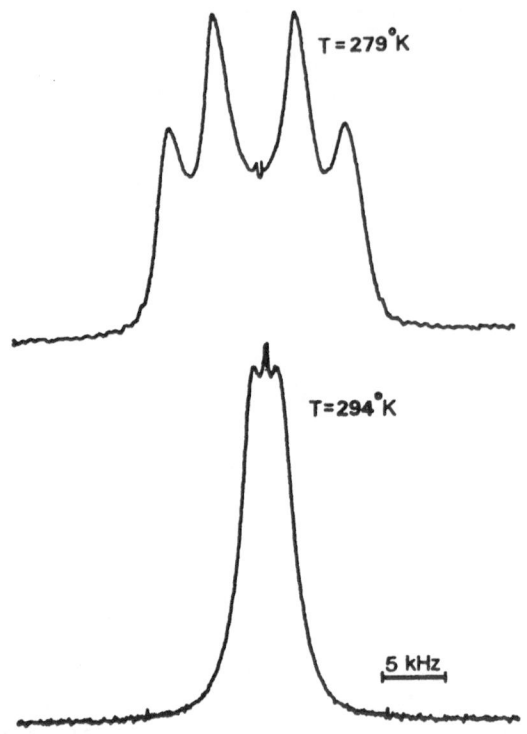

Fig. 4 ^{2}H-NMR spectral patterns of the cholesteric and blue
phases from MBBA-2',3',5',6'-d$_4$ with 70 wt% MBMBA.

phase may be controlled by the degree of biaxiality. Finally, we
show new ^{2}H-NMR spectral patterns for the blue phase.

REFERENCES

1.   J. W. Doane, in Magnetic Resonance of Transitions, edited by
        F. J. Owens, C. P. Poole, Jr., and H. A. Farach,
        Academic Press, New York, 1979, pp. 171-246.

2.   P. G. de Gennes, The Physics of Liquid Crystals, Oxford
        University Press, London, 1974.

3.   Z. Yaniv, N. A. P. Vaz, G. Chidichimo and J. W. Doane, Phys.
        Rev. Lett. 47, 46 (1981).

4. R. G. Priest and T. C. Lubensky, Phys. Rev. A 9, 893 (1974).

5. A. Wulf, J. Chem. Phys. 59, 1497, 6596 (1973).

6. S. A. Brazovskii and S. G. Dmitriev, Zh. Eksp. Theor. Fiz. 69, 979 (1975); Soviet Physics JETP 42, 497 (1976).

7. B. W. Van der Meer and G. Vertogen, Phys. Lett. 59A, 279 (1976).

8. H. Schroeder, in Liquid Crystals of One- and Two-Dimensional Order, edited by W. Helfrich and G. Heppke, Springer-Verlag, Berlin, 1980, p. 196.

9. D. W. Berreman and Terry J. Scheffer, Phys. Rev. A 5, 1397 (1972).

10. P. H. Keyes and C. C. Yang, J. Phys. (Paris), Colloq. 4, C3-376 (1979).

11. N. Vaz, M. Neubert and J. W. Doane, Mol. Cryst. Liq. Cryst. 68, 959 (1981).

12. D. Photinos and J. W. Doane, Mol. Cryst. Liq. Cryst. 76, 159 (1981).

13. G. Chidichimo, Z. Yaniv, N. A. P. Vaz and J. W. Doane, Phys. Rev. A 25, 1077 (1982).

14. N. A. P. Vaz, G. Chidichimo, Z. Yaniv and J. W. Doane, Phys. Rev. A (1982) (to appear).

15. T. Hashimoto, S. Ebisu, N. Inaba and H. Kawai, Polymer J. 13, 701 (1981).

16. A. Shetty, Ph.D. Thesis, Kent State University, Kent, Ohio (1981).

17. P. J. Collings and J. R. McColl, J. Chem. Phys. 69, 3371 (1978).

18. E. T. Samulski and Z. Luz, J. Chem. Phys. 73, 142 (1980).

DEUTERON MAGNETIC RELAXATION AND MOLECULAR DYNAMICS IN THERMO-TROPIC LIQUID CRYSTALS

Regitze R. Vold and Robert L. Vold

Department of Chemistry
University of California, San Diego
La Jolla, California 92093

INTRODUCTION

It has long been recognized that nuclear spin relaxation provides a convenient and powerful means for studying dynamic processes in ordered fluids. In this paper we discuss some recently developed NMR techniques which can provide novel, more detailed information about molecular motion in liquid crystals via the determination of individual spectral densities of motion. The potential of the techniques will be illustrated by deuterium relaxation data obtained in our laboratory and for reasons of brevity our discussion will be restricted to thermotropic nematic phases. Early work in this field (cf. reviews by Wade (1) and Doane (2)) was devoted mainly to measuring the interpreting overall spin-lattice relaxation rates of protons, taking advantage of the fact that numerous slow processes characteristic of the ordered medium possess strong Fourier components fluctuating near the Larmor frequency. A partial list of such processes includes the familiar fluctuations of nematic director orientation (3,4), translational self-diffusion (5,6), quasi-critical fluctuations of local order near phase transitions (7), and slow relaxation of local structures (8,9). Numerous refinements of the original model calcula-

561

tions have appeared, especially for order-director fluctuations
(ODF): detailed expressions are available for ODF with a high
frequency cutoff included in the mode expansion (10), for ODF with
the director aligned at an arbitrary angle with respect to the
magnetic field (11-13), for the combined effects of fast reorientation and ODF (9,12,14), including controversial cross-terms, and
for the effect of anisotropic elastic coefficients and viscosities
(15).

Major difficulties in testing the refined models arise because
the proton magnetic moment is so large that distinguishing intra-
from inter-molecular relaxation mechanisms is difficult. Field
cycling techniques (16), measurements of $T_{1\rho}$ (10,17,18), or the decay of dipolar order (18) permit information to be gathered about
the frequency dependence and mechanistic origin of proton relaxation rates, but they suffer from the fact that chemically distinct
protons generally show extensive spectral overlap. This has led
to several investigations of proton relaxation in selectively
deuterated liquid crystal molecules (17,19,20), for which considerable spectral simplication is achieved, as well as to studies of
the relaxation behavior of ^{13}C (21,22) and ^{2}H (17,19,23-26).

Deuterium is a particularly convenient spin probe for studies
of liquid crystal dynamics. It is not excessively expensive and
the sensitivity is adequate. The relaxation is dominated by a
single mechanism of intramolecular origin, and with current NMR
spectrometers measurements can be carried out in the 4 - 50 MHz
range - a region which is especially rich in dynamic effects. The
NMR spectra are often characterized by well-resolved quadrupolar
doublets (26,27) containing some dipolar fine structure such that
procedures (28,29) developed for analyzing relaxation of scalar
coupled systems may be used to analyze the more complex processes
which occur in ordered systems. These techniques basically aim at
determining individual spectral densities of motion, $J_q(\omega)$, rather
than overall correlation times, and have so far been applied

mostly to studies of solute relaxation and dynamics. Pioneering
work in this area has been done by Grant and coworkers who developed several schemes for analysis of relaxation behavior of both [13]C
(30-32) and protons (30,33), and by Jacobson et al. (34,35) who
measured ^{2}H spin lattice relaxation behavior of D_2O in lyotropic
liquid crystals.

The focus of our own work has been the development of pulsed
NMR techniques for measurements of deuterium relaxation rates in
nematic and smectic fluids, and so far we have concentrated our
efforts on studies of solute relaxation (36-39), but we emphasize
that many of the techniques developed in our laboratory are applicable to studies of the liquid crystal molecules themselves. Here
we summarize some recent results obtained for solutes in nematic
Merck Licristal Phase 5, and we present preliminary data for selectively deuterated derivatives of p-methoxybenzylidene-n-butylaniline (MBBA).

SPIN DYNAMICS

The NRM spectrum of a solute with one deuteron per molecule,
partially ordered in a nematic mesophase, consists of two well resolved lines. Using appropriate pulse sequences it is possible to
prepare this spin system in several different initial states, from
which one can monitor the relaxation of the spin system back to
thermal equilibrium. The rate of relaxation depends upon the nature of the initial excitation and Table 1 summarizes the information to be gained from several different kinds of pulse experiments.
The first four of these involve longitudinal (z) magnetization
only, and for a single deuteron the relaxation rates depend only
on two different spectral densities of motion, $J_1(\omega_o)$ and $J_2(2\omega_o)$.
Thus no matter what combination of techniques is employed, one
can determine at most two motionally significant parameters by
measuring longitudinal components of magnetization. Which combina-

Table 1. Pulse Sequences and Relaxation Rates for the Determination of Spectral Densities of Motion for a Single[a] Partially Ordered Spin I = 1.

Experiment	Pulse Sequence	Relaxation Rate	Reference
Non-selective Inversion-Recovery	$\pi - t_1 - \pi/2 - acq$	$J_1(\omega_0) + 4J_2(2\omega_0)$	34-36,38
Selective Inversion-Recovery[b]	$\pi_A - t_1 - \pi/2_A - acq$	$\frac{1}{2}J_1(\omega_0) + J_2(2\omega_0)$	34-36,38
Selective Inversion Recovery[b]	$\pi_A - t_1 - \pi/2_B - acq$	$4J_1(\omega_0) - J_2(2\omega_0)$	34-36,38
Quadrupolar Order	$\pi/2 - \tau - \pi/4 - t_1 - \pi/4 - acq$	$3J_1(\omega_0)$	40-42
1-Quantum Spin Echo	$\pi/2 - t_1/2 - \pi/2 - t_1/2 - acq$	$3J_0(0) + 3J_1(\omega_0) + 2J_2(2\omega_0)$	43,48
Selective 1Q Spin Echo	$\pi/2_A - t_1/2 - \pi_A - t_1/2 - acq$	$3J_0(0) + 3J_1(\omega_0) + 2J_2(2\omega_0)$	34-36,38
2-Quantum Spin Echo	$\pi/2 - \tau - \pi/2 - t_1/2 - \pi - t_1/2 - \pi/2 - acq$	$J_1(\omega_0) + 2J_2(2\omega_0)$	37,39,44-46

[a] If more than one spin is present the relaxation rates provide additional information about motional cross correlation terms.

[b] These decays are weakly biexponential and the table entries are initial decay rates.

tion of techniques is most suitable varies from system to system, and depends on such practical details as the available pulse power, degree of instrumental sophistication, and the relative magnitudes of relaxation rates and quadrupolar splittings.

The spectral density parameters appearing in Table 1 are Fourier transforms of auto correlation functions of second rank tensor components which appear in the relaxation Hamiltonian. In other words, that part of the relaxation Hamiltonian which connects spin states with $\Delta m = 1$ has a fluctuation spectrum described by $J_1(\omega_0)$ and the part connecting states with $\Delta m = 2$ is described by $J_2(2\omega_0)$. NMR relaxation measurements probe these fluctuation spectra at particular frequencies corresponding to spin energy level separations. It is important to note that this analysis involves no particular choice of motional model. It is therefore worthwhile to design relaxation experiments to measure, as accurately as possible, all the relevant spectral density parameters as a function of frequency and temperature: this provides the most stringent test possible by NMR of any proposed model of molecular motion.

For the case of a single deuteron, measurement of longitudinal relaxation behavior suffices to determine only two of the three relevant spectral densities. The spectral density at zero frequency, $J_o(0)$, contributes only to 1-quantum transverse relaxation rates and to measure it one must measure T_2 and then subtract the contributions to T_2 arising from $J_1(\omega_o)$ and $J_2(2\omega_o)$ (see Table 1). The determination of accurate T_2's of coupled spin systems is notoriously difficult (29,47,48), but recent developments in two-dimensional Fourier transform spectroscopy (49-52) are very promising in this regard.

Fig. 1 shows a two-dimensional spin echo spectrum of $CDCl_3$ in Phase 5 which was obtained using a $\pi/2$-$t_1/2$-α-$t_1/2$-t_2 pulse sequence (phase cycled to suppress many types (52) of pulse artifacts). Along the F_2 axis, which corresponds to the ordinary 1-quantum spectrum, only one of the two doublet components is shown. The sample temperature was 49.8°C, only 3°C below the N→I phase transition, and the experiment was performed shortly after the sample was placed in the field. The director was consequently not aligned uniformly throughout the sample, and the normal spectral lineshape (along F_2) is distorted. Along F_1 there are two peaks (43), the "Hahn" echo signal at $F_1 = 0$ and the "Carr-Purcell" echo signal at $F_1 = \nu_Q$, one half the quadrupolar splitting. The latter peak is responsible for echo modulation in one-dimensional versions of this experiment. Lineshape distortions due to director misalignment are even more apparent here than along F_2 because broadening due to external field inhomogeneity has been totally eliminated. The Hahn echo peak (which occurs with maximum intensity for a 90° refocussing pulse), is _not_ affected by quadrupolar splitting and is hence immune to misalignment distortion. Its width along F_1 provides a very accurate measure of T_2.

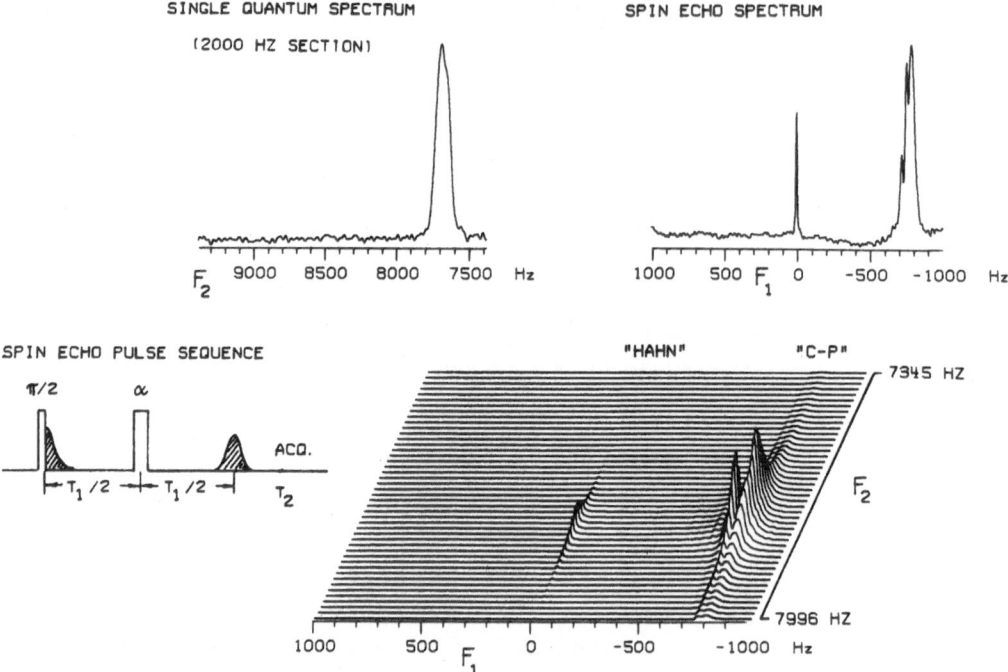

Fig. 1   Two-dimensional single quantum spin echo spectrum of
CDCL3 (10 mole %) in phase 5.

For systems with more than one deuteron additional terms appear
in the relaxation expressions, which account for pairwise cross
correlation in the motion of tensor components of electric field
gradients at each deuteron (53,54). These cross correlation terms
are entirely analogous to the dipolar cross correlation terms which
are useful in studies of relaxation and anisotropic molecular reorientation of scalar coupled spin systems (55,56), and the possibility of determining them in ordered fluids offers a promising new
source of motional information (37,57). Unfortunately, a combination of tailored spin-lattice relaxation experiments and ordinary
spin echo measurements of transverse relaxation rarely if ever
suffices to determine all the relevant spectral densities. It is

necessary in such cases to use more elaborate procedures, such as
the decay of multiple quantum coherence (44,45,49,58-60). The
transverse dephasing rate for the double quantum coherence of a
single deuteron as shown in Table 1 depends only on $J_1(\omega_o)$ and
$J_2(2\omega_o)$, but for systems containing more than one deuteron information about cross correlation terms may be obtained.

Fig. 2 shows spectra of acetonitrile-d_3 in Phase V, recently
obtained in our laboratory and described elsewhere in more detail
(54). At top left is the ordinary 1-quantum spectrum, showing
a (large) quadrupolar splitting as well as fine structure due to
deuteron-deuteron dipolar coupling. The other three traces are
slices of two-dimensional spectra, obtained using pulse sequences

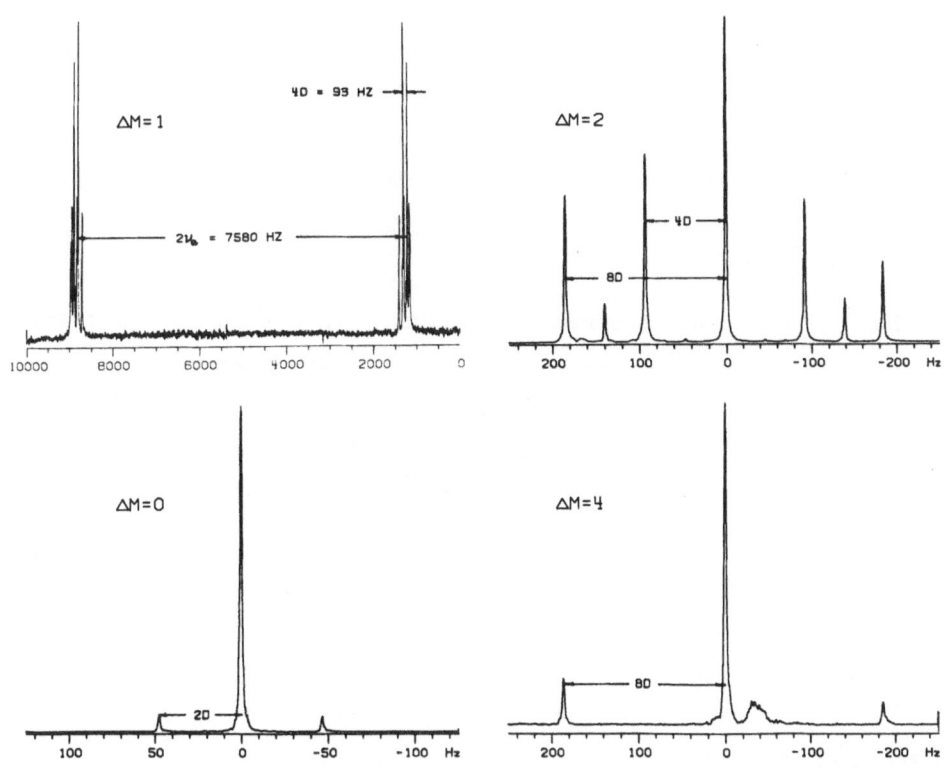

Fig. 2 Multiple quantum coherences for acetonitrile-d_3
        in phase 5.

(44,54) designed to selectively monitor the time evolution of 0-,
2- and 4-quantum coherences. In these experiments refocussing
pulses were used, so that the spectral linewidths correspond to true
transverse relaxation rates for the corresponding multiple quantum
coherence. These data, combined with spin lattice relaxation rates,
ordinarily suffice to overdetermine the set of relevant spectral
density parameters-- at least for systems of up to three deuterons,
which possibly represents the practical limit of the necessary
Redfield matrix element evaluations (39).

Multiple quantum relaxation experiments, although necessary to
determine all relevant spectral densities, tend to be both time consuming and expensive in terms of computer storage. It is fortunate-
ly possible to reliably determine both $J_1(\omega_0)$ and $J_2(2\omega_0)$ <u>without</u>
resorting either to such difficult procedures or even to a full
Redfield-type analysis of the relaxation equations. In particular,
the expressions for non-selective and selective spin-lattice relaxation rates given in Table 1 remain valid for multispin systems, including heteronuclear spins such as protons, as long as the system
may be effectively treated as a single group of chemically equivalent deuterons (42).

Multiple quantum coherence of order n is n-fold more sensitive
to static field inhomogeneity (49, 58) and even when refocussing
pulses are used, the measurement of long multiple quantum decay times
may suffer (57) from instabilities of the field frequency ratio in
much the same manner (61) as do 1-quantum decay rates. However,
it is possible to turn this enhanced sensitivity to field fluctuations to advantage, using multiple quantum decays in the presence
of purposely applied linear field gradients to measure anisotropic
translational diffusion (62) in ordered phases.

SOLUTE MOTION

An example of the type of information available from studies of deuterium relaxation of partially ordered solutes, we present relaxation data for deuterated p-diethynylbenzene (DEB) dissolved in nematic Phase V. A more complete account of this work will appear elsewhere (41). DEB, by virture of its elongated shape, is more highly ordered ($S_{zz} \sim 0.3$) than are smaller, more globular solutes. The acetylenic deuterons (see Fig. 3), whose principal field gradients lie along the long molecular (z) axis, therefore show a relatively large quadrupolar splitting (ca. 90 KHz) whereas the aromatic deuterons, which are oriented at ~60° to the z axis, have much smaller order parameters, $S \sim 0.006$.

Fig. 3 shows the results of measurements of R_1(nsl) and R_1(sl) for the acetylenic deuterons, as well as R_1(sum) and R_1(diff) for aromatic deuterons (obtained with Jeener-Broekaert pulse sequences (42,63)), as a function of temperature at three different Larmor frequencies. The general decrease of R_1(ns) for the acetylenic deuterons with increasing Larmor frequency is ascribable to contri-

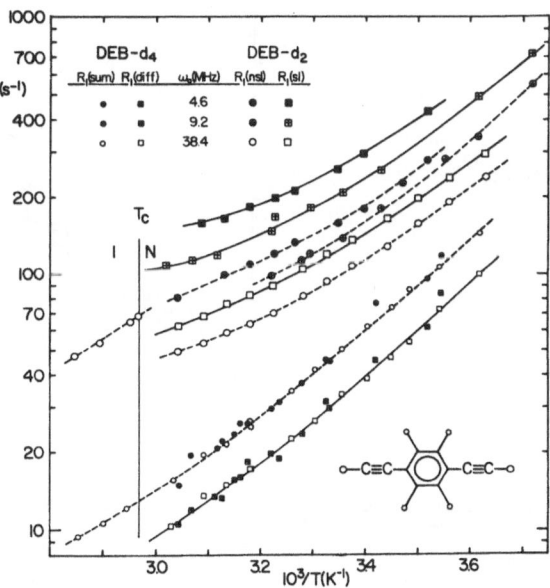

Fig. 3   Relaxation rates for deuterated derivatives of
         p-diethynylbenzene (DEB).

butions from order-director fluctuations, but the relaxation data
for the sum magnetizations alone are certainly not definitive in
this regard.

It is highly informative to use the data in Fig. 3 to derive
individual spectral density parameters -- fast reorientation contributes to both $J_1(\omega_o)$ and $J_2(2\omega_o)$ but ODF contributes only to
$J_1(\omega_o)$ (64). Fig. 4 shows such a decomposition, calculated from
the data in Fig. 3 according to the relations

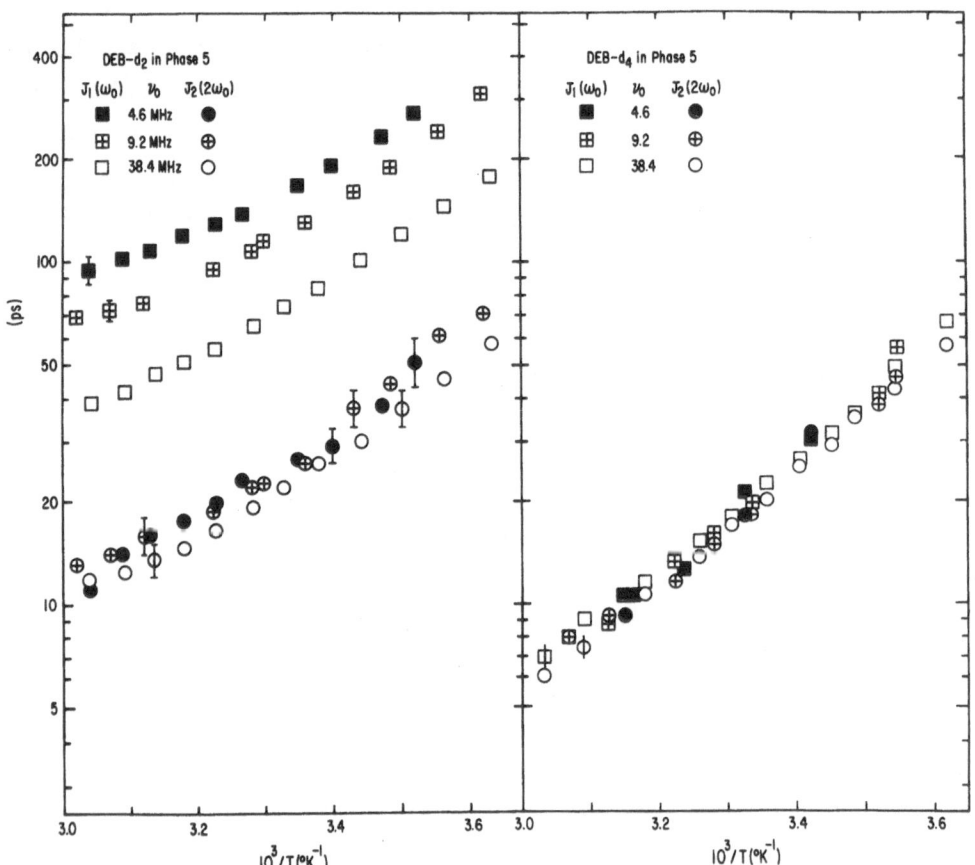

Fig. 4 Spectral density parameters for DEB derived from the
        data in Fig. 3. A full analysis of these data is given
        in Ref. (41).

$$J_1(\omega_o) = [4R_1(sl) - R_1(ns)]/9C \tag{1a}$$

$$J_2(2\omega_o) = [5R_1(ns) - R_1(sel)]/18C \tag{1b}$$

for acetylenic deuterons, and

$$J_1(\omega_o) = R_1(diff)/3C \tag{1c}$$

$$J_2(2\omega_o) = [3R_1(sum) - R_1(diff)]/12C \tag{1d}$$

for the aromatic deuterons, with

$$C = (3\pi^2/2)(e^2qQ/h)^2 \tag{1e}$$

Here we have used 198 KHz for the quadrupole coupling constant of the acetylenic deuterons and 183 KHz for aromatic deuterons.

It is immediately apparent from Fig. 4 that for acetylenic deuterons, virtually all of the frequency dependence of the spin lattice relaxation rates is carried by $J_1(\omega_o)$, and that $J_2(2\omega_o)$ is essentially frequency independent. Thus a reorientational correlation time τ_R can be extracted only from the $J_2(2\omega_o)$ values (after correcting for effects of static order (9)), and not directly from $J_1(\omega_o)$. The actual rotational contribution to $J_1(\omega_o)$ can be calculated using τ_R determined from $J_2(2\omega_o)$, and subtracted from the experimental values. The remainder we ascribe exclusively to ODF. It is roughly proportional to $\omega_o^{-1/2}$ and, as described more fully elsewhere (41), refinements such as a finite cutoff frequency (10) and a negative cross-term (9) between ODF and fast reorientation are needed to fit the data.

The spectral density parameters for aromatic deuterons of DEB behave quite differently from those of acetylenic deuterons. Since the order parameter for these deuterons is so small, ODF contributions are almost negligible and the corrections to both $J_1(\omega_o)$ and $J_2(2\omega_o)$ for static ordering are also unimportant.

The reorientational correlation time for aromatic deuterons ($\tau_R \approx 5J_2(2\omega_o)$) is much shorter than for acetylenic deuterons, because rotation about the long molecular axis is fast. If p-diethynyl-benzene is treated as a symmetric rotor, we find that the ratio $\tau_\perp/\tau_{||}$ is ca. 10 in both nematic and isotropic Phase 5 solutions (41).

SOLVENT DYNAMICS

The NMR spectra of deuterated liquid crystals are eminently well suited for study by the relaxation techniques developed for solute spins. Fig. 5 shows two recent examples from our laboratory:

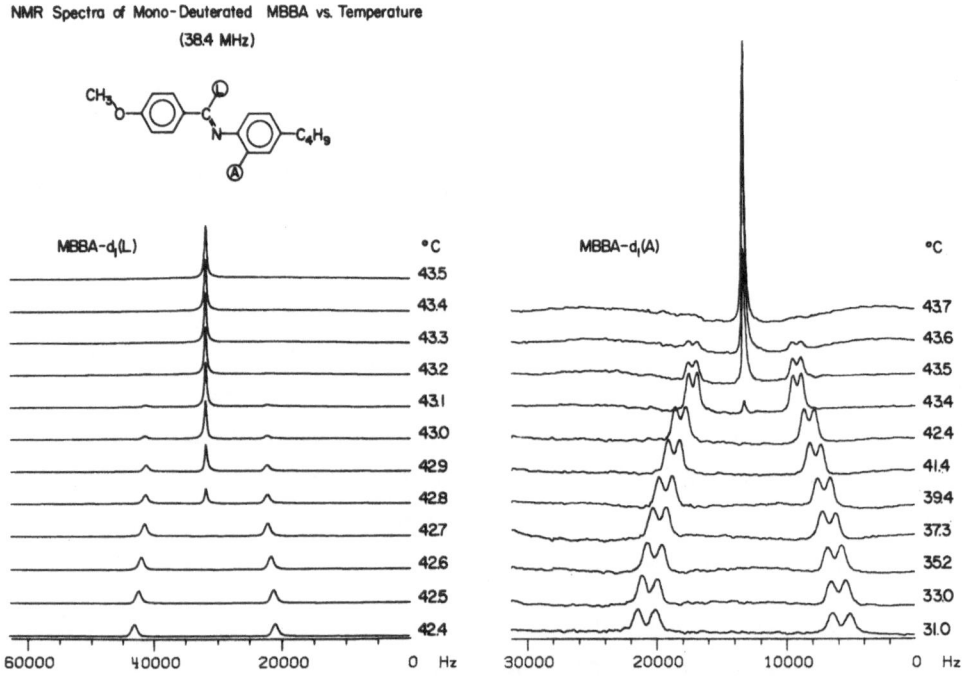

NMR Spectra of Mono-Deuterated MBBA vs. Temperature
(38.4 MHz)

Fig. 5   Spectra of mono-deuterated p-methoxybenzylidene-n-butylaniline (MBBA) near the nematic-isotropic phase boundary. Due to minor impurities there is a narrow biphasic region where nematic and isotropic MBBA coexist.

on the left spectra of p-methoxybenzylidene-n-butylaniline (MBBA)
deuterated at the linkage (L) position are shown as a function of
temperature, and on the right a similar set of spectra taken over
a wider temperature range of MBBA mono-deuterated in the aniline
(A) ring. In neither case was proton decoupling used, and the
aniline deuterons appear as a doublet of doublets due to dipolar
coupling to the ortho proton. In both cases, the nematic phase
spectra are further inhomogeneously broadened by many unresolved
proton couplings, so that considerable line narrowing takes place
in the isotropic phase.

Studies of relaxation near phase transitions require very
precise control of sample temperature, and this is especially
critical for liquid crystal systems because the temperature
dependence of the order parameter is quite strong near the
transition. The experiments were performed using a microprocessor-
based temperature controller capable of maintaining constant
temperature with $\pm 0.003°$, and for both liquid crystal samples in
Fig. 5 a narrow biphasic temperature region was observed, presumably because the samples contain minor (< ca. 2%) amounts of
impurities. It is interesting to note that the order parameter
for nematic MBBA remains essentially constant throughout this
biphasic region.

Fig. 6 shows preliminary spectral density data for linkage-
deuterated MBBA as a function of temperature. These data were
obtained using phase cycled Jeener-Broekeart sequences (42) at
ω_o = 38.4 MHz, a relatively high frequency for major contributions
from ODF. Possibly, most of the difference between $J_1(\omega_o)$ and
$J_2(2\omega_o)$ can be accounted for by static order parameter corrections,
but additional data at other Larmor frequencies are needed to
settle this question. Neither $J_1(\omega_o)$ nor $J_2(2\omega_o)$ show anomalous
behavior just below the nematic-isotropic phase transition, although the quality of these preliminary data must be improved

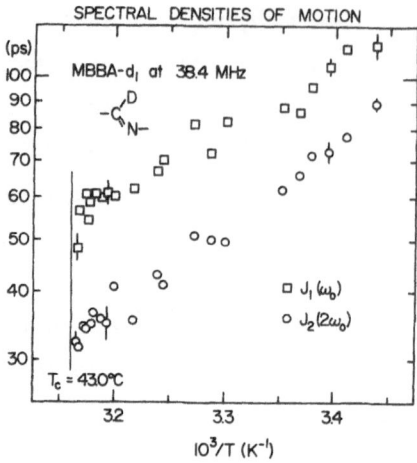

Fig. 6   Preliminary spin lattice relaxation data for linkage-
         deuterated MBBA at 38.4 MHz, obtained using the
         Jeener-Broekaert technique (63,42).

before ruling out pretransitional phenomena. Above the phase
transition, where the ^{2}H spectra show only a single line, we
cannot determine both spectral densities by methods described
here. However, combined measurements of both ^{2}H and ^{13}C relaxation rates (65) can yield this information, and preliminary data
(66) indicate that some as yet unidentified slow process influences
the deuteron relaxation for 20 - 30 degrees above the phase transition.

CONCLUSIONS

     Deuterium relaxation in liquid crystalline media can be very
informative about motional processes both of small, rigid solutes
and about the liquid crystal molecules themselves. Several new
types of pulsed NMR techniques have been developed for determining
individual spectral densities of motion rather than just overall
relaxation times and the information gained from these types of
experiments can be used to explore the molecular dynamics of

ordered phases. The limiting factor in applying these methods
is spectral resolution, but it is probable that they can be used to
advantage even in the case of perdeuterated liquid crystals, as
long as individual quadrupolar doublets are well separated.

ACKNOWLEDGMENTS

We wish to thank our collaborators over the last few years,
Geoffrey Bodenhausen, Harry W. Dickerson, David Jaffe, Raphael
Poupko, Lyndsie S. Selwyn, Steven W. Sparks and Nikolaus M.
Szeverenyi for their contributions to this work and the National
Science Foundation for generous support.

REFERENCES

1. C. G. Wade, Ann. Rev. Phys. Chem. 28, 47 (1977).

2. J. W. Doane, in Magnetic Resonance of Phase Transitions, (Eds.
        F. J. Owens, C. P. Poole, Jr. and H. A. Farach), Academic
        Press, New York, 1979, Ch. 4.

3. P. Pincus, Solid State Commun. 7, 415 (1969).

4. J. W. Doane and J. J. Visintainer, Phys. Rev. Lett. 23, 1421,
        (1969).

5. M. Vilfan, R. Blinc and V. Rutar, Solid State Commun. 11, 171
        (1975).

6. S. Zumar and M. Vilfan, Phys. Rev. A 17, 424 (1978).

7. B. Cabane, Adv. Molec. Relax. Proc. 3, 341 (1972).

8. C. F. Polnaszek and J. H. Freed, J. Phys. Chem. 79, 2283
        (1975).

9. J. H. Freed, J. Chem. Phys. 66, 4183 (1977).

10. J. W. Doane, C. E. Tarr and M. A. Nickerson, Phys. Rev. Lett.
        33, 620 (1974).

11. R. Blinc, NMR Basic Principles and Progress, 13, 97 (1976).

12. P. Ukleja, J. Pirs and J. W. Doane, Phys. Rev. A14, 414 (1976).

13. C. E. Tarr, F. Vosman and L. R. Whalley, J. Chem. Phys. $\underline{67}$,
        868 (1977).

14. P. Nordio and U. Segre, Gazz. Chim. Ital. $\underline{106}$, 431 (1976).

15. R. Blinc, M. Luzar, M. Vilfan and M. Burgar, J. Chem. Phys.
        $\underline{63}$, 3445 (1975).

16. V. Graf, F. Noack and M. Stohrer, Z. Naturforsch, $\underline{32a}$, 61
        (1977).

17. J. J. Visintainer, J. W. Doane, and D. L. Fishel, in $\underline{Liquid}$
        $\underline{Crystals}$ $\underline{3}$ (Eds. G. H. Brown and M. M. Labes), Gordon
        and Breach, New York, 1972, p. 329.

18. R. G. C. McElroy, R. T. Thompson, and M. M. Pintar, Phys. Rev.
        A $\underline{10}$, 403 (1974).

19. J. J. Visintainer, R. Y. Dong, E. Bock, E. Tomchuk, D. B.
        Dewey, A.-L. Kuo and C. G. Wade, J. Chem. Phys. $\underline{66}$,
        3343 (1977).

20. B. M. Fung, C. G. Wade, and R. W. Orwoll, J. Chem. Phys. $\underline{64}$,
        148 (1976).

21. M. Schwartz, P. E. Fagerness, C. H. Wang and D. M. Grant, J.
        Chem. Phys. $\underline{60}$, 5066 (1974).

22. H. Hutton, E. Bock, E. Tomchuk, and R. Y. Dong, J. Chem. Phys.,
        $\underline{68}$, 940 (1978).

23. V. Rutar, M. Vilfan, R. Blinc, and E. Bock, Mol. Phys. $\underline{35}$,
        721 (1978).

24. R. D. Orwoll, C. G. Wade and B. M. Fung, J. Chem. Phys. $\underline{63}$,
        986 (1975).

25. J. W. Emsley, J. C. Lindon and G. R. Luckhurst, Mol. Phys.
        $\underline{32}$, J187 (1976).

26. R. Y. Dong, J. Lewis, E. Tomchuk and E. Bock, J. Chem. Phys.
        $\underline{69}$, 5314 (1978).

27. S. Hsi, H. Zimmermann and Z. Luz, J. Chem. Phys. $\underline{69}$, 4126
        (1979).

28. L. G. Werbelow and D. M. Grant, Adv. Mag. Res. $\underline{9}$, 190 (1977).

29. R. L. Vold and R. R. Vold, Progr. NMR Spectr. $\underline{12}$, 79 (1978).

30. J. M. Courtieu, C. L. Mayne and D. M. Grant, J. Chem. Phys. 66, 2669 (1977).

31. L. G. Werbelow, D. M. Grant, E. P. Black and J. M. Courtieu, J. Chem. Phys. 69, 2407 (1978).

32. E. P. Black, J. M. Bernassau, C. L. Mayne and D. M. Grant, J. Chem. Phys. 76, 265 (1982).

33. J. Courtieu, N. T. Lai, C. L. Mayne, J. M. Bernassau and D. M. Grant, J. Chem. Phys. 76, 257 (1982).

34. J. P. Jacobsen, H. K. Bildsoe and K. Schaumburg, J. Magn. Reson. 23, 153 (1976).

35. J. P. Jacobsen and K. Schaumburg, J. Magn. Reson. 24, 173 (1976).

36. R. R. Vold and R. L. Vold, J. Chem. Phys. 66, 4018 (1977).

37. R. L. Vold, R. R. Vold, R. Poupko and G. Bodenhausen, J. Magn. Reson. 38, 141 (1980).

38. R. R. Vold and R. L. Vold and N. M. Szeverenyi, J. Phys. Chem., 85, 1934 (1981).

39. D. Jaffe, R. L. Vold and R. R. Vold, J. Magn. Reson. 46, 496 (1982).

40. H. W. Spiess, J. Chem. Phys. 72, 6755 (1980).

41. W. H. Dickerson, R. R. Vold and R. L. Vold, J. Phys. Chem. (in press).

42. R. L. Vold, W. H. Dickerson and R. R. Vold, J. Magn. Reson., 43, 213 (1981).

43. R. L. Vold and R. R. Vold, J. Magn. Reson. 42, 173 (1981).

44. G. Bodenhausen, R. L. Vold and R. R. Vold, J. Magn. Reson. 37, 93 (1980).

45. G. Bodenhausen, N. M. Szeverenyi, R. L. Vold and R. R. Vold, J. Am. Chem. Soc. 100, 6265 (1978).

46. R. Poupko, R. L. Vold and R. R. Vold, J. Phys. Chem. 84, 3444 (1980).

47. R. L. Vold and S. O. Chan, J. Chem. Phys. 53, 449 (1970).

49. W. P. Aue, E. Bartholdi and R. R. Ernst, J. Chem. Phys. <u>64</u>, 2229 (1976).

50. W. P. Aue, J. Karhan and R. R. Ernst, J. Chem. Phys. <u>64</u>, 4226 (1976).

51. G. Bodenhausen, R. Freeman, R. Niedermeyer and D. L. Turner, J. Magn. Reson. <u>26</u>, 133 (1977).

52. G. Bodenhausen, R. Freeman and D. L. Turner, J. Magn. Reson. <u>27</u>, 511 (1977).

53. R. Poupko, R. L. Vold and R. R. Vold, J. Magn. Reson. <u>34</u>, 67 (1979).

54. D. Jaffe, R. R. Vold and R. L. Vold, J. Magn. Reson. <u>46</u>, 475 (1982).

55. L. G. Werbelow and D. M. Grant, J. Chem. Phys. <u>63</u>, 544 (1975).

56. R. L. Vold and R. R. Vold and D. Canet, J. Chem. Phys. <u>66</u>, 1202 (1977).

57. D. Jaffe, R. L. Vold and R. R. Vold, J. Chem. Phys.

58. S. Vega and A. Pines, J. Chem. Phys. <u>66</u>, 5624 (1977).

59. A. Wokaun and R. R. Ernst, Mol. Phys. <u>36</u>, 317 (1978).

60. J. Tang and A. Pines, J. Chem. Phys. <u>72</u>, 3290 (1980).

61. A. Allerhand, Rev. Sci. Instr. <u>41</u>, 269 (1970).

62. J. F. Martin, L. S. Selwyn, R. R. Vold and R. L. Vold, J. Chem. Phys. <u>76</u>, 2632 (1982).

63. J. Jeener and P. Broekaert, Phys. Rev. <u>157</u>, 232 (1967).

64. J. W. Doane and D. L. Johnson, Chem. Phys. Lett. <u>6</u>, 291 (1970).

65. J. F. Martin, R. L. Vold and R. R. Vold, J. Magn. Reson. (in press).

66. S. W. Sparks, R. L. Vold and R. R. Vold, to be published.

AN EFFECTIVE PROTON HYPERFINE TENSOR FOR DI-TERTBUTYLNITROXIDE

Barney L. Bales and Ronney A. Dolin

Department of Physics and Astronomy
California State University, Northridge
Northridge, CA 91330

Robert N. Schwartz

Hughes Research Laboratories
3011 Malibu Canyon Blvd.
Malibu, CA 90265

ABSTRACT

Nitroxide spin labels are often used to probe the dynamics of liquid crystals and membranes. In analyzing the results, the EPR linewidth variation as a function of the label orientation enters as a parameter. This parameter comes on top of a number of other parameters, so it would be of interest to determine it from the results of a separate experiment. We report such a procedure in which the linewidth of the label is studied as a function of its alignment in a nematic liquid crystal. The alignment of the label is determined in the normal way by measuring line spacings in the nitrogen hyperfine pattern.

The linewidth is modeled as a convolution of a Lorentzian due to spin relaxation and a Gaussian due to unresolved hyperfine coupling. The Gaussian may be described by an effective hyperfine coupling a^e which may be measured directly from the EPR spectrum using a new method. Thus a^e vs. orientation is found. We then assume that a^e may be derived from an effective hyperfine tensor and use the measurements described above to find its principal values.

The procedure is demonstrated with DTBN dissolved in MBBA where remarkably consistent results are obtained. With tensor elements in hand, both the form and magnitude of the linewidth corrections may be made. In addition, these elements may be of use in studying the structure of the label.

I.   INTRODUCTION

Nitroxide spin labels are often used to probe the dynamics of liquid crystals and biological membranes.[1] Frequently, due to the highly fluid nature of these media, the label tumbles rapidly and anisotropically leading to sharp EPR lines whose positions may be analyzed to give the average orientation of the label.[1] In this so-called motional narrowing regime, theory[2] predicts Lorentzian shaped EPR lines whose linewidths, due to various spin relaxation mechanisms, lead to some understanding of the dynamics. On top of the Lorentzian linewidth, there is, almost invariably, an interesting, or perhaps annoying, Gaussian linewidth due to unresolved hyperfine coupling to protons in the label.[3-12] In most cases, in isotropic liquids, the Gaussian component can not be ignored in the determination of rotational correlation times and Heisenberg exchange frequencies: errors of 50% or more can arise.[3] In ordered fluids, the situation is potentially even worse because the anisotropic nature of the hyperfine coupling to the protons leads, in general, to a Gaussian broadening that depends on the orientation of the label with respect to the external magnetic field. This, of course, interferes with experimental testing of theories of the interesting Lorentzian linewidth variation with orientation that contains the dynamical information.[13-16]

The purpose of the present paper is to investigate the details of the orientational dependence of the Gaussian component of the linewidth for nitroxides using the spin label DTBN as an example. The basic strategy is to orient partially the label in a nematic liquid crystal, in this case MBBA, determine the orientation using the positions of the EPR lines in the nitrogen multiplet[2] measure by means of a new technique the Gaussian component of the linewidth,[17] and summarize the results in a usable way so that rigorous line-shape simulations may be carried out. Surprisingly, perhaps, we have found that the data are well reproduced by introducing the concept of an effective proton hyperfine coupling tensor. This is handy: the usual expressions for the orientational dependence of tensors predicts the Gaussian component as a function of orientation. Previously developed equations may then be applied to give the Lorentzian component.[3]

II. ORDER PARAMETER

The theory for the analysis of nitroxide line positions and linewidths is well developed so we merely summarize the necessary results here and refer the reader to the literature.[2] A molecular-fixed coordinate system with the x-axis along the NO bond and the z-axis along the nitrogen 2pπ orbital serves to define the orientation of a rigid nitroxide label like DTBN. See Figure 1. The nitrogen hyperfine coupling tensor and the g-tensor are both assumed to be diagonal in the above molecular frame, an assumption that seems to be reasonably justified from single crystal studies.[19] In the liquid crystal, in the motional narrowing regime, time averaged components of the nitrogen hyperfine coupling tensor and the g-tensor may be observed when the director is parallel with or perpendicular to the external field giving $a_{/\!/}^N$ and $a_{\perp}^N$ respectively as the line spacings and $g_{/\!/}$ and $g_{\perp}$ respectively as the g value. Following a development similar to reference 2,

$$a_{/\!/}^N = a^N + \frac{2}{3} S_{11}(a_{xx}^N - a_{yy}^N) + \frac{2}{3} S_{33}(a_{zz}^N - a_{yy}^N) , \qquad (1)$$

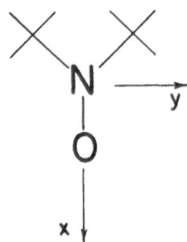

Fig. 1.   The nitroxide free radical DTBN with its molecular-
           fixed coordinate system. The eighteen protons in the
           tert-butyl groups interact with the unpaired electron
           to produce an unresolved hyperfine multiplet.

where a_{xx}^N, and a_{yy}^N and a_{zz}^N are the molecular-fixed diagonal elements of the nitrogen hyperfine coupling tensor and S_{11} (S_{33}) are
order parameters describing the time-averaged orientation of the
x-axis (z-axis) with respect to the director, which, in this case,
is also the direction of the external magnetic field.[2] The quantity a^N is the trace of the nitrogen hyperfine coupline tensor.
An expression similar to (1) holds for $g_{//}$ and expressions for $a_{\perp}^N$
and $g_{\perp}$ may be found in reference 2. Since the trace of a tensor
is invariant under rotation

$$a^N = \frac{1}{3} (a_{xx}^N + a_{yy}^N + a_{zz}^N) = \frac{1}{3} (a_{//}^N + 2a_{\perp}^N). \qquad (2)$$

Normally, one uses principal values of the magnetic tensors measured
with the nitroxide immobilized in low concentration in the solid
state, but there is a slight complication due to the well-known
fact that a^N and g are observed to depend on the dielectric constant of the medium in which the nitroxide resides,[20] thus
$a^N = K_N a_{solid}^N$ and $g = K_g g_{solid}$ where the constants K_N and K_g
depend on the medium but not on temperature. The effect is generally accepted to be due to a shift in unpaired-electron spin density
in the molecular framework[21] so it is plausible to assume that

each element of the magnetic tensor scales by the same factor[22], so an equation similar to (1) results

$$(3)$$

$$a^N_{//} = K_N \left\{ a^N_{solid} + \frac{2}{3} S_{11} (a^N_{xx} - a^N_{yy})_{solid} + \frac{2}{3} S_{33} (a^N_{zz} - a^N_{yy})_{solid} \right\}$$

Expressions similar to (2) and (3) hold for the g-tensor.

Inverting expressions similar to (3) and incorporating the corrections that account for the difference in the dielectric constant of the single crystal and the liquid crystal, gives

$$S_{11} = \frac{K_g^{-1}(g_{//} - g_{\perp})(a^N_{zz} - a^N_{yy})_{solid} - K_N^{-1}(a^N_{//} - a^N_{\perp})(g_{zz} - g_{yy})_{solid}}{\{(g_{xx} - g_{yy})(a^N_{zz} - a^N_{yy}) - (a^N_{xx} - a^N_{yy})(g_{zz} - g_{yy})\}_{solid}} \quad (4a)$$

$$S_{33} = \frac{-K_g^{-1}(g_{//} - g_{\perp})(a^N_{xx} - a^N_{yy})_{solid} + K_N^{-1}(a^N_{//} - a^N_{\perp})(g_{xx} - g_{yy})_{solid}}{\{(g_{xx} - g_{yy})(a^N_{zz} - a^N_{yy}) - (a^N_{xx} - a^N_{yy})(g_{zz} - g_{yy})\}_{solid}} \quad (4b)$$

$$S_{22} = -(S_{11} + S_{33}). \tag{4c}$$

DTBN, in first approximation, is a prolate spheroid with a length to width ratio of ~ 1.4, the long axis being very nearly the molecular y-axis. One expects such a molecule to be only slightly aligned in a nematic liquid crystal, with S_{22} positive.

III. THE EFFECTIVE HYPERFINE COUPLING CONSTANT

In the past, we have found it useful to define an equivalent hyperfine coupling constant

$$a^e = \left\{ \frac{4}{3} \alpha \sum_{j=1} I_j(I_j + 1) a_j^2 \right\}^{\frac{1}{2}}, \tag{5}$$

where a_j is the hyperfine coupling constant and I_j is the nuclear spin of the jth magnetic nucleus contributing to the unresolved hyperfine pattern. The constant α may be taken to be equal to 1.06. The reader is referred to reference 3 for details, but briefly, a^e is the value of the hyperfine coupling constant that n equivalent spin 1/2 particles would have in order to produce the same second moment as the actual hyperfine pattern. The second moment is important because it is directly related to the Gaussian linewidth.[3] The equivalent hyperfine coupling concept is very useful because computer simulation has shown that widely differing hyperfine patterns in detail lead to similar Gaussian broadening when a^e is similar.[3] This is important here because the simple nineteen line spectrum of DTBN in an isotropic liquid is expected to be very complicated in the nematic phase.

Using a recently developed technique in this laboratory,[17] we are able to directly measure a^e from the shape of the EPR line.

IV. THE EFFECTIVE HYPERFINE TENSOR

We postulate that, to an accuracy sufficient to predict the Gaussian linewidth orientational variation in ordered liquids, the effective hyperfine coupling is derived from a tensor $\underline{a}^e$ with principal axis system x,y,z. It would follow that an equation similar to (3) would hold.

$$a_{\parallel}^e = K_e \left\{ a^e + \frac{2}{3} S_{11}(a_{xx}^e - a_{yy}^e) + \frac{2}{3} S_{33}(a_{zz}^e - a_{yy}^e) \right\}, \qquad (6)$$

where the constant K_e would take account of dielectric constant differences from medium to medium, but in addition takes account of an interesting effect due to spin exchange which we discuss below.

We test (6) in the following way:

1) We measure $a_{//}^N$ and $g_{//}$ as a function of temperature as well as a^N and g in the isotropic range. Equations (4) then give S_{11} and S_{33} as a function of temperature.

2) We measure a^e in the isotropic temperature range of the liquid crystal and set that equal to $K_e a^e$ in (6).

3) Measurement of $a_{//}^e$ at two temperature, using (7) below, fixes the elements a_{xx}^e, a_{yy}^e, and a_{zz}^e; the overall fit tests (6).

4) As a test of consistency, $a_{\perp}^e$ is measured using a novel flow-technique; thus the invariance of the trace of $\underline{a}^e$ may be tested.

$$a^e = \frac{1}{3} (a_{xx}^e + a_{yy}^e + a_{zz}^e) = \frac{1}{3} (a_{//}^e + 2a_{\perp}^e) . \qquad (7)$$

Amazingly, the idea of an effective hyperfine tensor works very well, at least in the case of DTBN in MBBA.

V.   EXPERIMENTAL

The MBBA was purchased from 3M, the DTBN from Aldrich and both were used as received. Solutions of DTBN in MBBA were prepared volumetrically which lowered the transition temperature approximately 2-3° below 44°C, which was the transition temperature of the neat MBBA as received. In this work, temperature was merely a parameter so no effort was made to increase the transition temperature. For most experiments, the solutions were sealed in glass tubes after being subjected to several freeze-pump-thaw cycles to remove dissolved oxygen. The temperature was controlled to $\pm$ 1°C with a Varian nitrogen gas flow unit, the temperature being measured with thermocouple. An NMR gauss meter was rigged with a phase sensitive detector in order to record the NMR trace on the same spectral recording as the EPR using an X,Y,Y' recorder. An NMR trace before and after each EPR line allowed many measurements of the field scan rate as well as an accurate determination of the position of

each EPR line. The NMR frequency was measured with an HP 5300B
frequency counter. Careful insulation of the Hall probe, controlling the magnetic field, and recording each EPR line at least
three times led to the high precision quoted in Section VI. The
microwave frequency was measured with an HP 5342A counter. Progressive saturation curves were run and the data were taken well
below saturation. The modulation amplitude was maintained below
10% of the narrowest line in the study.

A somewhat novel flow-method was developed to measure the EPR
of DTBN in MBBA such that the director was perpendicular to the
external magnetic field.[18] To carry out these measurements, a 50
$\mu\ell$ glass pipet was connected with teflon tubing to form a closed
loop. The pipet was fixed in the microwave cavity with its axis
perpendicular to the external magnetic field. The tubing passed
through a peristaltic pump, which was variable in speed, which produced the fluid flow. The shear force would tend to align the
liquid crystal perpendicular to the field. As set up, the velocity
of the fluid could be varied from .012 cm/sec to 12 cm/sec and the
intention was to increase the flow rate until no further change in
the line positions occurred. As it happened, the lowest attainable flow rate produced the full shift in line position. Nitrogen
gas was bubbled through the solution for 20 minutes to degass the
sample and a slight positive pressure of the nitrogen gas was maintained to discourage the entrance of oxygen. Due to the geometry
of the set-up, g-values were not measured in the flow experiments.

The measurements of the Gaussian character of the EPR lines
described below require measurements of four points on the EPR
line as detailed in reference 17. These four points were measured
from the recorded spectra using the digitizing feature of a Servigor
281 interactive digital recorder interfaced with a minicomputer
which also carried out the calculations described in reference 17.

VI.   RESULTS

Fig. 2 shows the linewidth of DTBN vs temperature for various concentrations of the nitroxide in MBBA. As detailed below, the label becomes slightly aligned at temperatures below the isotropic-nematic transition temperature T_c which is 42°C for the samples in Fig. 2. The dramatic dropoff in linewidth as the temperature is lowered through T_c, as we shall see, may be almost entirely attributed to a decrease in a^e. One of the samples giving data in Fig. 2 was supercooled to 7°C, well below the normal freezing point of 18°C. At these low temperatures. It is noted that the linewidth experiences an increase with further temperature decrease, while the label continues to rotate rapidly and the ordering increases in a continuous fashion. Noteworthy is the fact that all of the samples

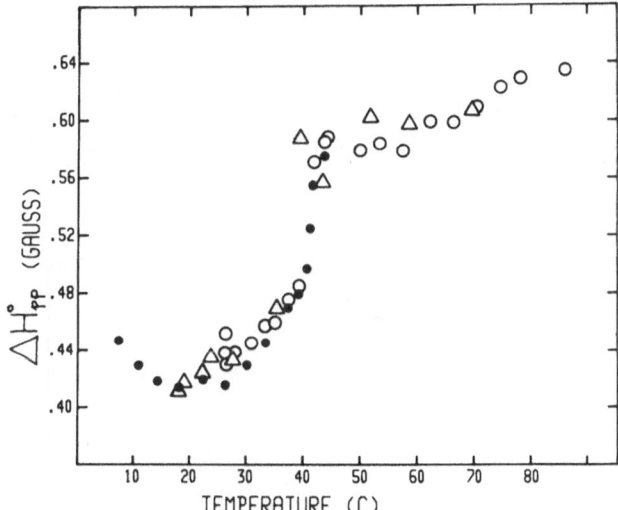

Fig. 2   The peak-to-peak linewidth of the $M_I = 0$ line of the EPR spectrum of DTBN for several low to intermediate concentrations in MBBA. Δ, 3.7 x 10^{-5} M; $\bullet$, 9.3 x 10^{-4} M; $\circ$, 4.8 x 10^{-3} M. The overall linewidths derived from the various samples are comparable and one sample, supercooled to 7°C shows an increase in linewidth with further decrease in temperature.

in Fig. 2 produce comparable linewidths, which, as we shall see, can be understood as a competition between two opposite effects brought on by spin exchange between pairs of DTBN molecules colliding in the fluid.

In the isotropic phase of MBBA, the nitrogen hyperfine coupling constant was found to be a^N = 15.500 $\pm$ .009 G and g = 2.00750 $\pm$ .0005 where the errors quoted are the standard deviations of six measurements in the same experimental setup. Thus, systematic errors, which are likely to be larger than the errors quoted above, especially for g, are not included, but these will tend to be unimportant in (4) since relative values are used in the computations. Using Libertini and Griffith[19] values a^N_{xx} = 7.59 G, a^N_{yy} = 5.95 G, a^N_{zz} = 31.78 G, g_{xx} = 2.00872, g_{yy} = 2.00616, and g_{zz} = 2.00270, yields a^N_{solid} = 15.11 G and g_{solid} = 2.00586 so K_N = 1.03 and K_g = 1.00.

In the nematic phase, careful measurements of a^N and g utilizing (4) and the data of the above paragraph yielded the elements of the ordering matrix as a function of temperature. The element S_{22} vs T is shown in Fig. 3 and is seen to be small and

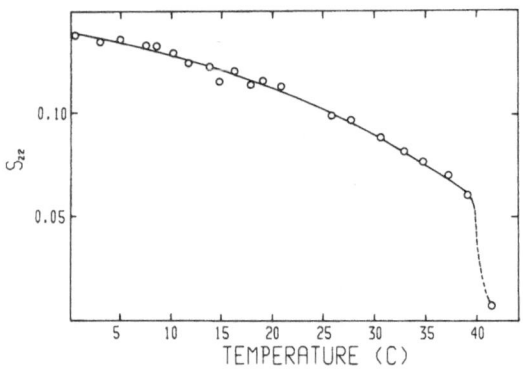

Fig. 3   The element of the ordering matrix describing the average orientation of the molecular y-axis, S_{22}, vs. temperature. The ordering is small and positive as expected from the shape of the DTBN molecule.

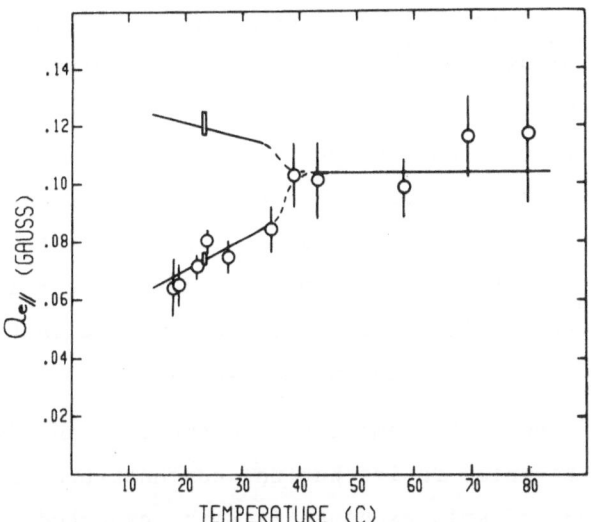

Fig. 4 The effective hyperfine coupling $a_{\parallel}^{e}$ vs. temperature for a
3.7 x 10^{-5} M solution of DTBN in MBBA. The points marked
with circles are measured 4 times from the same experimental
spectra and the error bars are the standard deviations.
The solid lines are plots of (6) with elements in Table 1.
The two points marked with rectangles at T = 23°C were derived from the flow alignment experiment as described in the
text. The heights of the rectangles are the estimated error
in the means derived from eleven measurements. The upper
rectangle is a measure of a_{1}^{e}.

positive, as anticipated. The scatter in the experimental points
is almost entirely due to imprecision in the measurement of g and
since this quantity is practically unimportant in the expression
for S_{33}, the scatter of data points in the plot of S_{33} vs T is a
factor of five less than that in Fig. 3. The solid line in Fig. 3
is a least squares fit of the data to a cubic with a coefficient
of correlation r = .996.

Figs. (4), (5), and (6) show the results of measurement of a^{e}
and $a_{\parallel}^{e}$ vs T for three, relative low concentrations of DTBN in MBBA.
These same three samples gave data in Fig. 2. The following points
are noteworthy:

1)   In Figs. (4) and (5) a^e is constant in the isotropic phase
within experimental error although there are two different complications in the two cases. The sample yielding Fig. 4 was so dilute
that the signal-to-noise was not sufficient to avoid relatively
large error bars. These error bars are the standard deviation in
four measurements on the same experimental spectrum at each temperature which shows that the measurement procedure itself is the precision limiting factor at low concentrations. In the isotropic phase
a^e = 0.105 G for the sample leading to Fig. 4.

On the other hand, the sample giving the data in Fig. 5 produced
excellent signal-to-noise ratios but the concentration is sufficient
that a small amount of spin exchange is to be expected. Spin exchange is known to lead to a collapse of hyperfine structure,[3] and,
in fact, in Fig. 5 we see that the expected decrease in a^e is realized: a^e = 0.085 G. Interestingly, the decrease in a^e offsets an
increase in the Lorentzian linewidth explaining the equality of overall linewidths for the various samples depicted in Fig. 2. Following
the reasoning further, a^e is not only expected to be smaller in
Fig. 5 than in Fig. 4 but also is expected to decrease at higher
temperatures. Perhaps a slight decrease is perceptible in Fig. 5
above 45°C but for the sake of judging the accuracy of (6), it may
be regarded as constant. We return to this interesting point in
Section VII. The error bars in Fig. 5 determined in the same way
as they were in Fig. 4 are of the order of the size of the symbols--
much smaller--due to the increased signal-to-noise ratio. Thus,
the random errors evident from the scatter of the points in Fig. 5
are due to experimental parameters and not the measurement process
that leads to a^e.

2)   In Fig. 6, we show the results derived from a sample supercooled to 7°C. Remarkable is the fact that in the temperature
range in which the total linewidth begins to increase with decreasing temperature, the effective coupling a^e continues to decrease.

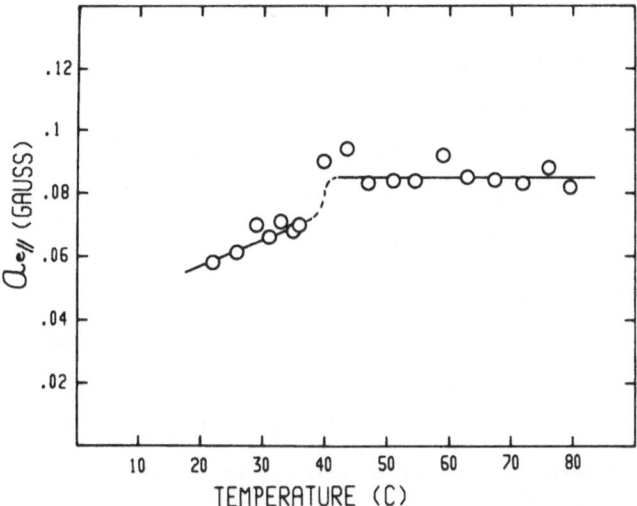

Fig. 5 The effective hyperfine coupling tensor $a^e_\parallel$ vs. temperature
for a 4.8 x 10^{-3} M solution of DTBN in MBBA. The solid line
is a plot of (6) using the elements of Table 1. The standard
deviation found by measuring the same spectrum several times
is of the order of the size of the symbols.

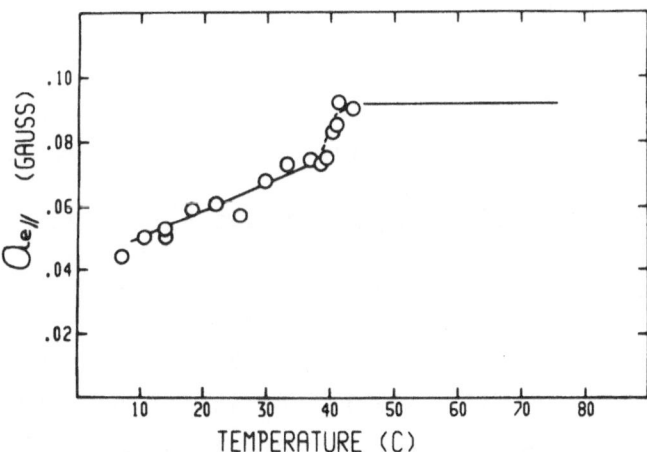

Fig. 6 The effective hyperfine coupling $a^e_\parallel$ vs. temperature for a
sample supercooled to 7°C. Note that $a^e_\parallel$ decreases monotonically with decreasing temperature while Fig. 2 shows that
the linewidth increases below 18°C. Concentration 9.3 x 10^{-4}
M. The solid line is a plot of (6) with elements given in
Table 1.

3) The solid lines in Fig. (4) - (6) are plots of (6) using the same fixed values for a_{xx}^e, a_{yy}^e, and a_{zz}^e given in Table 1; only K_e which we suppose to be temperature independent, is changed to account for partial proton hyperfine collapse due to spin exchange. The values of S_{11} and S_{33} used in (6) were taken from the smooth least squares fits such as the solid line in Fig. 3.

To present the results of the flow-alignment experiment, we return to Fig. 4. The lower and upper data points marked with rectangles at T = 23°C were measured with the liquid crystal at rest or flowing respectively. At the lowest attainable flow rate, the value of a^e was within experimental error of its value at all higher flow rates so eleven measurements at various flow rates were averaged to produce the datum represented by the higher rectangle in Fig. 4. The height of the rectangle is the estimated error in the mean. We should emphasize that the solid lines in Fig. 4 are a fit of (6) to the other data points in Fig. 4: the points from the flow experiment were left to fall where they might. The excellent fit is an encouraging test of the consistency of the idea of an effective tensor.

Table 1

Elements of the Effective Proton Hyperfine Tensor for DTBN[a]

Sample Concentration (M)	a_{xx}^e	a_{yy}^e	a_{zz}^e	$K_e a^e$	K_e
3.7×10^{-5}	−1.11	−0.13	+0.92	−.105	1.00
9.3×10^{-4}	−1.11	−0.13	+0.92	−.092	0.88
4.8×10^{-3}	−1.11	−0.13	+0.92	−.087	0.83

[a]Values in Gauss. Estimated error in tensor elements ± 10%; in a^e ± 5%. The sign of the tensor elements in based on the NMR assignment in reference 25.

VII. DISCUSSION

The problem that motivated the present work was that of putting the linewidth variation with orientation, due to unresolved hyperfine structure, on a firm quantitative footing. In the case of DTBN, the problem has been solved in a surprisingly handy, accurate way: the pertinent information is contained in the elements of the effective tensor given in Table 1. To compute the total linewidth ΔH^{o}_{pp}, one must convolute the Lorentzian linewidth ΔH^{L}_{pp} with the presently derived Gaussian component using the results of reference 3 as follows:

$$\Delta H^{o}_{pp} = \frac{1}{2} \Delta H^{L}_{pp} \left\{ 1 + [1 + 4n(a^{e}(\theta,\phi)/\Delta H^{L}_{pp})^{2}]^{\frac{1}{2}} \right\}, \qquad (8)$$

where $a^{e}(\theta,\phi)$ depends on the nitroxide orientation given by the polar angle θ and the azimuthal angle ϕ and would be computed in well-established ways[19] using the elements of Table 1.

Our methods to extract the Gaussian portion of a resonance line, of course, lead one to the more-important Lorentzian portion of the resonance line. Although it is not our primary concern here, we show the deconvolution of lines observed in the two orientations explored in the flow-orientation experiment in Table 2. This table serves to illustrate two points. First, the portion of the linewidth that is due to unresolved hyperfine structure is large--one must deal with it to accurately study dynamics. Second, at least in the present case, the orientational variation of the linewidth is dominated by the Gaussian portion: the Lorentzian width is constant within experimental error.

The elements in Table 1 are rather large and anisotropic. Actually, primitive calculations show that they are more or less what one would expect from electron-proton dipole couplings between an extended, unpaired electron in a 2pπ orbital and protons at distances typical of the DTBN molecule. The size of the couplings could explain why it has been necessary in the past to assume such

Table 2

Linewidth Components of DTBN in MBBA. T = 23°C.

	ΔH_{pp}^{O}	ΔH_{pp}^{G}	ΔH_{pp}^{L}	ΔH_{pp}^{O} (Calc)[a]
Director Parallel with Field	.485 ± 0.13	.324 ± .006	.27 ± .10	.486
Director Perpendicular to Field	.664 ± .034	.527 ± .057	.25 ± .06	.667

[a]Calculated from $\Delta H_{pp}^{O} = \frac{\Delta H_{pp}^{L}}{2} \{1 + \sqrt{1 + 4(\Delta H_{pp}^{G}/\Delta H_{pp}^{L})^2}\}$. Values in Gauss. Errors are standard deviations in eleven measurements.

large anisotropic linewidth components in deuterated nitroxide labels in order to fit solid state data.[23] A maximum variation of the linewidth predicted from the elements in Table 1 is ∼4.3 G and reducing this by the factor $I_H \gamma_H / I_D \gamma_D$ ∼ 3.7 gives about 1.2 G of anisotropy and could explain nicely the 1.4 G anisotropy that was needed to fit perdeuterated tempone data.[23] We caution that a different label is involved, but it would not be surprising if the results were similar.

The lack of a strong temperature dependence of a^e in Fig. 5 is very interesting since, for low viscosity liquids, one expects the collision frequence to vary as T/η, where η is the shear vis- cosity of the liquid.[4] The probability of spin exchange for DTBN upon collision has been found to be of the order unity in many liquids[4,6] so the spin exchange frequency ought to increase rapid- ly with temperature, leading to further collapse of the proton hyperfine structure at higher temperatures. This is not observed, but we note that recently, Berner and Kivelson[24] have observed a similar phenomenon using the concentration-dependent part of the linewidth of DTBN as the experimental observable. This interesting point seems worthy of further research.

REFERENCES

1. L. J. Berliner, Spin Labeling Theory and Applications, Academic Press, New York, 1976; Spin Labeling II, 1979.

2. See, for example, J. Seelig in Spin Labeling Theory and Applications, L. J. Berliner, Ed., Academic Press, New York, 1976, Chapter 10 and references therein.

3. B. L. Bales, J. Magn. Reson. 38, 193 (1980).

4. W. Plachy and D. Kivelson, J. Chem. Phys. 47, 3312 (1976).

5. G. Poggi and C. S. Johnson, Jr., J. Magn. Reson. 3, 436 (1970).

6. M. P. Eastman, R. G. Kooser, M. R. Das, and J. H. Freed, J. Chem. Phys. 51, 2690 (1969).

7. M. K. Ahn, J. Chem. Phys. 64, 134 (1976).

8. J. C. Lang, Jr., and J. H. Freed, J. Chem. Phys. 56, 4103 (1972).

9. C. C. Whisnant, S. Ferguson, and D. B. Chesnut, J. Phys. Chem. 78, 1410 (1974).

10. A. L. Kovarskii, A. M. Wasserman, and A. L. Buchachenko, J. Magn. Reson. 7, 225 (1972).

11. C. F. Polnaszek, Ph.D. thesis, Cornell University, 1975.

12. A. E. Stillman and R. N. Schwartz, J. Magn. Reson. 22, 269 (1976).

13. G. R. Luckhurst and A. Sanson, Mol. Phys. 24, 1297 (1972).

14. E. Meirovitch and J. H. Freed, J. Phys. Chem. 84, 2459 (1980) and references therein.

15. P. L. Nordio and P. Busolin, J. Chem. Phys. 55, 5485 (1971).

16. P. L. Nordio and U. Segre, Chem. Phys. 11, 57 (1975).

17. B. L. Bales, J. Magn. Reson. 48, 000 (1982).

18. A. T. Pudzianowski, A. F. Stillman, R. N. Schwartz, B. L. Bales, and E. S. Lesin, Mol. Cryst. Liq. Cryst. Letters 34, 33 (1976).

19. L. J. Libertini and O. H. Griffith, J. Chem. Phys. $\underline{53}$, 1359
     (1970).

20. B. R. Knauer and J. J. Napier, J. Am. Chem. Soc. $\underline{98}$, 4395

21. O. H. Griffith and P. C. Jost in Spin Labeling Theory and
     Applications, L. J. Berliner, Ed., Academic Press, New
     York, 1976. Chapter 12 and references therein.

22. W. L. Hubbell and H. M. McConnell, J. Am. Chem. Soc. $\underline{93}$, 314
     (1971).

23. J. S. Hwang, R. P. Mason, L. Hwang, J. H. Freed, J. Phys. Chem.
     $\underline{79}$, 489 (1975).

24. B. Berner and D. Kivelson, J. Phys. Chem. $\underline{83}$, 1406 (1979).

25. K. H. Hauser and J. C. Jochams, Mol. Phys. $\underline{10}$, 253 (1966).

ALKYL CHAIN FLEXIBILITY IN LIQUID CRYSTALS

Hirokazu Toriumi and Edward T. Samulski

Department of Chemistry and
Institute of Materials Science
University of Connecticut
Storrs, CT   06268

ABSTRACT

Several parameters for quantifying alkyl chain flexibility are
examined for free, unrestricted chains as a function of temperature.
The constraints imposed on alkyl chains in uniaxial liquid crystals
is modeled. The layer spacing in the smectic-A phases of the homologus series of terephthal-bis-(4n)-alkylanilines is computed in an
effort to estimate the degree of flexibility of the terminal alkyl
chains.

I.   INTRODUCTION

Alkyl chain flexibility impacts on the physics of polymers,
liquid crystals and membranes. It is particularly relevant to polymeric liquid crystals. In comb-like polymers, alkyl linkages control the degree of coupling of the dynamics of mesogenic pendant
groups to that of the main chain whereas, in linear polymers alkyl
spacers are frequently used to connect mesogenic monomers. Hence,
in the latter polymers the intrinsic properties of alkyl chains

determine the amount of orientational correlation that is transmitted to successive segments in the chain.

In Part II we contrast several ways of specifying the flexibility of "free" alkyl chains, ways that reflect their propensity for
propagating orientational information from one end of the chain to
the other. In Part III we model alkyl chains terminating the rigid
core of a mesogen in a uniaxial field imposed by neighboring molecules. We utilize a procedure developed to account for deuterium
nuclear magnetic resonance (DMR) data[1,2] and extend the model to
interpret observed oscillations in the layer spacings of smectic A
phases in the homologous series of terephthal-bis-(4n)-alkylanilines
(TBnA).[3]

II. FREE CHAINS

In this section we contrast different parameters that have
been proposed to quantify alkyl chain flexibility. These parameters
involve averages over all of the configurations $\{\phi\}$ of the chain.
They are computed exactly using the rotational isomeric state (RIS)
scheme[4] (non-bonded interactions between remote atoms are included).
The RIS variables, the alkane geometry, and the "united atom"[5]
characteristics are tabulated in Appendix. Configurational averages
of a conformation dependent function $f\{\phi\}$ for the alkyl chains are
given by

$$<f\{\phi\}> = \sum_{\{\phi\}} f\{\phi\}\exp(-E\{\phi\}/RT)/Z$$

where $Z = \sum_{\{\phi\}} \exp(-E\{\phi\}/RT).$

The energy of each configuration $E\{\phi\}$ includes the local RIS energies E_t and $E_{g^{\pm}}$, and non-bonded energies derived from a Lennard-
Jones 6-12 potential for all pairs of united atoms separated by

four or more bonds. As E{φ} is well-defined for alkanes in the RIS
scheme, the only variable in the computation of <f{φ}> is the temperature. So, in order to demonstrate how various configurationally averaged parameters depict chain flexibility, we examine them
for finite (up to 12 atoms) alkanes at selected temperatures.

Persistence Vector

For a chain with n bonds the end-to-end vector r connecting
the first and last atoms of the chain is

$$r = \sum_{i=1}^{n} l_i$$

where l_i is the ith bond vector. Consider the xyz-frame fixed to
l_1 (with the z-axis along l_1 and the y-axis in the plane defined by
l_1 and l_2; the x-axis completes a right-handed cartesian frame):

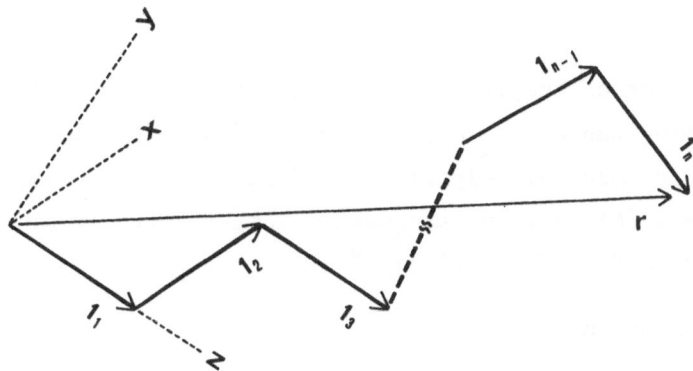

The persistence vector[6] a_n is defined by the configurational
average of r:

$$a_n \equiv <r> = \begin{bmatrix} <x> \\ <y> \\ <z> \end{bmatrix}$$

For a symmetric chain like the n-alkane <x> = 0; the chain extension
is specified by <y> and <z>. In Fig. 1 we illustrate a_n for n=2-11

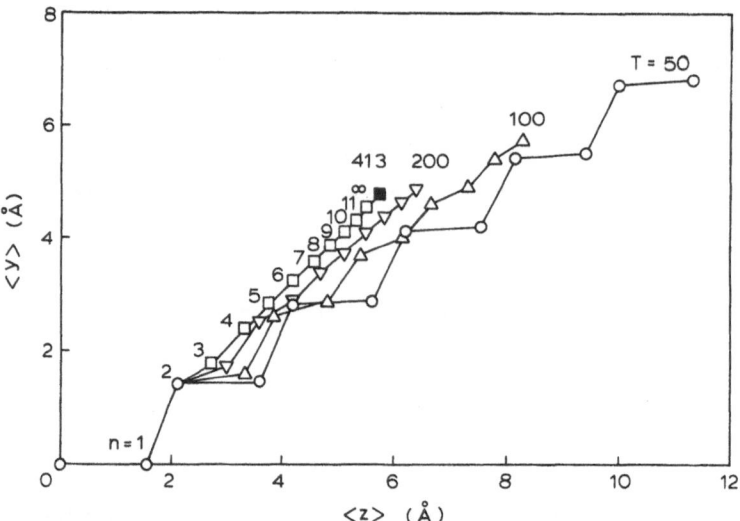

Fig. 1.   Persistence vectors of the alkyl chains having n bonds at
selected temperatures T (in Kelvin). A filled square represents a value for an infinite chain taken from ref. 6.

as a function of temperature. As the temperature is increased the
chain dimensions and a_n asymptotically approach limiting values; at
the melting point of polymethylene (413 K) $\lim_{n \to \infty} a_n$ gives $\langle z \rangle$ = 5.75Å
and $\langle y \rangle$ = 4.86Å (without excluded volume effects, i.e., without
remote nonbonded interactions).[6]

Bond Correlation

The persistence of orientational correlations down the chain
can be inferred from the configurationally averaged carbon–carbon
bond correlation function $\langle P_2^b \rangle_j$

$$\langle P_2^b \rangle_j = \langle P_2(\cos\alpha_{i,i+j}) \rangle = \frac{3}{2} \langle \cos^2\alpha_{i,i+j} \rangle - \frac{1}{2}$$

where $\cos\alpha_{i,i+j} = 1_i \cdot 1_{i+j} / |1_i| |1_{i+j}|$. In Fig. 2 we show $\langle P_2^b \rangle_j$
versus the number j at several temperatures. The oscillation as j
varies is caused by the approximately tetrahedral valence angles
between carbon–carbon bonds. The envelope of these oscillations is

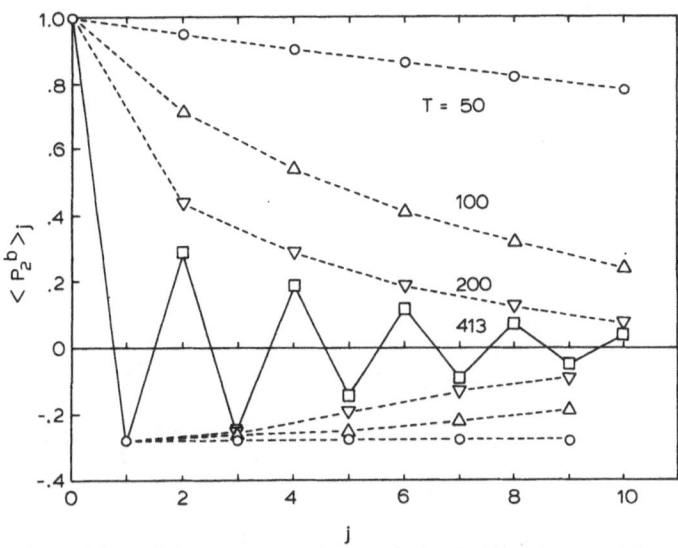

Fig. 2.   Bond correlations $<P_2^b>_j$ for pairs of C–C bonds in the
n–alkyl chain. The attenuation of the envelopes of the
oscillations shown by dashed lines with increasing temperature demonstrate the increase in chain flexibility;
the odd–even effect at 413K (the melting point of polyethylene) is illustrated by a solid line.

reduced dramatically with increasing chain flexibility (increasing

temperatures). The high–temperature values for $<P_2^b>_j$ are in good

agreement with the calculations of Baram and Gelbart for non–tetra-

hedral lattices.[7]

Chord Correlation

The oscillations induced in $<P_2^b>_j$ by the carbon–carbon valence

angle can be removed by rescaling and considering the orientational

correlations between chords c_i, the vectors connecting bond vectors

1_i and 1_{i+1}.

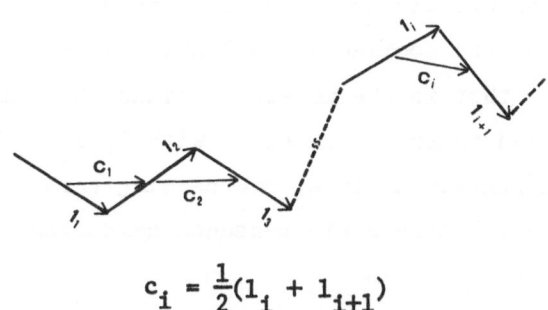

$$c_i = \frac{1}{2}(1_i + 1_{i+1})$$

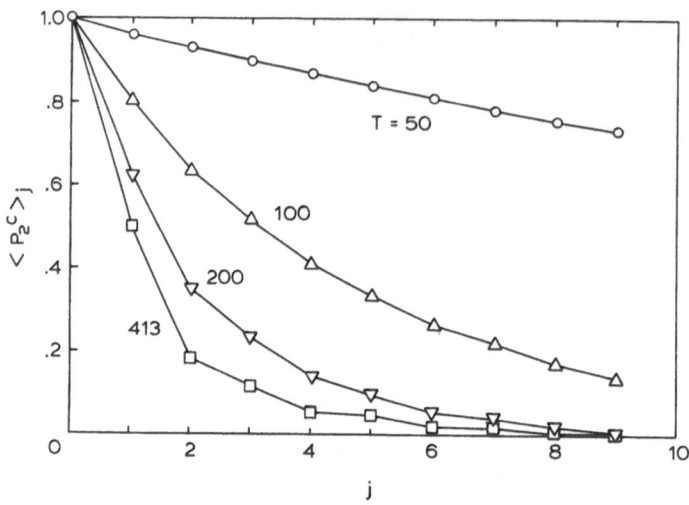

Fig. 3.   Chord correlation functions $\langle P_2^c \rangle_j$ for pairs of chords in
          the n-alkyl chain at selected temperatures T.

The chord correlation function $\langle P_2^c \rangle_j$ is given by

$$\langle P_2^c \rangle_j = \langle P_2(\cos \beta_{i,i+j}) \rangle$$

where $\cos\beta_{i,i+j} = c_i \cdot c_{i+j}/|c_i||c_{i+j}|$.   In the low-energy all-trans
conformation $\{\phi = 0\}$, all of the $\beta = 0$ and $\langle P_2^c \rangle_j = 1$ for all j. Fig.
3 shows that $\langle P_2^c \rangle_j$ decreases monotonically with increasing j at elevated temperatures.

Local Order Tensor

        Orientational order in nematic liquid crystals is given by an
ensemble average over the direction cosines relating the nematic
director to a molecular fixed 123-frame. By analogy we can specify
the orientational correlations from one end of the alkane to the
other by defining the configurational average of local order tensors
$\langle s \rangle_i$ relative to that in the molecular fixed 123-frame, S.   The
123-frame is fixed to the first chord with $3||c_1$, 2 bisecting the
C-C-C angle (perpendicular to c_1 and passing through the 2nd atom)
and the 1-axis completing a right-handed cartesian frame.

S is assumed to be diagonal in the 123-frame with $S_{11} = S_{22} = -\frac{1}{2}$ and $S_{33} = 1$. In a similar manner, local xyz-frames fixed to the ith methylene fragments are constructed. The configurationally averaged elements of the local order tensors $\langle s \rangle_i$ can be obtained from

$$\langle s_{\alpha\beta} \rangle_i = \langle \sum_{m,n}^{1,2,3} l_{m\alpha}^i S_{mn} l_{n\beta}^i \rangle \qquad (\alpha,\beta = x,y,z)$$

where $l_{m\alpha}^i$ is the direction cosine between the m-axis of the 123-frame and the α-axis of the ith xyz-frame (Note $\langle s \rangle_1 = S$).

Fig. 4 shows $\langle s_{zz} \rangle_i$ versus position i for several temperatures. At high temperatures $\langle s_{zz} \rangle_i$ oscillates between positive and negative values; the latter describe the tendency for the chords to order at right angles relative to c_1. Also shown at T=413 K is the "biaxiality" $\langle s_{xx} \rangle_i - \langle s_{yy} \rangle_i$ of the orientational order. Relative values of $\langle s_{zz} \rangle_i$ and the biaxiality can be extracted from deuterium NMR measurements of labeled chains.[1,2]

RIS Probabilities

The singlet probabilities, $P_{q;i}$, that bond i is in RIS q, and the doublet probabilities, $P_{qr;i}$, that bonds i and i+1 are in RIS q and r, respectively, also reflect chain flexibility. These quantities are shown for internal bonds (i=6) of dodecane in Fig. 5 as a function of temperature. The triplet "kink" probability $P_{g^{\pm}tg^{\mp}}$ is also shown.

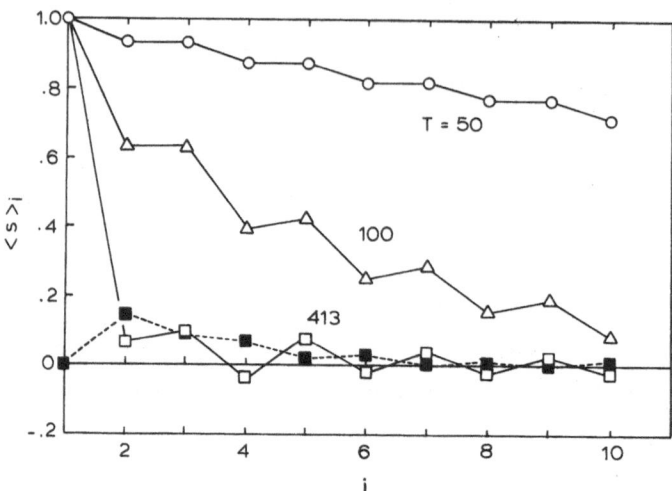

Fig. 4.   Configurationally averaged order parameters as a function
           of the chord number i. Open symbols illustrate the tem-
           perature dependence of the principal value $\langle s_{zz}\rangle_i$ and
           filled squares show the biaxiallity $\langle s_{xx}\rangle_i - \langle s_{yy}\rangle_i$ at
           413K.

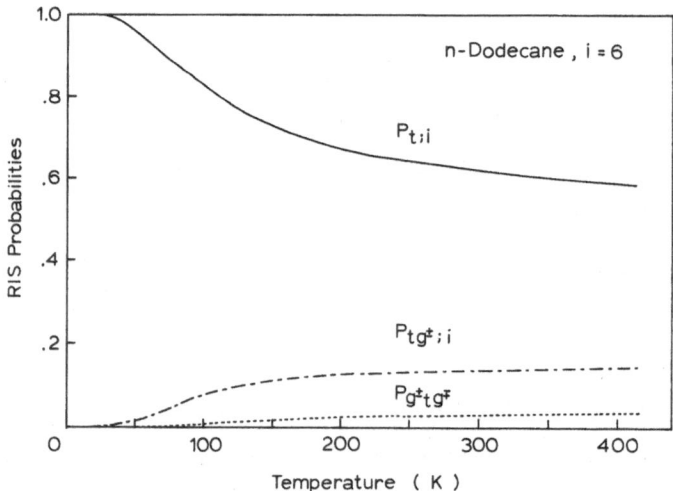

Fig. 5.   Temperature dependences of t, $tg^{\pm}$ and "kink" ($g^{\pm}tg^{\mp}$)
           probabilities for the center bond (i = 6) in n-dodecane.

III. RESTRICTED CHAINS

In this section we constrain the alkyl chains to the set of
configurations $\{\phi, r\}$ that fit within a cylinder of radius r. The
role of the cylinder is intended to mimic the uniaxial constraint
that the alkyl chain would experience in a nematic or smectic A
phase. We specifically attempt to model the behavior of the terminal alkyl chains in the homologus TBnA series (n=3-8). In particular, we determine the values of r which yield the layer spacings
d in the smectic A phases of TBnA reported by Kumar.[3] The idealized
assumption of an all-*trans* conformation for these mesogens while
indicating the proper pattern, overestimates the observed "even-odd"
increment for d with increasing n (Fig. 6).

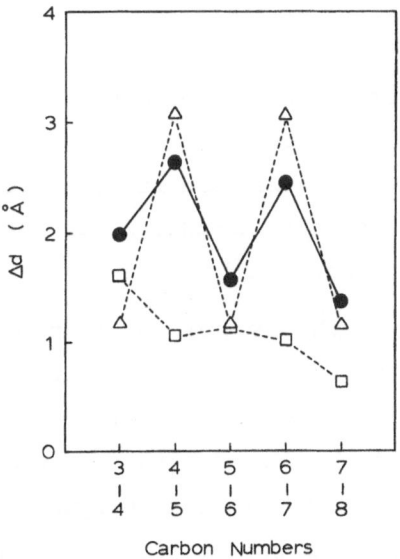

Fig. 6.   The change in smectic-A layer spacing Δd of TBnA at the
           smectic-C to -A transition temperature as a function of
           the number of carbon atoms in alkyl end chain. Closed
           circles are experimental values taken from ref. 3; open
           triangles and squares are calculated values for the
           all-*trans* chain and free chain, respectively.

The procedure utilized for specifying the orientation of the
mean-field cylinder relative to the TBnA molecular framework is
similar to that used in modeling deuterium NMR spectra of labeled
alkyl chains of mesogens:[2] The cylinder axis is assumed to be coincident with the axis of minor principal moment of inertia I_z of
each configuration of the TBnA molecule. (z is usually called the
"long molecular axis" of the mesogen.) This is essentially equivalent to assuming that repulsive (steric) interactions dominate the
mean-field and, each configuration $\{\phi,r\}$ adopts an orientation in
this mean-field that minimizes steric interactions with its surround-
ings. (Here we use Lennard-Jones interactions between the mesogen
atoms and the cylinder wall.) This restriction on the alkyl chain
flexibility caused by the cylinder is included in the configurational
average as an additional component of the energy for each configuration

$$E\{\phi,r\} = E\{\phi\} + \sum_{i} E_{i}.$$

The internal energy of the alkyl chain $E\{\phi\}$ is the same as that
used in part I; the sum is over the non-bonded interactions between
each united atom in the mesogen and the closest point on the cylinder
wall (a $CH_2 \ldots CH_2$ united atom Lennard-Jones potential is assumed for
each interaction; see ref. 1 and 2).

In the modeling of the smectic layer spacing d, we assume that
configurationally averaged values of the length of the TBnA mesogen
$<L\{\phi,r\}>_n$ is directly proportional to d. To compute $<L\{\phi,r\}>_n$ we
find the projection of the TBnA length on the minor principal inertia
axis (z-axis) for each configuration, weight it by the corresponding
energy $\exp(-E\{\phi,r\}/RT)$, and sum over all configurations; this procedure is carried out for several values of r. In practice, to
avoid a sum over the excessive number of configurations for the
higher homologs (n>5), we keep one alkyl chain fixed (the atoms are
located at their configurationally averaged coordinates with $r = \infty$,
i.e. a free chain) and compute the projection of the other alkyl

chain $A_z^n\{\phi,r\}$ on the z-axis. Additionally, we assume the contribution to $<L\{\phi,r\}>_n$ from the aromatic TBnA core is a constant, C_z.
This introduces marginal approximations as the principal z-axis
seldom deviates more than ~10° from the "para-axis" of the extended
(*trans*) core:

$$<L\{\phi,r\}>_n = C_z + 2 <A_z^n\{\phi,r\}>.$$

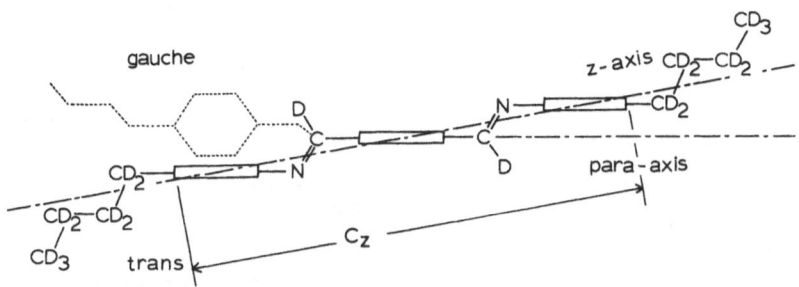

The calculated increment in the smectic A layer spacing as n
increases to n+1 in the TBnA series Δd is given by

$$\Delta d = 2(<A_z^{n+1}\{\phi,r\}> - <A_z^n\{\phi,r\}>).$$

For the unrestricted mesogen ($r = \infty$; free chains) the computed Δd
is shown in Fig. 6. Clearly the observed Δd corresponds to an
intermediate state of chain flexibility between that of the free
chain and the rigid all *trans* chain.

Using the procedures outlined and the assumption that,
$d \equiv <L\{\phi,r\}>_n$, the smectic layer spacing can be computed as a
function of r for the homologus series TBnA. In order to fit the
x-ray values of d or, equivalently, to infer the r-values required
for computing values of $<L\{\phi,r\}>_n$ commensurate with d, we must determine C_z by fixing a value of r for one member of the series. We
do this for TB4A by optimizing the cylinder radius to reproduce
reported quadrupolar splittings of the alkyl chain in the smectic A
phase of TBBA.[8] Such restricted chain model calculations have been
described before.[1,2] For TB4A we observed that the *gauche* core configuration (see diagram above) was needed to compute quadrupolar

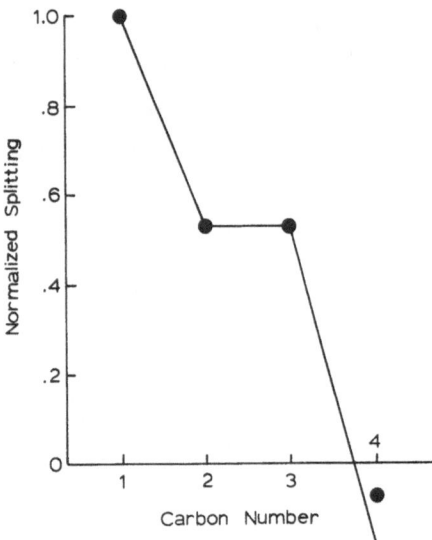

Fig. 7.   Normalized quadrupolar splittings for the smectic A phase
          of TB4A at 456K. Filled circles correspond to experimen-
          tal values taken from ref. 8 and the solid line to
          calculated values using r_o = 5,6 Å.

splittings in the nematic phase (*gauche*$^{\pm}$: *trans* ratio of ~2:1) but,
the *trans* core configuration only was necessary for the smectic A
phase. The optimized cylinder radius, r_o, which gives quadrupolar
splittings in agreement with the NMR results for the smectic A
phase of TB4A is r_o = 5.6 Å (Fig. 7). Having established the value
of r_o for TB4A we can use the layer spacing in TB4A to determine C_z
from d = C_z + 2 <A_z^4{φ,r_o}>. Now the r dependence of the calculated
values of <L{φ,r}>$_n$ can be intersected with the x-ray values[3] of d
in a plot of <L{φ,r}>$_n$ versus r. The intersections establish the
values of r that give the required restriction of the terminal
alkyl chains in the configurational average. The results are shown
in Fig. 8. These results suggest that the strength of the mean-
field, i.e., the value for r, exhibits an even-odd oscillation with
n in the TBnA series.

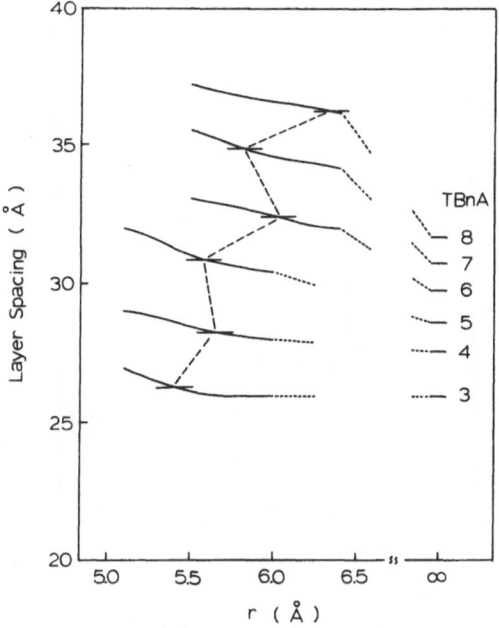

Fig. 8.   The computed dependence on r of the smectic A layer spacing (configurationally averaged molecular length) of TBnA
at the smectic-C to -A transition temperature. Short horizontal bars are experimental observations and a dashed
line represents the change in optimized value of r with n.

IV. CONCLUSIONS

The layer spacing d in the smectic A phase of the TBnA series
is computed from the configurationally averaged length of the TBnA
molecule. Restrictions imposed by the mean-field are approximated
by generating the configurations of the terminal alkyl chains in a
cylinder. The radius of the cylinder is a parameter in the calculation. It is optimized for TB4A to give agreement with deuterium
NMR data and the x-ray data under the assumption that d = $\langle L\{\phi,r\}\rangle$,
i.e., neglecting the orientational order and its modification of
the relationship between d and $\langle L\{\phi,r\}\rangle$. With these assumptions
the calculation yields values of r for other members of the TBnA
series. (Fig. 8).

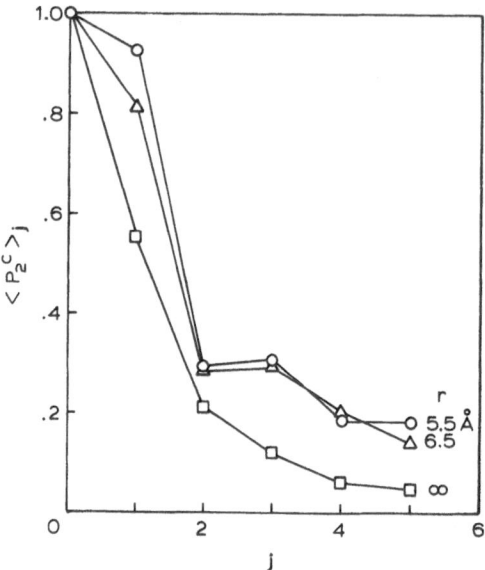

Fig. 9.   Chord correlation $<P_2^c>_j$ for the TB7A mesogen in smectic A
phase at 464 K as a function of r.

We can now contrast the alkyl chain flexibility in the smectic
A phase of TBnA with that determined for unrestricted chains in
part II. We show for example the computed chord correlation $<P_2^c>_j$
and the singlet probabilities $P_{t;i}$ for the TB7A mesogen as a function of the cylinder radius. $<P_2^c>_j$ decreases less rapidly and
undulates relative to that of the unrestricted alkyl chain with increasing j (Fig. 9). Roughly speaking, the inferred chain flexibility in the smectic-A phase at 464 K is comparable to that of a free
chain at about 200 K (contrast Fig. 9 with Fig. 3).

In the range of r-values that reproduce the x-ray layer spacing,
the probability of a *trans* RIS at a given bond shows a distinct alternation in the preference for a *trans* state followed by *gauche*
state relative to that computed for the free alkyl chain with r = ∞
(Fig. 10). This result is consistent with the oscillating incremen-
tal change in the lateral demisions of the mesogen that occurs when
the added C–C bond is an even or an odd member of the chain. Namely,

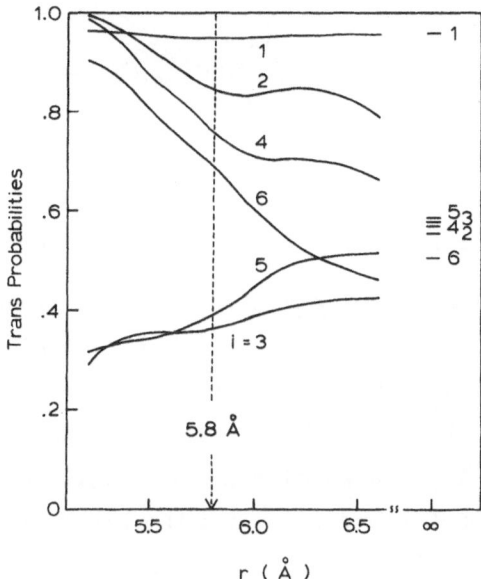

Fig. 10.   Dependence on r of the *trans* probability of bond i in the
TB7A mesogen in smectic A phase at 464 K.   The value of
r = 5.8 Å corrresponds to the optimized r value shown in
Fig. 8.

gauche (trans) conformers about the penultimate C-C bond increase
the width of the mesogen when the antepenultimate C-C bond exhibits
a preference for the *gauche (trans)* state.   These findings in the
context of this crude model calculation give new insights into the
character of terminal alkyl chains in thermotropic mesogens.

APPENDIX

The geometrical details of alkyl chains include the C-C and
C-D bond lengths: l_{CC} = 1.533 Å and l_{CD} = 1.090 Å; the carbon
valence angle: <CCC = 112.5°; and the angles between the deuterons,
<DCD = 109.0° and 109.47°, for methylene and terminal methyl groups
respectively.   The core geometry of TBnA, slightly simplified from
the X-ray data,[9] is shown below,

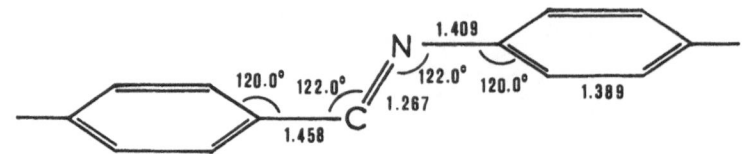

where the aromatic rings are assumed to possess hexagonal symmetry.

In the 3 RIS scheme for alkyl chains, the *gauche* energy $E_g\pm$ for $\phi_g\pm = \pm112.5°$ is assumed to be 0.4 kcal/mol higher than the *trans* energy E_t for $\phi_t = 0°$. One exception is the first dihedral angle in the alkyl chain attached to the aromatic ring in TBnA, where we assume 6 RIS with equal dihedral angle intervals of 60° (the first alkyl C-C-C plane is perpendicular to the plane of the aromatic ring for $\phi = 90°$).

The nonbonded energy for a pair of united atoms at a distance r is calculated by the Lennard-Jones 6-12 potential,

$$E_{NB} = A \ r^{-12} - B \ r^{-6}$$

where the parameters A and B are

$$B = \frac{\frac{3}{2} e \ (h/m^{1/2}) \ \alpha_1\alpha_2}{(\alpha_1/N_1)^{1/2} + (\alpha_2/N_2)^{1/2}}$$

$$A = B \ r_{min}^6/2 \qquad\qquad r_{min} = r_{w1} + r_{w2}$$

The atomic paralizability α, the effective number of outer-shell electrons, N, and the van der Waals radius r_w used are listed in Table I.

TABLE I

Data for evaluating nonbonded energy (ref. 5)

United Atoms	$\alpha(\overset{\circ}{A}^3)$	N	$r_w(\overset{\circ}{A})$
CH_2 (alkane)	1.77	7.0	1.95
CH_3 (alkane)	2.17	8.0	1.95
C (aromatic)	1.65	5.0	1.80
CH (aromatic)	2.07	6.0	1.90

Calculations of nonbonded energies are carried out for all pairs of united atoms separated by four or more bonds within the alkyl chain, and for TBnA the interaction of the end alkyl chain with the nearest aromatic ring is also included.

REFERENCES

1. E. T. Samulski, Ferroelectrics, 30, 83 (1980).

2. E. T. Samulski and R. Y. Dong, J. Chem. Phys., (in press).

3. S. Kumar, Phys. Rev. A23, 207 (1981).

4. P. J. Flory, "Statistical Mechanics of Chain Molecules", Interscience Publishers, New York (1969).

5. K. D. Gibson and H. A. Scheraga, Proc. Nat'l. Acad. Sci., 58, 420 (1967).

6. D. Y. Yoon and P. J. Flory, J. Chem. Phys., 61, 5366 (1974).

7. A. Baram and W. M. Gelbart, J. Chem. Phys., 66, 617 (1977).

8. B. Deloche, J. Charvolin, L. Liebert, and L. Strzelecki, J. de Physique, Colloque C1, 36, 21 (1975).

9. P. J. Doucet, J. P. Mormon, C. Chevalier, and A. Lifchitz, Acta Cryst., B33, 1701 (1977).

ORIENTATIONAL ORDER IN BINARY MIXTURES OF NEMATIC LIQUID CRYSTALS

P. Palffy-Muhoray[*], D. A. Dunmur, W. H. Miller

Dept. of Chemistry
University of Sheffield
Sheffield, U.K.

and

D. A. Balzarini

Dept. of Physics
University of British Columbia
Vancouver, B. C. Canada

ABSTRACT

Simple molecular field theory is used to examine the nematic-isotropic transition in binary mixtures of nematic liquid crystals. Using an extended Maier-Saupe model, the nematic-isotropic coexistence region can be determined for the entire range of concentrations; the extent of this region is determined by the ratio of transition temperatures and of the molar volumes of the constituents. The order parameters of the two nematic components can be obtained separately; these can be significantly different and lead to non-ideal behaviour in nematic mixtures. Predictions of the theory are compared with experiment, and new experimental results for two binary systems are presented.

---

[*]   currently on leave from Dept. of Physics, University of British Columbia, Vancouver, B. C. Canada

INTRODUCTION

Liquid crystalline mixtures, consisting of one or more mesogenic components, have been the subject of considerable interest in
the past in the attempt to produce room temperature mesomorphic
behaviour. More recently, interest in mixtures has increased because of their utility in electro-optic display applications. Considerable work in the theory of liquid crystalline mixtures has been
carried out (see for example[1]); statistical mechanical approaches
that have been used have been based on lattice models[2], hard particle theories[3] and molecular field theories[4]. The Maier-Saupe
(M-S) mean field theory[5] for pure nematogens has been extended to
binary mixtures (consisting of at least one nematogen) by Humphries,
James and Luckhurst[4,6] and Martire[7]. Although some characteristics
of mixtures were predicted by the above analyses, the problems of
determining the complete nematic-isotropic coexistence curve and
the order parameters of the constituents have remained unsolved.
In this paper we present a straightforward extension of the application of the M-S theory to mixtures. A result of the theory is a
simple method for the calculation of the nematic-isotropic coexistence region for any binary nematic mixture over the entire range of
composition and the order parameters of the constituents. The
coexistence region is determined by the ratio of transition temperatures and molar volumes of the pure components. The theory predicts
that, in general, a binary mixture will exhibit a two-phase region,
and it may also show azeotropic behaviour. The order parameters of
the two components usually differ, however they are connected by a
simple relation. The value of the order parameters along the coexistence curve is different from the M-S value and varies with
composition. The order parameters of pure materials can be obtained
from analysis of refractive index data[8,9]. In this paper we propose
a relation between the refractive indices of a binary mixture and
the order parameters of its components. Phase diagrams and refrac-

tive index data are presented for two binary systems; these results
are compared with predictions of the theory.

THEORY

Single Component Nematics

Before proceeding to mixtures, it is useful to consider a
one-component nematic liquid crystal consisting of N axially symmetric molecules confined to a volume V . The canonical configurational partition function Q for this system may be written in
terms of the single-particle pseudopotential $\varepsilon(\cos\theta)$ as

$$Q = \frac{1}{N!} \left[\frac{1}{4\pi} \int e^{\frac{-\varepsilon(\cos\theta)}{kT}} \ d^3\underset{\sim}{r} \ \sin\theta d\theta d\phi \right]^N \tag{1}$$

where the position of the particle is denoted by $\underset{\sim}{r}$, and the orientation of its symmetry axis is specified by the polar and azimuthal
angles θ and ϕ . We assume that the pseudopotential may be
written as

$$\varepsilon(\cos\theta) = \gamma^\circ - U\rho S P_2(\cos\theta) + \frac{1}{2} U\rho S^2 \tag{2}$$

where γ° is the single-particle pseudopotential in the isotropic
phase, containing entropic as well as energetic contributions of
the intermolecular interactions, U is the average strength of the
anisotropic pair interaction, ρ is the number density and the
nematic order parameter S is the average quantity

$$S = \langle P_2(\cos\theta) \rangle = \langle \frac{1}{2}(3\cos^2\theta - 1) \rangle.$$

The last term on the right hand side of eq. (2) arises from the
requirement that the average orientational potential energy per
molecule equals one half of the orientational energy per pair. In

pure liquid crystals, near the nematic-isotropic transition, the
density is frequently regarded as being nearly constant, hence
the volume dependence of the partition function is usually not considered explicitly. In mixtures, however, the number densities of
the individual components may vary significantly with temperature
and across phase boundaries; in order to account for the associated
entropy changes it is necessary to consider the volume dependence
of the free energy and of the partition function. On integration
over $\sim$ and ϕ, the Helmholtz free energy of the system becomes

$$F = -kT\ln Q = N\left\{\gamma^\circ + kT\ln\rho + \frac{1}{2}U\rho S^2 - kT\ln\int_0^1 e^{\frac{\rho USP_2(\cos\theta)}{kT}} d(\cos\theta)\right\} \qquad (3)$$

At constant density, the free energy is minimised if $\frac{\partial F}{\partial S} = 0$,
that is, if

$$S = \frac{\int P_2(\cos\theta)e^{\frac{\rho USP_2(\cos\theta)}{kT}} d(\cos\theta)}{\int e^{\rho USP_2(\cos\theta)} d(\cos\theta)} \qquad (4)$$

Eq. (4) is the M-S expression[10] for the order parameter, $S = 0$ is a
solution for all T , while for $\frac{\rho U}{kT} < 4.49$ non-zero solutions for
 S also exist. The temperature at which the transition from the
nematic to the isotropic phase occurs may be determined by equating
the chemical potentials of the two phases. Noting that at low
pressures $P = -\left(\frac{\partial F}{\partial V}\right)_{N,T} = -\frac{\rho \partial F}{V \partial \rho} \approx 0$, the chemical potential is

$$\mu = \left(\frac{\partial F}{\partial N}\right)_{V,T} = \gamma^\circ + kT\ln\rho + \frac{1}{2}U\rho S^2 - kT\ln\int e^{\frac{\rho USP_2(\cos\theta)}{kT}} d(\cos\theta) \qquad (5)$$

Letting $\mu_{isotropic} = \mu_{nematic}$ at the transition, eq. (5) gives

$$\frac{1}{2} U \rho S_c^2 = kT \ln \int e^{\frac{\rho U S_c}{kT} P_2(\cos\theta)} \, d(\cos\theta) \tag{6}$$

with the usual non-trivial solution[11] $S_c = 0.429$ and $\frac{\rho U}{kT_{NI}} = 4.54$

Binary Mixtures of Nematics

The preceeding formalism can be generalised to describe a two-component mixture, consisting of N_1 molecules of component 1 and N_2 molecules of component 2 . As suggested by the phase rule, we consider two coexisting liquid phases A and B . There are, respectively, N_{1_A} and N_{2_A} molecules of type 1 and 2 in phase A , and N_{1_B} and N_{2_B} in phase B ; $N_{1_A} + N_{1_B} = N_1$ and $N_{2_A} + N_{2_B} = N_2$. The pseudopotential of a molecule of component 1 in phase A is

$$\epsilon_A(\cos\theta_1) = \gamma_1^\circ - U_{11}\rho_{1_A} S_{1_A} P_2(\cos\theta_1) - U_{12}\rho_{2_A} S_{2_A} P_2(\cos\theta_1) \tag{7}$$

$$+ \frac{1}{2} U_{11}\rho_{1_A} S_{1_A}^2 + \frac{1}{2} U_{12}\rho_{2_A} S_{2_A} S_{1_A}$$

while for a molecule of component 2 in phase A it is

$$\epsilon_A(\cos\theta_2) = \gamma_2^\circ - U_{22}\rho_{2_A} S_{2_A} P_2(\cos\theta_2) - U_{21}\rho_{1_A} S_{1_A} P_2(\cos\theta_2) \tag{8}$$

$$+ \frac{1}{2} U_{22}\rho_{2_A} S_{2_A}^2 + \frac{1}{2} U_{21}\rho_{1_A} S_{1_A} S_{2_A}$$

Here U_{ij} is the mean anisotropic interaction energy between a molecule of component i and one of component j , $\rho_{iA} = \frac{N_{iA}}{V_A}$ where V_A is the volume of phase A and $S_{i_A} = \langle P_2(\cos\theta_i) \rangle$ is the order parameter of component i in the phase A . Replacing A by B in eqs. (7) and (8) gives the pseudopotentials for the two components in phase B ; the terms containing $\cos\theta$ in the pseudo-potentials are identical to those of Martire[12] .

We have assumed in the above that $\gamma^\circ_{1_A} = \gamma^\circ_{1_B} = \gamma^\circ_1$ and $\gamma^\circ_{2_A} = \gamma^\circ_{2_B} = \gamma^\circ_2$; explicitly, this means that the mean isotropic potential energy per molecule and the relative free volume per molecule for a given component is assumed to be the same in both phases. The free energy for the system is then $F = F_A + F_B$, where

$$
\begin{aligned}
F_A = N_{1_A} & \left\{ \gamma^\circ_1 + \tfrac{1}{2} U_{11} \rho_{1_A} S^2_{1_A} + \tfrac{1}{2} U_{12} \rho_{2_A} S_{2_A} S_{1_A} \right. \\
& \left. - kT\ln\frac{V_A}{N_{1_A}} \int e^{\frac{(U_{11}\rho_{1_A} S_{1_A} + U_{12}\rho_{2_A} S_{2_A})P_2(\cos\theta)}{kT}} \, d(\cos\theta) \right\}
\end{aligned}
$$

$$
\begin{aligned}
+ N_{2_A} & \left\{ \gamma^\circ_2 + \tfrac{1}{2} U_{22} \rho_{2_A} S^2_{2_A} + \tfrac{1}{2} U_{21} \rho_{1_A} S_{1_A} S_{2_A} \right. \\
& \left. - kT\ln\frac{V_A}{N_{2_A}} \int e^{\frac{(U_{22}\rho_{2_A} S_{2_A} + U_{21}\rho_{1_A} S_{1_A})P_2(\cos\theta)}{kT}} \, d(\cos\theta) \right\}
\end{aligned}
$$

$$
\tag{9}
$$

and replacing A by B in eq. (9) gives F_B . Instead of minimising the free energy with respect to the total volume as we had done for the single-component case, we make the usual simplifying assumption[12] that $V_A = N_{1_A} v_1 + N_{2_A} v_2$ and $V_B = N_{1_B} v_1 + N_{2_B} v_2$ where v_1 and v_2 are molecular volumes of components 1 and 2, equal to the reciprocal of the number density. The free energy can then be written as

$$F_A = V_A \left\{ \rho_{1_A} \gamma_1^o + \frac{1}{2} U_{11} \rho_{1_A}^2 S_{1_A}^2 + \frac{1}{2} U_{12} \rho_{1_A} \rho_{2_A} S_{1_A} S_{2_A} \right.$$

$$-kT\rho_{1_A} \ln \frac{V_A}{N_{1_A}} \int e^{\frac{(U_{11}\rho_{1_A} S_{1_A} + U_{12}\rho_{2_A} S_{2_A}) P_2(\cos\theta)}{kT}} \; d(\cos\theta)$$

(10)

$$+ \rho_{2_A} \gamma_2^o + \frac{1}{2} U_{22} \rho_{2_A}^2 S_{2_A}^2 + \frac{1}{2} U_{12} \rho_{1_A} \rho_{2_A} S_{1_A} S_{2_A}$$

$$\left. -kT\rho_{2_A} \ln \frac{V_A}{N_{2_A}} \int e^{\frac{(U_{22}\rho_{2_A} S_{2_A} + U_{21}\rho_{1_A} S_{1_A}) P_2(\cos\theta)}{kT}} \; d(\cos\theta) \right\}$$

and similarly for F_B. The variables which must now minimise F at a given temperature are S_{1_A}, S_{2_A}, S_{1_B}, S_{2_B} and N_{1_A} and N_{2_A}, since $N_{1_B} = N_1 - N_{1_A}$ and $N_{2_B} = N_2 - N_{2_A}$. It is straightforward to show that $\frac{\partial F}{\partial S_{1_A}} = \frac{\partial F}{\partial S_{2_A}} = 0$ if

$$S_{1_A} = \frac{\int P_2(\cos\theta) e^{\frac{(U_{11}\rho_{1_A} S_{1_A} + U_{12}\rho_{2_A} S_{2_A}) P_2(\cos\theta)}{kT}} d(\cos\theta)}{\int e^{\frac{(U_{11}\rho_{1_A} S_{1_A} + U_{12}\rho_{2_A} S_{2_A}) P_2(\cos\theta)}{kT}} d(\cos\theta)}$$

(11)

and

$$S_{2_A} = \frac{\int P_2(\cos\theta) e^{\frac{(U_{22}\rho_{2_A} S_{2_A} + U_{21}\rho_{1_A} S_{1_A}) P_2(\cos\theta)}{kT}} d(\cos\theta)}{\int e^{\frac{(U_{22}\rho_{2_A} S_{2_A} + U_{21}\rho_{1_A} S_{1_A}) P_2(\cos\theta)}{kT}} d(\cos\theta)}$$

(12)

and similarly for phase B . The integral equations of eqs. (11) and (12) for the order parameters S_{1_A} and S_{1_B} are coupled, and are difficult to solve. Minimisation of the free energy of eq. (10), in general, gives rise to six integral equations which need to be solved simultaneously; the equations for the chemical potentials are particularly complicated due to the particle number dependence of the pseudopotentials. However, if we make the usual assumption[4] that $U_{12} = U_{21} = \sqrt{U_{11}U_{22}}$,   then a considerable simplification may be effected by the following transformation.

We define the new variables σ_A and σ_B such that

$$\sigma_A = \sqrt{U_{11}} \; \rho_{1_A} S_{1_A} + \sqrt{U_{22}} \; \rho_{2_A} S_{2_A} \tag{13}$$

and

$$\sigma_B = \sqrt{U_{11}} \; \rho_{1_B} S_{1_B} + \sqrt{U_{22}} \; \rho_{2_B} S_{2_B} \tag{14}$$

The free energy can then be written as

$$F_A = V_A \left\{ \rho_{1_A} \gamma_1^\circ + \rho_{2_A} \gamma_2^\circ + \frac{1}{2}\sigma_A^2 - kT\rho_{1_A} \ln \frac{1}{\rho_{1_A}} \int e^{\sqrt{U_{11}}\sigma_A \frac{P_2(\cos\theta)}{kT}} d(\cos\theta) \right.$$

$$\left. - kT\rho_{2_A} \ln \frac{1}{\rho_{2_A}} \int e^{\sqrt{U_{22}}\sigma_A \frac{P_2(\cos\theta)}{kT}} d(\cos\theta) \right\}$$

$$\tag{15}$$

and similarly for F_B. Minimising F with respect to σ_A yields

$$\sigma_A = \rho_{1_A} \sqrt{U}_{11} \frac{\int P_2(\cos\theta) e^{\frac{\sqrt{U}_{11} \sigma_A \frac{P_2}{kT}(\cos\theta)}{}} d(\cos\theta)}{\int e^{\frac{\sqrt{U}_{11}\sigma_A \frac{P_2}{kT}(\cos\theta)}{}} d(\cos\theta)} +$$

$$+ \rho_{2_A} \sqrt{U}_{22} \frac{\int P_2(\cos\theta) e^{\frac{\sqrt{U}_{22} \sigma_A \frac{P_2}{kT}(\cos\theta)}{}} d(\cos\theta)}{\int e^{\frac{\sqrt{U}_{22} \sigma_A \frac{P_2}{kT}(\cos\theta)}{}} d(\cos\theta)} \qquad (16)$$

and similarly for σ_B. Eq. (16) is now a self-consistent equation for σ_A alone, which may be solved at a given temperature if the number densities are known. It is worth noting that elimination of σ_A between eqs. (16) and (13) recovers eqs. (11) and (12), the original expressions for S_{1_A} and S_{2_A}. The number of molecules in each phase is determined by minimising F with respect to N_{1_A} and N_{2_A}. Since $N_{1_A} + N_{1_B} = N_1$, $\frac{\partial F}{\partial N_{1_A}} = \frac{\partial F_A}{\partial N_{1_A}} - \frac{\partial F_B}{\partial N_{1_B}} = 0$, and F is minimised, as expected, if the chemical potentials of each component are equal in the two phases, that is, if $\mu_{1_A} = \mu_{1_B}$ and $\mu_{2_A} = \mu_{2_B}$. The evaluation of the chemical potentials is now straightforward, since although σ_A and σ_B are complicated functions of N_{1_A} and N_{2_A}, $\frac{\partial F}{\partial \sigma_A} = \frac{\partial F}{\partial \sigma_B} = 0$. Since $\rho_{1_A} = \frac{N_{1_A}}{N_{1_A}V_1 + N_{2_A}V_2}$ and $\rho_{2_A} = \frac{N_{2_A}}{N_{1_A}V_1 + N_{2_A}V_2}$ the chemical potentials for molecules of components 1 and 2 in phase A become, respectively,

$$\mu_{1_A} = \gamma_1^{\circ} + \frac{1}{2}\sigma_A^2 v_1 - kT\ln\frac{1}{\rho_{1_A}}\int e^{\sqrt{U}_{11}\,\sigma_A\frac{P_2}{kT}(\cos\theta)}\,d(\cos\theta) \qquad + kT\rho_{2_A}(v_2-v_1)$$

(17)

and

$$\mu_{2_A} = \gamma_2^{\circ} + \frac{1}{2}\sigma_A^2 v_2 - kT\ln\frac{1}{\rho_{2_A}}\int e^{\sqrt{U}_{22}\,\sigma_A\frac{P_2}{kT}(\cos\theta)}\,d(\cos\theta) \qquad + kT\rho_{1_A}(v_1-v_2)$$

(18)

and similarly for phase B . Expressing the number densities in terms of the volume fractions $y_{1_A} = \rho_{1_A}v_1$ and $y_{2_A} = \rho_{2_A}v_2$ and equating chemical potentials in phases A and B gives

$$\frac{1}{2}\frac{\sigma_A^2 v_1}{kT} - \ln\frac{1}{y_{1_A}}\int e^{\sqrt{U}_{11}\,\sigma_A\frac{P_2}{kT}(\cos\theta)}\,d(\cos\theta) + y_{2_A}\left(1 - \frac{v_1}{v_2}\right)$$

$$= \frac{1}{2}\frac{\sigma_B^2}{kT}v_1 - \ln\frac{1}{y_{1_B}}\int e^{\sqrt{U}_{11}\,\sigma_B\frac{P_2}{kT}(\cos\theta)}\,d(\cos\theta) + y_{2_B}\left(1 - \frac{v_1}{v_2}\right)$$

(19)

and

$$\frac{1}{2}\frac{\sigma_A^2}{kT} - \ln\frac{1}{y_{2_A}}\int e^{\sqrt{U}_{22}\,\sigma_A\frac{P_2}{kT}(\cos\theta)}\,d(\cos\theta) + y_{1_A}\left(1 - \frac{v_2}{v_1}\right)$$

(20)

$$= \frac{1}{2}\frac{\sigma_B^2 v_2}{kT} - \ln\frac{1}{y_{2_B}}\int e^{\sqrt{U}_{22}\,\sigma_B\frac{P_2}{kT}(\cos\theta)}\,d(\cos\theta) + y_{1_B}\left(1 - \frac{v_2}{v_1}\right)$$

From eq. (16) we get

$$\sigma_A = \frac{y_{1_A}\sqrt{U}_{11}}{v_1} \frac{\int P_2(\cos\theta)e^{\sqrt{U}_{11}\ \sigma_A \frac{P_2(\cos\theta)}{kT}}d(\cos\theta)}{\int e^{\sqrt{U}_{11}\ \sigma_A \frac{P_2(\cos\theta)}{kT}}d(\cos\theta)}$$

$$+ \frac{y_{2_A}\sqrt{U}_{22}}{v_2} \frac{\int P_2(\cos\theta)e^{\sqrt{U}_{22}\ \sigma_A \frac{P_2(\cos\theta)}{kT}}d(\cos\theta)}{\int e^{\sqrt{U}_{22}\ \sigma_A \frac{P_2(\cos\theta)}{kT}}d(\cos\theta)} \qquad (21)$$

and

$$\sigma_B = \frac{y_{1_B}\sqrt{U}_{11}}{v_1} \frac{\int P_2(\cos\theta)e^{\sqrt{U}_{11}\ \sigma_B \frac{P_2(\cos\theta)}{kT}}d(\cos\theta)}{\int e^{\sqrt{U}_{11}\ \sigma_B \frac{P_2(\cos\theta)}{kT}}d(\cos\theta)}$$

$$+ \frac{y_{2_B}\sqrt{U}_{22}}{v_2} \frac{\int P_2(\cos\theta)e^{\sqrt{U}_{22}\ \sigma_B \frac{P_2(\cos\theta)}{kT}}d(\cos\theta)}{\int e^{\sqrt{U}_{22}\ \sigma_B \frac{P_2(\cos\theta)}{kT}}d(\cos\theta)} \qquad (22)$$

The above four equations, eqs. (19) – (22) constitute the central result of the theory. At a given temperature, they may be solved simultaneously for σ_A, σ_B, y_{1_A} and y_{2_A} if U_{11} and U_{22} (or equivalently, the densities and transition temperatures of the pure components) are known. From σ_A and σ_B the four order parameters can be calculated via eqs. (11) and (12). Preliminary work suggests that at low temperatures two distinct nematic phases coexist, which become miscible above some temperature. At higher temperatures a nematic and an isotropic phase coexist, which become a single isotropic phase above some temperature. In general, the two components have different order parameters in each phase; in the present theory the degree of order is determined by the temperature and the transition temperatures and densities of the pure components. Results on coexisting binary nematic phases will be presented elsewhere; in this paper we restrict ourselves to coexisting nematic and isotropic phases.

Nematic – Isotropic Equilibrium

If phase B is assumed to be isotropic, then $\sigma_B = 0$. For simplicity, we set $y_{1_A} = a$ and $y_{1_B} = b$, that is, a and b are the volume fractions of component 1 in the nematic and isotropic phases respectively, then $y_{2_A} = 1-a$ and $y_{2_B} = 1-b$. We further let $\alpha = \dfrac{\sqrt{U}_{11} \, \sigma_A}{kT}$, and substitution of σ_A from eq. (21) into eqs. (19) and (20) gives

$$\frac{1}{2} \, \alpha \left\{ af'(\alpha) + (1-a)\frac{v_1}{v_2} \sqrt{\frac{U_{22}}{U_{11}}} \, f'\left(\sqrt{\frac{U_{22}}{U_{11}}} \, \alpha \right) \right\} - f(\alpha)$$

$$= \ln \left(\frac{b}{a}\right) + \left(1 - \frac{v_1}{v_2}\right)(a-b) \tag{23}$$

and

$$\frac{1}{2} \, \alpha \left\{ a \frac{v_2}{v_1} f'(\alpha) + (1-a) \sqrt{\frac{U_{22}}{U_{11}}} \, f'\left(\sqrt{\frac{U_{22}}{U_{11}}} \, \alpha \right) \right\} - f\left(\frac{U_{22}}{U_{11}} \, \alpha \right)$$

$$= \ln \left[\frac{1-b}{1-a}\right] + \left(1 - \frac{v_2}{v_1}\right)(b-a) \tag{24}$$

where $f(\alpha) = \ln \left\{ \begin{array}{c} \alpha P_2(\cos\theta) \\ e \ d(\cos\theta) \end{array} \right.$ and $f'(\alpha) = \dfrac{df(\alpha)}{d\alpha}$. Since

$\dfrac{\rho U}{kT_{NI}} = 4.54$ for both (pure) components, $\dfrac{U_{22}}{U_{11}} = \dfrac{T_{NI2}V_2}{T_{NI1}V_1}$. For a

given α , $f(\alpha)$ and $f'(\alpha)$ are easily evaluated, and if the ratio

of the transition temperatures $\dfrac{T_{NI1}}{T_{NI2}}$ and volumes $\dfrac{V_1}{V_2}$ are known,

the non-linear eqs. (23) and (24) may be solved simultaneously for

the volume fractions a and b . Knowing a , the temperature

corresponding to this value of α is obtained from eq. (22), rear-

ranged to give

$$\frac{T}{T_{NI1}} = \frac{4.54}{\alpha} \left\{ af'(\alpha) + (1-a) \frac{v_1}{v_2}\sqrt{\frac{U_{22}}{U_{11}}} \ f'\left(\sqrt{\frac{U_{22}}{U_{11}}} \ \alpha \right) \right\} \quad (25)$$

while the order parameters of the two components in the nematic

phase at this temperature are simply

$$S_1 = f'(\alpha) \qquad\qquad\qquad\qquad\qquad\qquad (26)$$

and

$$S_2 = f'\left(\sqrt{\frac{U_{22}}{U_{11}}} \ \alpha \right) \qquad\qquad\qquad\qquad (27)$$

Results of one set of calculations for the coexistence curve are

shown in Fig. 1; here $\dfrac{T_{NI1}}{T_{NI2}} = 1.5$ and $\dfrac{v_1}{v_2} = 1.5$. Typical behaviour,

as predicted by the theory, may be illustrated by considering a

mixture of fixed composition consisting of 40 mole per cent of

component 1 . At temperatures below $T_1 = 0.823 \ T_{NI1}$, a single

nematic phase exists. At $T = T_1$ an isotropic phase of composition

36% component 1 begins to form. As the temperature is increased

toward $T_2 = 0.838 \ T_{NI1}$ the amount of material in the isotropic phase

increases, and the composition of both phases becomes richer in

component 1 . At $T = T_2$, the nematic phase, of composition 44% of

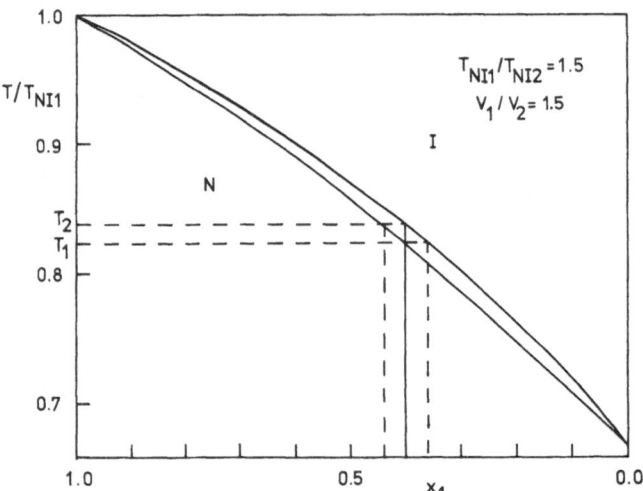

Fig. 1.   Nematic-isotropic phase diagram showing two phase region;
           x_1 is the mole fraction of component 1.

component 1 vanishes, and the entire mixture becomes isotropic.

The order parameters S_{c_1} and S_{c_2} of the two components along the

coexistence curve and the volume averaged order parameter

$<S_c> = aS_{c_1} + (1-a)S_{c_2}$ are shown in Fig. 2. It is interesting

to note that except for the pure components, the order parameters

S_{c_1} and S_{c_2} are significantly different from the M-S value of

$S_c = 0.429$. The temperature behaviour of the order parameters

for the mixture discussed above is shown in Fig. 3; $<S>$ again

denotes the volume averaged order parameter. For the purpose of

comparison, the M-S order parameter for a pure material is also

shown; the transition temperature was chosen to be equal to T_1 .

In the two-phase region $T_2 < T < T_1$ the nematic composition changes,

however the order parameters remain nearly constant. We have ob-

tained birefringence measurements for the binary mixtures of

cyano-biphenyls confirming this behaviour; these results will be

published elsewhere. Azeotropic behaviour is illustrated by the

system where $\frac{T_{NI1}}{T_{NI2}} = 0.99$ and $\frac{V_1}{V_2} = 0.25$ as shown in Fig.

4a. Since the two-phase region is quite narrow, we have plotted

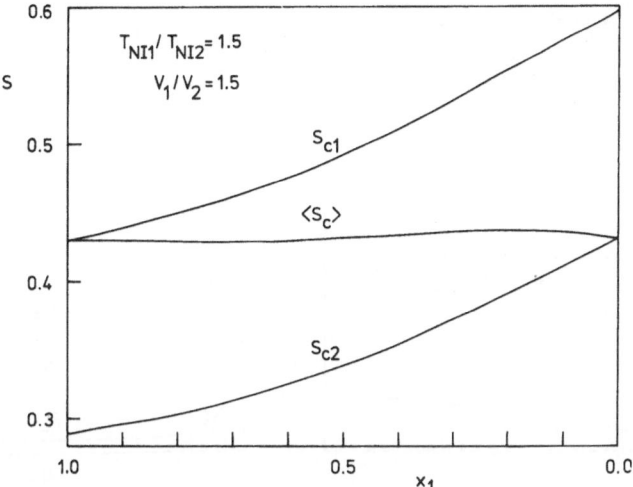

Fig. 2.   Volume averaged order parameter $\langle S_c \rangle$ and component order
parameters S_{c1} and S_{c2} evaluated along the coexistence
curve.

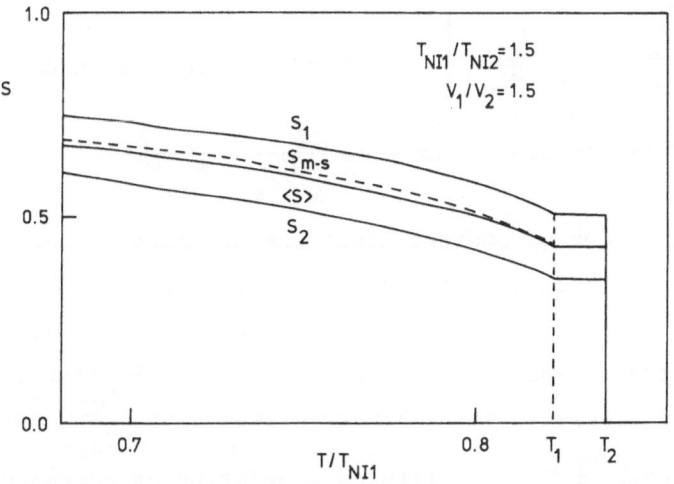

Fig. 3.   Order parameters as a function of reduced temperature:
the dashed line is the M-S order parameter.

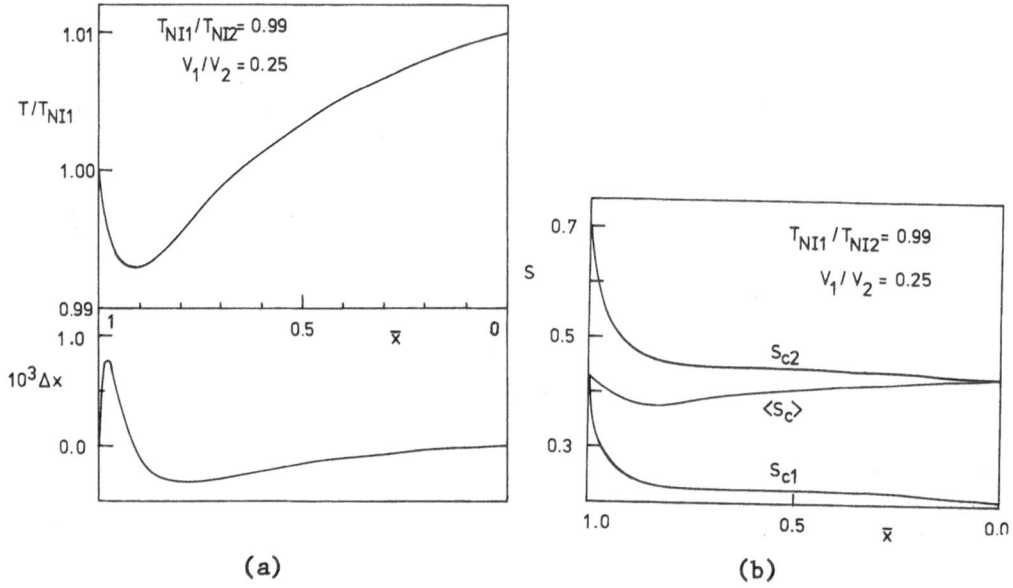

Fig. 4(a). Coexistence curve showing azeotropic behaviour: $\bar{x}$ is
the average mole fraction in the two phase region and Δx
is the extent of the two phase region.

4(b). Order parameters along the coexistence curve of an
azeotropic system.

here reduced temperature vs. average mole fraction

$$\bar{x} = \frac{1}{2} \left(\frac{N_{1_A}}{N_{1_A} + N_{2_A}} + \frac{N_{1_B}}{N_{1_B} + N_{2_B}} \right)$$ and the composition difference

between the two phases $$\Delta x = \left(\frac{N_{1_A}}{N_{1_A} + N_{2_A}} - \frac{N_{1_B}}{N_{1_B} + N_{2_B}} \right)$$

vs. $\bar{x}$. The order parameters along the coexistence curve are shown
in Fig. 4b; the volume averaged order parameter for the mixture
deviates considerably (11%) from the pure component value of 0.429
for low concentrations (13%) of component 2. Similar azeotropic
behaviour is predicted for a binary system with $\frac{T_{NI1}}{T_{NI2}} = 0.99$ as
above, but with $\frac{v_1}{v_2} = 4$. Mixtures consisting of components with
equal molar volumes but different transition temperatures also
exhibit two-phase regions. A portion of the coexistence curve for

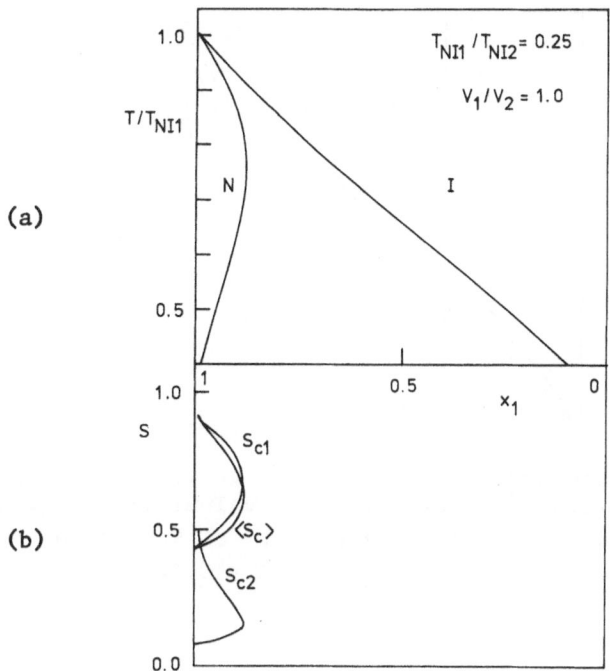

Fig. 5(a). Phase diagram of system exhibiting a restricted nematic
           range.
     5(b). Corresponding order parameters evaluated along the
           nematic phase boundary.

a system with $\dfrac{T_{NI1}}{T_{NI2}} = .25$ and $\dfrac{V_1}{V_2} = 1$ is shown in Fig. 5a; the corresponding order parameters along the coexistence curve are shown
in Fig. 5b. This system exhibits interesting behaviour in that a
nematic phase of a given composition (>82 % component 1) can
coexist with the isotropic phase at two different temperatures.
Similar behaviour has been predicted by Humphries and Luckhurst[13]
for mixtures of rods and spheres of equal volume.

     The relation between the order parameters of the two compo-
nents in the nematic phase of the binary mixture is given, in
parametric form, by eqs. (26) and (27). This relation is shown in
Fig. 6 for different values of $\dfrac{T_{NI1}V_1}{T_{NI2}V_2}$; negative order parameters
have been included for the sake of completeness.

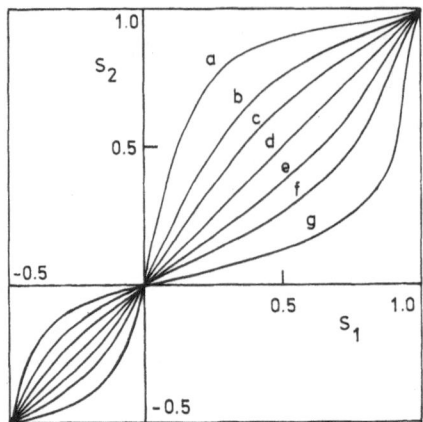

Fig. 6.   Curves showing the relationship between the component
order parameters for different values of $R = (T_{NI2}v_2/T_{NI1}v_1)$.   a) $R = 16$;   b) $R = 4$;   c) $R = 2$;   d) $R = 1$;
e) $R = 0.5$;   f) $R = 0.25$;   g) $R = 0.0625$.

COMPARISON WITH EXPERIMENT

     Although the phase behaviour of many binary liquid crystal
mixtures has been reported[1,14], the two phase region is rarely
observed and the predictions of statistical theories are difficult
to verify. The approach in this work has been to concentrate on
the order parameters of the components in the nematic mixtures. To
provide an experimental test of our theory we have analysed measurements of the refractive indices of two binary nematic systems, and
have obtained values for the volume averaged order parameter of
mixtures. For pure compounds the order parameter can be obtained
from refractive index measurements using the Haller extrapolation[8]
of the quantity

$$\left(\frac{n_e^2 - n_o^2}{\bar{n}^2 - 1}\right) = \left(\frac{\Delta\alpha}{\alpha}\right)S \tag{28}$$

where $\dfrac{\Delta\alpha}{\alpha}$ is the ratio of the polarisability anisotropy and the mean polarisability. This extrapolation is empirical, and assumes that the Vuks relation for the internal field is valid in aniso- tropic materials. Although there is no firm theoretical basis for the Haller extrapolation, we believe that it provides a practical method of comparing the relative order in different liquid crystals.

The principal refractive indices of component i may be written as

$$\left(\frac{n_e^2 - 1}{\bar{n}^2 + 2}\right)_i = \frac{\rho_i}{\varepsilon_o}\left(\alpha_i + 2\,\frac{\Delta\alpha_i}{3}\,S_i\right) \tag{29}$$

and

$$\left(\frac{n_o^2 - 1}{\bar{n}^2 + 2}\right)_i = \frac{\rho_i}{\varepsilon_o}\left(\alpha_i - \frac{\Delta\alpha_i}{3}\,S_i\right) \tag{30}$$

Using these results we obtain for the mixture refractive indices

$$\left(\frac{n_e^2 - n_o^2}{\bar{n}^2 - 1}\right)_{12} = y_1\left(\frac{\Delta\alpha_1}{\alpha_1}\right)S_1\left[y_1 + y_2\left(\frac{v_1\alpha_2}{v_2\alpha_1}\right)\right]^{-1} + y_2\left(\frac{\Delta\alpha_2}{\alpha_2}\right)S_2\left[y_2 + y_1\left(\frac{v_2\alpha_1}{v_1\alpha_2}\right)\right]^{-1}$$

$$\tag{31}$$

and if we assume that the mean polarisability is proportional to the molecular volume, then

$$\left(\frac{n_e^2 - n_o^2}{\bar{n}^2 - 1}\right)_{12} = y_1 \left(\frac{\Delta\alpha_1}{\alpha_1}\right) S_1 + y_2 \left(\frac{\Delta\alpha_2}{\alpha_2}\right) S_2 \qquad (32)$$

It is not possible to obtain component order parameters from refractive index measurements, but we may define an average order parameter for the mixture as

$$S_{12} = \left(\frac{\Delta\alpha}{\alpha}\right)_{12}^{-1} \left(\frac{n_e^2 - n_o^2}{\bar{n}^2 - 1}\right)_{12} \qquad (33)$$

If the polarisabilities of the pure components are unchanged in the mixture, then

$$\left(\frac{\Delta\alpha}{\alpha}\right)_{12} = y_1 \left(\frac{\Delta\alpha_1}{\alpha_1}\right) + y_2 \left(\frac{\Delta\alpha_2}{\alpha_2}\right) \qquad (34)$$

and the mixture order parameter is related to the component order parameters by

$$S_{12} = \left[y_1 \left(\frac{\Delta\alpha_1}{\alpha_1}\right) S_1 + y_2 \left(\frac{\Delta\alpha_2}{\alpha_2}\right) S_2 \right] \left(\frac{\Delta\alpha}{\alpha}\right)_{12}^{-1} \qquad (35)$$

For an ideal mixture in which the component order parameters are equal, the quantity $\left(\dfrac{n_e^2 - n_o^2}{\bar{n}^2 - 1}\right)_{12}$ is a linear function of volume fraction.

The phase diagram for the binary mixture of 44' n-pentyl cyanobiphenyl and 44' n-octyloxy cyanobiphenyl is shown in Fig. 7. The nematic-isotropic transition temperatures are approximately a linear function of mole fraction, but the birefringences (Fig. 8)

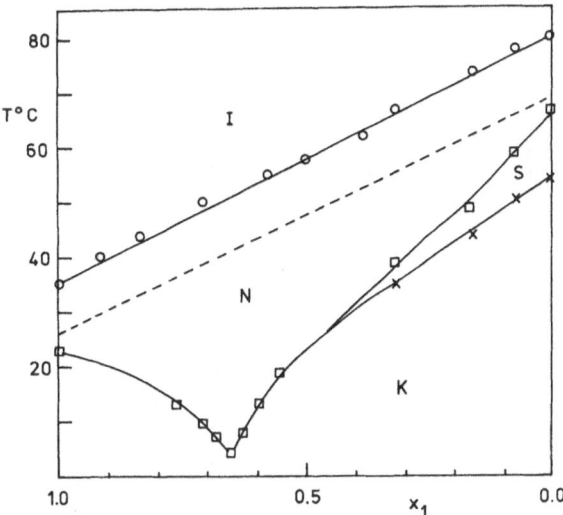

Fig. 7.   Phase diagram for the system 44' n-pentyl cyanobiphenyl
(mole fraction x_1) and 44' n-octyloxy cyanobiphenyl (I).
The dashed line corresponds to a constant reduced
temperature of 0.97.

of the mixtures at a constant reduced temperature of 0.97 lie consistently below the straight line joining the birefringences of the
pure components. Also plotted in Fig. 8 are the mixture order
parameters calculated using eq. (33). These show a significant
negative deviation from a linear dependence on composition. A
similar behaviour can be obtained from our theoretical calculations, as is shown in Fig. 4b. Although theory predicts that the
mixture order parameter and birefringence are functions of volume
fraction, we have plotted our results against mole fraction. This
is because the composition axis in phase diagrams is usually mole
fraction, and furthermore we do not have all the experimental data
on mixture densities to calculate volume fractions. For the
systems studied, the maximum difference between mole fraction and
volume fraction was 10%.

Mixtures of 44' n-pentyl cyanobiphenyl with 4 n-pentyl phenyl
4' n-pentyloxybenzoate have a smectic A phase, which is absent in

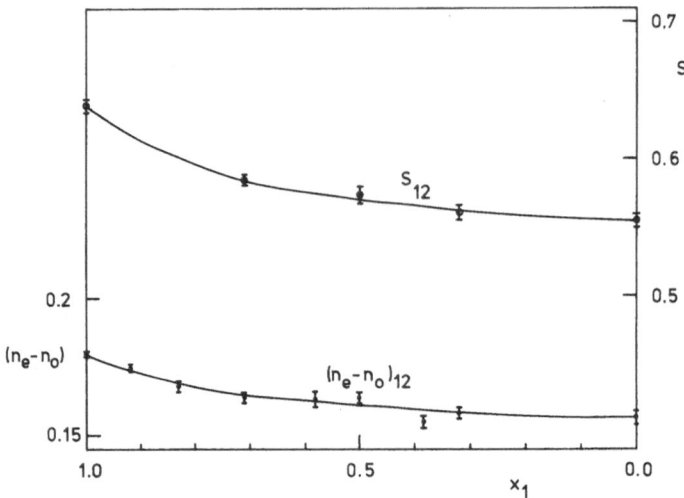

Fig. 8.   Birefringence ($n_e - n_o$) and mixture order parameter S_{12}
          for system I.

the pure components. The phase diagram for this system is shown in
Fig. 9, and birefringences of mixtures at constant reduced temperature of 0.98 are plotted in Fig. 10 along with the mixture order
parameters. We were able to locate the two-phase (nematic-isotropic)
region for this system, and its extent was approximately 0.2°C. At
the reduced temperature of 0.98 the system is smectic for the composition range X=0.4 to 0.7, and both the birefringence and order
parameter deviate substantially from ideal behaviour in the adjacent
nematic regions. 44' n-pentyl cyanobiphenyl has a relatively high
birefringence which is disproportionately reduced by addition of
the low birefringence benzoate ester. Values of the birefringence
in the smectic region were obtained by extrapolation of measurements
in the nematic phase, and are therefore subject to considerable uncertainty. Our derived results for the mixture order parameters
suggest that there is considerable disruption of order in the nematic region of the binary system. As the injected smectic phase is
approached there appears to be a significant increase in the mixture order parameter, which eventually stablizes the smectic phase.

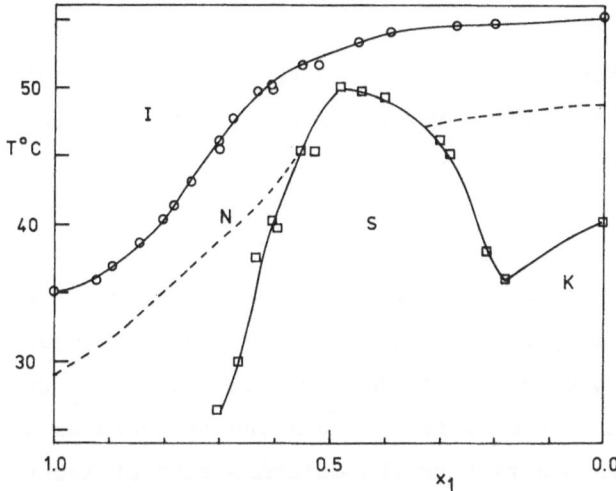

Fig. 9.   Phase diagram for the system 44' n-pentyl cyanobiphenyl
(mole fraction x_1) and 4 n-pentyl phenyl 4' n-pentyloxy
benzoate (II). The dashed line corresponds to a constant
reduced temperature of 0.98.

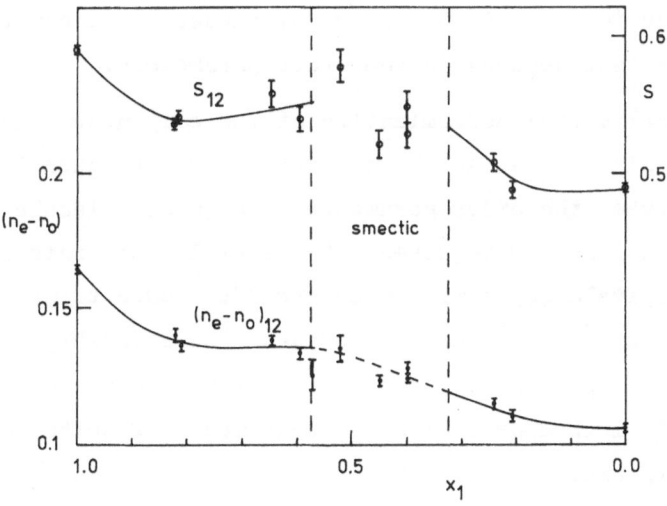

Fig. 10.   Birefringence $(n_e - n_o)$ and mixture order parameter S_{12}
for system II.

DISCUSSION

The application of mean field theory to binary mixtures of
nematics outlined above predicts that significant deviations from a
linear dependence nematic-isotropic transition temperatures on mole
fraction can be obtained. A two-phase nematic-isotropic region is
also predicted, and under certain conditions azeotropic behaviour
can arise. In our calculations it is assumed that the degree of
order in pure components is described by the M-S theory, and deviations from this behaviour in mixtures depend on the relative values
of the transition temperatures and molecular volumes of the two
components. In addition to the determination of isotropic-nematic
coexistence curves, the theory also yields the separate order parameters of the components in a mixture. These may be greater or less
than the M-S values for pure components, and the volume averaged
mixture order parameter also differs from the M-S value. This result
is of considerable experimental significance, since it suggests that
even for mixtures of mesogens which separately obey predictions of
M-S theory, there may be deviations from a linear dependence on
mole fraction for any mixture property (measured at constant reduced
temperature) that depends on the order parameters.

The experimental determination of the component order parameters in a mixture has not been carried out. Frequently the relationship between the order parameter of a guest molecule and a
liquid crystal host is required. For example, dichroic measurements
on liquid crystals doped with dyes are often used to probe the host
order parameter[15]. Our theory provides a relationship between the
component order parameters, and so it may be useful in providing a
more reliable method of analysing experimental results on two
component systems.

We have extended the relation between order parameters and
refractive indices in pure nematics to binary mixtures assuming the

Vuks internal field. The quantity $\left(\dfrac{n_e^2 - n_o^2}{\bar{n}^2 - 1}\right)$ is shown to be the volume average of the weighted component order parameters (eq. 31).

A detailed comparison of experimental results with the theory is difficult because of the assumption that the pure components of the binary mixtures separately obey predictions of the M-S theory. In practice this assumption is not fulfilled, and experimental deviations from 'ideal' behaviour are usually greater than those predicted by the theory. Experimental results for the two systems reported in this paper indicate that the birefringence of mixtures is lower than would be predicted from the birefringences of the pure components, as has been observed by other authors for different systems[16]. Such behaviour can be attributed to the reduction in the volume averaged order parameter for the mixture. Although the ratio of molar volumes of the systems studied experimentally is close to unity, it is possible that molecular association in one component[17] gives rise to an increased effective volume. As it stands, the theory cannot explain the more complex behaviour that is observed with some systems that exhibit injected smectic phases. Our measurements on one such system indicate that the order has a complex dependence on composition. Further experimental studies can be expected to provide valuable information on the local structure in these systems. The theory can be improved upon by allowing the isotropic part of the pseudopotential in the nematic phase to differ from that in the isotropic phase and also by explicitly including the effects of excluded volume. Inclusion of a term to allow for spatial order may successfully explain the formation of injected smectic phases.

ACKNOWLEDGEMENTS

We are grateful to the United Kingdom Ministry of Defense for the award of a Research Contract during the tenure of which this work was completed. Financial assistance from the U. K. Science and Engineering Research Council is also gratefully acknowledged. We would like to thank Dr. E. P. Raynes (R.S.R.E., Malvern, U.K.) for providing the phase diagram (Fig. 7), and also for many valuable discussions.

REFERENCES

1. H. Kelker and R. Hatz, <u>Handbook of Liquid Crystals</u>, Verlag Chemie, Weinheim, 1980.

2. M. A. Cotter, J. Chem. Phys. <u>66</u>, 1098 (1977).

3. M. A. Cotter and D. E. Martire, Mol. Cryst. Liq. Cryst. <u>7</u>, 295 (1969).

4. R. L. Humphries, P. G. James and G. R. Luckhurst, Symp. Faraday Soc. <u>5</u>, 107 (1971).

5. W. Maier and A. Saupe, Z. Naturforsch. <u>14a</u>, 882 (1959).

6. R. L. Humphries and G. R. Luckhurst, Chem. Phys. Lett. <u>23</u>, 567 (1973).

7. D. E. Martire, G. A. Oweimreen, G. I. Agren, S. G. Ryan and H. T. Peterson, J. Chem. Phys. <u>64</u>, 1456 (1976).

8. I. Haller, H. A. Huggins, R. Lilienthal and T. R. McGuire, J. Phys. Chem. <u>77</u>, 950 (1973).

9. P. Palffy-Muhoray and D. A. Balzarini, Can. J. Phys. <u>59</u>, 515 (1981).

10. M. A. Cotter, Mol. Cryst. Liq. Cryst. <u>39</u>, 173 (1977).

11. <u>Introduction to Liquid Crystals</u>, Ed. E. B. Priestley, P. W. Wojtowicz and P. Sheng, Plenum Press, New York, 1974.

12. <u>The Molecular Physics of Liquid Crystals</u>, Ed. G. R. Luckhurst and G. W. Grey, Academic Press, New York, 1977, p. 252.

13. R. L. Humphries and G. R. Luckhurst, Proc. Roy. Soc., <u>A352</u>, 41 (1976).

14. R. J. Cox, J. F. Johnson, A. C. Griffin and N. W. Buckley, Mol. Cryst. Liq. Cryst. <u>69</u>, 293 (1981).

15. M. A. Osman, L. Pietronero, T. J. Scheffer and H. R. Zeller,
         J. Chem. Phys. 74, 5377 (1981).

16. S. Denprayoonwong, P. Limcharoen, O. Phaovibul and I. M. Tang,
         Mol. Cryst. Liq. Cryst. 69, 313 (1981).

17. Hp. Schad and M. A. Osman, J. Chem. Phys. 75, 880 (1981).

THE EFFECT OF THE TRICRITICAL REGION ON THE SMECTIC A -

SMECTIC C TRANSITION

C. C. Huang and J. M. Viner

School of Physics and Astronomy
University of Minnesota
Minneapolis, MN 55455

ABSTRACT

So far all the available heat capacity data on the smectic A -
smectic C (SmA-SmC) transition of various liquid crystal compounds
show a mean-field jump at the transition temperature T_c with no
critical fluctuations above T_c. A Landau free energy including a
ψ^6 term has been proposed by the authors. The singular part of the
heat capacity derived from this free energy gives an excellent fit
to our heat capacity data of one liquid crystal compound. Our result suggests that the mean-field-like SmA-SmC transition is very
close to a mean-field tricritical point. Here we will explicitly
demonstrate that the crossover behavior between the mean-field and
the tricritical region is one of the major sources of discrepancy
among measured critical exponents associated with the SmC order
parameter.

INTRODUCTION

Since de Gennes[1] proposed the natures of the order in two important smectic phases (i.e., SmA and SmC) and suggested that the
nematic - smectic A (N-SmA) and SmA-SmC phase transitions may be

continuous and have helium-like critical exponents, there has been
much experimental and theoretical effort put into this subject.
Thus far the physical properties of the ordered phases have been
understood reasonably well from both light and x-ray scattering
studies.[2] On the contrary, our understanding of the current experimental situation for the N-SmA transition is confusing. Specifically there is still conflicting evidence concerning the values of
critical exponents and the interplay of two critical correlation
lengths.

Experimentally, the preparation of a well aligned and defect-
free SmC sample is somewhat more difficult than a SmA sample. Hence
there are less data on the SmA-SmC transition than on the N-SmA transition and, as with the N-SmA transition, the evidence which does
exist concerning the nature of the SmA-SmC transition is somewhat
conflicting also. Most measurements have been carried out on the
pretransitional phenomena of the tilt angle, the susceptibility and
the birefringence. The reported exponents vary from being mean-field
to helium-like or having values in between (Table I).

Our recent heat capacity study[12] on the SmA-SmC transition of
racemic 4-(2'-methyl-butyl)phenyl 4'-n-nonyloxybiphenyl-4-carboxy-
late (2M4P9OBC) (Fig. 1) has revealed an important property as to
the nature of the SmA-SmC transition. One can easily recognize
three prominent features associated with this heat capacity anomaly
(C_p). Firstly C_p is a linear function of temperature over a wide
temperature range from 136°C to 144°C above the transition temperature. Secondly there exists a very sharp jump in the high temperature side of the heat capacity anomaly. Actually the width between
90% and 10% of the heat capacity jump is 9×10^{-4} in reduced temperature which is very sharp. Finally, the full width at half height
of the heat capacity anomaly is very small also. Expressed in the
reduced temeprature scale, the width (t_o) is about 5.5×10^{-3}. The
first two features indicate that this SmA-SmC transition is mean-

TABLE I

Critical Exponents β and γ of SmA–SmC Transitions
in Various Liquid Crystal Compounds

Compound	β	γ	Ref.
A	0.40		3
A	0.34		4
B		1.0	5
A	0.55		6
A	0.51		7
C		1.28	8
D	0.47		9
B	0.36	1.3	10
A	0.32		11
Mean Field	0.5	1.0	
T.C.P.+	0.25	1.0	
XY–Model	0.35	1.3	

A: terephthal-bis-(4n)-butylaniline (TBBA)
B: undecylazoxymethylcinnamate
C: p-nonyloxy benzoate-p-butyloxy phenol
D: 4-n-pentyl-phenylthiol-4'-n-octyloxybenzoate (8̄S5)
+: Mean-field Tricritical Point

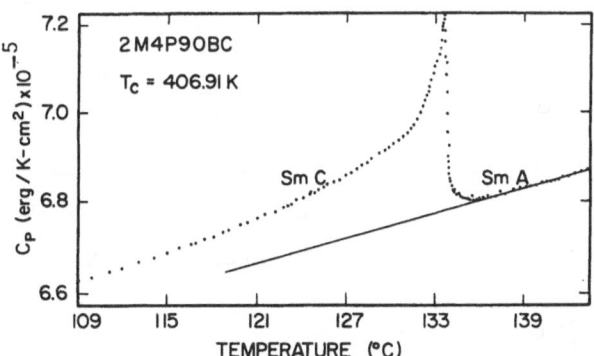

Fig. 1.   The total heat capacity per unit area (C$_p$) as a function
of temperature near the SmA–SmC transition of 2M4P9OBC.
The solid line is the best linear fit of the data with
temperature ranging from 135.6°C to 143.4°C. The measured liquid-crystal sample thickness was ∿0.17mm.

field like. However the smallness of t_o suggests that one should go beyond ordinary mean-field theory.

DISCUSSION OF EXTENDED MEAN FIELD MODEL

In the light of our experimental result we[12] suggested that the sixth power of the order parameter should be included in the singular part of the free energy.[13]

$$F_s = t|\Psi|^2 + b|\Psi|^4 + c|\Psi|^6. \tag{1}$$

Here $t = (T-T_c)/T_c$ and T_c is the SmA-SmC transition temperature. $\Psi = \psi e^{i\phi}$ is the SmC order parameter with ψ and ϕ being the tilt and azimuthal angles of the nematic director with respect to the normal of the smectic layer. c is a positive constant. For $b > 0$ ($b < 0$) we have a second-(first-) order transition. For $b = 0$ we have a Landau tricritical point. Thus far most SmA-SmC transitions have been shown by experiments to be continuous. Consequently we are interested in $b > 0$. Minimizing F_s with respect to $|\Psi|$ one obtains

$$(t + 2b|\Psi|^2 + 3c|\Psi|^4) \ |\Psi| \ = 0 \tag{2}$$

Then above and below T_c for a second order transition, $|\Psi|^2$ will be

$$|\Psi|^2 = \begin{cases} 0 & T > T_c & (3a) \\ \\ \dfrac{-b + b(1 + 3t'/t_o)^{1/2}}{3c} & T < T_c & (3b) \end{cases}$$

Here $t' = (T_c-T)/T_c$ and $t_o = b^2/c$. Substituting Eq. (3) into Eq. (1), one can calculate the heat capacity from $\Delta C = -T \ \partial^2 F/\partial T^2$. Then the singular part of the heat capacity becomes

$$\Delta C = \begin{cases} 0 & T > T_c & (4a) \\ \\ \dfrac{T}{2(3c)^{1/2}T_c^{3/2}} \ (T_m-T)^{-1/2} & T < T_c \end{cases}$$

where $T_m = T_c (1 + (t_o/3))$. Equation (4) results in a mean field
heat capacity jump at T_c being equal to $1/(2bT_c)$. Also in the reduced temperature scale the full-width at half height of the $(\Delta C/T)$
vs. T curve is t_o. The heat capacity expression obtained in Eq. (4)
gave an excellent fit to our measured data (Fig. 2). In addition,
from t_o ($=b^2/c$) and the heat capacity jump ($= 1/(2b\ T_c)$) we obtained
4.7 X 10^{-10} cm^3/erg and 3.9 X 10^{-17} cm^6/erg^2 for b and c, respectively. The coefficient of the power law fitting to ΔC for $T < T_c$
allowed us to determine c independently. This led to c equal to
3.8 X 10^{-17} cm^6/erg^2 which is in good agreement with c obtained
formerly. The excellent fit of Eq. (4) to our heat capacity data
with a small t_o means that the SmA-SmC transition is mean-field-
like but is also very close to the Landau tricritical point where
$t_o = 0$, i.e., b = 0. Although there exists several mean-field
models[14-16] for the SmA-SmC transition, all of them address the possibility of a continuous phase transition as well as the likelihood
of helium-like critical behavior. None of the existing theory provides any hint of the smallness in t_o. Our results should stimulate
further investigation as to the microscopic origin of the smallness

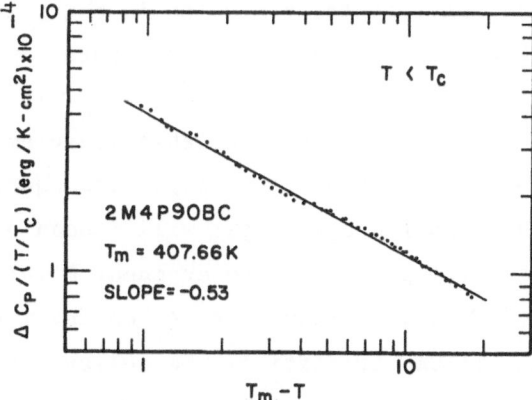

Fig. 2.   Log-log plot of $\Delta C_p (T/T_c)$ vs. $T_m - T$. See the text for
          the definition of T_m. The solid line with slope −0.53 is
          the best fit for fixed T_m.

of t_o. In the following some important consequences of our observation will be discussed.

From Eq. (3b), if $t' \ll t_o$ then $|\Psi|$ is proportional to $(t')^{1/2}$
which represents the critical behavior for a second-order mean-field
transition. But if $t' \gg t_o$ then $|\Psi|$ is proportional to $(t')^{1/4}$ which
is characteristic of a tricritical point. Therefore, the locus of
t_o's will separate the second-order transition region from the
tricritical region in a phase diagram.[17] In Fig. 3 we plot $\log_{10}$
$((1 + 3t'/t_o)^{1/2} - 1)$ vs. $\log_{10}(t'/t_o)$ and a gradual variation of
the slope which corresponds to twice the exponent β in the power
law representative of the order parameter (i.e., $|\Psi| \sim |t|^{\beta}$) is
clear. To examine the effective value of β in a given temperature
range (e.g., two decades in the reduced temperature scale), at
eight representive (t'/t_o) – values we have fitted the calculated
data (represented by solid dots in Fig. 3) one decade above and one
decade below the chosen point to the power law expression.

The effective values of β obtained in this way are also shown
in Fig. 3. The values of β clearly show a continuous variation
from being approximately 1/2 for $t'/t_o \ll 1$ to 1/4 for $t'/t_o \gg 1$. For
$t'/t_o \sim 1$, $\beta = 0.34$ which is the same as the helium-like exponent.
So far experimental heat capacity studies in the vicinity of SmA-SmC
transition have been carried out on several pure compounds[12,18] and
mixtures.[19] All available results show[12,17] $t_o \sim 10^{-3}$ which falls
in the temperature range for which reliable measurements of the
tilt angle can be obtained experimentally. Consequently, a power
law fitting to the measured tilt angle will depend on the temperature range of the fitting and lead to exponent β values ranging
from 0.25 to 0.5. This will result in misleading information as to
the nature of the SmA-SmC transition. We believe that this is the
major reason for the variation in the measured critical exponent β.

Equation (1) also allows us to calculate the susceptibility
explicitly $(\chi^{-1} = \partial^2 F/\partial |\Psi|^2)$.

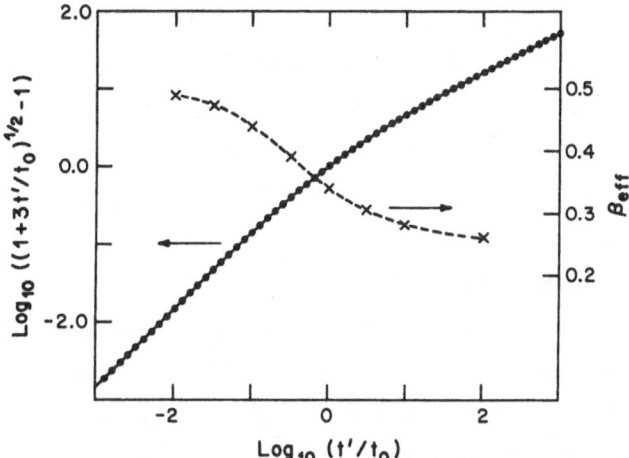

Fig. 3.   The solid line is the curve of $\log_{10}((1 + 3t'/t_o)^{1/2} - 1)$
vs. $\log_{10}(t'/t_o)$. The effective exponents β_{eff} are obtained by a power law fitting to a set of twenty datum-
points (represented by solid dots) in the immediate
neighborhood of the chosen (t'/t_o) -values (indicated by
crosses). The β_{eff} obtained in this way represents the
critical exponent one would get in fitting the order parameter data over two decades in the reduced temperature
scale. The dashed line is a continuous line drawn through
calculated values of β_{eff}.

$$\chi^{-1} = \begin{cases} 2t & T > T_c \\ 8t' + \frac{8}{3}t_o(1 - (1 + 3t'/t_o)^{1/2}) & T < T_c \end{cases}$$

For $T < T_c$, in the ordinary mean-field region $(t' \ll t_o)$ and the tricritical region $(t' \gg t_o)$, χ^{-1} becomes $4t'$ and $8t'$, respectively.
Hence, although χ^{-1} remains the same for both the ordinary mean-
field and tricritical cases in the region $T > T_c$, a clear crossover
occurs in the region $T < T_c$ from ordinary mean-field behavior
$(\chi^{-1} = 4t')$ to tricritical behavior $(\chi^{-1} = 8t')$. Unfortunately, it
is not easy to measure the susceptibility in the SmC phase.

SUMMARY

    From X-ray and light scattering, the bare correlation length
associated with the SmA-SmC transition[20] is found to be about the

same as the effective bare correlation length ($\xi_o = \xi_{o11}^{1/3}\xi_{o1}^{2/3}$)
associated with the N–SmA transition.[21] Here ξ_{o11} and ξ_{o1} are the
longitudinal and transverse bare correlation lengths for the SmA
order parameter with temperature far away from the N–SmA transition. Actually, all the bare correlation lengths are about the
size of a molecule. The distinguishing difference in the measured
critical behavior between the N–SmA and SmA–SmC transition originates again from the smallness in b or t_o. The mean-field heat
capacity jump can be expressed as $1/(2bT_c)$. Then the Ginzburg
criterion $\Delta T_o/T_c = k_B^2(32\pi^2(\Delta C)^2(\xi_o)^6)^{-1}$ will lead to a very small
critical region provided that the heat capacity jump (ΔC) is large
enough. Consequently, the smallness in b is not only responsible
for the closeness of the SmA–SmC transition to a Landau tricritical
point but is also responsible for the smallness of the critical
region for the SmA–SmC transition. However, because of the lack
of microscopic explanation of mean-field behavior for the SmA–SmC
transition with a small value for b, it would be interesting to see
if one can really find a helium-like SmA–SmC transition which is
permitted by the symmetry of the order parameter. Although the
susceptibility in the SmA phase will have critical exponents of 1
and 1.30 for the mean-field and helium cases respectively, experimentally it is not easy to measure the susceptibility.[5,8] Moreover,
in our heat capacity measurement of DOBAMBC[22] near its SmA–chiral
smectic C (SmA–SmC*) transition, we have found that the crossover
region determined by the Ginzburg criterion ($\Delta T_o/T_c$) is about
2×10^{-3}. This crossover between the mean-field and critical
region, similarly, will complicate the procedure of obtaining the
exponent γ from a power law fitting to the experimental data. Actually, Delaye[8] has obtained a distribution of values for exponent γ
ranging from 1.1 to 1.5 with a peak at 1.28 in different measurements of the same compound.

Because of the crossover behavior discussed here, a power-law
fitting to the tilt angle is not reliable for distinguishing a mean

field transition with a small b from a helium-like transition. Consequently, the heat capacity measurement is the best experimental tool to search for a helium-like SmA-SmC transition if there is any.

This work was supported by the U. S. Department of Energy under Contract No. DE-ACO2-79ER10461.

REFERENCES

1. P. G. de Gennes, Mol. Cryst. Liq. Cryst. 21, 49 (1973).

2. R. Schaetzing and J. D. Litster, Advances in Liquid Crystals, edited by G. H. Brown (Academic, New York, 1979), Vol. IV, p. 147.

3. R. A. Wise, D. H. Smith and J. W. Doane, Phys. Rev. A7, 1366 (1973).

4. P. J. Flanders, Appl. Phys. Lett. 28, 571 (1976).

5. M. Delaye and P. Keller, Phys. Rev. Lett. 37, 1065 (1976).

6. S. Meiboom and R. C. Hewitt, Phys. Rev. A15, 2444 (1977).

7. D. Guillon and A. Skoulios, J. Phys. (Paris) 38, 79 (1977).

8. M. Delaye, J. Phys. (Paris) 40, C3-350 (1979).

9. C. R. Safinya, M. Kaplan, J. Als-Nielsen, R. J. Birgeneua, D. Davidov, J. D. Litster, D. L. Johnson and M. E. Neubert, Phys. Rev. B21, 4149 (1980).

10. Y. Galerne, Phys. Rev. A24, 2284 (1981).

11. S. Kumar, Phys. Rev. 23, 3207 (1981).

12. C. C. Huang and J. M. Viner, Phys. Rev. A25, 3385 (1982).

13. According to Professor D. Litster, the validity of this free energy in the vicinity of a SmA-SmC transition has been demonstrated recently by the MIT group (i.e., J. D. Litster, R. J. Birgeneau and C. W. Garland) in light scattering, X-ray scattering and heat capacity studies carried out on one liquid crystal sample.

14. W. McMillan, Phys. Rev. A4, 1921 (1973).

15. A. Wulf, Phys. Rev. A11, 365 (1975).

16.   R. G. Priest, J. Phys. (Paris) $\underline{36}$, 437 (1975).

17.   C. C. Huang and S. C. Lien, Phys. Rev. Lett. $\underline{47}$, 1917 (1981).

18.   C. A. Schantz and D. L. Johnson, Phys. Rev. $\underline{A17}$, 1504 (1978).

19.   R. DeHoff, R. Biggers, D. Brisbin and D. L. Johnson, Phys.
        Rev. $\underline{A25}$, 472 (1982).

20.   J. D. Litster, private communication.

21.   J. D. Litster, et al., J. Phys. (Paris), $\underline{40}$, C3-339 (1979).

22.   S. C. Lien, J. M. Viner and C. C. Huang, submitted to Phys.
        Rev. DOBAMBC refers to p-decyloxybenzylidene-p-amino-2-
        methylbutyl cinnamate.

SPATIAL CORRELATIONS IN NEMATIC LIQUID CRYSTALS*

Chia-Wei Woo and Kean Feng[†]

Department of Physics
University of California, San Diego
La Jolla, CA 92093

and

Ping Sheng

Exxon Research and Engineering Company
Linden, NJ 07036

ABSTRACT

Liquid crystals are liquids. At such densities, spatial
correlations between molecules are expected to play an important
role in the determination of their collective and phase transition
properties. At short range, these correlations are quite anisotropic as a result of anisotropic intermolecular repulsions. We
recall a molecular potential model that takes into account anisotropic forces. We show how spatial correlations are accounted for
in the "orientationally averaged pair correlations" approximation
(OAPC) to affect the macroscopic properties of nematic liquid
crystals. We then perform a cell model calculation in which both
the molecular orientational distribution and the spatial distribution about lattice sites are determined self-consistently, by
solving coupled Euler-Lagrange equations which minimize the free
energy. Numerical work is carried out for a simple potential

without anisotropic terms. We are able to determine the potential
parameters which reproduce earlier OAPC results. Such an exercise
illustrates how the cell approximation works and provides us with a
theoretical framework for introducing anisotropic forces and pair
correlations into later work. It is hoped that the latter would
help remove some of the serious discrepancies between molecular
theories and experiment.

FORMULATION AND MODEL

We choose to describe our molecular theory of liquid crystals
in the following statistical mechanical formulation.

Begin with N molecules confined to a volume V, thus density
$\rho \equiv N/V$, under pressure P and temperature T. Assuming the molecules cylindrically symmetric, each will have its configuration
defined by a center-of-mass position vector $\vec{r}_i \equiv (x_i, y_i, z_i)$ and an
orientation (unit) vector $\hat{\Omega}_i \equiv (\theta_i, \phi_i)$.

Our focus will be on the general probability distribution func-
tion $P_N(1, 2, \ldots, N) \equiv P_N(\vec{r}_1, \hat{\Omega}_1, \vec{r}_2, \hat{\Omega}_2, \ldots, \vec{r}_N, \hat{\Omega}_N)$. If the interaction
between molecules is pairwise, represented by $v(i,j) \equiv v(\vec{r}_i, \hat{\Omega}_i, \vec{r}_j, \hat{\Omega}_j)$,
a functional $\mathfrak{I}$ of P_N can be constructed:

$$\mathfrak{I}\{P_N\} = \mathfrak{I}_0 + \int \left[\sum_{\substack{i<j \\ =1}}^{N} v(i,j) \right] \frac{P_N(1, 2, \ldots, N)}{N!} \, d\vec{r}_1 \, d\hat{\Omega}_1 \, d\vec{r}_2 \, d\hat{\Omega}_2 \ldots d\vec{r}_N \, d\hat{\Omega}_N$$

$$+ kT \int \frac{P_N(1, 2, \ldots, N)}{N!} \, \ell n \, \frac{P_N(1, 2, \ldots, N)}{N!} \, d\vec{r}_1 \, d\hat{\Omega}_1 \, d\vec{r}_2 \, d\hat{\Omega}_2 \ldots d\vec{r}_N \, d\hat{\Omega}_N \, , \quad (1)$$

so that it yields the equilibrium Helmholtz free energy F upon
minimization with respect to P_N. This can be shown readily by
varying P_N under the normalization constraint

$$\int P_N(1, 2, \ldots, N) \, d\vec{r}_1 \, d\hat{\Omega}_1 \, d\vec{r}_2 \, d\hat{\Omega}_2 \ldots d\vec{r}_N \, d\hat{\Omega}_N = N! \qquad (2)$$

The solution of the Euler–Lagrange equation leads immediately to the Boltzmann distribution:

$$P_N^0(1,2,\ldots,N) = \frac{N!}{Z} \exp\left[-\sum_{\substack{i<j \\ =1}}^{N} \frac{v(i,j)}{kT}\right], \tag{3}$$

where

$$Z = \int \exp\left[-\sum_{\substack{i<j \\ =1}}^{N} \frac{v(i,j)}{kT}\right] d\vec{r}_1 d\hat{\Omega}_1 d\vec{r}_2 d\hat{\Omega}_2 \ldots d\vec{r}_N d\hat{\Omega}_N . \tag{4}$$

These formally exact equations are of little practical use. Immediately upon the construction of a model $v(i,j)$, one seeks to retreat from the deluge of information contained in $P_N^0(1,2,\ldots,N)$ and focuses instead on its contractions: n–particle distribution functions $P_N^0(1,2,\ldots,n)$, $n \ll N$, such as the density (n=1) and the pair distribution function (n=2):

$$P_1^0(1) \equiv \rho f(\vec{r}_1,\hat{\Omega}_1) = \frac{N}{Z} \int \exp\left[-\sum_{\substack{i<j \\ =1}}^{N} \frac{v(i,j)}{kT}\right] d\vec{r}_2 d\hat{\Omega}_2 \ldots d\vec{r}_N d\hat{\Omega}_N$$

$$= N \int \frac{P_N^0(1,2,\ldots,N)}{N!} d\vec{r}_2 d\hat{\Omega}_2 \ldots d\vec{r}_N d\hat{\Omega}_N , \tag{5}$$

and

$$P_2^0(1,2) \equiv \rho^2 g(\vec{r}_1,\hat{\Omega}_1,\vec{r}_2,\hat{\Omega}_2) = N(N-1) \int \frac{P_N^0(1,2,\ldots,N)}{N!} d\vec{r}_3 d\hat{\Omega}_3 \ldots d\vec{r}_N d\hat{\Omega}_N . \tag{6}$$

Hierarchies of equations can be derived, by e.g. differentiating $P_n^0(1,2,\ldots,n)$ and expressing the results in increasingly higher-order P_n^0's (larger n's). They are then solved under closure approximations. At the same time, the free energy F can be cluster-expanded in P_n^0 and truncated at low orders. Such procedures are well known in statistical theories of classical liquids.[1]

We have found it more convenient to set up stages of approximations in a different way.[2] We return to the general probability distribution function, $P_N(1,2,\ldots,N)$, and approximate it as products of low-order distribution factors. Then we minimize the free energy functional $\mathcal{F}\{P_N\}$ with respect to these factors. For example, by writing

$$P_N(1, 2, \ldots, N) \approx N! \prod_{i=1}^{N} Q(\hat{\Omega}_i) \tag{7}$$

and minimizing $\mathcal{F}\{P_N\} \to \mathcal{F}\{Q\}$ with respect to Q, we obtain for $Q^0(\hat{\Omega}_1)$ the Maier-Saupe self-consistent equation and the mean field theory. To improve the theory, we go to a higher-level approximation and include spatial correlations, while ignoring their coupling with the orientational order:

$$P_N(1, 2, \ldots, N) \approx N! \left[\prod_{i=1}^{N} Q(\hat{\Omega}_i) \right] \Phi_N(\vec{r}_1, \vec{r}_2, \ldots, \vec{r}_N). \tag{8}$$

This results in a renormalized mean field theory, in which the orienting forces are modified by short-ranged pair distribution functions. The latter must, in turn, be solved for orientationally averaged intermolecular forces. Coupled self-consistent equations which minimize $\mathcal{F}\{P_N\} \to \mathcal{F}\{Q, \Phi\}$ with respect to Q and Φ give rise to what we call the "orientationally averaged pair correlations" theory - OAPC for short.

Now, both of these approximation schemes outlined above can be derived from the conventional formulation involving Eqs. (5) and (6). The mean field theory, Equation (7), is equivalent to solving the first BGKY (Born-Green-Kirkwood-Yvon) equation, with $P_2^0(1,2)$ approximated by $P_1^0(1)P_1^0(2)$. The OAPC theory, Equation (8), is equivalent to simultaneously solving the first and second BGKY equations, with $P_3^0(1,2,3)$ approximated by $P_2^0(1,2)P_2^0(2,3)P_2^0(3,1)/P_1^0(1)P_1^0(2)P_1^0(3)$, and with the second BGKY equation averaged over all orientations. In fact, this was how the OAPC theory was first constructed.[3] The difference between the present derivation and the conventional approach is that with the former we acquire a natural and rigorous method of evaluating the free energy. The cluster expansion procedure, which loses its validity at liquid

densities, is no longer necessary. The optimized Q and Φ, when substituted directly into Eq. (8) and then Eq. (1), yield an F which
is automatically consistent with our chosen level of approximation.

In all our work from 1979 and on, we have employed this new
formulation for our molecular theory of liquid crystals. It will
be the case in this paper as well.

What remains, then, is a specification of the model, i.e., the
pairwise interaction $v(i,j)$.

The interaction between a pair of cylindrically symmetric
molecules can be described by a potential of the form $v(i,j) =$
$v(r_{ij}, \hat{\Omega}_i \cdot \hat{\Omega}_j, \hat{\Omega}_i \cdot \hat{r}_{ij}, \hat{\Omega}_j \cdot \hat{r}_{ij}, \hat{\Omega}_i \times \hat{\Omega}_j \cdot \hat{r}_{ij})$. The last variable indicates chirality, and can be omitted from consideration
unless we deal with cholesterics. It is possible to expand v in
the other variables:

$$v(i,j) = v_0(r_{ij}) + v_2(r_{ij}) P_2(\hat{\Omega}_i \cdot \hat{\Omega}_j) + v_4(r_{ij}) P_4(\hat{\Omega}_i \cdot \hat{\Omega}_j) + \cdots$$

$$+ w_2(r_{ij}) [P_2(\hat{\Omega}_i \cdot \hat{r}_{ij}) + P_2(\hat{\Omega}_j \cdot \hat{r}_{ij})] + \cdots . \tag{9}$$

The first line gives us the familiar Kobayashi-McMillan potential, one that is frequently used by molecular theorists. It, however, does not contain anisotropic forces. The second line couples
spatial and orientational variables and represents the leading
anisotropic term. While the first line alone leads to reasonable
order parameter and volume changes at the clearing point T_{IN}, the
latent heat can be off from experiment by a factor of two. There
are other more serious discrepancies between theoretical results
and experimental data. First, the actual temperature dependence of
the order parameter σ_2 is much stronger than any molecular theory
can predict. Second, for MBBA the order parameter σ_4 turns negative
at temperatures near the clearing point. It is possible to include
higher order terms in the potential and thereby give the $P_4(\hat{\Omega}_i \cdot \hat{\Omega}_j)$

term something to negate, but that always seems to render the
nematic phase unstable.[4] Third, the experimental value of
d ℓn $T_{IN}(\rho)/d$ ℓn ρ] is always several times larger than those calculated. Hard rod models can produce high values for this para-
meter, but then the order parameters or some other property would
fall out of acceptable range.

We have felt that the inclusion of the anisotropic terms –
those in the second line of Eq. (9) – may account for some of these
discrepancies. A full-fledged OAPC formalism was set up and briefly
outlined in Ref. 5. For weak anisotropy, i.e., small $w_2(r)$, a
perturbation calculation was carried out; but no qualitative change
was observed.[5] This is not surprising since discrepancies as large
as those mentioned above cannot be mere perturbative effects. Also,
comparison of potential profiles obtained with Eq. (9) to those
obtained with rod models indicates that $w_2(r)$ can be at least as
strong as $v_2(r)$ even for short, spongy rods.[6] We are thus destined
to carry out numerical calculations for strongly anisotropic
potentials in the OAPC.

There are many parameters to be determined in Eq. (9). For
each set of these potential parameters, there are many _variational_
parameters to be determined in $P_N(1,2,...,N)$, Equation (8). In
order to simulate, say, MBBA, we need to determine all these
parameters self-consistently and show, at the same time, that the
free energies corresponding to these self-consistent solutions
imply a correct clearing temperature and correct order parameter
discontinuities. It is a horrendous numerical task – one that
cannot be accomplished without much preliminary analysis.

This has motivated us to consider a poor man's version of OAPC
– a cell model. Even the latter cannot be carried out without
first obtaining some insight from working with the first line of
Eq. (9). We present here a report on our successful effort to

reproduce the results of Ref. 2 using a cell model and a potential of the form

$$v(i,j) = v_0(r_{ij}) + v_2(r_{ij}) P_2(\hat{\Omega}_i \cdot \hat{\Omega}_j) + v_4(r_{ij}) P_4(\hat{\Omega}_i \cdot \hat{\Omega}_j) , \qquad (10)$$

in which the potential factors $v_{2\ell}(r)$ are softened in order to avoid mathematical divergences. In fact, we use Gaussians:

$$v_{2\ell}(r) = a_{2\ell} \, e^{-(\beta r)^2} , \qquad \ell = 0, 1, 2 . \qquad (11)$$

The coefficients $a_{2\ell}$ are to be fitted to experiment, in a way to be shown later.

SELF-CONSISTENT EQUATIONS

In the cell model, the general probability distribution function $P_N(1,2,\ldots,N)$ is approximated as a product of localized single particle functions. We further separate the spatial and orientational variables. Thus,

$$P_N(1, 2, \ldots, N) \approx N! \prod_{i=1}^{N} [Q(\hat{\Omega}_i) \, \phi(\vec{r}_i - \vec{R}_i)] , \qquad (12)$$

where $\{\vec{R}_i\}$ represents a set of lattice vectors consistent with the given density ρ. We shall take a simple rectangular lattice with lattice spacings d_x, d_y, and d_z so that $d_x d_y d_z = 1/\rho$. In the isotropic model, the molecular orientation is totally decoupled from its position, so $d_x = d_y = d_z = d$: a simple cubic structure. In later work when we introduce the anisotropic potential, $w_2(r_{ij})[P_2(\hat{\Omega}_i \cdot \hat{r}_{ij}) + P_2(\hat{\Omega}_j \cdot \hat{r}_{ij})]$, the preferred lattice structure for the nematic phase would undoubtedly be rectangular; i.e., $d_z > d_x = d_y$ for a director pointing in the z direction. It is probably worth mentioning that this should be the case also for certain orientationally ordered crystals. So, even for a probability function like Eq. (12), there would be coupling between position and orientation if there exist anisotropic forces.

Substituting Eq. (12) into Eq. (1), the free energy functional becomes

$$\mathfrak{F}\{P_N\} \approx \mathfrak{F}\{Q(\hat{\Omega}), \phi(\vec{r} - \vec{R})\}$$

$$= F_0 + \sum_{i<j} \int v(i,j) Q(\hat{\Omega}_i) Q(\hat{\Omega}_j) \prod_\ell \phi(\vec{r}_\ell - \vec{R}_\ell) d\hat{\Omega}_i d\hat{\Omega}_j d\vec{r}_1 d\vec{r}_2 \dots d\vec{r}_N$$

$$+ kT \sum_i \int Q(\hat{\Omega}_i) \ell n\, Q(\hat{\Omega}_i) d\hat{\Omega}_i + kT \int \prod_k \phi(\vec{r}_k - \vec{R}_k) \ell n \prod_\ell \phi(\vec{r}_\ell - \vec{R}_\ell) d\vec{r}_1 d\vec{r}_2 \dots d\vec{r}_N$$

$$= F_0 + \frac{1}{2} \sum_{i \neq j} \int \bar{v}(i,j) \phi(\vec{r}_i - \vec{R}_i) \phi(\vec{r}_j - \vec{R}_j) d\vec{r}_i d\vec{r}_j$$

$$+ kT \sum_i \int Q(\hat{\Omega}_i) \ell n\, Q(\hat{\Omega}_i) d\hat{\Omega}_i + kT \sum_\ell \int \phi(\vec{r}_\ell - \vec{R}_\ell) \ell n\, \phi(\vec{r}_\ell - \vec{R}_\ell) d\vec{r}_\ell \,,$$

$$\tag{13}$$

where

$$\bar{v}(i,j) = \bar{v}(j,i) = \int v(i,j) Q(\hat{\Omega}_i) Q(\hat{\Omega}_j) d\hat{\Omega}_i d\hat{\Omega}_j \quad.$$

$$\tag{14}$$

Note that $\bar{v}(i,j)$ is the same as the orientationally averaged pair potential defined in the OAPC formalism.[2] The normalization constraint, Equation (2), becomes:

$$\int Q(\hat{\Omega}) d\hat{\Omega} = 1 \,,$$

$$\tag{15}$$

and

$$\int \phi(\vec{r} - \vec{R}) d\vec{r} = 1 \quad.$$

$$\tag{16}$$

We next derive the Euler–Lagrange equations for $Q(\hat{\Omega})$ and $\phi(\vec{r} - \vec{R})$. Varying $\mathfrak{F}$ with respect to $\phi(\vec{r} - \vec{R})$ and introducing a Lagrange multiplier α_ϕ gives rise to:

$$0 = \frac{\delta \mathfrak{F}}{\delta \phi(\vec{r}_1 - \vec{R}_1)} = \sum_{i \neq 1} \int \bar{v}(i,1) \phi(\vec{r}_i - \vec{R}_i) d\vec{r}_i + kT[\ell n\, \phi(\vec{r}_1 - \vec{R}_1) + 1] + \alpha_\phi \,,$$

$$\tag{17}$$

and therefore

$$\phi(r_1 - R_1) = \frac{1}{Z_\phi} e^{-\frac{V(1)}{kT}} \,,$$

$$\tag{18}$$

where

$$V(1) = \sum_{i \neq 1} \int \bar{v}(i,1) \phi(\vec{r}_i - \vec{R}_i) d\vec{r}_i \,,$$

$$\tag{19}$$

$$Z_\phi = \int e^{-V(1)/kT} d\vec{r}_1$$

and   (20)

V(1) can be recognized readily as a mean field experienced by molecule 1 - a field that is obtained, first of all, by orientationally
averaging the interaction between molecules 1 and i, then spatially
averaged over all possible positions of molecule i, and finally
summed over all lattice sites $i \neq 1$. For our present model potential
we find, using addition theorem and assuming $Q(\hat{\Omega}) = Q(\hat{\Omega} \cdot \hat{n})$, where
$\hat{n}$ is the director,

$$\bar{v}(i,j) = v_0(r_{ij}) + v_2(r_{ij})\sigma_2^2 + v_4(r_{ij})\sigma_4^2 \equiv \bar{v}(r_{ij}|\sigma_2,\sigma_4) ,$$ (21)

with $$\sigma_{2l} = \int P_{2l}(\hat{\Omega} \cdot \hat{n})Q(\hat{\Omega})d\hat{\Omega} , \quad l = 1, 2,$$ (22)

and thus $$V(1) = \sum_{i \neq 1} \int \bar{v}(r_{i1}|\sigma_2,\sigma_4)\,\phi(\vec{r}_i - \vec{R}_i)d\vec{r}_i$$ (23)

Note that if the pairwise interaction contains anisotropic w_2
terms, $\bar{v}$ would be a function of $\vec{r}_{ij}$ rather than a central potential. This causes much complication. We merely circumvented it in
Ref. 5, and expect to deal with it properly in a paper to follow.

Varying $\mathfrak{J}$ with respect to $Q(\hat{\Omega})$ and introducing another Lagrange
multiplier α_Q gives rise to a second Euler-Lagrange equation:

$$0 = \frac{\delta\mathfrak{J}}{\delta Q(\hat{\Omega}_1)} = \sum_{i \neq 1} \int v(i,1)Q(\hat{\Omega}_i)\phi(\vec{r}_i - \vec{R}_i)\phi(\vec{r}_1 - \vec{R}_1)d\hat{\Omega}_i d\vec{r}_i d\vec{r}_1 + kT[\ln Q(\hat{\Omega}_1) + 1] + \alpha_Q .$$

(24)

For the present model and $Q(\hat{\Omega}) = Q(\hat{\Omega} \cdot \hat{n})$,

$$\int v(i,1)Q(\hat{\Omega}_i)\phi(\vec{r}_i - \vec{R}_i)\phi(\vec{r}_1 - \vec{R}_1)d\hat{\Omega}_i d\vec{r}_i d\vec{r}_1 = \gamma_0 + \gamma_2\sigma_2 P_2(\hat{\Omega}_1 \cdot \hat{n}) + \gamma_4\sigma_4 P_4(\hat{\Omega}_1 \cdot \hat{n}),$$

where   (25)

$$\gamma_{2l} = \int v_{2l}(r_{i1})\phi(\vec{r}_i - \vec{R}_i)\phi(\vec{r}_1 - \vec{R}_1)d\vec{r}_i d\vec{r}_1 \equiv \gamma_{2l}(\vec{R}_i, \vec{R}_1), \quad l = 0, 1, 2 .$$

(26)

Thus

$$Q(\hat{\Omega}_1) = \frac{1}{Z'_Q} e^{-\frac{1}{kT} \sum_{i \neq 1} [\gamma_0(\vec{R}_i, \vec{R}_1) + \sigma_2 \gamma_2(\vec{R}_i, \vec{R}_1) P_2(\hat{\Omega}_1 \cdot \hat{n}) + \sigma_4 \gamma_4(\vec{R}_i, \vec{R}_1) P_4(\hat{\Omega}_1 \cdot \hat{n})]} \qquad (27)$$

where

$$Z'_Q = \int e^{-\frac{1}{kT} \sum_{i \neq 1} [\gamma_0(\vec{R}_i, \vec{R}_1) + \sigma_2 \gamma_2(\vec{R}_i, \vec{R}_1) P_2(\hat{\Omega}_1 \cdot \hat{n}) + \sigma_4 \gamma_4(\vec{R}_i, \vec{R}_1) P_4(\hat{\Omega}_1 \cdot \hat{n})]} \, d\hat{\Omega}_1 . \qquad (28)$$

Equation (27) justifies our assumption that $Q(\hat{\Omega})$ depends only on the angle $\hat{\Omega} \cdot \hat{n}$. This would not be so obvious if anisotropic w_2 terms were included. There would be complicated additional terms in the exponential. One would have to rely on the symmetry of the lattice to prove the uniaxiality of the orientational order, as will be shown in a later paper.

Equations (18) and (27), along with their attending defining equations (20), (21), (22), (23), (26) and (28), form a self-consistency set. An attempt to solve these equations may, for example, begin with a constant $\phi(\vec{r}_1 - \vec{R}_1)$. Equations (26)–(28) give immediately Maier-Saupe's solution for $Q(\hat{\Omega}_1)$. Its substitution into the order parameters $\sigma_{2\ell}$, then the orientationally averaged potential $\bar{v}(i,j)$, and finally the mean field $V(1)$, yields a new spatial distribution function $\phi(\vec{r}_1 - \vec{R}_1)$ [through Eqs. (18) and (20)] which is surely not uniform. This leads naturally to successive iterations until both Q and ϕ converge.

REDUCTION OF EQUATIONS AND THE FREE ENERGY FORMULAS

In principle, one begins with assumed potential factors $v_{2\ell}(r)$. At and near the clearing point T_{IN}, the density has been measured. For example, for MBBA, $\rho = 0.002315$ $Å^{-3}$ at $T_{IN} = 318$ K. Its change across the first order transition is about 0.10–0.15%.[7,8,2] So we shall use the same density value for a range of temperatures about T_{IN}. Actually we are correct to use a fixed density. We deal with

a range of temperatures about T_{IN}^{*} = 317.2 K, the transition temperature for constant <u>density</u>[9,10] instead of constant <u>pressure</u>. (The
Helmholtz free energy that we use calls for fixing ρ and T. A
Maxwell construction must follow to convert it to Gibbs' potential
for P and T. See Refs. 9 and 10 for details.)

Using a simple cubic lattice with spacing d = ρ^{-3}, we may follow
the iterative process outlined in the last section and solve for Q
and ϕ at every temperature T. Free energies for the isotropic and
nematic phases, F_I and F_N, are then to be evaluated with the respective solutions. The theoretical transition temperature T_{IN}^{*} is determined, along with σ_2 and σ_4 at T_{IN}^{*}, when $F_I = F_N$.

There are a couple of practical matters to attend to. First,
the form of $\phi(\vec{r} - \vec{R})$, and next the formulas for F_I and F_N for such
a $\phi(\vec{r} - \vec{R})$.

We feel that for a simple cubic lattice, a spherically symmetric
$\phi(\vec{r} - \vec{R})$ would not be too bad. It is natural to expand $\phi(|\vec{r} - \vec{R}|)$
in a set of orthogonal functions. Since a lattice structure is used
here merely to simulate what should be a liquid structure, quantitative rigor is not high on our list of concerns. We shall take
only one term in the expansion - a Gaussian:

$$\phi(\vec{r} - \vec{R}) \approx \frac{\alpha^3}{\pi^{3/2}} e^{-\alpha^2 |\vec{r} - \vec{R}|^2} \equiv \phi(|\vec{r} - \vec{R}|) \ . \tag{29}$$

The solution for $\phi(\vec{r} - \vec{R})$ thus reduces to the determination of one
single parameter: α. (For a rectangular lattice, ϕ will be ellip-
soidal, and there will be two parameters.)

Other reductions are now possible. In particular, by going
over to relative coordinates

$$\begin{cases} \vec{r}_i + \vec{r}_1 = \vec{P} \ , \\ \vec{r}_i - \vec{r}_1 = \vec{Q} \ . \end{cases} \tag{30}$$

we find:

$$\sum_{i \neq 1} \gamma_{2\ell}(\vec{R}_i, \vec{R}_1) \equiv \sum_{i \neq 1} \int v_{2\ell}(r_{i1}) \phi(|\vec{r}_i - \vec{R}_i|) \phi(|\vec{r}_1 - \vec{R}_1|) d\vec{r}_i d\vec{r}_1$$

$$= \frac{\alpha^6}{\pi^3} \sum_{\{\vec{R}_i\}} \int v_{2\ell}(r_{i1}) e^{-\alpha^2 |\vec{r}_i - \vec{R}_i|^2} e^{-\alpha^2 |\vec{r}_1 - \vec{R}_1|^2} d\vec{r}_i d\vec{r}_1$$

$$= \frac{\alpha^6}{8\pi^3} \sum_{\{\vec{R}_i\}} \int v_{2\ell}(Q) e^{-\frac{\alpha^2}{2} |\vec{P} - (\vec{R}_i + \vec{R}_1)|^2} e^{-\frac{\alpha^2}{2} |\vec{Q} - (\vec{R}_i - \vec{R}_1)|^2} d\vec{P} d\vec{Q}$$

$$= \alpha^3 \left(\frac{1}{2\pi}\right)^{3/2} \sum_{\{\vec{R}_i\}} \int v_{2\ell}(Q) e^{-\frac{\alpha^2}{2} |\vec{Q} - (\vec{R}_i - \vec{R}_1)|^2} d\vec{Q}$$

$$\equiv G_{2\ell} \tag{31}$$

Or, taking Eq. (11) for $v_{2\ell}(Q)$,

$$G_{2\ell} = \frac{4\sqrt{2} \alpha^3 a_{2\ell}}{\pi^{3/2}} \int_0^\infty e^{-\left(\beta^2 + \frac{\alpha^2}{2}\right)(Q_x^2 + Q_y^2 + Q_z^2)} M(Q_x, Q_y, Q_z) dQ_x dQ_y dQ_z, \quad \ell = 0, 1, 2, \tag{32}$$

where

$$M(Q_x, Q_y, Q_z) = e^{-\frac{\alpha^2}{2} d^2} [\cosh(\alpha^2 Q_x d) + \cosh(\alpha^2 Q_y d) + \cosh(\alpha^2 Q_z d)]$$

$$+ 2e^{-\alpha^2 d^2} [\cosh(\alpha^2 Q_x d) \cosh(\alpha^2 Q_y d)$$

$$+ \cosh(\alpha^2 Q_y d) \cosh(\alpha^2 Q_z d) + \cosh(\alpha^2 Q_z d) \cosh(\alpha^2 Q_x d)]$$

$$+ 4e^{-\frac{3}{2}\alpha^2 d^2} [\cosh(\alpha^2 Q_x d) \cosh(\alpha^2 Q_y d) \cosh(\alpha^2 Q_z d)] \tag{33}$$

$$+ \cdots ,$$

in which contributions from three nearest neighbor shells are explicitly displayed.

Equations (27) and (28) now read:

$$Q(\hat{n} \cdot \hat{n}) = \frac{1}{Z_Q} e^{-\frac{1}{kT}[G_2 \sigma_2 P_2(\hat{n} \cdot \hat{n}) + G_4 \sigma_4 P_4(\hat{n} \cdot \hat{n})]} , \tag{34}$$

and
$$Z_Q = \int e^{-\frac{1}{kT}[G_2\sigma_2 P_2(\hat{\Omega}\cdot\hat{n}) + G_4\sigma_4 P_4(\hat{\Omega}\cdot\hat{n})]} d\hat{\Omega} , \tag{35}$$

which can be used to determine (σ_2,σ_4) self-consistently for given (a_2,a_4,β), T, and α.

In a similar manner, we find

$$V(1) = \sum_{i\neq 1} \int \overline{v}(Q|\sigma_2,\sigma_4)\, \phi\left(Q + (\vec{r}_1-\vec{R}_1) + (\vec{R}_1-\vec{R}_i)\right) d\vec{r}_i$$

$$= \frac{2\alpha^3}{\pi^{3/2}} \int_{-\infty}^{\infty} (a_0 + a_2\sigma_2^2 + a_4\sigma_4^2)\, e^{-\beta^2(Q_x^2+Q_y^2+Q_z^2)}\, L(u_x,u_y,u_z,Q_x,Q_y,Q_z)\, dQ_x dQ_y dQ_z ,$$

$$\equiv V(u_x,u_y,u_z) , \tag{36}$$

where

$$L(u_x,u_y,u_z,Q_x,Q_y,Q_z)$$

$$= e^{-\alpha^2[(u_x+Q_x)^2+(u_y+Q_y)^2+(u_z+Q_z)^2]} \times$$

$$\times \left\{ e^{-\alpha^2 d^2}[\cosh(2\alpha^2(u_x+Q_x)d) + \cosh(2\alpha^2(u_y+Q_y)d) + \cosh(2\alpha^2(u_z+Q_z)d)] \right.$$

$$+ 2e^{-2\alpha^2 d^2}[\cosh(2\alpha^2(u_x+Q_x)d)\cosh(2\alpha^2(u_y+Q_y)d)$$

$$+ \cosh(2\alpha^2(u_y+Q_y)d)\cosh(2\alpha^2(u_z+Q_z)d) + \cosh(2\alpha^2(u_z+Q_z)d)\cosh(2\alpha^2(u_x+Q_x)d)]$$

$$+ 4e^{-3\alpha^2 d^2}[\cosh(2\alpha^2(u_x+Q_x)d)\cosh(2\alpha^2(u_y+Q_y)d)\cosh(2\alpha^2(u_z+Q_z)d)]$$

$$\left. + \cdots \right\} , \tag{37}$$

and

$$\vec{u} \equiv \vec{r}_1 - \vec{R}_1 . \tag{38}$$

Since from Eq. (29)

$$\int \phi(|\vec{r}_1 - \vec{R}_1|)(x_1 - X_1)^2 d(\vec{r}_1 - \vec{R}_1) = \frac{1}{2\alpha^2} , \tag{39}$$

we deduce from Eqs. (18) and (36) the relation

$$\frac{1}{2a^2} = \frac{1}{Z_\phi} \int u_x^2 \, e^{-V(u_x, u_y, u_z)/kT} \, du_x \, du_y \, du_z \tag{40}$$

with

$$Z_\phi = \int e^{-V(u_x, u_y, u_z)/kT} \, du_x \, du_y \, du_z \;, \tag{41}$$

which can be used to determine α self-consistently for given (a_0, a_2, a_4, β), (d, T), and (σ_2, σ_4).

Our self-consistent equations thus reduce to the coupled set: Eqs. (34)-(35) with (32) and (22), and Eqs. (40)-(41) with (36)-(37).

Finally, by substituting these results into Eq. (13), we readily obtain free energy formulas for the isotropic and nematic phases:

$$F_I = F_0 - \frac{1}{2} N G_0^I - NkT \, \ell n \, 4\pi - NkT \, \ell n \, Z_\phi^I \;, \tag{42}$$

and

$$F_N = F_0 - \frac{1}{2} N G_0^N - NkT \, \ell n \, Z_Q^N - NkT \, \ell n \, Z_\phi^N - \frac{3}{2} N (G_2^N \sigma_2^2 + G_4^N \sigma_4^2) \;. \tag{43}$$

The superscripts I and N are to remind us that different values of $G_{2\ell}$, Z_Q, and Z_ϕ obtain for the solutions $(\alpha_I, 0, 0)$ and $(\alpha_N, \sigma_2, \sigma_4)$, which correspond to the two possible phases.

ACTUAL CALCULATION AND RESULTS

The scope of our present calculation is rather limited. We merely wish to find a set of potential parameters (a_0, a_2, a_4, β) which would allow us to reproduce in the present model the OAPC results[2]: $T_{IN}^* = 317.2$ K, $\sigma_2(T_{IN}^*) = 0.335$, and $\sigma_4(T_{IN}^*) = 0.052$. In fact we shall allow σ_4 to deviate by a few percent. It is recognized, of course, that our results will not be unique. We shall be content with gaining some familiarity with the new method and a

starting point for future calculations including anisotropic
repulsions.

The calculation begins with the recognition that any pair of
values for

$$
\begin{cases}
\xi_2 = \dfrac{G_2 \sigma_2}{kT} \\[2ex]
\xi_4 = \dfrac{G_4 \sigma_4}{kT} \ .
\end{cases}
\tag{44}
$$

would solve the self-consistent equations (34), (35), and (22).
Simply by evaluating σ_2 and σ_4 with Eq. (22), we identify the
solution (σ_2, σ_4) and the combinations G_2/kT and G_4/kT required for
yielding this solution. They are $G_2/kT = \xi_2/\sigma_2$ and $G_4/kT = \xi_4/\sigma_4$.
Table 1 shows a typical segment from a tabulation of $\xi_2, \xi_4, \sigma_2, \sigma_4$,
and the implied G_2/kT, G_4/kT, and G_2/G_4 values. It is an easy
matter to locate lines from different segments that contain the
desired σ_2 and σ_4. Each such line represents an acceptable set of
potential parameters (a_2, a_4). We shall presently discuss how one
such set will emerge as satisfying our other conditions.

Let us take one such line, say, the one shown in Table 1
marked with an asterick, and T = 317.2 K. Since $G_2/kT = 4.664$ and
$G_4/kT = 2.420$, we have $G_2 = 1,479.4$ kK and $G_4 = 767.6$ kK, with
$G_2/G_4 = 1.927$. Equation (32) implies that for fixed β and d,
$a_2/a_4 = 1.927$ for whatever input choice of $\alpha_N : \alpha_{in}$. Every choice of
α_{in} determines a_2 and a_4 uniquely. All that is missing in Eqs.
(36) is then the knowledge of a_0, which is needed for evaluating an
output $\alpha_N : \alpha_{out}$, with the help of Eq. (40). It is conceivable that
for every a_0 _some_ α_N can be found so that $\alpha_{in} = \alpha_{out}$. We will then
have obtained a model, (a_0, a_2, a_4, β), for which $(\sigma_2, \sigma_4, \alpha_N)$ satisfies
the coupled self-consistency, and $\sigma_2 = 0.335$ and $\sigma_4 = 0.056$ at
T = 317.2 K. Using the same model we can solve for α_I. The only
question that remains is whether the isotropic and nematic phases
would coexist, i.e., whether $F_I = F_N$. In general they would not. In

Table 1.   Solution of the self-consistency equation for orientational order.

$\xi_2 \equiv \dfrac{G_2 \sigma_2}{kT}$	$\xi_4 \equiv \dfrac{G_4 \sigma_4}{kT}$	σ_2	σ_4	$\dfrac{G_2}{kT}$	$\dfrac{G_4}{kT}$	$\dfrac{G_2}{G_4} \equiv \dfrac{a_2}{a_4}$
1.52	0.125	0.327	0.054	4.650	2.325	2.000
	0.130	0.326	0.053	4.656	2.456	1.896
	0.135	0.326	0.052	4.663	2.590	1.800
	0.140	0.326	0.051	4.669	2.729	1.711
	0.145	0.325	0.050	4.676	2.873	1.628
1.54	0.125	0.331	0.056	4.650	2.248	2.068
	0.130	0.331	0.055	4.657	2.374	1.962
	0.135	0.330	0.054	4.663	2.503	1.863
	0.140	0.330	0.053	4.670	2.636	1.772
	0.145	0.329	0.052	4.676	2.773	1.687
1.56	0.125	0.335	0.057	4.651	2.176	2.138
	0.130	0.335	0.057	4.657	2.296	2.029
*	0.135	0.335	0.056	4.664	2.420	1.927
	0.140	0.334	0.055	4.670	2.547	1.834
	0.145	0.334	0.054	4.677	2.678	1.746
1.58	0.125	0.340	0.059	4.651	2.107	2.208
	0.130	0.339	0.058	4.658	2.222	2.096
	0.135	0.339	0.058	4.664	2.341	1.992
	0.140	0.338	0.057	4.671	2.464	1.896
	0.145	0.338	0.056	4.677	2.589	1.806
1.60	0.125	0.344	0.061	4.652	2.042	2.279
	0.130	0.343	0.060	4.659	2.153	2.164
	0.135	0.343	0.060	4.665	2.267	2.058
	0.140	0.343	0.059	4.672	2.384	1.959
	0.145	0.342	0.058	4.678	2.505	1.867

Table 2. Solution of the self-consistency equation for spatial
          distribution.

a_0/a_2	a_0 (kK)	a_2 (kK)	a_4 (kK)	α_N (d^{-1})	α_I (d^{-1})	F_N-F_I (NkK)
1.497	355.9	237.8	123.4	1.534	1.400	-11.67
1.482	357.7	241.3	125.2	1.518	1.387	-6.47
1.472	360.0	244.6	126.9	1.500	1.376	-0.03
1.459	362.1	248.1	128.7	1.487	1.368	6.40
1.449	364.7	251.7	130.6	1.470	1.356	11.86

Table 3. Numerical results at various temperatures.

T (K)	σ_2	σ_4	α_N (d^{-1})	α_I (d^{-1})	F_N-F_I (NkK)
318.7	0.315	0.049	1.50	1.395	2.48
318.2	0.322	0.052	1.50	1.387	1.70
317.7	0.329	0.054	1.50	1.383	0.83
317.2	0.335	0.056	1.50	1.376	-0.03
316.7	0.340	0.058	1.50	1.372	-0.86
316.0	0.348	0.060	1.50	1.366	-1.95
315.4	0.354	0.062	1.50	1.361	-2.95

fact, for an arbitrarily selected line from Table 1, it is unlikely that some value of a_0 <u>must</u> exist so that its corresponding self-consistent α_N and α_I would yield $F_I = F_N$. We had to undergo a systematic search of Table 1 to locate the <u>right</u> line. It was done for a rather restricted choice of $\beta:\beta = 1/d$, one that requires the intermolecular potential to essentially vanish beyond third nearest neighbors.

Table 2 displays the effect of varying a_0 (or a_0/a_2) for the marked line, i.e., for $G_2/G_4 = a_2/a_4 = 1.927$. For each a_0/a_2, both α_N and α_I are listed, as is $F_N - F_I$. The third row approximately satisfies áll our conditions. In summary, at $\rho = 0.002315\text{Å}^{-3}$ or (d = 7.56 Å), the model

$$\begin{cases} a_0 = 360.0 \text{ kK} \\ a_2 = 244.6 \text{ kK} \\ a_4 = 126.9 \text{ kK} \\ \beta = 1/d = 0.132 \text{ Å}^{-1} \end{cases} \qquad (45)$$

would render a self-consistent solution

$$\begin{cases} \sigma_2 = 0.335 \\ \sigma_4 = 0.056 \\ \alpha_N = 1.500/d = 0.198 \text{ Å}^{-1} \\ \alpha_I = 1.376/d = 0.182 \text{ Å}^{-1} \end{cases} \qquad (46)$$

that places the constant-volume clearing temperature T_{IN}^{*} at 317.2 K.

To demonstrate that the last remark is indeed the case, and to obtain the temperature dependence of σ_2 and σ_4, we next take the model defined by Eq. (45) and solve the self-consistent equations at several other temperatures. Remember that the only useful lines are those with $G_2/G_4 = 1.927$, since we have no freedom left in the choice of (a_2, a_4). The results are given in Table 3. Note that the temperature dependence of σ_2 and σ_4 in the nematic range

$T < T_{IN}^{*}$ is very weak – exactly like the OAPC results reported in Ref. 2.

We have not bothered to carry out such calculations for other densities. An attempt was made to reproduce from this model the pair distribution function of Ref. 2. It can be shown from physical reasoning, or from calculating thermal averages of arbitrary two-particle operators, that the (isotropic) radial distribution function g(r) has the following form in our present model:

$$\rho g(r) \approx \frac{1}{4\pi} \int d\hat{\Omega} \left[\sum_{i \neq 1} \int \phi(|\vec{r}_1 - \vec{R}_1|) \, \phi(|\vec{r} + (\vec{r}_1 - \vec{R}_1) + (\vec{R}_1 - \vec{R}_j)|) d\vec{r}_1 \right] . \qquad (47)$$

The result of numerical calculation shows very soft regions of exclusion and mild peaks at neighboring sites: results which are quantitatively different from those of Ref. 2. This is not surprising since our present model potential does not contain repulsive cores.

Returning to the motivations stated in Section I, we have, insofar as possible, reproduced with our new method a model calculation that resembles the OAPC theory. As expected, we have not removed the serious discrepancies between molecular theory and experiments, but a theoretical framework that allows simpler ways for dealing with anisotropic repulsions has been established. We are at present making progress in numerical calculations for the latter, and expect to report our findings soon in a follow-up contribution.

REFERENCES

* Work supported in part by the National Science Foundation through Grant No. DMR-8008816.

† On leave from the Institute of Physics, Chinese Academy of Sciences, Beijing, China.

1. See, for example, T. L. Hill, <u>Introduction to Statistical Thermodynamics</u> (Addison-Wesley, 1960) and <u>Statistical Mechanics</u> (McGraw-Hill, 1956).

2. L. Feijoo, V. T. Rajan, and C.-W. Woo, Phys. Rev. A <u>19</u>, 1263 (1979).

3. V. T. Rajan and C.-W. Woo, Phys. Rev. <u>17</u>, 382 (1978).

4. P. J. Wojtowicz, in <u>Introduction to Liquid Crystals</u> (Plenum Press, 1975), pp. 56-58.

5. J. Shen, L. Lin, L. Yu, and C.-W. Woo, Mol. Cryst. Liq. Cryst. <u>70</u>, 1579 (1981).

6. D. Wu and C.-W. Woo (unpublished).

7. M. J. Press and A. S. Arrott, Phys. Rev. A <u>8</u>, 1459 (1973).

8. E. Gulari and B. Chu, J. Chem. Phys. <u>62</u>, 795 (1975).

9. Y. M. Shih, Y. R. Lin-Liu, and C.-W. Woo, Phys. Rev. A <u>14</u>, 1895 (1976); Y. M. Shih, H. M. Huang, and C.-W. Woo, Mol. Cryst. Liq. Cryst. (Letters) <u>34</u>, 7 (1976).

10. G. K. L. Wong and Y. R. Shen, Phys. Rev. A <u>10</u>, 1277 (1974).

MOLECULAR STRUCTURE AND ORDERING IN LIQUID CRYSTALS

J. Shashidhara Prasad and N. C. Shivaprakash

Department of Physics
University of Mysore
Mysore 570 006, India

ABSTRACT

In an earlier paper[1] we have discussed the variation of order parameter as a function of temperature, and the packing coefficient as a function of chain length for the homologous series of liquid crystals having symmetric and asymmetric molecules. Presently, we have studied the variation of thermal stability as a function of chain length for symmetric and asymmetric homologous series. The thermal stability increases as alkyl chain increases for symmetric homologous series. Clearly, they show exactly similar trend as observed for packing coefficients. We have also discussed the variation of order parameters with chain length at the transition temperatures for symmetric and asymmetric molecules. The order parameters have been theoretically evaluated for the two homologous series 4,4'-di-n-alkyloxyazoxybenzenes(symmetric) and p-(p-ethoxyphenylazo)phenyl alkanoates(asymmetric). The results obtained are in good agreement with the experimental data.

INTRODUCTION

The physical properties of liquid crystals such as thermal

stability, orientational order parameter, optical properties etc.,
play an important role in the design of liquid crystal devices.
These properties are strongly dependent on the molecular structure
viz., central rigid core, flexible end groups, existence of the
dipoles and so on. A proper understanding of these properties as
related to the molecular structures will simplify the choice and
tailoring of materials to suit any requirement.

In the past few decades lot of efforts have been made to investigate a quantitative relationship between the molecular structure
and phase transition behaviour for thermotropic liquid crystals.
It has been observed from several experimental data available in
the literature regarding the thermal stability and the order parameter that they are strongly dependent on the symmetry and asymmetry
of the molecules forming the mesogenic materials. In an earlier
paper[1] we have theoretically confirmed the observed changes in the
gradient of S factors of symmetric and asymmetric molecules. Presently, we have made an attempt to account for the observed variations of thermal stability with chain length for the homologous
series of symmetric and asymmetric molecules and discussed the variation of order parameter with chain length at transition temperature. We have also theoretically evaluated the order parameter for
symmetric (4,4'-di-n-alkyloxyazoxy benzenes) and asymmetric (p-(p-
ethoxyphenylazo)phenyl alkanoates) molecules and compared the results
with the experimental data.

THERMAL STABILITY

The eight homologous series we have studied are listed below:

  I   4-methyl-4'-alkylazoxybenzenes
  II 4-ethyl-4'-alkylazoxybenzenes
  III 4-4'-di-n-alkyloxyazoxybenzenes
  IV 4-4'-di-n-alkylazoxybenzenes
  V   4-methyl-4'-alkylazobenzenes

VI   4-ethyl-4'-alkylazobenzenes

VII   p-(p-ethoxyphenylazo)phenyl alkanoates

VIII   cholesteryl alkanoates.

The first four homologous series forms symmetric molecules and the last four series are asymmetric in nature. In the case of asymmetric molecules the alkyl chain is extended on only one side of the central moiety whereas in symmetric molecules the alkyl chain is extended symmetrically on both sides of the central rigid core. The homologous series I and II show the properties of symmetric molecules even though the alkyl chain is extended on only one side of the central moiety. This can be attributed to the fact that these compounds are composed of two-position isomers having equal probability of existence which in turn does not differentiate the existence of side chains of different length, thereby showing the properties of symmetric molecules.

The thermal stability data have been obtained from the literature[2]. The variations of thermal stability with chain length for the homologous series of symmetric (series I-IV) and asymmetric (series V-VIII) molecules are shown in Figs. 1 and 2 respectively. It is clear from the graphs that the thermal stability (clearing point) increases with chain length for the homologous series of symmetric molecules and the thermal stability (clearing point) decreases with chain length for the homologous series of asymmetric molecules. The trend is not well defined in the case of 4,4'-di-n-alkyloxyazoxy benzenes (Fig. 1(c)). The average thermal stability curve neither exhibits an increasing nor decreasing trend, but remains constant with increase of chain length. Clearly these curves show exactly a similar trend as observed for packing coefficients[1]. Thereby implying that an increase in packing coefficient increases the thermal stability and vice versa. The variations of packing coefficient with chain length are shown in Figs. 3 and 4 for comparison.

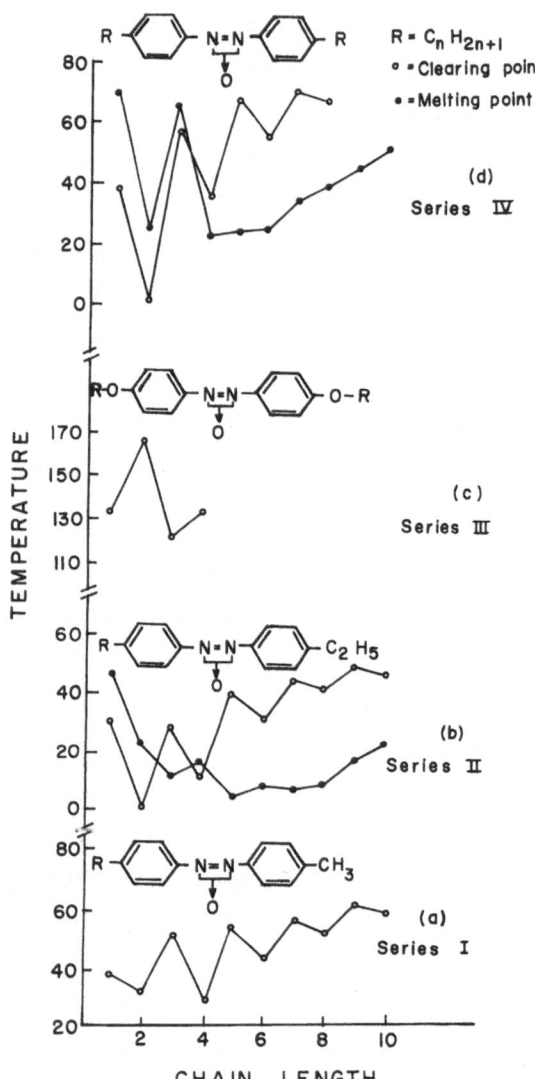

Fig. 1. Phase transition temperatures versus chain length for the
homologous series I to IV.

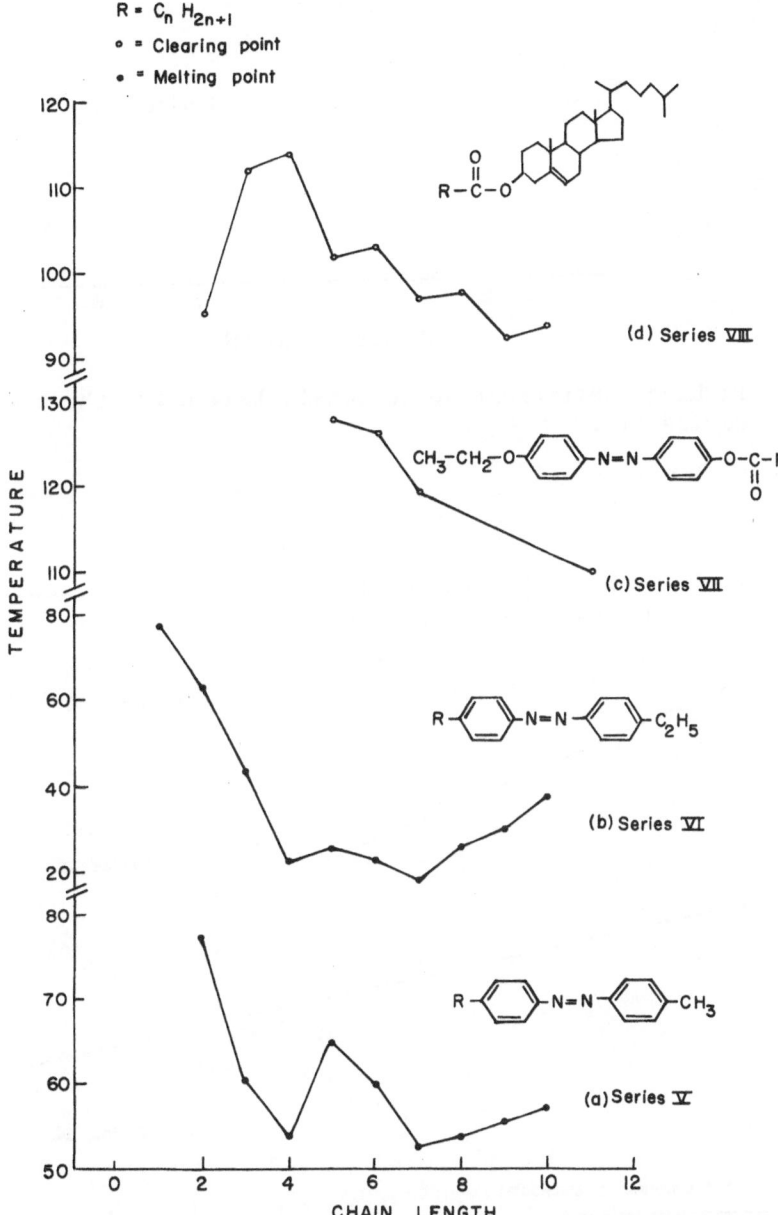

Fig. 2. Phase transition temperatures versus chain length for the homologous series V to VIII.

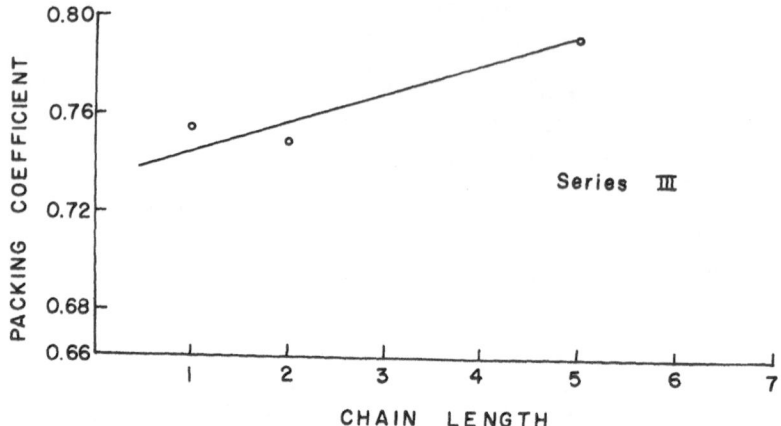

Fig. 3. Packing coefficient versus chain length for the homologous
series III.

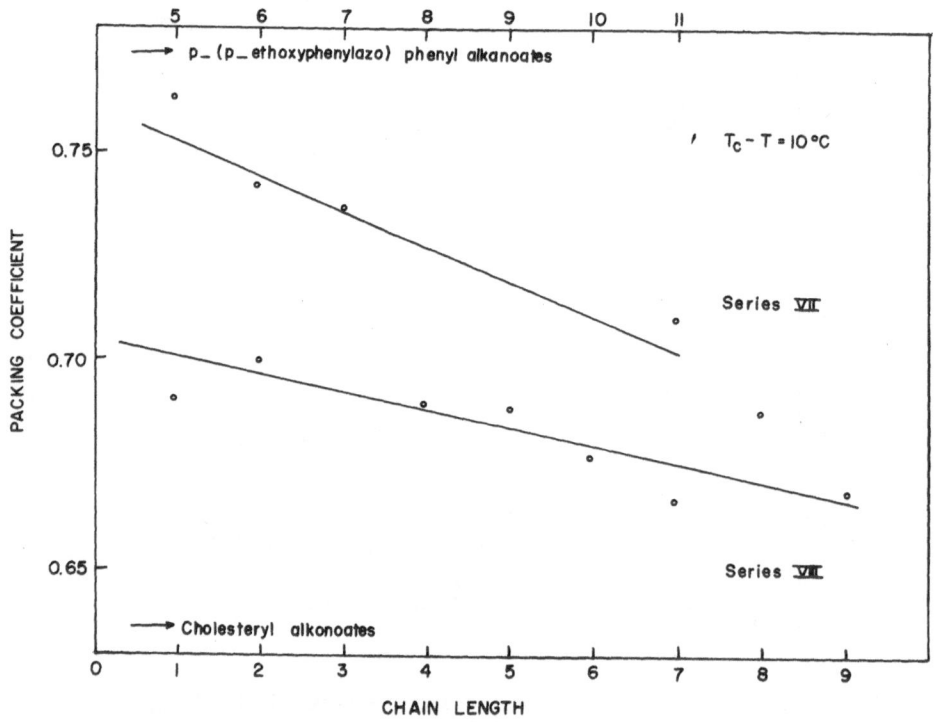

Fig. 4. Packing coefficient versus chain length for the homologous
series VII (upper curve) and series VIII (lower curve).

The variations of melting point with chain length are also plotted in Figs. 1(b), 1(d), 2(a) and 2(b). We observe that there are changes in slopes at some members of the series which indicates the onset of an additional or different mesomorphic phase. In the case of 4-methyl-4'-alkylazobenzenes only compounds with the heptyl and higher substituents were found to show monotropic nematic phase. The remaining compounds did not show this phase. This is quite evident from the change in the slope at the seventh member of the homologous series. Similarly the melting point curve of 4-ethyl-4'-alkylazobenzene series explicitly exhibits the above aspects. Thus we see that the melting point versus chain length curve reveals the onset of mesomorphic state[3] in any series as seen from the homologous series of Figs. 1(b) and 1(d) or the onset of an additional mesophase as seen from the homologous series plotted in Figs. 2(a) and 2(b). This is akin to the change in slope of packing coefficient versus chain length at the point of onset of new or additional phase.[4]

TRANSITION ORDER PARAMETER

The transition order parameter have been obtained from Marcelja[5] for the homologous series III and IV and from Watkins & Johnson[6] and Shivaprakash & Shashidhara Prasad[7] for the series VII. Figures 5 and 6 show the variations of order parameter with chain length at the transition temperature for the homologous series of symmetric (4,4'-di-n-alkyloxyazoxybenzenes and 4,4'-di-n-alkylazoxybenzenes) and asymmetric (p-(p-ethoxyphenylazo)phenyl alkanoates) molecules respectively. The transition order parameter decreases with chain length for symmetric molecules and increases with chain length for asymmetric molecules. This is in contrary to what is observed for the variation of order parameter with chain length at any given temperatures in the liquid crystalline phase. The curves are shown in Figs. 7 and 8 for the above homologous series. This feature can be explained as follows. The thermal stability increases with chain length for symmetric molecules i.e., higher homologs have higher

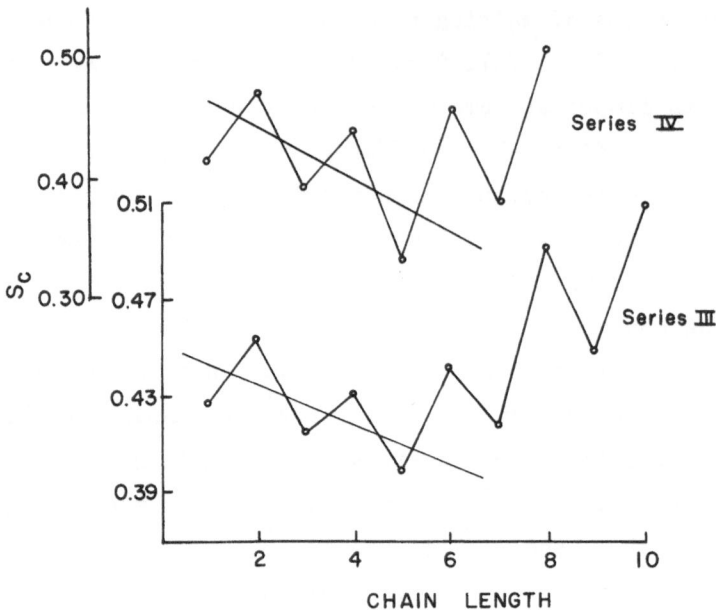

Fig. 5. Transition order parameter versus chain length for the
homologous series III (lower curve) and series IV (upper
curve).

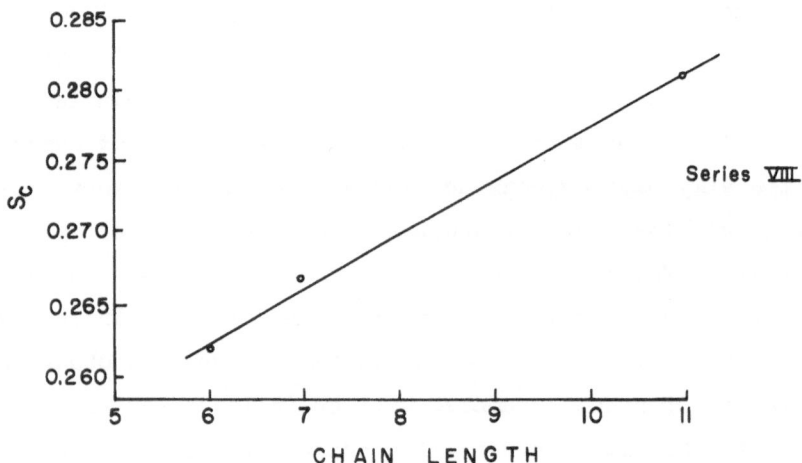

Fig. 6. Transition order parameter versus chain length for the
homologous series VII.

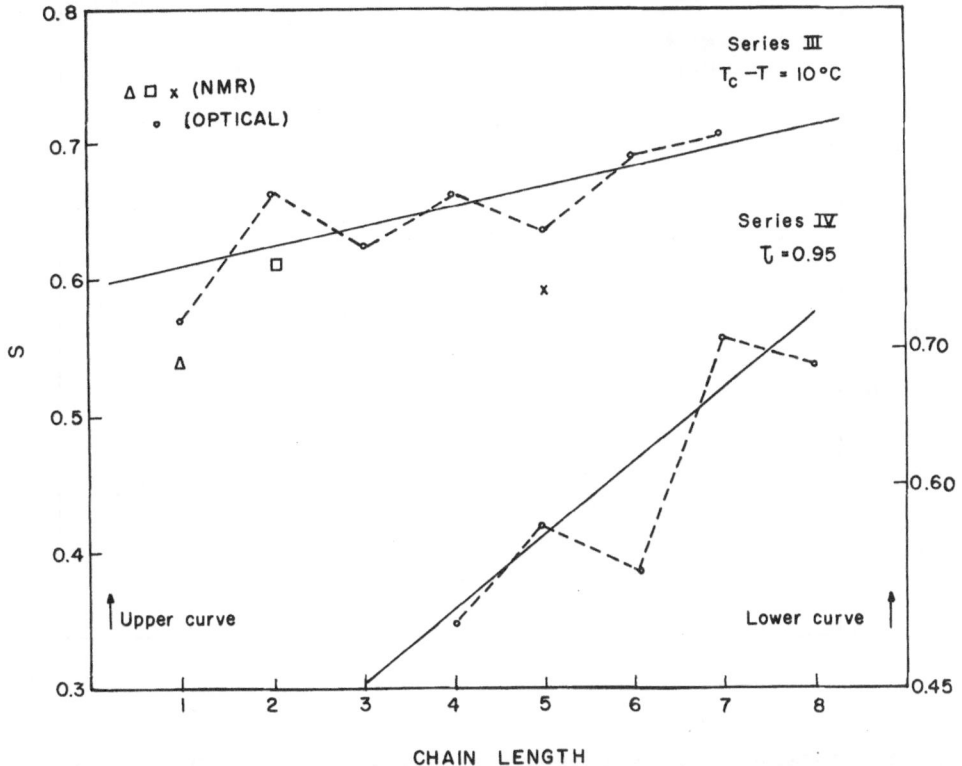

Fig. 7. Order parameter versus chain length for the homologous
series III (upper curve at $T_c - T = 10°C$) and Series IV
(lower curve, at τ_c (reduced temperatures) = 0.95;
$\tau_c = (T/T_{NI})(V/V_{NI})^2$).

transition temperatures. The higher the temperature, higher would
be the thermal energy. Thus higher thermal energy decreases the
transition order parameter for higher homologs. As such the transition order parameter decreases with chain length for symmetric
molecules, but at any other temperatures the order parameter increases with chain length similar to the packing coefficient behaviour. In the case of asymmetric molecules, the thermal energy decreases with chain length. So the lower thermal energies of higher
homologs at the transition results in increase of the transition
order parameter.

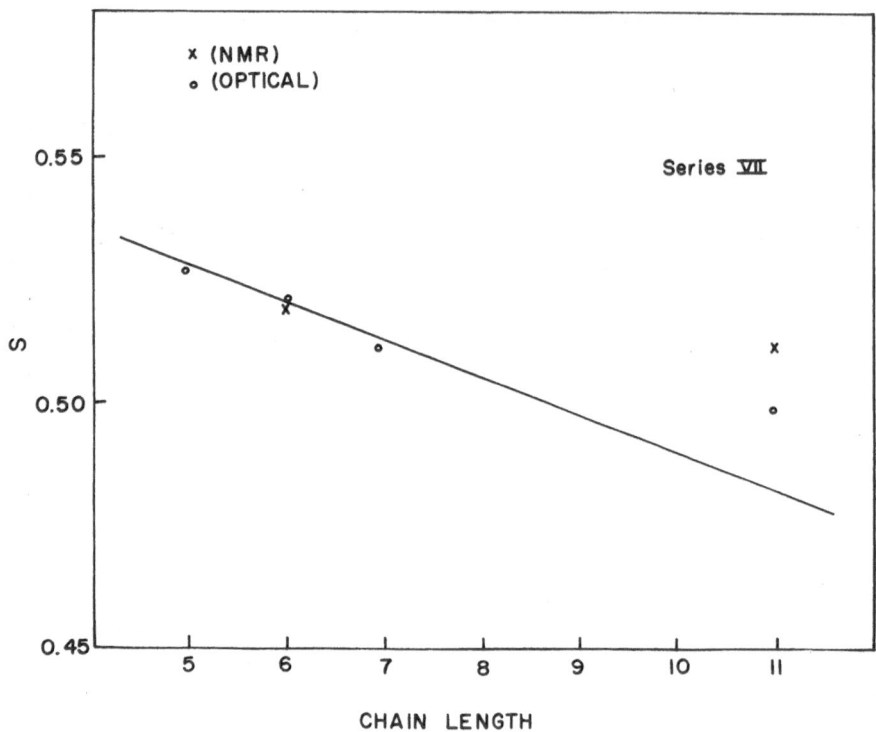

Fig. 8. Order parameter versus chain length for the homologous
series VII (at $T_c - T = 10°C$).

ORDER PARAMETER CALCULATION

The order parameters have been evaluated for the homologous
series 4,4'-di-n-alkyloxyazoxy benzenes (symmetric) and p-(p-
ethoxyphenylazo)phenyl alkanoates (asymmetric) using[1]

$$
S = \frac{\int_0^{\pi/2} P_2(\cos\theta) \exp\{\Gamma SP_2(\cos\theta)\} \sin\theta \, d\theta}{\int_0^{\pi/2} \exp\{\Gamma SP_2(\cos\theta)\} \sin\theta \, d\theta}
$$

where $\Gamma = \dfrac{(a-b)z}{kT} + \dfrac{5\pi}{32} (\Delta v)n$; $P_2(\cos\theta)$ the second order Legendre
polynomial, a the coefficient indicating the strength of the aniso-

tropic part of dispersion, b takes account of strength of the torque which depends on the distance between the centre of gravity of the rigid rod-like molecule and the geometric centre of the molecule, k Boltzmann constant, Z the coordination number of the molecules, n the concentration of molecules, Δv change in the volume. The geometrical data (length and breadth) have been obtained from the crystal structure analysis[8,9].

The calculated S values are plotted in Figs. 9 and 10 for the two homologous series. The theoretical curves are nearer to NMR data[6,7] as seen for the homologous series of Fig. 10 than the optical data[10] as seen for the series of Fig. 9, thereby suggesting that the NMR data is more precise and accurate as compared to the optical data.

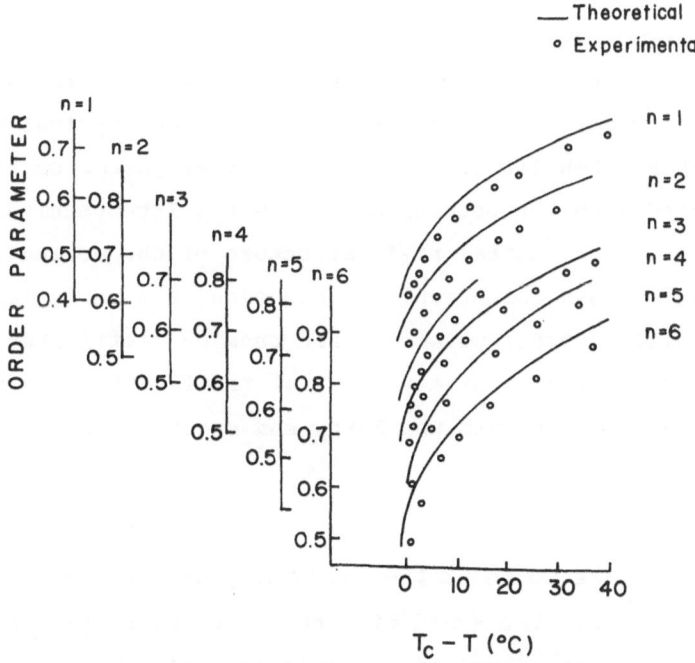

Fig. 9.   Order parrameter versus temperature for the homologous series III.

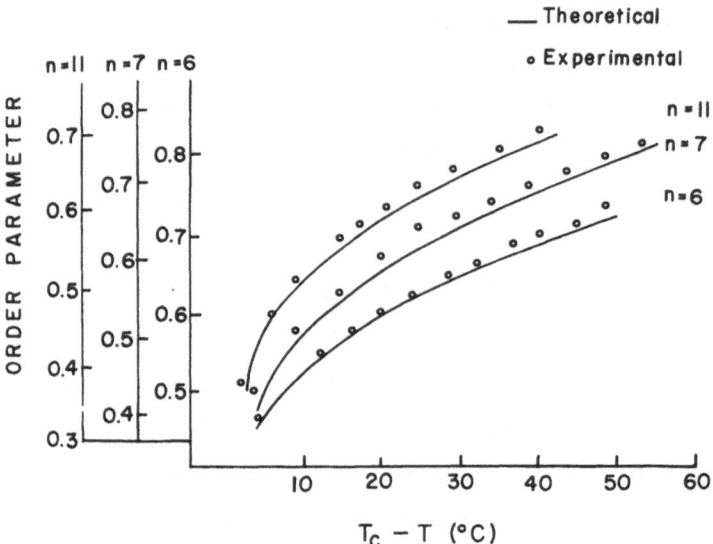

Fig. 10.   Order parameter versus temperature for the homologous
           series VII.

CONCLUSION

     We observe from our studies that the packing coefficient is a
meaningful parameter in describing the micro changes in the physical
properties of mesogenic molecular crystals.   The physical properties
such as thermal stability, orientational order parameter are inti-
mately related with the packing coefficient.   Irrespective of the
contribution of the central rigid structure of the molecule which determines the average anisotropic interaction, the variation of the
thermal stability, order parameter and packing coefficient depend
entirely on the symmetry or asymmetry of the flexible end groups,
which is in conformity with the X-ray and spectroscopic studies[11,12].

ACKNOWLEDGMENTS

     JSP would like to thank the University Grants Commission, India
for a Career award.   The award of a research fellowship to NCS by
the University Grants Commission, India is gratefully acknowledged.

REFERENCES

1. N. C. Shivaprakash and J. Shashidhara Prasad (paper presented
    at the 8th International Conference on Liquid Crystals
    held at Kyoto, Japan during Jun–Jul 1980) Mol. Cryst.
    Liq. Cryst. 74, 1815 (1981).

2. a) R. Dabrowski, K. Kenig, Z. Raszewski, J. Kedzierski and K.
        Sadowska, Mol. Cryst. Liq. 61, 61 (1980).

    b) H. Arnold, Z. Phys. Chem. (Leipzig) 226, 146 (1964).

    c) J. Van der Veen and W. H. de Jeu, J. Phys. Chem. 77, 2153
        (1973).

    d) D. Demus, H. Demus and H. Zaschke, Flussige Kristalle in
        Tableen (VEB Deutscher Verlag fur Grunstoffindustrie,
        Leipzig, 1974).

3. J. Shashidhara Prasad, Mol. Cryst. Liq. Cryst. 47, 115 (1978).

4. N. C. Shivaprakash, P. K. Rajalakshmi and J. Shashidhara
    Prasad, Mol. Cryst. Liq. Cryst. 51, 317 (1979).

    N. C. Shivaprakash, P. K. Rajalakshmi and J. Shashidhara
    Prasad, Mol. Cryst. Liq. Cryst. 60, 153 (1980).

    N. C. Shivaprakash, P. K. Rajalakshmi and J. Shashidhara
    Prasad, Liquid Crystals of one- and two- dimensional
    order, Eds. W. Helfrich and G. Heppke, Springer Series in
    Chemical Physics, Springer, Berlin, Heidelberg, New York,
    11, 72 (1980).

5. S. Marcelja, J. Chem. Phys. 60, 3599 (1974).

6. C. L. Watkins and C. S. Johnson, J. Phys. Chem. 75, 2452
    (1971).

7. N. C. Shivaprakash and J. Shashidhara Prasad, Mol. Cryst. Liq.
    Cryst. (Communicated).

8. J. Shashidhara Prasad, Acta Cryst. 'B' 35, 1404 (1979); 1407
    (1979).

9. A. J. Leadbetter and M. A. Mazid, Mol. Cryst. Liq. Cryst. 51,
    85 (1979).

10.   E. G. Hanson and Y. R. Shen, Mol. Cryst. Liq. Cryst. <u>36</u>, 193
            (1976).

11.   B. M. Craven and G. T. de Titta, JCS Perkin II, 814 (1976).

12.   N. C. Shivaprakash, P. K. Rajalakshmi and J. Shashidhara
            Prasad, Mol. Cryst. Liq. Cryst. <u>60</u>, 319 (1980).

      N. C. Shivaprakash, B. Narasimhamurthy and J. Shashidhara
            Prasad, Mol. Cryst. Liq. Cryst. <u>76</u>, 133, (1981).

NEW LIQUID CRYSTALS, POLYMERIC AND MONOMERIC, DERIVED FROM

BINAPHTHYL

A. J. East and B. C. Benicewicz

Celanese Research Company
Summit, New Jersey 07901

INTRODUCTION

In recent years, a large number of novel liquid crystal polymers have been described and the topic has been reviewed by several
authors[1,2]. One notable series of such mesogenic polymers consists
of stiff rod-like moieties connected by flexible in-chain units.
Examples of this type are the simple polyoxyethylene glycol polyesters derived from p-terphenyl-4,4"-dicarboxylic described by
Fayolle et al.[3], and the polyesterazines described by Roviello and
Sirigu[4], the polyester Schiff bases described by Blumstein et al.[5]
and the azoxydiphenol polyesters from the same group[6].

As part of a broad evaluation of thermotropic polymers at
Celanese Research Company in Summit, New Jersey, we examined homo-
polyesters of aliphatic dicarboxylic acids with the biaryl diol
(I), 6,6'-dihydroxy-2,2'-binaphthyl (DHBN).

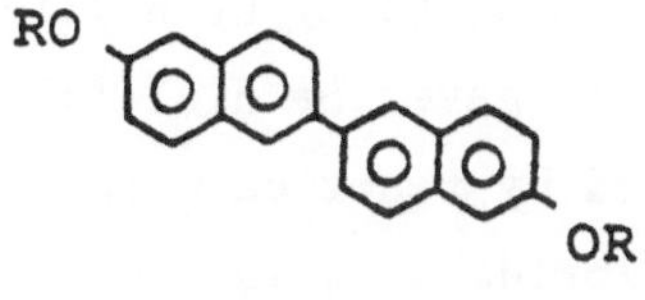

I	R = H
II	R = COCH$_3$
III	R = CH$_3$

We chose to work with the compound (I) because of all the
isomers of di-β-naphthol, this one had the most promise of yielding
thermotropic polymers. This was indeed found to be the case and
one of the raw materials for the polymer studies, the diacetate of
DHBN (II), was itself found to form a nematic liquid crystalline
phase. As a result, an homologous series of straight-chain aliphatic acid diesters of DHBN was prepared ranging from acetate up to
n-octanoate. This was probably an unusual state-of-affairs insofar
as the polymers came first. The liquid crystal properties of both
small molecules and simple homopolyesters could thus be compared.
This paper describes the synthesis and mesophase properties of these
materials. Greater emphasis has been placed upon the monomeric compounds since the complete identification of the polymer mesophases
is more complex and still under investigation.

SYNTHESIS

The biaryl diol (I) was synthesized by coupling 6-methoxy-2-
bromonaphthalene to give the dimethoxybinaphthyl (III). One early
attempt used the Ullman reaction but this was a complete failure.
More successful were coupling reactions based upon the Grignard
reagent derived from the bromomethoxynaphthalene. Several methods
are known involving the use of cupric salts (7) or thallous bromide
(8). The route chosen was that of Yamamato et al. (9) which involves reaction of the aryl halide with magnesium metal in the presence of a catalytic amount of [dipyridyl nickel II] chloride (10)
in tetrahydrofuran solution. The reaction proceeded exothermically
to give a good yield of the dimethoxy biaryl (III), which has been
previously isolated in low yield by reductive coupling of 2-bromo-6-
methoxynaphthalene[11]. Demethylation of (III) by boiling overnight
with hydrobromic-acetic acid mixture gave the diol (I).

The biaryl diol was esterified by reaction with a slight
excess of the appropriate acid chloride or acid anhydride in hot
pyridine. After working up, the crude material was repeatedly

recrystallized from hot anisole or xylene until the Differential
Scanning Calorimetry thermogram was constant. Mass spectroscopy,
infra-red and NMR were used to confirm the structure of the materials and HPLC showed their homogeneity.

The DHBN polyesters were made by a melt acidolysis reaction
between the DHBN diacetate (II) and the appropriate dicarboxylic
acid using a trace of magnesium metal as a catalyst[12]. This method
worked well for acids having eight or more carbon atoms. The C_6
and C_7 diacids gave less satisfactory polymers. The polymerization
temperature was in the 200–300°C range so that the polymer was
formed in its mesophase state. Analogous polymers have more frequently been made by solution reactions[4]. Initially, the melts
were clear, but rapidly became cloudy, then opaque, opalescent and
viscous as the molecular weight increased. A short vacuum cycle
completed the reaction. The higher molecular weight polymers were
fiber-forming.

CHARACTERIZATION OF MATERIALS AND LIQUID CRYSTALLINE PROPERTIES

All the materials investigated in this work were characterized
by Differential Scanning Calorimetry and polarizing microscopy.
Thermal measurements were made on a DuPont 1090 Thermal Analyzer
with a DSC cell in a nitrogen atmosphere at a heating rate of 20°C
per minute. Optical observations were made with a Leitz Ortholux
polarizing microscope equipped with a variable temperature Mettler
FP-5 hot stage apparatus.

The transitions, as recorded by the thermal analyzer, are reported in Table I and plotted as a function of the flexible chain
length in Figure 1. The first two members of the series exhibited
a nematic phase. In the third member of the series a smectic A
phase appeared in addition to the nematic and in the fourth member
there also appeared a smectic E phase. The nematic phase did not
appear in the sixth member or higher homologues and the smectic A

Table I
Transition Temperatures (°C) for
Monomeric Binaphthyl Liquid Crystals

$$C_nH_{2n+1}\overset{\overset{O}{\|}}{C}O\text{—(rings)—}O\overset{\overset{O}{\|}}{C}C_nH_{2n+1}$$

n	k		S_E		S_A		n		i
1	.	246	–	–	–	–	.	304	.
2	.	242	–	–	–	–	.	292	.
3	.	194	–	–	.	237	.	278	.
4	.	136	.	190	.	241	.	256	.
5	.	136	.	178	.	236	.	244	.
6	.	141	.	170	–	–	.	229	.
7	.	130	.	163	–	–	.	225	.

phase melted directly into an isotropic liquid. The phases were identified by observing the textures of the different phases using a polarizing microscope. The higher temperature smectic phase displayed a simple fan-shaped texture characteristic of a smectic A. In some preparations, textures with ellipses were observed. Also, in the higher homologues, slow cooling of the melts into the smectic phase permitted the observation of batonnets before coalescing into a fan-shaped texture.

On cooling the smectic A phase of the fourth and higher homologues, another smectic phase appeared which displayed a fan-shaped texture with concentric arcs. These striations are characteristic of a smectic E texture when formed from a smectic A. A mosaic structure was also observed with a different preparation.

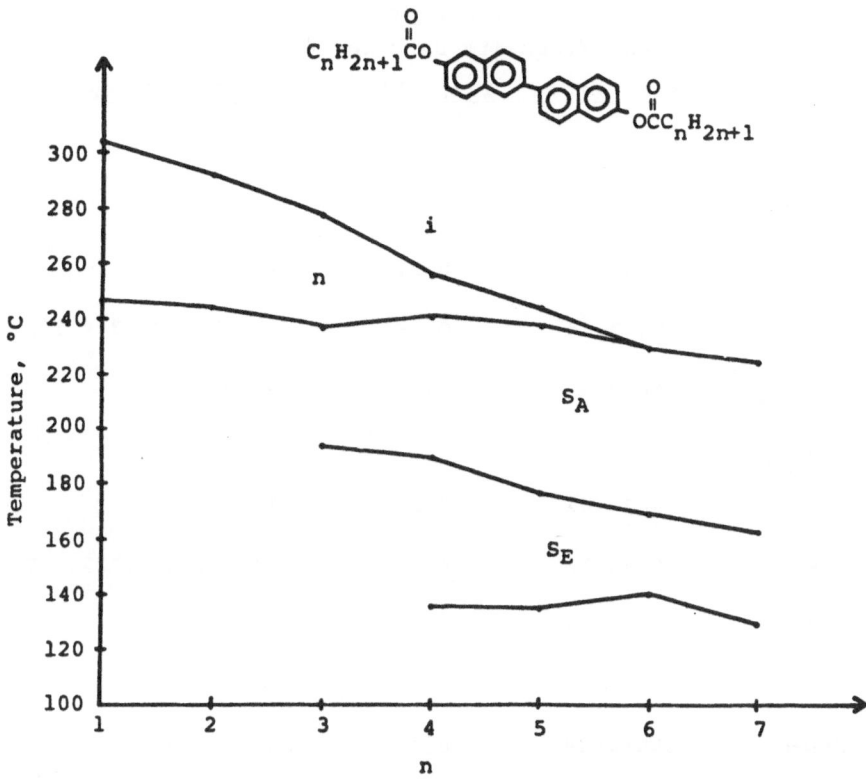

Fig. 1.   Plot of transition temperatures for the monomeric
          binaphthyl series.

The polymers made from aliphatic dicarboxylic acids and the
biaryl diol (I) were much harder to characterize than the low
molecular weight liquid crystal compounds. The data for these
polymers are given in Table II. The thermograms of some polymers,
as recorded by DSC, showed as many as five endothermic peaks which
were reproducible on subsequent heatings. An example is given in
Figure 2 for the polymer made from DHBN and tridecandioic acid.
Identifying the phases by optical microscopy was very difficult
because of the high viscosity of the polymer melts at lower temperatures. However, it was possible to obtain some information from
the microscopic observations. The final melting peak in all the
polymers investigated represented a melting from a nematic phase to

Table II
Thermal and Viscosity Data for DHBN Polymers

Polymer	Ref. No.	IV*, dl/g	DSC Endotherms, °C
DHBN-C_6	29058-10	0.30	305,390
DHBN-C_7	28858-25	0.30	216,393
DHBN-C_8	28858-9	0.90	278,412
DHBN-C_9	28858-14	0.62	202,230,282,344,380
DHBN-C_{10}	28858-6	0.50	257,375
DHBN-C_{11}	-	-	
DHBN-C_{12}	28858-42	0.89	217,326
DHBN-C_{13}	29058-12	0.61	221,237,279,327,337
DHBN-C_{14}	29059-27	1.51	225,312
DHBN-C_{15}	-	-	
DHBN-C_{16}	29059-28	0.83	203,218,302

*I.V.'s were determined at 0.1% concentration in pentafluorophenol.

an isotropic liquid. Therefore, it is possible to assign the transitions solid-nematic and nematic-isotropic to those endotherms for
polymers which only show two melting peaks. The polymer melt at
temperatures just below the nematic-isotropic transition was birefringent and fairly low in viscosity. If a very thin preparation
was made, the melt assumed a texture that is characteristic of low
molecular weight nematic liquid crystals. At lower temperatures,
the viscosity was greater and it was surmised that the high viscosity
prevents the molecules from rearranging and presenting textures similar to low molecular weight liquid crystals. It was not possible,
therefore, to identify the other transitions by the techniques used
in this investigation.

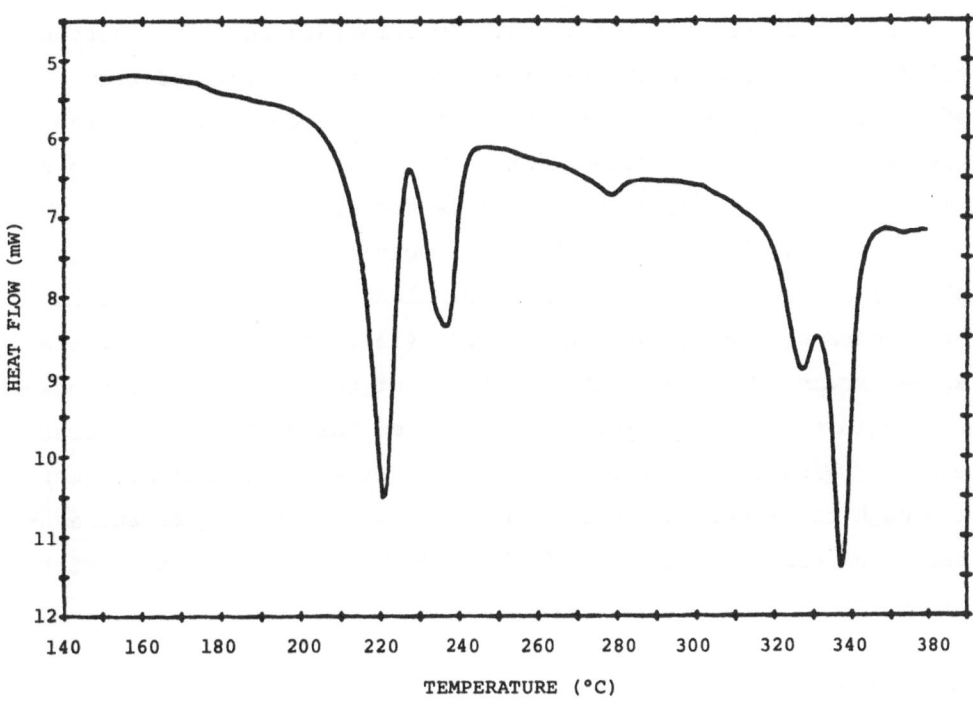

Fig. 2. Thermogram of DHBN-13 polymer showing multiple endotherms.

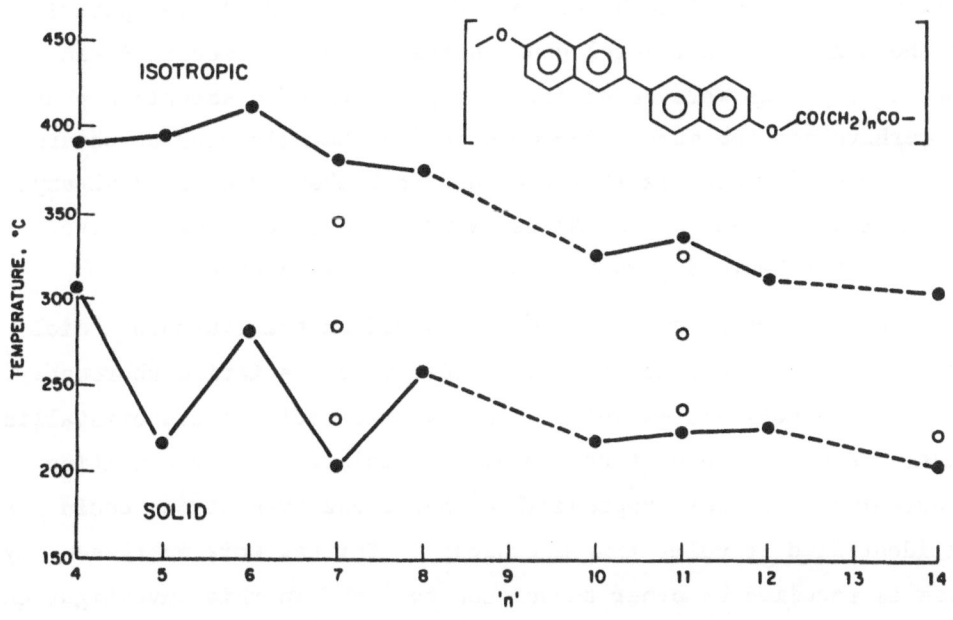

Fig. 3. Transition temperatures (°C) for DHBN polymers.

A plot of the polymer transition temperatures as a function of the flexible chain length is given in Figure 3. The odd-even effect is more pronounced in the solid-mesophase transition than in the nematic-isotropic transition. The anisotropic temperature range also becomes narrower as the flexible spacer increases in length which is typical with these types of polymers. A smectic phase (or phases) apparently forms with longer flexible spacers. This has been noted with other polymer systems (13). The longer flexible spacer seems to allow greater mobility of the rigid rod portions of the chain and, thus, layered alignment of these portions becomes easier. Future work on these polymer systems will include a more thorough characterization by other analytical techniques and synthesis of the DHBN-C_{11} and DHBN-C_{15} polymer when the appropriate dicarboxylic acids becomes available.

CONCLUSIONS

Two new series of liquid crystalline materials have been synthesized and identified based on 6,6'-dihydroxy-2,2'-bi-naphthyl. In the small molecule homologous series, nematic, smectic A and smectic E phases were detected. The presence of a smectic E phase is perhaps not too surprising considering the molecular structure. Many compounds which exhibit the smectic E phase contain aryl-aryl bonded rigid cores. As examples, many 4,4'-biphenyl derivatives and 4,4"-terphenyl derivatives exhibit smectic E phases.

The polymeric homologous series based on the binaphthyl diol were also liquid crystalline but harder to characterize thoroughly. Most of the polymers exhibited, solely, a nematic liquid crystalline phase. However, some of the polymers displayed multiple melting endotherms which were reproducible, but these transitions could not be identified by polarizing microscopy. Further work on these polymers is required by other techniques not used in this investigation such as viscosity measurements and x-ray analysis of oriented fibers.

ACKNOWLEDGEMENTS

The authors wish to thank Mr. Ben Morris for his expert assistance in obtaining the thermal measurements.

REFERENCES

1. R. W. Lenz and K. A. Feichtinger, Polymer Preprints 20(1),
        114-7 (1979).

2. J. Jin, S. Antoun, C. Ober and R. W. Lenz, Br. Polym. J.
        12(3), 132-46 (1980).

3. B. Fayolle, C. Noel, J. Billard, Journal de Physique 40(C3),
        Supplement #4, 485 (1979).

4. A. Roviello and A. Sirigu, J. Polym. Sci., Polym. Lett. 13,
        455 (1975).

5. A. Blumstein, K. N. Sivaramakrishnan, S. B. Clough and R. B.
        Blumstein, Mol. Cryst. Liq. Cryst. 49(8), 255-8 (1979).

6. S. Vilasagar and A. Blumstein, ibid. 56(8), 263-9 (1980).

7. M. S. Kharasch and O. Reimuth, "Grignard Reactions of Nonme-
        tallic Substances" (Constable:London) Chap. 5 (1954).

8. A. McKillop, L. F. Elsom and E. C. Taylor, Tetrahedron 26,
        4041-50 (1970).

9. T. Yamamato, Y. Hayashi and A. Yamamato, Bull. Chem. Soc.
        Japan 51(7), 2091-7 (1978).

10. M. Uchino, K. Asagi, A. Yamamato and S. Ikeda, J. Organo-
        metallic Chem. 84, 93-103 (1975).

11. K. W. Bentley, J. Chem. Soc. 2398-2402 (1955).

12. W. R. Sorensen and T. W. Campbell, "Preparative Methods in
        Polymer Chemistry", Ed. II (Interscience) 1968, p. 150.

13. L. Strzelecki and D. van Luyen, Eur. Polym. J. 16, 299 (1980).

IONENEOMERIC LIQUID CRYSTALS

Li-Ping Yu and Edward T. Samulski

Department of Chemistry and
Institute of Materials Science
University of Connecticut
Storrs, CT 06268 USA

ABSTRACT

A new class of liquid crystals based on salts of the 1,1-di-alkyl-4,4'-bipyridinium core are reported. Mesophase transition temperatures and stability ranges depend on both the nature of the anion and the length of the alkyl chain. Polymeric versions of these mesogens can be synthesized which retain electrochronic properties.

INTRODUCTION

There are examples of amphiphilic molecules (ion pairs) which exhibit thermotropic mesophases, e.g., neat soaps and, more relevant to this work, pyridinium salts.[1] For the most part, such mesogens are not symmetric--the molecules consist of a hydrophobic part and a polar site, usually localized at one extrema of the molecule. Here we explore a new class of symmetric liquid crystalline amphiphiles. They have a molecular shape characteristic of conventional mesogens, i.e., a rigid aromatic core terminated with flexible chains. These amphiphiles are based on bipyridyl (I) whose bipyri-

697

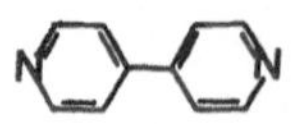

I

dinium salts (1,1--dialkyl-4,4'-bipyridinium dihalides; II) are
known as violagens. The violagens have been previously investigated in dilute aqeous solutions for electrochromic display applications[2,3]. The relevant phenomena is based on the reversable color
change that accompanys the oxidation-reduction reaction shown below.
In this report we examine some of the thermal properties of mesomorphic violagens.

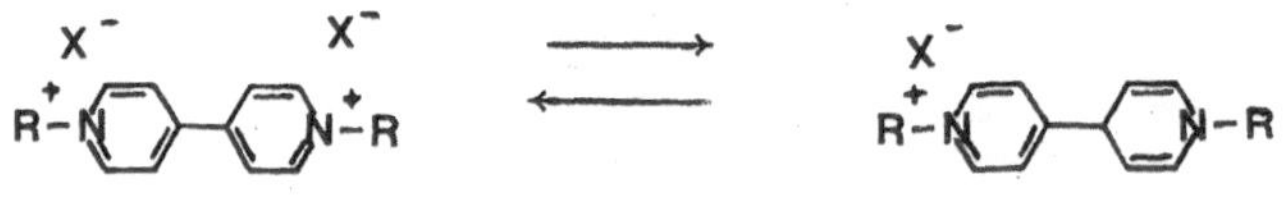

II

EXPERIMENTAL

Synthesis:

A.   1,1'dialkyl-4,4'-bipyridinium dibromide: the alkyl bromide (0.03 mole) was added to a solution of bipyridyl (0.01 mole in
25 ml of N-methyl pyrrolidone). After refluxing (120°C; 24 hrs)
the yellow crystalline solid dialkylated product was filtered from
the reaction mixture (0°C), washed with acetone, and recrystallized
from H_2O/acetone (15/85).

B.   Anion Exchange: The dihydroxide was prepared from the
dibromide prepared in A by adding freshly prepared Ag_2O (.006 mole)
to an aqueous solution of the dibromide (.005 mole), stirring 1 hr.,
and filtering. The filtrate was neutralized with an acid containing
the desired anion and the water removed (low pressure distillation)
to give the solid product which was recrystallized from ethanol/ace-
tone (10/90).

Mesophase Characterization

The mesophase transitions were characterized by DSC (DuPont
990 differential scanning calorimeter), polarizing optical micro-

scopy (Mettler variable temperature stage), and Thermal Optical
Analysis (integrated transmitted light intensity through crossed
polars versus temperature).

RESULTS AND DISCUSSION

The range of mesophase stability of the violagens is sensitive
to both the alkyl chain length and the anion used. For the latter
(for a fixed alkyl chain length) considerable variability was observed even when the anions were restricted to the halogens: $X^-=Cl^-$,
Br^-, and I^- (Fig. 1). Among these compounds thermal reversability
was poor; all three decomposed before exhibiting a stable isotropic
phase. More complex anions showed behavior ranging from an absence
of mesophase formation ($X^-=BF_4^-$) to reversable polymesomorphism
($X^-=CH_3SO_3^-$). The anion-dependent transitions are illustrated in
Fig. 2 for the diheptyl violagen. There is also evidence for solid-
solid transitions and/or super cooling of mesophases. As an example,

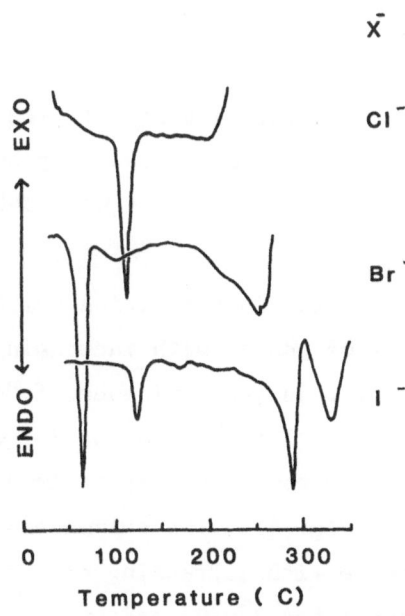

Fig. 1. DSC traces of n-decyl-violagen dihalides.

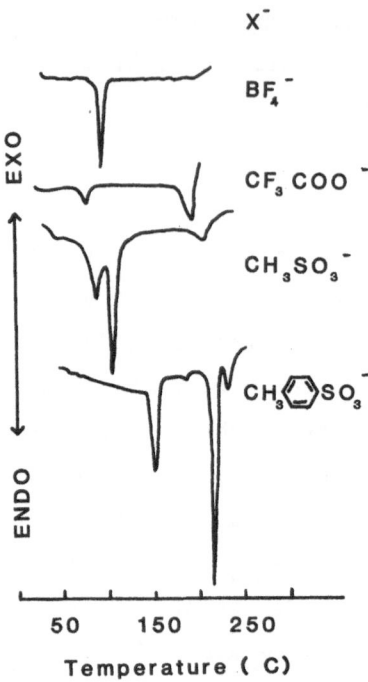

Fig. 2.   DSC traces of n-heptyl violagen salts with various anions
          (X⁻).

example, consider the stearate viologen ($X^-=CH_3SO_3^-$). It exhibits
low temperature transitions that appear only on the first heating;
the higher temperature transitions are well behaved and reproducible
(Fig. 3).

We anticipated changes in transition temperatures and/or
changes in the number of phases with increasing length of the flexible alkylchains. There is some evidence of "even-odd" variations
in the solid-liquid crystal transition as $R=-(CH_2)_x CH_3$ varies from
the pentyl to octyl (n=5 to 8) derivative in violagen series with
the anion $X^-=CH_3SO_3^-$ (Fig. 4). The polymesomorphic nature of these
materials does increase with increasing n; two mesophases are ob-
served for $5 \leq n < 7$, three for $7 \leq n \leq 9$, and evidence of a forth phase in
n=10 derivative.

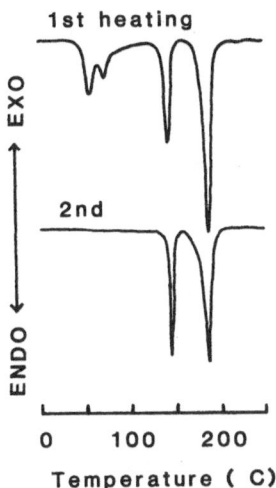

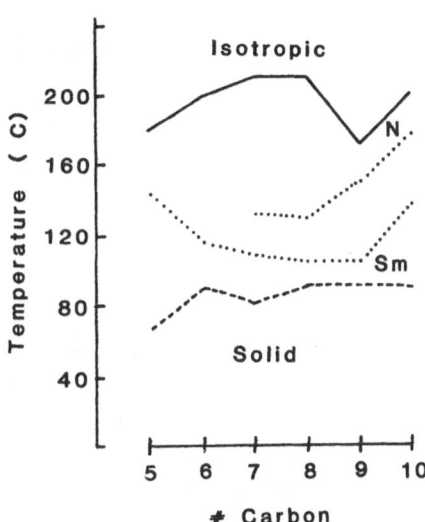

Fig. 3. DSC traces showing
monotropism/supercooling for
the stearate violagen
($X^- = -CH_3SO_3C$).

Fig. 4. Phase diagram derived
from DSC and microscopy
for n-alkyl violagens
as a function of the
number of carbon atoms
in the alkyl chain.

Generally, all of the observed mesophases for these materials
were very viscous by comparison with conventional thermotropics.
Examination of the textures via polarizing optical microscopy suggested that smectic phases were predominant.[4] However, cooling
from the less viscous isotropic state enabled the establishment of
what appeared to be a nematic texture in the higher homologs. The
textures of the high-temperature phases were sometimes further complicated by a simultaneous thermochromic transition, i.e., the pale
yellow melt would change reversably to deep purple at an elevated
temperature. This phenomemon, usually assoicated with a reversable
molecular rearrangment[5], is manifested in the proton NMR spectra of
the neat stearate violagen. Fig. 5 shows the stearate violagen NMR
spectrum as a function of temperature. The low resolution spectra
are characteristic of viscous anisotropic mesophases with incomplete averaging of dipolar couplings. The resolution increases as
one samples the higher temperature mesophases (refer to Fig. 3).

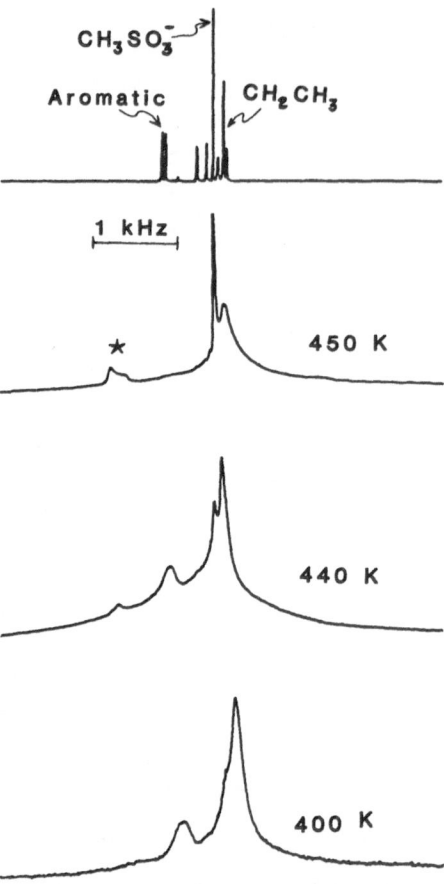

Fig. 5.   Proton NMR spectra of the stearate violagen
           ($X^- = -CH_3SO_3^-$) as a function of temperature. The top
           trace is obtained from a $CDCl_3$ solution of this violagen
           and some resonances are indicated; the remaining spectra
           are in neat phases of the violagen with the thermochromic
           transition accompanied by the appearance of new low-field
           resonances (indicated with *).

Simultaneously with the thermochronic transition, two very de-
shielded resonances appear (contrast with high resolution spectrum
of the isotropic violagen in $CDCl_3$) which undoubtedly reflects a
rearranged molecular structure.

       Lastly, the reaction of (I) with alkyl dihalides results in
ionene-like polymers:

The DSC studies of such polymers yielded ill-resolved and complex transition phenomena indicative of broad molecular weight distributions. There is, however, clear evidence from optical microscopy that the melts are liquid crystalline. And, when these polymers were cast (plasticized with ∿5% ethylene glycol), contiguous flexible films resulted which when sandwiched between conducting transparent electrodes, would change color from translucent yellow to opaque purple on application of a DC current.

During the course of our work on polymeric viologens we realized that there is considerable literature on these and related ionene polymers dealing with their electrical properties when doped with electron acceptors such as TCNQ[6], with their photochromic properties (color changes when exposed to light)[7], and as cationic redox polymers.[8] To our knowledge, however, the propensity of such polymers for exhibiting liquid crystalline phases has not been reported before. Lastly, both the monomeric and polymeric versions of these bipyridinium salts are very water soluble. Moreover, at low concentrations of H_2O (∿15%) birefingent gels are formed by these materials. Hence, it appears that ionenomeric liquid crystals encompass both liquid crystal types--lyotropic and thermotropic. As these materials retain photo-, electro- and thermochromic properties, they could be explored in conjunction with their mesomorphism for potential memory and display applications.

It should be noted that while these ionenomeric liquid crystals can be readily synthesized for further investigations, researchers should be aware that lower homologs are used as herbecides. Hence appropriate caution should be exercised when handeling these materials.

ACKNOWLEDGMENT

This work was supported in part by NIH funds (AM17497).

REFERENCES

1.   G. A. Knight and B. D. Shaw, J. Chem. Soc. 682 (1938).

2.   C. J. Schoot, J. J. Ponjee, H. T. van Dam, R. A. van Doorn and
          P. T. Bolwijn: Appl. Phys. Lett. 23, 64 (1973); T. Kawata,
          M. Yamamoto, M. Yamana, M. Tajima and T. Nakano; Japan.
          J. Appl. Phys. 14, 725 (1975).

3.   For a recent review see: C. L. Bird and A. T. Kuhn, Chem. Soc.
          Rev. 10, 49 (1981).

4.   J. Billard, private communication.

5.   J. H. Day, Chem. Rev. 62, 65 (1962).

6.   A. Rambaum, A. M. Hermann, F. E. Stewart and F. Gutman, J.
          Phys. Chem. 73, 513 (1969).

7.   M. S. Simon and P. T. Moore, J. Polym. Sci. 13, 1 (1975).

8.   A. Factor and G. E. Heinsohn, Polym. Lett. 9, 289 (1971).

MESOMORPHIC PROPERTIES OF AN HOMOLOGOUS SERIES OF ALKYL-TERMINATED
ENAMINE-KETONE CONTAINING LIQUID CRYSTALS

Brian C. Benicewicz[†], Samuel J. Huang,
Joseph A. Pavlisko[††] and Julian F. Johnson

Department of Chemistry and Institute of Materials
Science, University of Connecticut, Storrs, CT 06268

INTRODUCTION

In a recent paper, it was shown that compounds containing a
six-membered, hydrogen-bonded enamine-ketone ring were capable of
forming liquid crystalline phases.[1] These compounds are rigid rod
compounds that do not contain terminal polar or flexible aliphatic
substituents. The present study was conducted to investigate the
mesomorphic behavior of enamine-ketone containing compounds which
possess a more conventional structure, i.e., a rigid rod core with
para-substituted flexible aliphatic substituents. This work will
describe the synthesis and thermal properties of the homologous
series of bis[3-(p-n-alkylanilino)-2-butenoyl] benzenes of the
following general structure:

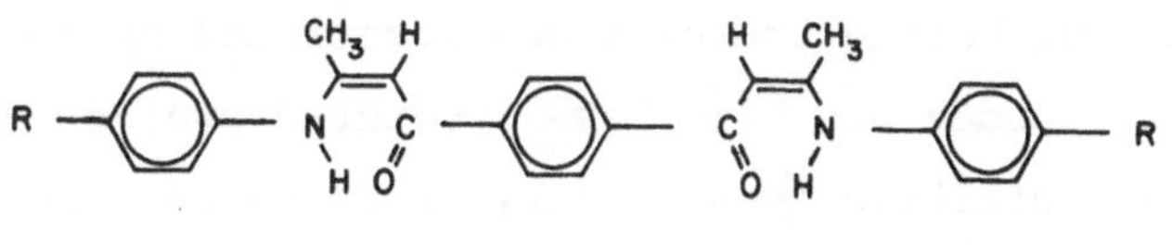

[†]Present address: Celanese Research Co., Summit, NJ
[††]Present address: Research Triangle Institute, Research
                     Triangle Park, NC

in order to further evaluate the effect of the enamine-ketone
moiety in liquid crystalline compounds.

EXPERIMENTAL

The enamine-ketone compounds, $C_1 \rightarrow C_8$, were prepared by the
reaction of terephthaloyldiacetone with the appropriate p-substituted anilines. The condensation was carried out in chloroform
with a small amount of acid catalyst.

$$R = C_1 \rightarrow C_8$$

The details of a typical reaction are as follows. To a 100 ml
flask equipped with a refluxing condenser were added in the following order: 1 mmole terephthaloyldiacetone, 2.7 mmole p-substituted
aniline, 25 ml chloroform, and two drops of concentrated hydrochloric acid. The mixture was refluxed for 22 hours. After removal
of the solvent with a rotary evaporator, the resulting yellow solid
was recrystallized several times from chloroform-methanol. The
yields were found to be 60-80%.

The enamine-ketone tautomer was identified by proton NMR
spectroscopy. Dudek and Holm have conducted proton NMR experiments
to study the equilibria in a variety of compounds with the enamine-
ketone structure in a variety of solvents.[2-4] These studies showed
that due to the appearance of a signal corresponding to a vinylic
hydrogen coupled with a low field signal which could only be assigned
to a hydrogen-bonded proton, the compounds must exist primarily in

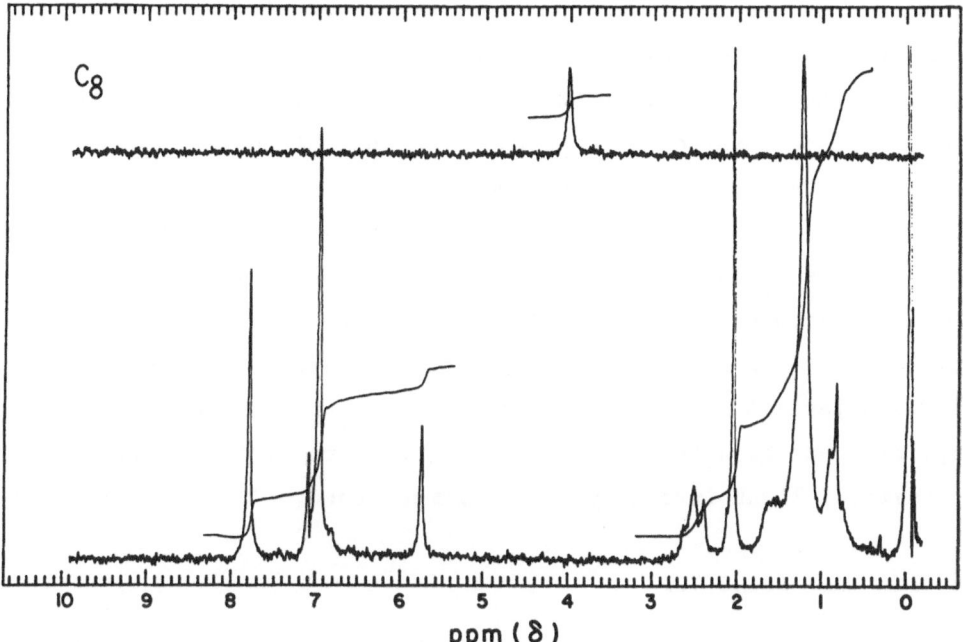

Fig. 1.   Proton NMR spectrum of bis [3-(p-n-octylanilino)-2-
butenoyl] benzene in chloroform downfield from TMS.
Upper scan is offset by 10 ppm.

a hydrogen-bonded chelate form. This was found to be true in almost
all solvents tested and solvents of differing polar character did
not noticeably affect the tautomeric equilibrium.   The preferential
existence of the enamine-ketone form was ascribed to greater stabilization through hydrogen-bonding and resonance. Fig. 1 shows the
proton NMR spectra of bis[3-(p-n-octylanilino)-2-butenoyl] benzene
where the upper scan is offset by 10 ppm. The appearance of a low
field signal at 14.0 ppm and a signal at 5.8 ppm corresponding to a
hydrogen-bonded and vinylic proton, respectively, agrees with the
enamine-ketone structure predominantly found by Dudek and Holm in
earlier studies.

Thermal measurements were performed on a duPont 990 Thermal
Analyzer equipped with a DTA/DSC cell. The samples were heated in
a nitrogen atmosphere at a rate of 5°C/minute. The transition

temperatures recorded were from solvent recrystallized samples which excluded any thermal history of the sample. The transition heats were obtained by measuring the area of the peaks with a planimeter and proper calibration with known standards. A thermal optical analyzer (TOA) was used to observe the textures of the phases which aided in the identification of the phase type and provided corroboration of the transition temperatures. The TOA consisted of a polarizing microscope equipped with a hot stage which allowed temperature and heating rate of the sample to be controlled. The transmitted polarized light was detected by a photodiode; the signal was amplified and transferred to a X-Y recorder to yield a plot of transmitted light intensity versus temperature.

RESULTS AND DISCUSSION

In this series the ethyl through octyl homologues were synthesized and investigated. The calorimetric data are presented in Table I. The nematic mesophase was the only mesophase detected and was present in all members of this series. The ethyl, butyl and hexyl homologues displayed solid polymorphism. The ethyl homologue displayed a solid→solid transition at 173°C which was immediately followed by an exotherm. The exotherm indicates that the molecules were probably rearranging to a more stable crystal form, followed at 184°C by melting into the nematic phase. Upon cooling and reheating the sample, the first peak increased in area, the exotherm disappeared and the second peak representing the solid→nematic transition decreased in size. Both peaks appeared at the same temperatures on second heating as on the first heating of the solvent recrystallized sample. It would appear that two crystal forms were present in the solvent recrystallized material whereas the melt crystallized sample produced only one crystal form.

The butyl homologue exhibited a very small endothermic DSC peak at approximately 107°C. A change in the transmitted light

Table I

Thermodynamic Data for the

bis[3-p-n-alkylanilino)-2-butenoyl] benzenes

Compound	Transition	$T,°C$	$\Delta H \times 10^4$,cal/mole	ΔS, cal/mole/°K
C_1	$K \rightarrow n$			
	$n \rightarrow i$			
C_2	$K_I \rightarrow K_{II}$	173	.211	4.73
	$K_{II} \rightarrow n$	184	.628	13.7
	$n \rightarrow i$	222	.039	.788
C_3	$K \rightarrow n$	173	1.12	25.2
	$n \rightarrow i$	224	.050	1.01
C_4	$K \rightarrow n$	150	.767	18.1
	$n \rightarrow i$	225	.035	.703
C_5	$K \rightarrow n$	150	.826	19.5
	$n \rightarrow i$	220	.040	.811
C_6	$K_I \rightarrow K_{II}$	134	.274	6.72
	$K_{II} \rightarrow n$	138	.510	12.4
	$n \rightarrow i$	198	.038	.807
C_7	$K \rightarrow n$	145	1.22	29.2
	$n \rightarrow i$	188	.044	.954
C_8	$K \rightarrow n$	157	1.38	32.1
	$n \rightarrow i$	175	.042	.937

intensity was also detected by the TOA apparatus; however, the
sample did not melt into a nematic phase until 150°C. The hexyl
homologue displayed a solid→solid transition at 134°C which is only
4°C before it melts into a nematic phase at 138°C.

The solid→nematic and nematic→isotropic transition temperatures for this series are plotted as a function of the alkyl tail
length in Fig. 2. There appears to be some evidence of an odd-even
alternation in the solid→nematic transition. There does not seem
to be any trace of this kind of alternation in the nematic→iso-
tropic transition. It should also be noted that the decreases in
the solid→nematic transition temperatures when advancing from the
propyl to butyl and pentyl to hexyl tails are greater than advanc-

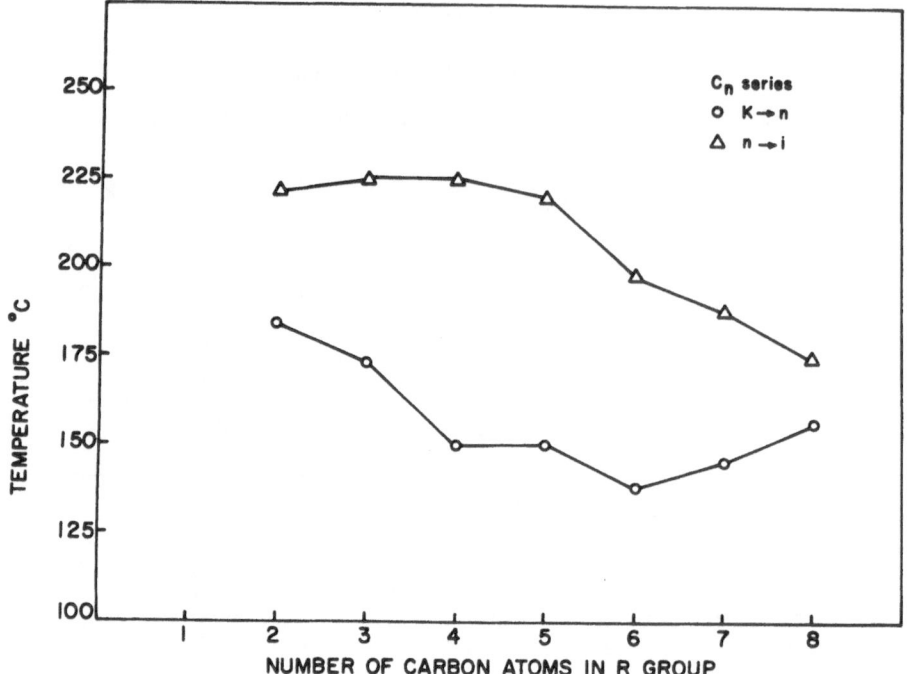

Fig. 2. Plot of transition temperatures versus the number of
carbon atoms in alkyl chain.

ing from the ethyl to propyl and butyl to pentyl tails. A curve
drawn through the solid→nematic transition temperatures of the even-
numbered carbon tail lengths falls below the curve created by drawing a line through the solid→nematic transition temperatures of the
odd-numbered carbon tail lengths. Although these trends are common
to liquid crystals when the flexible aliphatic chains are linked
directly to the aromatic core, they are usually found in the nema-
tic→isotropic transition.[5] The presence of the odd/even alternation
in the solid→nematic transition and lack of it in the nematic→iso-
tropic transition suggests that the aliphatic tails may play a
relatively less important role in the order of the nematic phase as
compared with other homologous series.

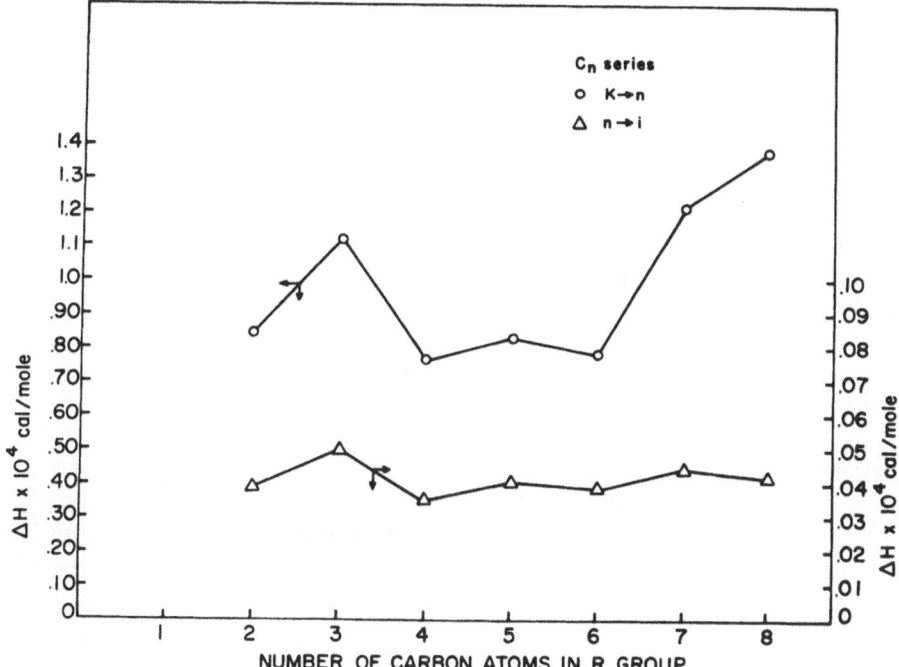

Fig. 3. Enthalpy change (ΔH) for the solid→nematic and
        nematic→isotropic transitions versus number of carbon
        atoms in alkyl chain.

In this series, an eight carbon chain tail length was the longest tail length synthesized. Fig. 2 shows that the nematic phase
stability range at the longer tail lengths decreases and, in the
case of the octyl homologue, is only 18°C wide. This convergence
is probably indicating the onset of the formation of a smectic
phase. In most homologous series exhibiting a smectic phase, the
increase in the smectic phase stability range coincides with a decrease in the nematic phase stability range. It is hypothesized
that the nonyl or decyl homologues would exhibit a smectic phase.

It is also instructive to look at the enthalpy (ΔH) and entropy
(ΔS) changes associated with these phase transitions. These values
are given in Table I and graphed in Figs. 3 and 4 as a function of
the aliphatic chain length. For the solid→nematic transition, both

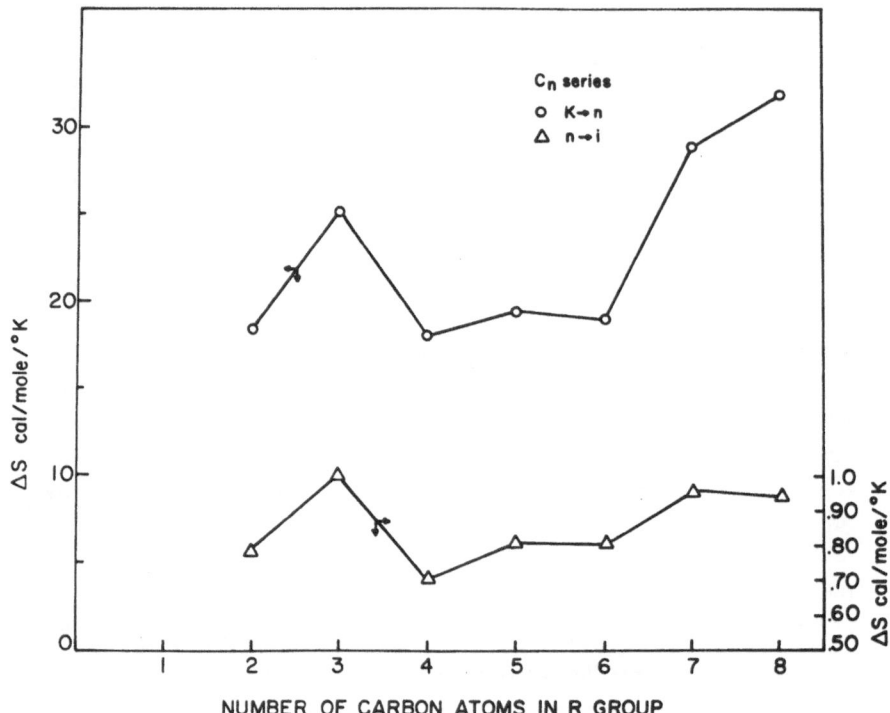

Fig. 4. Entropy change (ΔS) for the solid→nematic and nematic→
         isotropic transitions versus number of carbon atoms in
         alkyl chain.

ΔH and ΔS increase considerably when the aliphatic chain length
reaches seven and eight carbons long. This implies that the intermolecular forces between the aliphatic chains are becoming increasingly important. Since it is these forces that are important in
stabilizing the smectic mesophase, this is further evidence for the
onset of smectic mesophase formation.

    The alternation in the enthalpy and entropy changes for the
nematic→isotropic transition is very slight and appears to diminish
with increasing chain length, especially in the case of ΔH. If the
data are statistically analyzed and a least squares straight line
drawn through these points, the slope of this line would be close
to zero. These two observations seem to indicate that the terminal
aliphatic tails do not greatly affect the nematic mesophase order.

The compound bis[3-(p-1-methylbutylanilino)-2-butenoyl] benzene which contains branched aliphatic tails was synthesized, but did not exhibit any liquid crystalline properties. When heated, this compound melted from a crystalline solid directly to an isotropic liquid at 151°C. Studies on other compounds have shown that 1-methylation of alkyl tails had a larger effect than 2-, 3- or 4-methylation.[6] This is due to the steric interactions of the 1-methyl group with the ortho-hydrogen of the aromatic ring. This behavior demonstrates the sensitivity of liquid crystals to certain structural effects and the caution that must be used when making generalizations about homologous series.

CONCLUSIONS

A new structural class of liquid crystalline compounds containing the enamine-ketone moiety has been synthesized and investigated. These compounds exhibited nematic phases although some of the data indicate that smectic phases may be present if the homologous series were to be extended to longer terminal aliphatic chain lengths. The enthalpy and entropy data imply a low degree of interaction of the alkyl tails in the nematic phase. It appears, however, that the crystalline order of the solid phases is determined by the packing of the alkyl tails. Further work on this new class of liquid crystalline materials is being conducted to provide a better understanding of their mesophase properties and will be forthcoming.

ACKNOWLEDGEMENT

One of us (BCB) would like to gratefully acknowledge receipt of a UniRoyal Foundation Research Fellowship.

REFERENCES

1.   B. C. Benicewicz, S. J. Huang and J. F. Johnson, Mol. Cryst.
          Liq. Cryst., Letters 72, 9 (1981).

2.   G. O. Dudek and R. H. Holm, J. Am. Chem. Soc. 83, 2099 (1961).

3.   G. O. Dudek and R. H. Holm, J. Am. Chem. Soc. 83, 3914 (1961).

4.   G. O. Dudek and R. H. Holm, J. Am. Chem. Soc. 84, 2691 (1962).

5.   G. W. Gray, Molecular Structure and the Properties of Liquid
          Crystals, Academic Press:New York (1962)

6.   G. W. Gray and K. H. Harrison, Symposia of the Faraday Society
          5, 54 (1971).

7.   S. J. Huang, B. C. Benicewicz and J. A. Pavlisko, ACS
          Symposium Series, "Cyclopolymerization and Polymers with
          Ring-Chain Structures", 181st ACS National Meeting,
          Atlanta, GA, 1981, in press.

LIQUID CRYSTALLINE POLYMERS WITH AMPHIPHILIC AND NON AMPHIPHILIC SIDE CHAINS - EFFECT OF THE MAIN CHAIN ON THE PHASE BEHAVIOUR

H. Finkelmann, B. Lühmann, G. Rehage and H. Stevens

Institut für Physikalische Chemie
Technische Universität Clausthal
3392 Clausthal-Zellerfeld, FRG

INTRODUCTION

The most convenient classification of low molar mass liquid crystals (LC) refers to the constitution of the mesogenic molecules[1]: <u>Non amphiphilic LCs</u>, which are characterized by a rigid rod like molecular structure and <u>amphiphilic LCs</u>, which consist of molecules having a hydrophobic and a hydrophilic part. The non amphiphilic as well as the amphiphilic mesogenic molecules have become of interest as monomers for the synthesis of polymers, which have attached the mesogenic molecules to a polymer backbone as side chains (Fig. 1). The resulting macromolecular systems can exhibit liquid crystalline phases very similar to the low molar mass

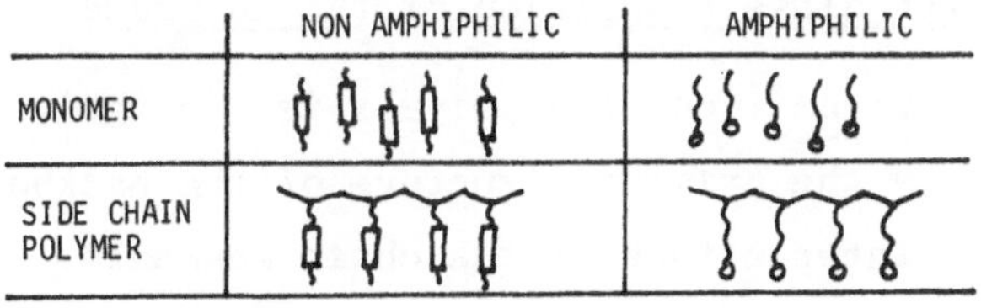

	NON AMPHIPHILIC	AMPHIPHILIC
MONOMER		
SIDE CHAIN POLYMER		

Fig. 1. Schem of <u>amphiphilic</u> and <u>non amphiphilic</u> monomers and side chain polymers (⊂⊃= rigid mesogenic group, o hydrophylic ⌇hydrophobic part of the amphiphile)

compounds. Due to the linkage of the mesogenic moieties to the polymer main chain, however, characteristic changes are expected in the phase behaviour as well as in the structure of the liquid crystalline phases. In this paper, at first we will describe, whether the phase behaviour and the structure of non amphiphilic LC side chain polymers depend on polymer specific properties like the glass transition and the degree of polymerization. Secondly the phase behaviour of a low molar mass amphiphilic LC will be compared with the phase behaviour of the corresponding amphiphilic side chain polymer.

NON AMPHIPHILIC SIDE CHAIN POLYMERS

Investigations of the past few years have proved that the LC state of the non amphiphilic side chain polymers can be varied by the chemical constitution of the mesogenic groups attached to the polymer backbone[2]. In principle the same systematics has been observed, which is known for conventional low molar mass LC. This systematics, however, is related to polymers, having a polymer backbone of the same chemical constitution and the same degree of polymerization. If, on the other hand, the liquid crystalline side chains remain, the influence of polymer specific properties on the liquid crystalline state can be observed. The glass transition and the degree of polymerization r of the polymer will be discussed in the following.

Influence of the glass transition of the LC phase

The glass transition of a polymer is - roughly described - determined i) by the primary structure of the backbone and ii) by intermolecular interactions of the chain segments. The primary structure describes the chemical constitution of the polymer main chain, which, on the other hand, directly influences its rigidity. Furthermore, the glass transition depends on the intermolecular interactions between the chain segments. For the LC state of the

LC side chain polymers it is of interest, whether the phase behaviour is more affected by a change in the rigidity of the polymer main chain, or whether the intermolecular interactions of the chain segments having the mesogenic units is more important.

The influence of the glass transition on the extent of the LC phase, due to a change of the rigidity of the polymer backbone, is demonstrated for three examples in Table 1. All polymers have the same mesogenic side group. These groups are linked to different polymer backbones, which differ in their flexibility. This is indicated in the glass transition temperatures T_g. The poly (methacrylate) exhibits the highest T_g. T_g is lowered, if the poly(methacrylate) main chain is exchanged by the poly(acrylate), which has a more flexible main chain because of the missing methyl group. The lowest T_g is observed for the poly(siloxane). Looking at the phase transitions liquid crystalline to isotropic, they are strongly influenced by the change of T_g: increasing flexibility of

Table 1. Phase behaviour of LC side chain polymers with the same mesogenic group linked to different polymer main chains (ΔT = extent of nematic phase).

CONSTANT MESOGENIC MOIETY : $-(CH_2)_2-O-\bigcirc-COO-\bigcirc-OCH_3$
FLEXIBLE SPACER

	PHASE TRANSITIONS (K)				ΔT
$\left[CH_2-\underset{\underset{COO-R}{|}}{\overset{\overset{CH_3}{|}}{C}} \right]$	g	369	n	394	i	25
$\left[CH_2-\underset{\underset{COO-R}{|}}{\overset{\overset{H}{|}}{C}} \right]$	g	320	n	350	i	30
$\left[O-\underset{\underset{CH_2-R}{|}}{\overset{\overset{CH_3}{|}}{Si}} \right]$	g	288	n	334	i	46

the main chain, which is indicated by the falling T_g, lowers the phase transition liquid crystalline to isotropic. The same effect was observed by Wendorff for LC poly(methacrylates)[3]. Even the chemical structure of the main chain remained and only the tacticity was changed, which describes the stereochemical arrangement of the monomer units. A change in the tacticity shifted T_g as well as the phase transition of the LC phase in the same direction.

These examples indicate that the flexibility of the main chain, which is determined by their chemical constitution, directly influences the _extent_ of the liquid crystalline state. With increasing rigidity of the main chain the motions of the mesogenic groups are more restricted. This can be noticed in an increase of the phase transition temperature liquid crystalline to isotropic.

As mentioned above, the flexibility of the polymer backbone is also determined by the interactions of the chain segments. For conventional polymers, these _interactions_ can be suppressed by adding low molar mass substances to the polymer. They cause a strong decrease of T_g (softening effect). If the softening effect also influences the liquid crystalline phase transitions was investigated for the LC side chain polymers. We therefore mixed a LC side chain polymer with a low molar mass LC of very similar constitution[4] (Fig. 2). For this mixture no severe distortion of the intermolecular interactions of the mesogenic groups can be assumed. Because of the very similar chemical structure the enthalpy of mixing should be approximately zero. Actually a strong shift of T_g is observed to lower temperatures with increasing amount of the low molar mass LC, indicating the softening effect on the polymer system. However, in contrast to the examples mentioned above, the phase transition liquid crystalline to isotropic is not affected by the shift of the glass transition. The clearing temperature T_c hardly depends on the composition as expected for nearly ideal mixtures of LC compounds of similar chemical constitution.

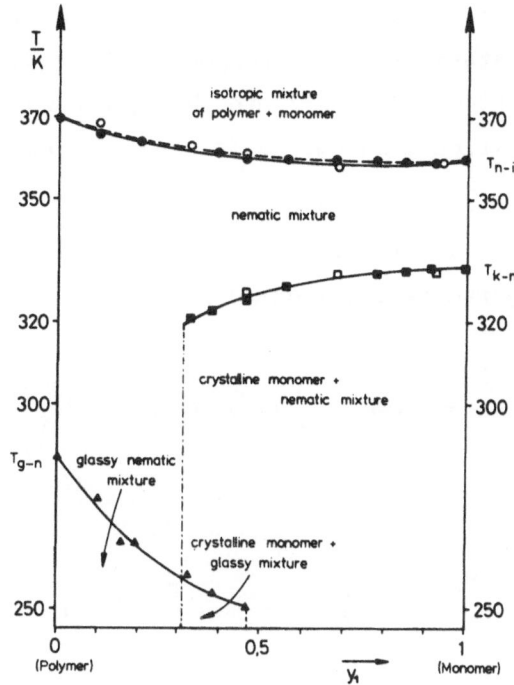

Fig. 2.   Phase diagram of mixtures of the monomer

$$CH_2=CH-CH_2-O-\bigcirc-COO-\bigcirc-OC_6H_{13}\quad\text{and the polymer}$$

$$CH_3-\underset{|}{\overset{|}{Si}}-(CH_2)_4-O-\bigcirc-COO-\bigcirc-OCH_3$$

These examples clearly indicate that the extent of the LC phase
of the polymer is mainly influenced by the flexibility of the polymer main chain determined by their chemical constitution. The flexibility of the backbone also influences the mobility of the mesogenic
side chains. This determines the extent of the LC phase. Intermolecular interactions of the main chain, which are also reflected
by T_g, essentially do not affect the temperature of the LC phase
transition.

Influence of the degree of polymerization on the LC phase

For conventional polymers it is well established that the glass
transition is also a function of the degree of polymerization r.

Following the results discussed above, consequently r will also
influence the LC state of the LC polymers. Here it is of interest,
in which way the LC state is affected by r, and which r is sufficient that an additional monomer unit does not change the LC phase
behaviour.

   To get a more detailed insight into these mechanism, we synthesized polydisperse oglio(siloxanes):

                                                                    (I)

   The oligomers were separated by gel permeation chromatography
(GPC) into monodisperse fractions of defined r. The phase behaviour
of these monodisperse oligomers was investigated by differential
scanning calorimetry (DSC) and polarizing microscopy. For the oligomers with m=3, the phase transition temperatures are shown as
function of r in Fig. 3. The monomer (r=1) and the dimer are not
liquid crystalline. Whereas for the monomer a rapid crystallization is observed at 338 K, the dimer can be supercooled and a transition into the glassy state is found at 255 K. An additional monomer unit produces for the trimer a nematic liquid crystalline phase,
exhibiting a clearing temperature of 285 K and a low temperature
glass transition at 270 K. The mesophase occurs in a temperature
interval of 15 K. With each additional monomer unit T_c steeply
increases and for r=10 the phase transition only differs by 3° from
the phase transition of the polymer with $\bar{r}$=95. The glass transition

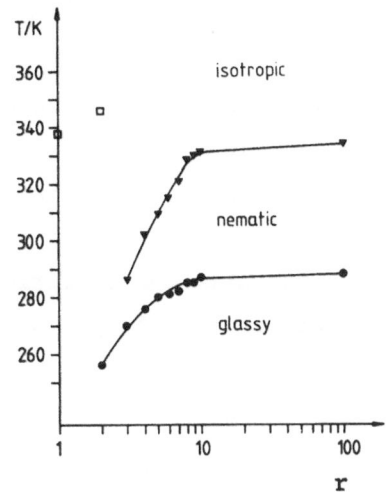

Fig. 3.   Phase behaviour of
          monodisperse oligomers
          1 as function of the
          degree of polymeriza-
          tion r   (m=3).

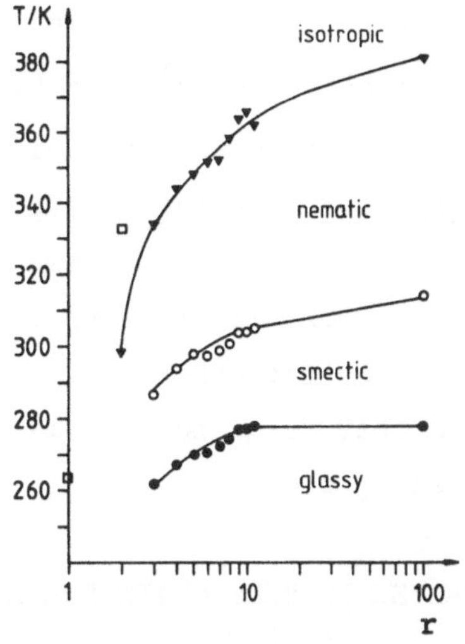

Fig. 4.   Phase behaviour of
          monodisperse oligomers
          1 as function of the
          degree of polymeriza-
          tion r   (m=6).

also shifts to higher temperatures with increasing r, but the difference between T_g of the dimer and T_g of the polymer ($\bar{r}$ = 95) is
only 18 K. It has to be noted that with increasing r the LC phase
is broadened from ΔT=15K for the trimer to ΔT=46 K for the polymer.

The same effect of the degree of polymerization on the extent
of the LC phase is also found for the oligomers with m=6 (Fig. 4).
The monomers is only crystalline. The dimer also crystallizes, but
by rapid cooling of the isotropic melt, the metastable phase transition isotropic to nematic can be detected. In this series the
trimer exhibits stable liquid crystalline phases for the first
time. Above the glass transition temperature (T_g=261) a smectic
phase exists, which is changed into a nematic phase at $T_{s,n}$=287 K;
T_c is 332 K. For these oligomers the phase transitions are also
shifted to higher temperatures with increasing r, but not as strong
as observed in the previous system. Furthermore it is notable that

the smectic phase is not broadened from r=3 to r=10, whereas the nematic phase is enlarged by 17 K.

These measurements for the first time prove that the LC state is <u>enlarged</u> with increasing degree of polymerization. The most drastic change in the phase transition temperatures, however, is observed for oligomers up to r≃10 for the polysiloxanes investigated. On the other hand, from these measurements it can be estimated that for r>100 the liquid crystalline state is no longer affected by r.

AMPHIPHILIC SIDE CHAIN POLYMERS

As described in the previous chapter, the linkage of non amphiphilic liquid crystalline molecules to a polymer backbone as side chains results in a broadening of the thermotropic liquid crystalline state and shifts the phase transitions towards higher temperatures. In comparison to these results it is of interest, whether the linkage of amphiphilic molecules to a polymer main chain also changes the lyotropic phase behaviour.

We therefore prepared a reactive low molar mass amphiphile II, consisting of 10-undecenoic acid as hydrophobic part of the molecule, and ethylenglycolether with eight glycol units as hydrophylic part[5]:

$$CH_2=CH-(CH_2)_8-COOH + CH_3-(O-CH_2-CH_2)_8-OH \longrightarrow$$
$$CH_2=CH-(CH_2)_8-COO - (CH_2-CH_2-O)_8-CH_3 \qquad II$$

In order to obtain the lyotropic liquid crystalline polymer, the monomeric amphiphiles II were attached as side chains to poly (methylhydrogensiloxane) in the presence of platinium catalyst:

$$
\cdots-\underset{\underset{H}{|}}{\overset{\overset{CH_3}{|}}{Si}}-O-\cdots + II \longrightarrow \cdots-\underset{\underset{CH_2-CH_2-(CH_2)_8-COO-(CH_2-CH_2-O)_8-CH_3}{|}}{\overset{\overset{CH_3}{|}}{Si}}-O- \qquad III
$$

The polymer has an average degree of polymerization of 95 . The phase behaviour of monomer II with water and of the corresponding

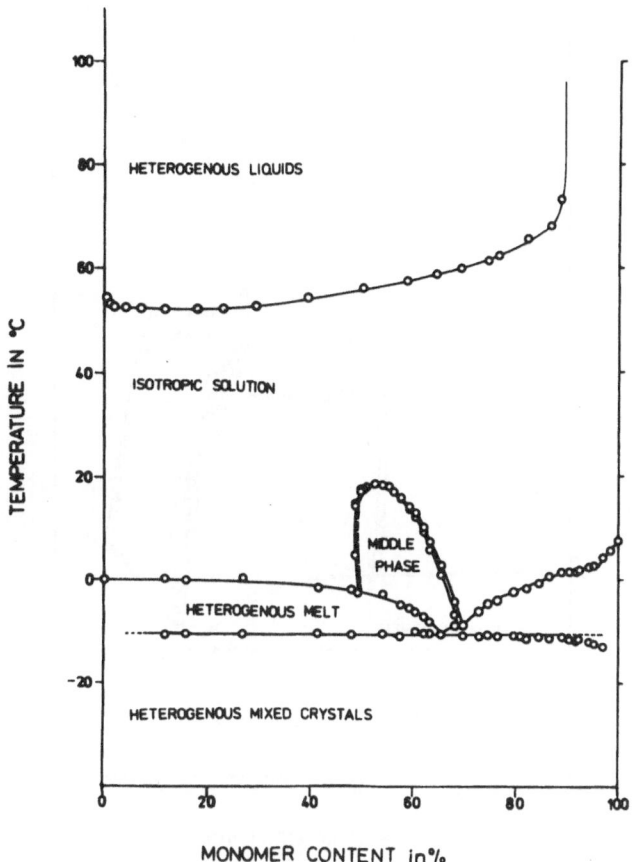

Fig. 5.   Phase diagram of the monomeric amphiphilic II and water.

polymer III with water was compared, using polarizing microscopy to
determine the phase boundaries and structure of the lyotropic LC
phases.

The monomer exhibits a liquid crystalline phase region in a
concentration range from 49% to 70% of II (Fig. 5). The microscopic texture was compared with textures described in literature[6] for
very similar systems, and the phase was identified as a middle phase
or M_1-phase of hexagonal packed micelles of indefinite length. The
phase boundary line between the liquid crystalline state and the

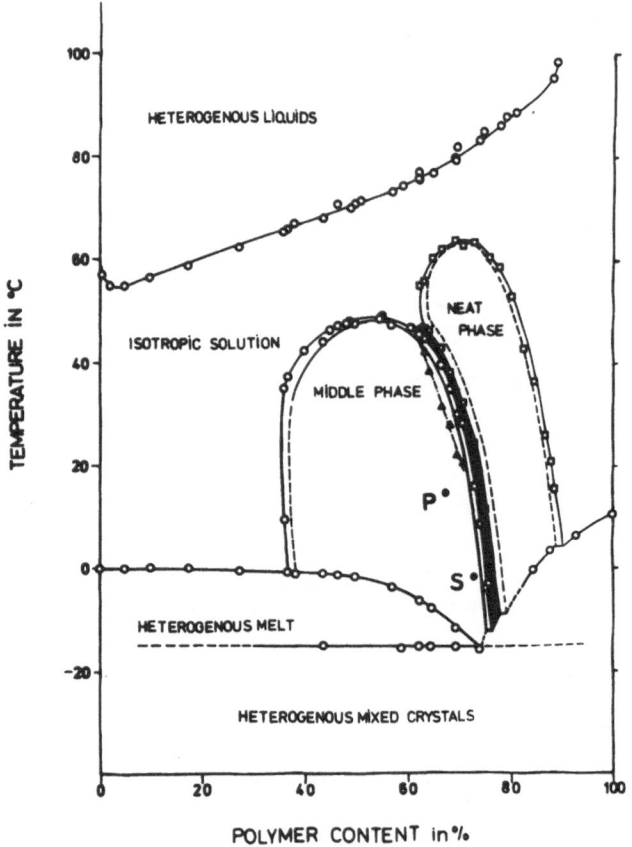

Fig. 6. Phase diagram of the polymeric amphiphile III and water.

isotropic solution has a maximum at 53% of II, which indicates the formation of a tripolyhydrate.

The phase diagram of the polymer III with water is shown in Fig. 6. In contrast to the monomer/water system, it is more complex and characterized by two different liquid crystalline phases, which are separated by a cubic phase (black area). At lower polymer concentrations, the polymer forms a M_1-phase with water, which is similar to the monomer system. At higher polymer concentrations, the liquid crystalline phase is a neat phase or G-phase of planar

micelles of indefinite length and width. The phase boundary line
of the G-phase to the isotropic solution has a concentration of 72%
polymer III at the maximum, which indicates the existence of a
1.5-polyhydrate.

Comparing the phase diagram of the monomer/water and the poly-
mer/water system, the mesomorphous regions are _enlarged_ for the
polymer/water system. This indicates that the polymer main chain,
which couples the monomeric surfactants, stabilize the mesophase.
This effect is analogous to the results obtained for the thermotropic, non amphiphilic side chain polymers and can also be explained
by the restriction of motions of the amphiphilic units, when they
are linked to the polymer backbone. The broadening of the lyotropic
phase, however, could also be explained by the lengthening of the
hydrophobic part of the amphiphilic monomer by one monomer unit of
the polymer main chain. Normally the miscibility gap is shifted to
lower temperatures, if the hydrophobic part of the amphiphile is
lengthened. This, however, is not observed for our monomer/water
and polymer/water system. The miscibility gap for the polymer/water
system is observed even at higher temperatures. This indicates
that the stabilization of the mesophase is more affected because of
the linkage of the amphiphiles of the main chain.

REFERENCES

1.   Ed. Ellis Horwood Series in Physical Chemistry: Liquid
          Crystals and Plastic Crystals, G. W. Gray, P. A. Winsor,
          Halsted, New York, 1974.
2.   H. Finkelmann in "Polymer Liquid Crystals", Ed. A. Cifferi, W.
          R. Krigbaum and R. B. Meyer, Academic Press Inc., New
          York, 1982.
3.   B. Hahn, J. H. Wendorff, M. Portugall, H. Ringsdorf, Colloid
          and Polymer Sci. 259, 875 (1981).

4.   H. Finkelmann, H. J. Kock and G. Rehage, Mol. Cryst. Liqu.
         Cryst. 000, 000 (1982) in press.

5.   H. Finkelmann, B. Lühmann and G. Rehage, Colloid and Polymer
         Sci. 56, 260 (1982).

6.   F. B. Rosevear, J. Amer. Oil Chem. Soc. 31, 628 (1954) and
         S. Friberg, Naturwissenschaften 64, 612 (1977).

THE FLUID MESOPHASES OF POLAR RODS

F. Hardouin, Nguyen Huu Tinh, M. F. Achard,
and A. M. Levelut*

Centre de Recherche Paul Pascal
Université de Bordeaux I
33405 Talence Cédex - France

* Laboratoire de Physique des Solides
Bât. 510
Université Paris-Sud
41504 Orsay - France

ABSTRACT

We report various unexpected polymorphisms of fluid mesophases revealed in pure or mixed cyano or nitro rod-like derivatives.

In addition to the smectic A polymorphism and the reentrant sequences we have recently found new mesophases in which a liquid-like short range ordering is kept although a long range two-dimensional periodic order is condensed. All these phases are due to the coexistence of two characteristic wavelengths one connected to the molecule and the other one to the length of antiparallel molecular pairs.

INTRODUCTION

During a few years we have investigated many new mesomorphic systems of polar compounds synthesized in our laboratory. These materials, rod-like shaped, involve usually three aromatic rings

and a strongly polar head (cyano or nitro). The general formula is
as follows:

R $-\langle 0 \rangle - X - \langle 0 \rangle - Y - \langle 0 \rangle -$ CN
or
NO_2

$R = C_nH_{2n+1}^-$, $C_nH_{2n+1}O^-$, $C_n H_{2n+1}COO^-$

$\left.\begin{matrix} X \\ Y \end{matrix}\right\} = - CH = CH-, - C \equiv C - , -CH = N-, -N = CH-,$ single bond,

$- COO -, -OCO-, - N = N-$.

Most of these compounds give rise to smectic A phases in which
an antiferroelectric order is favoured by the strong dipolar interactions. From a microscopic standpoint one can figure this arrangement as the formation of head to tail pairs of molecules. The stacking of the layers is then driven by two competitive periods, one
connected to the length of the molecular L(wave of density) and the
other one to the length of the antiparallel pair L' (wave of polarization). The thermal variation of the former one is weak, at the
opposite the latter one is likely to vary strongly from one phase
to the other as a function of the temperature. However the dependence of the wave of polarization on the specific polarity of each
molecule is not yet understood (Fig. 1).

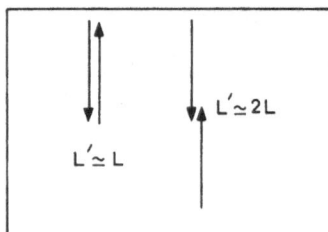

Fig. 1. Pair association with a more or less large overlap of the
molecules - (rod-like configuration).

In addition the experiments indicate most often that the system
has a ratio of overlapping $\frac{L}{L'}$ which is not an integer. But it has
been also shown in these fluid systems that this natural incommensurability is strained by a coupling which forces a commensuarate
lock-in of both waves such that they are able to produce simultaneously long range modulations. The consequences of these competi-
tions give rise to a rich polymorphism of "frustrated" smectics, so
called because they cannot fulfill the two conditions at the same
time.

THE SMECTIC A POLYMORPHISM

In our recent physico-chemical investigations we have discovered several phase transitions among different smectic A phases[1-5].
In terms of structure all these different phases have the features
which characterize a smectic A, i.e.: — a liquid-like order within
the layers — the molecules are perpendicular to the plane of the
layers (on average) — of course the phase is optically uniaxial.

The "bilayer" S_{A2} phase met in series with $R = C_n H_{2n+1}$,
$C_n H_{2n+1}$ O and $X = Y = -OCO -$ is clearly differentiated from the
other smectic A phases. It occurs in the S_{A2} a commensurate
lock-in of the two wavelengths with a ratio equal to 2 (Fig. 2).

At higher temperature in binary mixtures it may arise first a
"monolayer" smectic A (the layer spacing is one molecular length)[1-2]

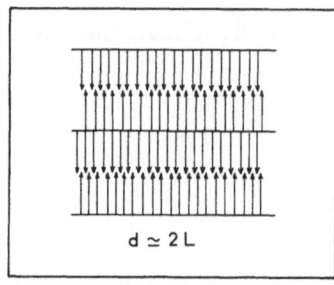

$d \simeq 2L$

Fig. 2. Schematic structure of the "bilayer" S_{A2}.

or a "partially bilayer" smectic A (the layer spacing is larger
than one molecular length and smaller than two)[3] or even both successively[4]. Since a first order transition has been evidenced between these two phases, we are led to distinguish more by indexing
S_{Ad} the "partially bilayer" phase and S_{A1}, of course, the "monolayer"
one. Nevertheless it happens that sometimes the evolution $S_{Ad} \rightarrow S_{A1}$
is obtained without any transition[6-7-8] like in the case of the
liquid-gas system around the critical point. Let us note that the
Prost's mean field theory gives a general interpretation of all
these $S_A - S_A$ transitions[9-10]. At last we stress that $S_{Ad} - S_{A2}$
phase transition has been revealed even in pure compounds.

FLUID BIDIMENSIONAL PHASES ($\tilde{S_A}$, $\tilde{S_C}$)

We have recently introduced the symbol $\tilde{S_A}$ to define a new
smectic phase which keeps a liquid like short range ordering although a long range two-dimensional periodicity is condensed[11,12].
Indeed the local order of this $\tilde{S_A}$ is same as in the S_{A2} but a long
range periodicity in a direction parallel to the layers gives a two
dimensional rectangular symmetry to this phase. It is observed as
in intermediate state between a S_{A1} phase and a S_{A2} phase and a
theorical explanation of this occurence in terms of escape from
incommensurability has been proposed by P. Barois et al.[13].

One can figure out this structure as a fluid "antiphase" with
periodic stacking faults in which the long molecular axis remains
normal to the layers. This $\tilde{S_A}$ phase has been discovered in a mixed
system of two cyanoderivatives:

$$R = C_n H_{2n+1} \qquad X = Y = -OCO-$$
$$n = 5 \text{ or } 6$$

$$R = C_n H_{2n+1} \qquad X = -CH = CH-, Y = -OCO-$$
$$n = 5$$

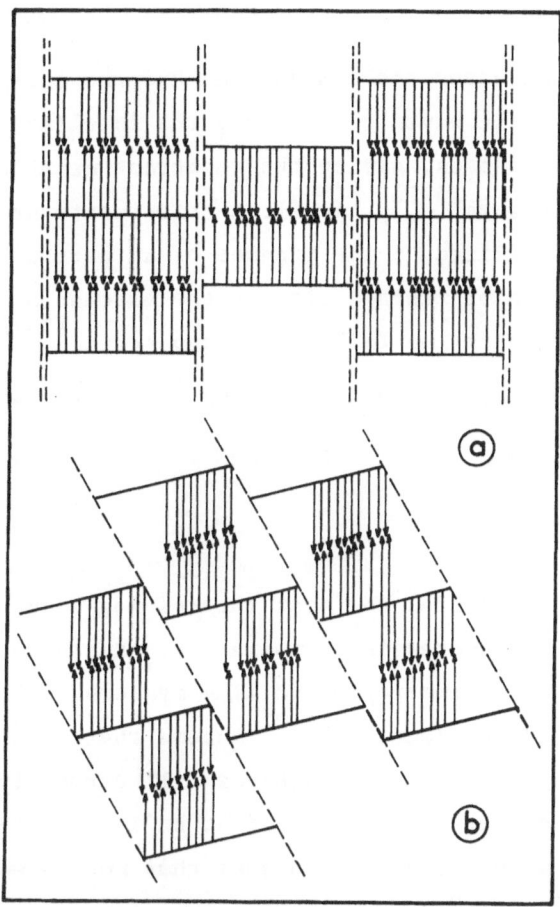

Fig. 3. a- Schematic dislocation model of the antiphase $S_A^{\sim}$ (two
        dimensional rectangular lattice).
     b- Schematic dislocation model of the serrated $S_C^{\sim}$ phase
        (two dimensional oblique lattice).
     ... periodic stacking faults.

Another novel fluid mesophase with a two-dimensional oblique
lattice has been found in series of nitro compounds: $R = C_nH_{2n+1}$,
$X = Y = - OCO - $. By analogy with the antiphase $S_A^{\sim}$ and to emphasize the fact that the "walls" are now tilted we propose the symbol:
$S_C^{\sim}$ (Fig. 3b). In the case $R = C_8H_{17}$, this "serrated" phase is
intermediate between a S_{Ad} and a S_{A2}^{14}.

REENTRANT SEQUENCES

In a number of cases the competition and the coexistence of the two characteristic waves can destroy the lamellar structure as the temperature decreases, giving the reentrant nematic phenomenon at atmospheric pressure[15-16]. At first this reentrant nematic phase has been stabilized at the expense of a "partially bilayer" S_{Ad} phase but it can be reached also at the expense of a "partially bilayer" smectic C with the following cyano series[17-18-19-20]:

$$R = C_n H_{2n+1} O \begin{cases} X = - COO- \;, \; Y = -CH = N - \; \text{or} \; Y = -N = CH - \\ X = - CH = CH - COO - \qquad Y = - CH = N- \end{cases}$$

CONCLUSION

The table I summarizes the most interesting sequences of transitions got upon decreasing temperature. Recent studies undertaken on these thermotropic rod-like systems with strong "antiferroelectric" interactions[21,22] assume that an even more complex polymorphism of fluid mesophases will certainly be defined. All domains in chemistry or physics might be interested in the understanding of such a number of phases that nobody would have expected five years ago.

TABLE 1

Various polymorphisms detected upon cooling in pure or binary systems of polar molecules. The techniques used to characterize the different phases are the optical microscopy, the differential scanning calorimetry and the X-ray diffraction.

I for isotropic Liquid; N for nematic; S_{A1}, S_{Ad}, S_{A2}, $S_{\tilde{A}}$, $S_{\tilde{C}}$, see text and references.

$$\underline{S_A - S_A \text{ polymorphism}}$$

$$I \rightarrow N \rightarrow S_{Ad} \rightarrow S_{A1} \rightarrow S_{A2}$$

$$\underline{\text{Polymorphisms including bidimensional fluid meosphases}}$$

$$I \rightarrow N \rightarrow S_{Ad} \rightarrow S_{A1} \rightarrow S_{\tilde{A}} \rightarrow S_{A2}$$

$$I \rightarrow N \rightarrow S_{Ad} \rightarrow S_{\tilde{C}} \rightarrow S_{A2}$$

$$I \rightarrow N \rightarrow S_{A1} \rightarrow S_{\tilde{C}}$$

$$\underline{\text{Reentrant polymorphisms}}$$

$$I \rightarrow N \rightarrow S_{Ad} \rightarrow N \rightarrow S_{A1}$$

$$I \rightarrow N \rightarrow S_{Ad} \rightarrow S_C \rightarrow N \rightarrow S_{A1}$$

REFERENCES

1. G. Sigaud, F. Hardouin, M. F. Achard and H. Gasparoux,
        J. Physique (Paris) Colloq. 40, C3-356 (1979).

2. F. Hardouin, A. M. Levelut, J. J. Benattar and G. Sigaud,
        Solid State Commun. 33, 337 (1980).

3. F. Hardouin, A. M. Levelut and G. Sigaud, J. Physique (Paris)
        42, 71 (1981).

4. A. M. Levelut, R. J. Tarento, F. Hardouin, M. F. Achard and
        G. Sigaud, Phys. Rev. A 24, 2180 (1981).

5. A. M. Levelut, B. Zaghloul and F. Hardouin, J. Physique
        (Paris) 43, L-83, 1982.

6. Advances in liquid crystal Research and Applications, H. T.
        Nguyen, G. Sigaud, M. F. Achard, H. Gasparoux and F.
        Hardouin, Pergamon Press Oxford, Budapest, 1980, p. 147.

7. Symmetries and Broken Symmetries in condensed matter physics,
        F. Hardouin, A. M. Levelut, G. Sigaud, M. F. Achard,
        H. T. Nguyen and H. Gasparoux, edited by N. Boccara,
        IDSET - Paris, 1981, p. 231.

8. P. E. Cladis, Mol. Cryst. Liq. Cryst. 67, 833 (1981).

9. J. Prost, J. Physique (Paris) 40, 518 (1979).

10. Liquid Crystal of one and two dimensional order, J. Prost,
        edited by H. W. Helfrich and G. Heppke, Springer, Berlin,
        1980, p. 125.

11. G. Sigaud, F. Hardouin, M. F. Achard and A. M. Levelut,
        J. Physique (Paris) 42, 107 (1981).

12. F. Hardouin, G. Sigaud, H. T. Nguyen and M. F. Achard,
        J. Physique (Paris) 42, L-63 (1981).

13. P. Barois, C. Coulon and J. Prost, J. Physique (Paris) 42,
        L-107 (1981).

14. F. Hardouin, H. T. Nguyen, M. F. Achard and A. M. Levelut,
        C. R. Acad. Sci. (to be published), J. Phys. (Paris) (to
        be published, english text).

15. See for example: G. Sigaud, H. T. Nguyen, F. Hardouin and
     H. Gasparoux, Mol. Cryst. Liq. Cryst <u>69</u>, 81 (1981)
     (Review with 58 references).

16. H. T. Nguyen, M. Joussot-Dubien and C. Destrade, Mol. Cryst.
     Liq. Cryst. <u>56</u>, 257 (1980).

17. W. Weissflog, G. Pelz, A. Wiegeleben and D. Demus, Mol. Cryst.
     Liq. Cryst. Lett. <u>56</u>, 295 (1980).

18. G. Pelz, S. Diele, A. Wiegeleben and D. Demus, Mol. Cryst.
     Liq. Cryst. Lett. <u>64</u>, 163 (1981).

19. H. T. Nguyen, F. Hardouin, C. Destrade and A. M. Levelut,
     J. Phys (Paris) <u>43</u>, L-33 (1982).

20. H. T. Nguyen, C. Destrade, H. Gasparoux, Mol. Cryst. Liq.
     Cryst. (to be published).

21. L. G. Benguigui and F. Hardouin, J. Phys. (Paris) <u>42</u>, L-111
     (1981).

22. L. G. Benguigui and F. Hardouin, J. Phys. (Paris) <u>42</u>, L-381
     (1981).

THE NEMATOGENEITY OF SOME HYDROCARBONS

R. Eidenschink, M. Roemer

E. Merck
Darmstadt, West Germany

F. V. Allan

EM Chemicals
Hawthorne, NY, 10532

ABSTRACT

Hitherto it has been tacitly assumed that the nematic character of an organic compound is strongly favored by moieties with
polar bonds in the direction of the longest molecular axis. It was
found, however, that hydrocarbons having a trans-4-alkylcyclohexyl
group attached to an aromatic ring exhibit wide nematic temperature
ranges. The transition temperatures of 4,4'-bis(trans-4-alkylcyclo-
hexyl)-biphenyls, 4-alkyl-4'-trans-4-alkylcyclohexyl-biphenyls and
4-trans-4-alkylcyclohexyl-alkylbenzenes are given. Because of their
low melting points, high solubilities in 4-trans-4-alkylcyclohexyl-
benzonitriles (PCH-compounds) and low viscosities, some of these
compounds are of great value in phases for fast switching electro-
optical devices.

INTRODUCTION

The discovery of the correlation of chemical structure with
the thermal stability of the nematic phase, which can be expressed

by the transition temperature from the nematic to the isotropic
state (T_{NI}), can still be regarded as the key for understanding
liquid crystalline behavior. To fulfill the requirement of a high
anisotropy of polarizability, which was assumed to be vital for a
high T_{NI}[1], aromatic systems with polar moieties at the end of, or
between, aromatic rings were the preferred subjects of synthetic
work. After the syntheses of some cyclohexane derivatives which
had distinctly higher T_{NI} values than the corresponding aromatic
compounds[2,3] repulsive interactions were recognized to play an
important role in stabilizing the nematic phase[4]. Also, the influence of polar groups, especially the cyano group, on the association of the rod-like molecules has been discussed[5] in this connection. Yet this did not explain the drastic changes in the T_{NI}
of nematic phases when a few groupings are interchanged in a molecule, the molecular shape being preserved[6]. So doubts arose as to
whether the theory of isotropic liquids, involving only the electrostatic, inductive, dispersive and repulsive interactions was adequate
for anisotropic liquids. Previously this has had common acceptance.
In 1980, one of the authors (R.E.) claimed that there are additional
specific interactions present in ordered liquids only. In this context, it seemed to be of interest to explore whether polar moieties
are required for nematic behavior. Furthermore, nematic hydrocarbons promised to have comparatively low viscosities, a property
urgently needed for nematic mixtures in electrooptical devices that
can be operated over a wide temperature range.

RESULTS AND DISCUSSION

As a first step in this direction of nonpolar nematics, the
cyano group of the 4-trans-4-alkylcyclohexyl-benzonitriles (1)[2] was
replaced by an n-alkyl group. A few of the resulting 4-trans-4-
alkylcyclohexyl-alkylbenzenes are listed in Table 1.

$$C_3H_7 - \langle \bigcirc \rangle - \langle \bigcirc \rangle - R \qquad (2)$$

Table I

Extrapolated clearing points ($^{\circ}$C) and viscosities
($mm^2 s^{-1}$ at 20°C)

R	Clp.	V
C_2H_5	-70	4.0
C_3H_7	-45	4.5
C_5H_{11}	-40	7.5

The clearing points of (2) are in the order of one hundred degrees below those of 4-trans-4-propylcyclohexyl-benzonitrile (PCH3). Their values, together with the viscosity values, were estimated by extrapolation from nematic mixtures of the PCH-type[8]. Despite their low practical clearing points, compounds (2) and their homologues are very valuable components for nematic phases that permit the operation of a twisted nematic cell at -40°C. In this context it should be mentioned that the alkoxy derivative (3) has a T_{NI} of 37°C[9]. This example

$$C_3H_7 - \langle \bigcirc \rangle - \langle \bigcirc \rangle - OC_2H_5 \qquad (3)$$

shows that molecules with one phenylcyclohexane unit can form enantiotropic nematic phases without the necessity for dimeric forms brought about by association[5]. Nevertheless, the general result[6] that the nematic state is promoted by polar bonds in the direction of the longest molecular axis of a sufficiently rigid molecule is also valid in this case.

The question which arose was: if by elongating the molecular shape an enantiotropic nematic phase could be obtained.   This was indeed the case as the examples in Table 2 show.

$$R-\langle \bigcirc \rangle-\langle \bigcirc \rangle-\langle \bigcirc \rangle-R' \qquad (4)$$

Table 2

Transition temperatures (^{o}C, C=crystalline,

S=smectic) of (4).

R	R'	C	S	N	I
C_3H_7	C_2H_5	66	134	166	
C_5H_{11}	H	58	81	98	
	CH_3	98	123	178	
	C_2H_5	34	146	164	
	C_4H_9	20	158	171	

The outstanding feature of the 4-trans-4-alkylcyclohexyl-4"-alkylbiphenyls of formula (4) is their high T_{NI} at a low value of viscosity, giving them a key role in forming broad range nematic mixtures.   In the case of R=C_5H_{11}, R"=C_2H_5 an extrapolated value of 20mm^{2}s^{-1} was measured.   Solubility in nematic matrices consisting of PCH-compounds is high[8] whereas solubility in 4'-alkylbiphenyl-4-carbonitriles is modest.

The above mentioned influences of polar groups on the T_{NI} is not so distinct in molecules with elongated hydrocarbon cores as the T_{NI} values for 4'-trans-4-pentylcyclohexyl-4-carbonitrile (5) and 4-methoxy-4'-trans-4-pentyl-biphenyl (6)[10] show.

$$C_5H_{11} - \langle \hexagon \rangle - \langle \bigcirc \rangle - \langle \bigcirc \rangle - CN \qquad (5)$$

C 95 N 219 I (°C)

$$C_5H_{11} - \langle \hexagon \rangle - \langle \bigcirc \rangle - \langle \bigcirc \rangle - OCH_3 \qquad (6)$$

C 80 N 165 I (°C)

The sequence of the rings in 4,4''-disubstituted hexahydro-p-ter-
phenyls is also known to have a huge impact on T_{NI} if the one
substituent is an alkyl and the other one a carbonitrile group[6].
As an example of this, the 1,4-bis(trans-4-alkyl)benzenes (7)

$$R - \langle \bigcirc \rangle - \langle \hexagon \rangle - \langle \bigcirc \rangle - R' \qquad (7)$$

are reported not to show any nematic phases at all[11]. Interestingly the terphenyls of type (8) also do not have nematic phases,
although it must be noted that the transition temperatures from the
smectic to the isotropic state are well above $200°C$[12].

$$R - \langle \bigcirc \rangle - \langle \bigcirc \rangle - \langle \bigcirc \rangle - R' \qquad (8)$$

The next step of our syntheses was the elongation of system
(4) by another 1,4-trans-cyclohexylene group leading to class (9).
As expected, the bis-(4-trans-4-alkylcyclohexyl)-biphenyls of (9)
have higher T_{NI} values than the corresponding compounds of type
(4). Compounds (9) have also an interesting variety of smectic
phases[13].

$$R - \langle \hexagon \rangle - \langle \bigcirc \rangle - \langle \bigcirc \rangle - \langle \hexagon \rangle - R' \qquad (9)$$

Table 3

Transition temperatures (^{o}C) of (9).

R	R'	C	S	S	S	N	I
C_3H_7	C_3H_7	155	210	220		325	
C_5H_{11}	H	68	181	184		190	
	CH_3	112	206	210		283	
	C_3H_7	61	232	251		311	
	C_5H_{11}	45	55	247	275	305	

The extrapolated viscosity of the pentyl propyl derivatives of (9) is only 42mm^{2}s^{-1} at 20°C. This common feature of the bisalkyl derivatives makes them valuable ingredients, used to raise the clearing point of nematic phases without elevating the viscosities too much; they also have good solubilities in PCH-compounds. The mono alkyl derivatives of (9) (R'=H) show distinctly higher bulk viscosities[8]; the reason for this is not clear.

Of course, the molecular building principle of combining 1,4-phenylene and 1,4-trans-cyclohexylene groups can be pursued further. Thus it was shown that 4,4'''-bis-(4-trans-pentylcyclo-hexyl)-p-quarterphenyl

C_5H_{11} —⬡—⬡—⬡—⬡—⬡—⬡— C_5H_{11} (10)

has a T_{NI} of 375°C. Solubilities of (10) in the liquid crystalline matrices tested turned out to be too low for a reasonable application in display technology, however.

In comparison to the few known examples of 4,4'''-dialkyl-p-quarterphenyls[14] and 4,4'''''-dialkyl-p-sexiphenyls[15], the hydro-genated systems (9) and (10), respectively, seemingly exhibit lower T_{NI} values. This can be attributed to the fact that because of the temperature dependence of the equilibrium between the equatorial-equatorial and axial-axial conformation in the cyclohexane rings, a noted deviation from a long shaped molecular form must be considered[6] for these elevated temperatures.

REFERENCES

1. W. Maier and A. Saupe, Z. Naturforsch. 13a, 564 (1958); 14a, 882 (1959); 15a, 287 (1960).

2. R. Eidenschink, D. Erdmann, J. Krause and L. Pohl, Angew. Chem. 89, 103 (1977).

3. R. Eidenschink, D. Erdmann, J. Krause and L. Pohl, Angew. Chem. 90, 133 (1978).

4. K. Tokita, K. Fujimura, S. Kondo and M. Takeda, Mol. Cryst. Liq. Cryst. 64, 171 (1981).

5. H. Schad and M. A. Osman, J. Chem. Phys. 75, 880 (1981).

6. R. Eidenschink, Kontakte (Merck) 1979 (1), 15; C.A. 92:21 614t.

7. Idem ibid. 1980 (3), 12; C.A. 94:148 560a.

8. G. Weber, F. Del Pino and L. Pohl, Proc. 10th Freiburger Arbeitstagung Fluessigkristalle (1980).

9. R. Eidenschink, J. Krause and L. Pohl, German Patent 26 36 684.

10. R. Eidenschink, D. Erdmann, J. Krause and L. Pohl, Deutsche Offenlegungsschrift 29 27 277.

11. H. Schubert, W. Schulze, H.-J. Deutscher, V. Uhlig and R. Kuppe, J. Phys. Colloq. 36, C 1 - 379 (1975).

12. H. Schubert, H. J. Lorenz, R. Hoffmann and F. Franke, Z. Chem. 6, 337 (1966).

13. W. H. de Jeu, in preparation.

14. D. Demus, H. Zaschke, Fluessige Kristalle in Tabellen, VEB Deutscher Verlag fuer Grundstoffindustrie, Leipzig 1974, p. 230.

15. Ibid., p. 232.

ACKNOWLEDGEMENT

This work has been supported by the Bundesministerium fuer Forschung und Technologie of the Federal Republic of Germany. The authors are solely responsible for the content.

SOME HETEROCYCLIC ANALOGUES OF BIPHENYL MESOGENS

D. J. Byron, D. Lacey and R. C. Wilson

Department of Physical Sciences, Trent Polytechnic
Burton Street, Nottingham NG1 4BU, England

INTRODUCTION

Extensive systematic studies of mesogens carrying six membered heterocyclic rings have not been undertaken. However, due to the success of the 4-n-alkyl- and 4-n-alkoxy-4'-cyanobiphenyls[1] in electro-optical 'field effect' displays of the twisted nematic type and the consequent interest in liquid crystals with a positive dielectric anisotropy, various heterocyclic analogues of the alkylcyanobiphenyls have been investigated. These studies have provided information on the effect of heterocyclic rings in promoting mesophase thermal stability. Thus, the isomeric 2,5-disubstituted cyano-alkylphenyl- and alkyl-cyanophenyl-pyrimidines have been reported by Scherrer and his co-workers[2] and certain of these compounds give rise to enantiotropic liquid crystals at relatively low temperatures. Nevertheless, the nematic mesophases of these substituted phenylpyrimidines have a higher thermal stability than the corresponding alkylcyanobiphenyls.[†] Hence, a 2-

---

[†] For example, the transition temperatures for 4'-n-pentyl-4-cyanobiphenyl[1] are C-N, 22.5°; N-I, 35°, whereas the corresponding values for 5-cyano-2-(4-n-pentylphenyl)pyrimidine[2] are C-N, 96°; N-I, 109°, and those for 5-n-pentyl-2-(4-cyanophenyl)pyrimidine[2] are C-N, 70.5-71°; N-I, [52°].

or 5-pyrimidyl ring promotes nematic thermal stability to a
greater extent than a phenyl ring. This finding is confirmed by
some related work by Zaschke et al[3] on phenylpyrimidines carrying
both alkyl and alkoxy substituents. These compounds are low
melting nematogens whereas the corresponding 4-n-alkyl-4'-n-
alkoxybiphenyls have no nematic mesophases and are only smectogenic.[4] Prior to the cited work of Scherrer and Zaschke and their
co-workers the only systematic studies undertaken had been on the
replacement of the central ring of a three ring mesogen[5] and these
had left undecided the effect of pyrimidyl relative to phenyl in
promoting nematic thermal stability.

COMPARISON OF PHENYL AND 2- AND 4-PYRIDYL RINGS IN MESOGENS

The effect of 2-, 3- and 4-pyridyl rings relative to phenyl
in promoting mesophase thermal stability has been investigated by
Young et al[6] and by Nash and Gray[7] but the results obtained were
conflicting and varied according to the system under investigation. In an attempt to obtain further information we have compared the properties of the 4-biphenylyl 4"-n-alkoxybenzoates
(Ia)[8] with those of the corresponding 4-(2'-pyridyl)phenyl and the
4-(4'-pyridyl)phenyl 4"-n-alkoxybenzoates (Ib and Ic)[9] in order to
determine the effect of replacing the terminal phenyl ring of the
biphenyl system by a 2- or a 4-pyridyl ring.

The transition temperatures for these three series are shown
plotted against the number of carbon atoms, n, in the n-alkyl
chain in Figs. 1, 2 and 3. It is apparent that although the
series differ only in the nature of the terminal substituent, Ar,
an alteration that causes no marked change in molecular size or

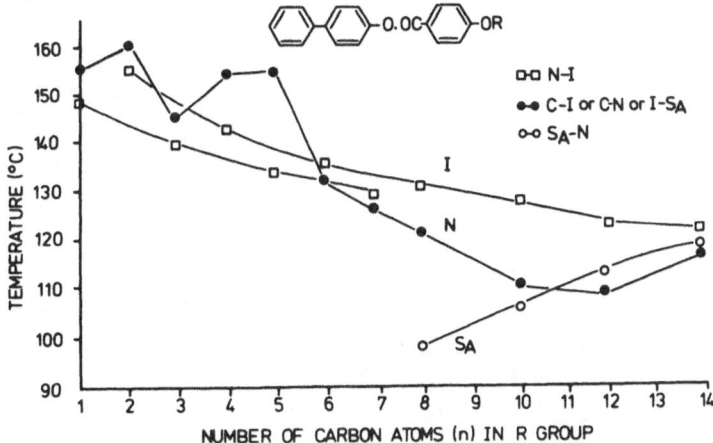

Fig. 1.   Plot of transition temperatures against alkyl chain
          length (n) for 4-biphenylyl 4"-n-alkoxybenzoates (Ia).

shape, nevertheless their liquid crystal behaviour is markedly
different. Thus when Ar=phenyl the compounds exhibit the most
commonly observed type of plot of transition temperature against
n, but when Ar=2-pyridyl the series is wholly nematogenic[†] and
when Ar=4-pyridyl nematic properties are confined to the early
members of the series, the later members showing only smectic
behaviour.

The three series may be compared by evaluation of the effect
of the terminal substituent on the thermal stability of the
smectic and the nematic phases and in Table I mean values of the
transition temperatures of certain comparable members of each
series are listed. The order of smectic (S_A) thermal stability is
Ar=4-pyridyl>phenyl>2-pyridyl, but the order of nematic thermal
stability is Ar=phenyl>2-pyridyl>4-pyridyl. Thus, the thermal stabilities of the S_A and the nematic phases are affected differently

---

[†] For 4-(2'-pyridyl)phenyl 4"-n-tetradecyloxybenzoate we have
estimated[10] the S_A-N transition temperature as 75° by extrapolation of values obtained for mixtures of the compound with the
corresponding 4-biphenylyl ester.

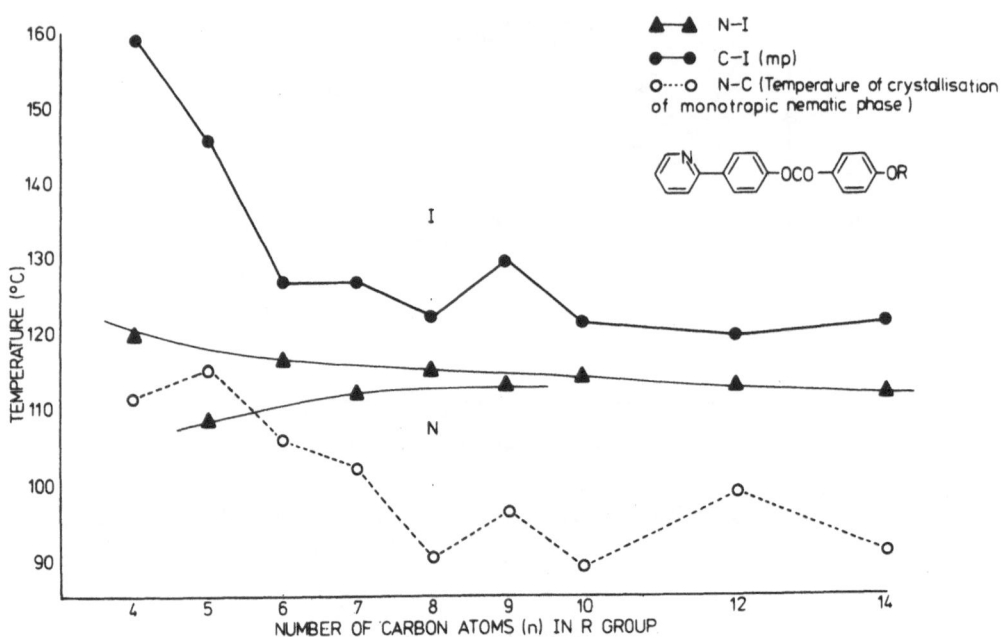

Fig. 2.   Plot of transition temperatures against alkyl chain
           length (n) for 4-(2'-pyridyl)phenyl 4"-n-alkoxybenzoates
           (Ib).

according to the position of the hetero-atom in the terminal ring.
When Ar=phenyl is replaced by Ar=4-pyridyl the smectic thermal
stability is increased by approximately 16.5° showing that the
presence of the terminal 4-pyridyl nitrogen atom slightly increases
the anisotropy of the molecular polarisability. In contrast, when
Ar=2-pyridyl, the lone pair of the nitrogen atom is laterally posi-
tioned, and the adverse dipole that this creates, possibly associated with the increase in molecular breadth resulting from a small
angle of twist between the 2-pyridyl ring and the benzene ring,
disrupts the lateral cohesive forces to such an extent that the
compounds do not show smectic phases. When Ar=2-pyridyl and Ar=4-
pyridyl, nematic thermal stability is substantially decreased
relative to Ar=phenyl, the decreases being 25.5° and 23.2°,
respectively.

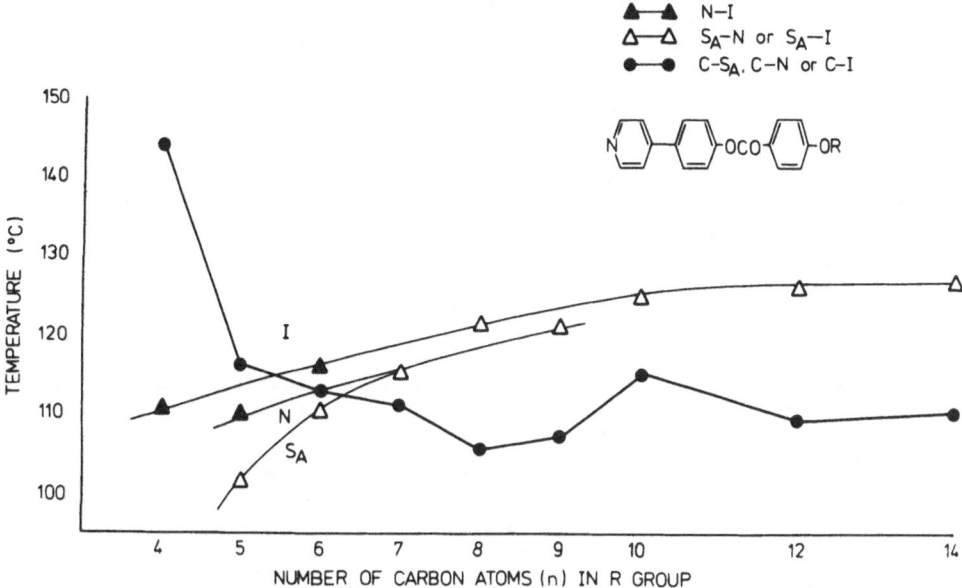

Fig. 3. Plot of transition temperatures against alkyl chain
         length (n) for 4-(4'-pyridyl)phenyl 4"-n-alkoxybenzoates
         (Ic).

Previous work on certain Schiff's bases (II) by Young et al[6]

RO—⟨◯⟩—⟨◯⟩—N=CHAr   ArCH=N—⟨◯⟩—⟨◯⟩—N=CHAr

(II)   (III)

and by Nash and Gray[7] is, in part, in agreement with these findings. For the thermal stability of the S_A phase (for R=n-octyl)
and the nematic phase (for R=CH$_3$) identical orders (Ar=4-pyridyl>
phenyl>2-pyridyl) were obtained. This order is the same as that
observed for the influence of the aryl group in promoting S_A
thermal stability of the esters discussed above, although it
conflicts with that obtained for the order of nematic thermal stability of these esters. For the dianils (III), Nash and Gray
obtained the order Ar=phenyl>4-phenyl and 2-pyridyl (neither of
the last two compounds being mesomorphic) for the influence of the

TABLE I

Mean m.p. and liquid crystal thermal stabilities ($^{\circ}$C) for

$$Ar - \underset{}{\bigcirc} - 0.0C - \underset{}{\bigcirc} - OC_nH_{2n+1}$$

Ar group	m.p. [a] (n=4-10,12,14)	Smectic A (n=8,10,12,14)	Nematic (n=4,5,6)
(phenyl)	127	108.1	137.7
(2-pyridyl)	130	–	114.5
(4-pyridyl)	114.5	124.6	112.3

[a] C-S, C-N or C-I transition

terminal substituents on nematic phase thermal stability. This order is very similar to that obtained for the esters discussed above. Nash and Gray suggest that in passing from the Schiff's bases (II) to the dianils (III) the inversion in the positions of phenyl and 4-pyridyl in the nematic thermal stability order may be due to large repulsive interactions between two 4-pyridyl rings placed end to end which will adversely affect the thermal stability of the nematic phase. It is possible that repulsion between nitrogen lone pairs of the heterocyclic rings and/or the oxygen atom of the alkoxy group may occur in the arrangement of the molecules in the nematic phases of the 4-(2'-pyridyl)phenyl and 4-(4'-pyridyl)phenyl 4"-n-alkoxybenzoates. This may be responsible for the relatively low thermal stability of these heterocyclic mesogens.

STERIC EFFECT OF THE NITROGEN ATOM OF 2-PHENYLPYRIDINE

The absence of smectic behaviour shown by the 4-(2'-pyridyl)-
phenyl 4"-n-alkoxybenzoates may be associated with the hetero-
nitrogen atom causing twisting about the bond connecting the
benzene ring to the 2-pyridyl ring. From studies of the influence
of 2- and 2'-substituents on the thermal stability of nematic
phases of biphenyl mesogens (eg. the Schiff's bases (IV) where
R=n-octyl-n-decyl and X=F, Cl, Br, I, Me, NO$_2$), Gray et al[11]
established that as the size of the 2-substituent, X, increases,
the mesophase thermal stability decreases, partly because the
molecule is (a) broadened by the substituent, but mainly because
the molecule is (b) thickened by the increase the substituent
causes in the average interplanar angle of the biphenyl system.
The overall steric effect [(a) + (b)] increases the separation of

RO—⟨◯⟩—CH = N—⟨◯⟩—⟨◯⟩—N = CH—⟨◯⟩—OR

(IV)

the long molecular axes and reduces the thermal stability of the
ordered arrangement of molecules in the liquid crystal, smectic
thermal stability being affected more than nematic thermal
stability.

In order to establish whether or not the hetero-nitrogen atom
of the 4-(2'-pyridyl)phenyl 4"-n-alkoxybenzoates(Ib) exerts a
steric effect we have compared[10] the liquid crystal behaviour of
this series with that of the corresponding 2'-chloro- and 2'-fluoro-
4-biphenylyl 4"-n-alkoxybenzoates (Vb and Vc). For the latter compounds, by analogy with the Schiff's base derivatives (IV), we

⟨◯⟩—⟨◯⟩—O.OC—⟨◯⟩—OR   (a) X = H
                          (b) X = Cl
(V)   (c) X = F

assumed that the 2'-halogeno-substituent exerts a steric effect
that causes a decrease in mesophase thermal stability relative to

the unsubstituted esters, the 4-biphenylyl 4"-n-alkoxybenzoates
(Va).

In Fig. 4, plots 3 and 4 show the N-I transition temperatures
of the 2'-fluoro- and 2'-chloro-4-biphenylyl 4"-n-alkoxybenzoates
(Vc and Vb) plotted against the number of C atoms in the n-alkoxy
chain. Also shown, for comparison, are a representative part of
the analogous plot for the 4-biphenylyl 4"-n-alkoxybenzoates (Va)
(Plot 1) and that for the 4-(2'pyridyl)phenyl 4"-n-alkoxybenzoates
(Ib) (Plot 2). As expected, it is clear that substitution of the
4-biphenylyl 4"-n-alkoxybenzoates by 2'-chloro- and 2'-fluoro-
substituents markedly lowers nematic thermal stability and almost
completely eliminates smectic properties. The extent of the substituent effect, which is in proportion to the size of the substituent (and is therefore a steric effect) is clear by inspection of
Fig. 4. For the 2'-chloro-compounds (Plot 4) only monotropic
nematic phases were observed and no mesophase transitions could be
obtained by supercooling the melts of the n-decyloxy and n-tetradecyloxy compounds. For the 2'-fluoro-compounds (Plot 3) the
early members also give monotropic nematic phases and only the
n-tetradecyloxy compound gives a smectic phase.

Table II lists comparisons of mean values for analogous members of the 4-biphenylyl 4"-n-alkoxybenzoates and (i) their 2'-
fluoro-derivatives, showing that the 2'-fluoro-substituent decreases
the thermal stability of both smectic and nematic mesophases by
37.5°, (ii) their 2'-chloro-derivatives, indicating that the 2'-
chloro-substituent lowers the thermal stability of the nematic
phase by 88.2°, and (iii) the 4-(2'-pyridyl)phenyl 4"-n-alkoxyben-
zoates, revealing that the presence of the hetero-nitrogen atom
decreases the thermal stability of the nematic phase by 16.6° and
that of the smectic phase by an estimated 41.5°. Corresponding
values for the effect of a 2-halogeno-substituent on the thermal
stability of the mesophases of the 4,4'-di-(p-n-alkoxybenzylidine-

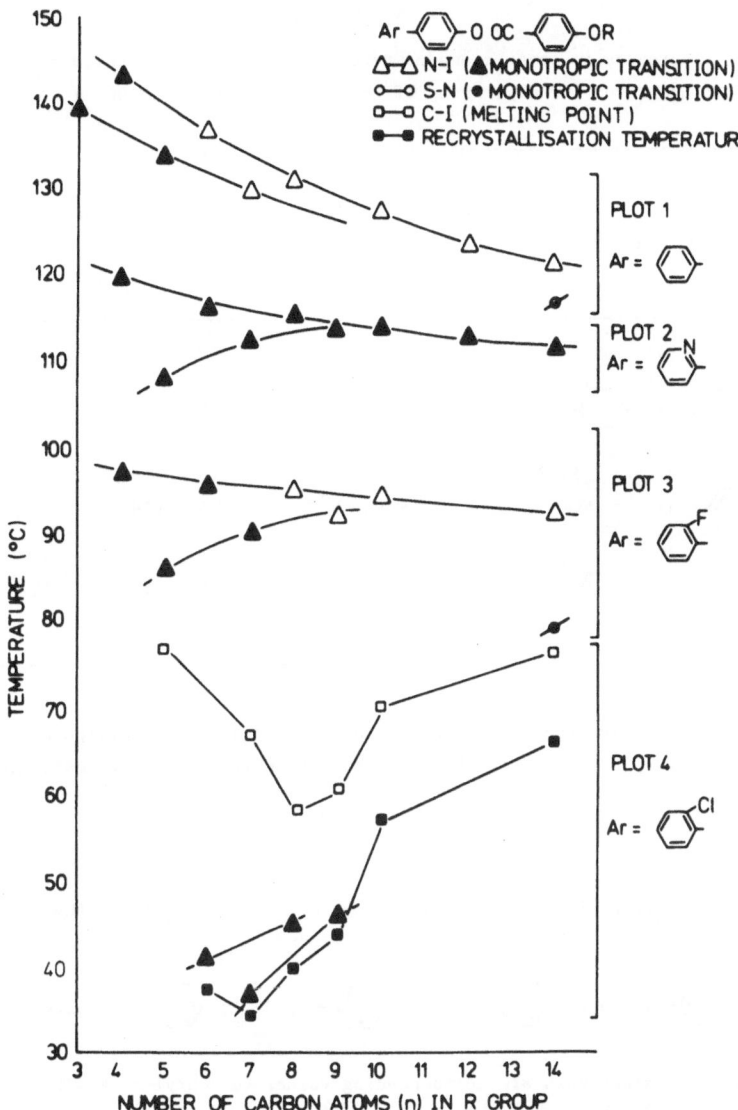

Fig. 4.   Plots of transition temperatures against alkyl chain
           length (n) for various 4-substituted phenyl 4"-n-alkoxy-
           benzoates (I and V).

amino)biphenyls (IV) are given in parentheses.   The results infer
that (a) the presence of the hetero-nitrogen atom at the 2-position
of the terminal ring of the 4-(2'-pyridyl)phenyl 4"-n-alkoxyben-

TABLE II

Decreases in smectic and nematic thermal stability caused by
changes of the aryl group of the series Ar——⟨ ⟩—— O.OC ——⟨ ⟩—— OR

| Change of Ar group | Decrease in liquid crystal thermal stability($^{\circ}$C) | |
	Smectic[a]	Nematic[b]
⟨ ⟩ to ⟨N⟩	41.5	16.6
⟨ ⟩ to ⟨F⟩	37.5(58.5)[c]	37.5(29.4)
⟨ ⟩ to ⟨Cl⟩	— (157.5)	88.2(84.1)

[a]Decreases are for the n-tetradecyl ethers. For 4-(2'-pyridyl)phenyl
4"-n-tetradecyloxybenzoate, the S_A-N transition temperature is estimated [10] as
75°.

[b]Decreases are derived from the following mean values ($^{\circ}$C) for the n-hexyl-n-nonyl
ethers:

Ar = ⟨ ⟩ , 130.6; ⟨N⟩ , 114; ⟨F⟩ , 93.1; ⟨Cl⟩ , 42.4.

[c]Values in parentheses are corresponding values (n-heptyl-n-decylethers) for
2-substituted 4,4'-di-(p-n-alkoxybenzylideneamino)biphenyls.

zoates causes an increase in separation of the long molecular axes
which reduces the thermal stability of the ordered molecular arrangement of the mesophases, and (b) this is due to a steric
effect similar to that which reduces mesophase thermal stability
of 2-halogenobiphenyl mesogens. As a hetero-nitrogen atom and its
lone pair is less bulky and hence has a smaller steric effect than

a benzenoid carbon atom bound to a halogeno-substituent it is to be expected that the decrease in nematic thermal stability caused by the hetero-nitrogen atom at the 2-position will be smaller than that caused by a 2-fluoro-substituent in the analogous biphenyl derivatives. Although the results are limited, the reverse trend is shown for the effect on smectic thermal stability. This may be due to repulsive interactions between the nitrogen lone pairs of adjacent molecules giving rise to an additional adverse effect on smectic thermal stability.

COMPARISON OF THE N-OXIDE AND CYANO-GROUPS AS MESOGENIC TERMINAL SUBSTITUENTS

The 4-n-alkyl- and 4-n-alkoxy-4'-cyanobiphenyls have proved very effective in electro-optical display devices of the twisted nematic type. The success of these compounds is due, in part, to the presence of the terminal strongly dipolar cyano-group. The N-oxide function has similarities to the cyano-group. For example, 4-phenylpyridine-N-oxide (VI) and 4-cyanobiphenyl (VII) have dipole

(VI)   (VII)

moments of $4.61D$[12] and $4.33D$,[13] respectively, and in each case the dipole is oriented along the major molecular axis. In order to explore whether or not this similarity extends to an analogous influence on liquid crystal behaviour we have investigated the N-oxides of certain of the 4-(4'-pyridyl)phenyl 4"-n-alkoxybenzo-ates(VIII)[14] and compared their mesomorphic behaviour with analog-ous members of the 4'-cyano-4-biphenylyl 4"-n-alkoxybenzoates (IXa) and the 4"-n-alkoxyphenyl 4'-cyanobiphenyl-4-carboxylates(IXb) re-ported by Coates and Gray.[15,16] A direct comparison of the N-oxide function and the cyano-group as mesogenic terminal substituents is thus possible.

RO—⟨ ⟩—CO.O—⟨ ⟩—⟨ ⟩—N→O   RO—⟨ ⟩—X—⟨ ⟩—⟨ ⟩—CN $\begin{array}{l}(a) X = CO.O \\ (b) X = O.OC\end{array}$

(VIII)   (IX)

The plot of transition temperatures against alkyl chain
length for the N-oxides of the 4-(4'-pyridyl)phenyl 4"-n-alkoxy-
benzoates is shown in Fig. 5. Certain of the high N-I transition
temperatures must be regarded as approximate as the compounds dark-
ened on exposure to sunlight and were subject to thermal decomposition under the conditions used to determine the transition temperatures. Two smectic mesophases were exhibited by some members
of the N-oxide series, but only the most thermally stable modification was identified, as smectic A. The general trends of the
transition temperature plots of the 4-(4'-pyridyl)phenyl 4"-n-
alkoxybenzoates (Fig. 3) and their N-oxides (Fig. 5) are very
similar and comparison of the two series reveals that the presence
of the N-oxide function increases both smectic and nematic thermal
stability by similar amounts (S_A by 73°; N by 71°). Alternatively,

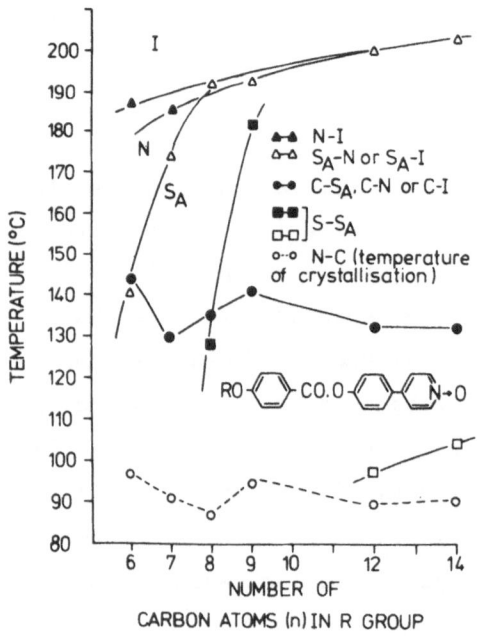

Fig. 5. Plot of transition
temperatures against alkyl
chain length (n) for N-
oxides of 4-(4'-pyridyl)-
phenyl 4"-n-alkoxybenzoates
(VIII).

the effect of the N-oxide function may be evaluated by comparision with the homologous series of 4-biphenylyl 4"-n-alkoxybenzoates (Ia).[8] This series gives rise to a plot of a different type but the comparison shows that the presence of the N-oxide function increases smectic thermal stability appreciably more than nematic thermal stability (S_A by 90°; N by 51°). The effect on liquid crystal thermal stability of the introduction of the cyano-group in the terminal position of the 4-biphenylyl 4"-n-alkoxybenzoates (Ia)[8] and the 4"-n-alkoxyphenyl biphenyl-4-carboxylates[17] to give the 4'-cyano-substituted esters (IXa and IXb)[15,16] is that the terminal cyano-group, which can conjugate with the aromatic system so increasing the overall axial molecular polarisability, enhances the thermal stability of both smectic and nematic mesophases (S_A by 101° for series (IXa); N by 107° for series (IXa) and 122° for series (IXb)). Although the data are limited, it appears that the general pattern of the plots of transition temperature against n for the unsubstituted esters[8,17] is not markedly altered by the introduction of the terminal cyano-group. However, both smectic and nematic properties are promoted and certain members[†] of both series (IXa and IXb) exhibit an S_B phase in addition to the S_A modification. Similar enhancement of smectic and of nematic thermal stability and the appearance of smectic polymorphism occurs when the terminal hetero-nitrogen atom of the 4-(4'pyridyl)phenyl 4"-n-alkoxybenzoates is converted into the corresponding N-oxide.

Clearly, the N-oxide function and the cyano-group resemble each other in their effects on liquid crystal thermal stability. Both substituents enhance mesophase thermal stability when situated in a terminal position in a mesogen, although the cyano-group promotes nematic thermal stability more than the N-oxide function, which, conversely, substantially enhances smectic properties.

[†] See footnote on p. 230 of reference 14.

The compounds that provided the data from which these comparisons were made have mesophases of rather high thermal stability and we felt that it would be worthwhile to compare the behaviour of the N-oxide function and the cyano-group as terminal
substituents in mesogens that give rise to liquid crystals of much
lower thermal stability. Appropriate reference compounds of this
type are the 4-n-alkoxy-4'-cyanobiphenyls (X).[1] For comparison
with these compounds we investigated an analogous series of pyri-
dine-N-oxides, namely the N-oxides of the 4-(4'-n-alkoxyphenyl)-
pyridines (XI).[18] In both (X) and (XI) the powerful dipole
moments due to the N-oxide function and the cyano-group are
oriented along the major molecular axis.

RO—⟨圖⟩—⟨圖⟩—C≡N   RO—⟨圖⟩—⟨圖⟩—N→O
        (X)   (XI)

The N-oxides (XII) were not sensitive to light and were not
subject to decomposition when heated to determine their transition
temperatures. The transition temperatures for the N-oxides (XII)
are shown plotted against the number of carbon atoms, n, in the
n-alkyl chain in Fig. 6, and, for comparison, the plot, for the
analogous 4-n-alkoxy-4'-cyanobiphenyls (X)[1] is shown in Fig. 7.
In contrast with the cyanobiphenyls (X) the members of the N-oxide
series show no nematic properties and give rise to smectic A
phases only. For the first few members these smectic phases are
monotropic but the S_A-I transition temperatures then increase
rapidly as n increases. Comparison of the plots shown in Figs. 6
and 7 shows that the first two members of the series of N-oxides
have a lower liquid crystal thermal stability (S_A-I transitions)
than the analogous 4-n-alkoxy-4'-cyanobiphenyls (N-I transitions).
However, the gradients of the S_A-I curves for the N-oxides are
substantially greater than for the cyanobiphenyls and the later
members of the N-oxide series all have higher transition temperatures than their cyano-analogues. The mean values for S_A thermal

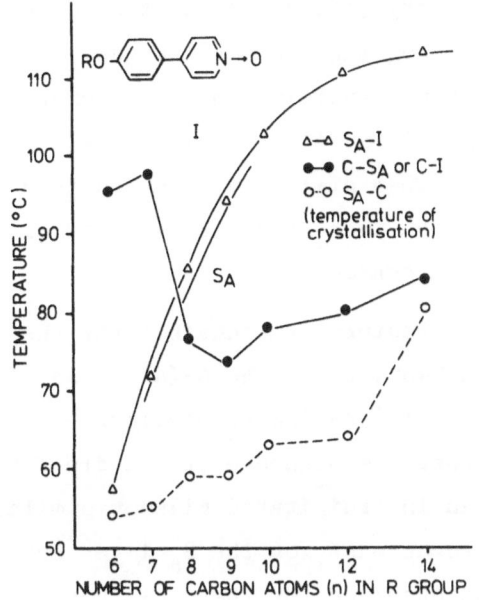

Fig. 6. Plot of transition temperatures against alkyl chain length (n) for N-oxides of 4-(4'-n-alkoxy-phenyl)pyridines(XI).

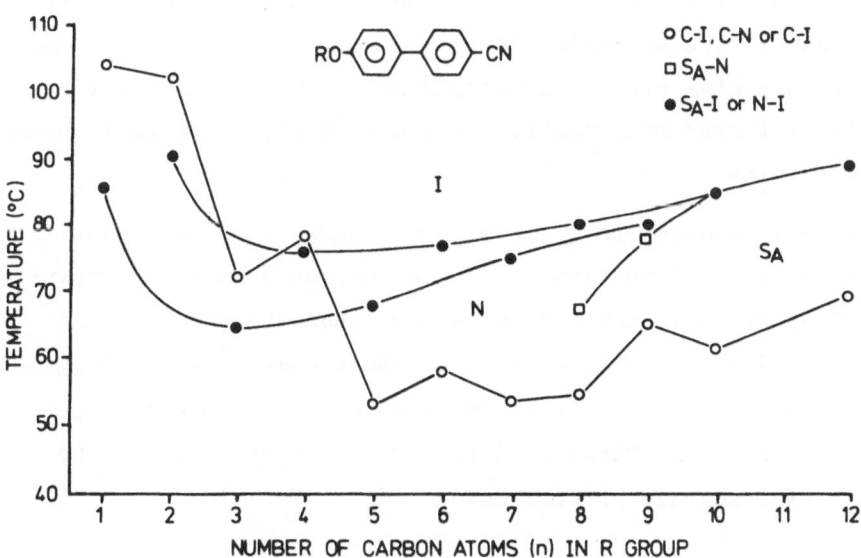

Fig. 7. Plot of transition temperatures against alkyl chain length (n) for 4-n-alkoxy-4'-cyanobiphenyls(X)

stability of corresponding members (n=8,9,10,12) of the two series
are 98.3° for the N-oxides and 79.5° for the cyano-compounds. As
the N-oxides of the 4-(4'-n-alkoxyphenyl)pyridines have no nematic
properties a comparison with the 4-n-alkoxy-4'-cyanobiphenyls
cannot be made, but it is clear that when similarly situated in a
terminal position, the N-oxide function increases smectic thermal
stability substantially more than the cyano-group.

There are no obvious structural features responsible for the
pronounced smectic character of the N-oxides of the 4-(4'-pyridyl)
phenyl 4"-n-alkoxybenzoates and the 4-(4'-n-alkoxyphenyl)pyri-
dines.[18] We have suggested [14,18] that the presence of the dipolar
N-oxide function may give rise to an interdigitated bilayer similar
to that which has been proposed[19] to account for the S_A layer spacings of 1.5 - 1.6 molecular lengths observed for the 4-n-alkoxy-
and 4-n-alkyl-4'-cyanobiphenyls.[1] It has been suggested that the
presence of the strongly dipolar terminal cyano-group in these
compounds may be the cause of the bilayer arrangement and that
this interdigitated structure may enhance smectic thermal stability. Hence, the strongly dipolar N-oxide function in a terminal
position may give rise to a bilayer arrangement which is the
structural factor responsible for the markedly smectogenic nature
of the N-oxides.

It may be noted that the 4-(4'-n-alkoxyphenyl)pyridines
themselves are not mesogenic. These compounds melt and recrystallise at appreciably lower temperatures than the corresponding
N-oxides. However, a much smaller dipole moment[†] acts along the
long molecular axis of these compounds and no terminal 'substituent' is present. These factors may be responsible for the
absence of mesogenic properties.

[†] 4-Phenylpyridine has a dipole moment of 2.55D.[20]

REFERENCES

1. G. W. Gray, J. Phys. (Paris) <u>36</u>, Cl, 337 (1975); G. W. Gray,
    K. J. Harrison, and J. A. Nash, J. Chem. Soc. Chem.
    Comun. 431 (1974); G. W. Gray and A. Mosley, J. Chem.
    Soc. Perkin II, 97 (1976).

2. A. Boller, M. Cereghetti, M. Schadt and H. Scherrer, Mol.
    Cryst. Liq. Cryst. <u>42</u>, 215 (1977).

3. H. Zaschke, H. Schubert, F. Kuschel, F. Dinger and D. Demus,
    DDR Patent, 95,892 (1973); H. Zaschke, J. Prakt. Chem.
    <u>317</u>, 617 (1975).

4. D. Demus, L. Richter, D. E. Rurup, H. Sackman and H. Schubert,
    J. Phys. (Paris), <u>36</u>, Cl, 349 (1975); D. Demus, H. Demus
    and H. Zaschke, Flussige Kristalle in Tabellen (VEB
    Deutscher Verlag für Grundstoffindustrie, Leipzig,
    1976).

5. H. Schubert and H. Zaschke, J. Prakt. Chem. <u>312</u>, 494 (1970);
    H. Schubert, Wiss Z. Univ. Halle XIX, '70M, H.5, S.1.

6. W. R. Young, I. Haller and L. Williams, <u>Liquid Crystals and</u>
    <u>Ordered Fluids</u>, J. F. Johnson and R. S. Porter, eds.,
    Plenum Press, New York, 1970, p. 383.

7. J. A. Nash and G. W. Gray, Mol. Cryst. Liq. Cryst. <u>25</u>, 299
    (1974).

8. D. J. Byron, D. Lacey and R. C. Wilson, Mol. Cryst. Liq.
    Cryst. <u>45</u>, 267 (1978). See also, B. K. Sadashiva and G.
    S. R. Subba Rao, Curr. Sci. <u>44</u>, 222, (1975); B. K.
    Sadashiva, Mol. Cryst. Liq. Cryst. <u>55</u>, 135 (1979).

9. D. J. Byron, D. Lacey and R. C. Wilson, Mol. Cryst. Liq.
    Cryst. <u>62</u>, 103 (1980).

10. D. J. Byron, D. Lacey and R. C. Wilson, Mol. Cryst. Liq.
    Cryst. <u>73</u>, 273 (1981).

11. For example, see the following reviews for the original
    references: G. W. Gray, <u>Liquid Crystals and Plastic</u>
    <u>Crystals</u>, G. W. Gray and P. A. Winsor, eds., Ellis

Horwood, Chichester, 1974, 1, p. 130, et seq; G. W. Gray, Advances in Liquid Crystals, G. H. Brown, ed., Academic Press, London and New York, 1976, 2, p. 45 et seq.

12. A. N. Sharpe and S. Walker, J. Chem. Soc. 4522 (1961).

13. K. B. Everad, L. Kumar, and L. E. Sutton, J. Chem. Soc. 2807 (1951).

14. D. J. Byron, D. Lacey and R. C. Wilson, Mol. Cryst. Liq. Cryst. 75, 225 (1981).

15. D. Coates and G. W. Gray, Mol. Cryst. Liq. Cryst. 37, 249 (1976).

16. D. Coates and G. W. Gray, Mol. Cryst. Liq. Cryst. 31, 275 (1975).

17. D. J. Byron, D. Lacey, and R. C. Wilson, Mol. Cryst. Liq. Cryst. 51, 265 (1979).

18. D. J. Byron, D. Lacey and R. C. Wilson, Mol. Cryst. Liq. Cryst. 76, 253 (1981).

19. G. W. Gray, Proc. Int. Conf. Liquid Cryst., 5th, 1974; J. E. Lydon and C. J. Coakley, Proc. Int. Conf. Liquid Cryst., 5th, 1974; J. Phys. (Paris), 36, C1, 45 (1975); G. W. Gray and J. E. Lydon, Nature (London), 252, 221 (1974).

20. A. N. Sharpe and S. Walker, J. Chem. Soc. 2974 (1961).

ALIGNMENT OF LIQUID CRYSTAL MOLECULES ON VARIOUS SURFACES:

MYTHS, THEORIES, FACTS

Joseph A. Castellano

Stanford Resources, Inc., P. O. Box 20324
San Jose, CA   95160

INTRODUCTION

Over 70 years ago, Mauguin[1] produced an aligned liquid crystal layer by sandwiching p-azoxyanisole between two glass plates, the surfaces of which had been rubbed unidirectionally with paper. Ever since that classic experiment, scientists have been attempting to study and understand the mechanism of rubbing and the microscopic nature of the liquid crystal/surface interaction.

Because many of these studies were carried out sporadically and over such a long period of time, it is not surprising that a good deal of confusion has arisen regarding the relation between the results obtained and the nature of the experiments performed. In this paper, many of the methods used in past studies are categorized and the myths, facts and theories are identified. In addition, results of some recent work on the alignment of cyanobiphenyl liquid crystals on silicon monoxide films deposited and treated in various ways will be presented.

Two terms which will be mentioned often in this paper are HOMOGENEOUS and HOMEOTROPIC alignment. Homogeneous alignment results when liquid crystal molecules lie parallel to a surface; it becomes

uniform when all the molecules at the surface point in the same
direction. In homeotropic alignment, the molecules are oriented
with their long axes perpendicular to the surface. Another term
used is the tilt angle which is defined as the angle between the
long molecular axis and the surface.

The problem with approaching a subject of this nature is that
many different materials have been used as the alignment substrate
or "alignment layer" including both organic and inorganic substances
and mixtures of the two. The techniques used to form these materials
on the surfaces of such things as glass, ceramics and organic
polymers include: dip coating, spraying, roller coating, offset
printing, sputtering, high-vacuum sublimation (evaporation) and
even chemical reaction at the surface. Moreover, a number of
techniques have been used to create preferred molecular directionality
and/or tilt of the molecules on these alignment layer surfaces
including: rubbing, high-speed buffing, mechanical abrasion, chemical
milling, ion milling, high vacuum sublimation at oblique angles and
laser milling. And finally, the type of materials used to create
these latter effects has included such substances as polyesters,
cellulose, cellulose acetate, nylon, alumina, diamond, silicon
monoxide, magnesium fluoride, silicon carbide and the list goes on.
In too many cases, the type of material used is not even identified,
merely being referred to as a piece of "cloth". Thus it becomes
clear that one not only has a scientific problem but a logistic
one as well.

DISCUSSION

In reviewing and organizing these techniques, materials and
effects, it is convenient to separate them into two major categories:
(A) organic alignment layer materials and (B) inorganic films and
substrates. Within each category we can then identify the following:

1.   Specific alignment layer material

2.   Technique for forming alignment layer

3.   Method and specific material used to create
      directionality and/or molecular tilt

4.   Observed result(s) - THE FACTS

5.   Proposed mechanisms - THE THEORIES

6.   Assumptions based on hearsay - THE MYTHS

To simplify our discussion, we will refer to the use of glass substrates exclusively.

A.   Organic Films and Materials

The first known report of surface alignment of liquid crystals
was provided by Mauguin[1] in 1911 when he rubbed a raw glass plate
with a piece of paper and obtained uniform homogeneous alignment of
p-azoxyanisole.   This technique was then used by a number of
researchers who fabricated cells for examination by electric and
magnetic fields.[2]   Chatelain,[3] theorized that the orientation
resulted from dipole interactions between an ordered layer of adsorbed
fatty impurities and the nematic molecules.   However, he did not rule
out alteration of the substrate surface itself.

From these reports and other papers, the myth developed that
"rubbing always produces homogeneous alignment".   Creagh and Kmetz[4]
found, however, that lecithin, a surface active agent, produced
homeotropic alignment of liquid crystals such as MBBA
(p-methoxybenzylidene-p'-n-butylaniline), regardless of whether the
surface was rubber or not.   The use of lecithin as a homeotropic
aligning agent has also been reported by Hiltrop and Stegemeyer.[5]
They proposed a mechanism in which the lecithin is visualized as
forming a brush-like film on the surface with the long alkyl chains
pointing out of the surface.   Holes of molecular dimensions in this
film are penetrated by the nematic molecules which become anchored
with their long axes protruding out of this monomolecular layer.
Elastic interaction between these anchored molecules and adjacent

molecules in the bulk cause the homeotropic alignment.

In a related study, Kahn[6] found that a quaternary ammonium salt containing silicon and having a long alkyl chain such as N,N-dimethyl-N-octadecyl-3-aminopropyltimethoxysilyl chloride produced homeotropic alignment of MBBA. In these cases the surface active agent has an ionic "head" and long, non-polar "tail". A mechanism which supports the observed facts would say that the "head" becomes anchored to the substrate with the non-polar "tail" protruding from it. The alkyl chain of the MBBA molecule would then be attracted to the tail and the long axes of the MBBA molecules would then be perpendicular to the substrate surface (ergo, homeotropic alignment) as shown in Figure 1. The attractive force between the head of the aligning layer material and the glass surface may be so great that even rubbing may not be sufficient to disrupt it.

Another myth which has developed over the years is that uniform homogeneous alignment is caused by grooves in the organic alignment layer material as a result of the rubbing action. Although this is certainly a plausible mechanism, it is by no means the only mechanism.

It is widely known, although little has been published, that rubber polymer films of such polymers as polyvinyl alcohol, poly-imides, polyesters, polysiloxanes, etc. will produce uniform homo-geneous alignment and very small tilt angles of most liquid crystals which come into contact with them. One of the major problems with much of this work has been the lack of consistency. In some cases, the rubbing is done with diamond paste, while in others a "soft" organic material is used. In the diamond rub case, microgrooving of the surface coating is easier to visualize; it might be classified as grinding by the definition of Adamson.[7] The mechanism in this case would then involve the molecules lying in the grooves with their long axes parallel to the groove direction.

In most cases, however, the polymer film is rubbed or buffed with another organic polymer material, often described as the "cloth".

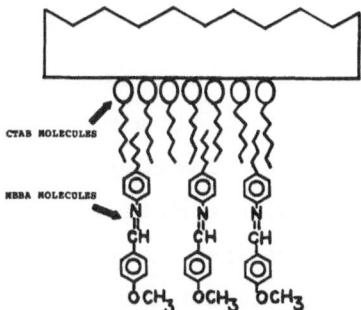

Fig. 1   Proposed homeotropic alignment mechanism of MBBA on CTAB
         treated surface.
         CTAB = Cetyltrimethylammonium Bromide
         MBBA = p-Methoxybenzylidene-p'-Butylaniline

Because of the proliferation of the synthetic fibers industry during
the past 40 years, which has apparently been ignored by workers in
this field, cloth can have a variety of chemical structures. It is
the nature of the rub material as well as the surface coating that
determines the microscopic structure of the treated surface layer.
According to Adamson,[7] the action of rubbing or buffing, sometimes
called polishing, produces very high localized heating which results
in melting of one of the polymer materials. The higher melting
material causes melting of the lower melting material. Thus, if
one rubs a layer of polyvinyl alcohol (melting point 200 degrees C)
with a Dacron polyester cloth (melting point 260 degrees C), the
polyvinyl alcohol layer will melt and its long molecular chains
will reorient themselves in the direction of the rubbing; this reorientation is then "frozen-in" when the material cools down. The
orientation of polymer molecules in films by external forces is
well known[8] and well studied by X-ray diffraction and other
techniques. The alignment and tilt of liquid crystal molecules on
such treated surfaces then becomes a complex combination of geometrical (steric) factors and Van der Waal's interactions between the

oriented polymer and liquid crystal molecules.

The mechanism proposed above is an alternative to the grooving
theory. It can also be used to explain the Mauguin case (rubbing
glass with paper) as follows: The high local temperatures created
during the rubbing of glass, a high melting material, with paper
(cellulose) results in a transfer of the melted cellulose material
to the glass with the long polymer chains oriented in the rub
direction. The cellulose molecules become anchored to the surface
via hydrogen bonds. The liquid crystal molecules then become
aligned to these oriented chains. It is an experimental fact that
you cannot destroy the alignment layer with ordinary solvents but
you can destroy it by heating above 250 degrees C (above the melting
point of cellulose).

B.   Inorganic Films and Substrates

I.   Oblique Evaporation Method. One of the most widely used
techniques for the alignment of liquid crystal materials on surfaces
was the oblique evaporation of various silicon oxides. The first
report of the use of this technique was made by Janning.[9] It
involved depositing thin silicon monoxide films, 100 Angstrom units
or less, onto substrates at an angle of approximately 85 degrees to
the normal (Figure 2, Ψ = 85 degrees).

The angular deposit causes the film to grow in a preferred
direction. Liquid crystals applied to such a surface, become
aligned in the direction of film growth. That is, the molecular
director points toward the direction of evaporation (in the Y-Z
plane of Figure 2).

In Janning's experiment both the front and rear plates were
treated in the same way and the plates were then assembled into a
twisted-nematic field effect device using MBBA as the liquid
crystal material.

Guyon and his coworkers[10] studied the technique further, being

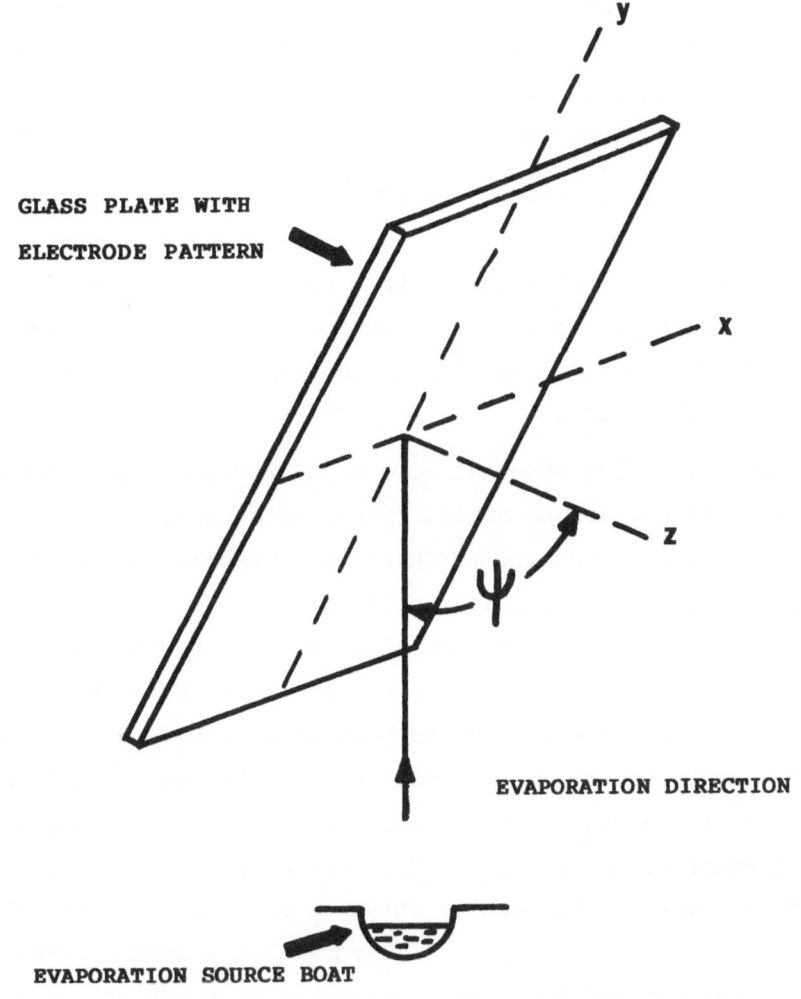

Fig. 2 Geometry of the oblique evaporation technique.

particularly concerned with the angle of incidence, Ψ, of the beam
of evaporating material. Films of gold or silicon monoxide evaporated at normal incidence ($\Psi = 0$) gave non-uniform, homogeneous
alignment. When the incidence angle was greater than 45 degrees
the alignment was uniform and homogeneous. In this case the
director is along the direction X, perpendicular to the plane of

incidence of the beam (Figure 2). Then, remarkably, when the
incidence angle was greater than 72-75 degrees, the alignment was
uniform and homogeneous but the director was oriented in the Y-Z
plane! Figure 3 shows the direction of alignment for each case.

Electron micrographs of the films prepared by evaporation at
Ψ = 80 degrees showed a definite columnar growth structure. No such
defined structure is apparent to this or other observers in the
electron micrographs of the films evaporated at Ψ = 70 degrees,
although Guyon claims to see "channels roughly perpendicular to the
direction of the beam".

In a similar but somewhat more sophisticated study, Goodman
and co-workers evaporated silicon monoxide at angles of Ψ = 60 and
83 degrees and examined the surfaces with transmission electron
microscopy at various glancing angles of the electron beam. At 83
degrees, the columnar growth, which is again observed, appears as
needle-like crystals which protrude from the surface at an angle of
35-40 degrees with the glass substrate (Figure 4). On the basis of
elastic energy factors, they explained the observed results and
proposed that the molecular director orients itself parallel to the
long direction of the columns. Unfortunately, this explanation is
not consistent with the results obtained by Crossland, et. al.,[12]
who found that the tilt angles of cyanobiphenyl liquid crystals
varied between 28 and 40 degrees depending upon the molecular chain
length.

It is widely known that films deposited at glancing and oblique
angles of incidence often adopt a morphology which has been described
as consisting of tilting columns of the evaporant material pointing
towards the evaporation source. The angle between these columns and
the substrate is difficult to measure accurately, and no systematic
comparison between the tilt angle of the nematic director and the
column angle in the aligning film has yet been reported. The
hypothesis that the nematic director is aligned along the column

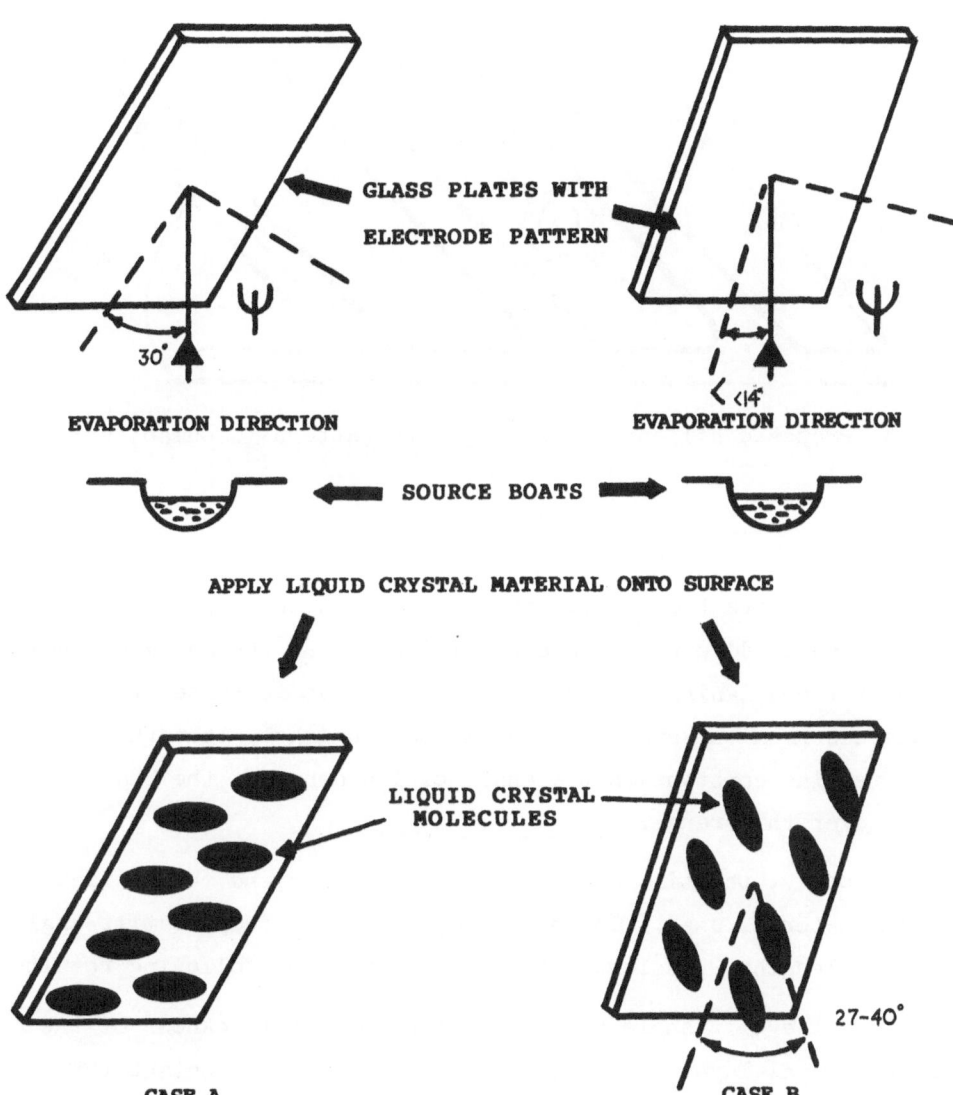

Fig. 3 Alignment of liquid crystal molecules on slope evaporated silicon monoxide.

EVAPORATION PERFORMED AT PSI = 83 DEGREES

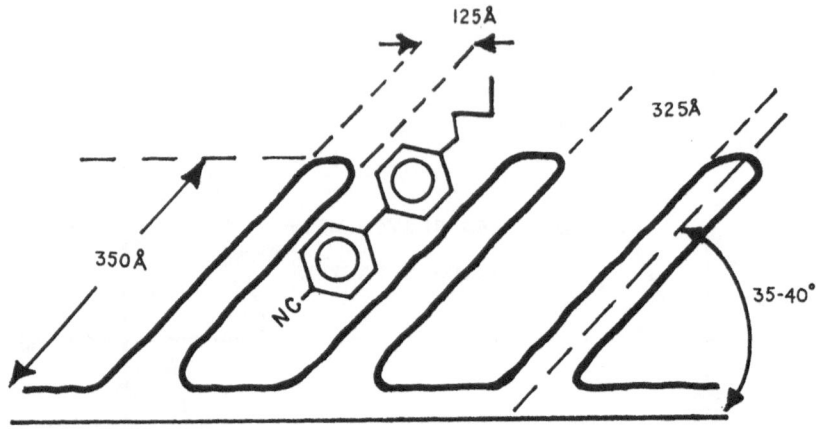

Fig. 4 Proposed silicon monoxide growth pattern (Goodman, et. al.)

direction appears too simplified for at least two reasons. First,
the tilt angle does not change significantly as the angle of incidence of the aligning film of silicon monoxide increases from 5 to
14 degrees from the substrate and is most unlikely that the column
angle remains constant as the angle of incidence of the beam
varies over this range.

Secondly, the tilt angle was found[12] to depend on the liquid
crystal material used. In their comprehensive study of tilt angle
vs chain length, Crossland, et. al. obtained the following results:

1.   The tilt angle was independent of the thickness of the
liquid crystal layer up to 100 microns and it was constant throughout
the nematic temperature range of the homologous series of 4-n-alkoxy-
and 4-n-alkyl-4'-cyanobiphenyls studied.

2.   The tilt angle did not vary by more than the experimental
error of 2-3 degrees with the angle of incidence of the silicon
monoxide films deposited at Ψ = 85, 82.5 & 76 degrees.

3.   For surfaces treated with SiO evaporated at $\Psi = 85$ degrees, the tilt angle was found to be a function of the nematic liquid crystal material being aligned:

(a) If the cyanobiphenyl compound has an unbranched alkyl group, the tilt angle increases approximately linearly with the alkyl chain length.

(b) If the compound has an unbranched alkoxy chain, the tilt angle is several degrees higher than that observed with the comparable alkyl derivative of similar chain length.

(c) The tilt angles observed with the alkoxy derivatives increase with increasing chain length, but show an alternation or "odd-even" effect.

4.   The tilt angle depends on the evaporant used to create the alignment. If carbon is used, variations of tilt angle with chain length are very small.

These observations suggest that forces other than those due merely to surface topography also operate. The odd-even effect observed for the alkoxy derivatives is very similar to that exhibited by the nematic-isotropic transition temperatures of these materials[13] which have been discussed in terms of the effects of increasing aliphatic chain length on the anisotropy of molecular polarizability. This, in turn, controls the attractive-dispersive forces acting between aliphatic chains. The chains may have less affinity for the silicon monoxide surface than they have for themselves, and therefore they exhibit a tendency to adopt a configuration maximizing their own interaction.[12] This would result in the molecules' tendency to bend away from the surface as the chain length increased.

Changing the alignment material will alter the balance between the tendency of the chains to align normal to the surface, and their tendency to associate with it (i.e., lie flat on it). It may be that the increase in tilt angle is not observed when carbon layers

are used because the chains have more of a tendency to associate
with carbon surfaces than with silicon monoxide.

II. Other Evaporation Techniques

A 90 degree rotation of the molecular alignment with respect to
the film incidence occurs when the angle between the incidence
direction and the substrate exceeds approximately 18 degrees (i.e.
Ψ < 72 degrees, Case A in Figure 3). This change is accompanied by
a decrease in the tilt to zero.

The myth has developed that these evaporation angles produce a
topographical structure with very narrow, parallel channels into
which the liquid crystal molecules lie. The question is where is
the evidence for the existence of such channels? In many attempts
to observe these channels, we and others (14, for example) have
been unsuccessful.

We have found that the physical and chemical properties of
evaporated films of silicon monoxide change as function of deposition rate, incidence angle, chamber pressure and post evaporation temperature. This suggests that the stoichiometry of the
film also changes as a function of these parameters. Hollinger,
et. al.[23] have found this to be the case.

Thus, the mechanism of uniform alignment of the liquid crystal
molecules on these evaporated films is still a matter of speculation.
We submit that the alignment is likely to be more a function of the
molecular interaction between the silicon monoxide and liquid
crystalline compounds than any topographical features. We theorize
that the Ψ < 75 degree films of silicon monoxide produce highly
ordered hexagonal structures which associate with the aromatic rings
of the liquid crystal compounds (e.g., via Van der Waal's forces)
to produce the uniform, non-tilted alignment.

III. Rubbed Films

Since films evaporated with Ψ < degrees exhibit no tilt, twisted nematic field effect cells made using this technique exhibit a high degree of reverse tilt.[15]   This occurs because the molecules can turn either to the right or to the left when an electric field is applied and the molecules begin to turn in the direction of the field.   This results in the activated segments taking on a "patchy" appearance.

Raynes[15] discovered two solutions to this problem.   One uses a higher percentage of chiral additive in the nematic liquid crystal mixture; however, this results in a higher threshold voltage.   The second uses unidirectional rubbing of one or both surfaces with paper.   The rubbing is performed in a direction perpendicular to the direction of evaporation.   Using cyanobiphenyl type compounds, tilt angles of 1 to 3 degrees can be obtained with this technique. No specific mechanism is suggested by Raynes other than to say that a reorientation of the surface occurs.

A number of other workers have discussed the techniques of rubbing or buffing the surface of a dielectric film to obtain a resulting alignment layer which produces low tilt angles. Ristango[16] rubbed a sputter deposited film of silicon dioxide with a piece of cotton cloth (cellulose). The silicon dioxide was deposited normal to the surface and the twisted nematic cells were filled with cyanobiphenyl type liquid crystals. The cells exhibited uniform, homogeneous alignment and a low, but unmeasured tilt angle. Similar results were reported by workers at Sanyo Electric Co.[17] and by Gurtler and Casey,[18] both using silicon oxide films rubbed with various soft and hard materials.

The uniform homogeneous alignment of liquid crystal molecules on surfaces which have been purposely abraded or mechanically grooved has been the subject of several papers. Berreman[19] produced parallel alignment of p-azoxyanisole and MBBA on soda lime glass and quartz by lapping the surface with a 1.0 micron diamond paste mixture. Electron micrographs of the surface proved the existence of grooves.

However, no grooves were observed when the surface was rubbed with paper or various fabrics. The tilt was always zero in the mechanically grooved cases, as long as the surfaces were clean. Berreman also found that when large grooves were used, the grooves acted as capillaries and the flow of liquid crystal material along the grooves produced a uniform homogeneous alignment.

Flanders and his co-workers[20] obtained uniform homogeneous alignment of MBBA on surfaces of silicon dioxide into which a grating type pattern was plasma etched. In a related study, Little et. al.[21] produced grooves in the surfaces of indium and tin oxide films by ion-beam milling in an Argon atmosphere.

As a result of this work on mechanically and otherwise abrading the surfaces of vacuum deposited films, the myth has developed that "rubbing hard surfaces (i.e., silicon monoxide, glass, quartz) with anything produces grooves which align the liquid crystal material". Apparently many workers have interpreted the results to apply to all cases.

We have obtained uniform homogeneous alignment with tilt angles of 1 to 3 degrees by rubbing silicon monoxide films, which had been vacuum deposited on soda lime glass and indium-tin oxide coated substrates at incidence angles of $\Psi < 75$ degrees (including $\Psi = 0$ degree), with cellulose acetate fabric on a high speed buffing wheel. The same result was obtained whether or not the film was heated at 525 degrees Centigrade; the effects of the rubbing survives these high temperatures. Scanning electron micrographs of the surface before and after rubbing and after the high temperature treatment showed no signs of grooves or other abrasions (Figure 5). Using Auger spectroscopy we examined the rubbed and unrubbed films before and after heat treatment; no evidence of a retained organic film or additional carbon (above the background) was found. On the other hand, an evaporated silicon monoxide film rubbed with 0.5 micron diamond paste showed abrasion grooves when examined by scanning

UNRUBBED   **CELLULOSE ACETATE RUBBED**

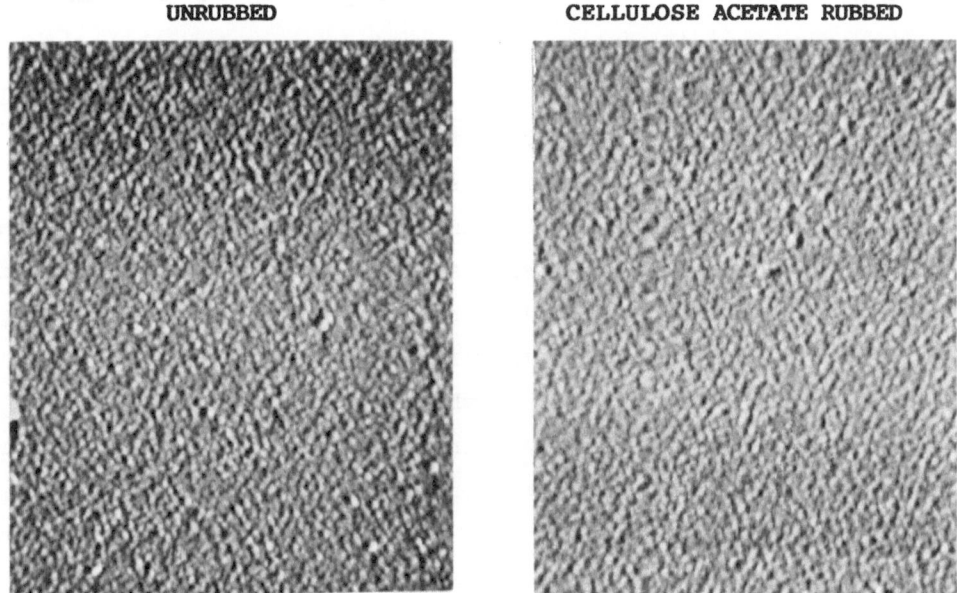

CELLULOSE ACETATE RUBBED & FIRED AT 500 DEGREES CENTIGRADE

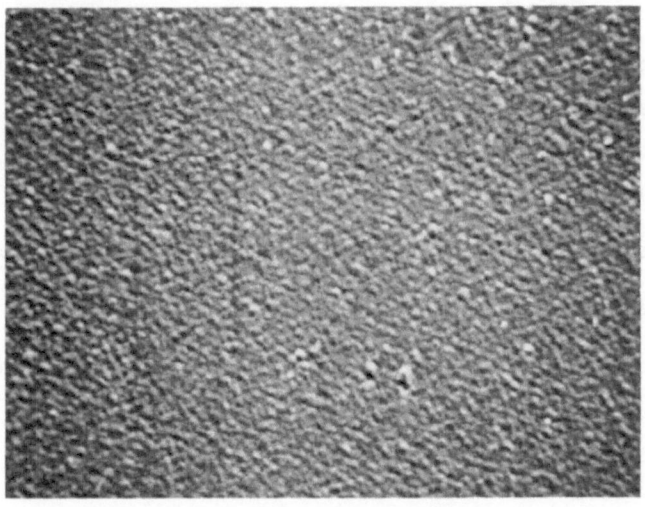

Fig. 5   Scanning electron micrographs of silicon monoxide treated
         surfaces (Magnification = 8500X).
         Evaporation performed at PSI = 60 degrees.

electron microscopy; the film gave zero tilt and uniform homogeneous alignment of a mixture of cyanobiphenyl liquid crystal compounds.

Thus, we see no evidence for grooves in silicon monoxide films rubbed with soft organic materials. Several mechanisms are possible. One proposes that the rubbing or buffing action causes an actual lossening and spreading of microcrystallites of the deposited film across the surface orienting them in the direction of and tangential to the rub head or buffing wheel. Another, suggests that a charging of the surface is produced during the rubbing and that these "trapped charges" produce alignment of the liquid crystal molecules. The difficulty with this mechanism is that it fails to explain the tilt and it assumes that the charges do not dissipate at the high temperatures used during the heat treatment. Whatever the mechanism is, it is certainly not due to grooves. All the work we have performed and other experiments reported to date with actual grooved surfaces yield zero tilt.

CONCLUSIONS

The alignment and tilt of liquid crystal molecules on surfaces treated in various ways is a complex phenomenon; not one, but many mechanisms are involved. The mechanism of alignment and tilt are determined by:

1. The technique used to coat the substrate surface.

2. The nature of the surface coating material.

3. The method and nature of material used to orient the surface material. Cloth must be defined.

4. The nature of the liquid crystal material.

ACKNOWLEDGEMENT

We would like to express our deep appreciation to the Fairchild Camera & Instrument Corporation, Mt. View, California and Conic

Semiconductor, Ltd., Hong Kong, for their financial support. Many helpful suggestions of the late Dr. T. W. Nakagawa are also grate- fully acknowledged.

REFERENCES

1.   C. Mauguin, Bull. Soc. fr. Min. 34, 71 (1911).

2.   G. Friedel, Ann. Physique 18, 273 (1922).

3.   P. Chatelain, Bull. Soc. fr. Min. Crist. 66, 105 (1943).

4.   L. Creagh, A. Kmetz, Mol. Cryst. Liq. Cryst. 24, 59 (1973).

5.   K. Hiltrop, H. Stegemeyer, Mol Cryst. Liq. Cryst. 49, 61 (1978).

6.   F. J. Kahn, Appl. Phys. Letters 22(8), 386 (1973).

7.   A. Adamson, Physical Chemistry of Surfaces, 3rd ed., John Wiley
         & Sons, NY, 1976, 246-7.

8.   A. V. Tobolsky, Properties and Structure of Polymers, John Wiley
         & Sons, NY, 1960.

9.   J. Janning, Appl. Phys. Letters 21(4), 173 (1972).

10. E. Guyon, P. Pieranski, M. Boix, Lett. Appl. & Eng. Sci. 1, 19
         (1973); Appl. Phys. Letters 25, 479 (1974).

11. L. Goodman, J. McGinn, C. Anderson, F. Digeronimo, Proc. S.I.D.
         18(1), 11 (1977).

12. W. Crossland, J. Morissy, B. Needham, J. Phys. D:Appl. Phys.
         9, 2001 (1976); J. Phys. D:Appl. Phys. 10, L175 (1977).

13. G. Gray, A. Mosley, J. Chem. Soc. Perkin II, 97 (1976).

14. M. Kuwahara, H. Yayama, H. Onnagawa, K. Miyashita, Oyo Butsuri,
         50(11), 1147 (1981)(Japan); CA 96:14105c.

15. P. Raynes, Electronic Letters 10(9), 141 (1974).

16. C. Ristango, private communication. Work at Micro Display
         Systems, Inc.

17. Sanyo Electric Co., Japan 53,090,959 (1978).

18. R. Gurtler, J. Casey, Mol. Cryst. Liq. Cryst. 35, 275 (1976).

19. D. Berreman, Mol. Cryst. Liq. Cryst. 23, 215 (1973); Phys. Rev.
         Letters 28, 1683 (1972); also private communication.

20. D. Flanders, D. Shaver, H. Smith, Appl. Phys. Letters 32(10),
        597 (1978).

21. M. Little, H. Garvin, Y-S. Lee, France 2,308,675 (1976);
        CA 87:P125391f.

22. H. Birecki, F. J. Kahn, The Physics and Chemistry of Liquid
        Crystal Devices, ed. G. Sprokel, Plenum Press, NY, 1980,
        115-23.

23. G. Hollinger, Y. Jugnet, T. Minh Duc, Solid State Comm. 22(5),
        277 (1977); G. Farquhar, Diss. Abst. Int. B, 39(4), 1842
        (1978).

DEVELOPMENT OF DUAL-FREQUENCY ADDRESSABLE LIQUID CRYSTALS

R. L. Hubbard, J. C. H. Liang, and K. R. Koehler/Beran

Tektronix 50/426
P. O. Box 500
Beaverton, OR 97075

ABSTRACT

The effects of several classes of liquid crystal components were evaluated with respect to the frequency dependence of their dielectric constants in mixtures. Empirical rules are suggested for the use of these components to produce dual-frequency addressable displays that can be operated at elevated temperatures with modest high-frequency drive.

INTRODUCTION

A few liquid crystal materials have been reported that have unusually low dielectric relaxation[1-4]. Some of these were briefly available commercially in sample sizes. A low enough dielectric relaxation frequency allows one to drive the material "on" at a low frequency as well as drive the material "off" at a frequency higher than the relaxation frequency. Since the rise time of liquid crystals is inversely proportional to the square of the electric field, the rise times can be shorteneed to just a few milliseconds at higher drive voltages. The fall times can also be reduced to a few milliseconds by using a high frequency drive[5]. Until just recently[6]

781

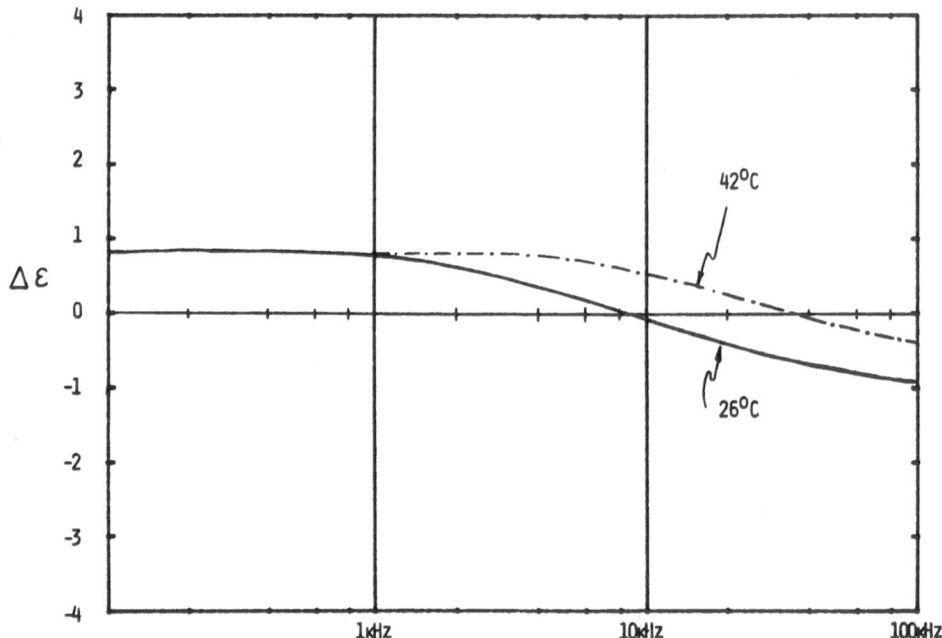

Figure 1. Temperature dependence of the dielectric anisotropy of
           ZLI-1085.

this two-frequency drive approach was the only practical way to have
liquid crystal devices with response speeds in the millisecond
range.   The disadvantages of this method include the problems associated with the high frequencies and higher drive voltages needed
as well as the temperature dependence of the dielectric relaxation
point. The first materials developed for this use had cross-over
frequencies (f_c) of around 10 kHz[7]. This required the high frequency
drive to be about 20 kHz and 50 V RMS at room temperature (see Fig.
1). In a device however, the current this produced would tend to
heat the cell somewhat which would shift the cross-over frequency
higher and therefore the drive frequency would have to be moved up
to 50 kHz. This would produce even higher heat and eventually the
cell could not be switched "off" at all.

    These dual-frequency addressable liquid crystal materials
might be more useful in devices then, if either the temperature

dependence could be reduced or the cross-over frequency moved down to lower frequencies where the heat produced would be minimal. The purpose of this investigation was to develop materials with lower cross-over frequency and incidentally lower birefringence for wider viewing angle.

EXPERIMENTAL

The components used for this study were either obtained from EM Labs, Hoffman La Roche or synthesized at Tektronix Laboratories by Jason Liang. Components of the series 1XXX are from EM while the series 2XXX were synthesized here. Eutectic mixtures were calculated using the usual Schroeder-Van Laar equations and correcting to weight percent. Thermal data was collected with a Perkin-Elmer DSC-2 and checked with polarized microscopy. Synthesized components were identified by proton and carbon FT-NMR as well as Infrared spectroscopy. Components were purified with a Waters preparative liquid chromatograph using hexane, ethyl acetate and THF followed by recrystallization. Refractive indicies were measured with an Abbe refractometer with oriented surfaces.

The dielectric constants were obtained using a Princeton Applied Research lock-in amplifier and preamp to measure the capacitance of a 255 micron thick layer of the liquid crystal parallel and perpendicular to a 12 kGauss magnetic field. The temperature was 26 C. Lowering the magnetic field had no appreciable effect on the capacitance and the results were reproducible to within 2%. The parallel and perpendicular dielectric constants were calculated and plotted on a Tektronix 4051.

APPROACH

The dielectric relaxation frequency of liquid crystals has been described as being related to the rotation around the short molecular axis of the molecule[4,8]. Even though the choices of components for

this study were made with this in mind the approach used here was
mostly empirical. The intent was to find structural features that
were conducive to the desired properties and to optimize those properties within certain limits.

Our experiments began with the intent to use commercially
available esters with cyclohexyl rings to reduce birefringence. Components with the structure I (where at least one of the rings X, Y
or Z is <u>trans</u>-cyclohexyl)

$$R - X - Y - \underset{\underset{O}{\|}}{C} - O - Z - R' \qquad (I)$$

have not only lower birefringence than completely aromatic esters
but also exhibit inherently lower dielectric relaxation frequencies
as well as higher clearing points. Components with the structure II

$$R - X - \underset{\underset{O}{\|}}{C} - O - Y - R' \qquad (II)$$

will be used to reduce viscosity. Components with the structure III

$$ (III) $$

will be used to increase the parallel dielectric constant and lower
the cross-over frequency.

We found that the use of components with cyano substituents
perpendicular to the long molecular axis did not have a strong effect on either the cross-over frequency or the negative anisotropy
but did cause serious problems with the thermal stability of the
final mixtures made with them. The component S-1461:

C_3H_7 —⬡— ⬡— $C - O$ — (CN) — C_4H_9 (S-1461)

‖
O

which includes one cyclohexyl ring had better dielectric properties
than S-1014:

C_5H_{11} —⬡—⬡— $C - O$ — (CN) — C_7H_{15} (S-1014)

‖
O

but since structures I have inherently larger perpendicular dielectric constants than completely aromatic esters we were able to dis-
pense with such components as S-1461 and S-1014 entirely. Other
structures with strong perpendicular dipoles have been studied by
several groups[9-11] and they have also been found to reduce the
thermal stabilities of nematic phases due mostly to steric forces.

BASIC EUTECTIC COMPONENTS

The components that were selected to make up the bulk of the
eutectic base were chosen from structures Ia-If. We were not able
to compare complete series of homologous components since several
components had either several irreproducible solid phases, small
unstable nematic phases, or produced solutions instead of eutectics.
For some of these components no percentage could be found that was
compatible in mixtures. Table 1 lists the thermal data for structures I. It should be noted that some of our data do not correspond to the published data furnished by the manufacturer. In several cases we found that the reported melting point was in fact a
transition from a smectic to nematic phase. As mentioned above

TABLE 1[1]: BASE EUTECTIC COMPONENTS

$$R-X-Y-CO_2-Z-R'$$

	X	Y	Z	R	R'	mp	np	cp	Component No.
Ia	CO_2	Ph	Ph	C_5H_{11}	C_3H_7	38		53	2006
				C_5H_{11}	C_7H_{15}	38		61	2009
				C_5H_{11}	C_5H_{11}	37		57	2003
				C_5H_{11}	C_4H_9	37		48	2004
				C_3H_7	C_5H_{11}	55		59	2005
Ib	Ph	Ph	Ph	C_5H_{11}	C_5H_{11}	95		176	1011
Ic	Cy	Ph	Ph	C_3H_7	C_3H_7	71	89	186	1222
				C_3H_7	C_4H_9	90		173	2021
				C_3H_7	C_5H_{11}	82		171	2019
				C_3H_7	C_7H_{15}	58	99	162	2022
				C_4H_9	C_5H_{11}	77	115	167	2028
				C_5H_{11}	C_3H_7	72	92	179	2014
				C_5H_{11}	C_4H_9	73	129	169	2017
				C_5H_{11}	C_5H_{11}	(72)	(137)	170	2010
				C_5H_{11}	C_6H_{13}	60	136	160	2027
				C_5H_{11}	C_7H_{15}	(45)	139	161	2011
				C_5H_{11}	$C_5H_{11}O$	27	147	187	2025
Id	Cy	Ph	Cy	C_2H_5	C_3H_7	39	94	134	1232
				C_3H_7	C_3H_7	(92)		159	1224
				C_4H_9	C_3H_7			155	1273
				C_5H_{11}	C_3H_7	67		154	1223
Ie	Cy	Cy	Cy	C_5H_{11}	C_5H_{11}	43		184	2029
If	Cy	Cy	Ph	C_5H_{11}	C_5H_{11}	172		187	2032

1   Ph = phenyl ; Cy = cyclohexyl

there were cases where repeated thermal cycling produced anomalous
results. It was necessary for us to use only materials that had
reproducible nematic phases or in some cases reproducible smectic
phases but in all cases the nematic point was used for the eutectic
mixture calculations rather than the melting point.

The first class studied was Ia which had the advantage of low
viscosity but produced relatively high cross-overs and high birefringence. There did not seem to be a significant relationship between alkyl chain length and cross-over frequency. Several of these
components were used as a base to evaluate single components of
other families. A typical experiment was to add a component from
class Ic to three components of class Ia and observe the change in
dielectric behavior. All of the results are of a comparative nature
rather than absolute. To evaluate individual components would require a variable temperature measurement system which we did not
originally have available. As will be shown however, it was possible to observe general trends and produce the desired properties.

There was not much of an effect when one of the ester groups
was replaced with a phenyl ring (Ib) as shown in Figure 2. This
was not pursued since it was of more interest to use cyclohexyl
rings.

The effect of substituting a cyclohexyl ring in the ester (Ic)
was significant as seen in Fig. 2. As was expected the clearing
point of the mixture was raised about 20C and the birefringence was
lowered from 0.17 to 0.14. The cross-over frequency was lowered
from 20kHz to 15kHz. Again there was no significant effect found
from changing the alkyl groups. When two class Ic components were
added the cross-over dropped to 10kHz (Fig. 3). When the last
phenyl ring was changed to cyclohexyl there was no significant effect on the dielectric curve (Fig. 4). Roughly the same result is
obtained when all three rings are cyclohexyl. When the first two
rings are cyclohexyl and the last is phenyl there is a dramatic

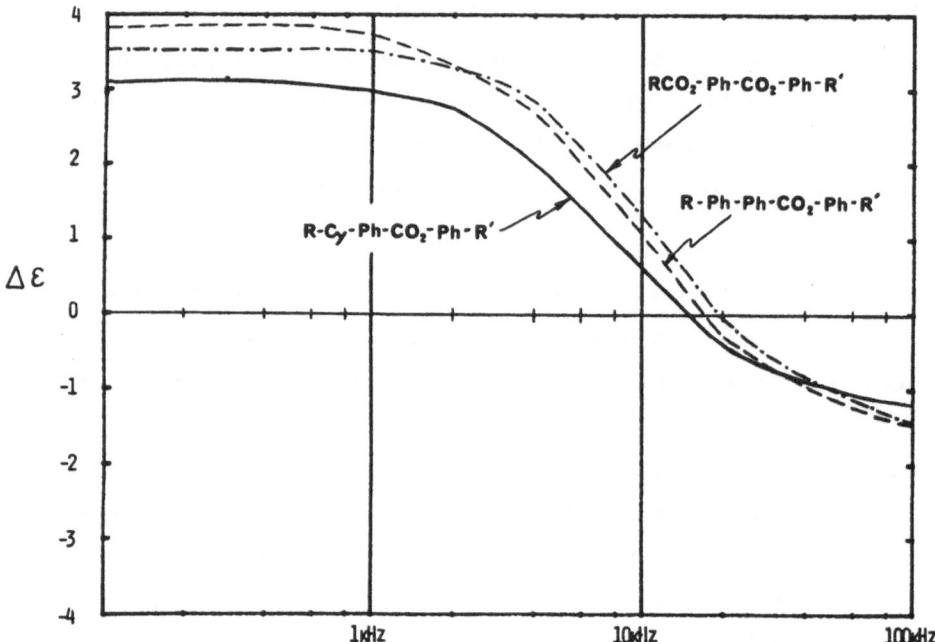

Fig. 2. Effect on dielectric behavior of replacement of carboxy
         group with unsaturated or saturated ring.

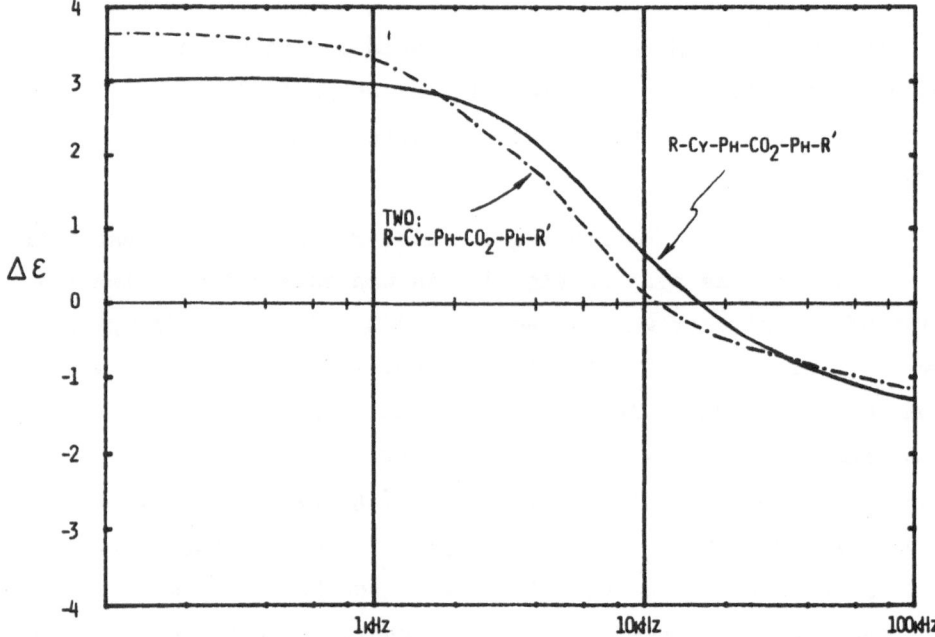

Fig. 3. Synergistic effect obtained with ester homologues.

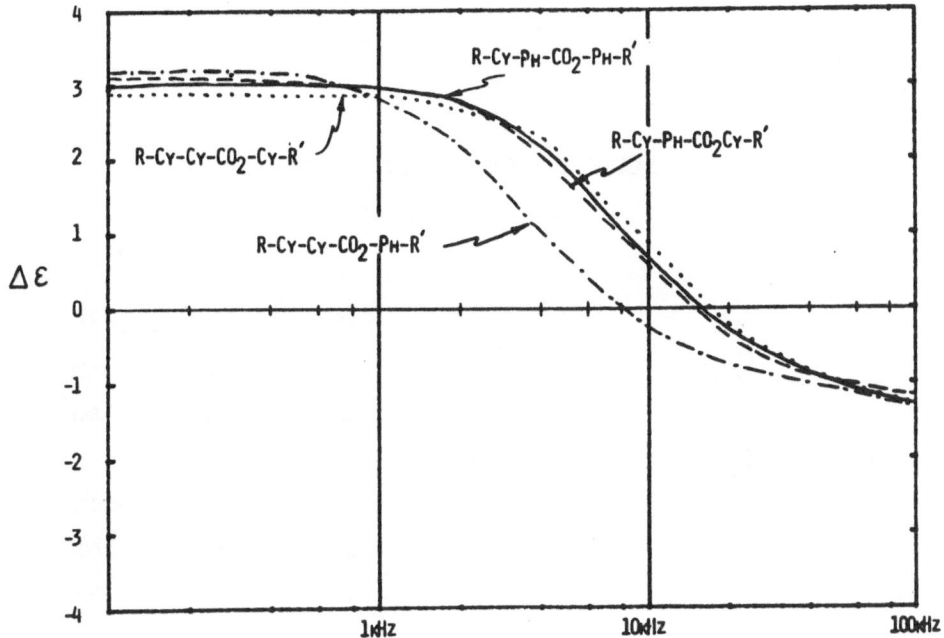

Fig. 4. Effects obtained by varying the position of cyclohexyl and phenyl rings.

drop in the cross-over frequency. When only class Ic-If structures are used in the base eutectic the birefringence is lowered to 0.12 and the cross-over is the lowest. Unfortunately, the temperature range is shifted so high that it is necessary to add dopants to lower the melting point to below room temperature as will be discussed later.

CYANO ADDITIVES

The addition of a component with a strong positive dielectric anisotropy such as structure III lowers the cross-over frequency as well as increases the positive anisotropy at low frequencies. Table 2 lists the structures and the thermal properties of these components.

TABLE 2[1]: CYANO ADDITIVES

R-X-Y-Ph-CO$_2$-Ph-CN

	R	X	Y	mp	cp	Component No.
IIIa	C_4H_9	-	-	66	(42)	1500[2]
	C_5H_{11}			60	(56)	1530[2]
	C_7H_{15}			43	56	1540[2]
IIIb	C_3H_7	CO_2	-	95	97	2001
	C_5H_{11}			66	68	2002
	C_7H_{15}			61	86	2007
IIIc	C_5H_{11}	Cy	-	82	222	2016
	C_5H_{11}	Ph	-	97	221	2026
IIId	H	Ph	CO_2			2035
	C_3H_7			109	257	2033
	C_4H_9			124	192	2008
	C_5H_{11}			76	119	2012
	C_6H_{13}					2039
	C_7H_{15}			56	109	2037
	C_2H_5O				153	2038
	C_3H_7O			78	144	2040
	C_4H_9O			122		2024
	$C_5H_{11}O$					2041
	$C_6H_{13}O$			127	247	1339
IIIe	C_5H_{11}	Cy	CO_2	80	217	2020

1 Ph = phenyl ; Cy = cyclohexyl

2 From Hoffman LaRoche

All of the eutectic mixtures discussed above included 10% or less of one of the structure III components (IIId, R=butyl, X=Ph, Y=CO) for comparison purposes. It was found that IIId components were the best of the cyano additivies for reducing the cross-over frequency and that generally the longer the molecule and the more extended the unsaturation, the lower the cross-over tended to be. Fig. 5 shows the effect of extending the length of the rigid nucleus of the cyano additive (IIIa to IIId) and Figs. 6, 7 and 8 show the effect of lengthening the alkyl chain in these families. As shown in Fig. 9, the replacement of an ester functionality with a ring (IIIc) decreases the cross-over as well as slightly lowers the over-all anisotropy. The replacement of the first phenyl ring in IIId with a cyclohexyl ring shortens the conjugation and lowers the cross-over somewhat.

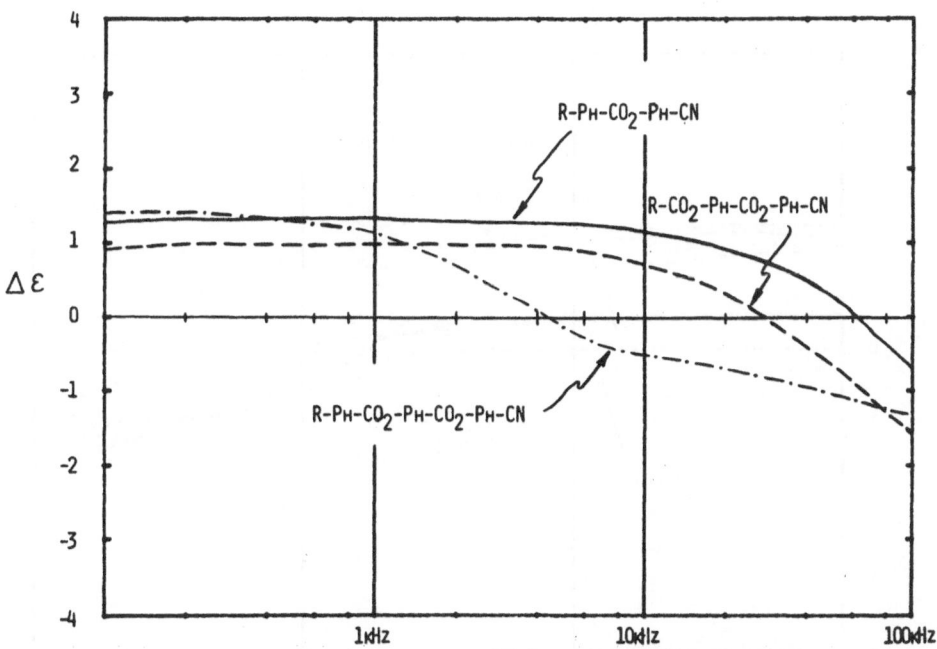

Fig. 5. Effect obtained by lengthening the congugation of cyano additives.

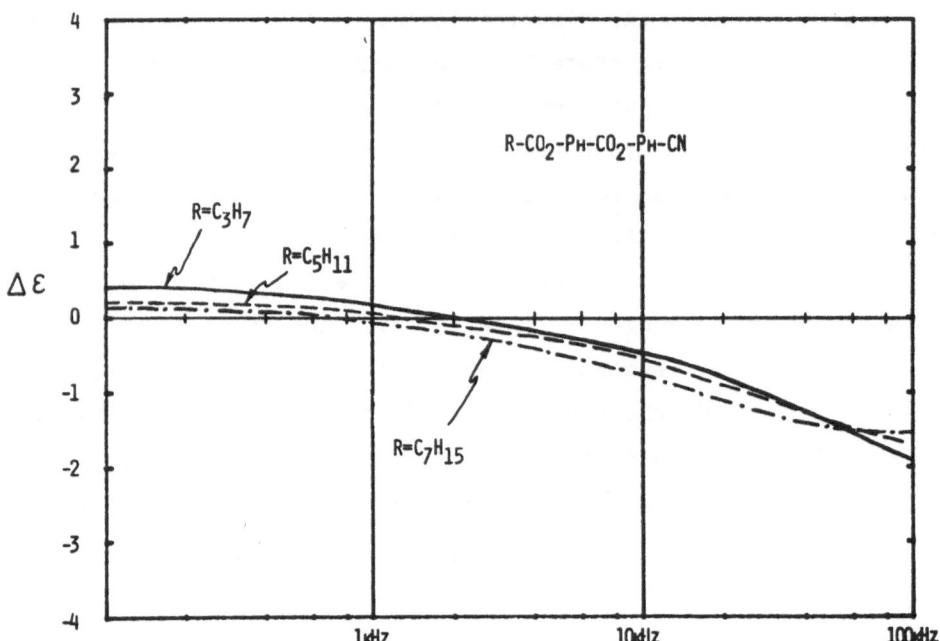

Fig. 6. Effect obtained by lengthening alkyl chain in cyano
additive IIIb.

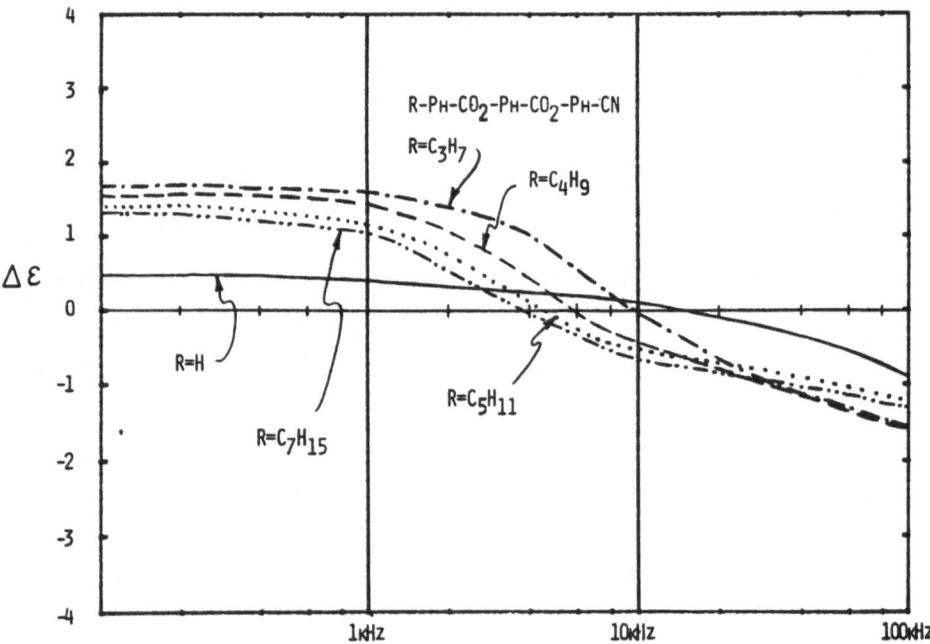

Fig. 7. Effect obtained by lengthening alkyl chain in cyano
additive IIId.

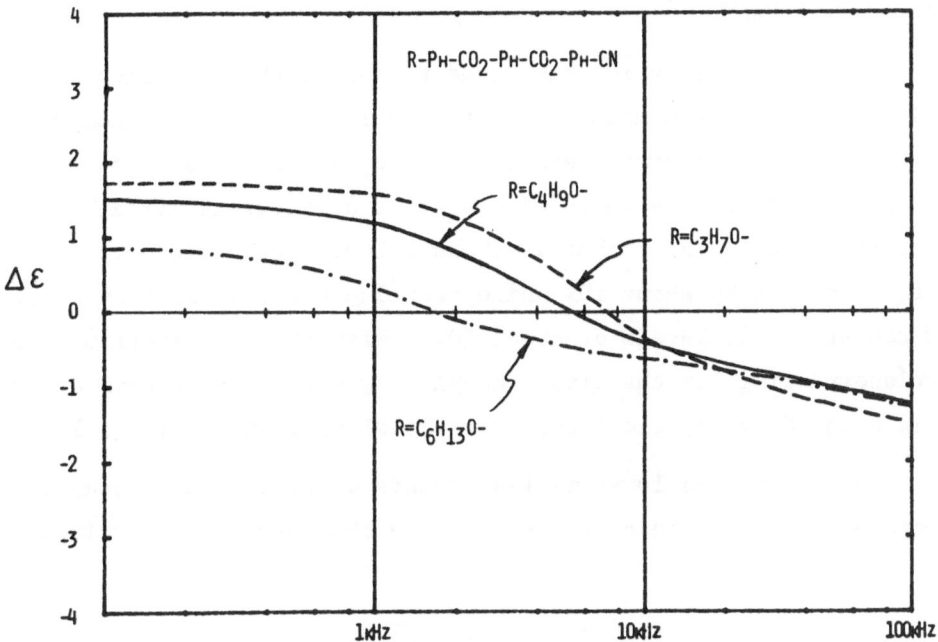

Fig. 8. Effect obtained by use of homologous alkoxy substituents in IIId.

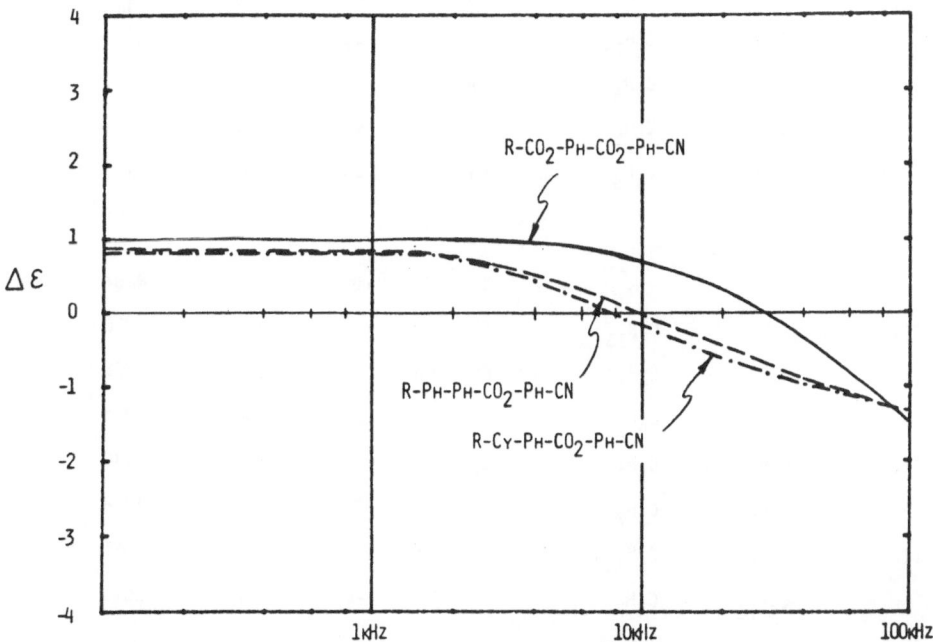

Fig. 9. Effect obtained by replacement of carboxy group with ring.

VISCOSITY DOPANTS

It was necessary in many cases to reduce the viscosity and melting point of the mixtures by the addition of smaller components with appropriate thermal behavior (Table 3). The choice of dopants was determined on the basis of the least deleterious effect on cross-over frequency. The biphenyl ethers S-1009 and S-1010 only increased the cross-over by about 10% while reducing the viscosity by a factor of two at dopant levels of about 10%. With the substitution of a cyclohexyl ring for the first phenyl ring (cyclohexylphenyl ethers S-1476 and S-1477), the cross-over was slightly worse (Fig. 10).

To obtain even lower melting points we evaluated several ester components (II). The best results were obtained with a cyclohexyl

TABLE 3[1]:   VISCOSITY DOPANTS

R-X-Ph-O-R

	R	R´	X	mp	cp	Component No.
	C_5H_{11}	C_6H_{13}	Ph	82	84	1009
	C_5H_{11}	C_2H_5	Ph	72	81	1010
	C_3H_7	C_2H_5	Cy	41	(37)	1476
	C_3H_7	C_4H_9	Cy	35	(32)	1477

$R-X-CO_2-Y-R´$

	R	R´	X	Y	mp	cp	Name
IIa	C_3H_7	C_5H_{11}	Ph	Ph			2036
IIb	C_3H_7	C_5H_{11}	Cy	Ph	17	35	2023
	C_5H_{11}	C_5H_{11}			33	46	2013
IIc	C_3H_7	C_5H_{11}	Cy	Cy	(50)	65	2030
	C_5H_{11}	C_3H_7			22	51	2018
IId	C_3H_7	C_5H_{11}	Ph	Cy		16	2031
	C_5H_{11}	C_3H_7			-5	22	2015

1   Ph = phenyl ; Cy = cyclohexyl

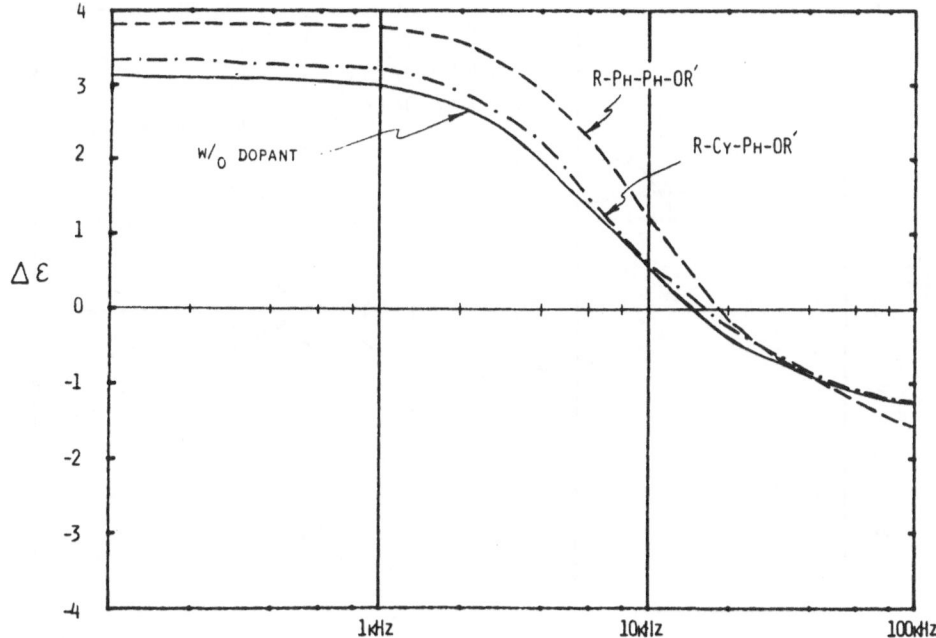

Fig. 10. Effect of ether viscosity dopants.

acid ring and either a phenyl (IIb) or a cyclohexyl alcohol (IIc)
ring as shown in Fig. 11. The viscosities were lowered to about
50cp and the melting points were lowered below zero. Fortunately
it was possible to add several of these components to mixtures without increasing the cross-over more than would be expected from only
one additive (Fig. 12).

DISCUSSION

This empirical study has shown that it is possible to produce
materials with useful dielectric properties that might have value
for dual-frequency addressed displays. One of the advantages of
these mixtures was that the low cross-over frequencies (1-5 kHz)
and high clearing points (130-150 C) resulted in materials that
could be run continuously for days at 40 C with a 2 kHz high frequency drive and response times of less than a millisecond. It is

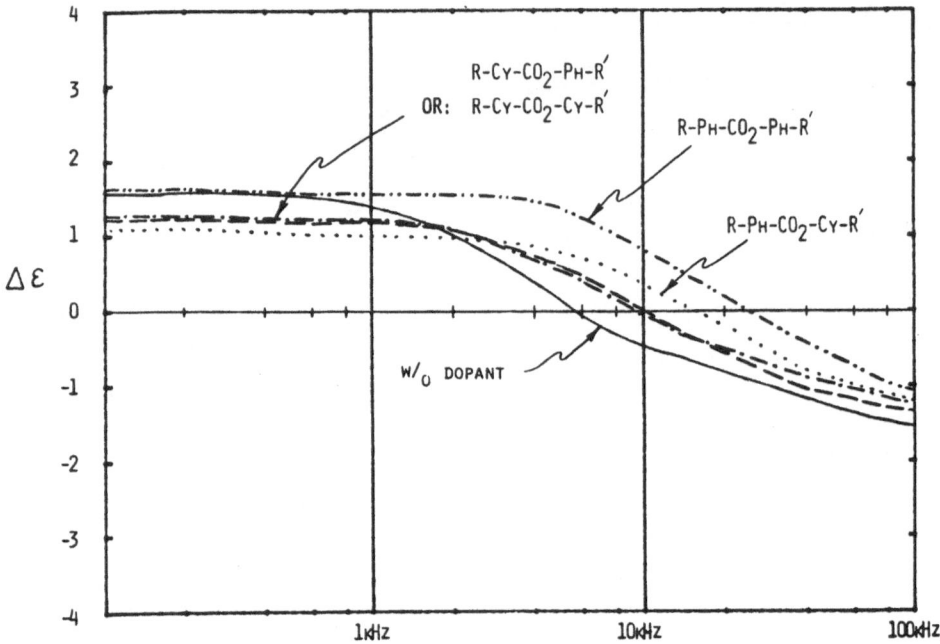

Fig. 11.   Effect of ester viscosity dopants.

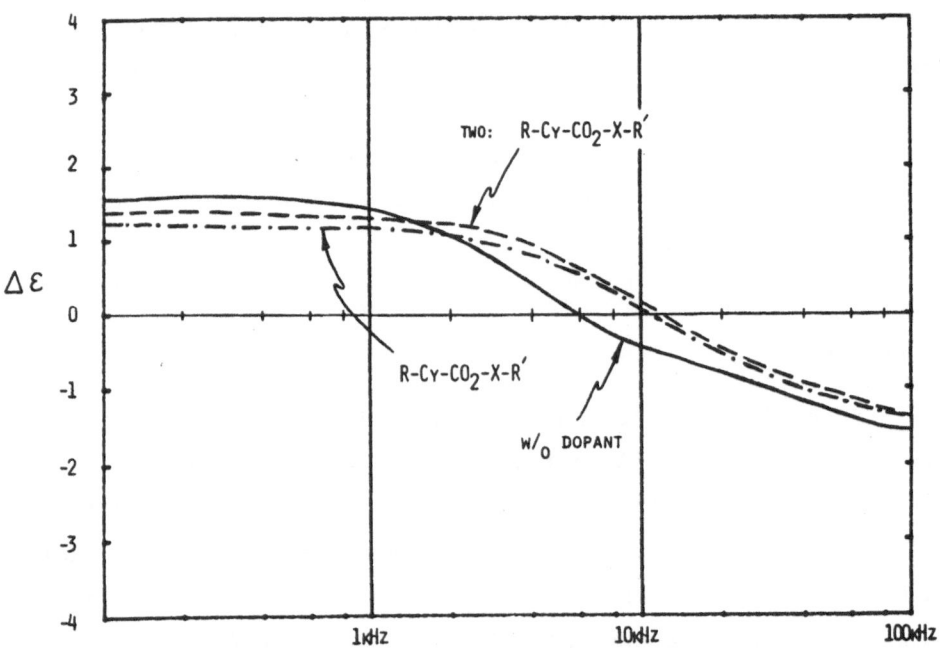

Fig. 12.   Synergistic effect of two viscosity dopants.

also useful to be able to modify the dielectric properties to suit specific applications.

Despite limited time and resources we were able to produce some useful materials to study. It would be preferable to have dielectric constant data on individual components and over a wide temperature range. We might then be able to determine more accurately what specific structural features contribute to the parallel and perpendicular dielectric constants and thus develop a theoretical basis for the design of improved materials.

ACKNOWLEDGEMENTS

All of the dielectric, viscosity and birefringence measurements were made by Rickey Koehler/Beran. Phil Bos designed the dielectric constant apparatus that Rickey built. The DSC data were provided by Hal Frame and Cindy Savage of the Analytical Service Lab. Some of our starting materials for synthesis were provided by Virgil Hedgecoth of Electronic Materials Corp.

REFERENCES

1. M. Schadt, J. Chem. Phys. 56, (4), 1494 (1972).

2. W. H. DeJeu, et al., Phys. Lett. 39A (5), 355 (1972).

3. G. Baur, A. Steib, G. Meier, Liquid Crystals & Ordered Fluids 3, 465, PLenum Press (1973).

4. W. H. DeJue and Th. W. Lathouwers, Mol. Cryst. Liq. Cryst. 26, 225 (1974).

5. C. S. Bak, K. Ko and M. M. Labes, J. App. Phys. 46 (1), 1 (1975).

6. R. L. Hubbard and P. J. Bos, IEEE Trans. Elect. Dev. Ed. 28 (6), 723 (1981).

7. W1, ZLI-518, ZLI-1085 (E. Merck, Darmstadt).

8. L. Bata and G. Molnar, Chem. Phys. Lett. 33 (3), 535 (1975).

9.   T. Gunjima, Y. Nakagawa and Y. Masuda, Paper presented at the
     5th International Liquid Crystal Conference, 1974,
     Stockholm, Sweden.

10. D. J. Byran, D. Lacey, and R. C. Wilson, Mol. Cryst. Liq.
     Cryst. 73, 273 (1981).

11. K. Takei, S. Kanbe and Shionozaki, U. S. Patent 4,287,085
     (1980).

DIELECTRIC STUDIES OF MONOESTER AND DIESTER NEMATOGENS

M. F. Bone and A. H. Price

Edward Davies Chemical Laboratories, University College of
Wales, Aberystwyth SY23 1NE, UK

M. G. Clark and D. G. McDonnell

Royal Signals and Radar Establishment, Malvern
Worcestershire WR14 3PS, UK

ABSTRACT

The permittivities of several monoester nematogens have been
studied in the nematic and isotropic phases at frequencies up to
18 GHz. The dielectric relaxations observed in the nematic phase
are in agreement with theoretical predictions for a rigid nematogen.
Values for the nematic order parameter and the angle between the
"long axis" of the molecule and its electric dipole are estimated
from the data, and the definition of the "long axis" in the molecular
theory of nematics is discussed. Two distinct relaxation frequencies
are observed in the isotropic phase of substituted phenyl benzoates,
probably due to anisotropic rotational molecular diffusion indicative
of residual nematic-like short-range order. This interpretation is
supported by quantitative comparison with nematic phase data, and by
data on a benzene solution and the non-nematic phenyl benzoate
parent. Data on cyclohexane carboxylate esters reveals interesting
similarities and contrasts with the benzoate esters. Recent developments in liquid crystal materials for two-frequency addressed

devices are considered in the context of the data on monoesters
together with studies of dielectric relaxation in monoester/diester
mixtures.

1.   INTRODUCTION

It is well known[1] that in general a rigid dipolar nematogen may
show up to four dielectric relaxations in its nematic phase, corresponding to the contributions to the permittivity components parallel
($\varepsilon_{//}$) and perpendicular ($\varepsilon_{\perp}$) to the nematic director $(\underset{\sim}{n})$ of the
electric dipole components along ("longitudinal") and perpendicular
("transverse") to a molecular "long axis" which is statistically
distributed about $\underset{\sim}{n}$. These dielectric relaxations are of interest
both from the point of view of the molecular physics involved, and
because in some materials the slowest of these relaxations lies at
audio frequencies and thus can be exploited in liquid crystal devices
by addressing the device either sequentially or simultaneously with
frequencies both below and above this relaxation.[2]   In this paper we
shall be concerned with both these related yet diverse aspects.

The phenomenon of low-frequency dielectric relaxation in the
nematic phase is found in largish molecules, notably, but certainly
not exclusively, the diester materials in which three saturated
and/or unsaturated rings are linked by two ester bridges. Motivated
by possible applications, the low frequency dielectric properties of
nematic mixtures containing diesters have been studied for a number
of years, and monoester materials, such as substituted phenyl
benzoates, identified as components of interest for formulating
mixtures with the desired blend of dielectric and other properties.[3]

The interpretation of material properties, such as permittivity,
in terms of the molecular theory of nematics rests on the concept of
a molecular "long axis" which is statistically distributed about the
director,[4] the distribution being characterized by the nematic order
parameter

$$S = \frac{1}{2}\langle 3\cos^2\theta - 1 \rangle \qquad (1.1)$$

where θ is the angle between the molecular long axis and $\underline{n}$, together with corresponding averages of the higher even order Legendre polynomials. As emphasized by Frank,[5] the definition of the long axis, although usually taken for granted, is not obvious and indeed appears impossible if only static properties are considered. We show in this paper that the molecular long axis can be properly defined when a dynamical property such as dielectric relaxation is considered. Indeed, dielectric relaxation data can be used to obtain experimental information on the direction of the long axis in the molecular frame, and also to estimate S without the need for tricky and potentially inaccurate normalization such as the extrapolation to absolute zero required in the determination of S from refractive index data.

After outlining our experimental procedures in Sec. 2, we give in Sec. 3 permittivity data at frequencies up to 18 GHz on several monoester nematogens in their nematic and isotropic phases. The interpretation of these data is discussed in Sec. 4. In order to set these studies in the context of applications, we outline in Sec. 5 some of the design strategies and constraints involved in practical mixtures such as those formulated recently at RSRE,[3] and discuss permittivity data on diester containing mixtures in the frequency range up to 500 kHz. On this last topic we are deliberately selective both for reasons of space and because more comprehensive information will be published elsewhere.

2. EXPERIMENTAL PROCEDURES

Measurements of the real component of the permittivity of diesters and diester/monoester mixtures at frequencies up to 500 kHz were made using thin homeotropically-aligned (lecithin) liquid crystal cells as described previously.[6] The capacitance of the cell is measured empty and full within the frequency range 150 Hz to

550 kHz by use of a Hewlett-Packard 4800A Vector Impedance Meter. For determination of $\varepsilon_\perp$ (which in all cases was independent of frequency in the range of 150 Hz to 550 kHz) a magnetic field was applied in the plane of the cell and the measured capacitance extrapolated to infinite magnetic field by an extrapolation procedure. The magnetic field was routinely left on and oriented perpendicular to the cell to reinforce the lecithin-induced homeotropic alignment for measurements of $\varepsilon_{//}$, although the alignment was sufficiently good that extrapolation was not required. The temperature of the cell could be controlled to within 0.1°C over the range 0-200°C. The error in the permittivities measured by this method is estimated as about 3%.

Measurements on monoesters at frequencies up to 150 MHz were made with a Wayne Kerr B201 bridge (100 kHz to 3MHz) and a Boonton 250A RX-meter (0.5 MHz upwards), using a specially designed rectangular cell[7] temperature controlled to $\pm$ 0.1°C. Corrections for lead effects followed the method of O'Konski and Edwards.[8] Alignment in the nematic phase was achieved by application of a 0.2 T magnetic field, which for the 1 mm cell gap used is sufficient to obviate the need for extrapolation to infinite field at our experimental accuracy of 5%. An iterative procedure with visually-determined initial estimates was used to fit to each loss peak in the experimental data the Fuoss-Kirkwood equation

$$\cosh^{-1}(\varepsilon''_m / \varepsilon'') = \alpha \ln |\omega\tau| \qquad (2.1)$$

where ε'' is the imaginary part of the permittivity and ε''_m its maximum value, ω the angular frequency, τ the relaxation time, and α an empirical parameter normally lying between zero and unity, $\alpha = 1$ corresponding to a simple Debye-like relaxation.

Measurements on selected data were extended up to 18 GHz using slotted line[9] (200 to 1800 MHz) and sweep frequency[10] (4 to 18 GHz) techniques. In both cases the cell was constructed from a length of coaxial line operating in the TEM mode. A magnetic solenoid wound

round the cell was used to align the nematic director perpendicular
to the measuring field. In the sweep frequency apparatus an electric
aligning field could be applied between the inner and outer conductors.
Both cells were temperature controlled to $\pm$ 0.1°C. Data were processed by a similar procedure to that described above. As shown in
Sec. 3, the additional data produced only slight changes in the
estimated dielectric parameters.

Refractive indices for 633 nm HeNe laser radiation were determined by direct measurement on an extended range Abbé refractometer
using lecithin-induced alignment in the nematic phase. Densities
were measured by use of 2 cm^3 pycnometer of conventional design
calibrated with deionised water. For both measurements temperatures
were controlled to $\pm$ 0.1°C.

3. RESULTS

3.1 Glossary of Acronyms

All alkyl substituents are n-alkyl; phase transition temperatures
are given in °C.

ME35 = $C_3H_7 \cdot C_6H_4 \cdot COO \cdot C_6H_4 \cdot C_5H_{11}$	K 14 N 19.9 I
ME55 = $C_5H_{11} \cdot C_6H_4 \cdot COO \cdot C_6H_4 \cdot C_5H_{11}$	K 35 (N 26) I
H1 = 1:1 molar mixture of ME35 and ME55	K 12.5 N 21.8 I
DE35 = $C_3H_7 \cdot$ cyclo-$C_6H_{10} \cdot COC \cdot C_6H_4 \cdot C_5H_{11}$	K 30.5 N 37 I
DE55 = $C_5H_{11} \cdot$ cyclo-$C_6H_{10} \cdot COO \cdot C_6H_4 \cdot C_5H_{11}$	K 35.5 N 48 I
H2 = 1:1 molar mixture of DE35 and DE55	K 5.5 N 42 I
FDE35 = $C_3H_7 \cdot$ cyclo-$C_6H_{10} \cdot COO \cdot 2'-FC_6H_3 \cdot C_5H_{11}$	K 36 (N 26.5) I
FDE55 = $C_5H_{11} \cdot$ cyclo-$C_6H_{10} \cdot COO \cdot 2'-FC_6H_3 \cdot C_5H_{11}$	K 17.5 N 36.5 I
H3 = 1:1 molar mixture of FDE35 and FDE55	K 16 N 32 I

DiE1 = $C_3H_7 \cdot$ cyclo$-C_6H_{10} \cdot COO \cdot 3' - C_6H_3Cl \cdot COO \cdot C_6H_4 \cdot C_5H_{11}$ K 51 N 140 I

DiE1(3) = 28% DiE1 + 72% H1 by weight   K 5 N 46 I

DiE1(8) = 27% DiE1 + 73% H3 by weight   K 12 N 52 I

DiE2 = $C_3H_7 \cdot C_6H_4 \cdot COO \cdot 3' - C_6H_3Cl \cdot COO \cdot C_6H_4 \cdot C_5H_{11}$ K 55 N 136 I

DiE2(1) = 31.5% DiE2 + 68.5% H1 by weight   K 4 N 49 I

DiE2(3) = 12.2% DiE2 + 87.8% H1 by weight   K 11 N 32 I

DiE2(4) = 46.7% DiE2 + 53.3% H1 by weight   K 17 N 64 I

3.2 Monoester Materials

The experimentally measured dielectric relaxation parameters
for the monoesters and monester mixtures studied are collected in
Table I. Figures 1 to 6 show examples of the data obtained for
nematic and isotropic phases. Densities and refractive indices have
been measured as functions of temperature for most of the materials
in Table I, and are used to interpret the dielectric data as appropriate.

3.3 Diesters and Diester/Monoester Mixtures

Figure 8 shows the experimentally measured phase diagram of
DiE2 + H1; the other diester/monoester mixtures studied have very
similar phase diagrams. Figures 9 and 10 show permittivity data up
to 500 kHz for selected mixtures. As mentioned earlier, more
comprehensive information on this data will be published elsewhere.

4.   INTERPRETATION OF MONOESTER DATA

4.1 Theory

The physical effects that we wish to emphasize can be adequately
demonstrated by using the expressions for permittivities of nematics
given by Maier and Meier.[11] This simple approach has advantages for

Table I Measured Dielectric Relaxation Parameters of Monoesters and Monoester Mixtures

Material and Phase	T/K	ε'_o	ε''_m	α	ε'_∞	f_m/MHz
ME35	286.9	4.61	0.54	0.95	3.47	0.96
		3.47	0.34	0.84	2.66	240
nematic	288.1	4.57	0.50	0.97	3.54	1.1
parallel		3.54	0.35	0.82	2.69	260
	289.8	4.57	0.46	0.90	3.54	1.4
		3.54	0.365	0.78	2.61	280
	291.1	4.53	0.43	0.90	3.57	1.7
		3.57	0.37	0.78	2.62	290
ME35	286.9	4.37	0.73	0.81	2.56	240
nematic	288.1	4.36	0.73	0.82	2.56	260
perpendicular	289.8	4.34	0.72	0.78	2.48	280
	291.1 (a)	4.33	0.73	0.87	2.65	290
	291.1 (b)	4.33	0.076	1.0	4.18	15
		4.18	0.73	0.87	2.50	275
	294.1	4.36	0.20	1.0	3.96	8.34
		3.96	0.63	0.80	2.37	314
ME35	296.9	4.32	0.20	1.0	3.93	8.79
		3.93	0.62	0.79	2.37	336
isotropic	302.6 (a)	4.26	0.19	1.0	3.88	13.2
		3.88	0.61	0.80	2.36	429
	302.6 (b)	4.26	0.24	1.0	3.78	14.0
		3.78	0.65	0.99	2.46	384
	311.6	4.15	0.17	1.0	3.81	15.2
		3.81	0.55	0.75	2.35	515
ME35 1.55 mol dm^{-3} in benzene	294.0	3.27	0.07	1.0	3.13	100
		3.13	0.27	0.985	2.58	910
Phenyl benzoate 1.55 mol dm^{-3} in benzene	294.0	3.10	0.315	0.87	2.38	3900
ME55	302.0	4.05	0.195	1.0	3.66	11.5
		3.66	0.45	1.0	2.76	235
isotropic	308.5	3.98	0.20	1.0	3.58	15.1
		3.58	0.425	1.0	2.72	300
	314.0	3.92	0.205	1.0	3.51	19
		3.51	0.415	1.0	2.68	360
H1	285.3	4.6	0.56	1.0	3.48	0.65
		3.48	0.30	0.78	2.71	207
nematic	290.0	4.54	0.495	1.03	3.58	1.06
		3.58	0.32	0.79	2.77	253
parallel	293.2	4.43	0.415	1.0	3.60	1.35
		3.60	0.38	0.70	2.51	357

(continued)

Table I (Continued)

Material and Phase	T/K	$\mathcal{E}'_0$	$\mathcal{E}''_m$	α	$\mathcal{E}'_\infty$	f_m/MHz
H1	285.3					
		4.38	0.70	0.79	2.61	207
nematic	290.0					
		4.32	0.69	0.79	2.57	253
perpendicular	293.2					
		4.26	0.68	0.70	2.32	357
H1	296.8	4.308	0.235	1.0	3.86	8.75
		3.86	0.52	0.95	2.76	230
isotropic	304.0	4.21	0.22	1.0	3.77	15.5
		3.77	0.48	1.0	2.81	280
	312.0	4.12	0.22	1.0	3.68	18.0
		3.68	0.46	1.0	2.76	350
DE55 nematic	312.5					
		3.19	0.34	0.66	2.16	4150
perpendicular	316.0					
		3.14	0.367	0.76	2.17	4250
DE55 isotropic	325.5	2.95	0.315	0.84	2.20	4330

Notes

(i)   For a relaxation with maximum loss $\mathcal{E}''_m$ occurring at a
        frequency f_m, $\mathcal{E}'_0$ and $\mathcal{E}'_\infty$ are the values of $\mathcal{E}'$ well below
        and well above f_m, and α is the Fuoss-Kirkwood parameter.
        Note that $\mathcal{E}'_0 - \mathcal{E}'_\infty = 2\mathcal{E}''_m/\alpha$.

(ii)   Where two entries are given for the same temperature that
        denoted (a) is based on data taken up to 150 MHz, and that
        denoted (b) on data taken up to 900 MHz (ME35 isotropic)
        or 18 GHz (ME35 nematic). See also text.

clarity of exposition, but the approximation that the internal field
corrections are isotropic, although likely to be accurate for monoesters with their small dielectric anisotropies, may be less desirable in general.

The expressions given by Maier and Meier for the principle
permittivities in the nematic phase are:

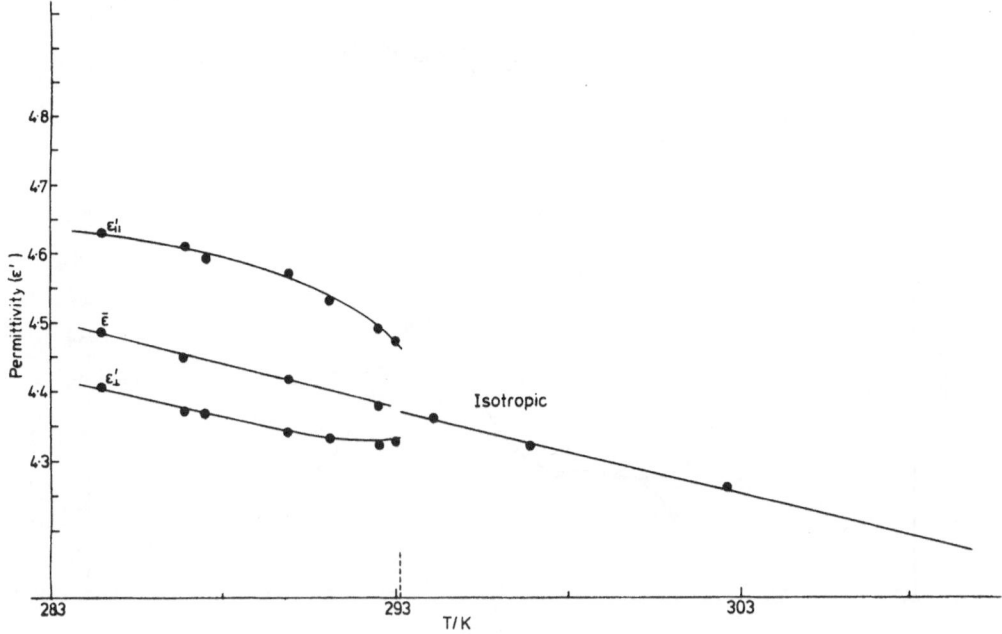

Fig. 1 Temperature dependence of the permittivities of ME 35 at 100 kHz.

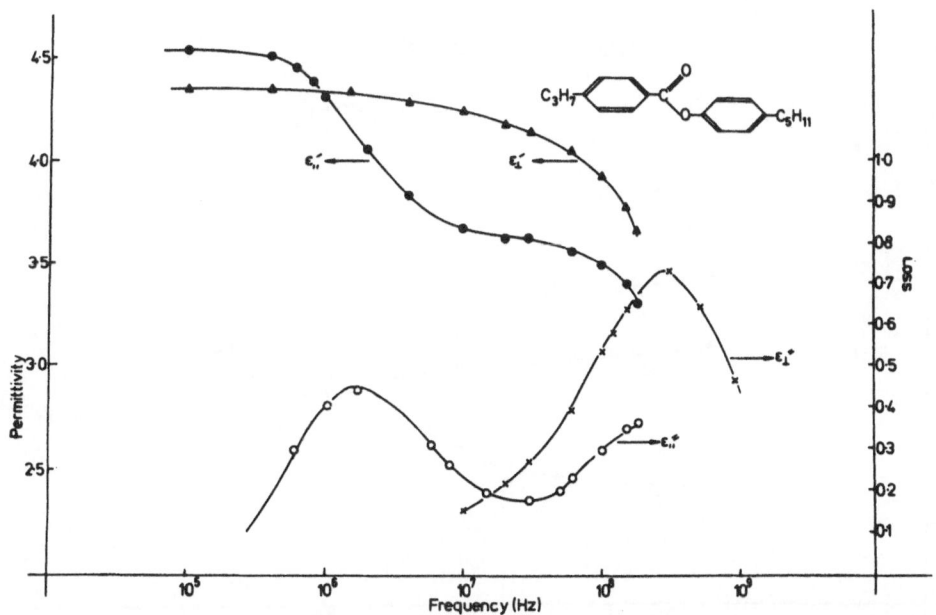

Fig. 2 Permittivity and loss curves for ME35 in its nematic phase at 18°C.

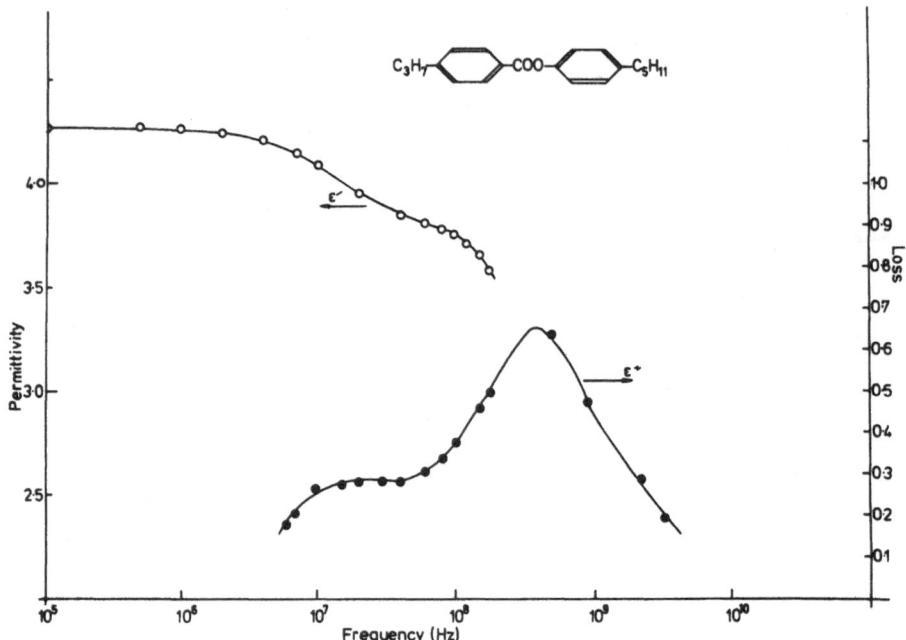

Fig. 3   Permittivity and loss curves for ME35 in its isotropic
         phase at 29°C.

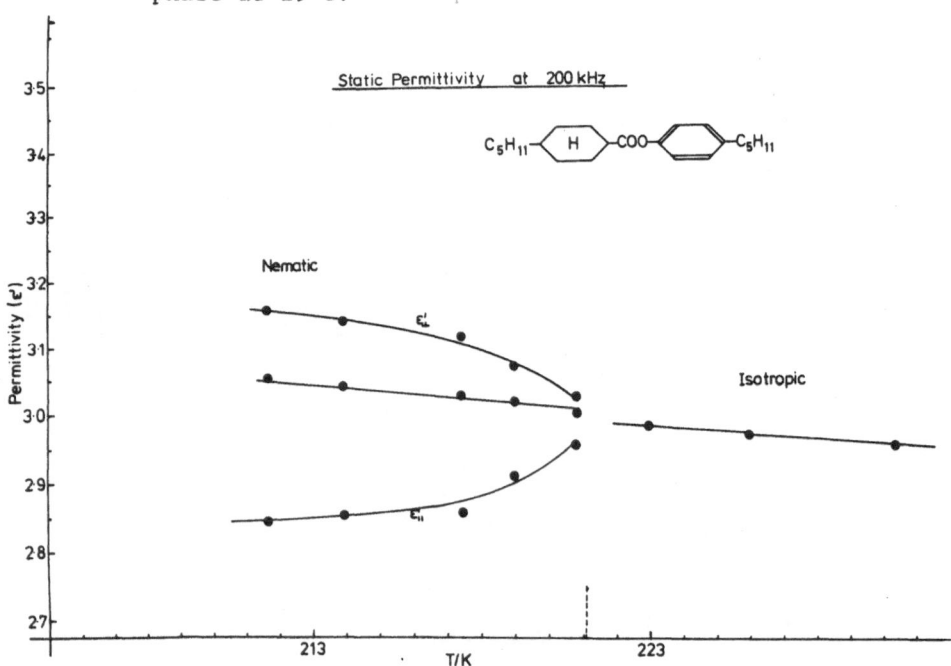

Fig. 4   Temperature dependence of the permittivities of DE55
         at 200 kHz.

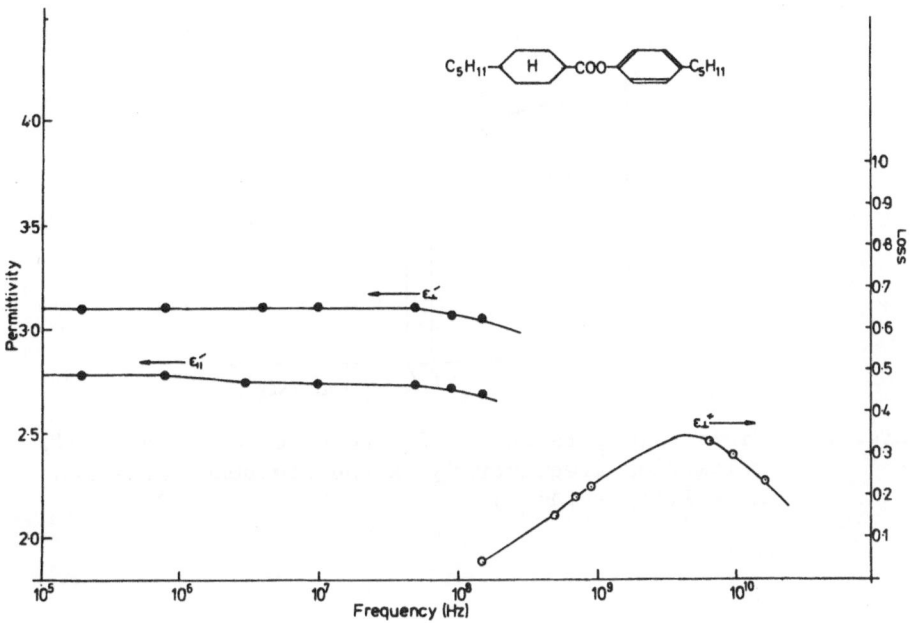

Fig. 5   Permittivity and loss curves for DE55 in its nematic phase
at 39.5 °C.

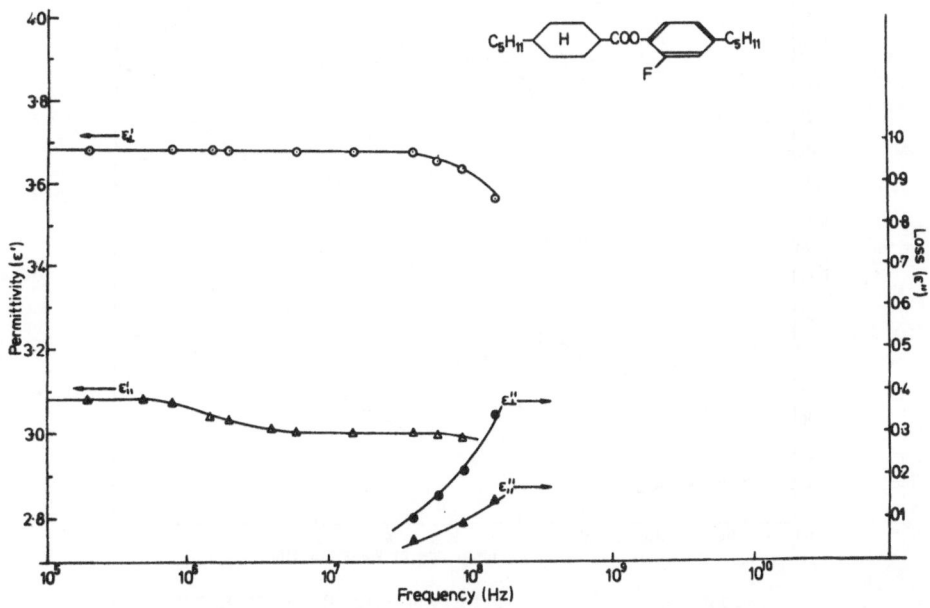

Fig. 6   Permittivity and loss curves for FDE55 in its nematic phase
at 25°C.

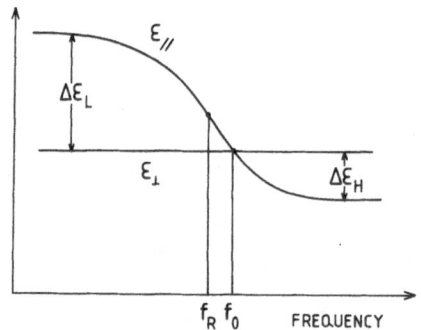

Fig. 7   Dielectric parameters of a two-frequency material.   The
          relaxation frequency f_R is the frequency at which
          $\Delta\varepsilon = \frac{1}{2}(\Delta\varepsilon_L - |\Delta\varepsilon_H|)$.

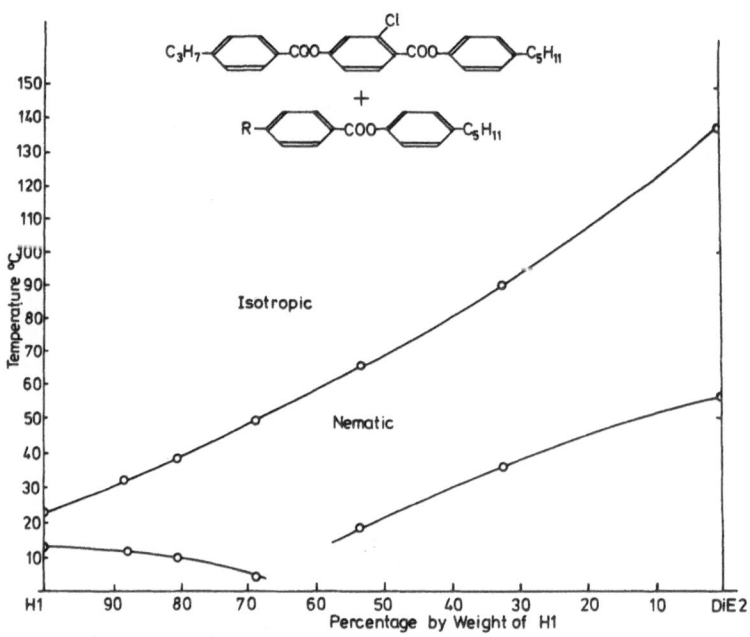

Fig. 8   Phase diagram for H1 + DiE2.

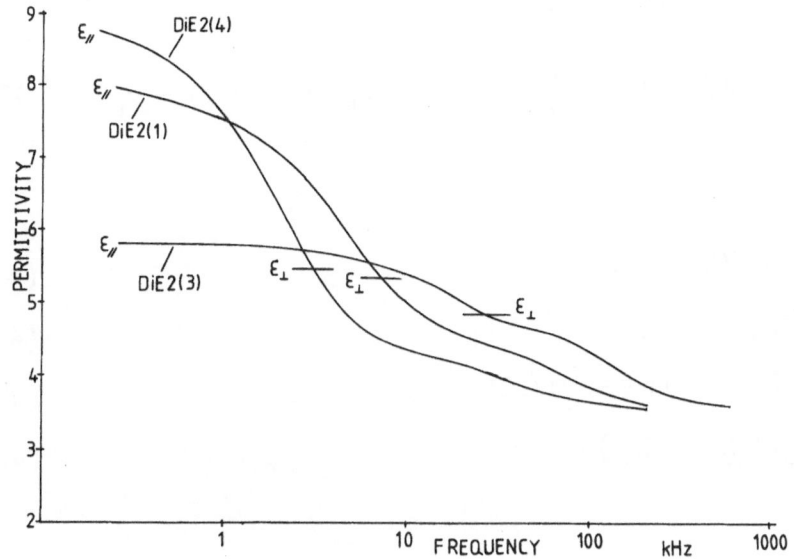

Fig. 9   Low frequency permittivities at 5°C of diester/monoester
mixture DiE2 + H1.

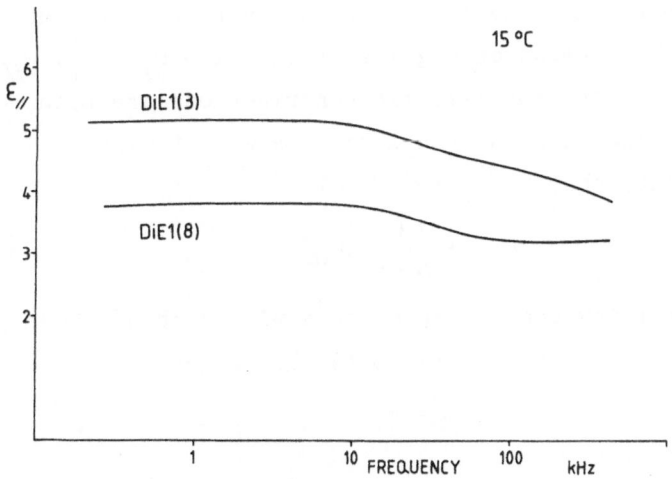

Fig. 10   Frequency dependence of $\varepsilon_{||}$ at 15°C of diester/monoester
mixtures DiE1(3) and DiE1(8).

$$\varepsilon_{\parallel} = 1 + (NhF/\varepsilon_o)\{\bar{\alpha} + \tfrac{2}{3}\Delta\alpha S + (F/3k_B T)[\mu_{\ell}^2(1 + 2S) + \mu_t^2(1-S)]\}$$

$$(4.1)$$

$$\varepsilon_{\perp} = 1 + (NhF/\varepsilon_o)\{\bar{\alpha} - \tfrac{1}{3}\Delta\alpha S + (F/3k_B T)[\mu_{\ell}^2(1-S) + \mu_t^2(1+\tfrac{1}{2}S)]\}$$

$$(4.2)$$

where

$$h = 3\bar{\varepsilon}/(2\bar{\varepsilon} + 1) \tag{4.3}$$

$$F = 1/(1 - \bar{\alpha}f) \tag{4.4}$$

$$f = (2N/3\varepsilon_o)[(\bar{\varepsilon} - 1)/(2\bar{\varepsilon} + 1)] \tag{4.5}$$

$$\bar{\varepsilon} = \tfrac{1}{3}(\varepsilon_{\parallel} + 2\varepsilon_{\perp}) \tag{4.6}$$

In eqns. (4.1) to (4.6) ε_o is the permittivity of free space, S is the order parameter given by eqn. (1.1), μ_{ℓ} and μ_t are the components of the molecular electric dipole along and transverse to the long axis, $\bar{\alpha}$ and $\Delta\alpha$ are the spherical mean and the anisotropy of the molecular polarizability, and N is the number of molecules per unit volume. Suppose that the contributions of μ_{ℓ} to $\varepsilon_{\parallel}$ and $\varepsilon_{\perp}$, and of μ_t to $\varepsilon_{\parallel}$ and $\varepsilon_{\perp}$, relax at angular frequencies $\omega_{\parallel}^{\ell}$, $\omega_{\perp}^{\ell}$, $\omega_{\parallel}^{t}$, and $\omega_{\perp}^{t}$, respectively. Since the last two contributions are both relaxed by rotational diffusion about the long axis we anticipate (in agreement with experiment) that

$$\omega_{\parallel}^{t} = \omega_{\perp}^{t} \tag{4.7}$$

If the permittivity changes associated with each of these relaxations is calculated from eqns. (4.1) to (4.6), we get:

$$\Delta \varepsilon_{/\!/}'^{\ell} = 2(\varepsilon_m'') {}_{/\!/}^{\ell} / \alpha_{/\!/}^{\ell}$$

$$= 3[(\varepsilon_{/\!/\infty} - 1)/(\bar{\varepsilon}_\infty + 2)](h_1 F_1 - h_2 F_2) +$$

$$+ [N/3\varepsilon_o k_B T](h_1 F_1^2 - h_2 F_2^2) \mu_t^2 (1 - S) +$$

$$+ [N/3\varepsilon_o k_B T] h_1 F_1^2 \mu_\ell^2 (1 + 2S) \tag{4.8}$$

$$\Delta \varepsilon_{/\!/}'^{t} = 2(\varepsilon_m'') {}_{/\!/}^{t} / \alpha_{/\!/}^{t}$$

$$= 3[(\varepsilon_{/\!/\infty} - 1)/(\bar{\varepsilon}_\infty + 2)](h_2 F_2 - h_\infty F_\infty) + \tag{4.9}$$

$$+ [N/3\varepsilon_o k_B T] h_2 F_2^2 \mu_t^2 (1 - S)$$

$$\Delta \varepsilon_{\perp}'^{\ell} = 2(\varepsilon_m'')_{\perp}^{\ell} / \alpha_{\perp}^{\ell}$$

$$= 3[(\varepsilon_{\perp\infty} - 1)/(\bar{\varepsilon}_\infty + 2)](h_1 F_1 - h_2 F_2) + \tag{4.10}$$

$$+ [N/3\varepsilon_o k_B T](h_1 F_1^2 - h_2 F_2^2) \mu_t^2 (1 + \tfrac{1}{2}S) +$$

$$+ [N/3\varepsilon_o k_B T] h_1 F_1^2 \mu_\ell^2 (1 - S)$$

$$\Delta \varepsilon_{\perp}'^{t} = 2(\varepsilon_m'')_{\perp}^{t} / \alpha_{\perp}^{t} \tag{4.11}$$

$$= 3[(\varepsilon_{\perp\infty} - 1)/(\bar{\varepsilon}_\infty + 2)](h_2 F_2 - h_\infty F_\infty) +$$

$$+ [N/3\varepsilon_o k_B T] h_2 F_2^2 \mu_t^2 (1 + \tfrac{1}{2}S)$$

where ε_m'' is the maximum loss associated with the relaxation, α is the Fuoss-Kirkwood factor [eqn. (2.1)],

$$\varepsilon_{\parallel\infty} = n_e^2 \tag{4.12}$$

$$\varepsilon_{\perp\infty} = n_o^2 \tag{4.13}$$

$$\bar{\varepsilon}_\infty = \tfrac{1}{3}(n_e^2 + 2n_o^2) \tag{4.14}$$

n_o and n_e being the ordinary and extraordinary refractive indices, and

$$hF = \bar{\varepsilon}(\bar{\varepsilon}_\infty + 2)/(2\bar{\varepsilon} + \bar{\varepsilon}_\infty) \tag{4.15}$$

$$hF^2 = \bar{\varepsilon}(2\bar{\varepsilon} + 1)(\bar{\varepsilon}_\infty + 2)^2/3(2\bar{\varepsilon} + \bar{\varepsilon}_\infty)^2 \tag{4.16}$$

where in eqns. (4.15) and (4.16) the mean permittivity $\bar{\varepsilon}$ is taken at angular frequencies ω_1, ω_2 or ω_∞ such that

$$\omega_1 < \omega_{\parallel}^\ell < \omega_\perp^\ell < \omega_2 < \omega_{\parallel}^t = \omega_\perp^t < \omega_\infty \tag{4.17}$$

and we ignore the small relaxations of dipole contributions arising from molecular inertia terms.

Note that when S = 0, i.e. an isotropic fluid, $\Delta\varepsilon_{\parallel}'^\ell + \Delta\varepsilon_{\parallel}'^t$ and $\Delta\varepsilon_{\perp}'^\ell + \Delta\varepsilon_{\perp}'^t$ are both equal, the sum of either eqn. (4.8) + eqn. (4.9), or eqn. (4.10) + eqn. (4.11), reducing to the well-known Onsager equation.

The four equations (4.8) to (4.11) contain only three unknowns, μ_ℓ, μ_t, and S. Thus from the dielectric relaxation data at any one temperature it is possible to determine the effective molecular electric dipole moment μ $[=(\mu_\ell^2 + \mu_t^2)^{1/2}]$, the angle β $[=\tan^{-1}(\mu_t/\mu_\ell)]$ which the long axis makes with the dipole, and the order parameter S, and even, if all four relaxations are observed, to test the adequacy of the theory by checking the consistency of eqns. (4.8) to (4.11). The redundancy in eqns. (4.8) to (4.11) arises because we have assumed that the molecular orientational distribution function depends only on the angle between the long axis and the director. The most general assumption consistent with uniaxial nematic symmetry would introduce an extra order parameter, leading to the four equations containing four unknowns.

The wider implications of these results are worthy of comment. Let (ϕ θ Ψ) be the Euler angles describing the rotation from laboratory axes to the frame associated with a particular molecule. Consider any static property of the nematic described by a property tensor, here taken for convenience as a spherical tensor, $<T_{kq}>$, where the conical brackets denote averaging over the molecular distribution. Then if the property tensor in the molecular frame is T'_{kp}:

$$\langle T_{kq} \rangle = \delta_{oq} T'_{ko} S_k \qquad (4.18)$$

where, consistently with the theory above, the molecular distribution is assumed to depend only on θ, and the order parameter S_k is given by the generalization of eqn. (1.1)

$$S_k = \langle P_k(\cos \theta) \rangle \qquad (4.19)$$

Writing

$$T'_{kp} = A_k C_{kp}(\beta \alpha) \qquad (4.20)$$

where C_{kp} are the modified spherical harmonics and (α β γ) are the Euler angles describing the orientation of $\underset{\sim}{T}'$ in the molecular frame, eqn. (4.18) becomes

$$\langle T_{kq} \rangle = \delta_{oq} A_k P_k(\cos \beta) S_k \qquad (4.21)$$

Experimental measurements yield the quantity $<T_{ko}>$. Thus it is only possible to determine S_k by extrapolating $<T_{ko}>$ to absolute zero, where the assumption that $S_k(T = 0)$ is unity yields a value for $A_k P_k(\cos \beta)$ provided this quantity is independent of temperature (both explicitly and implicitly through dependence on the order

parameters). Even then the angle between the long axis and the chosen
molecular frame cannot be determined without a priori knowledge of
A_k, thus raising questions as to whether the long axis, and by impli-
cation S_k, is defined independently of the particular static property
considered.[5] An analogous argument applies if the molecular distribution is allowed to have the most general angular dependence consistent with nematic symmetry, namely on θ and Ψ.

However, the arguments leading to eqns. (4.8) to (4.11) take
the long axis as that axis about which the rotational diffusion
correlation time is shortest. This corresponds to the relaxation
time

$$\tau^t = 1/\omega_{//}{}^t = 1/\omega_{\perp}{}^t$$

in our approximation. The choice of axis is consistent with the
averaging over Euler angles leading to (4.18). Thus for all static
properties, and dynamic properties with time scales longer than those
for which molecular inertia becomes important, (which includes the
properties of greatest practical interest) the long axis may be defined as the largest-eigenvalue principal axis of the rotational
diffusion tensor[12]

$$\underset{\sim}{D}^{rot} = \int_0^\infty dt \langle \underset{\sim}{\omega}(0)\underset{\sim}{\omega}(t)\rangle \tag{4.22}$$

where $\underset{\sim}{\omega}$ is the molecular angular velocity. Clearly since the
molecular reorientational motion is diffusive over the timescales of
interest, the choice of the principal axes of inertia, which is
sometimes suggested for defining the long axis, is inappropriate.
From the definition we see that the long axis is defined in the
molecular frame by the local dynamical structure of the particular
fluid, and that its orientation relative to the molecular geometry
can and should be determined experimentally.

4.2 Benzoate Esters

The temperature dependence of the low frequency (100 kHz)
permittivities of the benzoate esters (e.g. Fig. 1) is typical of
simple nematogens, with both $\bar{\varepsilon}'$ in the nematic phase and ε' in the
isotropic phase having negative temperature coefficients and negli-
gible discontinuity at the clearing point. The kind of association
exemplified by polar nematogens such as the cyanobiphenyls is thus
clearly absent here, as might be expected since the role of esters
when used with cyanobiphenyls in mixtures for multiplexed LCDs
includes breaking up that association.

In the nematic phases of the benzoate esters three of the four
predicted relaxations can be clearly resolved (see Table I and Fig.
2). This is sufficient to apply eqns. (4.8) to (4.11). Table II
gives the values calculated for the angle β between the long axis
and the molecular dipole

$$\beta = \tan^{-1}(\mu_t/\mu_\ell) \qquad (4.23)$$

together with values for S and the effective dipole moment μ.

The values obtained for the dipole moment are in acceptable
agreement with the value of 5.4×10^{-30} C m predicted from group moment
calculations[13] bearing in mind the limitations of both estimates.
The group moment calculation suggests that the molecular electric
dipole is directed along the C=O double bond. Although knowledge of
β alone does not define the direction of the long axis unambiguously,
it is possible that the long axis is directed along the bisector of
the O=C-O angle. The variations with temperature of μ and β in
Table II are not significant. The values obtained for S are reasonable, and as T → T_c become close to the values predicted by Maier-
Saupe mean field theory.

The calculations also yield estimates of $\Delta\varepsilon_\perp'^\ell$ for the unresolved
relaxation. The values predicted are rather small and, in agreement

with experiment, would be difficult for us to resolve, particularly
if broadened or close to the transverse dipole relaxation. Within
the experimental and fitting uncertainty, the value $\Delta\varepsilon\rfloor^{\ell} = 0.344$
predicted for ME35 at 291.1 K is in agreement with the value of 0.15
obtained by fitting two loss peaks to data taken up to 18 GHz.

In the isotropic phase the benzoate esters show two clearly
resolved relaxations (Table I and Fig. 3). These relaxations may be
quantitatively understood if it is assumed that the rotational diffusion tensor is sufficiently anisotropic for distinct relaxation
times to be associated with the longitudinal and transverse dipole
components. From Table II it is seen that μ and β deduced from
isotropic phase data on this assumption agree well with those obtained from the nematic phase, particularly bearing in mind the experimental errors, and that the phases do have different long range order.
Table III gives values of the anisotropy of the rotational diffusion
tensor $D_{/\!/}^{rot}/D_{\perp}^{rot}$ (assumed to be axially symmetric) obtained from the
equation[12]

$$\frac{D_{/\!/}^{rot}}{D_{\perp}^{rot}} = 2\left(\frac{\varepsilon_o'^t}{\varepsilon_o'^{\ell}}\right)\left(\frac{2\varepsilon_o''^{\ell} + \varepsilon_{\infty}'^{\ell}}{2\varepsilon_o'^t + \varepsilon_{\infty}'^t}\right)\left(\frac{\omega^t}{\omega^{\ell}}\right) - 1 \qquad (4.24)$$

where ω^{ℓ} and ω^t are the angular frequencies of the lower and upper
frequency relaxations, respectively, and we have assumed an Onsager
relationship between the dielectric relaxation time and the dipole
correlation time. From comparison of the (a) and (b) data sets at
302.6 K in ME35 it is clear that the estimates of $D_{/\!/}^{rot}/D_{\perp}^{rot}$ are
fairly sensitive to experimental error. Nevertheless, they are indicative of considerable anisotropy, with nematic-like local structure
persisting well into the isotropic phase.

Perhaps even more surprisingly, the alkyl benzoate ester appears
to enjoy a nematic-like local environment even in 1.55 mol dm^{-3}
solution in benzene, although the relaxation frequencies are both

Table II. Molecular and Fluid Parameters Calculated from Dielectric
           Data

Material and Phase	T/K	T/T_c	S	β/deg	$\mu/10^{-30}$ C m
ME35	286.9	0.979	0.56	58	6.59
	288.1	0.983	0.53	59	6.55
nematic	289.8	0.989	0.49	59	6.74
	291.1	0.994	0.42	58	6.57
ME35	294.1	1.004	0.0	65	6.63
	296.9	1.013	0.0	65	6.61
isotropic	302.6	1.033	0.0	65	6.62
	311.6	1.063	0.0	65	6.59
ME35 in C_6H_6	294.0	—	0.0	64	5.64
$C_6H_6COOC_6H_6$ in C_6H_6	294.0	—	0.0	—	6.00
ME55	302.0	1.010	0.0	56	5.60
isotropic	308.5	1.032	0.0	56	5.63
	314.0	1.050	0.0	55	5.67
H1	285.3	0.968	0.58	58	6.60
nematic	290.0	0.984	0.54	59	6.57
	293.2	0.995	0.41	62	7.06
H1	296.8	1.007	0.0	58	5.92
isotropic	304.0	1.031	0.0	56	5.71
	312.0	1.058	0.0	55	5.75
DE55 nematic	312.5	0.974	0.61	82	5.13
DE55 isotropic	325.5	1.014	0.0		5.05
FDE55 nematic	298.0	0.963	0.56	80	6.12

raised and $D_{//}^{rot}/D_{\perp}^{rot}$ is less (Table III), as might be expected. On
the other hand unsubstituted phenyl benzoate in 1.55 mol dm^{-1} solution
in benzene shows only a single relaxation at much higher frequency
(Table I). This may be understood by supposing that the structure
of the phenyl benzoate/benzene mixture can be thought of as a liquid
of benzene rings in which a proportion of pairs are bridged by ester
groups. The addition of C_3 and C_5 alkyl terminal substituents to the
bridged benzenes will clearly disrupt this structure, and thus allows
the latent nematicity of ME35 to manifest itself.

Something must be said concerning the possibility of internal
reorientations involving the dipole-bearing ester linkage, since we
have so far treated the monoesters as rigid molecules. An internal
reorientation which changed only part of either μ_ℓ or μ_t would give
an additional loss peak at higher frequency than the corresponding
rigid molecule relaxation.[12] There is no evidence for this in the
nematic phases studied. If the two loss peaks observed in the
isotropic phases of ME35, ME55, and H1 were interpreted as a single
rigid molecule reorientation ($\underset{\sim}{D}^{rot}$ isotropic) coupled with an internal
reorientation involving the dipole-bearing ester linkage, then the
lower relaxation frequency would be assigned to the rigid molecule
reorientation. This appears unrealistically low in view of the
transverse dipole relaxation frequencies observed in the nematic
phases of ME35 and H1. Further, it is difficult to see why the same
effects should not be observed in the phenyl benzoate solution if
internal reorientation involving the ester linkage were the cause.
Thus there is no evidence for internal reorientations increasing the
number of loss peaks, although we cannot exclude the possibility of
internal processes which increase observed relaxation frequencies
without increasing the number of peaks. Such an effect would not
affect the calculations leading to the numbers in Table II, but the
values of $D_{//}^{rot}/D_{\perp}^{rot}$ in Table III could be affected since, for example,
increasing ω^t would cause $D_{//}^{rot}/D_{\perp}^{rot}$ to be overestimated.

Table III.   Anisotropy of the Rotational Diffusion Tensor Estimated
             from Relaxation Data on Isotropic Phases

	T/K	T/T_c	$D_{\parallel}^{rot}/D_{\perp}^{rot}$
ME35	294.1	1.004	83
	296.9	1.013	84
	302.6(a)	1.033	71
	302.6(b)	1.033	59
	311.6	1.063	75
ME55	302.0	1.010	42
	308.5	1.032	41
	314.0	1.050	39
H1	296.8	1.007	55
	304.0	1.031	37
	312.0	1.058	40
ME35 in C_6H_6	294.0	—	18

4.3 Cyclohexane Carboxylate Esters

The temperature dependence of the low frequency (200 kHz) permittivities of the DE and FDE cyclohexane carboxylate esters is again
typical of simple nematogens (Fig. 4) with both $\bar{\varepsilon}'$ in the nematic
phase and ε' in the isotropic phase having negative temperature coefficients and negligible discontinuity at the clearing point. Thus
the kind of association exemplified by cyanobiphenyls is clearly also
absent in these materials.

However, the frequency dependence of ε in these materials is at
first sight very different to that of the benzoate esters. In both
the nematic and isotropic phases very little frequency dependence is
seen up to 10^8 Hz, when the prominent relaxation associated with the
transverse component of the electric dipole begins to appear. In

DE55 this relaxation has been precisely located at about 4 GHz (see Table I) by use of the slotted line and sweep frequency techniques. Experiments to locate the relaxation precisely in the fluorine-substituted esters are in progress, but the indications from the data up to 150 MHz are that it is at a similar frequency.

This rather more than ten-fold increase over the corresponding relaxation frequency in the substituted phenyl benzoate esters is surprising. It would not be expected from the literature, where very similar relaxation frequencies are reported for e.g. cyclohexylbromide and bromobenzene, although in our measurement of unsubstituted phenyl benzoate in benzene (Table I) we do observe a relaxation frequency of about 4 GHz. One can obviously speculate that the greater flexibility conferred by the cyclohexane ring may be involved, perhaps allowing an internal reorientation which increases the relaxation frequency.

Of more immediate practical interest (Sec. 5) is the relaxation of the contribution of μ_ℓ to $\varepsilon_{//}$, which in the substituted phenyl benzoate esters occurred with significant intensity at about 1 MHz. This feature is very weak in the cyclohexane esters, although we have succeeded in detecting it in the 1 to 3 MHz region in DE55, FDE55, and H3 in their nematic phases; Figures 5 and 6 show examples. There is a hint (unsubstantiatable at our experimental accuracy) of this relaxation in some of the isotropic data. Unfortunately, even in the nematic phase the intensity of the loss is too low for the relaxation frequency to be accurately determined. However, it seems clear that, in contrast with the transverse dipole relaxations, the longitudinal dipole relaxation of $\varepsilon_{//}$ occurs at a similar frequency in both benzoate and cyclohexane carboxylate esters. The low intensity of the relaxation in the cyclohexane esters implies that the long axis is almost perpendicular to the electric dipole. Since, for the unfluorinated DE materials at least, group moment calculations suggest that the electric dipole lies in the same direction relative to the molecular frame, i.e. along the C=O double bond, as in the benzoate esters,

the angle of the long axis to the molecular frame differs by about
30° between the two types of ester, emphasizing that the long axis is
very much a property of the molecule in its fluid environment rather
than of the molecule in isolation.

It is possible to analyse the permittivity data along the lines
developed in Sec. 4.1 without the high-frequency data required to
locate the transverse dipole relaxations precisely. When the longi-
tudinal dipole relaxation of $\varepsilon_{//}$ can be observed, $\Delta\varepsilon'^{\ell}_{//}$ can be obtained
from $\varepsilon'_{//}$ vs frequency (Figs. 5 and 6) and, since $\Delta\varepsilon'^{\ell}_{\perp}$ will clearly be
very small, the values of $\Delta\varepsilon'^{t}_{//}$ and $\Delta\varepsilon'^{t}_{\perp}$ can be taken as

$$\Delta\varepsilon'^{t}_{//} = \varepsilon'_{//} - \Delta\varepsilon'^{\ell}_{//} - n_e^2 \qquad (4.25)$$

$$\Delta\varepsilon'^{t}_{\perp} = \varepsilon'_{\perp} - n_o^2 \qquad (4.26)$$

where $\varepsilon'_{//}$ and $\varepsilon'_{\perp}$ are the low frequency ("static") permittivities. The
results of analyses of this kind on DE55 and FDE55 are included in
Table II, together with an estimate of the dipole moment of DE55 in
its isotropic phase. For comparison, group moment calculations[13]
predict a dipole moment of 5.4×10^{-30} C m for DE55; that for the fluo-
rinated ester will depend on the angle between the plane of the fluo-
rinated ring and the plane of the ester linkage. Taking the group
moment due to the F substituent as 5.0×10^{-30} C m, the value of
6.12×10^{-30} C m for μ gives a value of 85.5° for the angle between the
plane of the F-substituted benzene ring and the plane of the COO ester
group, where the angle is measured from the trans configuration with
the fluorine opposed to the C=O double bond.

Since we have estimated only β, the direction of the long axis
is not located unambiguously, and one cannot say that the similar
values for β obtained for the DE and FDE materials imply that the long
axis has a similar orientation in both. However, if it is assumed
that the long axis has the same orientation relative to the ester
group in both DE55 and FDE55, then by straightforward vector geometry
it can be shown that the only possible directions are

$$\underset{\sim}{\ell} = (0.80 \ - 0.30 \ 0.51)$$

$$(4.27)$$

or

$$\underset{\sim}{\ell} = (0.75 \ - 0.59 \ 0.31)$$

$$(4.28)$$

where we take right-handed axes with Ox bisecting the O=C-O angle, Oy in the O=C-O plane on the C=O side, and Oz perpendicular to the O=C-O plane. Note also that if we had only the ME and FDE data, it would be tempting to assume that the only difference between the benzoate and F-substituted cyclohexane carboxylate esters was that the fluorine dipole largely cancelled the longitudinal component of the ester dipole, with the long axis remaining virtually parallel to Ox. Clearly, such an assumption is called into question by the DE data.

The picture emerging from our study that the benzoate and cyclohexane carboxylate esters have similar dipole moments but different long axes is consistent with the change in static dielectric anisotropy from positive to negative in passing from the benzoate to the cyclohexane esters.

If we neglect $\Delta\varepsilon'^{\ell}_{//}$ in eqn. (4.25), which is equivalent to taking $\beta = 90°$, then S and μ may be calculated solely from the static permittivity data and refractive index data. Unfortunately S is rather sensitive to this approximation and is underestimated, but the value of μ obtained for DE55, 5.1×10^{-30} C m, is independent of temperature, within experimental error, over the nematic and isotropic range examined ($T/T_c = 0.969$ to 1.025), and agrees well with the values obtained from more exact analysis (Table II) and from group moments. Evidently an analogous procedure could be applied if μ_t were negligible, since $\Delta\varepsilon'^{\ell}_{//}$ and $\Delta\varepsilon'^{\ell}_{\perp}$ could then be estimated from static permittivity data and refractive index data. However, since molecules for which such an approximation is applicable may have quite large positive dielectric anisotropies the warning in Sec. 4.1

concerning our use of isotropic internal field corrections should be heeded.

Finally, we remark that since both the benzoate and the cyclo-hexane carboxylate esters have at best rather short nematic ranges, measurements of S by methods which involve extrapolations or experimental quantities to absolute zero (e.g. the refractive index method) are inaccurate.

5.   DIESTER/MONOESTER MIXTURES

5.1 Two-frequency Materials for Practical Applications

Liquid crystal materials in which the longitudinal dipole relaxation of $\varepsilon_{//}$ occurs in the audio frequencies are currently required for two applications,[14] namely fast-switched twisted nematic shutters and two-frequency multiplexed twisted nematic matrix displays. The materials required have the property that, as a result of the relaxation, the dielectric anisotropy changes sign at a "cross-over frequency" f_o. The cross-over frequency will only be equal to the relaxation frequency f_R if the low frequency and high frequency dielectric anisotropies $\Delta\varepsilon_L$ and $\Delta\varepsilon_H$ (see Fig. 7) are equal in magnitude. The applications involve addressing the device with both low and high frequency signals, either sequentially in shutters or simultaneously in multiplexing.

In addition to the standard material requirements of wide phase range, lack of colour, purity, stability, low conductivity, and safety, there are for each application specific desirable combinations of refractive indices, frequency-dependent permittivities, and visco-elastic properties. The interplay of these factors is quite complex,[15] so that rather than attempting to formulate a single mixture for all applications it is better to establish mixture systems in which the properties for each specific application can be met as closely as possible. Although the compositions of practical mixtures are more complex than those that will be considered here, simple diester/monoester mixtures exemplify many of the principles involved.

If two or more rather similar compounds are mixed, a single low
frequency relaxation which shows no evidence of broadening is observed. This is exemplified in Table I by the data on mixture Hl,
where the Fuoss-Kirkwood α parameter for the low frequency relaxation
remains close to unity, and is equally true for mixtures of similar
diesters. The properties of such mixtures are effectively those of
a single average component. However, when two very different single
compounds or mixtures are mixed, two distinct low frequency relaxations are observed, each associated with one of the components
although modified by the other. This effect has been disclosed in
the literature for monoester/diester mixtures,[16] azoxy/diester and
azoxy/lateral-CN-substituted-monoester mixtures,[17] and cyanobiphenyl/
cyanoterphenyl mixtures.[18]

A successful design philosophy for two-frequency mixtures thus
consists in combining two components A and B, each of which may
itself be a mixture. The function of component A is to provide the
audio frequency relaxation, while that of B is to enable other
properties of the material to be modified while retaining the desired
dielectric parameters. Thus A has f_R in the range desired for f_o,
together with sufficiently large $\Delta\epsilon_L$ and $\Delta\epsilon_H$ and sufficient miscibility range with B. Component B has $\Delta\epsilon$ close to zero or negative,
and should desirably show no perceptible relaxation in the specified
range of drive frequencies. By use of this approach mixtures have
been formulated which combine low f_o with relatively large dielectric
anisotropies and viscosities less than 100 cP;[3] an example is shown
in Table IV.

5.2 Discussion of Data

Although miscibility can sometimes be a problem when mixing
molecules of different sizes the diester + monoester system is not
troublesome in this respect. The phase diagram of DiE2 + Hl, shown
in Fig. 8, is typical of our results.

Table IV. An Example of a Recently Developed Two-Frequency Material

```
Phase range:
        melting point                          supercools
        clearing point                         101 °C
Dielectric properties at 20 °C:
        Δε at 100 Hz                           +2.95
        Δε at 40 kHz                           -1.8
        ε⊥                                     5.25
        cross-over frequency f_o               5.1 kHz
Birefringence (20 °C) Δn^D                     0.12
Viscosity (20 °C)                              98 cP
```

This mixture system is also convenient for studying the influence of each component on the other relaxation since both DiE2 and H1 have strong low frequency relaxations which both lie below 500 kHz at accessible temperatures. Thus Fig. 9 shows data for three different mixtures at 5°C.

However, it should be recognised that there is a difficulty in comparing the relaxation frequencies of even single compounds. Quite general theoretical considerations suggest that f_R can be written approximately as

$$f_R \approx f_D(T)/g(T/T_c) \qquad (5.1)$$

where f_D is comparable to the Debye relaxation frequency of an isotropic liquid. Thus neither absolute temperature T nor reduced temperature T/T_c emerge unambiguously as a basis for comparison. Variants such as viscosity for T, or order parameter for T/T_c, suffer the same problem. That the importance of neither factor can be overlooked is evidenced by the observation that diesters such as DiE1 and DiE2, which have unusually low f_R on the basis of equal-temperature comparisons with simpler 2-ring compounds, have similar relaxation frequencies to the simpler compounds when compared at the same reduced temperature.

The problem of a basis for comparison is much greater in the
case of mixtures such as DiE2 + H1. Suppose that eqn. (5.1) can be
applied separately to the relaxation associated with each component:
is T_c to be taken as T_c of the mixture, or T_c of the component, or
some intermediate value? The possibility of local structure effects
which strongly differentiate the environments of component A and
component B molecules in the mixture, thus preserving the identities
of the two low frequency relaxations, clearly has to be taken
seriously, particularly in view of the effects we have observed for
monoesters in their isotropic phases and benzene solution (Sec. 4.2).
Higher frequency studies of mixtures such as DiE2 + H1 with a view
to applying the analysis employed in Sec. 4 to each set of relaxations
are currently in progress, and may well elucidate some of these
questions.

A practical point concerning the use of benzoate monoesters is
that the presence of the monoester relaxation tends to increase the
temperature dependence of $\Delta\varepsilon_H$ for frequencies well above the cross-
over frequency, since both relaxation frequencies vary rapidly with
temperature. For convenience in temperature-compensating the high
frequency drive voltages in a device it may be preferable to have
a single unbroadened relaxation. The low frequency relaxation is
actually favourable in this respect as it characteristically is close
to the simple Debye form:

$$\varepsilon(\omega) = \varepsilon'_\infty + (\varepsilon'_0 - \varepsilon'_\infty)/(1 + i\omega\tau_R) \qquad (5.2)$$

as evidenced by a Fuoss-Kirkwood α close to unity if ε'' is fitted,
or

$$(\partial\varepsilon''/\partial\log f)_{f=f_R} = -1.1513(\varepsilon'_0 - \varepsilon'_\infty) \qquad (5.3)$$

if ε' is measured as a function of frequency f. Hence the relative
ease with which the two low frequency relaxations are resolved in
Fig. 9.

Figure 10, which shows $\varepsilon_{//}$ against frequency for the same concentration of DiEl in mixtures H1 and H3, encapsulates some of the
important features of monoester/diester mixtures. The diester
relaxation has virtually the same frequency and magnitude in both
mixtures, although the curve for DiEl(8) is shifted down by about
the same amount as the 1.4 units by which $\varepsilon_{//}$ of H3 at 15°C is less
than $\varepsilon_{//}$ of H1. In DiEl(3) the higher frequency end of the diester
relaxation overlaps with the beginning of the rather obvious H1
relaxation, whereas the almost negligible H3 relaxation leaves $\varepsilon_{//}$
for DiEl(8) virtually constant above the diester relaxation. Thus,
whereas benzoate monoesters are preferable for probing the interactions which occur in diester/monoester mixtures, the DE and FDE
cyclohexane carboxylate esters have an advantage in practical
mixtures.

REFERENCES

1. G. Meier in <u>Dielectric</u> <u>and</u> <u>Related</u> <u>Molecular</u> <u>Processes</u>, ed.
        M. Davies, The Chemical Society, London, 1975, Vol. 2,
        pp. 183-97.

2. H. K. Bücher, R. T. Klingbiel, and J. P. VanMeter, Appl. Phys.
        Lett. <u>25</u>, 186 (1974).

3. D. G. McDonnell and R. A. Smith, unpublished work.

4. W. H. de Jeu, <u>Physical</u> <u>Properties</u> <u>of</u> <u>Liquid</u> <u>Crystalline</u>
        <u>Materials</u>, Gordon and Beach, New York, 1980.

5. F. C. Frank, <u>Proc.</u> <u>Internat.</u> <u>Conf.</u> <u>Liq.</u> <u>Cryst.</u> <u>Bangalore</u> <u>1979</u>,
        ed. S. Chandrasekhar, Heyden, London, 1980, pp. 1-6.

6. M. G. Clark, E. P. Raynes, R. A. Smith, and R. J. A. Tough, J.
        Phys. D, Appl. Phys. <u>13</u>, 2151 (1980).

7. A. Buka, P. G. Owen, and A. H. Price, Mol. Cryst. Liq. Cryst.
        <u>51</u>, 273 (1979).

8. C. T. O'Konski and A. Edwards, Rev. Sci. Instr. <u>39</u>, 1456 (1968).

9. T. W. Dakin and C. N. Works, J. Appl. Phys. <u>18</u>, 789 (1947);
        G. Williams, J. Phys. Chem. <u>63</u>, 534 (1959).

10.   A. H. Price and G. H. Wegdam, J. Phys. E 10, 479 (1977).

11.   W. Maier and G. Meier, Z. Naturforsch. 16a, 262 (1961).

12.   C. J. F. Böttcher and P. Bordewijk, Theory of Electric
      Polarization, 2nd Ed., Elsevier, Amsterdam, 1978,
      Vol. 2, Chaps. 11 and 15.

13.   R. J. W. Le Fevre and A. Suridaram, J. Chem. Soc. 1962, 3904;
      N. E. Hill, W. E. Vaughan, A. H. Price, and M. Davies,
      Dielectric Properties and Molecular Behaviour, Reinhold-
      vanNostrand, London, 1969.

14.   M. G. Clark, Microelectronics J. 12, 26-32 (1981).

15.   M. G. Clark and K. J. Harrison, SID Internat. Symp. Digest 12,
      82 (1981); M. G. Clark and I. A. Shanks, SID Internat.
      Symp. Digest 13, 172 (1982).

16.   L. Bata and G. Molnar, Chem. Phys. Lett. 33, 535 (1975).

17.   M. I. Barnik, L. M. Blinov, A. V. Ivashchenko, and
      N. M. Shtykov, Kristallografiya 24, 811 (1979) [Engl.
      transl.: Sov. Phys. Cryst. 24, 463]

18.   H. R. Zeller, Phys. Rev. A 23, 1434 (1981).

INTERMOLECULAR GUEST-HOST INTERACTIONS AND THE OPTICAL ORDER

PARAMETER OF PLEOCHROIC DYES

F. C. Saunders, L. Wright, and M. G. Clark

Royal Signals and Radar Establishment, Malvern
Worcestershire WR14 3PS, UK

ABSTRACT

The optical order parameters S_{op} of anisotropic dye molecules
in various nematic hosts have been studied as a function of temperature, in order to compare different guest-host systems over the
same range of reduced temperature. Two anthraquinone dyes, one of
which has hydrogen bonding and the other not, have been studied in
cyano and non-cyano two-ring nematic host mixtures, with and without cyano and non-cyano high clearing point additives. Extrapolation of S_{op} to absolute zero is attempted in order to determine
experimentally the angle between the transition moment and the
"long axis" of the molecule. It is found that the temperature
dependence of S_{op} has a different shape to that of the host order
parameter, the reason perhaps being the rather strong deviations
from cylindrical symmetry of the anthraquinone molecule. The
comparison of reduced temperature curves of S_{op} reveals two
effects: (i) the hydrogen-bonding dye is more sensitive to the
chemical nature of the host, presumably due to varying competition
between intramolecular and intermolecular interactions; (ii) although
the host order parameter was unchanged by addition of the high

831

clearing point additives studied, the presence of such materials
in a mixture lowers S_{op} significantly, indicating some kind of
local structure effect, possibly preferential solvation of the dye
by the larger molecule. The shift in clearing point on adding dye
broadly correlates with direct evidence on the degree to which each
host orders the dye. We speculate on a possible role for dipole/
induced-dipole interactions in guest-host interactions.

1.   INTRODUCTION AND THEORY

The use of anisotropic dye molecules as solutes in liquid
crystal hosts is now well established, and various liquid crystal
devices based on this guest-host effect have been developed.[1]
However, the practical application of these devices is constrained
by availability of suitable dyes. Thus there is continuing pressure
for improvements in the properties of anisotropic dyes, including in
particular properties such as solubility and order parameter (defined
below) which depend on the guest-host interactions. Additionally,
it appears to us a matter of some interest that relatively asymmetric
molecules such as anthraquinone dyes should perform so well. With
these points in mind we have undertaken an investigation of the
relationship between dye performance and guest-host interactions by
studying in detail two selected anthraquinone dyes in a variety of
hosts. The first results of this study are presented here.

A figure of merit for the absorption anisotropy of a dye
aligned in a nematic host is the "optical order parameter" S_{op},
defined in terms of the experimentally measured absorbances for
light polarized parallel and perpendicular to the nematic
director, $A_{//}$ and $A_{\perp}$, by

$$S_{op} = (A_{//} - A_{\perp})/(A_{//} + 2A_{\perp}) \qquad (1.1)$$

Evidently $-\frac{1}{2} < S_{op} < 1$, with $S_{op} = 0$ corresponding to isotropy and $S_{op} = 1$ to perfect alignment. At a molecular level S_{op} is the order parameter of the transition moment of the absorption:

$$S_{op} = \tfrac{1}{2}\left\langle 3(\underset{\sim}{n} \cdot \underset{\sim}{\mu}/\mu)^2 - 1 \right\rangle \tag{1.2}$$

where $\underset{\sim}{n}$ is the nematic director and $\underset{\sim}{\mu}$ the transition moment. By use of angular momentum theory it is readily shown that S_{op} can be written

$$S_{op} = \tfrac{1}{2}(3\cos^2\beta - 1)S_G \tag{1.3}$$

where β is the angle between $\underset{\sim}{\mu}$ and the long axis $\underset{\sim}{\ell}$ of the dye in the fluid ($\underset{\sim}{\ell}$ being defined as the largest-eigenvalue principal axis of the rotational diffusion tensor $\underset{\sim}{D}_G^{rot}$ of the dye molecule[2]), and S_G is the guest order parameter

$$S_G = \left\langle \tfrac{1}{2}(3\cos^2\theta - 1) \right\rangle \tag{1.4}$$

θ being the angle between $\underset{\sim}{\ell}$ and $\underset{\sim}{n}$. In the case of a degenerate transition eqn (1.3) becomes[3]

$$S_{op} = (\Delta\alpha/3\bar{\alpha})S_G \tag{1.5}$$

where

$$\Delta\alpha = \tfrac{1}{2}(2\alpha_{33} - \alpha_{22} - \alpha_{11}) \tag{1.6}$$

$$\bar{\alpha} = \tfrac{1}{3}(\alpha_{11} + \alpha_{22} + \alpha_{33}) \tag{1.7}$$

the α_{ii} being the absorption probabilities for light polarized parallel to the principal axes of $\underset{\sim}{D}_G^{rot}$ with the 3-axis parallel to $\underset{\sim}{\ell}$

In both eqn (1.3) and (1.5) we have assumed, as is commonly done, that the distribution function of the dye molecules, f, is a function of θ only, $f(\theta)$. This is strictly true only for cylindrically symmetric molecules. For the most general distribution function consistent with nematic symmetry, $f(\theta,\psi)$, where (ϕ θ ψ) are the Euler angles describing the rotation of the laboratory frame into the molecular frame (principal axes of $\underset{\sim}{D}_G^{rot}$), eqn (1.3) becomes

$$S_{op} = \tfrac{1}{2}(3\cos^2\beta - 1)\left\langle\tfrac{1}{2}(3\cos^2\theta - 1)\right\rangle +$$

$$\tfrac{1}{2}\sqrt{3}\sin^2\beta\,\cos 2\gamma\left\langle\tfrac{1}{2}\sqrt{3}\sin^2\theta\,\cos 2\psi\right\rangle \qquad (1.8)$$

where (β γ) are the angular polar coordinates of $\underset{\sim}{\mu}$ in the molecular frame, and we have used the symmetry of $f(\theta,\psi)$:

$$f(\theta,\psi) = f(\pi - \theta, \psi) = f(\theta, -\psi) = f(\pi - \theta, -\psi) \qquad (1.9)$$

However, we shall concentrate mainly on the simplest form, eqn (1.3), although we shall present and discuss evidence which possibly suggests that this form may not be fully adequate to describe an anthraquinone dye in a nematic host.

Equation (1.3) may be rewritten as a product of three factors

$$S_{op} = \tfrac{1}{2}(3\cos^2\beta - 1)\cdot(S_G/S_H)\cdot S_H \qquad (1.10)$$

where S_H is the analogue of eqn (1.4) for the host nematic molecules, and can be determined experimentally from, for example, refractive indices measured as a function of temperature. Thus S_H measures

the intrinsic alignment of the host, which is a function of temperature having broadly the form given by Maier-Saupe mean field
theory. The first factor in eqn (1.10) is a measure of the alignment of the transition moment relative to the long axis. The
angle β would appear unlikely to be strongly temperature dependent,
although if strong temperature dependence of the absorption wavelength were observed this might be an indication to the contrary.
Clearly any deviation of β from zero will reduce S_{op}.

The factor S_G/S_H measures the orientational fluctuations of
the guest molecule relative to those of the host. At finite
temperatures the guest molecule may show greater or lesser ordering
depending on its relative size and shape and its interactions with
the host. Thus S_G/S_H may be greater or less than unity. As $T \to 0$
we assume that both S_H and S_G tend to unity, implying that

$$\lim_{T \to 0} (S_G/S_H) = 1 \qquad (1.11)$$

and thus

$$\lim_{T \to 0} S_{op} = \tfrac{1}{2}(3\cos^2 \beta - 1) \qquad (1.12)$$

From this discussion it is seen that each of the factors in
eqn (1.10) may be estimated experimentally. The host order
parameter S_H may be determined, for example, from refractive index
measurements. By measuring S_{op} as a function of temperature and
extrapolating to absolute zero, β may be obtained from eqn (1.12).
This at least partly defines and determines experimentally the
direction of the long axis, since the direction of the transition
moment can be defined, and at least in principle calculated, in
arbitrarily chosen molecular axes by use of quantum mechanics.
Finally, using the measured S_{op}, S_G/S_H may be calculated from
eqn (1.10).

This approach is the basis of our study reported here, but, as will be seen below, it is by no means straightforward to carry through in practice. Two dyes out of the very large number that have been examined in our laboratory have been selected for this study in depth. Both are proprietary anthraquinone materials, one having hydrogen-bonding substituents, and the other having substituents which although polarizable have no hydrogen-bonding character. The dyes have been studied in polar and broadly non-polar two-ring nematic hosts with and without a variety of high clearing point additives.

2.   EXPERIMENTAL PROCEDURES

Measurements of optical order parameter, S_{op}, for the lowest energy electronic transition were made using a Perkin Elmer model 554 spectrometer. A temperature-controlled enclosure built into the sample chamber of the spectrometer allowed both a reference cell and the sample cell to be heated or cooled between $0°C$ and $60°C$. This temperature control was achieved using nitrogen initially cooled in a Red-point Associates vortex tube (with pre-cooling by passage through dry ice to attain temperatures less than $10°C$), and then passed over a heating element controlled by a CRL model 405 temperature controller. The temperature at the sensor was maintained to $0.1°C$ by the controller, but at the cell the temperature, although constant to $0.1°C$, was sometimes slightly lower, and so a separate thermocouple was attached to the cell to increase the accuracy of readings. A slight temperature gradient was found across the cell, normally less than $0.5°C$. Also mounted in the spectrometer sample holder were two HN32 polarizers, one for each beam, whose polarization axes had been adjusted to be vertical.

The cells used for these measurements were made using 3 mm thick glass whose surfaces had been treated with polyvinyl alcohol and rubbed to produce a low tilt homogeneous alignment on both surfaces. The two glass plates were assembled antiparallel so that the tilt of about 3° was uniform across the cell. The cells were spaced with either 12 μm or 25 μm thick mylar, and variations in thickness across the cell minimized by adjusting pressure on the edges of the glass to produce the minimum number of interference fringes across the cell when viewed in monochromatic sodium light. The cells were mounted in square mounts with carefully machined right angle corners, filled with the liquid crystal mixture under test, and mounted in the spectrometer temperature enclosure so that the alignment direction was parallel to the vertically polarized light. The sample cell contained the liquid crystal host and dye mixture while the reference contained the same host material without dye.

A base line correction had previously been carried out without the cells in the sample chamber. This compensates for any difference in the polarization of light present in the two beams of the spectrometer. The monochromator was then taken to a wavelength a long way from the dye absorption, so that any difference in the measured absorbance of the cells could be seen. These differences were normally only due to contamination of the outer surfaces of the cells, which were then thoroughly cleaned with a solvent and tissue until there was zero absorbance difference between the cells. With the cell mounted with its alignment direction parallel to the polarization direction of the beam the absorbance $A_{//}$ of the cell over an appropriate spectral range was scanned; special care was taken to read the absorbance value at the peak absorption wavelength from the spectrometer's digital display as this is more accurate than trying to read the result off the trace afterwards. The cells were then both rotated so that the alignment directions were perpendicular to the polarization direction and the process repeated to obtain $A_{\perp}$. Both the cells must be rotated to eliminate any possible differences in reflection off the glass to liquid crystal interfaces due to the birefringence of the liquid crystal. After changing the measurement temperature the cells were left for approximately 20 minutes to reach thermal equilibrium.

We found that no problems were introduced by using cells with a slight tilt. If the rubbed plates were assembled in the parallel configuration to produce zero tilt in the centre of the layer the results obtained were identical. Much more important was that there should be no twist in the cells due to the rubbing directions not being parallel to the sides of the glass, this was tested and found to be less than two or three degrees across the cell.

Measurements of refractive indices for sodium D light were made by use of an extended range Abbé refractometer with lecithin-induced homeotropic alignment of the nematic. Identical results were obtained for host mixtures with and without dye.

Clearing points were determined using a Mettler hot stage and polarizing microscope. The centre of the observed clearing range was recorded as T_c. The effect of dissolved dye on the host clearing point was measured by taking the change in the centre of the clearing range. All dyed mixtures nominally contained either 1% by weight of dye A (see Sec. 3.1) or 2% of dye B, although it is not easy to weigh out these small quantities of dye with great accuracy.

3.   RESULTS AND DISCUSSION

3.1 Glossary of Acronyms

All alkyl substituents are n-alkyl.

(i)   2-ring nematogens

K15 = $C_5H_{11} \cdot C_6H_4 \cdot C_6H_4 \cdot CN$
K21 = $C_7H_{15} \cdot C_6H_4 \cdot C_6H_4 \cdot CN$

E1 = BDH Chemicals proprietary eutectic mixture of K15 and K21

1695 = E Merck proprietary mixture of $C_nH_{2n+1} \cdot$cyclo-$C_6H_{10} \cdot$cyclo-$C_6H_{10} \cdot CN$ compounds

FDE = Mixture of C_nH_{2n+1}cyclo-$C_6H_{10} \cdot COO \cdot 2' - FC_6H_3 \cdot C_5H_{11}$ compounds

PCH302 = $C_3H_7 \cdot$cyclo-$C_6H_{10} \cdot C_6H_4 \cdot OC_2H_5$

PCH304 = $C_3H_7 \cdot$cyclo-$C_6H_{10} \cdot C_6H_4 \cdot OC_4H_9$

35F1BCO = $C_3H_7 \cdot$bicyclo$(2,2,2)$-$C_8H_{14} \cdot COO \cdot 2' - FC_6H_3 \cdot C_5H_{11}$

55F1BCO = $C_5H_{11} \cdot$bicyclo$(2,2,2)$-$C_8H_{14} \cdot COO \cdot 2' - FC_6H_3 \cdot C_5H_{11}$

F1BCO = Mixture of 35F1BCO and 55F1BCO

(ii) <u>high clearing point additives</u>

T15 = $C_5H_{11} \cdot C_6H_4 \cdot C_6H_4 \cdot C_6H_4 \cdot CN$

BICH5 = $C_5H_{11} \cdot$cyclo-$C_6H_{10} \cdot C_6H_4 \cdot C_6H_4 \cdot CN$

CHE = $C_2H_5 \cdot$cyclo-$C_6H_{10} \cdot COO \cdot C_6H_4 \cdot C_6H_4 \cdot CN$

BICH52 = $C_5H_{11} \cdot$cyclo-$C_6H_{10} \cdot C_6H_4 \cdot C_6H_4 \cdot C_2H_5$

HD35 = $C_3H_7 \cdot$cyclo-$C_6H_{10} \cdot C_6H_4 \cdot OOC \cdot$cyclo-$C_6H_{10} \cdot C_5H_{11}$

BB21 = $C_2H_5 \cdot C_6H_4 \cdot C_6H_4 \cdot C_6H_4 \cdot C_6H_4 \cdot CH_3$

CBC53 = $C_5H_{11} \cdot$cyclo-$C_6H_{10} \cdot C_6H_4 \cdot C_6H_4 \cdot$cyclo-$C_6H_{10} \cdot C_3H_7$

(iii) <u>mixtures containing high clearing point additives</u>

E43 = BDH Chemicals proprietary eutectic mixture of cyano-biphenyl compounds with high clearing point additive

1132 = E Merck proprietary mixture of three cyano-phenylcyclohexane (PCH) compounds and one biphenyl-cyclohexane (BICH) compound

(iv) <u>dyes</u>

A = proprietary anthraquinone dye with substituents having strong hydrogen-bonding character

B = proprietary anthraquinone dye with polarizable substituents which do not have any hydrogen-bonding character

3.2   Summary of Data

Table I summarizes the various mixtures that have been studied and the data that have been taken for each; S_{op} denotes the optical order parameter and $S_{\Delta n}$ the host order parameter as determined from refractive index measurements. The extrapolation to absolute zero which is required to extract order parameters from refractive index data was performed by plotting $\log[(n_e^2 - n_o^2)/(\bar{n}^2 - 1)]$ against $\log[1 - (T/T_c)]$, where $\bar{n}^2 = \frac{1}{3}(n_e^2 + 2n_o^2)$. The extrapolation of S_{op} to absolute zero is discussed in Sec. 3.3. We estimate that experimental errors in the measured order parameters are no greater than 0.005. Thus all calculations have been performed to 3 decimal places to control rounding errors, although data are only quoted to 2 places here to avoid any illusion of excessive accuracy.

Clearly it is not possible to reproduce here in full all the data taken. Figures 1 to 4 and Table II give results on systems which, in the light of exploratory work, were studied with greater completeness, and which exemplify the points we wish to make in this paper.

3.3   Extrapolation of S_{op} to Absolute Zero

Assuming that the molecular orientational distribution function $f = f(\theta)$, then if some experimentally measured static property of a liquid crystal $<U>$ is a measure of S, they are related by

$$\langle U \rangle = A P_2(\cos \beta) S \tag{3.1}$$

where

$$P_2(\cos \beta) = \tfrac{1}{2}(3\cos^2 \beta - 1) \tag{3.2}$$

Table I. Summary of Experimental Data

Mixture (% by wt)	Phase transitions (°C)	$\frac{10^3 \Delta T}{T_c}$ A	B	S_{op} vs T A	B	Extrapolated S_{op} (T = 0) A	B	$S_{\Delta n}$ vs T
E1	N37.3I	+1.6	+8.1	X	X			X
E1 + 10% T15	N55.5I	+0.3	+6.1	X	X			
E1 + 15% BICH5	N58.7I	-2.1	+2.1	X	X			X
E1 + 10% CHE	N52.4I	0	+5.5	X	X			
E1 + 15% BICH52	N55I		+2.1	X	X			X
E1 + 10% HD35	N51.3I	0	+4.0	X	X			
E1 + 10% BB21	N62I			X	X			
E1 + 10% CBC53	N62.5I			X				
E1 + 15% BICH52 + 10%CBC53	N90I				X			
E43	K-10N79I	+2.8	+9.9	X	X	1.0	1.0	X
1695	S_G6S_H13N72I	-2.6	-0.9	X	X	1.12	0.97	X
FDE	N36.1I				X			
FDE + 10% CBC53	N58I	+1.5	+5.1	X	X	0.93	1.09	X
50%PCH302 + 50%PCH304	N33I				X			
35%PCH302 + 35%PCH304 + 20% BICH52 + 10%CBC53	N82I		+2.0		X			
1132	K-6 N69.7I	+5.8	+8.2	X	X	1.0	0.99	X
F1BCO	N59.6I	-1.5	+0.6	X	X	1.05	1.04	X
F1BCO + 10% CHE	N73.4I	-1.7	-3.8	X	X	1.07	1.00	X
F1BCO + 15% BICH52	N71.5I	-1.2	+0.9	X	X	1.04	1.04	X
F1BCO + 10% CBC53	N80.6I	-2.8	+1.4	X	X	1.08	1.06	X

X = order parameter measured as a function of T/T_c

and $AP_2(\cos \beta)$ is determined by extrapolation to T = 0. As detailed in Sec. 1, S_{op} is unusual in that A = 1 [cf. eqn (1.3)], so that extrapolation of S_{op} to T = 0 should yield direct information on the angle of the long axis to the transition moment. It is also unusual in that whereas one generally only has at best an approximate a priori value for $AP_2(\cos \beta)$, both eqn (1.3), and the more general eqn (1.8), imply that

$$S_{op}(T = 0) \leqslant 1 \qquad (3.3)$$

Thus in contrast to the usual situation one has a quantitative
test of whether the rather long extrapolation to absolute zero has
been carried out successfully.

A number of extrapolations were performed by plotting $\log S_{op}$
against $\log(1 - T/T_c)$, with results recorded in Table I. In all
cases the experimental data extended down to at least $T = 0.85T_c$,
the lowest being some of the F1BCO mixtures, which were measured
down to $0.78T_c$. Nevertheless, it is seen that many of the extrapolations yield values of $S_{op}(0)$ greater than unity, in violation of
eqn (3.3). In order to confirm these observations great care was
taken with experimental procedure and additional data points were
collected. Further, the values of $S_{op}(0)$ reported are for straight
lines fitted by eye with a bias towards $S_{op}(0) = 1$, although in
practice the data gave little or no latitude to exercise this bias.

The reason for these results is that in general the S_{op} against
T curves have a different shape to the curves of $S_{\Delta n}$ against T.
Since the extrapolation to $T = 0$ is inevitably a long one, its
success is dependent on using an extrapolating function which has
the correct form. Thus in Figs. 1 and 2 we see that for E43 and
1132 the S_{op} curves have the same shape as the $S_{\Delta n}$ curves and even
though the absolute values of S_{op} at higher T/T_c are rather
different for the different curves they all extrapolate back to
$S_{op}(0) = 1$ to within experimental error. In the case of FDE + 10%
CBC53 (Fig. 3) the curve for dye A appears to parallel the curve
for $S_{\Delta n}$, and one might be confident that the extrapolation to 0.93
at absolute zero represented a confirmed example of $\beta \neq 0$ were it
not for the fact that the curve for dye B extrapolates to 1.09.
However, a much clearer example of the differences in shape is
provided by 1695 (Fig. 4), where, although the curves for both A
and B lie above $S_{\Delta n}$, one extrapolates to 1.12 and the other to
0.97. Thus it seems clear that S_{op} has a temperature dependence
at lower temperatures, inaccessible to experiment, which is not

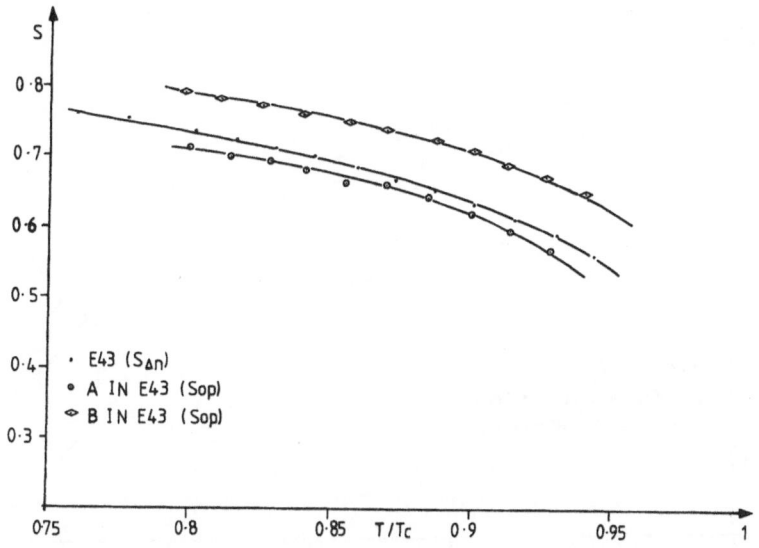

Fig. 1   Reduced temperature curves of refractive index order parameter $S_{\Delta n}$ for E43, and optical order parameters S_{op} of dyes A and B in E43.

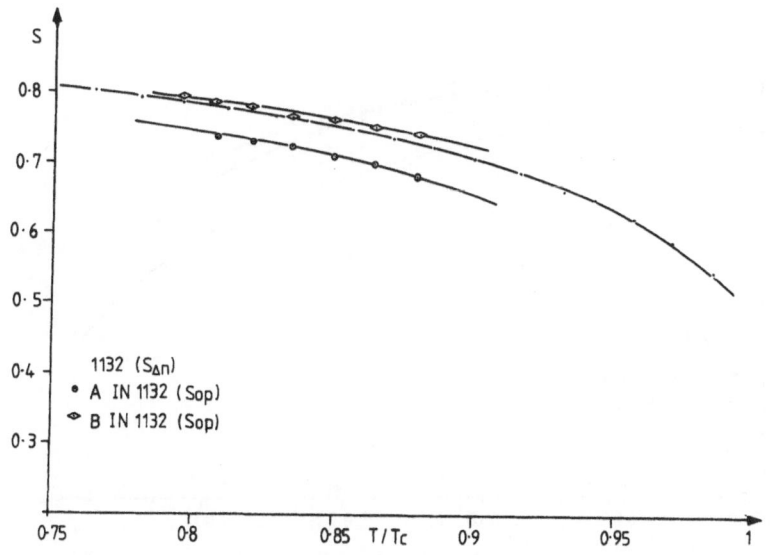

Fig. 2   Reduced temperature curves of $S_{\Delta n}$, S_{op}^{A}, and S_{op}^{B} for 1132.

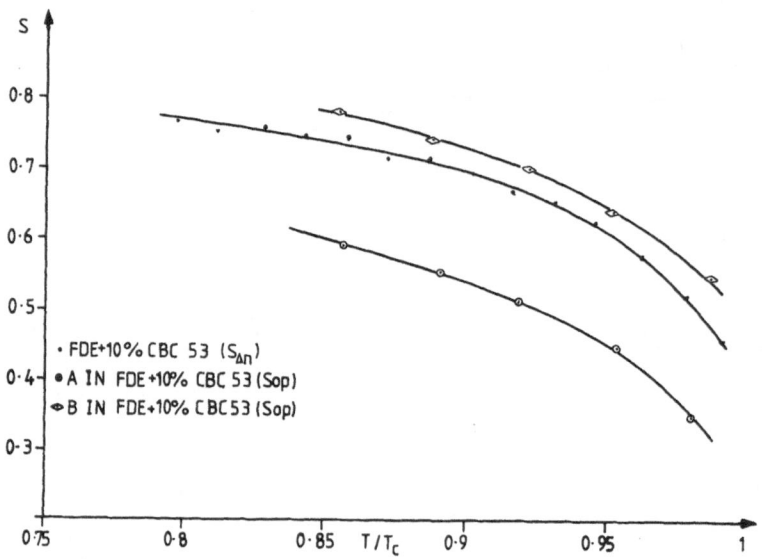

Fig. 3   Reduced temperature curves of $S_{\Delta n}$, S_{op}^{A}, and S_{op}^{B} for FDE + 10% CBC53.

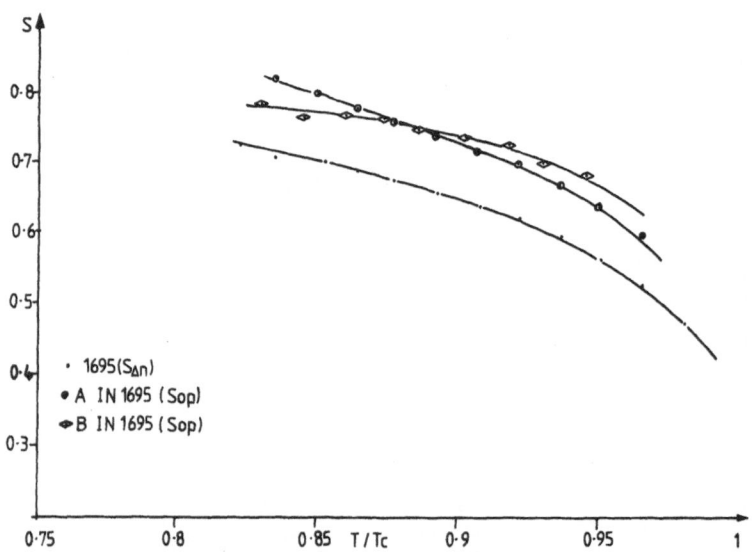

Fig. 4   Reduced temperature curves of $S_{\Delta n}$, S_{op}^{A}, and S_{op}^{B} for 1695.

reproduced correctly by functional forms which are apparently successful for extrapolation of $S_{\Delta n}$.

This interpretation was confirmed by attempts to perform the extrapolation numerically using a recently-developed method[4] based on a function of Maier-Saupe form which has proved convenient for extrapolation of refractive index data.[5] This function invariably proved too "stiff", consistently yielding values of $S_{op}(0)$ which were greater than unity but were in proportion to the hand-extrapolated values.

Further evidence was provided by examining the temperature variations of S_{op}/S_H for various mixtures (e.g. Table II). Over the range accessible to experiment, these turn out to be quite strong and lacking in any obvious consistent rationalism. For example, S_{op}/S_H for dye A in F1BCO + 10% CBC53 goes from 0.68 at $T = 0.95T_c$ to 0.87 at $T = 0.78T_c$, whereas for dye B S_{op}/S_H goes from 0.98 to 1.02 over the same range in the same host.

There are at least two possible reasons for the observed effect. First, β may acquire temperature dependence, or S_G/S_H may acquire unexpected additional temperature dependence, through the effects of thermal fluctuations: vibrations, librations, and possibly internal reorientations of the molecules. However, Osman et al.[6] were unable to adduce any evidence that such effects can be large in anthraquinone dyes. Second, anthraquinone dyes are certainly not cylindrically symmetrical so that eqn (1.8) may be more appropriate than eqn (1.3), although the two become identical if $\beta = 0$. Provided $\beta \neq 0$, mean field calculations of the two order parameters involved[7] suggest that additional temperature dependence of the form observed could arise in this way. Certainly one might anticipate that anthraquinone guest molecules might be more likely to demonstrate effects due to their molecular distribution depending on θ and Ψ than the more rod-like nematogens.

Table II.　Values of S_{op}/S_H for Various Mixtures

T/T_c	E1			E1 + 15% BICH5		E1 + 15% BICH52	
	s_H	s^A_{op}/s_H	s^B_{op}/s_H	s^A_{op}/s_H	s^B_{op}/s_H	s^A_{op}/s_H	s^B_{op}/s_H
0.95	0.59	1.00	1.15		1.09	0.91	1.11
0.925	0.64	0.99	1.12	0.93	1.07	0.91	1.09
0.90	0.67	0.99	1.10	0.93	1.06	0.92	1.08
0.875	0.70	1.01	1.09	0.94	1.05	0.93	1.07
0.85	0.72	1.01	1.08	0.95	1.05		1.06

T/T_c	F1BCO			F1BCO + 10% CHE		F1BCO + 10% CBC53	
	s_H	s^A_{op}/s_H	s^B_{op}/s_H	s^A_{op}/s_H	s^B_{op}/s_H	s^A_{op}/s_H	s^B_{op}/s_H
0.95	0.62	0.81	1.00	0.72	0.95	0.68	0.98
0.925	0.66	0.83	1.00	0.77	0.96	0.74	0.98
0.90	0.69	0.85	1.01	0.80	0.97	0.77	0.99
0.875	0.72	0.87	1.01	0.82	0.97	0.80	1.00
0.85	0.74	0.89	1.01	0.84	0.98	0.81	1.01

Notes

(i)　Addition of the high clearing point additive alone produced no
measurable change in $S_H(T/T_c)$.

(ii)　S_H is determined by the refractive index method; identical values
at the same T/T_c were obtained with and without dye in the mixture.

Whatever the cause, in view of the difficulties encountered
we use for the remainder of this paper S_{op}/S_H where we would have
preferred S_G/S_H. These difficulties must surely also act as a
warning concerning other order parameters obtained by extrapolation,
such as $S_{\Delta n}$, where a test such as that provided by eqn (3.3) is
not available.

3.4 Relative Host Sensitivity of S_{op}

An observation which recurred throughout our studies was that the dye A was much more sensitive to the chemical nature of the host than dye B. This is illustrated in Fig. 5, where the curves for S_{op}^A and S_{op}^B from Figs. 1 to 4 have been collected together. It is seen that the curves for B are clustered so that as functions of T/T_c they are almost coincident, whereas those for A are scattered. There is also a consistent tendency for the A curves to fall more rapidly at high T/T_c than the B curves. These features, particularly the latter, were evident also during our attempts to fit $S_{op}(T)$ data numerically as discussed in Sec. 3.3.

We attribute these observations to the hydrogen-bonding nature of A. The dye A has both O–H and N–H functions which can indulge in intramolecular H-bonding with the ketonic oxygens.

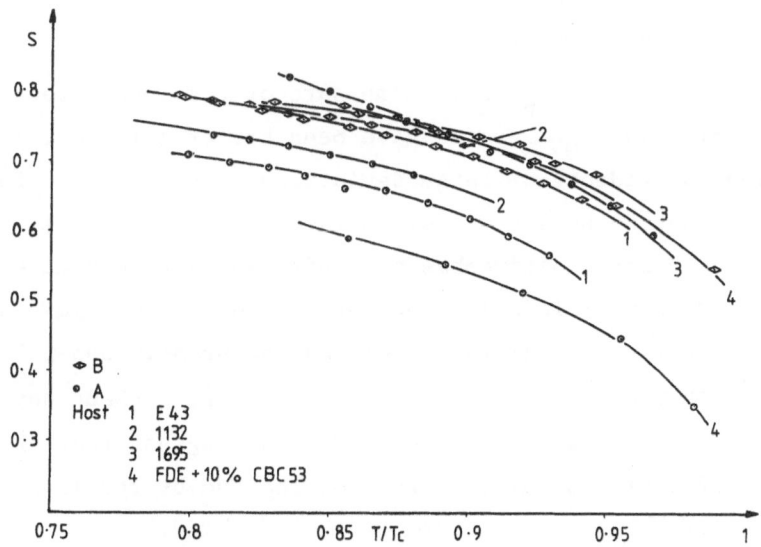

Fig. 5   Superimposed reduced temperature curves of S_{op}^A and S_{op}^B collected from Figs. 1 to 4.

Each of these H-bonding groups will be influenced by features of
the solvent molecules such as polar substituents (e.g. -CN), ester
bridges, and perhaps even electrostatic screening by conjugated
systems such as benzene rings. Thus the extent of intramolecular,
and perhaps intermolecular, H-bonding will vary from host to host.
The dye B has no H-bonding functions and thus remains characteristically more invariant.

3.5 Effect of High-Clearing Point Additives

Table II illustrates the effect of high clearing point
additives. We were unable to detect any change in $S_H(T/T_c)$, as
determined by the refractive index method, due to the addition of
either the high clearing point additives or the dye. The latter
would certainly be expected to have no effect, but it is worth
noting that the high clearing point additives are present in quite
large concentrations. It appears, however, that at least in the
cases we have studied in detail, once the change in T_c is accounted
for the high clearing point additive is not affecting the long
range order of the nematic.

From eqn (1.10), S_{op}/S_H is a measure of the local ordering of
the guest molecule; S_G/S_H would have been better (cf. Sec. 3.3)
but we have no evidence of large deviations of β from zero for the
transitions that we have studied. Evidently from Table II, the
high clearing point additive has a significant detrimental effect
on the dye ordering. This is true for both dye A and dye B and
for both the polar host El ($\Delta\epsilon \gg 0$) and the broadly non-polar
($\Delta\epsilon \sim -1$)[8] FlBCO host. Since this effect occurs without any
detectable change in the long range order we suggest that it is
indicative of preferential solvation of the dye by the larger
molecule.

3.6 ΔT_c as a Measure of Guest-Host Interactions

From Table I it is seen that the fairly low concentrations

of dye used (1% A, 2% B) nevertheless produce significant shifts in T_c. Although it would be unwise to interpret our crude data in detail, there is clear evidence that the activity coefficient of the dye in the nematic phase can be either greater or less than that in the isotropic phase. As might be anticipated from theory,[9] there is a broad (but imperfect) correlation between our measurements of ΔT_c and the degree of ordering of the dye, more positive or less negative $\Delta T_c/T_c$ corresponding to increased ordering of the dye. With our definition (Sec. 2), ΔT_c for a mole fraction x_G of dye is related to the activity coefficients γ_N^∞ and γ_I^∞ in the nematic and isotropic phases by[9]

$$\Delta T_c/T_c = \tfrac{1}{2}[(\gamma_I^\infty/\gamma_N^\infty) - (\gamma_N^\infty/\gamma_I^\infty)](R/\Delta S_{NI})x_G \qquad (3.4)$$

where ΔS_{NI} is the entropy change at the nematic to isotropic transition of the pure host, and we should recall that smaller $\gamma_N^\infty/\gamma_I^\infty$ implies greater guest compatibility with the nematic phase of the host and vice versa. Correlations between the activity coefficients and the ordering of the guest by the host would be anticipated, although their manifestation in ΔT_c may be modified by variations in ΔS_{NI}.

The higher order parameter dye B has a more positive (or less negative) $\Delta T_c/T_c$ than dye A in the same host. In the E1-based hosts both dyes have less positive (more negative) $\Delta T_c/T_c$ when high clearing point additives are present. The picture is more confused for the F1BCO-based hosts, but for these hosts both dyes uniformly show less positive or more negative $\Delta T_c/T_c$ than in the E1-based hosts, reflecting the effect, clearly seen in Table II, that both dyes are better ordered in the highly polar $\Delta \varepsilon \gg 0$ hosts.

4.   CONCLUDING DISCUSSION

In this paper we have attempted to elucidate guest/host interactions in nematic solutions by obtaining, from experimental studies
of S_{op} and $S_{\Delta n}$ as functions of reduced temperature T/T_c, estimates
of the three factors which contribute to S_{op} [eqn (1.10)]. The
difficulties that have been encountered in attempting to determine
the angle β between the long axis and the transition moment by
extrapolating S_{op} to absolute zero, are in themselves informative,
and, in view of the fact that S_{op} is one of the few experimental
measures of order which does not involve an experimentally-
determined normalization [A in eqn (3.1) is unity], are worthy of
further study.

Despite these problems the value of studying dye order
parameters as functions of temperature, and of comparing the S_{op}
against T/T_c curves of different dye/host mixtures, has been clearly
demonstrated. It is difficult to see how the effects of H-bonding
and high clearing point additives could have been clearly identified
from studies at one fixed temperature, since variations in S_{op} are
then also heavily influenced by the changes in clearing point.

The picture which emerges is that the behaviour of a dye
molecule aligned in a nematic host is quite complex and subtle.
The temperature variation of the dye order parameter appears to
have a different form to that of the host, possibly because the
rather asymmetric shape of the anthraquinone molecules leads to
their distribution function depending on Ψ as well as θ. Significant local structure effects are observed, both associated with
hydrogen bonding and because the dye molecule appears sensitive to
the presence of high clearing point additives in the host, even
though these additives have no perceptible affect on the long
range order (comparing at equal T/T_c) as measured by S_H.

It is interesting to speculate on the intermolecular interactions which give rise to these effects. In addition to the steric

repulsions and dispersive attractions it may be that the
dipole/induced-dipole force between the dipolar nematic host
molecules and the highly polarizable dye guest are significant.
Such an interaction would be expected to favour aligning the
transition moment of the lowest electronic transition (the one that
we have studied) antiparallel to the permanent dipole of the host,
thus accounting for the relatively high ordering of anthraquinone
dyes in highly positive ($\Delta\varepsilon \gg 0$) hosts where the host dipole is
longitudinal and thus broadly parallel to the director. If such a
mechanism is significant some correlation with the presence or
absence of other low energy transitions would be expected.

Whatever the mechanism of alignment, it is worth emphasizing
our view[2] that the long axis is a dynamical property of the
molecule in the fluid, and is determined by the local structure of
the guest molecule and its neighbours; it is not a property of the
molecule alone. In the light of this, and the effects elucidated
here, the observation of Osman et al.[6] that the angle between $\underset{\sim}{\mu}$
and an arbitrary axis in the anthraquinone molecule does not
correlate well with S_{op}, is to be expected.

ACKNOWLEDGEMENTS

We thank R. J. A. Tough and M. J. Bradshaw for applying their
numerical extrapolation technique to our data, and E. P. Raynes
for numerous helpful discussions.

REFERENCES

1.   T. Uchida and M. Wada, Mol. Cryst. Liq. Cryst., 63, 19 (1981).
2.   M. F. Bone, A. H. Price, M. G. Clark, and D. G. McDonnell,
         Proc. 4th Internat. Symp. Liq. Cryst. and Ordered Fluids,
         (1982).

3.   M. G. Clark, Displays, 1, 17 (1979).

4.   R. J. A. Tough, to be published.

5.   M. J. Bradshaw and R. J. A. Tough, unpublished work.

6.   M. A. Osman, L. Pietronero, T. J. Scheffer, and H. R. Zeller,
        J. Chem. Phys., 74, 5377 (1981).

7.   G. R. Luckhurst, The Molecular Physics of Liquid Crystals,
        G. R. Luckhurst and G. W. Gray, Eds., Academic Press,
        London, 1979, Chap. 4.

8.   J. Constant and E. P. Raynes, Mol. Cryst. Liq. Cryst., 70, 105
        (1981).

9.   D. E. Martire, The Molecular Physics of Liquid Crystals,
        G. R. Luckhurst and G. W. Gray, Eds., Academic Press,
        London, 1979, Chaps. 10 and 11.

LIQUID CRYSTAL ALIGNMENT ON SUBSTRATE SURFACES, A COLLECTIVE

PHENOMENON

H. Birecki

Hewlett-Packard Company
Solid State Laboratory
1501 Page Mill Road
Palo Alto, CA 94304

ABSTRACT

Alignment of cyano biphenyl materials on substrates coated with
fluorocarbon polymer has been studied as a function of temperature.
A discontinous change of alignment between homogenous and homeotropic states has been observed in the nematic phase by a change in
the birefringence of the bulk of the sample. The alignment change
is reversible and exhibits supercooling and superheating behavior
typical of a first order phase transition.

The dependence of the transitional behaviour on the hydrocarbon chain length of the liquid crystal indicates that the transition
is driven by polar interactions similar to those responsible for
formation of lipid bilayers. The difference in surface energy between homogeneous and homeotropic alignment is on the order of 0.1
dynes/cm close to the smectic-nematic phase transition temperature.

These experiments show an example of alignment of liquid crystals at bounding surfaces being governed by collective interactions
of molecules with the substrates.

INTRODUCTION

The interactions of liquid crystals with solid substrates govern the behaviour of many liquid crystal devices. There have been numerous studies examining the orientation of various liquid crystals on many types of substrates[1]. It has been found that liquid crystal orientation can be controlled by surface topography[2] as well as by the chemical nature of the surface. In some instances the orientation of the nematic director has been found to be slightly temperature dependent. Such dependence has been explained by interaction of the bulk order parameter with an anisotropic surface[3].

The only observation of an orientational phase transition at the substrate was made by Ryschenkov and Kléman[4]. In that case the alignment of MBBA on substrates coated with products of decomposition from burnt paper changed continuously from tilted to homeotropic as the temperature was raised. This is similar to what happens at a free MBBA surface[5]. Very close to the isotropic transition the alignment abruptly became tilted again. The authors argue that this transition is due to the appearance of the isotropic phase of MBBA at the substrate interface because of impurities.

In this paper a clear observation of a structural phase transition at the substrate interface is presented. Studies of such effects are not only technologically important but are also important steps in understanding the physics of interactions at surfaces.

SUMMARY OF THE EXPERIMENTAL FINDINGS

In the experiments described below, the alignment of cyano biphenyl liquid crystals on surfaces coated with pentadecyl-fluoro-octo-methacrylate (PDFOM) polymer is studied as a function of temperature. It has been found that mixtures of octyl-cyanobiphenyl (8CB) and octyl-oxy-cyanobiphenyl (8OCB) exhibit a first order phase transition of the alignment. At high temperatures the alignment of

the director of the nematic phase is close to being parallel to the
substrate surface. As the temperature is lowered the alignment becomes perpendicular to the interface in a discontinuous fashion. The
transition is reversible and exhibits supercooling and superheating
effects typical of a first order phase transition.

Figure 1 shows the transition temperatures as function of weight
% concentration of 8OCB. For concentrations less than about 35%, the
transition to the smectic phase occurs in the supercooling region and
the surface alignment transition cannot be observed unless the liquid
crystal is polled with an electric field strong enough to reorient
the boundary layer of liquid crystal close to the substrate. The
transition back to the 'parallel' alignment is observed upon heating.
The width of the superheating-supercooling temperature range decreases as the concentration of 8OCB increases. To within experimental error pure 8OCB does not exhibit any hysteresis.

The transition is also present when pure heptyl-oxy-cyanobi-
phenyl (7OCB) is used in the experiment. It exhibits a large (5 deg.
C.) hysteresis of the alignment. Neither hexyl-oxy-cyanobiphenyl
(6OCB) nor nonyl-oxy-cyanobiphenyl (9OCB) exhibit transitional behaviour on studied surfaces. The alignment of 9OCB is homeotropic
throughout the nematic phase and the alignment of 6OCB is 'parallel'.
No experiments on the behaviour of 6OCB in electric field were conducted. This dependence of alignment on the length of the aliphatic
chain indicates that polar interactions similar to those which cause
formation of lipid bilayers are responsible for the above transition.

MATERIALS, SURFACES AND SAMPLE PREPARATION

The liquid crystal materials used in these experiments were obtained from E. Merck (British Drug House) and used without further
purification. The pentadecyl-fluoro-octo-methacrylate polymer layers
were prepared by coating FC723 surface modifier commercially available from 3M company. Thickness of the fluoropolymer layer was about

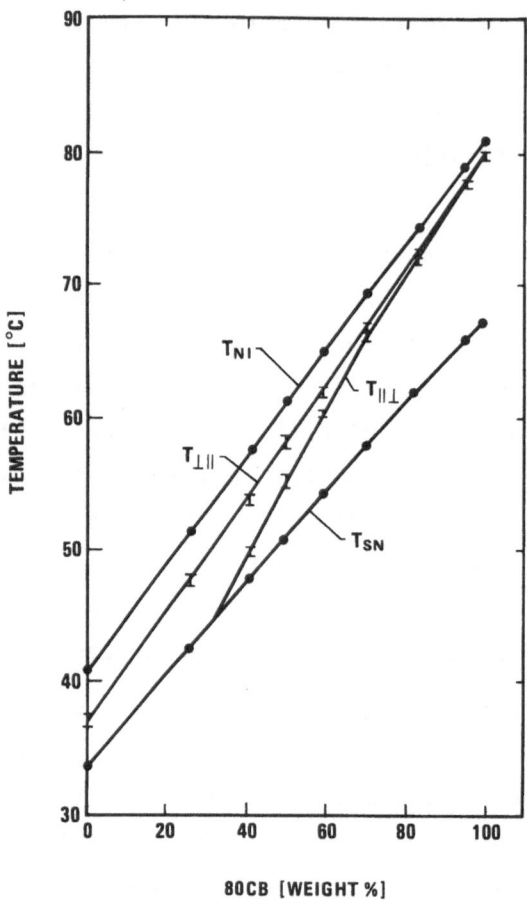

Figure 1. Transition temperatures vs. concentration of 8OCB in the
          8CB-8OCB mixtures. T_{NI} is for the nematic to isotropic
          transition, T_{SN} is for the smectic to nematic transition,
          $T_{\parallel\perp}$ is for transition from 'parallel' to homeotropic
          alignment and $T_{\perp\parallel}$ is for the reverse transition. The
          error bars indicate the uncertainties due to sample
          surface inhomogeneities.

500 Å. Substrates treated for parallel alignment of liquid crystals were coated with a 1000 Å thick layer of polyimide.

    Fused silica plates, crown glass lenses and fused silica plates
coated with conductive InSnO were used as substrates. The choice of
substrate has some influence on the alignment transition temperatures.

This dependence has not yet been investigated. However changes as large as 1 degree C. have been observed.

Samples were prepared in cells of varying thickness. The thickness variation was achieved either by using a convex lens as one of the substrates, or by making a wedge sample with spacers at one end only. Cells were filled with liquid crystals at temperatures above the nematic to isotropic transition. This was done to eliminate flow alignment problems. In many instances the alignment transition temperature varied across the cell in a manner directly correlated with cell filling patterns. This may be due to chromatographic effects of impurities attaching to substrate surfaces during cell filling. In addition to these impurity effects other irregularities of surface properties, which may relate to cleaning procedures and handling are responsible for fairly large error bars in the determination of transition temperatures.

The absolute sample temperature was measured with an accuracy of 0.2 deg.C. using a thermistor. Temperature gradients were less than 0.1 deg.C. as indicated by smectic-nematic and nematic-isotropic transitions.

EXPERIMENTAL OBSERVATIONS

In order to observe the alignment transitions the sample is placed between two crossed polarizers and the transmission of normally incident monochromatic light (λ = 543 nm) from a mercury lamp is observed. The structural transition is evidenced by an abrupt change in effective birefringence of the sample.

At high temperatures one observes a set of dark and bright fringes with very high contrast. The fringes follow contours of equal cell thickness. Initially as temperature is lowered and birefringence increases the fringes move toward thinner parts of the cell. Their appearance is highly insensitive to small (< 20 deg.) changes in observation angle. Such fringes are easily understood if one considers the liquid crystal sample to be composed of small domains in which the director has various azimuthal angles β (see figure 2) but the same tilt angle α which can vary along the cell normal. In such a case the transmission, t, of light incident along the sample normal is given by

$$t = 1/2*<\sin^2 2\beta>*\{1-\cos[2\Pi*d*(<n_e>-n_\perp)/\lambda]\} \tag{1}$$

where d is the sample thickness, $n_\perp$ is the ordinary index of refraction $n_e = n_\parallel/\{1+[(n_\parallel/n_\perp)^2-1]*\cos^2\alpha\}^{\frac{1}{2}}$ is the extraordinatory refractive index, $n_\parallel$ is the principal extraordinary refractive index, the averaging of α is done over the cell thickness and averaging of β is done over the domains in the sample area under consideration.

The decoupling of α and β is only possible if there is no twist of the director (change of β) between the sample substrates.

The insensitivity of the fringe pattern to changes in observation angle indicates that the director is almost parallel to the

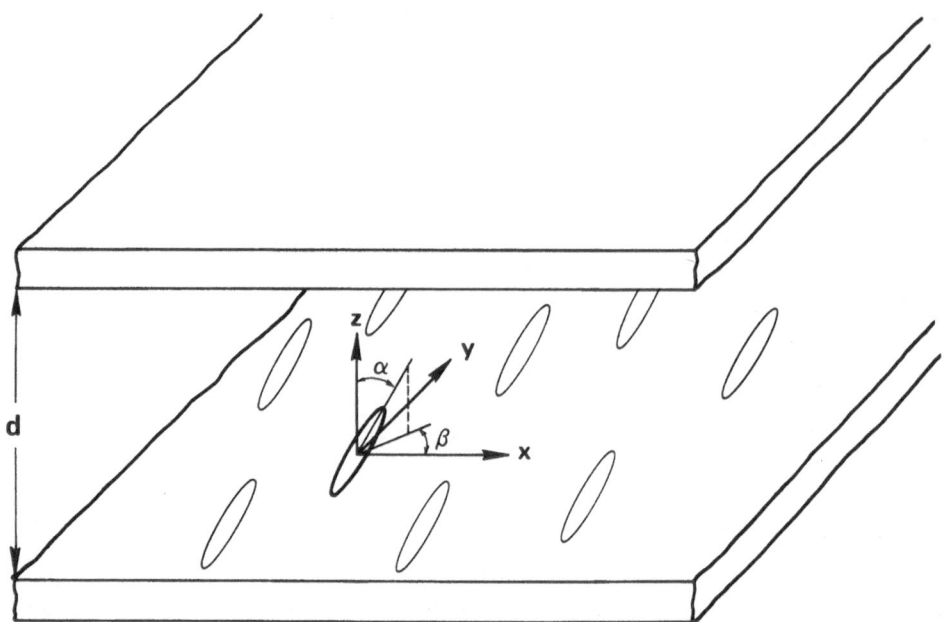

Figure 2.   Liquid Crystal Director Geometry.

substrates. The high contrast of the fringes shows that there is very little twist of director between substrates.

Fringes with high contrast are visible at small cell thicknesses. As the cell thickness increases the contrast decreases until the fringes disappear. The disappearance of fringes is related to appearance of surface disclinations in the sample. These disclinations are observed under a microscope and are discussed later.

When the temperature is lowered below $T_{||\perp}$ the fringe pattern changes significantly. If one substrate is coated with polyimide for parallel alignment then the fringe period roughly doubles indicating decrease in the effective birefringence by a factor of about 2. In addition the fringe pattern becomes very sensitive to the observation angle. If both substrates are coated with PDFOM then fringes disappear and the sample becomes dark between crossed polarizers. Such appearance is consistent with liquid crystal alignment becoming homeotropic on the fluoropolymer coated surface. Figure 3 shows a typical fringe pattern in a sample undergoing alignment transition.

The behaviour of liquid crystal has also been observed under a polarizing microscope. Large temperature gradients allow these observations to be used only in a qualitative manner. These observations confirm that in the range of sample thicknesses where high

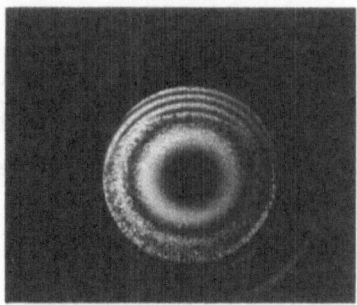

Figure 3. Typical appearance of sample undergoing the alignment transition. The fringe period doubling can easily be seen in the upper part of the sample.

contrast fringes are visible the liquid crystal undergoes no twist
between the substrates. The patterns seen under the microscope between crossed polarizers are those typical of a conically aligned
nematic. A network of strength 1 disclinations is connected by regions of liquid crystal where total extinction of transmitted light
is possible only when polarizers are crossed. The dark 'brushes'
emanating from the disclinations rotate with the rotation of polarizers.

As sample thickness increases the strength 1 bulk disclinations
are gradually replaced by surface disclinations. These form loops
or terminate in strength 1/2 disclinations at both ends. The liquid crystal adjusts to the surface conditions throughout the thickness of the sample and considerable twist is present in the bulk.
This becomes particularly evident when one observes the behaviour
of the dark 'brushes' of the strength 1/2 disclinations. Upon rotation of the polarizer which is on the side of the sample where the
disclination is anchored, the 'brush' rotates with the polarizer. If
the other polarizer is rotated then the 'brush' remains stationary,
only the contrast changes.

An interesting effect occurs at a fraction of a degree below
the nematic to isotropic transition. At that temperature all the
surface disclinations become unstable. The opened ones collapse
into strength 1 bulk disclinations and the loop types do likewise
or vanish. All the strength 1 disclinations decouple from their
anchoring points at the surfaces and the sharp features associated
with surface irregularities disappear as well. Such decoupling of
surface disclinations from the bounding surfaces has been previously
discussed by R. B. Meyer[6].

The surface alignment transition to the homeotropic state can
be clearly seen under the microscope. Loops of surface disclinations form predominantly around small surface irregularities and
strength 1 disclinations. As they grow in size the dark 'brushes'

emanating from bulk disclinations do not change and cross the surface disclinations in a continuous fashion. This confirms the lack
of twist in the thin parts of the cell. If both cell substrates
are coated with PDFOM, a conoscopic cross appears below the transition temperature in the focal plane of the microscope objective.
The cross does not move as the cell is rotated indicating true homeotropic alignment of the liquid crystal.

ANCHORING ENERGY

When one of the substrates is coated with polyimide to assure
parallel alignment, the transition to the homeotropic state is suppressed in the region of very small cell thicknesses. This is due
to the fact that a homeotropic alignment on one substrate and parallel on the other requires energy to support the bend and splay distortions of the nematic. This energy per unit area is given by[7]

$$e = \Pi^2 K/8d \qquad\qquad (2)$$

where K is the elastic constant assumed to be equal for both splay
and bend. For d small enough this energy becomes larger than the
difference in the energy due to the surface alignment and the transition is suppressed. This suppression is clearly visible under the
microscope when a long focal length convex lens is used as one of
the substrates. Close to the nematic to smectic transition this
suppression occurs for cell thickness of about 0.1 microns for the
liquid crystal materials studied. Thus assuming K to be 10^{-6} dynes
one can estimate the anchoring energies on the fluoropolymer coated
surface to be on the order of 0.1 dynes/cm close to the nematic to
smectic transition temperature.

DIRECTOR TILT AT THE SURFACE

The angle that the nematic director makes with the substrate
surface is an important parameter which can be used to determine surface interactions. Its measurement in the case of samples studied

in the above experiments is not a simple task as the surfaces do
not have a preferred direction. Thus conoscopic techniques cannot
be used.

An attempt to measure this tilt was made for two samples. For
a mixture of 50% 80CB with 50% 8CB and for pure 8CB. The measurement relied on determining the diameter of several fringes in the
light transmission pattern when a crown glass convex lens with radius
of curvature of 258 mm was used as one of the substrates. A flat
fused silica disc was used for the other substrate. Both substrates
were coated with PDFOM. The director tilt was established from the
data using equation (1). The principal refractive indices were measured using an Abbé refractometer. Sample thickness was determined
from the radius of curvature of the lens and the fringe diameter.

The measurement resulted in cosα=.24 (α=76 deg) and cosα=.21
(α=78 deg) at 56.4 and 57.8 deg.C. respectively for the mixture,
and cosα=.12 (α=83 deg) for the pure 8CB at 34.4 deg.C. The smaller
tilt of the director away from the substrate surface for pure 8CB
correlated with the smaller cell thickness at which the surface disclinations appeared. This would be understandable as the energy of
a bulk disclination perpendicular to the substrates decreases with
decreasing α. However, as the elastic constants of these materials
are unknown, no inference can be drawn from this correlation.

The major source of error in the above tilt measurement is the
uncertainty of the absolute temperature in the measurement of birefringence using the Abbé refractometer. The birefringence varies
rapidly with temperature and errors on the order of 0.5 deg.C. have
large impact on the tilt determination. Clearly a direct method of
tilt measurement is needed.

DISCUSSION

It is usually accepted that the alignment of liquid crystal on
various substrates is defined by the appropriate binding of a mono-

molecular layer of liquid crystal at the interface. The order induced by the surface in that monolayer is propagated into the bulk
of the liquid crystal. Such order at the surface is not identical
to the mesophase nature of liquid crystals and exists even in the
isotropic phase[8]. The binding of the monomolecular layer is usually considered in terms of single molecules interacting with the
substrate.

As the experiments described above indicate, the interaction
between the substrate and the liquid crystal are governed by coop-
erative phenomena. The dependence of the surface alignment characteristics on the aliphatic chain length in the 6OCB-9OCB series indicates that the polar interactions are driving the observed surface
phase transitions. The longer the aliphatic chain the more clearly
one can divide the molecule into a polar head, which tends to be
driven into the medium with high dielectric constant, and a nonpolar
group.

A structural phase transition at the interface driven by such
interactions has been studied theoretically by Parsons[9]. His model
for a free surface is based on an interplay of Van der Vaals forces,
which promote parallel alignment, and polar interactions which favor
homeotropic alignment. This model predicts a continuous transition
from homeotropic to tilted alignment. Since the observed transition
is first order, Parsons' model is inapplicable. Work is currently
in progress to explain observed phenomena in terms of a modified
model for polar forces which allows for a first order phase transition.

An alternative model explaining the observed behaviour would
be one that assumes two specific binding sites on the molecule where
it can attach to the substrate. The binding energies in such a model
have to depend on the nematic order parameter. Since the studied
fluoropolymer surface is highly inert, such an explanation is rather
unlikely.

ACKNOWLEDGMENTS

I gratefully acknowledge the help of Dr. Steven L. Naberhuis in the preparation of the liquid crystal mixtures and many fruitful discussions with him, Dr. Hsia Choong and other members of our laboratory.

REFERENCES

1. P. Datta, G. Kaganowicz, A. W. Levine, J. Coll. Int. Sci. <u>82</u>, 167 (1981) and references therein; D. Armitage, J. Appl. Phys. <u>51</u>, 2552 (1980) and references therein.

2. D. W. Berreman, Phys. Rev. Lett. <u>28</u>, 1683 (1972); M. Nakamura and M. Ura, J. Appl. Phys. <u>52</u>, 210 (1981) and references therein.

3. H. Mada, Mol. Cryst. Liq. Cryst. <u>51</u>, 43 (1979).

4. G. Ryschenkov and M. Kléman, J. Chem. Phys. <u>64</u>, 404 (1976).

5. M. A. Bouchiat and D. Langevin-Cruchon, Phys. Lett. <u>34A</u>, 331 (1971).

6. R. B. Meyer, Sol. State. Comm. <u>12</u>, 585 (1973).

7. J. E. Proust and L. Ter-Minassian-Saraga, Coll. Pol. Sci. <u>254</u>, 492 (1976).

8. K. Miyano, Phys. Rev. Lett. <u>43</u>, 51 (1979).

9. J. D. Parsons, Phys. Rev. Lett. <u>41</u>, 877 (1978).

ALIGNMENT OF LIQUID CRYSTALS IN THE VICINITY

OF THE SmA-N TRANSITION

G. J. Sprokel

IBM Research Laboratory
San Jose, California 95193

N. J. Chou

IBM Research Laboratory
Yorktown, New York 10598

ABSTRACT

$\varepsilon(T)$ and $\varepsilon(H)$ data were obtained for a number of cells having SiO_2 aligning layers, e-beam deposited at oblique angles. The cells were filled with a mixture of cyanobiphenyls which has a narrow nematic phase between its SmA and I phases. In general, the cell alignment depends on the angle of deposition as was found for pure nematics, except that the smectic phase dominates. Cybotactic domains determine the average tilt at small angles of evaporation. At larger deposition angles, the smectic configuration persists throughout the entire nematic region.

Surface induced alignment of liquid crystals which have only a nematic phase has been well documented. The topic of this paper is the alignment of liquid crystal mixtures which have a SmA-N transition, in particular such mixtures which have a narrow nematic range. The aligning layers are e-beam evaporated quartz films deposited at

oblique angles between the substrate surface and the source. The
cell alignment is obtained by comparing permittivity-temperature
curves in the vicinity of the transitions for the cell itself with
those for the cell held in a magnetic field directed either parallel
with or perpendicular to the surface. In the nematic phase, $\varepsilon(H)$
curves are determined and compared with curves calculated from continuum theory. The transitions and the resulting cell textures are
examined with a polarizing microscope equipped with a temperature-
controlled chamber. The results are summarized here; experimental
details and a discussion follow.

At very shallow deposition angles (5°), the aligning layer has
little effect. The SmA and N phases are essentially randomly
oriented.

At somewhat larger deposition angles (10-15°), the aligning
layer begins to affect the cell texture. In most cells, a fine
focal conic texture appears. The nematic phase is in general
tilted, but uniformly oriented.

At still higher deposition angles (20-50°), a well-developed
focal conic texture appears. The axis of the hyperbola are perpendicular to the plane of incidence of the evaporation beam. The
nematic phase is aligned uniformly parallel with the surface, free
of disclinations and shows strong extinction if the plane of incidence of the evaporation coincides with the polarizer axis.

Finally, at still higher deposition angles, the characteristic
fan texture appears. The nematic phase remains uniformly parallel.

If the nematic range is small, retention of the smectic order
well into the nematic phase is apparent under the microscope. For
well-ordered cells, the smectic structure dominates the whole nema-
tic phase. $\varepsilon(T)$ curves - and $\varepsilon(H)$ curves in the nematic phase -
depend on the thermal history of the sample.

AUTOMATED ϵ AND ρ MEASUREMENTS

Figure 1 illustrates the principle of the measurement discussed in detail elsewhere.[1] Cells are made by etching the electrode pattern using conventional photolithography. Next, the aligning layer is deposited and the substrates are joined using epoxy-coated mylar gaskets. Typical cell spacing is 10 μm. The electrode material is Cr/Au or Cr/Al. For cells intended for microscopy, one of the substrates is ITO. ITO cells are not suitable to determine the absolute values of ϵ and ρ; the series resistance of the ITO film can introduce an error of 10%. Although this resistance could be determined for a particular cell, the method is time consuming. All-metal cells have negligible loss up to 10 MHz.

The cell is suspended inside a double-walled thermostat. The temperature is controlled to about 0.05°C. The assembly is placed between the poles of an NMR magnet such that the cell is oriented either parallel with the magnetic field or perpendicular to it. The cell temperature is measured, independent of the control sensor, by a thermistor cemented to it.

The measuring circuit is a negative feedback loop. This arrangement effectively cancels the stray capacitance which is the largest source of systematic error in the measurement. Considering the cell as a parallel RC circuit, it follows from circuit theory:

$$X_1 = \frac{X_2 + Tg\phi}{1 - X_2 Tg\phi}$$

$$R_1 = R_2 \cdot \frac{V_i}{V_o} \cdot \sqrt{\frac{1 + X_1^2}{1 + X_2^2}}$$

$$C_1 = \frac{X_1}{\delta R_1} .$$

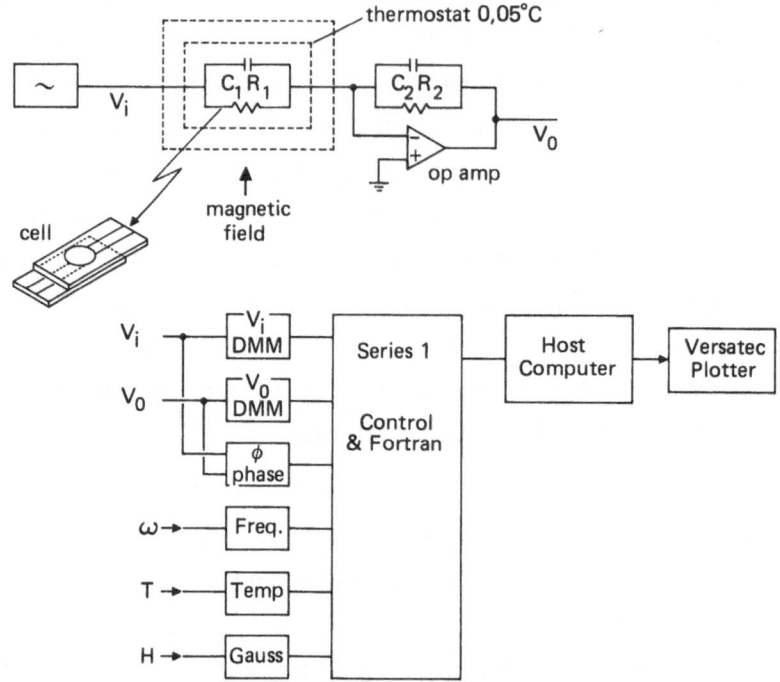

Fig. 1.   Automated ε, ρ Measurements for L. C. Cells.

R_2 and C_2 are standard parts, the values are chosen such that the
phase lag ϕ is always well below $\pi/2$. An IBM Series 1 computer
controls the data flow and performs the initial calculations. From
the input data V_1, V_0, ϕ and ν, the program calculates R_1 and C_1
for the filled cell and compares these data with the stored data
for the empty cell to obtain ε and ρ. At the end of a run, ε, ρ, T
and H are transmitted to a host computer for further calculations
and a graph of $\varepsilon(T)$ or $\varepsilon(H)$ is prepared. All curves in this paper
are computer plots. The machine can be used from subaudiofrequen-
cies to about 100 kHz. All data in this paper were obtained at 1
kHz. The instrument is checked by comparing data for an RC network
with those from a commercial LCR meter (HP 4274A). Data agree to 1%.

ALIGNING LAYER DEPOSITION

Aligning layers are thin films of SiO_2 deposited by e-beam evaporation of quartz slugs at about 10^{-7} Torr. The vacuum chamber is pumped by a cryogenic pump enhanced by a differential ion pump. There is no organic material anywhere in the system. Samples are mounted in a fixture which can be rotated along a horizontal axis to select any angle between the source and the substrate surface. This angle is here referred to as the "angle of deposition". It plays a major role in determining the alignment of the cell. The refractive index of the film is measured by ellipsometry. The index is typical 1.45 for deposition of more than 20°, but decreases at smaller deposition angles. The film thickness is about 300Å.

RESULTS

A.   Cr/Al-ITO Cells

Figures 2 and 3 show $\varepsilon(T)$ and $\varepsilon(H)$ curves for six cells differing only in the angle of deposition of the aligning layer. The angles were varied from 0° to about 70°. All cells have Cr/Al and ITO electrodes to permit microscopic examination of the cell textures. All cells are filled with a commercial mixture of cyanobiphenyls available from BDH as catalog number S2. The mixture contains 50% K24, 39% K30 and 11% M30.

All $\varepsilon(T)$ curves were obtained by slowly cooling the sample from isotropic to nematic to smectic. The cooling rate is 0.1-0.2°C per minute. The sequence was kept the same: the newly-filled cell is cooled without a field, heated to isotropic and cooled with the magnetic field directed parallel with the cell surface, then heated again to isotropic and cooled with the magnetic field perpendicular to the surface (note the exception for the 6.5° sample; this does not affect the discussion). The field is about 14 kGauss.

The graphs in Figs. 2 and 3 are arranged in the order of the angle of deposition. It is seen that the aligning effect of the

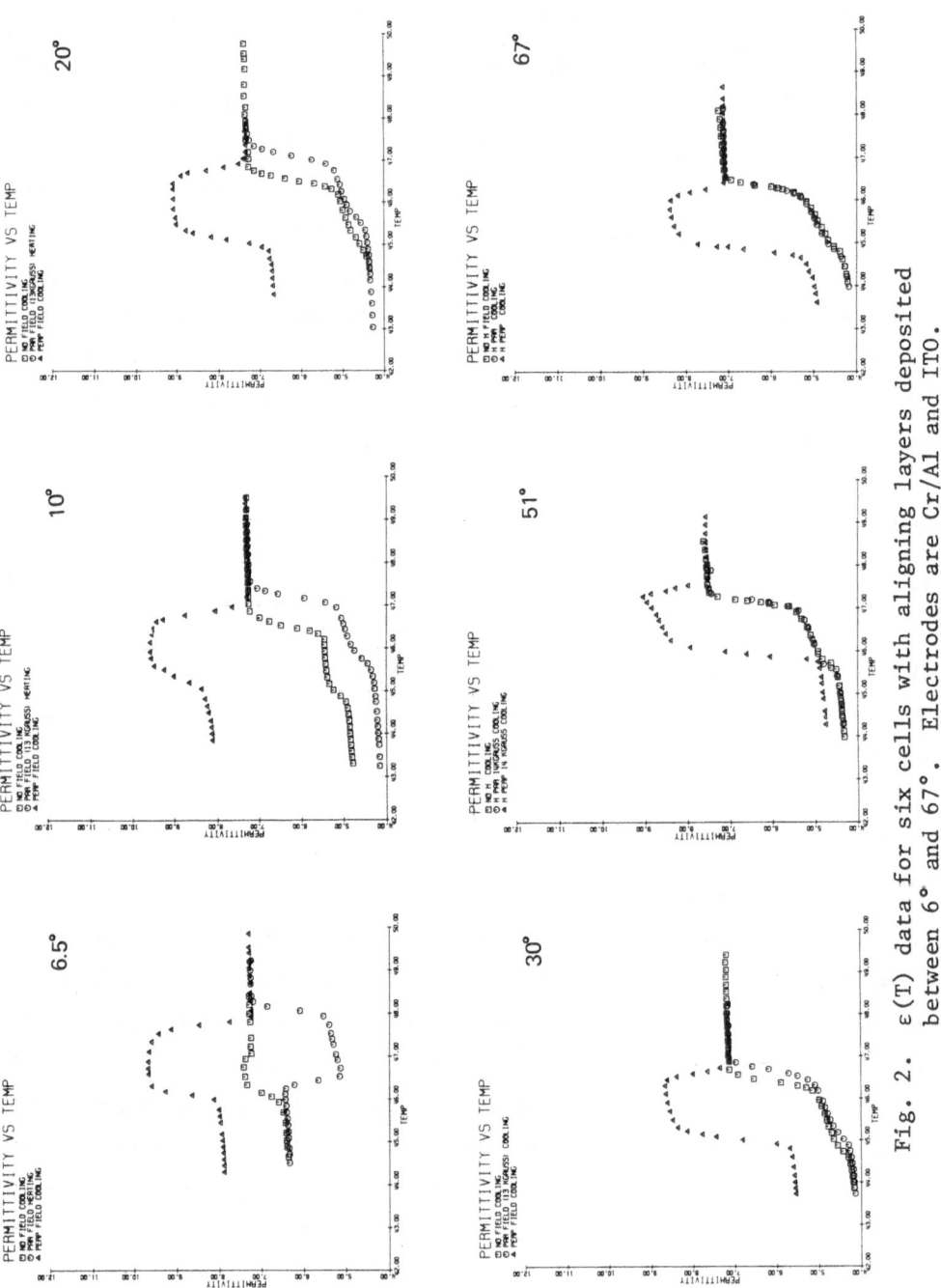

Fig. 2. ε(T) data for six cells with aligning layers deposited
between 6° and 67°. Electrodes are Cr/Al and ITO.

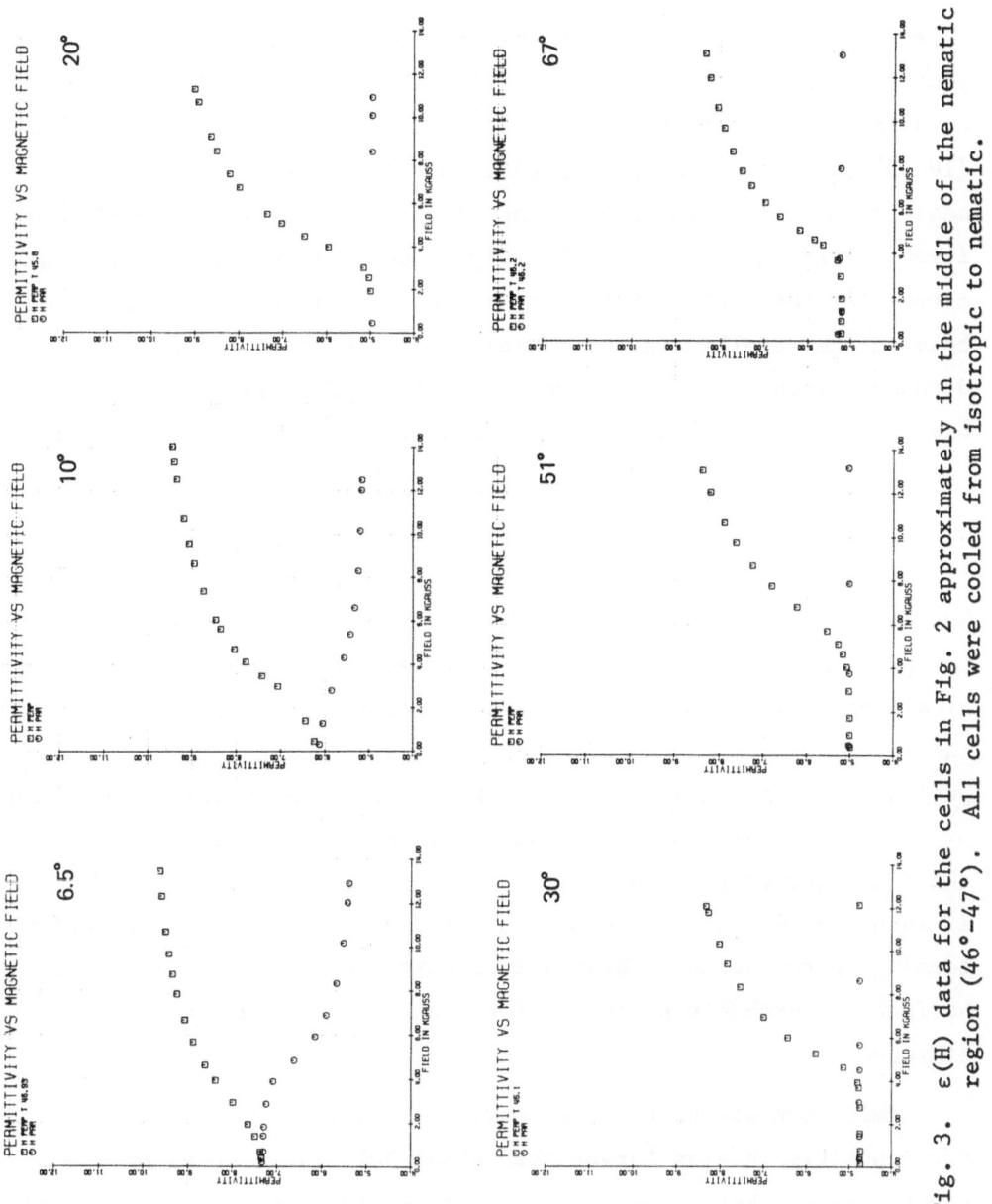

Fig. 3. ε(H) data for the cells in Fig. 2 approximately in the middle of the nematic region (46°-47°). All cells were cooled from isotropic to nematic.

deposited film increases rapidly with the angle of deposition and
reaches its maximum at about 30°. At very small angles (6°), the
permittivity of the cell does not change at the I→N transition.
There is no orientation in the nematic phase. At the N→SmA transition, ε decreases but remains fairly high. There is little ordering
even in the smectic phase. As the angle of deposition is increased
(10-15°), the cells tend to become parallel aligned while in the
nematic phase. There is a further decrease in ε at the N→SmA transition. The trend continues for higher deposition angles. The ε(T)
curves for the cell-as-filled approach the ε(T) curve for the cell
held in the parallel magnetic field. At around 30-40° angle of
deposition, the two ε(T) curves, no-field and parallel-field, coincide. The aligning film has attained its maximum effect.

Referring now to the cooling curves with the magnetic field
oriented perpendicular to the surface and thus opposing the surface-
induced alignment, the cell tends to align in the field direction,
but as the effect of the aligning layer increases, it is less able
to do so; ε in the nematic phase decreases as the angle of deposition increases. Pretransitional ordering in the neighborhood of
the N→SmA transition is rather pronounced. In the nematic phase, ε
begins to decrease well above the transition. As a larger fraction
of the cell becomes more smectic-like, the orientation effect of
the field decreases. For well-aligned cells, ε in the smectic phase
is much smaller than for poorly-ordered cells. The smectic regions
forming in the nematic phase are oriented more parallel with the
surface and probably propagate from the surface (see section on
microscopy).

Referring now to the ε(H) curves in Fig. 3, it is seen that
for deposition angles larger than about 30°, the samples show a
sharp Freedericksz transition and are essentially parallel aligned.
At about 10° angle of deposition, the sample appears tilted. At
very low angles, the sample is a mosaic rather than a uniformly

Angle of Deposition 10°

Angle of Deposition 20°

Angle of Deposition 40°

Angle of Deposition 70°

Evaporation Direction ⟶

Fig. 4.   Texture of cells in the smectic phase. The angle of depo-
          sition was varied from 10° to 70°. Between 20° and 50°
          angle of deposition, all focal conics are perpendicular
          to the plane of incidence of the evaporation beam.
          Magnification 250x.

aligned body. As will be shown, such $\varepsilon(H)$ curves cannot be described by continuum theory.

   <u>Cell Textures.</u> Figure 4 shows the textures of newly-filled
cells in the SmA phase. Ordered smectic phases become first apparent at about 10° angle of deposition, but well-developed focal
conics require higher angles. Between 20° and 50° angle of deposition, the cell texture is very uniform; all focal conics align with
their hyperbola orthogonal to the plane of incidence of the evaporation. At still higher angles, this directionality disappears

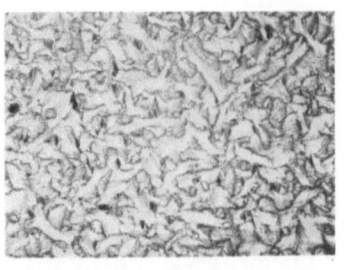

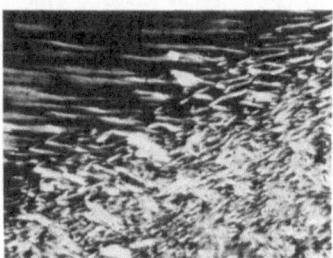

Fig. 5. Aligning effect of cooling a fan-texture sample in a
         parallel magnetic field.

again, a fan texture develops, but the axis of the fans are random
in the plane. It was found that the fan texture can be realigned
by cooling the sample in a magnetic field parallel with the surface,
but orthogonal to the direction of evaporation. Figure 5 shows
this sequence. The newly-filled cell shows the fan texture, but
the cell cooled in the field shows well-aligned focal conics. If
the cell is heated slowly, the uniform alignment breaks up at the
SmA→N transition. Figure 6 is included here to show that the
aligned focal conic texture persists throughout the entire nematic

region and disappears only in the isotropic phase. The deposition angle was 30°. The top row in Fig. 6 shows the texture in the smectic phase between crossed polarizers. The sample is aligned parallel with the polarizers and at 45°. The middle row shows the cell in the nematic phase for the two positions, parallel and 45°. It is clear that part of the cell remains smectic-like. The bottom photograph was taken at the I→N transition as the sample was cooled. The regions which are already nematic, enclosed by disclination loops, show the focal conic texture. In between, the sample is isotropic and uniformly black.

B.   Cr/Au Cells

Figure 7 shows $\varepsilon(T)$ and $\varepsilon(H)$ curves for cells with Cr/Au electrodes and SiO_2 aligning films. The graphs are arranged in ascending order of the deposition angle. All data were taken in the sequence described above. The similarity of the curves in Fig. 7 and those in Figs. 2 and 3 will be apparent. Thus the aligning effect of the SiO_2 film does not depend on the substrate on which it is deposited. Again, it is seen that for very low deposition angles (7°), there is little alignment in the N or Sm phases. At around 20° angle of deposition, the $\varepsilon(T)$ curve for the cell by it-self approaches the $\varepsilon(T)$ curve for the cell in a parallel magnetic field, while at around 30° angle of deposition, the two $\varepsilon(T)$ curves are nearly identical.

The cells were made to compare experimental $\varepsilon(H)$ curves with continuum theory. As was pointed out, the ε data for ITO cells are subject to a systematic error caused by the series resistance of the ITO film. Calculation of the $\varepsilon(H)$ function from continuum theory was discussed in detail by H. Gruler et al.[2] and is now avail-able in many textbooks.[3,4] Here we use the numerical method dis-cussed before.[5] To reiterate briefly, ε is obtained by numerical integration of Eq. (1):

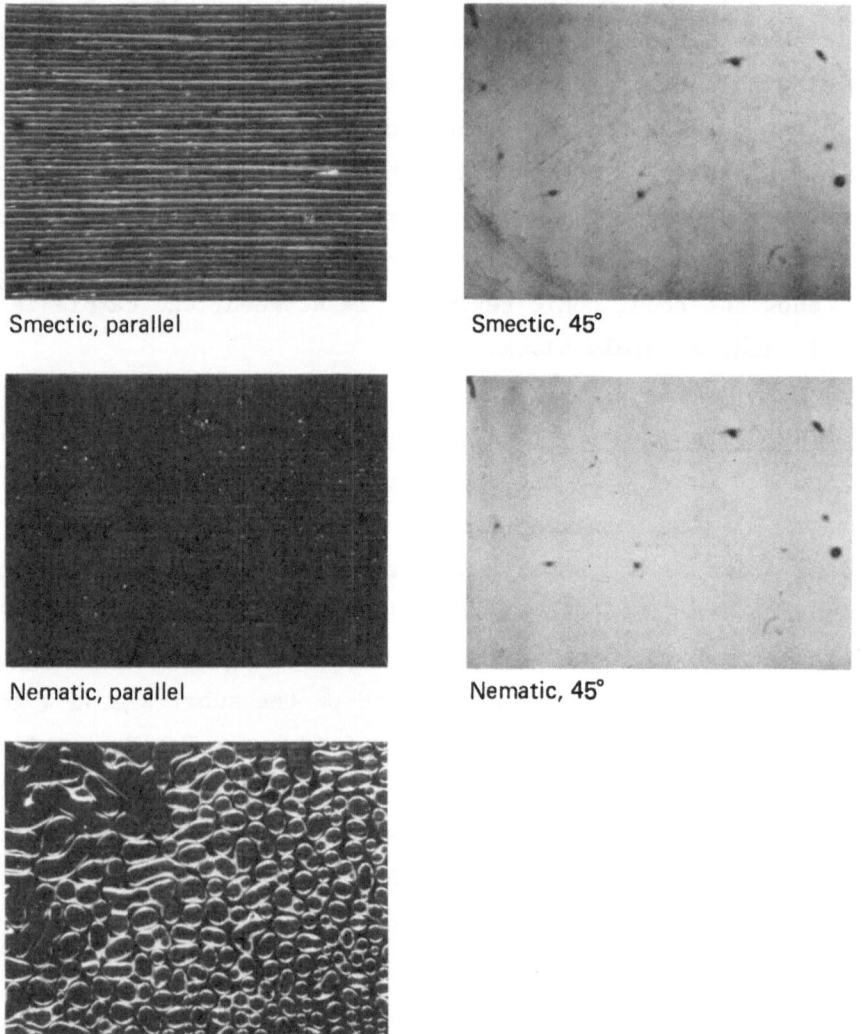

Smectic, parallel   Smectic, 45°

Nematic, parallel   Nematic, 45°

I→N Transition

Fig. 6.   Top row: aligned focal conic texture of a cell with an
           aligning layer deposited at 30°. Smectic phase; parallel
           with polarizer and at 45°. Middle row: nematic phase,
           parallel with the polarizer and at 45°. Much of the
           focal conic texture remains. Bottom row: cell at the I→N
           transition. Most of the cell is already nematic. In the
           nematic phase, the smectic aligned focal conic texture is
           already apparent.

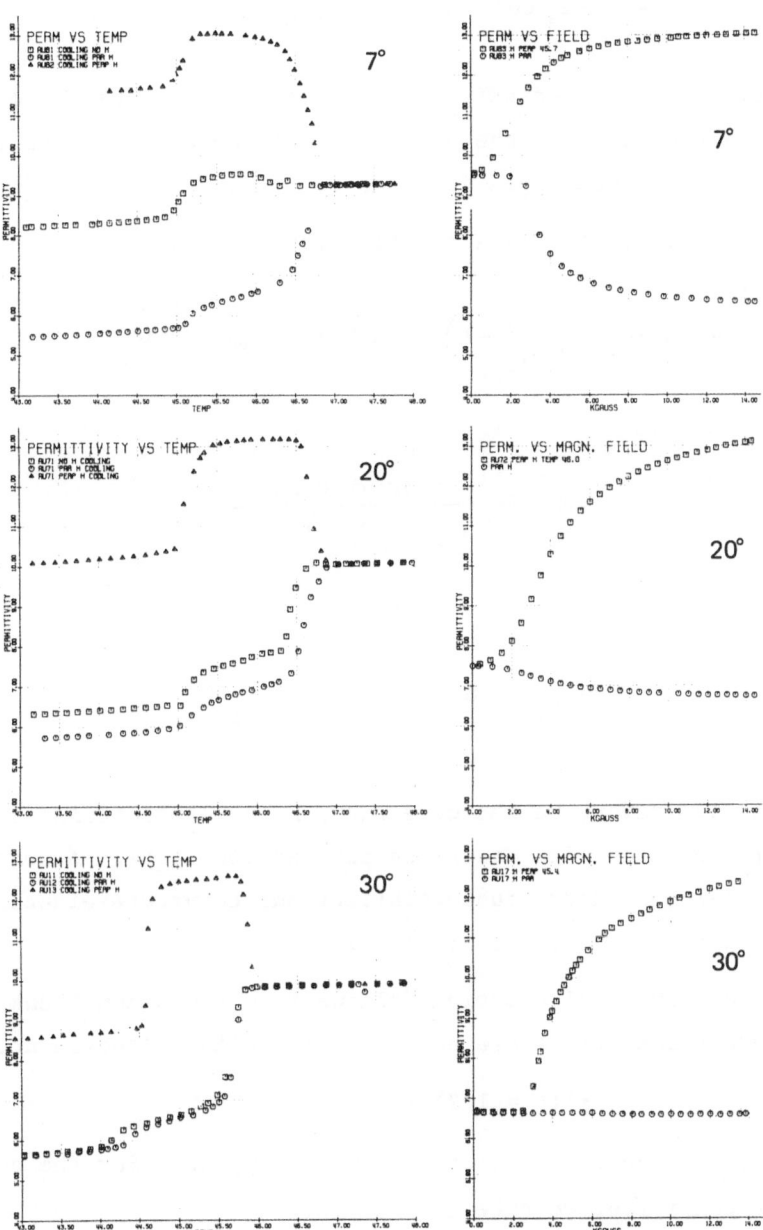

Fig. 7. $\varepsilon(T)$ and $\varepsilon(H)$ curves for cells with Cr/Au electrodes and SiO$_2$ aligning layers deposited at 7°, 20°, and 30°. Note the similarity with Figs. 2 and 3.

$$\varepsilon^{-1} = \int_0^1 (\varepsilon_1 \cos^2 \phi(\zeta) + \varepsilon_2 \sin^2 \phi(\zeta))^{-1} d\zeta . \tag{1}$$

ε_1 and ε_2 are the principals of the permittivity tensor. $\phi(\zeta)$ is the tilt angle taken as the angle between the director and the x axis. ζ is the normalized distance to the surface: $\zeta = \frac{z}{d}$.

The function $\phi(\zeta)$ is a solution of Eq. (2):

$$\phi'' + \phi'^2 \cdot \left(\frac{k_{33}}{k_{11}} - 1 \right) \cdot g(\phi) \pm A \cdot g(\phi) = 0$$

where

$$g(\phi) = \frac{\sin \phi \ \cos \phi}{1 + \sin^2 \phi \left(\frac{k_{33}}{k_{11}} - 1 \right)}$$

$$A = \frac{\Delta \chi}{k_{11}} \cdot d^2 \cdot H^2 = \pi^2 \cdot \left(\frac{H}{H_t} \right)^2 .$$

The plus sign refers to the case where $\vec{H}$ is perpendicular to the surface; the minus sign describes the parallel case. A contains the material constants, susceptibility and curvature elasticity, the cell dimension and the field.

The differential equation follows from the Oseen-Frank theory. Two boundary conditions are needed, but one is self-evident:

$$\phi'(\zeta = 1/2) = 0$$

Otherwise a plane of disclinations would appear. For the second boundary condition, we take

$$\phi(\zeta = 0) = \phi_0 .$$

ϕ_0 is a constant, independent of the field, but otherwise unrestrained. It is the purpose of the computer program to determine

ε_1, ε_2, the ratio $\dfrac{K_{33}}{K_{11}}$ and ϕ_0 such that the calculated curve coin-

cides with the experimental data. The special case $\phi_0 = 0$ is discussed in textbooks and referred to as the Freedericksz transition.
For this special case, Eq. (2) can be rewritten as an elliptic
integral and series expansions for this type of integral are known.
This has been the approach taken in most papers. The computer
program used here solves Eq. (2) using the Runge-Katta scheme as
modified by Gill[6] and then integrates Eq. (1) using the Simpson
technique. Starting values can be obtained from the experimental
results; the program then iterates until the calculated and experimental curves coincide. If agreement is not possible, it is concluded that the cell cannot be described by the continuum theory
approach.

Results of these calculations are shown in Figs. 8, 9 and 10
for the $\varepsilon(H)$ curves in Fig. 7.

Figure 10 (deposition angle was 30°) has the form of a classical Freedericksz transition. The tilt angle $\phi_0 = 0$. The threshold
is 2.9 kGauss. Best fit was obtained with $k_{33}/k_{11} = 1.2$. However, ε
is not a strong function of k_{33}/k_{11}. See also G. Baur.[7]

Figure 9 (deposition angle 20°) was calculated with $\phi_0 = 20.7°$.
Good agreement is obtained for both branches of the $\varepsilon(H)$ function.
While agreement with theory for the $\phi_0 = 0$ case has been discussed
before, the data in Fig. 9 show for the first time that the theory
applies equally well to uniformly tilted cells.

Figure 8 (deposition angle 7°) was calculated with $\phi_0 = 42.7°$,
which gives good agreement at $H=0$ and at high fields, but very poor
agreement between 0-5 kGauss. The branch for parallel H suggests
that part of the cell is in effect perpendicularly aligned and has
a threshold of about 3 kGauss. Completely random orientation corresponds to an average tilt angle of 35.3°. Thus part of the cell
must be oriented nearly perpendicular. This would agree with the

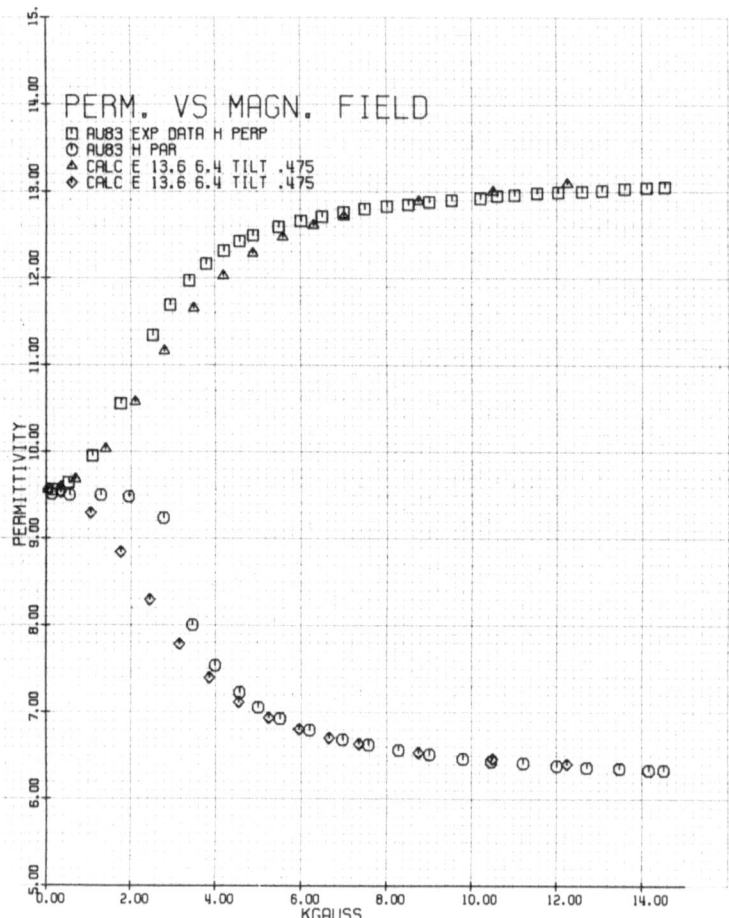

Fig. 8.   Calculated and experimental ε(H) curves for the cell in
          Fig. 7, top row. Tilt angle 42.7°. The cell is not
          uniform; the calculated and experimental curves do not
          agree between 0-5 kGauss.

increase in ε in the middle of the nematic region (see the ε(T)
part of Fig. 7). The cells are best described as randomly oriented
with smectic-like regions in which the orientation is nearly
perpendicular to the surface.

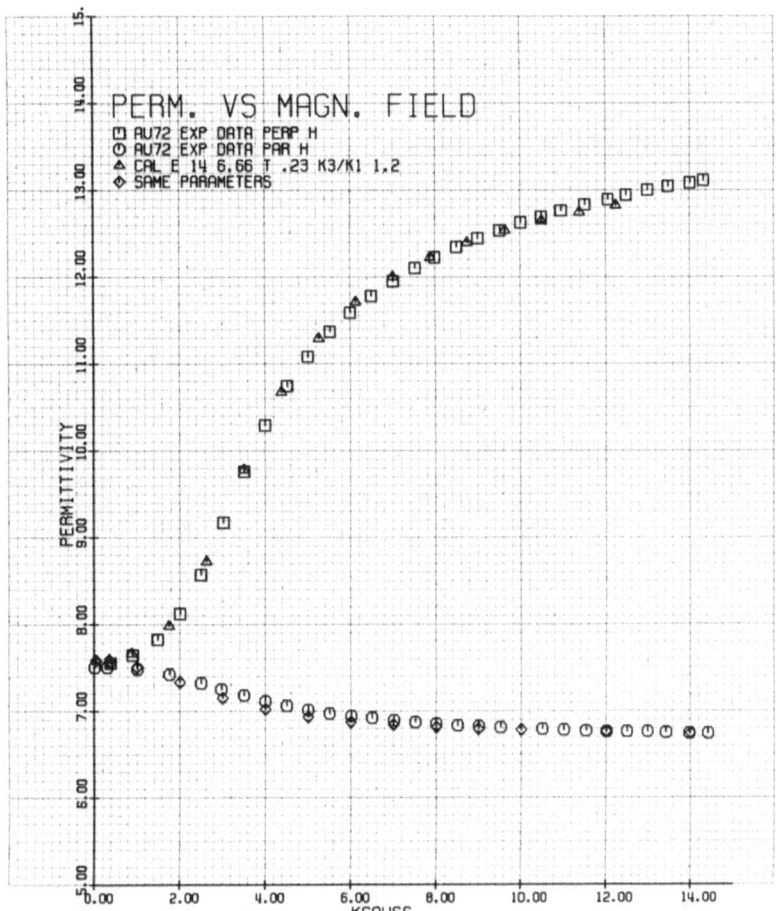

Fig. 9.   Calculated and experimental ε(H) curves for the second
          cell in Fig. 7, middle row. Tilt angle 20.7°. Good
          agreement for both ε(H) branches. The cell is uniformly
          tilted.

COOLING AND HEATING ε(T) CURVES

    For uniformly parallel aligned cells, the ε(T) curve without a
field coincides with the curve for the sample held in a parallel
oriented field. For tilted cells, the knee in the ε(T) curve
occurs about 0.2°C lower; for randomly oriented cells, there may be
no knee at all. Thus it seems appropriate to take the knee in the

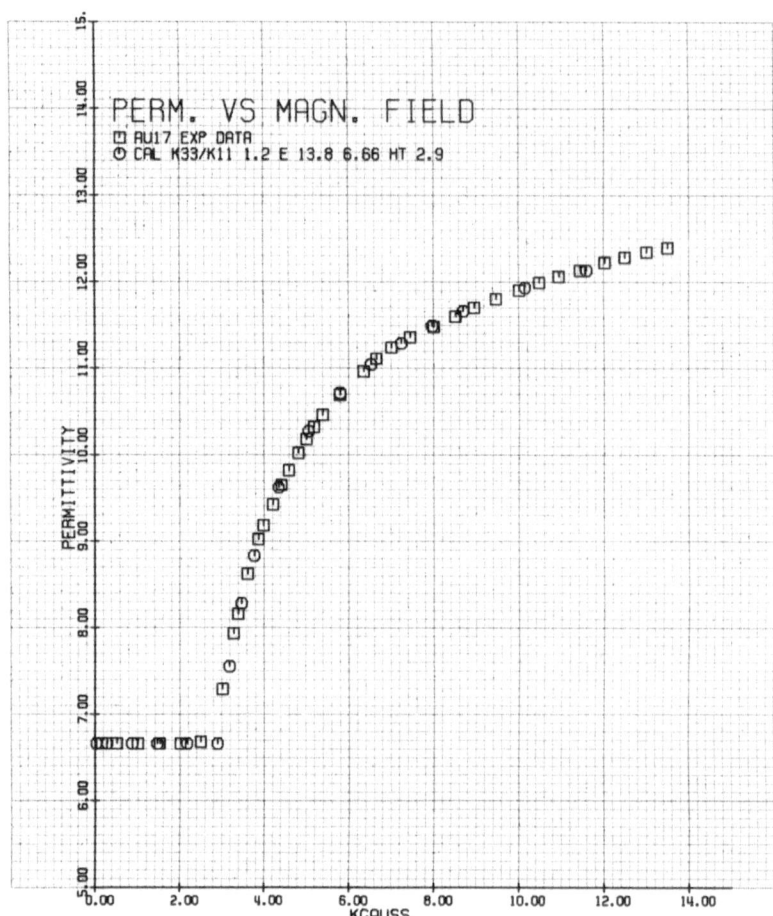

Fig. 10.   Calculated and experimental ε(H) curves for the third
           cell in Fig. 7, bottom row.   Tilt angle 0°; classical
           Freedericksz transition.

ε(T) curve for the sample held in a parallel field as the onset of
the I→N transition.   The end of the transition is 0.3-0.5°C below
this.   The rate of cooling (0.1-0.2°C per minute) is slow enough to
keep the sample in temperature equilibrium even during the transition.

    All previous curves were obtained by cooling the sample.   Fig-
ure 11 exemplifies a cooling-heating sequence.   The sample used is

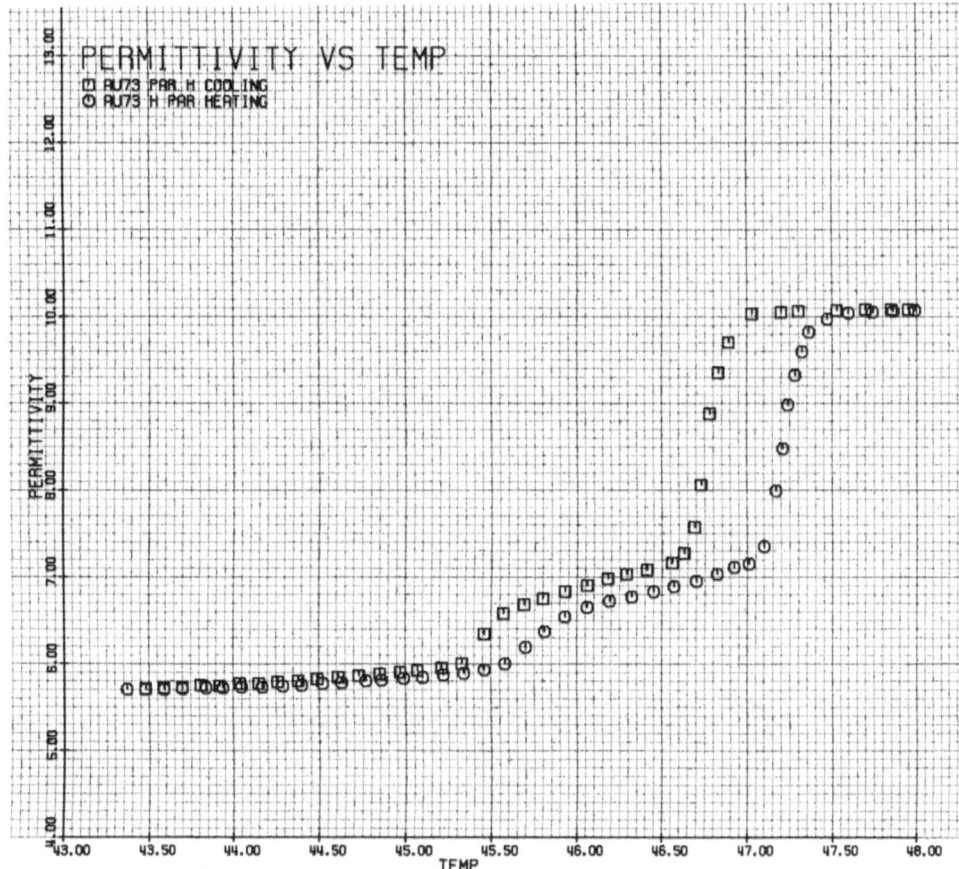

Fig. 11.   Cooling-heating sequence in parallel H field for the 20.7° tilt cell (second cell in Fig. 7).

the 20.7° tilt sample from Figs. 9 and 7. The sequence is: cooling from isotropic to smectic in a parallel field followed by heating back to isotropic in the field. The onset of the I→N transition during cooling occurs at the same temperature as the N→I transition during heating, but in each case the complete realignment takes about 0.4°C. This implies that the realignment occurs in the next phase, e.g., during heating realignment occurs in the isotropic phase as defined by the cooling curve. The same phenomena occurs at the SmA→N transition, although this is not very clear from Fig. 11.

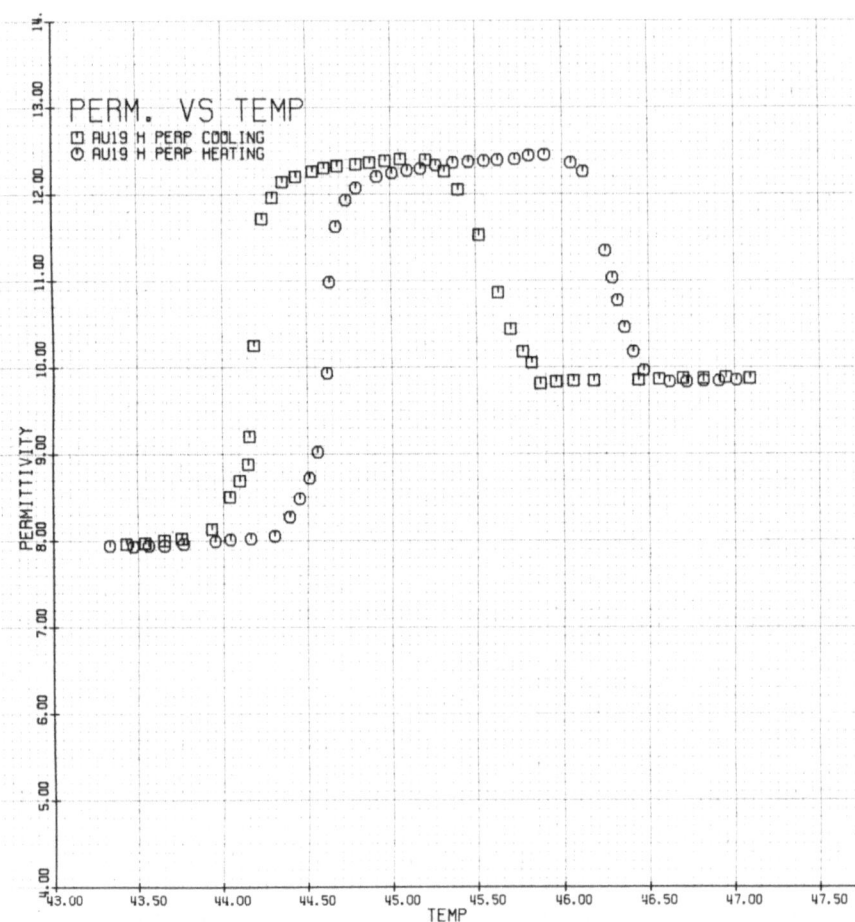

Fig. 12. Cooling–heating sequence in a perpendicular H field.

Field and surface induced alignment cooperate and the net change is
small. Figure 12 shows a cooling–heating sequence for a sample
with zero tilt held in a magnetic field oriented perpendicular to
the surface. Again, the onset of the N→SmA transition during cooling occurs at the same temperature as the SmA→N transition during
heating, but the transition takes 0.4–0.5°C. The loop is reproducible and not an artifact of the instrumentation. For this sample,
the temperature was kept constant at the transition and only fluctuations were observed, not a steady drift.

The difference between the heating and cooling curves is attributed to the retention of the smectic phase throughout the nematic phase, as was illustrated in Fig. 6. For this reason, only cooling curves were used throughout this paper.

DISCUSSION

The results reported here for smectic mixtures which have a narrow nematic range agree in principle with earlier work.

At high evaporation angles (>30°), the cell aligns parallel with the surface (i.e., 0° tilt) and orthogonal to the plane of incidence of the evaporation beam. This is in agreement with Raynes et al.,[10] Scheffer et al.[8] and earlier work.

At intermediate angles, the sample is tilted approximately toward the source. At 20° deposition, the tilt is about 21° (Fig. 9). This is in the range indicated by Raynes.

At still lower angles (6°-7°), the observed tilt is 45° (Fig. 8), but the sample is no longer uniform. Scheffer found 25° for CHBA at 6° angle of deposition. Crossland et al.[9] found 42° for octyloxycyanobiphenyl, a compound similar to the ones used here. However, these authors did not report ε(H) curves or discuss sample uniformity.

The data in Figs. 11 and 12 and the photomicrographs in Fig. 6 clearly demonstrate smectic-like ordering in the nematic and isotropic phases. Earlier evidence of smectic ordering was reviewed by Chandrasekhar.[4] More recently, Coles et al.[11] interpreted their light-scattering data in the isotropic phase of dodecylcyanobiphenyl, as caused by cybotactic domains (1979). Achard et al.[12] found that the N→I transition entropy is affected by a smectic phase at lower temperature. However, our data suggest that the cybotactic ordering does not reach equilibrium in the time span of several minutes and perhaps much longer. At any temperature in the nematic phase, the sample is more smectic-like if it is heated from the smectic phase as compared to cooled from the isotropic phase. If the sample is carefully cooled, one can obtain a series of $\varepsilon(H)$ curves showing that the threshold increases as the temperature decreases. The temperature dependence of the splay and bend constants are being measured, the data are as yet incomplete.

ACKNOWLEDGMENTS

We want to express our gratitude for the help offered by the Research Computing Center; in particular R. W. Martin for programming the Series 1, Wes Christensen for APL and W. E. Rudge for graphics. The interface hardware was made by Central Scientific Services. We want to thank W. Schillinger and his group.

REFERENCES

1. G. J. Sprokel, Abstract 311, ECS Meeting, Los Angeles, 1979.

2. H. Gruler, T. J. Scheffer, G. Meier, Zeitschrift f. Naturforschung 27a, 966 (1972).

3. P. G. de Gennes, The Physics of Liquid Crystals, Claredon Press (1974).

4. S. Chandrasekhar, Liquid Crystals, Cambridge, University Press, (1977).

5. G. J. Sprokel, Mol. Cryst. Liq. 42, 233 (1977).

6. B. Carnahan, H. A. Luther, J. O. Wilkes, Applied Numerical Methods, Wiley (1969).

7. G. Baur, Optical Characteristics of Liquid Crystal in: Physics and Chemistry of Liquid Crystal Devices, IBM Symposia Series, G. J. Sprokel, ed., Plenum Press (1979).

8. T. J. Scheffer, J. Nehring, JAP 48, 1783 (1977).

9. W. A. Crossland, J. H. Morissy and B. Needham, J. Phys. D., Appl. Phys., G2001 (1976).

10.   E. P. Raynes, D. K. Rowell and I. Shanks, Mol. Cryst. Liq.
         Cryst. Lett. <u>34</u>, (4) 105 (1976).

11.   H. J. Coles and C. Strazielle, Mol. Cryst. Liq. Cryst. (L) <u>49</u>,
         259 (1979).

12.   M. F. Achard, G. Sigand and F. Hardonin, <u>Liquid Crystals of
         One- and Two Dimensional Order</u>, W. Helfrich, ed., Springer
         (1980).

NEMATIC-SUBSTRATE INTERACTION AND THE BOUNDARY LAYER PHASE TRANSITION

Ping Sheng

Corporate Research Science Laboratories, Exxon Research &
Engineering Company, P. O. Box 45, Linden, NJ 07036

I. INTRODUCTION

Substrate alignment of liquid crystals is an essential element
in the functioning of a liquid crystal display device. While the
primary role of the treated substrate is to anchor the nematic
director at the solid-liquid crystal interface in some pre-designed
manner, it has been pointed out in an earlier work[1] that if a similar
anchoring effect exists for the orientational order parameter S, then
the substrate would also induce a boundary layer, extending about
1000 Å from the substrate, within which the orientational order can
be significantly different from that of the bulk. In particular,
this would mean that the liquid crystal molecules in the vicinity of
the substrate can possess a finite degree of orientational ordering
even when the bulk is isotropic -- a property whose experimental
utilization led eventually to the observation of this surface-induced
effect.[2] It is the purpose of the present paper to examine the
implications of substrate alignment from the viewpoint of a general
substrate-nematic interaction. By considering the alignment action
as arising from an anisotropic substrate potential, it is shown that
the resulting boundary layer temperature variation can provide an

excellent account of the experimental results on substrate-induced
birefringence, with values of surface aligning potential deduced as
a by-product of theoretical fit to the data. The theory furthermore
predicts the existence of a novel phase transition in the regime of
weak substrate aligning strength. Since this phase transition involves only the boundary layer, it will be labeled the "boundary
layer phase transition".

In what ensues, the form of the substrate-nematic interaction is
considered in Section II. The calculation of substrate-induced
birefringence and its comparison with experimental results are
presented in Section III. In Section IV we describe the nature of
the boundary layer phase transition and the condition for its
occurrence.

II. SUBSTRATE POTENTIAL

Consider a sample of nematic liquid crystal bounded on one
side by a substrate. The solid-liquid crystal interface is defined
as $z=o$, and the sample is assumed to be uniform in the x and y
directions. The substrate is treated such that the nematic molecules
in its immediate vicinity experience an aligning force along some
fixed spatial direction $\hat{n}$. Since the aligning action is assumed to
be the same for $\hat{n}$ and $-\hat{n}$, the nematic-substrate interaction potential
ν can be expressed as[3]

$$\nu(\theta,z) = - G\delta(z) \sum_{n=1}^{\infty} a_{2n} P_{2n} (\cos \theta), \qquad (1)$$

where θ is the angle between the long-axis of a nematic molecule and
$\hat{n}$, z is the distance from the substrate, G is a constant denoting the
strength of the potential, P_{2n} denotes the Legendre polynomial of
order 2n, and a_{2n}'s are the series expansion coefficients, with
$a_2 = 1$. In Eq. (1), we have incorporated the short-range nature of

the substrate potential by the delta function $\delta(z)$. If we now truncate $\nu(\theta,z)$ to the leading term of the series and average the resulting potential over the many molecules within a small spatial region, then the macroscopic potential takes the form

$$V = \langle\nu(\theta,z)\rangle = -G\delta(z)\langle P_2(\cos\theta)\rangle = -G\delta(z)S. \tag{2}$$

Here $S \equiv \langle P_2(\cos\theta)\rangle$ is the orientational order parameter of the nematic liquid crystal.

III. SUBSTRATE-INDUCED BOUNDARY LAYER

To study the implications of the substrate potential, Eq. (2), we start with the Landau-deGennes free energy of the system:[4]

$$\frac{\Phi}{A} = \int_0^\infty \phi \, dz - \frac{G}{A} S_o \quad , \tag{3}$$

$$\phi = f(S) + L \left(\frac{dS}{dz}\right)^2 \quad , \tag{4}$$

$$f(S) = a(T-T^*) S^2 + BS^3 + CS^4 \quad , \tag{5}$$

where Φ is the free energy, A is the area of the planar sample, S_o denotes the value of S at $z=o$, ϕ is the Landau-deGennes free energy density, T is the temperature, and a, T^*, B, C, L are material parameters. Since $S(z)$ is determined by the condition of minimum free energy, we first minimize Φ variationally with respect to $S(z)$, treating S_o as fixed. This results in the Euler equation, which can be integrated once to yield

$$\xi_o^2 \left(\frac{dS}{dz}\right)^2 = F(S) + K \quad , \tag{6}$$

Here $F(S) \equiv f(S)/aT_K^o$, $\xi_o = (L/aT_K^o)^{1/2}$ is the correlation length, and

T_K^o is the bulk phase transition temperature. The integration constant K is determined by the condition that at $z \to \infty$, $dS/dz = 0$. That means

$$\xi_o^2 \left(\frac{dS}{dz}\right)^2 = F(S) - F(S_b) \quad , \tag{7}$$

where S_b is the bulk value of the order parameter. Since we want to study surface-induced phenomena, the relevant part of the free-energy is that of the boundary layer, which can be obtained by subtracting out the bulk free energy:

$$\frac{\Phi_{BL}}{A\xi_o a T_K^o} = \int_0^\infty [F(S) - F(S_b) + \xi_o^2 \left(\frac{dS}{dz}\right)^2] \, dz - g \, S_o$$

$$= 2\int_{S_b}^{S_o} \sqrt{F(S) - F(S_b)} \, dS - g \, S_o \quad , \tag{8}$$

where $g \equiv G/A\xi_o a T_K^o$ is the dimensionless parameter characterizing the substrate aligning potential, and we have used Eq. (7) to convert the z integration to S integration. The equilibrium value of S_o is given by the condition $d\Phi_{BL}/dS_o = 0$, or

$$F(S_o) = F(S_b) + \frac{g^2}{4} \quad , \tag{9}$$

with S_b found by the stipulation that

$$F(S_b) = \text{absolute minimum of } F(S). \tag{10}$$

In case Eq. (9) has multiple roots, the correct S_o value is that which gives the lowest value of Φ_{BL} as expressed by Eq. (8). Once S_o is obtained, the functional profile of the boundary layer, $S(z)$, can be determined by the integral of Eq. (7), or

$$\frac{z}{\xi_0} = \int_{S(z)}^{S_o} \frac{dS}{\sqrt{F(S) - F(S_b)}} \tag{11}$$

Equations (9), (10), and (11) have been solved numerically and the substrate-induced boundary layer profiles for g = 0.02 are plotted in Fig. 1 at various temperatures. In the calculation, we have used the measured values[5] of a=0.065 J/cm^3K, B=-0.53 J/cm^3, C=0.98 J/cm^3, T*=307.14K for the nematic liquid crystal 4-pentyl-4'-cyanobiphenyl (PCB), and the estimated value of L=4.5X10^{-14} J/cm. From the figure we see that at T> T_K^o, the boundary layer has finite values of the order parameter S. Since the isotropic bulk cannot produce any optical birefringence, the phase retardation measured at temperature T>T_K^o must be given by[6]

$$\rho = 4\pi C \frac{\xi_0}{\lambda} (\sin \psi) I \quad , \tag{12}$$

$$I = \int_0^\infty S \frac{dz}{\xi_0} = \int_0^{S_o} \frac{dS \ (a \ T_K^o)^{1/2}}{\sqrt{a(T-T^*)+BS+CS^2}} \quad ,$$

where C is a proportionality constant relating the difference, Δn, in the indices of refraction perpendicular and parallel to the optical axis, and the orientational order parameter S (Δn = CS), λ is the wavelength of light, and ψ is the angle between the light ray and the optical axis. The $\sin \psi$ in Eq. (12) arises from the $\sin^2 \psi$ factor of the phase retardation formula, multiplied by 1/$\sin \psi$ due to the lengthening of the light path when the light ray is not perpendicular to the substrate. A factor of 2 is also included in Eq. (12) to account for the two surfaces of a liquid crystal cell.

Theoretical calculation of the phase retardation and the comparison with experimental results are shown in Figs. 2. We have

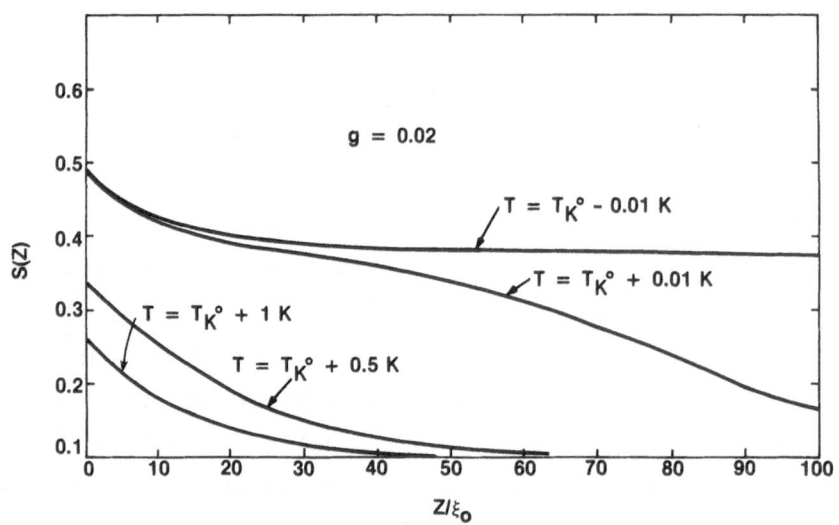

Fig. 1   Calculated boundary layer profile for PCB at various
         temperatures. The two top curves correspond to the
         boundary layer just before and just after the bulk nematic-
         isotropic phase transition. The substrate potential
         strength g is noted in the figure.

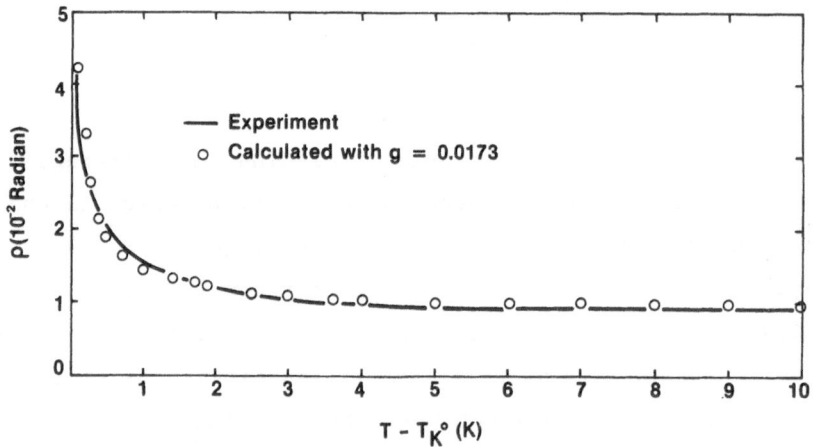

Fig. 2   Theoretical fit to the data (from Ref. 2) on wall-induced
         birefringence. The best fit is obtained with g = 0.0173.
         The baseline is treated as an adjustable parameter in the
         fit.

used the measured values of[7] C=0.319 for PCB, ψ=45° for the homeotropic sample, ξ_0 = 6Å, and λ=6328Å for the He-Ne laser used in the
experiment.[2] Since the experimental data inherently contains some
amount of birefringence from the glass substrate,[2] we have fitted
data with an adjustable constant added to Eq. (12). From the figure
we see that not only is the theory in excellent agreement with
experiment, but also, as a by-product, the values of the substrate
potential can be deduced. For PCB, the fitted values of g = 0.0173
translates into a substrate alignment potential of G/A = 0.24 erg/cm^2.

IV. BOUNDARY LAYER PHASE TRANSITION

Suppose now the alignment potential can be made very weak. In
Fig. 3 we show the calculated phase retardation as a function of
temperature for g = 0.008. There is a discontinuous jump, occurring
at $T = T_K^o + 0.049$ K, that is related to a phase transition in the
boundary layer. In Fig. 4 we show the S(z) profile at $T = T_K^o$ and
$T = T_K^o + 0.049$K. It is seen that whereas at $T = T_K^o$ the bulk undergoes a first-order phase transition, the value of S_o stays fixed.
However, at $T = T_K^o + 0.049$K the S_o has an abrupt jump, signaling a
boundary-layer phase transition. The physics of the boundary-layer
transition can be understood as follows. The layer of molecules at
the nematic-substrate interface experiences two types of forces:
the elastic force, represented by the L(dS/dz)^2 term of the free
energy, and the substrate aligning force. At the boundary layer
phase transition, there is a tradeoff between the two types of
energies: the gain in the substrate potential energy resulting from
the drop in S_o is offset by the lowering of the elastic energy.
Figure 5 shows the variation of S_o as a function of temperature for
various values of the substrate potentials. It is clear that the
boundary layer transition occurs only in the range of 0.0056 =
$g_o < g < g_c$ = 0.011. For $g < g_o$, the transition in S_o occurs
simultaneously with the bulk transition, whereas at $g > g_c$ S_o exhibits

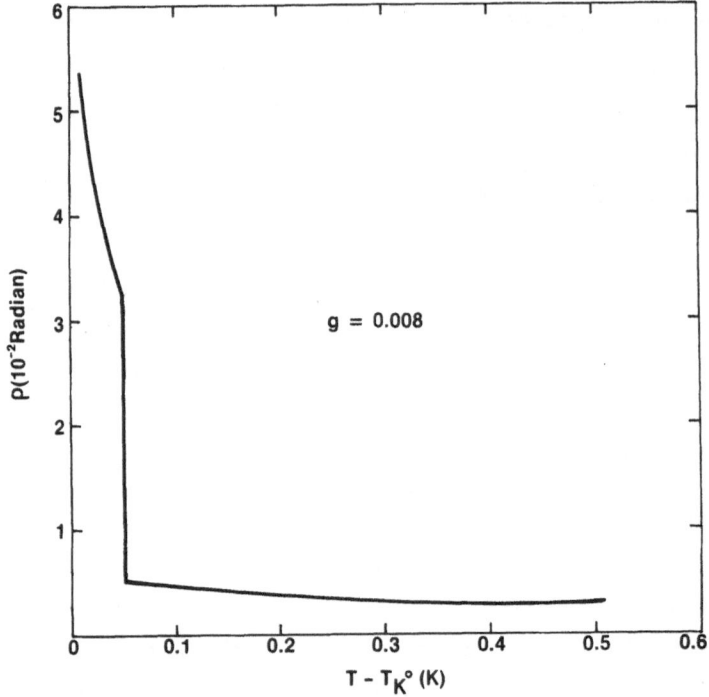

Fig. 3   Calculated birefringence for PCB on a substrate with
          substrate potential g = 0.008. The discontinuous drop
          at T ≅ T$_K^o$ + 0.05K is caused by a boundary-layer phase
          transition.

a continuous temperature variation. For PCB, the calculated values
of g_o and g_c translates into $G_o/A \simeq 0.075$ erg/cm^2 and $G_c/A \simeq 0.15$
erg/cm^2.

V.   CONCLUSION

       Properties of nematic liquid crystals aligned by a short-range,
arbitrary strength substrate potential are examined in the framework
of Landau-deGennes theory. It is shown that the substrate potential,
which can arise from surface treatment of liquid crystal display
cells, induces a boundary layer, extending about 1000 Å from the
substrate, within which the orientational order parameter value can

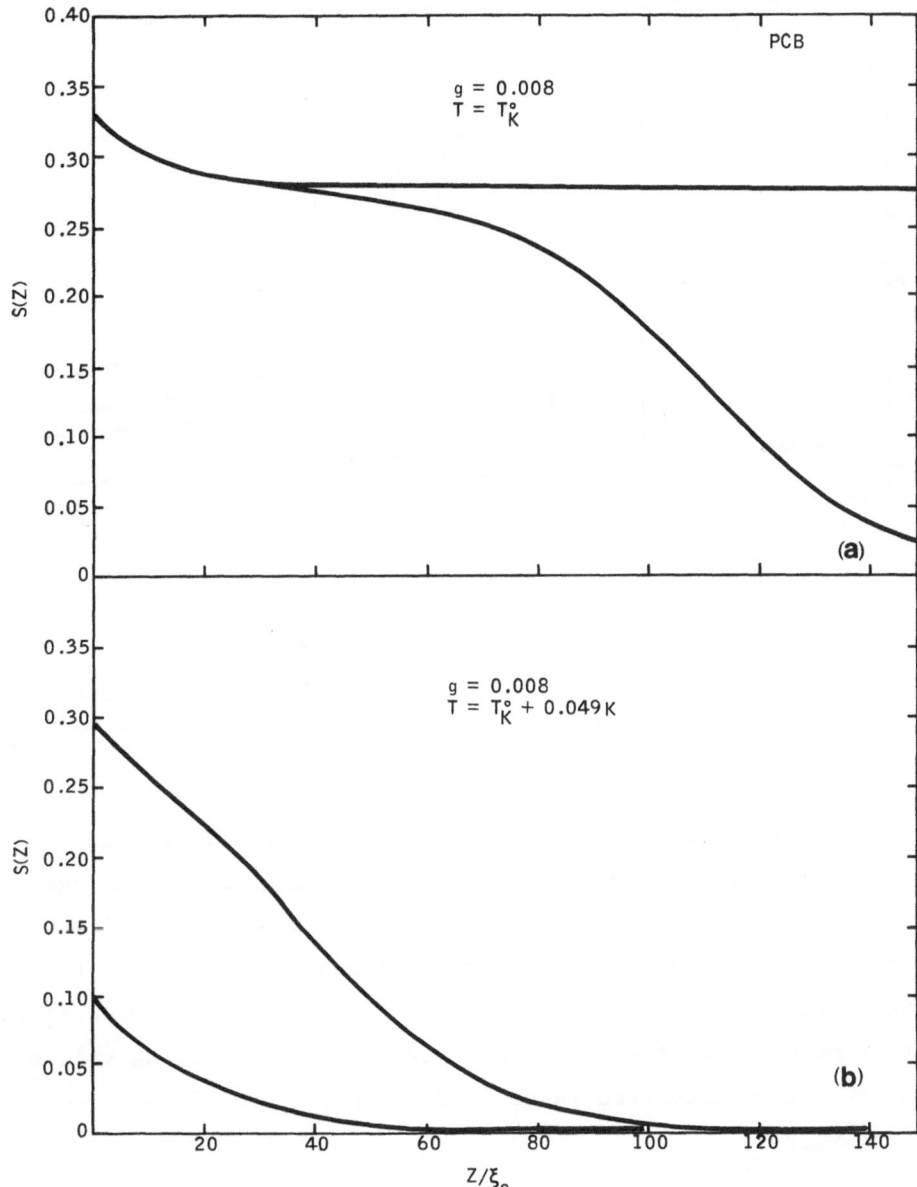

Fig. 4   (a) Boundary layer at $T = T_K^o$. The two curves correspond
          to S(z) just before and just after the bulk transition.
          (b) Boundary layer profiles just before and just after a
          boundary layer phase transition at $T = T_K^o + 0.049K$.

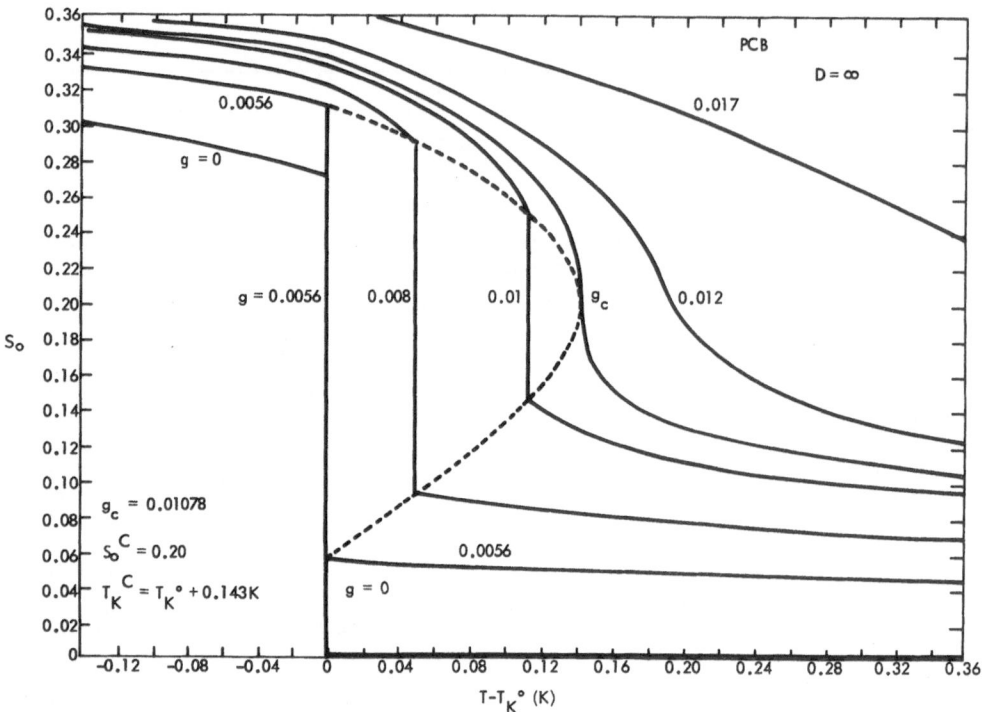

Fig. 5 S_O plotted as a function of temperature for various
          substrate potential values. There is a critical point
          at g = 0.0108, $T = T_K^O + 0.143K$.

differ significantly from that of the bulk. Analysis of the data on
substrate-induced birefringence not only produces excellent agreement between the theory and experiment, but also yields value of the
substrate aligning potential. A novel boundary-layer phase transition, which occurs only in a range of weak substrate potentials,
is also predicted. For PCB, the lower and upper limiting values of
this range are calculated to be ~0.075 erg/cm^2 and ~0.15 erg/cm^2,
respectively.

The author wishes to thank K. Miyano for providing the optical
birefringence data in digital form.

REFERENCES

1. Ping Sheng, Phys. Rev. Lett. 37, 1059 (1976).

2. K. Miyano, Phys. Rev. Lett. 43, 51 (1979).

3. See, for example, P. Sheng and P. J. Wojtowicz, Phys. Rev. A14, 1883 (1976).

4. P. G. de Gennes, Mol. Cryst. Liq. Cryst. 12, 193 (1971).

5. H. J. Coles, Mol. Cryst. Liq. Cryst. 49, 67 (1978).

6. M. Born and E. Wolf, (MacMillan, New York, 1964), p. 699.

7. R. G. Horn, J. Phys. (Paris) 39, 105 (1978).

OPTICAL PROPERTIES OF THE BLUE PHASE OF CHOLESTERIC LIQUID CRYSTALS

J. H. Flack, P. P. Crooker, D. L. Johnson, and S. Long

Department of Physics and Astronomy
University of Hawaii
Honolulu, Hawaii 96822

INTRODUCTION

Although the blue phase -- a stable liquid crystal phase appearing over a small temperature region between the cholesteric and isotropic phases -- was first observed long ago, only in the last few years has an understanding of its structure begun. It is now known[1] that the blue phase only occurs in short pitch cholesterics and that it usually appears in two forms, BPI and BPII, in the sequence Ch - BPI - BPII - Iso. A third phase, BPIII, has also been reported between BPII and the isotropic phase.[2] Both theory[3] and experiment[4,5] indicate that the symmetry of these phases generally is cubic, however, the exact nature of the structure is still under investigation.

The fact that the blue phase has never been observed in nematics, nor in cholesterics with selective reflection wavelength λ_c < 700 nm is significant. Brazovskii, et al.[6] have argued that the essential features of the isotropic-cholesteric transition which lead to the appearance of an intermediate phase are due to: (1) There is a translational symmetry breaking in which the system goes from a periodic lower phase, characterized by wavevector

$\vec{k}_i$, to a uniform upper phase in which the direction of $\vec{k}_i$ is un-
specified and in which the free energy is degenerate on a sphere
in k-space. (2) The lower phase has an unusually long periodic
length P. As the transition temperature is approached the diverg-
ing correlation length ξ can satisfy $\xi \sim P$, at which point the transition becomes qualitatively different. Put briefly, the blue
phase appearance is closely connected with the periodicity of the
lower phase. One therefore expects the cholesteric pitch length
P to have a profound effect on the presence and behavior of the
blue phase.

A good way to test these ideas is to experimentally vary the
pitch of a cholesteric and observe the resulting blue phase behavior. The cholesteric pitch is easily lengthed by adding nematic material or by adding cholesteric material of opposite
handedness; however this procedure alters more than the pitch if
the two compounds are chemically different. A better method is to
mix chiral and racemic versions of the same compound. These mixtures have a continuously adjustable pitch with little, if any,
variation in their chemical properties. Hence the effect of pitch
and pitch alone can be isolated.

We present here a review of some recent experiments investigating blue phase behavior as a function of cholesteric pitch. In
mixtures of chiral CB15 and nematic E9 we have looked at both the
wavelength and polarization of the Bragg reflections with a view
to determining the symmetry of the blue phase structure. In
another study using mixtures of chiral and racemic material we have
determined the phase diagram as a function of pitch.

OPTICAL PROPERTIES OF CB15 - E9 MIXTURES

Chiral CB15 and nematic E9 are cyanobiphenyl compounds[7] which
have received recent attention by other researchers.[8] Mixtures of
these materials form striking blue phase platelet textures (Fig. 1).

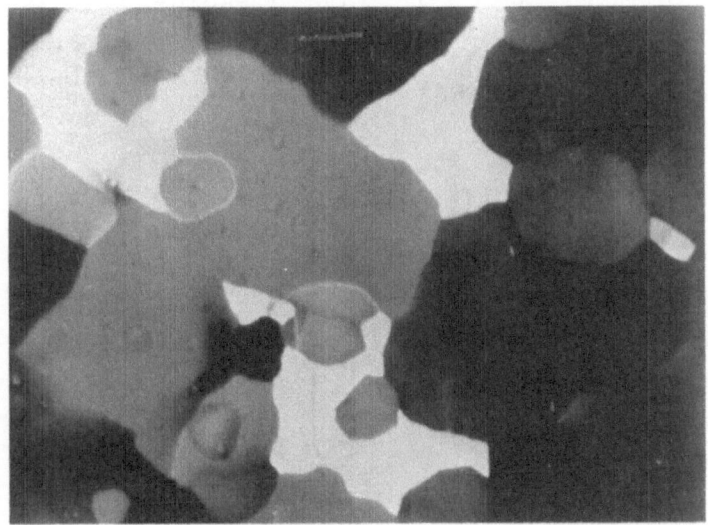

Fig. 1 Blue phase platelet texture. Distinct red, yellow, and blue
        areas corresponding to definite Bragg planes are shown.
        Other non-Bragg orientations are dull green or black.

When illuminated with white light, only a few vivid colors are observed; these colors are clearly a signature of the blue phase's internal structure. The wavelengths associated with the colors are
easily measurable in the microscope and spectrometer arrangement of
Fig. 2. The incident white light is first polarized by the polarizer,
then passed through a Berek prism to the sample. The Berek prism
acts like an achromatic half-wave plate and essentially preserves the
type of polarization. Reflected light from the sample is then
analyzed and passed out of the microscope into a recording spectrometer.

Fig. 3 shows the reflected wavelengths[5] as a function of
temperature for three different mixtures, each characterized by
its cholesteric selective reflection λ_c. Clearly there are three
blue phase manifestations. BPI occurs for all pitches below

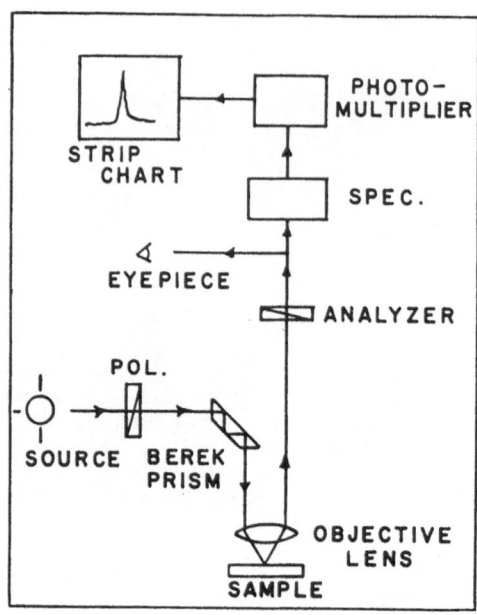

Fig. 2   Apparatus for detection of blue phase wavelengths. The
           apparatus below the spectrometer is incorporated in an
           incident light microscope. Quarter wave retarders are
           added at the polarizer and analyzer positions if circu-
           lar polarization is desired.

~ 700 nm and has a decreasing wavelength with temperature. BPIIB
appears when λ_c > 540 nm and does not exhibit a BPI-BPII wavelength gap. For λ_c ~ 540 nm BPIIB coexists with BPIIA, which <u>does</u>
exhibit a gap, and below 540 nm only BPIIA is seen. The ratios of
the wavelengths, after correcting for refractive index dispersion,
is 1 : $1/\sqrt{2}$: $1/\sqrt{3}$: $1/\sqrt{4}$, starting from the uppermost wavelength.
These wavelength ratios satisfy the condition for Bragg scattering
from a body centered cubic (bcc) or simple cubic (sc) structure.
This particular experiment will not, however, distinguish between
the two.

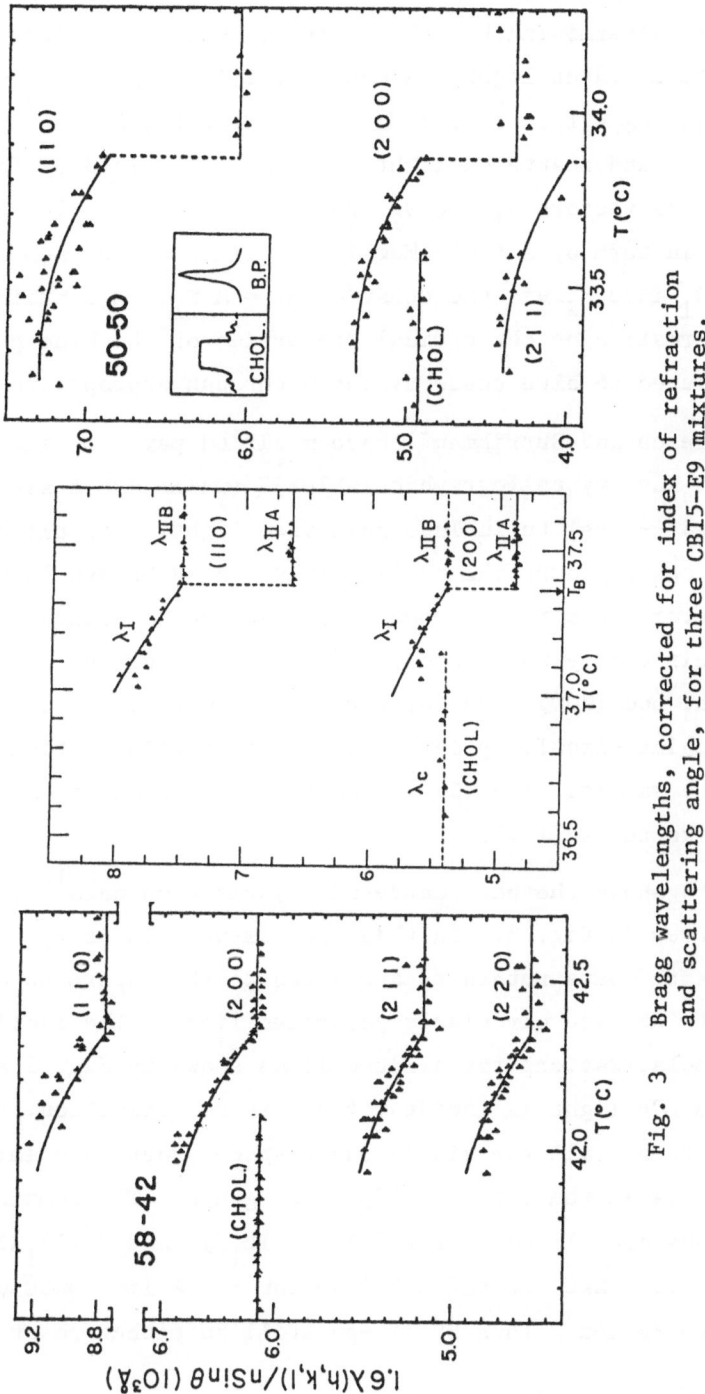

Fig. 3   Bragg wavelengths, corrected for index of refraction and scattering angle, for three CBI5-E9 mixtures.

Further information on the structure of the blue phase can be obtained by determining how the Bragg scattering affects polarization of the incident light. A convenient language for describing this polarizing effect is the Mueller matrix formalism.[9] Briefly, the incident and scattered light beams are described by four component Stokes vectors S_1 and S_2, respectively. The blue phase is described in turn by a 4 x 4 Mueller matrix M which mathematically scatters S_1 into S_2 via the equation $S_2 = M \, S_1$. M contains the desired information on the optical properties of the blue phase and can be related to blue phase symmetry through appropriate theories.

Hornreich and Shtrikman[10] have modified parts of the International X-Ray Crystallographic Tables[11] (which are designed for unpolarized x-rays) to include polarized light. For all the chiral sc space groups the sc(111) line should behave like a simple mirror, that is, it should reverse the handedness of any incident circularly polarized light. Of the four chiral bcc space groups, the bcc (200) line for three of them reflects back only one particular circular polarization if that polarization is incident on the sample. The (200) line for the fourth bcc space group behaves like the sc(111).

Fig. 4 shows the backscattered polarization data[12] for some of the curves in Fig. 3. In this experiment polarizing and analyzing quarter-wave retarders are added to the apparatus of Fig. 2 in order to utilize circularly polarized light. The incident/detected polarizations for all the lines shown in Fig. 3 are RCP/RCP. When LCP light is incident there is no reflection. Thus, according to HS, all the simple cubic space groups are immediately ruled out, as is the bcc 0^5(I432) space group. The remaining space groups are all bcc : T^3(I23), T^5(I2$_1$3), and 0^8(I4$_1$32). Note, however, that the sc(111) line for BPIIA lies outside the observation region. Thus BPIIA may still be either sc or bcc.

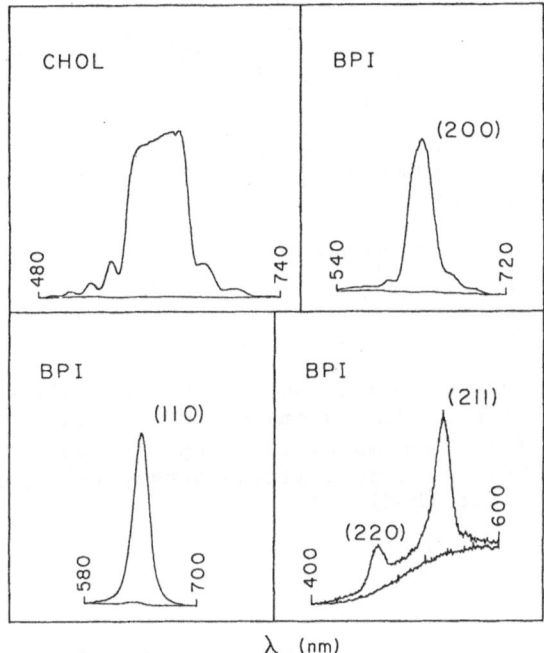

Fig. 4 Lineshapes for circular polarized light in CB15-E9 mixtures. The incident/detected polarizations for the upper curves are RCP/RCP; for the lower curves LCP/LCP.

In backscattering, certain Mueller matrix elements become zero as a result of the scattering symmetry. Changing the scattering angle allows these same elements to be nonzero and additional information is gained. Fig. 5 shows our experimental setup for determining the Mueller matrices for a Bragg scattering angle of 45°. Monochromator MC is first adjusted to the peak of the Bragg line of interest, then fixed polarizer P_1 and adjustable compensator C_1 admit light of specific polarization S_1 to the right angle prism, on which is mounted the sample. After reflecting from the sample, the light then passes through continuously rotating compensator C_2 and fixed analyzer P_2. As C_2 rotates, its angular position and the corresponding intensity is stored in the

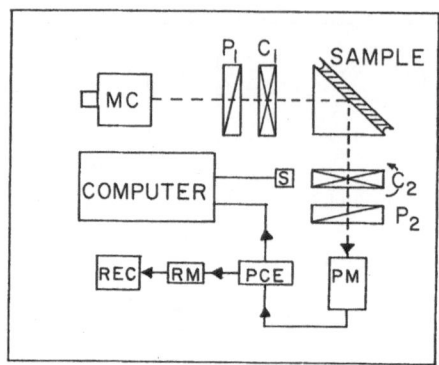

Fig. 5   Apparatus for measuring the Mueller matrices of the
            Bragg lines. Monochromater (MC), polarizer (P), compen-
            sator (C), photomultiplier (PM), ratemeter (RM), recor-
            der (REC), angular position sensor (S), photon counting
            electronics (PCE).

computer, which then reconstructs the scattered Stokes vector S_2.
Repeating the process for 16 input and output Stokes vectors, the
Mueller matrix is then calculated. The temperature of the sample
is next changed, the monochromator reset to the Bragg peak, and a
new matrix obtained.

The Mueller matrix for the BPI-BPIIA (110) Bragg line (we use
bcc notation henceforth) is shown versus temperature in Fig. 6.
This matrix has been normalized by dividing each element by M_{11};
such a matrix shows polarization information only. Although this
transition shows a wavelength gap, no significant manifestation of
the jump is seen in the Mueller matrix. We have also measured
Mueller matrices for the BPI-BPIIB (110) and (200) lines and again
we see no temperature dependence throughout the blue phase region.
The conclusion is that there is no qualitative change in the structure, although the strength of the scattering, and hence of the
order, may be changing.

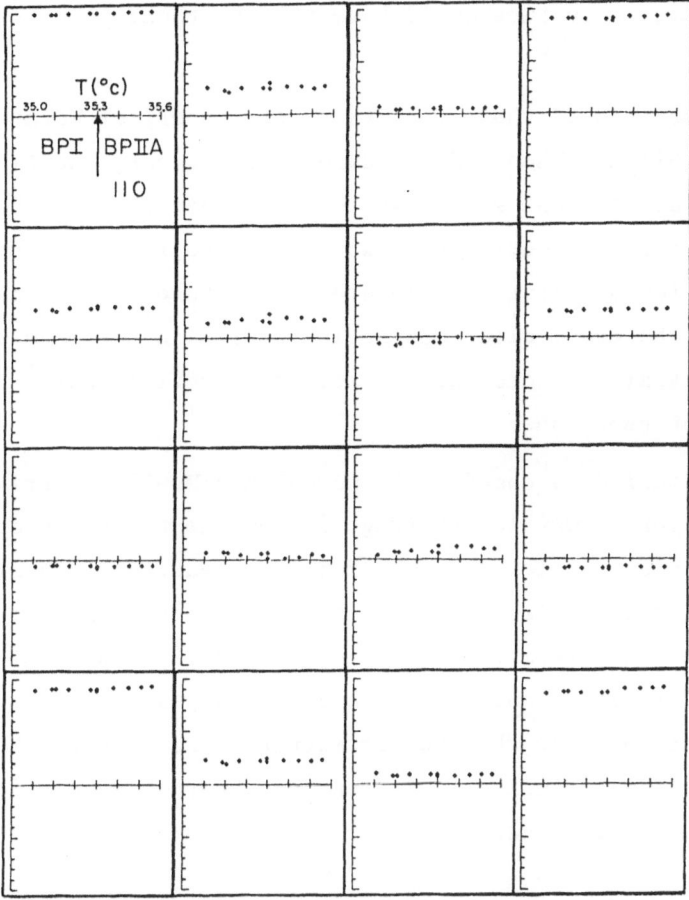

Fig. 6   Typical normalized Mueller matrix versus temperature for
the bcc (200) line.

Further analysis of the matrices reveals that the blue phase
behaves like a reflecting elliptical polarizer with the ellipse
having axial ratio ~ .7 and a tilt angle dependent on the orientation imparted to the sample by the surfaces. Thus any theory of

the blue phase structure would have to take these results into
account.

A comparison of these results to the theory of Hornreich and
Shtrikman (HS) has been made elsewhere[13], however, questions arise
because HS assume weak scattering which is not the case here. A
rigorous theory for our results would have to first incorporate
the actual (or at least a postulated) structure for the blue phase,
then solve the exact electromagnetic problem in the manner used by
Berreman, et al. for cholesteric selective reflections.[14] To date
this has not been done.

Additional data showing the normalized Mueller matrix of the
BPIIB line versus wavelength (Fig. 7) reveals the behavior of the
polarization at wavelengths away from the peak. In the simplest
approximation the reflected polarization ellipse tends to flatten
as one moves away from the peak; in addition, a continual rotation
of the major axis as one passes through the peak is observed. On
the wings of the line the Mueller matrix is no longer interpretable
as an elliptical polarizer. As yet, no theory has been developed
which addresses this data.

PHASE DIAGRAMS FOR CHIRAL-RACEMIC MIXTURES

CB15 and E9 are chemically different compounds; consequently
the thermodynamic properties of mixtures of them will be a function
of concentration irrespective of pitch variation. We have, however, obtained both chiral and nonchiral forms of the same liquid
crystal mixture.[15] The chiral mixture contains the nematic cyanobiphenyls M24(18%), M30(17.6%), and M36(4.4%) to which has been
added 60% chiral CB15. The nonchiral mixture is identical except
that it contains racemic CB15. Combination of these chiral and
nonchiral mixtures results in a new mixture with the same concentrations of the nematics and CB15, but with a chirality which is
now continuously adjustable.

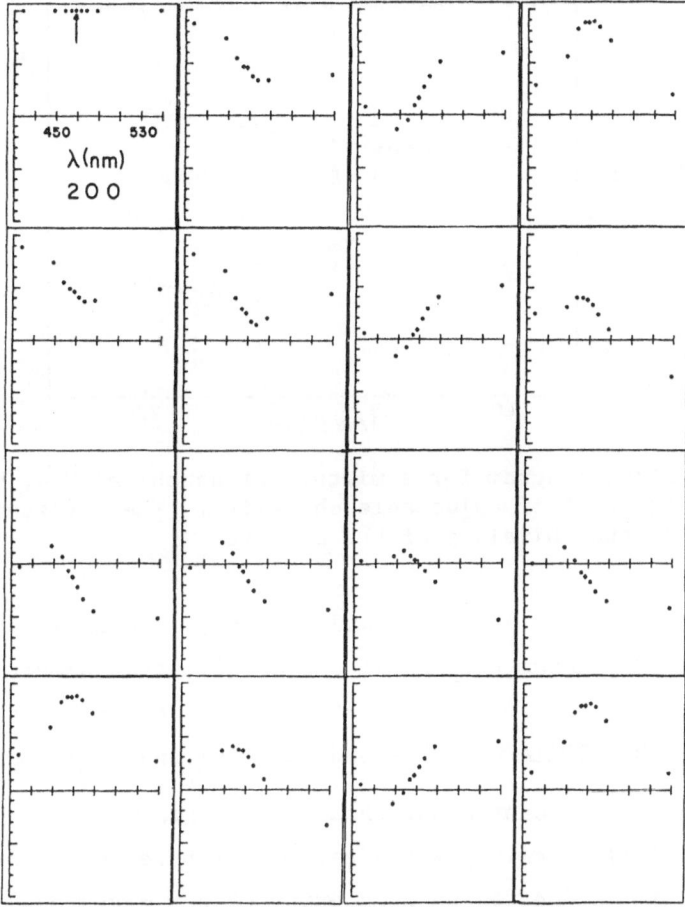

Fig. 7   Typical normalized Mueller matrix versus wavelength for
the bcc (200) line.

Fig. 8 shows the phase diagram for this mixture. The independent variable is the chirality, given by 1/nP, where n is the
average refractive index and P is the pitch in the cholesteric
phase; nP is just the cholesteric selective reflection wavelength.
At low chirality (np > 700 nm) the isotropic transforms directly
to the cholesteric. At higher chirality (np < 700 nm) both blue

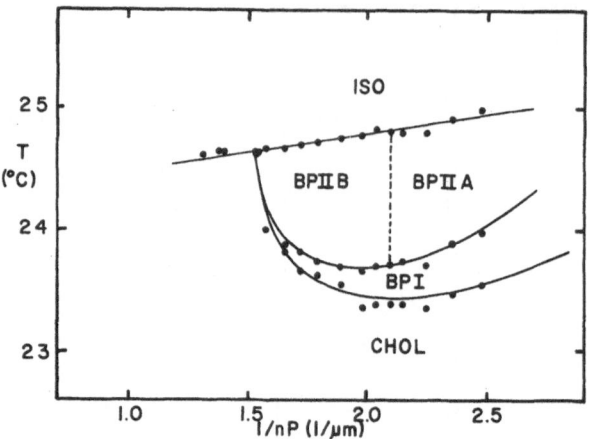

Fig. 8   Phase diagram for a mixture of nonchiral M24, M30, M36,
          and CB15 of adjustable chirality. The horizontal axis
          is the chirality of the mixture.

phases, BPI and BPII, appear simultaneously, the BPII appearing as
BPIIA. At nP ≃ 480nm both BPIIA and BPIIB appear simultaneously,
in a manner similar to the CB15-E9 mixture. As the chirality increases further BPIIA vanishes and only BPI and BPIIB survive.

It should be pointed out that this diagram is somewhat different from chiral-racemic phase diagrams for pure compounds.[16]
Probably this is due to the complexity of our mixtures. Also the
diagram is quite unlike diagrams predicted theoretically by
Hornreich and Shtrikman,[17] who so far have only treated single
component systems possessing one or two Bragg lines. It is therefore premature to expect good agreement with our experiment.

ACKNOWLEDGEMENTS

The authors would like to thank R. Hornreich and S. Shtrikman
for extensive correspondence and C. F. Hayes for useful discussions. This work was supported by a grant from Research Corporation and by National Science Foundation Grant No. DMR-8022855

REFERENCES

1. H. Stegemeyer and K. Bergmann, in Liquid Crystals of One- and
    Two-Dimensional Order, W. Helfrich and G. Heppke, ed.
    (Springer Verlag, Berlin 1980); K. Bergmann and H.
    Stegemeyer, Ber. Bunsenges. Phys. Chem. 82, 1309 (1978);
    Z. Naturforsch. 349, 251 (1979).

2. M. Marcus, J. Phys. (Paris) 42, 61 (1981).

3. A. Saupe, Mol. Cryst. Liq. Cryst. 1, 59 (1969); S. Alexander,
    Proc. of the Colloque Pierre Curie, ESPC Paris 1-5, 1980,
    in press; R. M. Hornreich and S. Shtrikman, J. Phys.
    (Paris) 41, 335 (1980); H. Kleinert and K. Maki, Fortschr.
    Phys. 29, 219 (1981); S. Meiboom, J. P. Sethna, P. W.
    Anderson, and W. F. Brinkman, Phys. Rev. Lett. 46, 1216
    (1981); P. L. Finn and P. E. Cladis, to be published.

4. S. Meiboom and M. Sammon, Phys. Rev. Lett. 44, 882 (1980);
    Phys. Rev. A24, 468 (1981).

5. D. L. Johnson, J. H. Flack and P. P. Crooker, Phys. Rev.
    Lett. 44, 882 (1981).

6. S. A. Brazovskii, Sov. Phys. - JETP 41, 85 (1975); S. A.
    Brazovskii and S. G. Dmitriev, Sov. Phys. JETP 42, 497
    (1976); S. A. Brazovskii and V. M. Filev, Sov. Phys.
    JETP 48, 573 (1978).

7. CB15 and E9 are British Drug House, Inc. designations.

8. B. B. Rao, J. Her, and J. T. Ho, to be published; M. Marcus,
    to be published.

9. W. A. Shurcliff, Polarized Light (Harvard University Press,
    Cambridge 1962).

10. R. M. Hornreich and S. Shtrikman, Phys. Lett. 82A, 354 (1981).

11. International Tables for X-Ray Crystallography, Vol. 1
     (Kynoch Press, Birmingham, 1952).

12. J. H. Flack and P. P. Crooker, Phys. Lett. 82A, 247 (1981).

13. J. H. Flack, P. P. Crooker, and R. C. Svoboda, to be pub-
        lished.

14. D. W. Berreman and T. J. Scheffer, Mol. Cryst. Liq. Cryst. $\underline{11}$,
        395 (1970).

15. We are indebted to D. G. McDonnell for supplying us with this
        mixture. M24, M30 and M36 are trade names of British
        Drug House, Inc.

16. M. Marcus and J. W. Goodby, to be published.

17. R. M. Hornreich and S. Shtrikman, Phys. Rev. A$\underline{24}$, 635 (1981).

LATTICE PARAMETERS IN BLUE PHASE MIXTURES

Jih Her and John T. Ho

Department of Physics and Astronomy
State University of New York at Buffalo
Amherst, New York 14260

ABSTRACT

The blue phases BPI and BPII in two sets of mixtures of cholesteric and nematic materials have been studied using optical Bragg
reflection. A critical cholesteric pitch exists for each set of
mixtures separating two types of behavior of the BPII lattice parameters. The extent of lattice contraction at the BPI-BPII transition in short-pitch mixtures appears to be universal. There is no
simple correlation between the blue phase lattice parameters and
the cholesteric pitch.

The phenomenon of the cholesteric blue phases has attracted a
lot of experimental[1-7] and theoretical[8-12] attention recently. The
two commonly observed blue phases (BPI and BPII in order of increasing temperature) are found to have structures which are consistent
with either body-centered cubic (bcc) or simple cubic symmetry.
Agreements to date between theory and experiment are mostly qualitative. Detailed information about the lattice parameters of the
blue phases and their relation with the cholesteric pitch in various
systems is highly desirable. We report here optical Bragg reflec-

tion measurements on two sets of blue phase mixtures. The results
in both systems show the existence of a critical cholesteric pitch
separating two types of behavior of the BPII lattice parameters.
In addition, the blue phase lattice parameters do not appear to
have a simple correlation with the cholesteric pitch.

   The samples studied were mixtures of the chiral biphenyl CB15
with the nematic mixture E9, and of the non-sterol chiral mixture
TM59 with the nematic mixture E5,[13] in various concentrations by
volume. The CB15/E9 mixtures have been studied extensively[3,4,7]
and some of our results have been reported.[7] The TM59/E5 mixtures
were chosen as an additional system to study partly because the
molecular structures of TM59 and CB15 are somewhat different and
partly because, unlike the CB15/E9 mixtures, the two constituents
have similar isotropic transition temperatures (53°C and 50°C respectively). Samples were found between glass slides separated by
a 25-μm spacer. The sample temperature was controlled to a stability of better than 0.01°C. Bragg reflection spectra in the backscattering geometry were measured using a halogen-tungsten light
source and a grating monochromator scanned by a stepping motor. The
spectra were stored digitally in a signal averager. To correct for
the intrinsic dispersion in the optical system, a reflection spectrum was also taken with each sample in the isotropic phase and used
to normalize all the cholesteric and blue phase spectra. The peak
reflection wavelength was determined with a relative accuracy of
1 nm.

   The reflection spectra of the short-pitch CB15/E9 samples are
illustrated in Fig. 1 for the (50:50) mixture in the vicinity of
the cholesteric peak. Upon heating, the broad flat-topped cholesteric peak at λ_c persists up to 33.85°C, above which the sharp BPI
peak appears. The peak wavelength λ_I decreases with increasing temperature. Above 34.26°C, the sample transforms into the BPII phase,
with the appearance of two peaks at λ_{II} and λ'_{II} in decreasing wave-

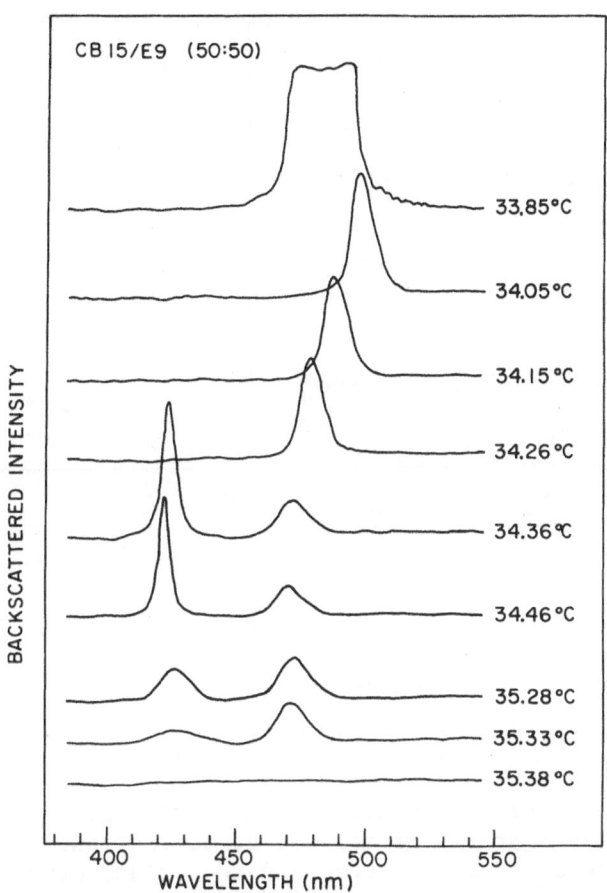

Fig. 1. Temperature dependence of reflection spectra between 380
         and 550 nm in a CB15/E9 (50:50) mixture.

length. The transition from λ_I to λ_{II} is continuous. The λ_{II} and
λ_{II}' wavelengths are not consistent with a simple cubic structure,
but may be indicative of the coexistence of two interpenetrating
lattices. The change in peak intensity upon further heating is
probably related to the increasing presence of the isotropic phase
in the two-phase region. Above 35.33°C, the sample is entirely in
the isotropic phase. The λ_I and λ_{II}' reflections have been identified with the (200) lattice parameters if the bcc structure is
assumed.[3]

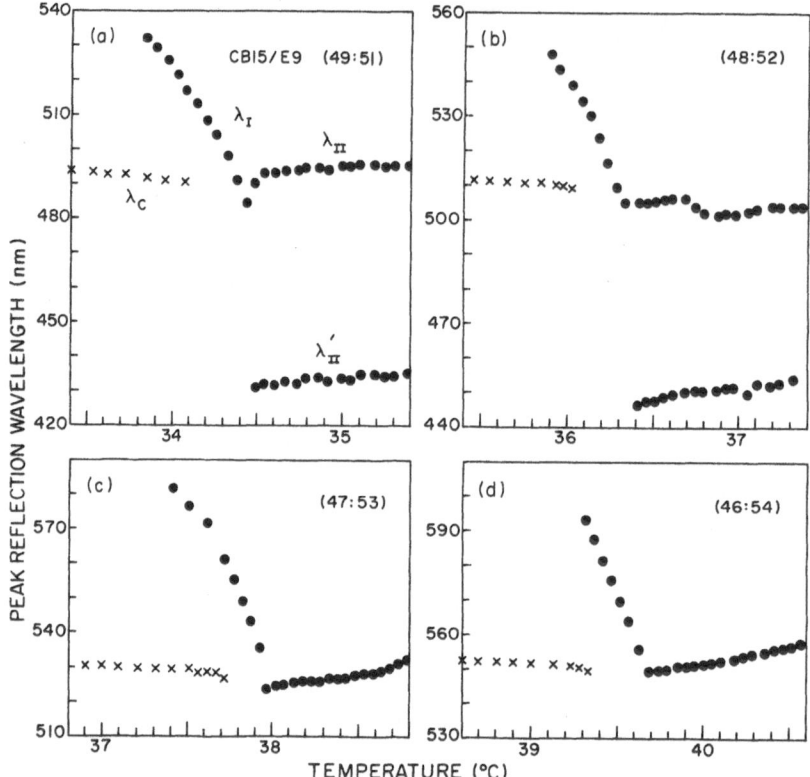

Fig. 2.   Temperature dependence of peak reflection wavelength in
         CB15/E9 mixtures of various concentrations.   The crosses
         denote cholesteric peaks and the points denote blue phase
         peaks, including supercooled BPI peaks.

The effect of increasing the E9 content and increasing λ_c on
the behavior of the lattice parameters in the CB15/E9 mixtures is
shown in Fig. 2.   It can be seen that the coexistence of the λ_{II}
and λ_{II}' peaks persists up to 52% of E9.   Beyond that, λ_{II}' disappears, and the BPI–BPII transition is characterized only by a
continuous change from λ_I to λ_{II}.   It should be mentioned that,
while the coexistence of λ_{II} and λ_{II}' was seen in all of our short-
pitch CB15/E9 mixtures, it was observed in only one concentration
by Flack and Crooker.[4]   Another observation is that the minimum
value of λ_c is typically almost, but not exactly, equal to the
value of the $\lambda_{II}(200)$ peak at the BPI–BPII transition.[7]   It may be

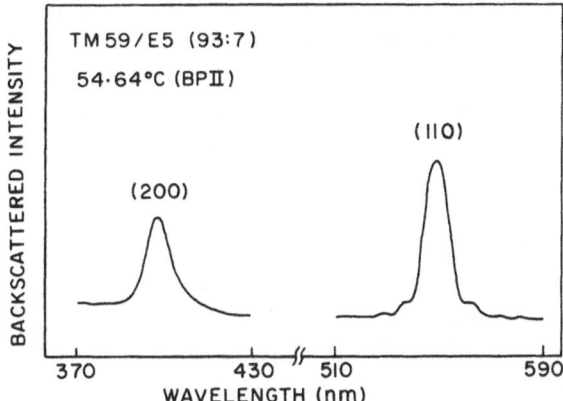

Fig. 3. Two harmonic reflections of a TM59/E5 (93:7) mixture in the BPII phase at 54.64°C.

argued, however, that the close proximity of the two wavelengths may be more intriguing and significant than their small differences.

To test the generality of the results in the CB15/E9 mixtures and to avoid the problem of the large difference between the isotropic temperatures of CB15 and E9 (-30°C and 82°C, respectively), we have studied mixtures of TM59 and E5, with 0 to 8% by volume of E5. Since these mixtures have not been previously examined, the blue phase structures have to be first determined. Because of the short cholesteric pitch in these mixtures, only two harmonic peaks are observed for each blue phase structure. An example is shown in Fig. 3 for the mixture with 7% of E5 in the BPII phase, in which two λ_{II} peaks are obtained and identified as the (110) and (200) reflections if the bcc structure is assumed. The identification is based on the fact that the backscattered Bragg peaks should occur at $\lambda(h\kappa\ell) = 2nd/(h^2 + \kappa^2 + \ell^2)^{1/2}$, where d is the lattice parameter, n is the average refractive index, and ($h\kappa\ell$) are the Miller indices. This is illustrated for the (96:4) mixture in Fig. 4, where plots of λ/n versus $(h^2 \times \kappa^2 + \ell^2)^{-1/2}$ yield straight lines passing through the origin for all three pairs of representative λ_I, λ_{II} and λ'_{II} peaks.[14]

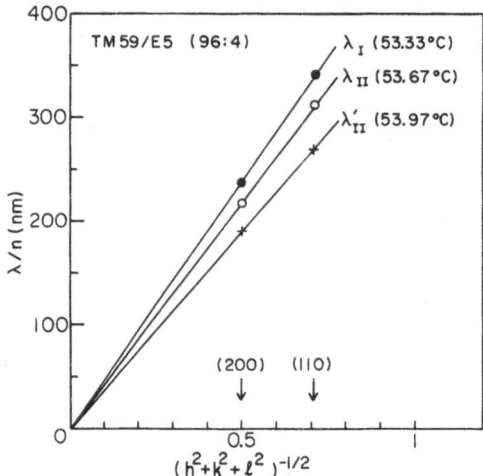

Fig. 4. Peak reflection wavelength λ normalized by the average refractive index n versus $(h^2 + \kappa^2 + \ell^2)^{-1/2}$, where $(h\kappa\ell)$ are the Miller indices of the first two harmonics of the bcc structure, for three representative pairs of peaks at various temperatures in a TM59/E5 (96:4) mixture.

The effect of increasing the E5 content and increasing λ_c on the behavior of the lattice parameters in the TM59/E5 mixtures is shown in Fig. 5. For mixtures with up to 5% of E5, there is again a coexistence of two BPII lattice parameters λ_{II} and λ'_{II}, with the former transforming continuously from λ_I. This coexistence is observed in most, but not all, of the short-pitch samples. While there are general qualitative similarities in the behavior of the CB15/E9 and TM59/E5 mixtures, there are also some important differences. The critical cholesteric pitch λ_c separating the two types of BPII behavior in the TM59/E5 mixtures is about 400 nm, which is considerably shorter than the critical λ_c in the CB15/E9 mixtures. Another difference is that the minimum value of λ_c in the TM59/E5 mixtures is consistently and significatnly (up to 11%) larger than the value of $\lambda_{II}(200)$ at the BPI-BPII transition. This implies there might not be any simple correlation between $\lambda_{II}(200)$ and λ_c. This observation, however, is not inconsistent with the qualitative

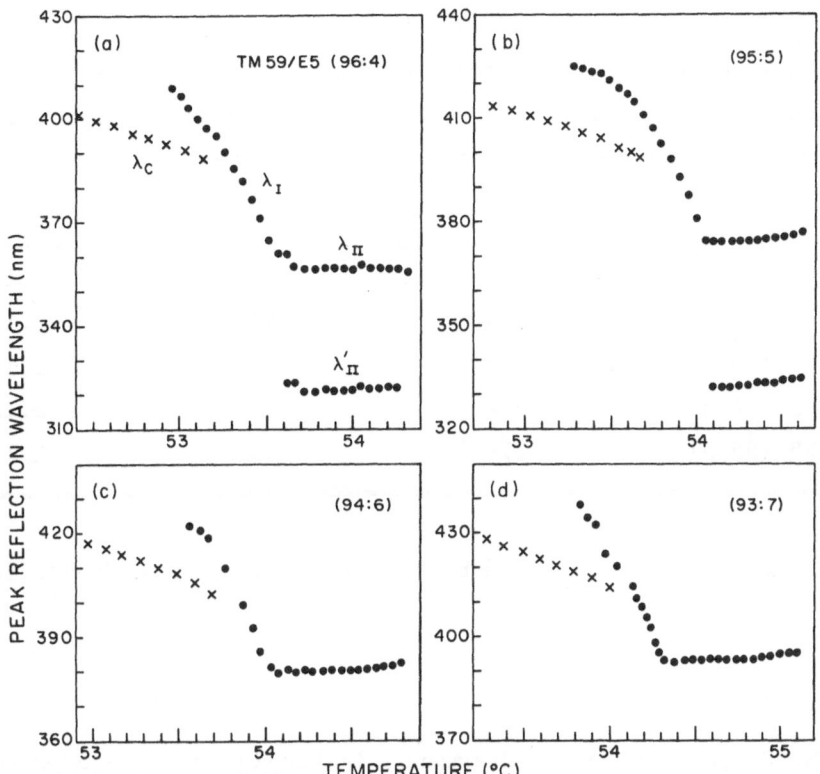

Fig. 5. Temperature dependence of peak reflection wavelength in
TM59/E5 mixtures of various concentrations. The crosses
denote cholesteric peaks and the points denote blue phase
peaks, including supercooled BPI peaks.

Landau-theory result including harmonics that $\lambda_{II}(110)$ is red-
shifted with respect to λ_c.[8]

    One striking feature emerging from both sets of data is the
apparent universality in the extent of the lattice contraction from
λ_I to λ'_{II} at the BPI-BPII transition. Table I shows the values of
λ_I, λ'_{II}, and λ_I/λ'_{II} at the BPI-BPII transition obtained in our
short-pitch mixtures. These results show that the lattice con-
tracts at the transition by about 13% in all the mixtures studied.
This universality has also been found to extend to other blue phase
systems that have been studied,[7] and is possibly significant.

TABLE I

Ratio of the lattice parameter λ_I in the BPI phase to the
lattice parameter λ'_{II} in the BPII phase at the BPI-BPII transition.

Mixture	λ_I (nm)	λ'_{II} (nm)	λ_I/λ'_{II}
CB15/E9 (50:50)	465	411	1.13
CB15/E9 (49:51)	484	430	1.13
CB15/E9 (48:52)	506	448	1.13
TM59/E5 (96:4)	362	324	1.12
TM59/E5 (95:5)	374	332	1.13

This work was supported by the National Science Foundation
under Grant No. DMR-8019985.

REFERENCES

1. H. Stegemeyer and K. Bergmann, in Liquid Crystals of One- and
      Two-Dimensional Order, W. Helfrich and G. Heppke, editors
      Springer-Verlag, Berlin, 1980.

2. S. Meiboom and M. Sammon, Phys. Rev. Lett. 44, 882 (1980);
      Phys. Rev. A 24, 468 (1981).

3. D. L. Johnson, J. H. Flack, and P. P. Crooker, Phys. Rev.
      Lett. 45, 641 (1980).

4. J. H. Flack and P. P. Crooker, Phys. Lett. 82A, 247 (1981).

5. M. Marcus, J. Phys. (Paris) 42, 61 (1981).

6. P. H. Keyes, A. J. Nicastro, and E. M. McKinnon, Mol. Cryst.
      Liq. Cryst. 67, 59 (1981).

7. J. Her, B. B. Rao, and J. T. Ho, Phys. Rev. A 24, 3272 (1981).

8. R. M. Hornreich and S. Shtrikman, J. Phys. (Paris) 41, 335
      (1980); Phys. Rev. A 24, 635 (1981).

9. S. Meiboom, J. P. Sethna, P. W. Anderson, and W. F. Brinkman,
      Phys. Rev. A 46, 1216 (1981).

10. H. Kleinert and K. Maki, Fortschr. Phys. <u>29</u>, 219 (1981).

11. S. Alexander, in Proceedings of the Colloque Pierre Curie 1980, to be published.

12. P. L. Finn and P. E. Cladis, to be published.

13. These are PDH Chemicals designations.

14. These peaks are also consistent with the (100) and (110) reflections if the simple cubic symmetry is assumed.

BLUE PHASE STRUCTURE ANALYSIS BY OPTICAL BRAGG DIFFRACTIONS

Dwight W. Berreman

Bell Laboratories
Murray Hill, New Jersey   07974

ABSTRACT

Hornreich and Shtrikman have derived selection rules for polarization and intensity of light that is Bragg-diffracted in the
back-reflection direction by cubic crystalline models of the cholesteric blue phases. These rules help to identify the space groups
of the various phases. We go beyond space-group determination and
compute reflection of both normally and obliquely incident light of
various polarizations for specific models of order parameter and
orientation within each unit cell. Our optical computations are
done with a 4X4 matrix technique that was developed to study the
optical properties of stratified anisotropic media. The new features in our use of the technique are to average the optical dielectric tensor over many intermediate planes per cycle parallel to
each set of Bragg planes, and to Fourier analyze the variations of
these average tensors. We ignore scattering by transverse variations
of the dielectric tensor. From the simplest Landau models of cubic
blue phases we predict reflection from only one or two sets of Bragg
planes but more reflections have been observed. Meiboom, et al.
have proposed models in which the refractive index is isotropic

along disclination tubes and uniaxial elsewhere. The more complex
reflection spectra that we have computed from one of these models
with O^8 symmetry agree well with experimental data for blue phase
I. Data for blue phase II are consistent with O^2 symmetry.

INTRODUCTION

This paper gives the method and results of computation of
Bragg reflection intensities for some specific structural models of
cholesteric blue phases having O^2 simple cubic symmetry and O^5 and
O^8 body centered cubic symmetries. The method is readily extended
to any model and any symmetry, and should help to determine details
of the structure of blue phases.

OPTICAL DIELECTRIC TENSORS OF CHOLESTERICS

The local optical dielectric tensor $\underline{\varepsilon}_1$ in any nonmagnetic
medium with negligible intrinsic optical activity, such as a cholesteric liquid crystal, may be written as a scalar term, ε_s, plus the
product of another scalar, ε_t times a traceless, symmetric tensor,
$\underline{Q}$. $\underline{Q}$, ε_s and ε_t may all depend on position in the medium. For
structures of cubic symmetry it is only necessary to specify one
diagonal term and one off-diagonal term in $\underline{Q}$ to completely describe
this tensor.

A cholesteric material in the phase with focal-conic or Grand-
jean texture, which we shall call the "planar phase", has the tensor
$\underline{Q}$ defined by the following elements, wherein c is 360° divided by
the unit cell dimension, a_o, and the coordinates a, b and c are
along the cubic axes.

$$6 \cdot Q_{aa} = 1 + 3\cos(c), \qquad 6 \cdot Q_{bc} = 0,$$

$$6 \cdot Q_{bb} = 1 + 3\cos(c-180°), \quad 6 \cdot Q_{ca} = 0,$$

$$6 \cdot Q_{cc} = -2, \qquad \text{and} \quad 6 \cdot Q_{ab} = 3\cos(c-90°).$$

A constant diagonal tensor of zero trace may be noted in $\underline{Q}$ for this "planar" phase which is absent in cubic phases. Its optical effects are minor if the optical anisotropy is small. The terms ε_s and ε_t are independent of position in this phase, except near disclinations.

LANDAU MODELS OF THE BLUE PHASES

In the simplest cubic Landau models of the blue phases the scalar terms ε_s and ε_t are independent of position. In Table 1 we give functions for Q_{aa} and Q_{bc} for simple Landau models with 0^2, 0^5 and 0^8 symmetries.[1,2] The other seven tensor elements, shown at the top of Table 1, can be obtained by changing a to b, b to c and c to a once or twice; that is, by cycling the coordinates. The tensors $\underline{Q}$ have been normalized in such a way as to prevent the ellipsoid representing the optical dielectric tensor, which is the square of the refractive index, from deviating from a sphere of radius ε_s by more than $2\varepsilon_t/3$ in any direction. That is, the major axis of $\underline{Q}$ is at most 2/3. The normalization constant may be either positive or negative, and its magnitude depends on the sign. Only the 0^8 models seemed to be promising with either sign, and we designate them as 0^8+ and 0^8-.

Landau models yield biaxial optical properties (three unlike eigenvalues of $\underline{\varepsilon}$ or $\underline{Q}$) at most positions in the cubic unit cell. The characteristic equation for the eigenvalues of $\underline{Q}$ is

$$\underline{Q} - \lambda \underline{\underline{1}} = \underline{0}$$

Since the sum of the diagonal terms in $\underline{Q}$ is zero ($\underline{Q}$ is traceless), the characteristic equation has the form

$$\lambda^3 + u\lambda + v = 0.$$

This equation has two roots of one sign and one of the other with the exception of isolated points where one or three roots are zero. The sum of the roots is zero. The Landau models of appropriate

TABLE 1

$\underline{Q}$ for four Landau models with chiral cubic symmetry.

$$\underline{\underline{Q}} = \begin{vmatrix} Q_{aa} & Q_{ab} & Q_{ca} \\ Q_{ab} & Q_{bb} & Q_{bc} \\ Q_{ca} & Q_{bc} & Q_{cc} \end{vmatrix}$$

For O^2 symmetry;

$12 \cdot Q_{aa} = \cos a \cos b + \cos c \cos a - 2 \cos b \cos c$
$\qquad\qquad\quad + 3 (\cos c - \cos b)$

and

$12 \cdot Q_{bc} = \backslash |\overline{2} \sin a (\cos c - \cos b) + \sin b \sin c - 3 \sin a$

For O^5 symmetry;

$6 \cdot Q_{aa} = -\cos a \cos b - \cos c \cos a + 2 \cos b \cos c$

and

$6 \cdot Q_{bc} = -\backslash |\overline{2} \sin a (\cos c - \cos b) - \sin b \sin c$

For O^8 symmetry;

$C \cdot Q_{aa} = \sin c \cos a + \sin a \cos b - 2 \sin b \cos c$

and

$C \cdot Q_{bc} = \backslash |\overline{2} (\cos a \cos b - \sin a \sin c) - \cos b \cos c$

where C=6 for O^8+ and C=-3 (approximately) for O^8-.

sign give an array of rather narrow tubular regions where there are one negative and two positive roots. The tensors from Landau theory give the minimal number of Bragg reflections consistent with the specified symmetries. As mentioned by Hornreich and Shtrikman[3], higher harmonics will produce additional reflecting planes in the back-reflection direction.

UNIAXIAL MODELS OF THE BLUE PHASE

Meiboom, Sethna, Anderson and Brinkman (MSAB) recently described alternatives to the Landau models that have only uniaxial
or isotropic (null) $\underline{Q}$ tensors everywhere.[5] These models give many
higher harmonics of appreciable amplitude which would give additional Bragg reflections. The MSAB models may be developed from
the Landau models in the following way, although it is not necessary
to start from Landau models.

In the ordinary cholesteric phase the optical dielectric
tensor is almost perfectly uniaxial, so that the secular equation
has one position and two equal negative roots. Regions with one
negative and two positive roots in the Landau models of the blue
phases are more like the "discotic" than the cholesteric phase. It
is necessary to introduce a network of disclination lines in order
to fill a space of cubic symmetry with uniaxial substance. A
suitable and reasonable network of disclination lines may be placed
at the loci of lines where there are two equal positive roots of
the secular equation for the Landau model. In the MSAB model a
network of tubes of fixed radius is centered on these lines. Directors in the rest of space may at first be set in the direction of
the eigenvector corresponding to the most positive root. This
forms a continuously varying array of directors but it is not the
configuration of minimal energy. In order to show the order of
magnitude of the dependence of Bragg reflection intensities on
details of the model we have computed Bragg reflections from this
unrelaxed uniaxial model.

The last step in the development of the MSAB model is to minimize the Oseen-Frank energy of the director array. This is done by
allowing the director array to rotate to a configuration of minimum
energy.[5-7] The result depends somewhat on the diameter of the
tubes and on the ratio of the cell dimension to the helicity of the
cholesteric. Consequently, the tube diameter and helicity are also

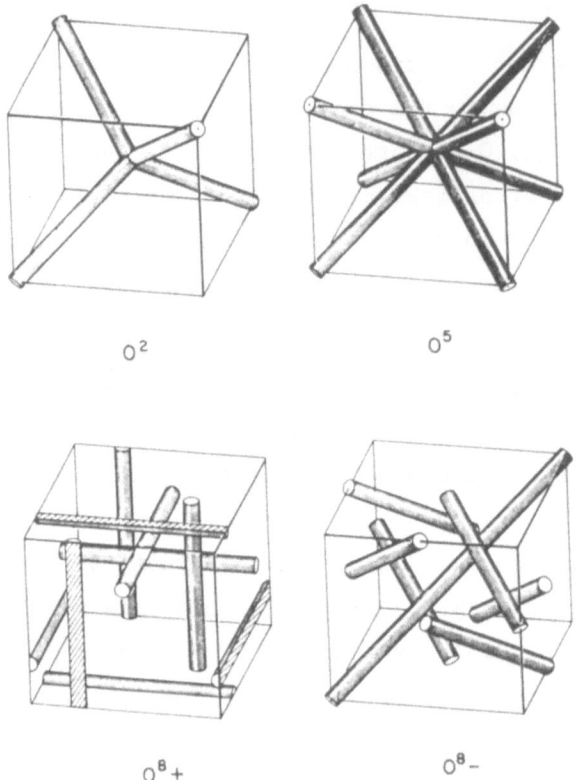

O^2 O^5

O^8+ O^8-

Fig. 1.   Configuration of tubes surrounding disclinations in cubic
          unit cells of O^2, O^5, O^8+ and O^8- symmetry.

adjusted to minimize energy. The resulting tube diameters are
somewhat smaller than the varying diameters of the tubes containing
two positive roots in the Landau theories. We used the same tube
diameters in the unrelaxed models as in the MSAB models.

    Figure 1 shows the loci of these disclination lines or tubes
in unit cells of O^2, O^5, O^8+ and O^8- symmetry. Although all the
tubes illustrated are straight, there is actually a slight spiral
in the tubes for O^8+ symmetry as they thread past near neighbors.
This spiraling is not sufficient to have an appreciable effect on
the Bragg spectra at the precision we intend to use; it is not
required by the symmetry; and it has also been ignored by Meiboom,

et al.[5-7] The tubes for 0^8+ are located where the directors are
uniaxial and parallel to the corresponding lines for the 0^8- Landau
model, and vice-versa. This is a consequence of the fact that both
types of lines are loci of degenerate principal axes.

OPTICAL COMPUTATIONS

The computation of reflection by stratified anisotropic media
can be done with a 4X4 matrix technique.[4,8,9] (This matrix technique does not involve the 4X4 Mueller matrices mentioned in Ref.
14, but it can predict them.) Consider a beam of plane waves
impinging at an angle Θ from the normal to a set of Bragg planes
designated by their Miller indices, (h k 1). Because of local
variations in refractive index these "plane" waves will actually
have values of the electromagnetic field components that vary
somewhat in the plane of a wave-front. However, if simultaneous
Bragg scattering of a single planar incident wave into more than
one direction in one crystalline domain can be neglected, the
electromagnetic waves may be treated as if the variation, parallel
to the Bragg planes, of any field component is purely of the form
$e^{ik_x x}$. Here, k_x is the component of the propagation vector parallel
to the Bragg planes. Rectangular coordinates are chosen so that z
is normal to the Bragg planes, x is parallel to them in the plane
of optical propagation and y is parallel to the Bragg planes and
normal to the propagation. The succeeding equations are simplified
by making the unit of length in the x, y, z coordinate system equal
to the Bragg spacing;

$$B = 1 = a_o / \sqrt{(h^2+k^2+1^2)}.$$

The next step in the computations is to average the tensor $\underline{\underline{\epsilon}}_1$
over parallel planes at each of numerous levels of the periodic
structure in the (h k 1) direction. These averages describe a
plane stratified medium with equivalent reflective and transmis-
sive properties for that particular set of Bragg planes. The

average tensor may be designated $\underline{\underline{\epsilon}} = \underline{\underline{\epsilon}}$ (h,k,l,z). From there on
the computation may proceed using the 4X4 matrix technique.[4,8,9]

The 4X4 matrix propagation equation for electromagnetic waves
in stratified media is[8]

$$\frac{\partial \Psi(z)}{\partial z} = i \underline{\underline{\Delta}} \Psi(z)$$

where

$$\Psi = \begin{vmatrix} E_x \\ H_y \\ E_y \\ -H_x \end{vmatrix}$$

is a "vector" composed of four independent electromagnetic field
components and

$$\underline{\underline{\Delta}} = \begin{vmatrix} \dfrac{-k_x \epsilon_{zx}}{\epsilon_{zz}} & 1 - \dfrac{k_x^2}{\epsilon_{zz}} & \dfrac{-k_x \epsilon_{yz}}{\epsilon_{zz}} & 0 \\[4mm] \epsilon_{xx} - \dfrac{\epsilon_{zx}^2}{\epsilon_{zz}} & \dfrac{-k_x \epsilon_{zx}}{\epsilon_{zz}} & \epsilon_{xy} - \dfrac{\epsilon_{zx}\epsilon_{yz}}{\epsilon_{zz}} & 0 \\[4mm] 0 & 0 & 0 & 1 \\[4mm] \epsilon_{xy} - \dfrac{\epsilon_{zx}\epsilon_{yz}}{\epsilon_{zz}} & \dfrac{-k_x \epsilon_{yz}}{\epsilon_{zz}} & \epsilon_{yy} - k_x^2 - \dfrac{\epsilon_{yz}^2}{\epsilon_{zz}} & 0 \end{vmatrix}$$

In these expressions the unit of length is chosen so that the ratio
the optical frequency of the light to the velocity of light in
vacuum is unity and k_x is the projection of the propagation vector
parallel to the layers in the same units. We have also moved the
imaginary numbers around so that the expressions are superficially

different from those in Ref. 8. Note that for normal incidence, k_x is zero.

More insight into the optical properties may be obtained by Fourier analyzing the spatial variation of mean values of $\underline{\varepsilon}$ over a cycle normal to the Bragg planes. The optical properties of a periodic stratified medium near a Bragg reflection are almost identical to those of a medium with sinusoidal variation, wherein the amplitudes and phases of the varying components of the optical dielectric tensor are determined by Fourier analysis.[10] There is a complex 3X3 tensor of such numbers, $\underline{F}$, and a complex scalar term, f, which come from such analysis. In the case of non-cubic media, such as the ordinary cholesteric phase, we may also need a constant traceless tensor, $\underline{T}$, to describe the most general sinusoidally varying material as

$$\underline{\varepsilon}(h,k,l,z) = \varepsilon_s (1 + f)\underline{1} + \varepsilon_t (\underline{T} + \underline{F}).$$

The scalar f plays the same role as the scalar "structure factor" in X-ray diffraction. The tensor $\underline{F}$ plays a similar role but is generally negligible for x-rays. We will find that it is f rather than $\underline{F}$ that is usually negligible for cholesterics.

Fortunately the main features of $\underline{F}$ can be specified with one real and four complex numbers, rather than nine complex numbers, because it is symmetric and traceless, and the phase of the element F_{xx} may be set at zero unless small effects of the azimuth of the incident beam are considered. We have chosen to avoid complex numbers in $\underline{F}$ by defining it in terms of amplitudes and phase angles (in degrees). For a uniaxial cholesteric in the planar phase the elements of $\underline{F}$ may be read off from the elements of $\underline{Q}$. They may be defined by amplitudes and phase angles (parenthesized) as follows.

$$F_{xx} = 0.5 \ (0), \qquad F_{yz} = 0,$$

$$F_{yy} - 0.5 \ (180°), \qquad F_{zx} = 0,$$

$$F_{zz} = 0, \quad \text{and} \quad F_{xy} - 0.5 \ (-90°).$$

The nonzero elements of $\underline{\underline{T}}$ are

$$T_{xx} = T_{yy} = 1/3 \quad \text{and} \quad T_{zz} = -2/3$$

for the planar phase.

The elements of $\underline{\underline{F}}$ for various models of cubic blue phases are listed in similar fashion in Tables 2–5. $\underline{\underline{T}}$ is 0 for cubic blue phases. In Table 2 we present structure tensors, $\underline{\underline{F}}$, or Fourier analyses of variations normal to low-index Bragg planes, of the components of Q tensors averaged parallel to those planes in three models with 0^2 symmetry. The entries under F_{ii} are in the order F_{xx}, F_{yy} and F_{zz}. Those under F_{jk} are F_{yz}, F_{zx} and F_{xy}. Tables 3 to 5 show such data for 0^5, 0^8+ and 0^8- symmetry. We also did some

TABLE 2

Structure tensors, $\underline{\underline{F}}$, for 0^2 symmetry.

Model:	Simple Landau				Unrelaxed uniaxial; tube fraction =.00475				MSAB, $k_{11}=2k_{22}=k_{33}$; same; a_0 / pitch =0.64			
(h k l)	F_{ii}	ϕ_{ii}	F_{jk}	ϕ_{jk}	F_{ii}	ϕ_{ii}	F_{jk}	ϕ_{jk}	F_{ii}	ϕ_{ii}	F_{jk}	ϕ_{jk}
(1 0 0)	.25	0°	0		.2791	0°	0		.2886	0°	0	
	.25	180°	0		.2793	180°	.0007	81°	.2885	180°	0	
	0		.25	-90°	.0005	-113°	.2817	-90°	0		.2727	-90°
(1 1 0)	.0833	0°	0		.1040	0°	.0005	180°	.1043	0°	.0008	-2°
	.0833	180°	0		.1105	180°	.0006	-177°	.1114	180°	.0009	-41°
	0		.0833	-90°	.0078	0°	.1122	-90°	.0071	2°	.1102	-90°
(1 1 1)	0		0		.0172	0°	0		.0112	0°	0	
	0		0		.0165	-2°	0		.0123	-3°	0	
	0		0		.0351	180°	0		.0235	179°	.0005	66°
(2 0 0)	0		0		.0145	0°	0		.0116	0°	0	
	0		0		.0145	-2°	0		.0116	0°	0	
	0		0		.0296	178°	0		.0232	180°	0	
(2 1 0)	0		0		.0185	0°	.0147	180°	.0245	0°	.0162	1°
	0		0		.0008	176°	.0179	-99°	.0092	179°	.0170	91°
	0		0		.0177	-177°	.0103	-89°	.0153	-179°	.0151	-90°
(2 1 1)	0		0		.0049	0°	.0206	-161°	.0044	0°	.0202	2°
	0		0		.0047	-43°	.0212	-74°	.0041	-90°	.0207	86°
	0		0		.0100	160°	.0005	-94°	.0058	137°	.0022	-50°
(2 2 0)	0		0		.0123	0°	0		.0154	0°	0	
	0		0		.0125	179°	0		.0152	180°	0	
	0		0		.0014	-1°	.0145	-89°	0		.0181	-88°

TABLE 3

Structure tensors, $\underline{\underline{F}}$, for O^5 symmetry.

Model:	Simple Landau				Unrelaxed uniaxial; tube fraction =.00925				MSAB, $k_{11}=2k_{22}=k_{33}$; same; a_o / pitch =1.00			
(h k l)	F_{ii}	Φ_{ii}	F_{jk}	Φ_{jk}	F_{ii}	Φ_{ii}	F_{jk}	Φ_{jk}	F_{ii}	Φ_{ii}	F_{jk}	Φ_{jk}
(1 0 0)	0		0		0		0		0		0	
(1 1 0)	.1667 .1667 0	0 180°	0 0 .1667	-90°	.2133 .2203 .0065	0 180° 1°	0 0 .2190	-90°	.2080 .2076 .0004	0 180° -175°	.0012 .0018 .2257	-66° 4° -90°
(1 1 1)	0		0		0		0		0		0	
(2 0 0)	0 0 0		0 0 0		.0235 .0240 .0483	0 0 180°	0 0 0		.0084 .0084 .0169	0 0 180°	0 0 0	
(2 1 0)	0		0		0		0		0		0	
(2 1 1)	0 0 0		0 0 0		.0080 .0064 .0093	0 -95° 141°	.0405 .0401 .0045	176° -91° -52°	.0027 .0034 .0058	0 -33° 162°	.0362 .0382 .0002	38° 113° -64°
(2 2 0)	0 0 0		0 0 0		.0336 .0329 .0011	0 180° 166°	0 0 .0364	-90°	.0452 .0468 .0017	0 -180° -4°	.0006 .0016 .0523	-63° 3° -89°

TABLE 4

Structure tensors, $\underline{\underline{F}}$, for O^8+ symmetry.

Model:	Simple Landau				Unrelaxed uniaxial; tube fraction =.015				MSAB, $k_{11}=2k_{22}=k_{33}$; same; a_o / pitch =0.70			
(h k l)	F_{ii}	Φ_{ii}	F_{jk}	Φ_{jk}	F_{ii}	Φ_{ii}	F_{jk}	Φ_{jk}	F_{ii}	Φ_{ii}	F_{jk}	Φ_{jk}
(1 0 0)	0		0		0		0		0		0	
(1 1 0)	.1667 .1667 0	0 180°	0 0 .1667	-90°	.2122 .1963 .0159	0 180° 180°	0 0 .2169	-90°	.2155 .2109 .0047	0 180° -166°	.0014 .0016 .2084	-53° 14° -90°
(1 1 1)	0		0		0		0		0		0	
(2 0 0)	0 0 0		0 0 0		.0419 .0419 0	0 180°	0 0 .0294	-90°	.0289 .0289 0	0 180°	0 0 .0229	-92°
(2 1 0)	0		0		0		0		0		0	
(2 1 1)	0 0 0		0 0 0		.0491 .0488 .0014	0 -179° 106°	.0352 .0343 .0467	-42° 54° -90°	.0548 .0518 .0064	0 -174° 122°	.0326 .0309 .0513	139° -116° -89°
(2 2 0)	0 0 0		0 0 0		.0051 .0040 .0014	0 -178° 177°	0 0 .0066	-92°	.0066 .0101 .0036	0 180° -2°	.0009 0 .0073	5° -90°

TABLE 5

Structure tensors, $\underline{F}$, for O^8- symmetry.

(h k l)	Simple Landau F_{ii}	Φ_{ii}	F_{jk}	Φ_{jk}	Unrelaxed uniaxial; tube fraction =.010 F_{ii}	Φ_{ii}	F_{jk}	Φ_{jk}	MSAB, $k_{11}=2k_{22}=k_{33}$; same; a_o / pitch =0.80 F_{ii}	Φ_{ii}	F_{jk}	Φ_{jk}
(1 0 0)	0		0		0		0		0		0	
(1 1 0)	.3333	0	0		.1974	0	0		.1803	0	.0010	-83°
	.3333	180°	0		.1637	180°	0		.1712	180°	.0011	27°
	0		.3333	-90°	.0337	180°	.1624	-90°	.0090	180°	.1672	-90°
(1 1 1)	0		0		0		0		0		0	
(2 0 0)	0		0		.1688	0	0		.1747	0	0	
	0		0		.1688	180°	0		.1747	180°	0	
	0		0		0		.1672	-90°	0		.1784	-90°
(2 1 0)	0		0		0		0		0		0	
(2 1 1)	0		0		.0596	0	.0055	39°	.0567	0	.0059	-114°
	0		0		.0622	168°	.0062	-17°	.0614	165°	.0036	132°
	0		0		.0129	-84°	.0631	-95°	.0161	-81°	.0584	-98°
(2 2 0)	0		0		.0140	0	0		.0102	0	0	
	0		0		.0165	180°	0		.0088	180°	0	
	0		0		.0024	0	.0122	-90°	.0014	178°	.0071	-90°

computations assuming that the scalar f was of a magnitude that might be expected if the tubes in the MSAB model were less dense than the rest, using the ratio of densities for isotropic and cholesteric phases of cholesteryl nonanoate at the transition temperature.[5,11] We found that even for that material, which has very small optical anisotropy, the contribution of f was negligible in comparison to $\underline{F}$ in the low-order Bragg reflections studied. In many cases it was exactly zero, as indicated by Hornreich and Shtrikman.[3] For more anisotropic media the relative contribution of f would be even smaller. For brevity we have not tabulated those results here.

In Table 6 we show computed peak Bragg reflection intensities for circularly polarized light reflected straight back and for such light reflected at 45° from normal. (The 45° angle is measured from the z-axis within the liquid crystal or in an overlying medium, such as a prism, having the same value of ϵ_s. Without a prism, refraction at the air-to-sample interface would not allow such a large angle except in a transmission geometry.) These computations

TABLE 6

Normalized Bragg reflectances for circularly polarized light.

(h k l) Mult.	Pol. Sym.	Simple Landau				Unrelaxed uniaxial				MSAB			
		rt.	left	rt. 45°	left 45°	rt.	left	rt. 45°	left 45°	rt.	left	rt. 45°	left 45°
(1 0 0) 6	O^2	0	.154	.010	.335	0	.193	.012	.421	0	.193	.014	.423
(1 1 0) 12	O^5	0	.017	.001	.037	0	.030	.002	.065	0	.029	.002	.064
	O^8_+	0	.069	.004	.149	0	.117	.009	.255	0	.115	.007	.251
	O^8_+	0	.069	.004	.149	0	.109	.004	.238	0	.109	.007	.239
	O^8_-	0	.273	.018	.590	0	.072	.003	.161	0	.072	.004	.158
(1 1 1) 8	O^2	0	0	0	0	0	0	.001	.001	0	0	.001	.001
(2 0 0) 6	O^5	0	0	0	0	0	0	0	0	0	0	.001	.001
	O^8_+	0	0	0	0	0	0	.002	.002	0	0	0	0
	O^8_+	0	0	0	0	0	.003	.001	.007	0	.002	0	.004
	O^8_-	0	0	0	0	0	.069	.005	.152	0	.077	.005	.167
(2 1 0) 24	O^2	0	0	0	0	0	0	0	.001	0	.001	0	.002
(2 1 1) 24	O^5	0	0	0	0	0	0	0	.001	0	0	.001	0
	O^8_+	0	0	0	0	0	0.	.002	.003	0	.002	.002	.001
	O^8_+	0	0	0	0	0	.006	.003	.013	0	.007	.001	.017
	O^8_-	0	0	0	0	0	.009	.001	.021	0	.008	.001	.018
(2 2 0) 12	O^5	0	0	0	0	0	0	0	.001	0	.001	0	.001
	O^8_+	0	0	0	0	0	.003	0	.007	0	.006	0	.012
	O^8_+	0	0	0	0	0	0	0	0	0	0	0	0
	O^8_-	0	0	0	0	0	0	0	.001	0	0	0	0
Planar phase 2		0	.614	.040	1.341								

are based on the data in Tables 2-5. They are normalized for a medium in which the product of the number of Bragg layers and the ratio ϵ_t/ϵ_s is unity. The actual reflectance is proportional to the square of the number of correlated planes and to $\underline{\underline{T}}^2$ or $(\epsilon_t/\epsilon_s)^2$, so long as the reflectance is considerably less than unity. (Specifically, we computed the results for ten layers with the ϵ ratio equal to .01, and multiplied the results by 100. This kept the computed reflectances small enough to avoid significant "primary extinction", or loss of incident beam intensity by Bragg diffraction.) At the bottom of Table 6 we give the Bragg reflection intensities for the cholesteric planar phase for comparison with the other values, assuming small optical anisotropy in order to avoid problems mentioned above from the constant diagonal terms in $\underline{Q}$.

Comparison of Tables 2-5 with Table 6 will reveal the trend of the relation between average tensor variation and reflection inten-

sity. Certain Bragg planes listed in Table 6 give oblique reflection but no back-reflection. The 4X4 matrix for back-reflection
only contains off-diagonal components of $\underline{\underline{\varepsilon}}$ with z in their subscripts in second order. When incidence is oblique, they appear in
first order.[9] When non-zero components with z in their subscripts
appear in the expansions of $\underline{\underline{F}}$ shown in Tables 2-5, there will
usually be oblique Bragg reflection even if back-reflection is
forbidden.

Off-diagonal terms in $\underline{\underline{F}}$ contribute to preferential reflection
of light circularly polarized in one sense. In a few cases the
reflectance for right and for left circularly polarized light are
alike, as with the (2 0 0) planes with O^5 symmetry. This occurs
when there is modulation of ε_{zz} and when off-diagonal (rotation)
terms are small or absent. It should also be noted that the selection rule that exists in some cases against Bragg reflection of
circularly polarized light of one sense but not the other is not
valid when the direction of reflection is oblique.

The peak relative intensity of Bragg reflection caused by a
number, n, of parallel, coherently correlated Bragg planes is proportional to n^2 so long as the relative intensity is considerably
less than unity. Under the same conditions the breadth of the
reflection band is inversely proportional to n. Consequently the
area under a rocking curve for a monochromatic beam incident on a
thin monocrystal is proportional to n times the reflectance of a
single cycle. The intensity of light diffracted in one direction
from a beam of white light incident on a "powder" of randomly
oriented domains is also proportional to n times the reflectance of
a single cycle, but the intensity is also proportional to the multiplicity of equivalent planes in the crystal (6 to 48 in cubic
phases, 2 in the planar phase) in that case. Multiplicities are
listed on Table 6 under the Miller indices of the planes.

For blue light, cholesteryl nonanoate in the planar cholesteric phase near the temperature of transition to blue phase I has
ε_s = 2.207 and ε_t = -0.039.[12] If these numbers do not change much
on transition to the blue phase, about 57 parallel layers would be
required to give the peak reflection intensities shown in Table 6.
(Of course, the number greater than unity for the planar phase and
other numbers above about 0.2 in Table 6 would be inaccurate for so
many planes.) Other cholesterol compounds have similar values of
ε. Meiboom and Sammon have reported measurements of residual transmission through randomly oriented samples of such cholesterics.[13]
The dips in transmission versus wavelength as the wavelength becomes
shorter than that for normal (straight-back) Bragg reflection should
be roughly proportional to the product of the multiplicities and
the reflectances in Table 6. Figure 2 shows such products of backscattering intensities and multiplicities, computed for four MSAB
models for circularly polarized light, relative to the intensity to
be expected from the planar phase. The abscissa is the ratio of

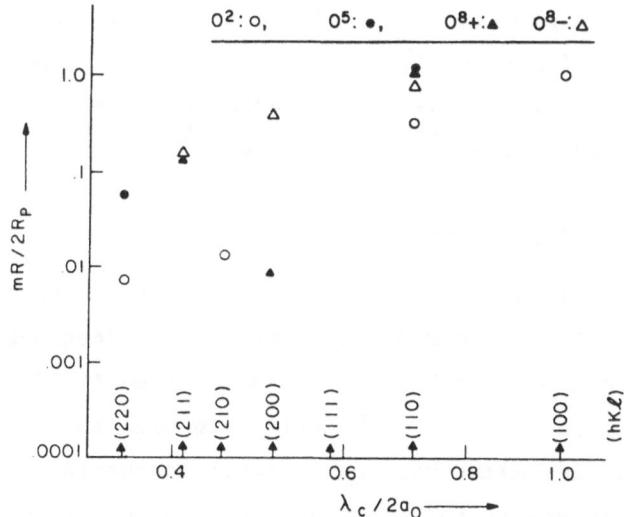

Fig. 2.   Log-log plot of multiplicity (m) times reflectance (R)
          relative to that for the planar phase, versus relative
          wavelength for four MSAB models of the blue phase in
          straight-back reflection.

the wavelength of light in the liquid crystal, λ_c, to twice the
unit cell dimension, a_o. The plot is on a log-log scale so that
experimental data plotted on a similar scale against wavelength in
any units can be laid on top of the plot for comparison without
renormalization.

In the case of a thick, very uniform monocrystalline sample the
width of the "Darwin" reflection band is proportional to the ampli-
tudes in $\underline{\underline{F}}$ and to $\varepsilon_t/\varepsilon_s$. The total number of planes is not important, so long as it is large, because "primary extinction" then
limits penetration of the beam into the sample. In this case the
numbers in Table 6 are proportional to the square of the intensity
of light reflected from a parallel incident beam of "white" light,
or to the square of the integrated intensity under the "rocking
curve" for monochromatic incident light.

Most chiral nematics other than cholesterol compounds have
positive rather than negative optical anisotropy and ε_t is much
larger. They give much stronger reflections in the blue phase with
the same number of planes, and broader Darwin reflection bands if
there are enough planes to give nearly total reflection. Diffraction data from such more anisotropic media have been reported by
Johnson, Flack and Crooker[14] and by Marcus.[15]

COMPARISON WITH AVAILABLE DATA

Although available data are meager, reflections have been
reported that must come from planes that the simplest Landau models
do not allow. Meiboom and Sammon[13] have observed two Bragg reflections above the "fundamental" reflections in blue phase I of
cholesteryl-chloride mixed with cholesteryl-nonanoate. The steps
in transmitted intensity that they observed agree very well with
the (1 1 0), (2 0 0) and (2 1 1) line strengths computed from the
MSAB model with O^8- symmetry. They do not agree at all with the
other three symmetries investigated or with the Landau model of O^8-

symmetry. The data are not of sufficient precision to distinguish between the unrelaxed and the relaxed MSAB models. The two steps in scattering intensity observed by Meiboom and Sammon for blue phase II are consistent with an MSAB model of O^2 symmetry but the evidence is insufficient to be conclusive.

Measurements of Darwin reflection bandwidths by Marcus[15] show a band from the planar cholesteric phase about four times as broad as the broadest band in blue phase I at the transition temperature. Blue phase II has a band about twice as broad as blue phase I at the transition. If ϵ_t does not change much across the phase transitions then, in order to fit Marcus' data, the reflection in Table 6 for blue phase II should be 1/4 as intense as for the planar phase, and that for blue phase I should be 1/16 as intense. The intensities for (1 0 0) planes of simple cubic O^2 symmetry computed with any of the models are about 1/4 that for the planar phase. Thus any of the O^2 models agree with Marcus' observed bandwidth. Marcus also suggests that blue phase II is simple cubic on the basis of other evidence.[16] Although the O^8- MSAB model comes closest of any model tested to agreeing with Marcus' observations, it gives about 1/9 rather than 1/16 the intensity of the planar phase. Perhaps the model is inaccurate or perhaps the value of ϵ_t changes abruptly at the phase transition.

Johnson, Flack and Crooker[14] show an incompletely resolved Darwin band in blue phase I which they attribute to (1 1 0) reflection. It appears to be of about the same width as the bands measured by Marcus. They also show reflections at wavelengths appropriate for (2 0 0), (2 1 1) and (2 2 0) planes. The fact that they see the (2 2 0) line suggests that if the MSAB model of O^8- symmetry is close to accurate, then their intensities are proportional to the first power of $\underline{\underline{F}}$. This means that they must have nearly total reflection even in that band.

The reflection intensities computed from specific models are consistent with the selection rules listed by Shtrikman, et al.[3] for normally incident light. However, specific model computations reveal very large variations among "strong" and "weak" reflections and give new information about obliquely incident light. The tables of Shtrikman, et al. omitted the (2 2 0) planes for O^2 symmetry, which we find to give stronger reflections than O^8. Pure symmetry arguments could not have predicted the differences between O^8+ and O^8-.

CONCLUSIONS

We believe that subjecting any specific model of a cholesteric blue phase to optical analysis of the type described will provide a very rigorous test of its validity. The data that we have found in the literature are very limited and no one liquid crystal has been exhaustively studied. Marcus' Darwin bandwidth data and the scattering intensities observed by Meiboom and Sammon are consistent with any of the models of O^2 symmetry for blue phase II. Available data suggest that blue phase I is at least very similar to the O^8- MSAB model but we have not made an exhaustive study of the possible models or symmetries. More detailed diffraction data for both phases will be very useful.

ACKNOWLEDGEMENT

We wish to thank S. Meiboom, M. Sammon and M. Marcus for supplying useful information bearing on this paper that has not been acknowledged elsewhere.

REFERENCES

1. The form of the O^5 tensor for Landau theory was worked out for us by S. Alexander and that for O^8 by W. F. Brinkman (private communications).

2. The 0^5 tensor was also published by R. M. Hornreich and S. Shtrikman, "Proc. Garmish Partenkirchen Conf. Liq. Cryst." W. Helfrich and G. Heppke, eds., Springer, New York, 1980, pgs. 185ff.

3. R. M. Hornreich and S. Shtrikman, Phys. Lett. A 82A, 354 (1981).

4. D. O. Smith, Opt. Acta 12, 13 (1965).

5. S. Meiboom, M. Sammon and W. F. Brinkman, "A Lattice of Disclinations: The Structure of the Blue Phase of Cholesteric Liquid Crystals" (Phys. Rev. A, in press).

6. M. Sammon, "Numerical Three Dimensional Relaxation of Liquid Crystal Director Fields", Mol. Cryst. Liq. Cryst. 89, 305 (1982).

7. S. Meiboom, J. P. Sethna, P. W. Anderson and W. F. Brinkman, Phys. Rev. Lett. 46, 1216 (1981).

8. D. W. Berreman and T. J. Scheffer, Mol. Cryst. and Liq. Cryst. 11, 395 (1970).

9. D. W. Berreman, Mol. Cryst. and Liq. Cryst. 22, 175 (1973) and Jour. Opt. Soc. Am. 63, 1374 (1973).

10. D. W. Berreman, Phys. Rev. B 14, 4313 (1976): Eqs. 18-25 show how this result is derived.

11. D. Demus, H.-G. Hahn and F. Kuschel, Mol. Cryst and Liq. Cryst. 44, 61 (1978).

12. B. Boettcher and G. Graber, Mol. Cryst. and Liq. Cryst. 14, 1 (1971).

13. S. Meiboom and M. Sammon, Phys. Rev. Lett. 44, 882 (1980) and Phys. Rev. A 24, 468 (1981).

14. D. L. Johnson, J. H. Flack and P. P. Crooker, Phys. Rev. Letters 45, 641 (1980).

15. M. Marcus, "Relative Order Parameters of Cholesteric and Blue Phases", Phys. Rev. A25, 2276 (1982).

16. M. Marcus, "Crystallography of 'Blue' Phase I and II", Phys. Rev. A25, 2272 (1982).

MESOMORPHIC POLYMERS AS THREE COMPONENT SYSTEMS

F. Cser

Research Institute for Plastics

H-1950. Budapest, Hungary

ABSTRACT

The nature of mesomorphic polymers is proposed to be explained
by the properties of a three component thermodynamic system, where
the components are not independent, they are coupled. The components of this system are the polymeric main chain, the rigid core
in the side chain and the flexible chain (spacer) between the
former ones. The spacer behaves as a plasticizer of the main chain,
it effects the phase transition of the rigid core as well as it
decouples the thermodynamic systems of the two previous components
from each other in the measure of its increasing length and
flexibility.

The effect of the spacer on the thermodynamic parameters of
the main chain as well as of the rigid core is investigated by expressing its length as molar fraction using data obtained on polymers built up from vinylic monomers substituted by a long aliphatic
chain as well as on different chemical class of mesomorphic substances.

The thermodynamic properties of both systems are led back to
the packing density of the monomeric units. The packing of the

molecules with relatively short spacer is determined by the rigid core, where the phase transition temperatures decrease with the increasing length of the spacer. From a critical length the packing is determined by the spacer, where the phase transition temperatures increase with the increasing length of the spacer. Nematic, smectic A, B and C structures are formed in the liquid part of the "eutectic" melt of the spacer and the rigid core, the other smectic phases are part of the solidus of the rigid core. A rigid chain increases the phase transition temperatures, it decreases the liquidity of the mesomorphic system.

The glass transition temperature and the clearing point of the mesomorphic polymers decrease with increasing length of the flexible chain segment between polymeric main chain and rigid core. The plasticized polymer chain shows the same effect as the flexible chain itself.

INTRODUCTION

There are many ways to form mesomorphic systems. Some sorts of molecules have thermotropic mesomorphic states. These molecules contain a rigid aromatic or polynuclear cycloaliphatic core connected to aliphatic chains (1). Some sorts of molecules form amphiphillic mesomorphic states (2). These molecules have an ionic head connected to an aliphatic chain. Some rod-like molecules form lyotropic states (3,4). Here the rod-like molecule is rigid and it is dispersed in a flexible fluid. Some polymers have thermotropic mesomorphic states, too. These polymers maintain aromatic rigid cores either in the main chain connected by flexible chain segments, or in the side chain of a periodic main chain connected by a flexible segments to the main chain.

Although the different types of mesomorphic systems seem to be very different, we can find one common feature in their build up: they have at least two well definable parts, a rigid core and a flexible surrounding.

The hard core may be a series of aromatic rings connected by rigid chemical groups (e.g. azo-, azoxy-, azomethine-, carboxilate-, vynilene-, etc.), it may be a polycondensed cycloaliphatic ring system (e.g. cholestane-), it may be a rigid helix (e.g. alpha helix of proteins, DNA, RNA, etc.) or it may be the coordination sphere of the ionic heads of amphiphyllic materials. The soft medium consists of either a paraffinic chain, polyoxyaliphatic chain or it may be a small molecular izotropic solvent.

The two parts of the molecules in the thermotropic systems (rigid core and flexible chain segments) form two thermodynamic subsystems which are not fully independent, they are coupled. The amphiphyllic and lyotropic systems may be also divided in two thermodynamic subsystems. The two subsystems have different inner temperatures as the intermolecular forces between the molecular parts are very different. The general structure of mesomorphic systems is represented schematically in Figure 1.

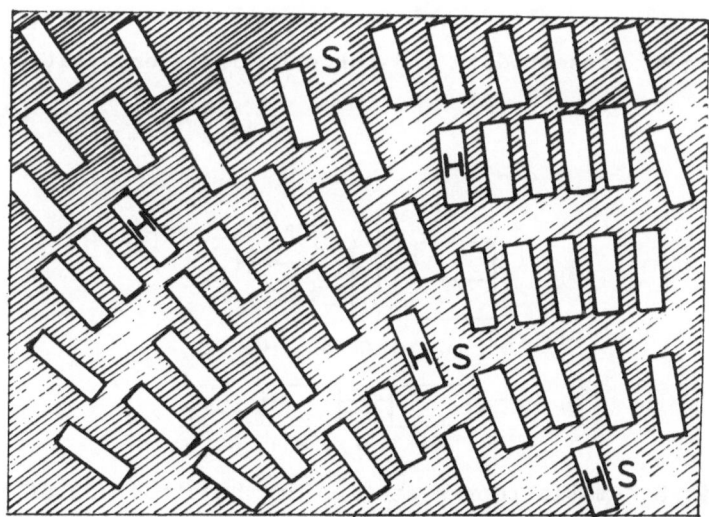

Fig. 1 Representation of the common features in mesomorphic
        systems. Hard cores (H) and soft medium (S) form
        two-component thermodynamic systems.

In the case of polymers with mesogenic groups (hard cores) in the main chain (polyazomethines, polyaramides, poly-Shiff bases, aromatic polyesters, etc.) the thermodynamic system may also be regarded as two coupled subsystems: the thermodynamic system of the hard cores and the thermodynamic system of the connecting flexible chain segments, the so-called "spacers".

In polymers with mesogenic groups in the side chain there are three different parts of the system with basically different inner forces: the thermodynamic system of the main chain, the thermodynamic system of the hard cores and the thermodynamic system of the flexible segments connecting the two previous ones. The situation is presented on Figure 2.

This paper deals with the more general case, i.e. with polymers with a mesogenic group in the side chain (case A in Figure 2). There are only a small number of homologous series of mesomorphic polymers where the monomeric units differ only in the length of the flexible segment.

This means the data for investigating the problem directly are too few. The problem should be divided in two parts. First we are going to present data and concept for interpretation of the

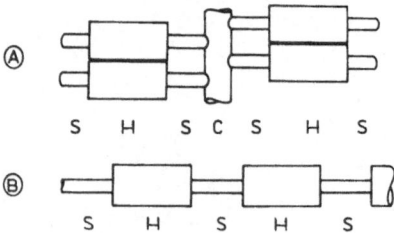

Fig. 2 Schematic representation of the structure of meso-
        morphic polymers by hard cores (H, C) and soft
        segments (S). The mesogenic groups are either in
        the side chains (A) or in the main chain (B).

effect of the flexible side chain to the main chain here after
we are going to do the same for the interaction of the soft and
hard parts of the molecules.

Interaction of the Main Chain and the Soft Side Chain

Many papers (5-9) have dealt with the effect of the aliphatic
side chains on the properties of comb-like polymers with different
main chains. It was established, that the glass transition temperature (T_g) of the polymers decreases with increasing length of the
soft side chain. At a given chain length characteristic to the type
of the main chain (acrylates, vinylesters, α-olefines etc.) the
crystallization of the side chains appears and the capability of
the soft segment to be crystallized becomes characteristic with increasing chain length. The T_g can also be detected in many cases
and the two parts of the polymers appear as a two phase system (6).

The usual way to investigate the interaction of two components
in a system is the representation of thermodynamic data as a function of the mole fraction of the components. It seems to be convenient to represent the transition temperatures of comb-like polymers as a function of the mole fraction of the soft segment. Eqn.
1 gives a formula for calculating the mole fraction of the soft
segments in the series of n-alkyl acrylates:

$$x_{CH_2} = \frac{n \times 14,027 + 1,008}{n \times 14,027 + 1,008 + 71,057} \qquad \text{Eqn. 1}$$

where n is the number of carbon atoms in the side chain.

In this equation the terminal methyl groups were considered
an isomorphous part of the chain in spite of the conception of
Flory (10) who regarded this group as nonisomorphous with the
interchain CH_2 groups. Kitaigorodskii (11) pointed out, that the
terminal groups determine the layer stacking in long chain paraffinic crystals, but they have no effect to the packing of the

chains within the layer. This means, the terminal methyl groups
are isomorphous with the interchain methyl groups. Figure 3 represents the melting point of n-alkanes as a function of mole
fraction of the interchain methyl groups with respect to the
terminal groups. The data were extracted from a handbook (12).
In the Figure we drew the straight line by the conception of Flory
using the melting enthalpy of a CH_2 group extrapolated from the
melting enthalpies of n-paraffines. The experimental points show
that the terminal methyl groups are rather isomorphous with the
interchain methyl groups, as they do not follow the Flory equation.

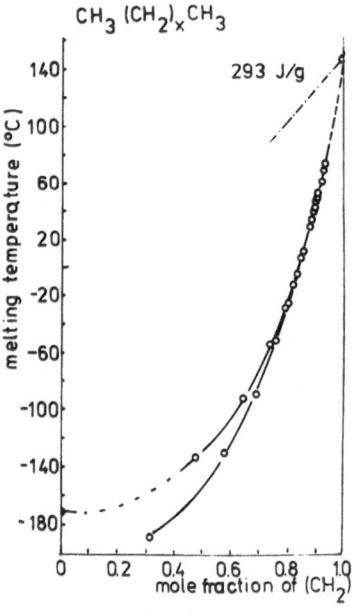

Fig. 3 Melting points of n-paraffines as a function of mole
        fraction of the methylene groups with respect to the
        terminal methyl groups. The straight line is calculat-
        ed by the Flory equation.

The transition points (T_g, melting point, T_m) of many comb-like polymers are represented on Figure 4 as a function of the mole fraction of CH_2 groups in the side chain with respect to the rest of the monomeric units. In this Figure the terminal methyl groups were separated from the interchain methylene groups and they were added to the rest of the monomeric units, as Reimschuessel (8) did.

The T_g decreases by increasing fraction of methylene groups up to x_{CH_2} = 0.6. Below this methylene content side chain crystallinity could not be detected. Above this concentration of methylene groups the melting point of the side chain crystals increases with increasing length of the side chain. The points scatter around a straight line which intersects the ordinate at the melting point of polymethylene. The melting points seem to be independent of the quality of the head of the monomer in this representation. The

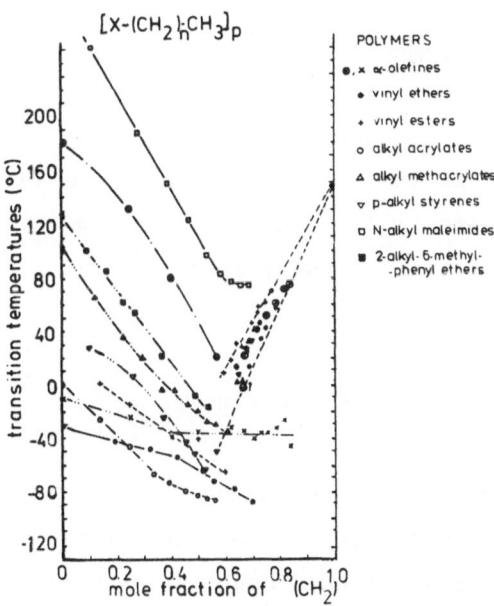

Fig. 4 Transition temperatures of some polymers with long paraffinic chains in their side chains as a function of the mole fraction of methylene groups in the side chain.

melting enthalpy of the methyleneic groups calculated from the
slope of the straight line was found to be 300 J/g which value
corresponds to the melting enthalpy of polyethylene with orth-
orhombic celle (290 J/g) (12). We can conclude that the main
chain with a given amount of soft segments behaves as a simple
liquid which solves the crystals formed from the paraffinic side
chains.

On the left side of Figure 4 the melting points of the crystals
of polyoleffines and the glass transition temperatures of the other
polymers including poly α-olefines (x) can be seen. If we regard
the transition temperatures of ply α-olefines only, they show an
eutectic-like behaviour. The glass transition temperatures are
approximately independent of the chain, the melting point curves of
the crystalline side chains and of the crystalline main chain
intersect at the glass transition temperature.

We have many possibilities to analyse the interaction of the
two parts of the monomeric units. Reimschuessel (8) regarded the
situation as a critical phenomenon. He defined a critical T_g, the
glass transition temperature of polymers with the greatest, number
of carbon atoms in the side chains where the side chain crystallization is yet absent (T_g^o). The integral form of the differential
equation (eqn. 2) fitted the measured points of many polymers well
(see Figure 5).

$$\frac{dT_g}{dM} = - K M (T_g - T_g^o) \qquad\qquad \text{Eqn. 2}$$

where K is a constant depending on the quality of the monomeric
units, M is the molecular mass of the monomer. In the integral
form of eqn. 2 an additional constant characteristic the quality
of the monomer also appears.

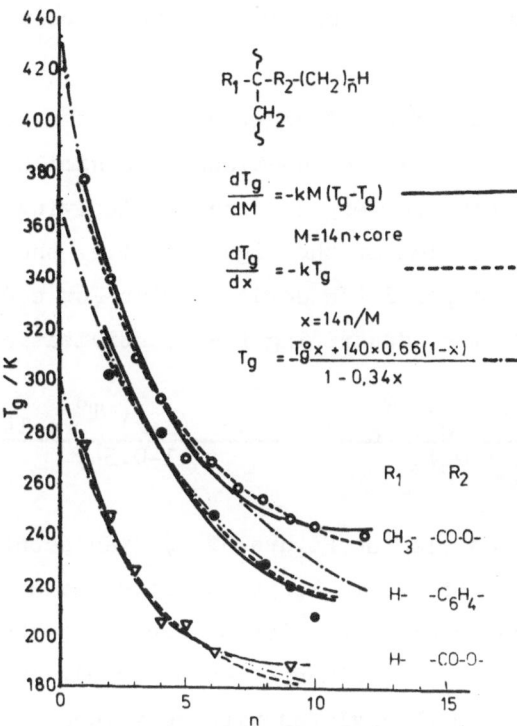

Fig. 5 Glass transition temperatures of some comb-like polymers
as a function of the length of the paraffinic side
chain.

We tried an other exponential equation on the base of the
mole fraction of the soft segments:

$$T_g = T_g^0 \quad \exp(-kx) \qquad \text{Eqn. 3}$$

where k and T_g^0 are constants and x is the mole fraction of methylene groups in the side chains including also the terminal ones.
The equation could also be fitted to the measured points with a
very good correlation coefficient. These curves are also represented on Figure 5. It is interesting, that T_g^0 was found to be
independent of the quality of the monomer with a value scattered
around 140 K ($\pm$ 5K). In literature this temperature often is

appears as the glass transition temperature of polyethylene
(13-15).

Now we have two different equations with the same correlation
coefficient (0.99!). The two equations have similar forms but
their physical meaning is very different. Applying the conclusion
drawn from the right side of the Figure 5, i.e. the main chain is
a simple liquid we applied the Gordon-Taylor-Fox equation. This
is the simplest form of describing the plasticization of polymers
(eqn. 4)

$$T_g = \frac{\alpha_1 \phi_1 T_g 1 + \alpha_2 \phi_2 T_g 2}{\alpha_1 \phi_1 + \alpha_2 \phi_2} = \frac{0.66 \ (1-x) \ T_g^\circ + 140x}{1-0.34x} \qquad \text{Eqn. 4}$$

For the curves given also on Figure 5. ϕ_2 was substituted by x,
$\alpha_1/\alpha_2 = 0.66$ and T_g^2 (infinite length of side chain) by 140 which
was the mean value obtained by fitting eqn. 3 to the experiments.
A very good fit to the experiments was observed. The glass transition temperature of methacrylates with very long side chains do not
fit the equation. These results are not in contradiction with the
previous conclusions, we can conclude now, that one part of the
soft segments behaves as plasticizer of the main chain. This means,
the thermodynamic system of comb-like polymers can really be regarded as the interaction of two semi-independent thermodynamic
subsystems. The polymethylene side chains form crystals interacting with the main chain in the form of low molecular crystalline plasticizers. We investigated many phase diagrams of an
amorphous polymer and a crystalline plasticizer (16-18) and their
forms were similar to those in Figure 4.

Interaction of the Hard Cores and the Soft Segments

For this investigations we chose the phase transition data
collected in the book of Demus et al (1). The way presented above
was followed and the phase transition temperatures were represented as a function of the mole fraction of methylenic groups of

the soft segments. Here only a small number of the extracted examples will be presented. Many similar diagrams can also be constructed from homologous series where the number of their members are sufficiently large.

Figure 6 shows the phase transition temperatures of p-alkoxy-benzoic acids as function of the mole fraction of the soft segment. In the upper part of the Figure the same data are shown as a function of the number of carbon atoms in the soft segment. This is a representation generally applied in the relevant literature.

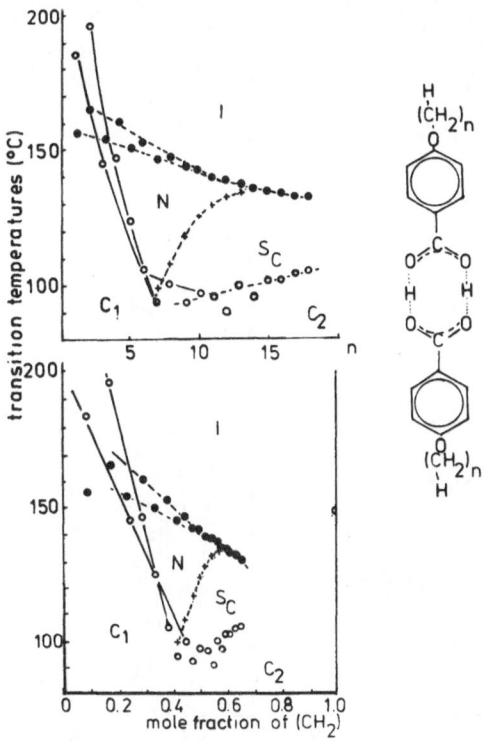

Fig. 6 Transition temperatures of p-alkoxy benzoic acids as a function of the length of the soft chain segments. Upper part shows the representation used in the literature. The lower part shows the representation as a function of the mole fraction of methylene groups with respect to the rigid core. o = melting points, ● = clearing points, + = transition from S_C to N. N = nematic, S = smectic, C = Crystalline, I = isotropic.

Linear change of the transition temperatures can be observed in the former case. The only exception is the transition temperature from S_C to N.

There are two hypothetic straight lines for the clearing points due to the well known even-odd effect. The line of the odd-homologous is continued for the members of longer chain length.

Figure 7 displays the transition point as a function of the mole fractions for two other series. The even-odd effect was al-

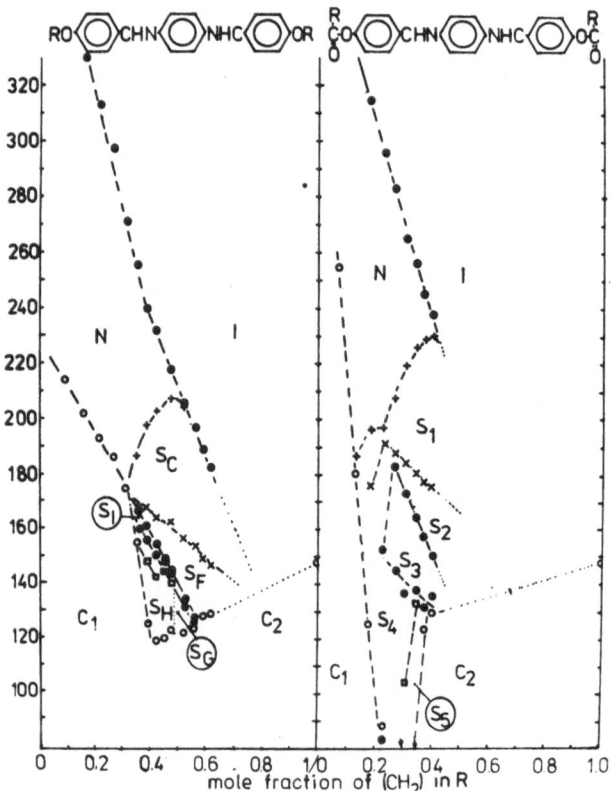

Fig. 7   Transition temperatures of bis-p-alkoxy- and bis-p-alkyl-carboxy-phenyl-p-phenylene azomethines as a function of the mole fraction of CH_2 groups in the soft segments. o = melting point, ● = clearing point, □, ◑, ◕, x, + = mesomorphic phase transition temperatures, C= crystalline, S = smectic, N = nematic, I = isotropic.

ready vanished in these compounds. Many straight lines may be
drawn in the system on the left side of the Figure. Wax-like smectic states are present at the compounds belonging to this series.
They can be found below the extrapolated line of melting points.
Neat-like smectic states are situated above the extrapolated line
of melting points (smectic C and F). The curve separating the nematic and the smectic C states has a form of a solubility curve of
two liquids.

Figure 8 shows the transition points of a homologous series
of p-alkoxy-phenyl-p-alkoxy-benzoate. We can see here, that the
two alkyl chains in the opposite parts of the molecules have the
same effect, although, they are not equivalent. The length of the
alkyl chains effects the transition points, it is not possible to
construct a universal diagram for the members with different chain
length in the two side of the molecules but with the same mole fraction of CH_2 groups. The intermolecular forces of the rigid cores
are much less in this series than in the previous one, since the
transition temperatures are also lower here.

The sloppe of the straight lines fitting the clearing points
best is less when the clearing points are smaller. If we extrapolate the straight lines of the clearing points of Figures 6-8
to greatest value of x_{CH_2} ($x \rightarrow 1$) the clearing point of the hypothetic mesomorphic state of polymethylene may be found. According
to the diagrams presented here, this temperature lies in the
vicinity of 80°C. By chance, there is a transition in the temperature range near 80°C of semi-crystalline polyethylene with a medium
line of a NMR signal (19).

Let us consider now the melting points. The melting points
are decreasing from the side of either aromatic or of the aliphatic
parts of the diagram with increasing mole fraction of the other component. We claim, there are two basic crystalline forms before or
after the minimum of the melting points in each series. They are

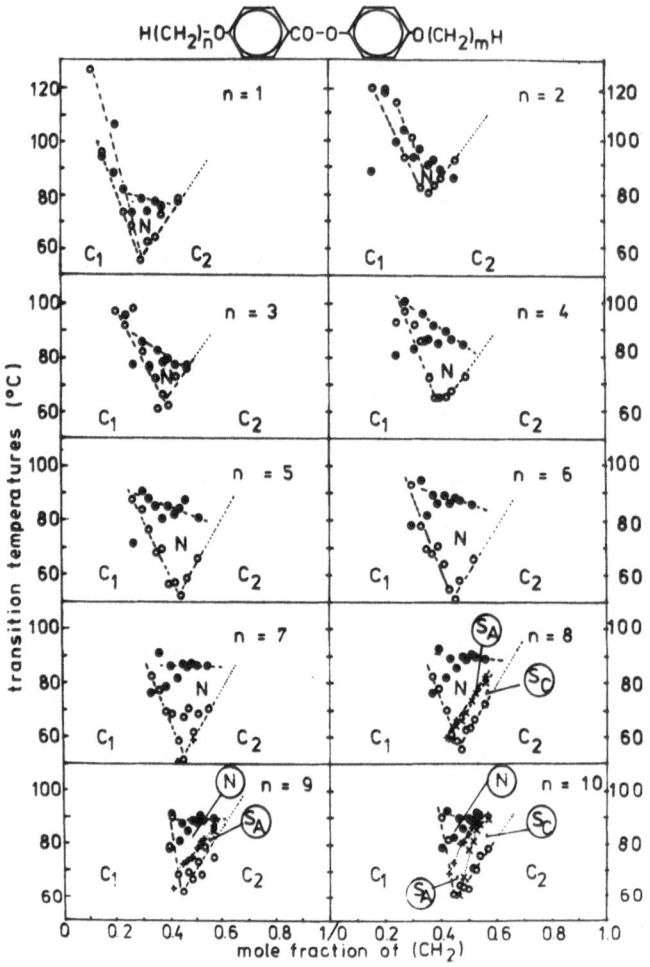

Fig. 8 Transition temperatures of p-alkoxy-phenyl-p-alkoxy-
       benzoates as a function of the mole fraction of CH_2
       groups in the soft segments. o = melting point, ● =
       clearing point, + = $S_C \rightarrow S_A$ transition, = S_A --N
       transition.

indicated by C_1 or C_2. In the crystals of C_2 the packing of molecules is determined by the packing of the aliphatic side chains as
it is evident from many of X-ray works done on substances with
longer paraffinic chains (21). The packing in C_1 is determined by

the aromatic cores. Many of X-ray works done on precursors of mesomorphic substances indicated a loose packing of side chains with
increased mobility when the aliphatic chains were short.

Figure 9 represents the shape of p-alkoxy benzoic acids using
van der Waals radii of 0.18, 0.12 and 0.15 nm-s for C, H and O atoms
respectively. The cross section of the different parts of the
molecules differ in each possible conformation of the molecules at
least in 0.03×10^{-18} m^2. If the packing is determined by the

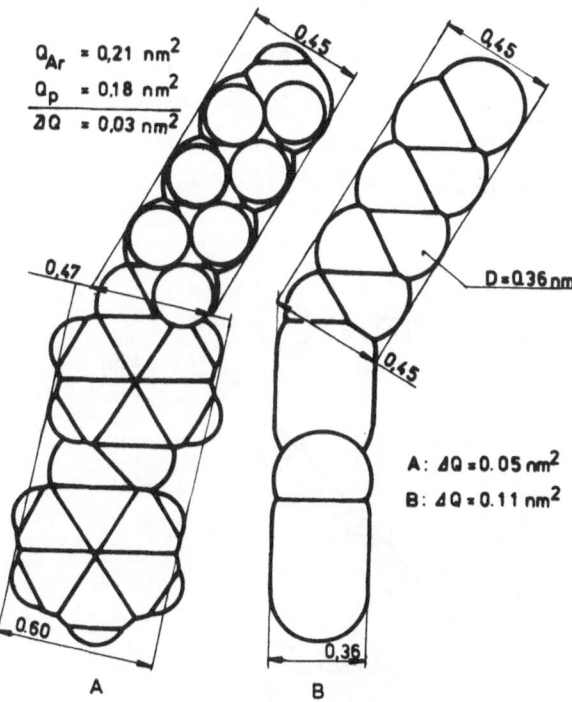

Fig. 9   van-der Waals shapes of molecules of p-alkoxy-benzoic acids. A = planar molecule, B = tilted molecule.
Q = cross section. Ar and p in subscript mean aromatic and paraffinic respectively.

aromatic core a free volume arises in the crystal with increasing
length of the aliphatic chain. This causes decreasing contact energy
with respect to one carbon atom and it causes increasing entropy of
the crystal too. Concrete data (22) of measured entropy change in
the melting processes of p-alkoxy benzoic acids are represented in
Figure 10. Linear entropy change as a function of increasing chain
length may be found only at greater chain length. The entropy de-
clines from the linear function of short chain lengths. The pre-
cisity of the measurements is not sufficient to indicate the same
effect at the greater chain lengths.

      Figure 11 contains analogous data for the homologous series
of p-methoxy-benzal-p-alkoxy anilides (1,22). This homologous
series has more or less similar properties.

      In the study of more than 20 series we did not find any be-
having very different as compared to the systems presented above,
thus this type of behavior seems to be general in thermotropic
mesomorphic substances.

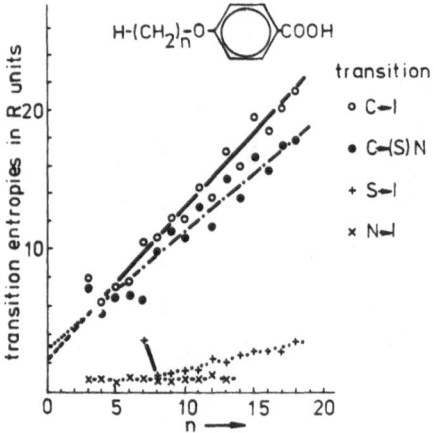

Fig. 10   Transition entropies of p-alkoxy-benzoic acids as a
          function of the number of carbon atoms in the soft
          segments.

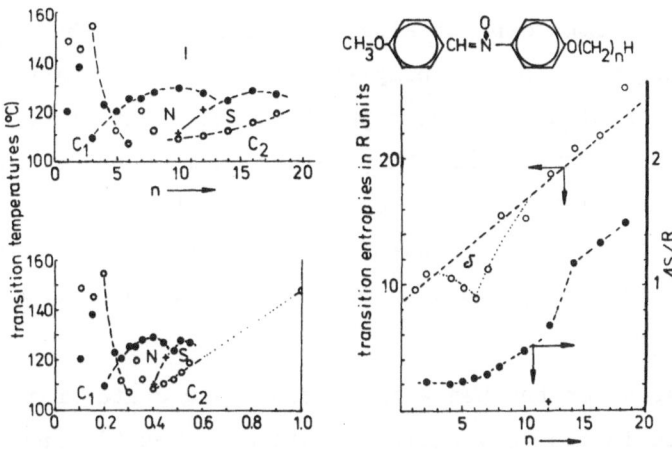

Fig. 11   Transition temperatures and entropies of p-methoxy-
          benzal-p-alkoxy-anilid oxides as a function of the
          length of the alkyl chain. o = melting, ● = clearing.
          + = S → N transition.

Let us now consider the mesomorphisms from the point of thermodynamics. As discussed above, the melting points of homologous
series have a minimal value at a given length of the soft segment.
There are two basic crystalline forms (C_1 and C_2) for the short
and for the long chains respectively. In this respect, the mesomorphic materials have many similarities to the eutectics. In the
crystals of mesogenic materials the two thermodynamics subsystems
are separated in the space, although, they are chemically coupled.
This is the situation in smectic states, too. The wax-like mesomorphic states are formed in the area C_1, the neat-like mesomorphic
states are situated in the liquid melt of the semi-eutectic melt.
The position of the mesomorphic states is consistent with their
fluidities.

The clearing points are nearly on a straight line. The melting point of some compounds is greater than the clearing point. In
this case the mesomorphic states are monotropic. The monotropic

mesomorphic states can not always be demonstrated by experiments. The clearing point of the polymethylene can be assumed by the extrapolation of the clearing point to the greatest mole fractions of methylene groups. This temperature is much below the melting point of the polymethylene, consequently, the absence of its experimental detection is evident. The experiments (relaxational, NMR, etc.) indicate some kind of transitions in this temperature range.

The area for the neat-like mesomorphic state in the diagrams are divided in two very different states: nematic and smectic states. The phase transition between them seemed to be a second order process in many experiments (23). Regarding the nature of the two phases this transition is the separation (or it is the mixing) of the two subsystems. In the smectic A (or smectic C) state the molecules are arranged in layers in which the centre of molecules are not fixed by symmetry. The hard and the soft segments form consecutive layers, the parts of the system are isolated. In the nematic state the layered structure is not present, the parts of the molecules are completely mixed. In cybotactic nematic phases some separation of the hard and soft segments is proposed by Chistyakov (24) and de Vries (25). The form of the curves separating the two phases (smectic and nematic) have also a form of the curves separating the homogeneous and heterogeneous phases in the phase diagram of two nonmiscible liquids with positive heat of mixing. The mixing of aromatic compounds with aliphatic compounds has positive heat of mixing according to the theories and the experiments (26).

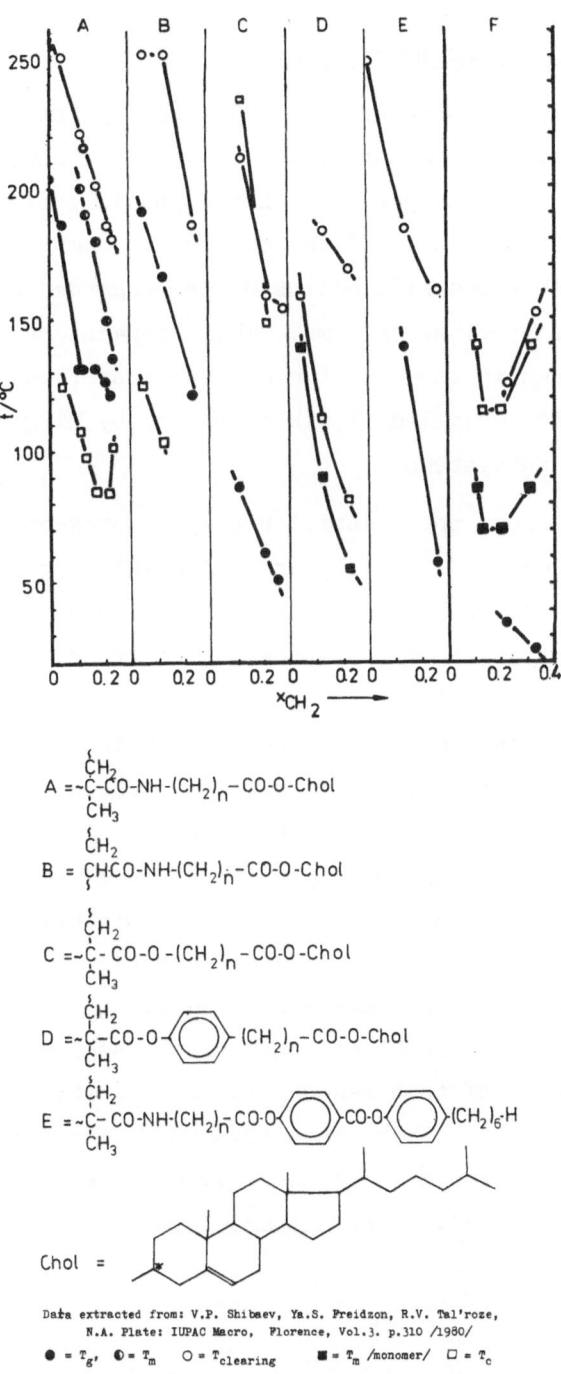

A = $\sim\underset{\underset{\displaystyle CH_3}{|}}{\overset{\overset{\displaystyle CH_2}{|}}{C}}$-CO-NH-$(CH_2)_n$-CO-O-Chol

B = $\underset{\xi}{\overset{\overset{\displaystyle CH_2}{\xi}}{C}}$HCO-NH-$(CH_2)_n$-CO-O-Chol

C = $\sim\underset{\underset{\displaystyle CH_3}{|}}{\overset{\overset{\displaystyle CH_2}{\xi}}{C}}$-CO-O-$(CH_2)_n$-CO-O-Chol

D = $\sim\underset{\underset{\displaystyle CH_3}{|}}{\overset{\overset{\displaystyle CH_2}{\xi}}{C}}$-CO-O-⟨◯⟩-$(CH_2)_n$-CO-O-Chol

E = $\sim\underset{\underset{\displaystyle CH_3}{|}}{\overset{\overset{\displaystyle CH_2}{\xi}}{C}}$-CO-NH-$(CH_2)_n$-CO-O-⟨◯⟩-CO-O-⟨◯⟩-$(CH_2)_6$-H

Chol =

Data extracted from: V.P. Shibaev, Ya.S. Freidzon, R.V. Tal'roze,
N.A. Plate: IUPAC Macro, Florence, Vol.3. p.310 /1980/

● = T_g,   ◑ = T_m ○ = $T_{clearing}$ ■ = T_m /monomer/   □ = T_c
/monomer/

Fig. 12   Transition temperatures of some mesomorphic polymers
with mesogene groups in the side chain.

Interactions in Mesomorphic Polymers

The data from literature fit well our hypothesis, that the
thermotropic mesomorphic substances as well the comb-like polymers
can be regarded as two component thermodynamic systems, where both
of the components are part of the molecule. Comb-like polymers
with hard cores in the side chains may be regarded as the combination of the two previous systems. The interactions here are even
more complex than in the pure two component systems, as one part of
the soft segments - called flexible spacers by Ringsdorf (27) - is
a part of both subsystems.

Figure 12 shows the transition temperatures of homologous
series of some mesomorphic polymers. A great decrease in the glass
transition temperatures and in the clearing point can be observed
with increasing length of the spacer. The transition temperatures
at x = 0, where there is no spacer, is extremely high. This indicates the direct interaction of the main chain and the hard core.
This type of macromolecule has a helical structure and the macromolecule has a rod-like form which is the origin of its anisotropy
(28). The anisotropy of this type of polymers vanishes parallel
to the chemical decomposition of the macromolecule, which is above
250°C in the case of polymers with a cholesteric group as a hard
core (29). The hard cores are connected to the main chain directly,
there is a relatively short and nonflexible segment between two hard
cores. The length of this segment is shorter (0.25 nm) than the
average diameter of the hard core (0.5 nm). The hard cores are in
a highly repulsive position in the polymer molecules. The macromolecules are collective thermodynamic systems (see Figure 13).

The collective nature of the helices decreases with increasing amounts of flexible chains between the main chain and the hard cores. The diameter of the soft chain is shorter than the diameter of the hard cores, it has more space around the macromolecule, the

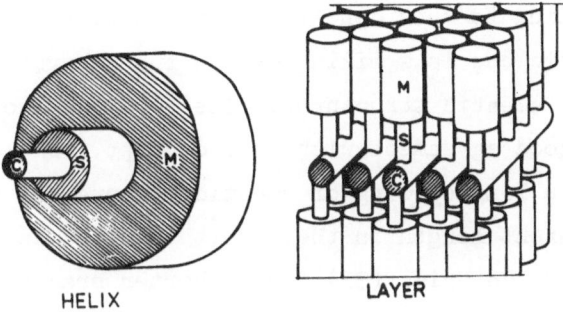

HELIX   LAYER

Fig. 13   Schematic representation of the extreme structures with short (helix) or long (layer) flexible spacer of mesomorphic polymers.

hard cores are yet not in repulsive position. The thermodynamic systems of the hard core and the main chain will therefore be decoupled (27) by increasing the length of the soft segment. As it was shown above, the soft segment acts also as a plasticizer of the main chain, which means, the main chain will also be a part of the soft segment with a length of the segment above its critical value (number of carbon atoms exceeds 6-7). In this case, smectic A structures were detected in comb-like polymers with cholesteryl group as a hard core according to the miscibility rule suggested by Sackmann and Demus (30). The clearing points of these smectic polymers were also high with respect to the clearing point of compounds with cholesterin as a hard core (31).

CONCLUSIONS

We considered the mesomorphic systems as two component thermodynamic systems. The reconsideration of the thermodynamic data of
thermotropic mesomorphic substances supported this hypothesis. We
also considered the comb-like polymers as two component thermodynamic systems. The data were also in good agreement with this hypothesis. According to this concept the transition from smectic
states to nematic one is a mixing of the two subsystems which were
isolated in the smectic structures. The X-ray data of Diele (32)
are in a very good agreement with this concept. Diele found a
linear change in long spacing of smectic mixtures of two components
with very different length in their soft segments as a function of
the composition. de Vries (33) found shorter spacings in all
smectic state compounds with oblique layers than the length of the
molecule. He stated: all the oblique layers should be tilted
ones due to this observation. The shortening of the molecules
in mesomorphic state is evident according to this hypothesis, since
the hard cores form close packed crystalline or liquid two dimensional layers in the smectic state, but the soft segments fill the
volume between two layers of hard cores as a liquid. They are not
in the fully extended form. The X-ray analysis of crystalline TBBA
(34) and of many other compounds (20,21) showed rotating end of
alkoxy chain. The motion of the soft segments in the smectic state
is more dominant, i.e. it has plastic (or even liquid) nature.

In conclusion the monotropic clearing point of polymethylene
can be defined as a result of the extrapolation of the clearing
points to the polymethylene end of the diagrams. There is a relaxational transition of semicrystalline polyethylene near to the
extrapolated temperature the nature of which is consistent with a
nematic state.

ACKNOWLEDGEMENT

The author is indebted to Miss Judit Horváth for her kind cooperation in collecting data for this study.

REFERENCES

1. D. Demus, H. Demus and H. Zaschke, Flussige Kristallen in Tabellen. Deutsh. Vrlg. Grundstoffindustrie, Leipzig (1974).

2. P. A. Winsor, in. Liquid Crystals & Plastic Crystals, Ed, G. W. Gray, P. A. Winsor, John Wiley, N.Y., Vol. 1, (1974) p. 199.

3. S. P. Papkov, V. G. Kulitshichin, Liquid Crystalline State of Polymers, Russ. Chimiya, Moscow (1977).

4. M. Panar and L. F. Beste, Macromolecules, 10, 401 (1979).

5. N. A. Platé and V. P. Shibaev, J. Polym. Sci. Macromol. Rev. 8, 117 (1974).

6. E. F. Jordan Jr., D. W. Feldeisen and A. N. Wrigley, J. Polym. Sci. A-1, 9, 1835 (1971).

7. N. A. Platé and V. P. Shibaev, Comb-Like Polymers and Liquid Crystals, Russ. Chimiya, Moscow (1980).

8. H. K. Reimschuessel, J. Polym. Sci. Polym. Chem. Ed., 17, 2447 (1979).

9. J. M. Barrales-Rienda and J. I. González de la Campa, J. Macromol. Sci. - Phys. B18, 625 (1980).

10. P. J. Flory, Principles of Polymer Chemistry, Cornell University Press, N.Y. (1953).

11. A. I. Kitaigorodskii, Organic Chemical Crystallography, Consut. Bureau, N.Y. (1961).

12. J. Brandrup and E. H. Immergut, Eds., Polymer Handbook. Interscience Pub., N.Y. (1975).

13. C. A. Sperati and H. W. Starkweather, Jr., Adv. Polym. Sci. 2, 465 (1961).

14. A. Nishioka and M. Watanaba, J. Polym. Sci. <u>24</u>, 298 (1975).

15. J. L. O'Toole, Modern Plastics Encylopedia, <u>45</u>, 48 (1968).

16. G. Hardy, F. Cser, G. Kovács, N. Fedorova and G. Samay,
    Acta Chim. Acad. Sci. Hung. <u>79</u>, 143 (1973).

17. F. Cser, K. Nyitray and G. Hardy, in <u>Mesomorphic Order in
    Polymers and Polymerization in Liquid Crystalline Media</u>,
    Ed. A. Blumstein, ACS. Symp. Ser -No. 74 Washington,
    D.C. (1978) p. 95.

18. F. Cser, K. Nyitrai and G. Hardy, J. Varga, G. Menczel, J.
    Polym. Sci. Symp. <u>69</u>, 91 (1981).

19. R. Kitamura and F. Horii, NMR Approach to the Phase
    Structure of Linear Polyethylene in <u>Advances in Polymer
    Science</u>, Vol. <u>26</u>, Springer Vrlg., Berlin (1978) p. 139.

20. R. F. Bryan, Proc. of Pre-Congress Symposium on Organic
    Crystal Chemistry, Poznan, p. 105 (1978).

21. R. F. Bryan, R. W. Miller and M. S. Shen, Abstr. Amer. Cryst.
    Assocn. <u>5</u>, 37 (1977).
    B. M. Graven and G. De Titta, J. Chem. Soc. Perkin Trans II.
    814 (1976).

22. E. M. Barrall II and J. F. Johnson, in <u>Liquid Crystals Plastic
    Crystals</u>, Ed. G. W. Gray and P. A. Winsor, John Wiley,
    N.Y., Vol 2, p. 245 (1974).

23. P. G. de Gennes, <u>The Physics of Liquid Crystals</u>, University
    Press, Oxford (1974).

24. V. I. Chistyakov and W. Chaikowsky, Mol. Cryst. Liqu. Cryst.
    <u>7</u>, 269 (1966).

25. A. de Vries, J. Phys. <u>36</u>, Col C1 (1975).

26. C. A. Cruz, J. W. Barlow and D. R. Paul, Macromolecules <u>12</u>,
    726 (1979).

27. H. Finkelmann, H. Ringsdorf, W. Siol and J. H. Wendorff, in
    <u>Mesomorphic Order in Polymers and Polymerization in
    Liquid Crystalline Media</u>, Ed. A. Blumstein, ACS. Symp.
    Ser. No. 74, Washington, D.C. (1978). p. 22.

28. F. Cser, J. Phys. <u>40</u>, Colloque C3-459 (1979).

29. F. Cser, K. Nyitrai and G. Hardy, in <u>Advances in Liquid
    Crystal Research and Applications</u>, Ed. L. Bata, Akadémia-
    Pergamon, Budapest-Oxford (1981) p. 845.

30. H. Sackmann and D. Demus, Mol. Cryst. Liqu. Cryst. <u>21</u>, 239
    (1973).

31. G. Hardy, F. Cser, K. Nyitrai and K. Kiss, this book.

32. S. Diele (Halle) personal communication, to be published.

33. A. de Vries, in <u>Advances in Liquid Crystal Research and
    Applications</u>, Ed. L. Bata, Akadémia-Pergamon,
    Budapest-Oxford, (1981) p. 71.

34. J. Docet, A. M. Levelut and M. Lambert, Acta Cryst. <u>B33</u>,
    1710 (1977).

COLLOIDAL CRYSTALS AND GLASSES*

P. M. Chaikin and P. A. Pincus

Department of Physics, University of California
Los Angeles, California 90024

ABSTRACT

We have investigated some of the aspects of the solidification
of charged polystyrene spheres (polyballs) both theoretically and
experimentally. We review the treatment of the interparticle interaction and test its validity with elastic constant measurements.
The crystal structure is then calculated as a function of the ratio
of the screening length to the particle separation, and the unusual
melting properties are explored. When polydispersed systems of
spheres are made the result is often a colloidal glass. We present
some elementary calculations related to the occurrence of the glass
phase and compare these results to the experimental observations.

Monodisperse latices made from charged polystyrene spheres
provide a unique system for the study of many basic problems in
condensed matter physics. The interparticle interaction is electrostatic in origin, spherically symmetric, controllable through the
electrolyte concentration and potentially completely understandable.

*Research supported by the Office of Naval Research under grant
#N00014-76-C-1078.

The fact that under suitable conditions the latices are seen to
crystallize is evidence of the importance of the interactions. The
opalescence which results from Bragg scattering of visible light
makes the crystalline state even more accessible for detailed study.
With a known and controllable interaction we should be able to calculate the crystal structure, the melting transition, and the possible existence of glass phases. The latter case is particularly
intriguing in the simplifying case of working with a spherically
symmetric potential.

There has been considerable previous work on these colloidal
crystals[1-8]. Theoretically they have been treated as interacting
via a screened coulomb potential.[5] In some cases numerical calculations have been done using the nonlinear Boltzmann-Poisson
equation.[6-8] The order-disorder transition has been computer and
experimentally observed[8] and the crystal structure predicted and
observed.[9,10] The shear modulus has also been calculated and
measured.[11-15]

In the present paper we will review some of the work that we
and other authors have done and comment additionally on the interactions, the melting transition and the formation of glasses. We
find that the nonlinear Boltzmann-Poisson (BP) equation can be well
represented by its linearized and much simpler form, the Debye-
Huckel equation, where the effective charge in the DH model is not
necessarily the charge on the sphere. We perform the exact numerical
calculations in a spherical model to obtain the "renormalized
charge" as a function of the actual charge. The crystal structure
is then calculated and we find FCC for large screening and BCC for
slight screening. We find that any phase diagram which involves
temperature has an anomalous "reentrant" behavior due to the temperature dependence of the dielectric constant of the solvent. In
connection with the formation of glass phases, we find that a considerable difference in the charges of the spheres ($\sim$ a factor of

three) is necessary to form a glass rather than a crystal.

The polystyrene spheres have sulfonate groups on their surface with an areal density of ~ 1 electronic charge/$(4 \times 10^3 \text{ A}^2)$. The solvated spheres with their surrounding compensating proton distribution can be described by the Boltzmann-Poisson equation:

$$\nabla^2 \Phi = \rho_0 e^{-\frac{e\Phi}{kT}}$$

(1)

which is a self-consistent mean field treatment of the proton distribution and the electrostatic potential. Being mean field it ignores dynamics, thermodynamic fluctuations and additional correlations, as well as specific interactions. Even with these limitations, this equation has not been analytically solved in three dimensions. Most commonly it is linearized by expanding the exponential in a power series in the charge density, or more accurately in $e\phi/kT$, and keeping only the first term. The result is the Debye-Huckel (DH) approximation which has the well behaved solution:

$$\Phi = \frac{Ze\,e^{-Kr}}{\epsilon r} \qquad K^2 = \frac{4\pi e Z e n}{\epsilon kT} \;.$$

(2)

Z is the charge on a sphere, ϵ is the dielectric constant of the solvent and n is the density of the polystyrene spheres. This is the screened coulomb or Yukawa potential which has a convenient short range analytic form.

The question which must be addressed is the validity of the DH approximation for the samples of interest. It must be compared with the exact numerical solutions to the BP equation. If the colloid is crystalline then a Wigner-Seitz unit cell can be associated with each polystyrene sphere as shown in Fig. 1a. Since this is related by translational symmetry to every other unit cell in the crystal, it must be charge neutral to preserve the charge

neutrality of the entire sample. At certain points on the surface
of the cell the electric field must be zero from symmetry. Since
the unit cells of cubic structures are polyhedra which resemble
spheres we follow the spirit of the calculation of Wigner and Seitz
in the band structure of Sodium[17] and approximate the unit cell as
being a sphere (Fig. 1b). In this case the charge neutrality requires that the electric field is zero at the surface of the sphere.

For the spherical Wigner-Seitz cell the solution is straightforward. The electric field is zero at the cell boundary and the
charge density is some finite value just inside this surface. The
potential and charge density can thus be computed from one numerical
integral from the outer surface to the radius of the latex particle.[8]
In Fig. 2 we have plotted the proton density from the numerical
computation using the BP equation. Also shown is the solution to
the DH equation with the same number of protons in the cell.

In considering the spherical unit cell model it becomes apparent
that the Debye-Huckel approximation must be appropriate for the
interactions. Since each cell is charge neutral, the electric
field is zero at the cell surface and the potential can be taken as
zero at this radius. Moving inside the cell the potential increases
slowly. Thus for some distance in from the surface of the cell the

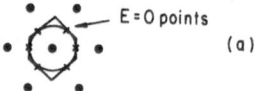

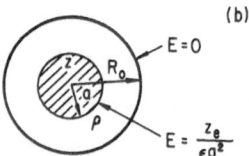

Fig. 1   (a) Wigner-Seitz cell in crystal. (b) Spherical Wigner-
          Seitz cell used in numerical solution for the Boltzmann-
          Poisson equation.

conditions are correct for the linearization of the BP equation.
However, since the charge cloud toward the center of the cell can
be highly distorted, the effective charge which comes into the DH
equation near the cell surface may be quite different from the
actual charge dissociated from the sphere. The correct effective
charge for DH is obtained by setting the charge density at the
surface equal to the exact value obtained from the BP equation.
The DH approximation with the calculated effective charge is shown
in Fig. 2 for comparison with the exact and conventional DH
treatments.

In Fig. 3 the effective charge per sphere is plotted as a
function of the actual charge per sphere from the calculations
described above. At low values of charge per sphere, the effective charge is the actual charge. At higher charge the effective
charge saturates. The saturation value of the effective charge is

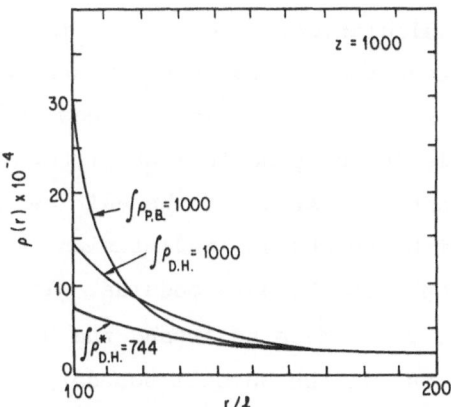

Fig. 2   Charge density of counter charge surrounding a colloidal
          particle. The exact calculation from the Boltzmann-
          Poisson equation is compared with the Debye-Huckel
          approximation using the actual counter charge and the
          effective charge. The Berne length 1 is about 7 Angstroms.

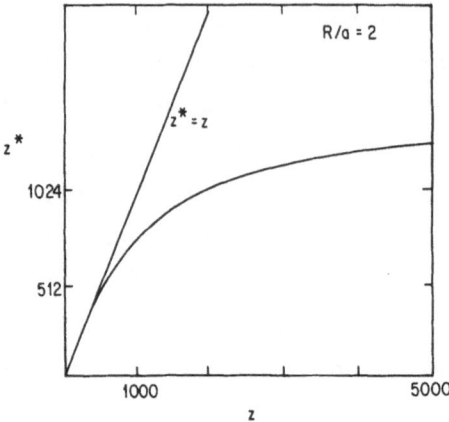

Fig. 3   The effective charge to be used in a Debye-Huckel
         approximation as a function of the actual charge of the
         colloidal particle for a volume fraction of $\sim 12\%$.

set by the condition that the chemical potential for the protons
must be constant throughout the system. As the real charge increases the screening length decreases. Roughly the potential at
the radius of the polyball cannot be larger than the chemical
potential at infinity: $Z^*e^2/a \sim k_B T \ln(Z^* R^3/1^3) \sim 27 k_B T$, where
$1 \sim 1$ angstrom is the thermal wavelength of the proton.

The numerical calculations also provide insight into the interactions and give direct results for such properties as the osmotic
pressure and bulk modulus. If we wish to know the effective force
between two polyballs in the spherical WS picture, the only way it
can enter is to change the radius of the unit cell. The energy
change associated with the change in dimensions, when repeated
throughout the crystal, is the bulk modulus. In the present treatment, the only force which acts to expand the WS cell is the osmotic
pressure of the protons at the WS cell boundary. Thus the osmotic
pressure Π and bulk modulus are simply related to the exactly
solved proton density p at the boundary by: $\Pi = p k_B T$. It is
interesting to note that in this picture the force between the
polyballs is entirely due to the pressure of the proton gas which

entropically always wants to expand. The description in terms of
osmotic pressure is another way of calculating the interparticle
interaction and the two ways must yield equivalent results, they
are not separate forces which add.

The interaction between the particles as given by the Boltzmann-
Poisson equation is always repulsive, as suggested by the osmotic
pressure argument above. If we imagine two polyballs which attract,
then the polyballs and their compensating charge must be contained
in some finite region of space. Since this is charge neutral there
are points outside this region with zero electric field. It takes
no work to move protons into this region, while the additional phase
space increases the entropy. The proton gas will thus expand, reduce
the screening between the polyballs, and cause the polyballs to
increase their separation. Unlike quantum systems, it is not
possible to have a bound state resulting from the interactions of
charged, counterion compensated spheres. The latex made of these
charged spheres will always want to expand to fill the solvent
volume unless other interactions, such as van der Waals or dispersion forces play an important role.

In order to experimentally test the theoretical prediction that
the DH approximation is appropriate with a renormalized charge, it
is necessary to measure the interparticle interaction strength.
Elastic moduli measurements provide this information most
directly.[11-15] The bulk modulus was measured by gravitational
compression in reference 18. Shear modulus measurements are more
relevant for studying the solid phases. In Fig. 4 we show the
measured shear modulus of a 0.109 micron sample with 2% volume
fraction as a function of electrolyte concentration.[15] The
completely deionized value can be used to determine the effective
charge per sphere. The charge per sphere was found to be ∿300 from
the shear modulus, 600 from the conductivity, and 4500 from the
conductometric titration.[16] The solid curve in Fig. 4 is the

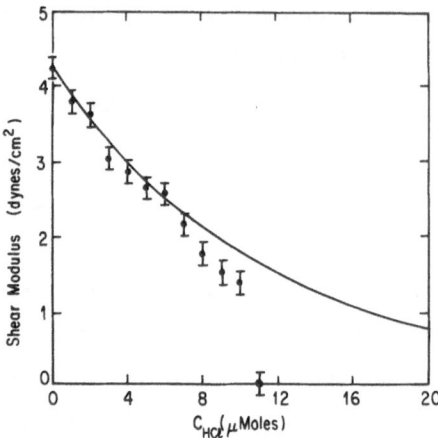

Fig. 4   Shear Modulus of a colloidal crystal of .109 micron spheres
         with 1% volume fraction as a function of HCl concentration.
         The points are experimental results. The line is the
         prediction of Debye-Huckel theory with the charge of the
         sphere set by the measurement at zero electrolyte con-
         centration. Note the first order transition at 11
         micromolar.

prediction of DH theory with the effective charge calculated from
the deionized sample. The agreement is quite reasonable and shows
that the interaction is adequately described by DH with an effective
charge which does not change appreciably with electrolyte concentration. The latter result has also been obtained from numerical
evaluation of the BP equation in the presence of electrolyte.[8]

What is particularly interesting in Fig. 4 is the sudden drop
in the shear modulus at an electrolyte concentration of 1.1 micro-
molar HCl. We take this as the first order melting transition of
the colloidal crystal. From studies of this and other samples the
drop in the shear modulus occurs at the same concentration of
electrolyte as does the disappearance of opalescence.

If the interparticle forces are known, it should be possible
to calculate the crystal structure of the solid phase. With the
short range potential of equation 2 it is possible to directly

perform the Madelung sum (with either a programmable calculator or
a computer) to find the energy of different configurations and
compare them. For fixed density (n) the ratio of the Madelung sums:

$$E = \frac{N}{2} \sum_i \frac{z_e^2 e^{-K r_i}}{\epsilon \, r_i} \quad = \quad \frac{N}{2} \frac{z^2 e^2 f_1 (K a_s)}{\epsilon \, a_s} \tag{3}$$

for different structures is a function of only one variable, Ka_s,
where $a_s = n^{-1/3}$ and f_1 is a specific sum for each lattice[10]. For a
short range repulsive interactions we expect that FCC with near
neighbors that are more numerous but further away than in the BCC
structure, will be the more stable. Since the ground state of the
classical Wigner crystal ($Ka_s = 0$) is BCC,[19] there must be a crossover.
In Figure 5 we show the difference in relative energies between BCC,
FCC, and HCP as a function of Ka_s.[20] We see that HCP is never favored,
BCC is stable for $Ka_s < 1.72$, FCC for $Ka_s > 1.72$, and that in the region
of interest the fractional difference between BCC and FCC is quite
small. Experimentally only the BCC and FCC phases have been observed
and most careful studies are in the low Ka_s regime and report BCC.

In the calculations above we have completely ignored entropy
considerations. To see which phase is stable we should compare the
Free energies, $F = E - TS$. If the entropy of one phase is higher, then
it will usually be the preferred phase at high temperatures. In the
present case there is a very interesting and important observation
that must be made. The internal energy E is strongly dependent on
temperature through the temperature dependence of the dielectric
constant of water. To first approximation the dielectric constant
results from free dipoles and therefore varies as $1/T$. Within this
approximation all of the energies in the problem increase linearly
with T. The partition function is then independent of T. Actually
water has a dielectric constant which varies as $1/T^{1+d}$ with $d \sim .5$.
Thus the energy terms vary more rapidly with temperature than the
entropy term and we have the interesting possibility of melting upon

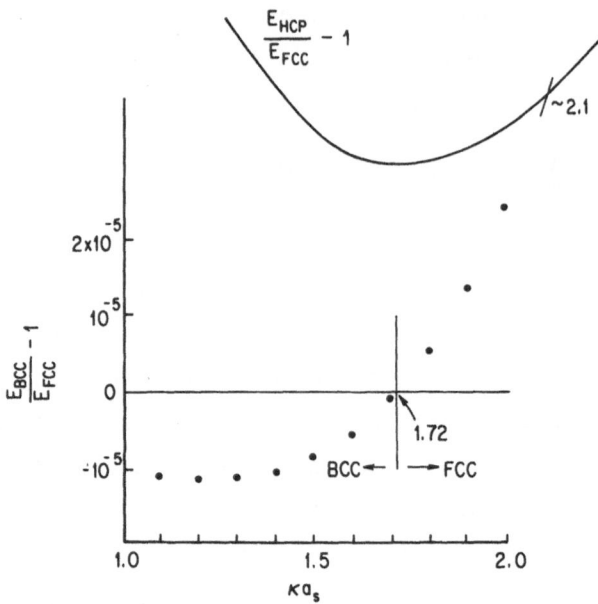

Fig. 5   Relative energies of BCC, FCC and HCP phases as a function
          of the inverse screening length times a characteristic
          interparticle distance.

cooling. If we were to include only nearest neighbor interactions
the form of the free energy difference would be:

$$\Delta F \sim T\left[A\, T^{d}e^{-B\, T^{d/2}} - \Delta s\right] \tag{4}$$

Such an expression predicts a transition to a low entropy state on
either heating or cooling - reentrant melting in the case of the
liquid solid transition.

While one of the most intriguing aspects of these latices are
their monodispersity, another interesting question is the state which
results from well characterised mixtures of spheres with different
charges and diameters. At the present we are only able to make
mixtures of commercially prepared spheres and we therefore vary the
charge and the diameter at the same time. However, at the particle
concentrations at which we are working the diameter should not matter.

Typically the interparticle distance is 5 times the radius so we are
far from a hard sphere transition. If we mix two latices which
individually form crystals there are several possibilities: the
particles can phase separate into two crystalline regions, they can
form a well defined compound analogous to an intermetallic, or they
can form a glass. In our investigations we have found only the latter
two.

Two sets of experiments have been performed.[15,16] In the first
we used mixtures of .109 micron spheres with .220 micron spheres.[15]
The second set used .109 and .089 particles. The particle concentrations were equal in the two starting colloidal crystals so the
particle concentration was the fixed parameter. The deionized latices
were mixed and then measured. In the case of the .109/.89 micron
mixtures the samples always exhibited strong Bragg scattering and a
shear modulus indicating crystalline structure and probably a
substitutional alloy. The .109/.220 mixtures with fractional concentrations more than 5% from either end of the phase diagram showed
no Bragg scattering or opalescence when examined with a laser beam
or under a metallurgical microscope. The latter samples did however
produce strong resonances in our shear modulus apparatus and their
shear modulus is shown in Figure 6.[15] Plotted as the solid line in
this figure is the most elementary mean field calculation for the
modulus using the measurements of the colloidal crystals to obtain
the effective charge per sphere.

The absence of long range order as probed by Bragg scattering
in a sample with a finite shear modulus serves as an operational
definition of a glass. It is perhaps unusual to find a glass in a
system with spherically symmetric interactions and a well defined
bimodal distribution of charges. The samples probably represent a
quenched condensed glass. Some evidence for this comes from the
observation that samples which have been left undisturbed for periods
of up to a year show the formation of small crystals. While we have

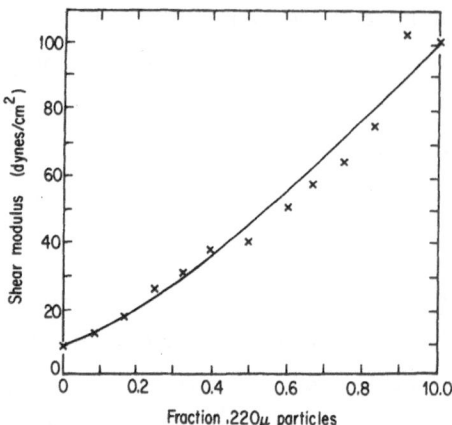

Fig. 6   Shear modulus for a series of mixtures of .109 and .22
          micron spheres. The opalescence vanishes for ∿5%
          concentration from either end.

not yet been able to study the structure in these crystallites in
detail, we have taken a sample from one of the crystallites for
electron micrography. We have found that the crystallite contained
∿ the same number of spheres of each size (the same as the initial
mixture) this suggests that a compound has been formed.

   If we make a mixture with fraction f_1 of spheres of charge Q_1
and f_2 of spheres of charge Q_2 it would appear as a well defined
problem to calculate at what stage we should expect the loss of long
range order and the presence of the glass phase. While we have not
been able to come close to solving this problem exactly we have a mean
field like calculation for obtaining an approximate solution. We
can imagine the loss of long range order in the glass phase as
similar to the melting of a crystal by molecular motions which
produce an rms displacement which is some sizable fraction of the
unit cell dimensions. A little algebra gives the deviation of the
interparticle spacing for particles interacting via a Yukawa

potential as:

$$\left\langle \frac{\Delta r}{r} \right\rangle = \langle \delta \rangle = f_1 f_2 (Q_2/Q_1 - 1 + Q_1/Q_2 + 1)/(Ka_s + 1)$$ (5)

The Lindemann criteria for melting is that $\langle \delta \rangle \sim 0.2$. Applying the same criteria for the glass we find from equation 6 that the ratio Q_1/Q_2 must be less than $\sim .3$ for a 50% mixture of spheres with $Ka_s = 1-2$. Experimentally we have results for the two sets of samples which indicate that the critical ratio for glass formation in a 50%-50% mixture is between .25 and .6 since the latter remained crystalline while the former was a glass.

Finally we note that there is a good deal of interesting hydro-dyanmics to be studied in the colloidal crystals. They are highly nonlinear in their response to shear. At low shear the crystals respond elastically. As the shear is increased the response is plastic and it is possible to observe the motion of dislocations as the crystal begins to yield. At a shear stress which exceeds the (low) yield stress of the solid the crystal flows. Recent studies by Clark and Ackerson[21] indicate that higher shear melts the crystal in an unusual transition in which two dimensional order is retained in sheets which slip over one another. We have also observed unusual behavior in moderate oscillatory shear in a tube.[22] In this case the colloidal crystal forms a periodic array of well oriented crystalline regions along the length of the tube. Experiments indicate that this is a hydrodynamic instability induced and probed by the colloidal crystal.

In conclusion we have shown that charged colloids present a unique system for studying a wide variety of fundamental problems in solid state physics. The interparticle interactions can be directly probed by shear modulus measurements and tend to confirm the idea that the screened coulomb interaction is appropriate as long as a renormalized effective charge is used rather than the actual charge on the surface of the sphere. The structure calculation seems to be in agreement with experiment, but a careful study of structure as a function of the screening length is needed to fully test our understanding. There remains a great amount of work to do in the areas of glass and compound formation and this could be facilitated by being able to make spheres with a quasi-continuous range of charges. The interesting hydrodynamics which can be studied with these strongly interacting latices is also just beginning.

We would like to acknowledge useful discussions with W. Dozier, H. Lindsay and P. Goldbart.

REFERENCES

1.   H. J. Van Den Hul and J. Vanderhoff, J.C.I.S 28, 336 (1968).

2.   P. A. Hiltner and I. M. Krieger, J. Phys. Chem. 73, 2386 (1969).

3.   D. W. Schaefer and B. J. Ackerson, Phys. Rev. Letters 35, 1448 (1975).

4.   R. Williams and R. S. Crandall, Phys. Lett. 48A, 225 (1974).

5.   E. J. W. Verwey and J. T. G. Overbeek, Theory of Stability of Lyophobic Crystals, Elsevier, Amsterdam, 1948.

6.   I. Snook and W. Van Megen, Faraday Transactions II 72, 216 (1976).

7.   T. Ohtsuki, S. Mitaku and K. Okano, Japan. J. Appl. Phys. 17, 627 (1978) and T. Ohtsuki, A. Kishimoto, S. Mitaku and K. Okano, Japan. J. Appl. Phys. 20, 509 (1980).

8.   P. Grant, P. Pincus, D. Hone, G. Morales, S. Alexander and P. M. Chaikin, to be published.

9.   J. Madeirose Silva and B. J. Mokross, Solid State Commun. 33, 493 (1980) and Phys. Rev. B21, 2972 (1980) and G. L. Hall, Phys. Rev. B24, 2881 (1981).

10. P. M. Chaikin, P. Pincus, S. Alexander and D. Hone, to be published in J. Colloid and Interface Science.

11. S. Mitaku, T. Ohtsuki, K. Enari, A. Kishimoto and K. Okano, J. Jap. Appl. Phys. 17, 305 (1978).

12. E. Dubois-Violette, P. Pieranski, F. Rothen and L. Strzlecki, J. de Physique 41, 369 (1980).

13. W. B. Russel and D. W. Benzing, JCIS 83, 163 (1981).

14. J. F. Joanny, J. Colloid Interface Sci. 71, 622 (1979).

15. H. M. Lindsay and P. M. Chaikin, to be published in J. Chem. Phys.

16. H. M. Lindsay, private communication.

17. E. Wigner and F. Seitz, Phys. Rev. 43, 804 (1933) and 46, 509 (1934).

18. R. S. Crandall and R. Williams, Science 198, 293 (1977).

19. L. L. Foldy, Phys. Rev. B3, 3472 (1971).

20. We would like to thank Mr. P. Goldbart for calculating the HCP Madelung sum.

21. N. A. Clark and B. J. Ackerson, Phys. Rev. Letters 46, 123 (1981).

22. W. D. Dozier and P. M. Chaikin, to be published in J. de Physique.

SELF DIFFUSION CONSTANT AND VISCOSITY OF CHARGED POLYSTYRENE
COLLOIDS*

B. Dozier, H. M. Lindsay and P. M. Chaikin

Department of Physics
University of California
Los Angeles, California 90024

and

H. Hervet and L. Leger

College de France
Paris, France

ABSTRACT

We have used the "Forced Rayleigh scattering" technique with
spiropyran dyed polystyrene spheres in aqueous suspension, to measure the self diffusion as a function of the strength of interaction
between the particles. The latter quantity was modified both by
adding electrolyte and by changing the particle concentration, as
well as by using spheres with different diameters and charges. We
find that the diffusion constant decreases monotonically from the
Stokes value as the repulsive interaction is increased, until the
interactions are sufficiently strong to form a colloidal crystal.
At this point there is a first order transition indicated by the
decrease of the diffusion constant by ∿ four orders of magnitude

*Research supported by ONR under grant #N00014-76-C-1078

and the presence of Bragg reflections. These measurements are
compared with viscosity measurements done in the liquid and solid
states, with quasielastic light scattering measurements and with
models for diffusion and viscosity in interacting systems.

Latices formed from aqueous suspensions of charged monodisperse
polystyrene spheres appear to be ideal model systems to test our
understanding of the solid and liquid states of matter. The interaction between these particles can be well characterized and described
by a screened coulomb potential. This short range spherically symmetric potential is also experimentally controllable by changing
the electrolyte concentration which determines the screening length.
With such a system it then becomes possible to calculate and inter-
relate the properties of the interacting particles as they form different phases and to study the phase transitions.

There has been a great deal of work on these systems, particularly in the ordered phase[1-7]. Due to the purely repulsive coulomb
potential the particles will form the classical analog of a Wigner
crystal with the particles arranged on a lattice with long range
order. The typical lattice spacing studied is ∿0.5 micron so that
the crystals Bragg scatter visible light and thus appear opalescent[1-3]. Using monochromatic light from a small laser the crystal
structure is readily obtainable from the position of the Bragg spots.
The existence of the ordered phase has also been shown by a variety
of elastic constant experiments which not only indicate the existence
of a finite shear modulus characteristic of the solid, but also provide a measurement of the strength of the interparticle interaction[5-7]. The mechanical properties of the liquid state have been
investigated by fewer authors although the viscosity has been measured after melting[6].

A more detailed study of the liquid state has been done with
the technique of quasi-elastic light scattering (QELS)[8-11] where
the monodisperse nature of the spheres makes them ideal candidates

for the diffusion studies in the highly screened non-interacting
state. Both QELS and classical light scattering studies have shown
that colloid samples with large ratios of particle separation to
particle diameter show liquid like structure factors in the wave-
vector (q) dependence of the total scattered intensity and the effective diffusion constant. The structure factor goes to a constant as the interparticle interactions are reduced by increasing
the temperature or adding electrolyte.

We have performed experiments treating the mechanical and
dynamic properties in both the solid and liquid phases. Two different light scattering experiments were performed: QELS and "Forced
Rayleigh Scattering" (FRS)[12]. The QELS experiment measures diffusion processes which are related to both single particle and collective motions. The FRS experiment, using labelled particles, measures the self-diffusion directly. Our results on the QELS experiments are very similar to what has been previously obtained[8-11].
We will therefore report only the FRS experiments. The shear modulus[5-7] measured in the solid phase gives us knowledge of the interaction strength as a function of electrolyte concentration which
can be extended into the liquid state. The measured viscosity can
then be related to the elastic constant and the self-diffusion constant to obtain a more complete picture of the liquid state.

The polystyrene latex samples were purchased as 10% nominal
suspensions from Dow Diagnostics[13]. The concentrations were checked
by weight measurements of dried samples, by Bragg scattering from
the crystalline state, and QELS from the liquid state. The samples
were diluted in deionized water to the appropriate concentrations
and then deionized over a bed of ion exchange resin for periods of
several days to several months. The colloids formed from the .109
micron spheres would begin to crystallize within an hour and were
completely crystallized within 1-2 days. The colloids made with
the 0.038 micron carboxylate modified spheres (up to a 10% volume

fraction) were never observed to form a solid by any measurement we performed, although they were allowed to deionize several months.

The charge of the spheres was determined by conductivity measurements and conductometric titration for both sets of particle sizes. The measured charge was 600 electronic charges for 0.109 microns and 1600 charges for the 0.038 micron spheres. However, for the .109 micron spheres shear modulus measurements in the completely deionized solid provided the most useful measurement of the effective charge to be used in a Debye-Huckel approximation to the potential.[7]

The measurement of the self-diffusion constant by FRS is new to these systems. The technique is more fully described in ref. 12. The 350 nm line of a Argon ion laser is split and then recombined on a colloidal sample. The (adjustable) path length difference between the beams results in the formation of an interference pattern and a sinusoidal intensity variation. The colloidal particles contain a photochromic dye which changes its absorption in the red when excited by the UV. The sinusoidal intensity of UV thus produces an absorption grating in the visible which can easily be probed by diffracting a beam from a He-Ne laser. The intensity of the diffracted spot decays as the absorption grating fades. If the lifetime of the excited dye is sufficiently long then the decay is governed by the diffusion of the dyed spheres from the dark to the light regions. For the interference spacing used ($\sim$5 microns) this displacement of a sphere over a large distance, past neighboring spheres, corresponds directly to a self-diffusion process.

The dye that we used was spiropyran. The spiropyran was dissolved in xylene, and then mixed with a master suspension of polystyrene spheres for a period of 1-2 days. The latex was then fractionally distilled under vacuum, centrifuged, micron filtered and ion exchange resin was added. The excited state lifetime of the spiropyran was $\sim$8 minutes. Thus the smallest diffusion constant that could be measured was $\sim 10^{-12}$ cm^2/sec.

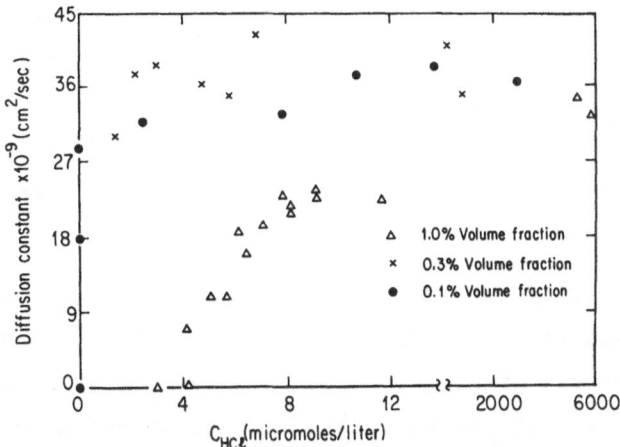

Figure 1. Self-diffusion constant of 0.109 micron polystyrene
spheres as a function of HCl concentration for several
volume fractions. Note the break in the abscissa to show
the limiting values at high electrolyte concentration.

In figure 1 we show the self-diffusion constant (D_s) of a
series of latices prepared with 0.109 micron spheres as a function
of the concentration of added HCl. At high electrolyte concentration the screened interaction is weak and the curves all approach
the Stokes limit (kT/6πnr, where n is the solvent viscosity). At
lower electrolyte concentration the interaction effects are important and D_s decreases. Note that this result is in sharp contrast
to the results of a QELS experiment which at long wavelength would
show a large increase in D_{eff}, due to the increased osmotic pressure
or equivalently, to the strong reduction of the structure factor at
low q.[8-11]

The only sample shown that solidified in the cuvette was the
1% volume fraction. The diffusion constant of this sample was less
than we could measure $(D_s < 10^{-12})$, and the sample was opalescent.
Upon adding HCl, the opalescence dissappeared abruptly and D_s jumped
by ∿4 orders of magnitude, indicating the first order transition to
the liquid state. A similar first order transition can be seen in
the drop of the shear modulus as the transition is approached from
the solid phase.[7]

Although contamination from the cuvette and the atmosphere was sufficient to melt the crystals of lower volume fraction, it was possible to observe solid regions (opalescence) immediately after the sample was added to the cuvette or upon addition of ion exchange resin. In all cases the diffusion constant in the solid phase was $<10^{-12}$ cm^2/sec.

The viscosity was measured with an Ostwald viscometer which uses the drain time of a fluid through a capillary under a quasi-constant pressure gradient. Under the conditions of the experiment the shear stress at the tube wall was ~25 dynes/cm^2 which is probably higher than the yield stress of the solid. Thus the solid "shear melted" and flowed so that no transition was observed in the viscosity as the sample melted with added HCl.[14]

In figure 2 we have plotted the measured viscosity as a function of the electrolyte concentration for the 0.109 micron spheres at a volume fraction of 1%. For a highly screened interaction the viscosity approaches that of pure water at the measurement tempera-

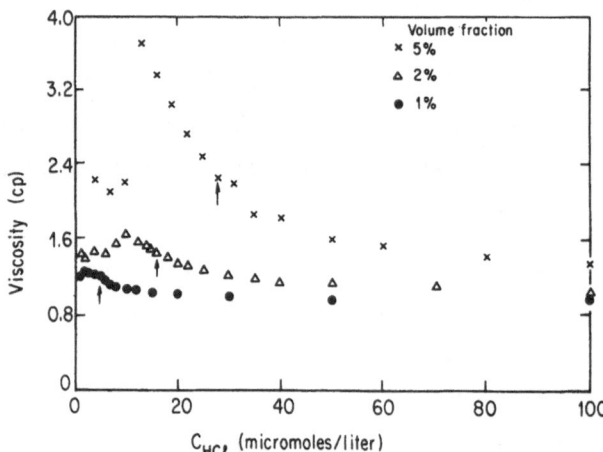

Figure 2. Viscosity of suspensions of 0.109 micron spheres as a
function HCl concentration. Arrows indicate the concentration corresponding to the melting transitions.

ture of $\sim$20C. As the interactions increase the viscosity increases.
Note that the viscosity (n) near the melting point is considerably
higher than what could be expected from the Einstein relation for a
suspension ($n=n_0(1+2.5Q)$, where Q is the volume fraction). A characteristic of these measurements is that the viscosity starts at a
large value in the solid, decreases as HCl is added and then jumps
back to a high value before again decreasing continuously as the
colloid passes into the liquid state. This behavior is reminiscent
of the series of transitions observed by Clark et al. upon application of increasing shear.[14] At low shear the solid is rigid. At
higher shear planes slip past one another, and at higher shear the
solid completely melts to the liquid.

In order to treat the polyballs as a simple interacting liquid
it is necessary to separate the contribution of the water from the
observed properties. The model then consists of two interpenetrating fluids interacting with themselves and additionally interacting
with each other through a simple Stokes drag. Assuming that the
Stokes drag is sufficiently strong that the velocity fields of the
two fluids are the same, we find that the viscosity is just the sum
of the two viscosities, or:

$$n=n_1+n_2 \tag{1}$$

where n_1 is the viscosity of water and n_2 is the viscosity of the
polyball liquid.

The self-diffusion constant is more subtle. However, mobility
considerations suggest that the diffusion constants add reciprocally:

$$D_2^{-1}=D_{s1}^{-1} + D_{s2}^{-1} \tag{2}$$

where D_{s1} is the self-diffusion constant of the spheres in water
(given by Stokes) and D_{s2} is the diffusion constant of the spheres
in the hypothetical polyball liquid.

Although a rigorous treatment of the viscosity of the polyball
liquid must involve a calculation of the velocity-velocity correla-

tion function in the presence of the screened Coulomb interaction,
we attempted a more naive argument as a first guess. Both the
shear modulus and the viscosity involve the transfer of momentum
between layers under shear stress. In the former case the response
is a displacement gradient while in the latter there is a velocity
gradient. The relationship between the two is set by the characteristic time for strain relaxation.[15] In the solid this is long, the
creep is small and the response is predominantly elastic. In the
liquid the strain relaxation occurs on a time scale given by the
molecular motions and the interparticle spacing. For a conventional
liquid it would be of the order of the interparticle spacing (a)
divided by the sound velocity. For the polyball liquid in the
presence of the water it is given by the time it takes to diffuse
the distance between spheres due to the collisions with the water,
$t = a^2/D_{s1}$.

The viscosity is then simply proportional to the product of
the shear modulus, G, times t:

$$n_2 = BGt \tag{3}$$

From our studies of the shear modulus in the solid phase we find
that G is well described by the Debye-Huckel approximation in the
region of parameter space of the present experiments.[7] We thus
feel confident that we can extrapolate this dependence into the
liquid state to evaluate G.

In figure 3 we show the prediction of equation 3 plotted vs.
the measured polyball liquid viscosity given by n_2 in equation 1.
The quantity which is varying is the electrolyte concentration.
The agreement is quite good. The fact that the coefficient B which
scales equation 3 to the observed viscosity is approximately 0.01
indicates that the distance required for strain relaxation is of
the order of 0.1 interparticle spacings, approximately equivalent
to the rms displacement used in a Lindemann criteria for melting.
In more conventional liquids, such as water, the coefficient B is

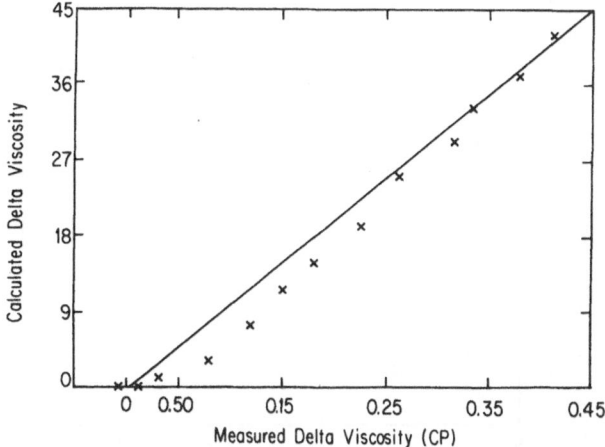

Figure 3. The calculated vs. measured viscosity for 0.109 micron, 2% sample, using equation 3 with B=0.01 The parameter being varied is the HCl concentration.

also ∿0.01. Equation 3 has been used previously for highly viscous and viscoelastic materials where t is the crossover time between elastic and viscous behavior.[15] In the present material there is no observed elasticity in the liquid phase up to much higher frequencies than given by 1/t. In the present case t depends on Stokes diffusion, rather than the interparticle forces themselves, as would be the viscoelastic case.

The diffusion constant for a simple liquid is related to the liquid viscosity by a Stokes-like equation:

$$D_s = k_B T / Anr \qquad (4)$$

where r is an effective radius defined by concentration ($r = r_s / Q^{-1/3}$, r_s is the polyball radius) and A is a constant between 1 and 6π.[16] If we determine the diffusion constant and viscosity of the polyball liquid by separating the water contributions we can see whether the inverse scaling is correct.

In figure 4 we plot the reciprocal of the total diffusion constant measured by FRS and the total viscosity as a function of volume fraction for 0.038 micron spheres. Note that the two curves have very similar shapes. Subtracting their intercepts to obtain

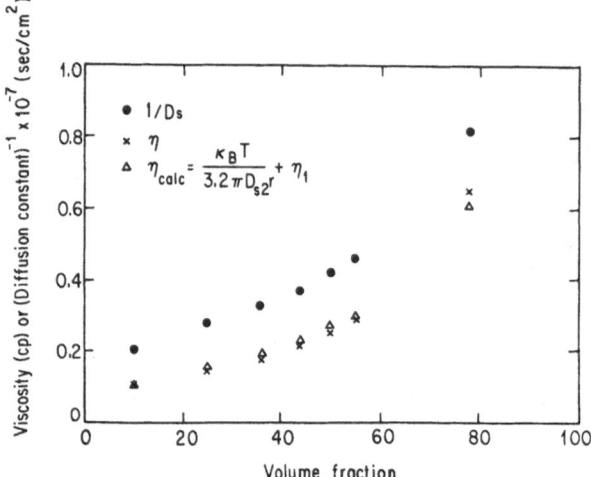

Figure 4. Viscosity and reciprocal self-diffusion constant as a
          function of volume fraction for 0.038 micron spheres.
          The *'s are the viscosity calculated from the diffusion
          constant and equation 5.

n_2 and D_{s2}^{-1} for the polyball liquid the curves are seen to be $\sim$
proportional. Also plotted is the viscosity calculated from the
diffusion constant using:

$$n_{calc} = kT/AD_{s2}r + n_1 \tag{5}$$

The coefficient A is found to be $\sim 3.7\pi$. For more conventional
liquids measurements indicate $A = 2\pi - 4\pi$.

The equations 3 and 4 relate the properties of the liquid
phase to the shear properties of the solid phase and through it to
the interaction potential between the particles. The advantage of
the present treatment is that a single measurement in the solid
phase, the shear modulus with no added electrolyte serves to characterize both the phases by measuring the effective charge per
sphere in a Debye-Huckel approximation. Less satisfactory is the
use of an indirect measurement of the effective charge through a
conductivity measurement.

While we have tried other treatments of the self-diffusion constant and viscosity as suggested by the works of other authors[6,16], the results of equations 3 and 4 have provided the best fit, and also some insight into the nature of the interacting fluid.

In conclusion, we have measured the viscosity and the diffusion constant of strongly interacting charged polystyrene spheres in colloidal suspension. We are able to observe a first order solid-liquid transition. We find that the polyball liquid behaves as a conventional liquid and therefore many of the properties of the liquid state can be explored using our knowledge of and ability to control the interparticle interaction.

We would like to acknowledge many useful discussions with Prof. P. Pincus.

REFERENCES

1.   T. Alfrey Jr., E. B. Bradford and J. W. Vanderhoff, J. Opt. Soc. Am. 44, 603, (1954).

2.   P. A. Hiltner and I. M. Krieger, J. Phys. Chem. 73, 2386 (1969).

3.   R. Crandall and R. S. Williams, Phys. Lett. 48A, 225 (1974).

4.   I. Snook and W. Van Megen, Faraday Transactions II, 72, 216 (1976).

5.   P. Pieranski, E. Dubois-Violette, F. Rothen and L. Strzelecki, J. de Physique, 41, 369 (1980).

6.   S. Mitaku, T. Ohtsuke and K. Okano, Jap. J. Ap. Phys. 19, 439 (1980), ibid. 17, 627 (1978), S. Mitaku, T. Ohtsuki, E. Enari, A. Kishimoto and K. Okano, J. Jap. Ap. Phys. 17, 305 (1978).

7.   H. M. Lindsay and P. M. Chaikin, to be published in J. Chem. Phys.

8.   D. W. Schaefer and B. J. Ackerson, Phys. Rev. Letters, 35, 1448 (1975).

9. J. C. Brown, P. N. Pusey, J. W. Goodwin and R. H. Ottewill, J. Phys. A., $\underline{8}$, 664 (1975).

10. P. N. Pusey, J. Phys. A. $\underline{11}$, 119 (1978).

11. D. W. Schaefer, J. Chem. Phys. $\underline{66}$, 3980 (1977).

12. D. W. Pohl, S. E. Scharz, V. Irniger, Phys. Rev. Letters $\underline{32}$, 31 (1973).

13. Dow Diagnostics, P. O. Box 68511, Indianapolis, Indiana 46268.

14. N. A. Clark and B. J. Ackerson, Phys. Rev. Letters $\underline{46}$, 123 (1981).

15. Landau and Lifshitz, Elasticity, Pergamon Press, Oxford, 1979.

16. H. J. V. Tyrrell, Diffusion and Heat Flow in Liquids, Butterworth and Co., London, 1961.

INFLUENCE OF MOLECULAR CONFORMATION ON THE HELICAL TWISTING POWER

OF TERPENES IN NEMATIC LIQUID CRYSTALS

Paul R. Gerber

Central Research Units
F. Hoffmann-La Roche & Company
Limited Company
4002 Basel, Switzerland

ABSTRACT

Measurements of the helical twisting power (HTP) $(cP)^{-1}$ of various terpenes dissolved in the nematic mixture RO-TN-403 are presented. Here c is the concentration of terpenes and P is the pitch of the induced cholesteric structure. It is attempted to correlate the measurements with quantities calculated from molecular conformations of the terpenes. The conformations and their relative strain energies are obtained by modelling the molecules with a molecular mechanics programm. Starting from these conformations various quantities are calculated which are candidates for a description of the HTP. Because symmetry arguments give only partial conditions for the loci of zeros in configurational space even in simplest cases, the choice of usefull functions for the HTP has to be guided by intuition and comparison with experiments. The work presented here is a step towards a solution of this rather difficult task.

1. INTRODUCTION

Much work has been devoted to studying the induction of a cholesteric state in a nematic solvent by adding chiral guest molecules. In the low guest-concentration limit the helical twisting power (HTP) h characterizes this phenomenon. It is defined by the limit

$$h = \lim_{c \to o} (cP)^{-1} \qquad\qquad (1.1)$$

where c is the concentration of the chiral additive and P is the pitch of the cholesteric state, i.e. the distance along the helical axis over which the director turns by a full angle 2π. Obviously h is characterized by the chiral molecule and also by the nematic host and depends on temperature. Cholesteric states of a given pitch are usefull in many applications and so the determination of h has been a subject of interest for some time[1-5]. Several theoretical investigations have been devoted to this subject. Some of these[6-8] start from the principal electrostatic interactions of charged particles and end up with expressions, which cannot usually be evaluated for systems of interest. Most often these complex expressions are condensed in parameters which are then subject to empirical determination. A second approach restricts itself to symmetry considerations of chirally substituted achiral molecules[9].

This approach led to a verification of sum rules[10-12] but fails to make quantitative predictions. Furthermore many qualitative models with plausible heuristic argumentation were proposed usually to cope with particular systems[13-15].

The aim of the present work is, firstly, to present several measurements of h with its temperature coefficient for a set of chiral molecules in a commercial liquid crystal mixture. Secondly it is attempted to produce values of h, starting from the molecular conformation of the chiral additives.

2. MEASUREMENTS

As nematic host we used the commercial mixture RO-TN-403 from F. Hoffmann-La Roche with a clearing temperature of $81.5^\circ C$[16]. As chiral dopands we added several terpene molecules and two steroids in concentrations of roughly 3 wt%. The cholesteric pitches of the resulting mixtures were measured by the Grandjean-Cano method in wedge shaped cells. The temperature was varied between 5 and $65^\circ C$. A quadratic polynomial of the temperature was fitted to the data of inverse pitch at various temperatures. Dividing by the concentration c one obtains to a good approximation the HTP in the form

$$h = \frac{1}{cP} = A + B(T - 25^\circ C) + C \ (T - 25^\circ C)^2. \qquad (2.1)$$

In Table I the coefficients A and B are listed together with the concentration used in the measurements and with the clearing-temperature depression coefficient

$$\tau = [T_c(c) - T_c(o)]/c. \qquad (2.2)$$

Pitch values larger than 80μm could not be measured. The corresponding entries in Table I are nevertheless important for comparison with the calculations of the next section. In this set of measurements we have not varied the nematic host, mainly because it appeared from the literature and from previous own experience that large variations in the helical twisting power of a single chiral molecule in various nematic solvents are rather the exception than the rule.

The steroids with their longish shape show a more pronounced HTP than the terpenes which are of more spheric shape and are expected to have a rather low degree of orientational order. Thus, it seems that a good orientational order of the chiral additive is a prerequisite for a large twisting power. Sign and value of h are however, quite unpredictable. Most systems show a decrease of the absolute value of h with increasing temperature.

TABLE 1. Helical twisting power h(2.1) and its temperature
coefficient $\frac{dh}{dT}$, of several molecules in the nematic
mixture RO-TN-403. Also shown are the relative clearing
temperature shifts (2.2) and the concentration of the
actually measured solution. The last column contains
calculated H-values (3.5) as described in the text.
Measured values for T = 25°C.

M O L E C U L E	STRUCTURE	$h\left[\frac{gr}{Mol\cdot\mu m}\right]$	$100\cdot\frac{dh}{dT}$ $\left[\frac{gr}{Mol\cdot\mu m\cdot K}\right]$	$\tau\left[K\frac{gr}{gr}\right]$	c[wt %]	H[arb]
1R-(+)-cis-pinane [20]		-1.97	.06	- 99	3.321	15
1S-(-)-trans-pinane [20]		———	———	- 200	3.250	- 13
1S-(-)-α-pinene [20]		———	———	- 256	3.201	12
1S-(-)-β-pinene [20]		3.98	- .24	- 322	3.255	39
(+)-3-carene [20]		-1.82	-1.00	- 308	3.150	- 24
(+)-camphor [20]		———	———	- 466	3.196	4
(-)-fenchone [20]		-3.69	1.51	- 420	3.474	22
(+)-camphene [20]		———	———	- 280	3.068	- 2
R-(+)-pulegone [20]		-1.84	2.56	- 391	3.225	- 15
5β-estra-3,17,dione [21]		20.16	-10.93	- 300	2.964	116
5α-estra-3,6,7-trione [22]		-26.44	2.78	- 234	2.988	- 118

3. CALCULATIONS

Theoretical treatments of the cholesteric phase have been
presented in many papers. In rigorous derivations of h which start
from basic intermolecular interactions, the expressions for the
pitch are of such complex nature that there is little hope to
arrive at a meaningfull evaluation with the data available for
realistic systems. For this reason we attempted an alternative
more heuristic approach. Based on the experience that the HTP
appears in most cases to be mainly determined by the chiral guest
molecule, and less influenced by the nature of the nematic host, we
tried to find relatively simple expressions, obeying the correct
symmetry relations which would correlate with h as measured in the
previous section. Such an approach cannot aim at a true physical
explanation but rather at obtaining a manageable model with usefull
predictive power.

In our approach we assumed steric effects i.e. the shape of
the molecules to be of greatest importance. Furthermore it appeared
plausible to us that the degree of orientation of the molecule with
respect to the nematic director $\underset{\sim}{n}$ would have to play an important
role. The simplest characterization of this degree of order is
given by the orientation tensor

$$O_{ij} \equiv \langle (\underset{\sim}{n} \cdot \underset{\sim}{e}_i)\,(\underset{\sim}{n} \cdot \underset{\sim}{e}_j) \rangle, \qquad\qquad (3.1)$$

where $\underset{\sim}{e}_i$ are the basis vectors of an orthogonal coordinate frame
attached to the chiral molecule. The brackets indicate averaging
over the molecular orientations. The second quantity of importance
must account for the chirarlity of the molecule. Given an axis
through the origin of the molecular frame along a unit vector $\underset{\sim}{a}$ and
two atoms at positions $\underset{\sim}{r}_1$ and $\underset{\sim}{r}_2$ a quantity describing chiral
properties of this arrangement is proportional to[15]

$$(\underset{\sim}{a} \cdot (\underset{\sim}{r}_1 - \underset{\sim}{r}_2))\,(\underset{\sim}{r}_1 \times \underset{\sim}{r}_2 \cdot \underset{\sim}{a}) \qquad\qquad (3.2)$$

In order to make use of this expression in our problem we have first to find a suitable definition of the origin. This was done by requiring the equality

$$\sum_a g_a \vec{x}_a = 0 \tag{3.3}$$

where the sum runs over all atoms of the molecule. The weight factors g_i were taken proportional to the atomic volumes as determined by van der Waals radii and bond radii in a calotte model in accordance with our assumption of the primary importance of steric interactions. The values are listed in Table II. The expression (3.2) can now be extended to a molecular quantity by forming the tensor

$$D_{ij} = \sum_{a,b} (\vec{e}_i \cdot (\vec{x}_a - \vec{x}_b))(\vec{x}_a \times \vec{x}_b \cdot \vec{e}_j) \; G_{ab}(|\vec{x}_a - \vec{x}_b|). \tag{3.4}$$

TABLE II

Values for Van der Waals radii and bond radii in Å as used to calculate the weights g which are determined by the volume of the Atom in a calotte model.

A T O M	Van der Waals radius	bond radius	weight g
H —	1.20	.37	5.2
⟩C⟨	1.65	.77	5.6
⟩C =	1.65	.67 (=)	8.2
O =	1.40	.55	8.6

The weight $G_{ab}(|\underset{\sim}{x}_a - \underset{\sim}{x}_b|)$, a scalar quantity under coordinate transformations, leaves a lot of freedom in this expression which can later on be utilized to render our final quantity

$$H = \sum_{ij} O_{ij} D_{ji} \qquad\qquad (3.5)$$

suitable for a description of the HTP.

For this expression only the symmetric part D^S of D is relevant because O is by definition a symmetric tensor. The pseudotensor D is traceless as can be seen from $(\underset{\sim}{x}_a - \underset{\sim}{x}_b) \cdot (\underset{\sim}{x}_a \times \underset{\sim}{x}_b) = 0$. Hence the equation $\underset{\sim}{x} \cdot \underset{\sim}{D}^S \cdot \underset{\sim}{x} = 0$ defines a zero cone which separates the directions of eigenvectors of positive and negative signs. While inversion symmetry of the molecule obviously leads to D = 0, two perpendicular symmetry planes do not impose $D^S = 0$. The two planes merely constitute the zero-cone of D^S which then implies H = 0. Similarly one obtains H = 0 for a single mirror plane because this plane always belongs together with its normal to the zero cone of D^S. Thus, three distinct symmetry planes are required to lead to $D^S = 0$ by symmetry.

The weight functions $G_{ab}(x)$ must be viewed as containing the information of mutual orientation action between the atomic pair, a,b and the nematic host. We have chosen them to be of the form

$$G_{ab}(x) = g_a \cdot g_b \cdot f(x) \qquad\qquad (3.6)$$

with a still arbitrary function f(x). This form omits the possibility of taking into account the nature of the bonding between atoms a and b and may thus represent the case of pure steric interaction with the nematic host.

To have f(x) = const. in (3.6) is not usefull because D vanishes in this case owing to (3.3). Other simple functions are e.g. power laws $f(x) \sim x^\gamma$. More structured functions introduce in addition lengths scales which may offer the possibility to account for the nature of the host.

An important part of the numerical calculations is the determination of molecular conformations which are needed in order to evaluate the expressions (3.4). We obtained those from molecular mechanics calculations. A rough structure was obtained by using a model-builder program of Wipke and coworkers[17] which could be operated with a graphic input. The structure, thus obtained, was afterwards refined by using the MMI molecular mechanics program of Allinger et al.[18,19] which utilizes a well tested interatomic force field. A problem with such calculations is that there is some uncertainty whether the algorithm has reached the conformation corresponding to the absolute minimum of strain energies or only a secondary relative minimum. Furthermore, relatively flexible molecules may have several states of low strain energy which are comparably populated by thermal excitation. This possibility may provide a mechanism for explaining a strong temperature dependence of h as it is occasionally observed. The effect of temperature would then be to provide varying occupation probability ratios for two conformations with very different values of H (3.5). However, in our present attempt with expressions of the form (3.4) we have avoided such complications by restricting ourselves to molecular structures with fairly unambiguous minimal energy conformation.

In order to evaluate the expression for H (3.5) we needed values for the ordering tensor O_{ij}. A simple and, most probably, quite reasonable way is to utilize the second moment tensor

$$S_{ij} = \sum_a g_a \, r_{ai} \, r_{aj} \tag{3.7}$$

by approximating

$$O \approx S/ \, \mathrm{trace}(S). \tag{3.8}$$

A more complicated but perhaps more consistent approximation might be to utilize a normalized form of the tensor

$$T_{ij} = \sum_{a,b} r_{ai} \, r_{bj} \, G_{ab}(|\underset{\sim}{x}_a - \underset{\sim}{x}_b|), \quad O \approx T/\mathrm{trace}(T), \tag{3.9}$$

with the same function G as in (3.4). Because of (3.3) T becomes proportional to S for the choice (3.6) with $f(x) \equiv 1$.

4. NUMERICAL RESULTS

We have performed calculations of H (3.5) with the expressions (3.4) and (3.6) using approximate order tensors of the form (3.8) and (3.9). The function $f(x)$ was taken to be a power law

$$f(x) = x^{\gamma}. \qquad (4.1)$$

The exponent was varied between -2 and 2.

To discuss the findings it is easiest to use the Eigenvectors of O as the basis vectors of the molecular frame. Then only the diagonal elements D_{ii} are of importance. Let the largest Eigenvalue of O correspond to the direction of coordinate axis number one, the smallest to axis number three. A typical finding was then that $|D_{11}|$ was smaller by a factor of three to ten then $|D_{22}|$ and $|D_{33}|$ which, hence, were almost of equal magnitude and opposite sign. This finding was encountered for the majority of choices of γ (4.1) and order parameter approximations. As a consequence the final quantity H was a small number obtained from a difference of large numbers. This fact is further exemplified by the exact result

$$\sum_{ij} S_{ij} D_{ji} = 0 \quad \text{for} \quad \gamma = 2 \qquad (4.2)$$

which can easily be derived with the aid of (3.3). But (4.2) is one of our possible choices.

These observations show that quite small changes in the values assumed for O may lead to substantial changes in the values obtained for H. This result is rather unsatisfactory and, perhaps illustrates the difficulty observed in the past of obtaining a clear understanding of basic mechanisms leading to induced cholesteric twisting. Despite of this difficulty we have attempted to produce

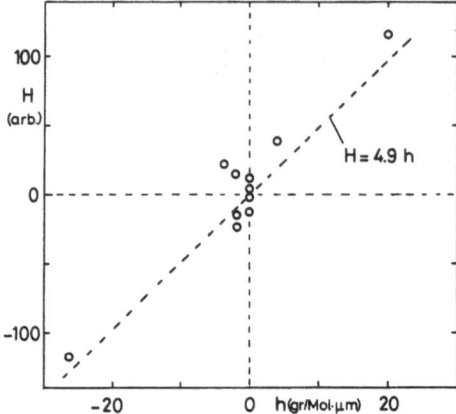

Figure 1. Calculated values of H in arbitrary units versus helical
twisting power h of the molecules of Table I as measured
at 25°C.

values of H which correlate reasonably with the measured h-values.
Taking the second moment tensor to calculate the order tensor
(equation 3.8) and an exponent $\gamma = \frac{1}{2}$ (equation 4.2) to evaluate D
(3.4) with (3.6) we obtained for H the values listed in Table I.
The correlation with the measured h-values can be seen in Figure 1.
Obviously this choice for calculating O and D is able to provide
large H values of opposite sign to the two steroids whereas all the
terpenes have comparatively small values, which agrees with the
experiments. For the terpenes themselves the model seems to be
only partially suited to produce the correct sequence of H-values,
in particular for 1R-(+)-Cis-pinane and (-)-Fenchone the signs of H
are opposed to the measured ones. For the rest of the terpenes the
sequence is quite reasonable.

In conclusion we may say that calculating a quantity of the
nature of H appears to be a delicate task because quite small
changes in conformation or in the assumed order parameter values
often alter the final result dramatically. For a good description
of the HTP, if it is at all possible along the lines presented
here, accurate conformations and information on the state of order

are mandatory before a decision on suitable values for the free parameters and functions is possible.

ACKNOWLEDGEMENT

The author thanks Dr. A. Boller for synthesizing the steroids of Ref. 21 and 22 and Drs. R. Geiger, H. Braun and P. A. Straub for discussions and advice in the molecular mechanics calculations.

REFERENCES

1. H. Finkelmann and H. Stegemeyer, Ber. Bunsenges. Phys. Chem. 82, 1302 (1978).

2. J.-P. Berthault, J. Billard and J. Jaques, C. R. Acad. Sc. Series C, 284, 155 (1977).

3. M. Hilbert and G. Soladie, Mol. Cryst. Liq. Cryst. 64, (Letters) 211 (1981).

4. Z. M. Elasvili, T. S. Piliasvili, G. S. Chilaya and K. G. Japaridze, Z. Chem. 19, 453 (1979).

5. G. Heppke, H. Marschall, P. Nürnberg, T. Oesterreicher and G. Scherowsky, Chem. Ber. 114, 2501 (1981).

6. W. J. A. Goossens, Mol. Cryst. Liq. Cryst. 12, 237 (1970).

7. B. W. Van der Meer, G. Vertogen, A. J. Dekker and J. G. J. Ypma, J. Chem. Phys. 65, 3935 (1976).

8. V. B. Magalinskii and N. A. Glushkov, Sov. Phys. Crystallogr. 25, 634 (1980).

9. E. Ruch and A. Schönhofer, Theor. Chim. Acta 10, 91 (1968), Theor. Chim. Acta 19, 225 (1970).

10. W. J. Richter and E. H. Korte, Ber. Bunsenges Phys. Chem. 82, 812 (1978).

11. W. J. Richter, H. Heggemeier, H. J. Krabbe, E. H. Korte and B. Schrader, Ber. Bunsenges. Phys. Chem. 84, 200 (1980).

12. E. H. Korte, P. Chingduang and W. J. Richter, Ber. Bunsenges. Phys. Chem. 84, 45 (1980).

13. G. Gattarelli, B. Samori, C. Stremmenos and G. Torre, Tetrahedron 37, 395 (1981).

14. G. S. Chilaya, Z. M. Elashvili, L. N. Lisetski, T. S. Piliash- vili and K. D. Vinokur, Mol. Cryst. Liq. Cryst. 74, 261 (1981).

15. B. W. Van der Meer and G. Vertogen, Z. Naturforschung 34a, 1359 (1979).

16. M. Schadt and F. Müller, IEEE Transactions Ed15, 1125 (1978).

17. W. T. Wipke, J. Verbalis and P. Grind, Princeton 1972 (unpub- lished).

18. D. H. Wertz and N. L. Allinger, Tetrahedron 30, 1579 (1974).

19. N. L. Allinger, J. T. Sprague and T. Liljefors, J. Amer. Chem. Soc. 96, 5100 (1974).

20. These compounds were obtained from Fluka AG., Chemical Factory, Buchs, Switzerland.

21. R. E. Counsell, Tetrahedron 15, 202 (1961).

22. R. Gardi and C. Pedrali, Steroids 2, 387 (1963).

VIBRATIONAL SPECTRA OF LIQUID CRYSTALS – XIII: CRYSTALLIZATION
KINETICS STUDY OF 4, OCTYLOXY, 4', CYANO BIPHENYL BY RAPID RAMAN
SPECTROSCOPY

Bernard J. Bulkin and James M. Sloan

Department of Chemistry
Polytechnic Institute of New York
Brooklyn, New York   11201

There has been considerable discussion over the years of the
role of metastable crystalline phases in the thermodynamics of
liquid crystals. Many studies have attempted to draw information
about the liquid crystalline phases from a study of the metastable
crystalline phases. Particularly active in this regard has been
studies using infrared and Raman spectroscopy.[1]

In this paper we use Raman spectroscopy to examine a new aspect
of these metastable crystalline phases -- the kinetics of the phase
transition from metastable to stable crystalline phase. In previous
papers[2,3] we have commented in detail on the pretransition effects
observed in the phase transition of stable crystalline phases to
nematic liquid crystals, and this pretransition phenomenon, in which
disorder is seen in the crystalline phase, has also been studied by
others.[4,5] The metastable phase freezes out some of the positional
disorder of a liquid crystalline phase. Study of the transition
from metastable to stable crystalline phase thus represents a way
of studying the reverse of the pretransition effects we have
previously studied.

EXPERIMENTAL

The technique used to follow the crystallization process is rapid scanning Raman spectroscopy. In this experiment, the gratings of a double monochromator are oscillated through a small angle to scan the spectrum. The oscillation is effected by a motor turning in one direction only. The apparatus is essentially that which has been described in the literature[6] now interfaced to and controlled directly by a computer.

To produce the metastable crystalline phase, samples of 4, cyano 4' octyloxy biphenyl (8OCB) are heated into the nematic phase (75°) and quenched. This is done in a melting point (1.1 mm od) capillary which is a typical sample cell for Raman spectroscopy. DSC results on such samples indicate that the metastable form does not survive to melt and give a smectic phase. Rather it begins to convert to the stable crystalline form about 15° below the normal transition temperature for crystal-smectic. It is this conversion metastable-stable crystal in the temperature range 7-12° below the c-Sm transition, which has been studied.

RESULTS

In this paper, we concentrate on results for one key region of the 8OCB spectrum, the CN stretching region. This is seen between 2200 and 2250 cm^{-1}. We have already reported[7] that the metastable phase shows a single band in this region, with wavenumber maximum at 2225 cm^{-1}. In the stable crystal, two distinct maxima are seen, a strong band at 2228 cm^{-1}, and a weaker band at 2217 cm^{-1}.

The results of following this transition at 44°C by rapid scanning Raman spectroscopy are shown in Figure 1. In Fig. 1, the spectrum is scanned from 2205 to 2245 cm^{-1} repeatedly. Each scan takes 12 sec. The spectrum is sampled at 0.8 cm^{-1} intervals although the results have been displayed as continuous curves.

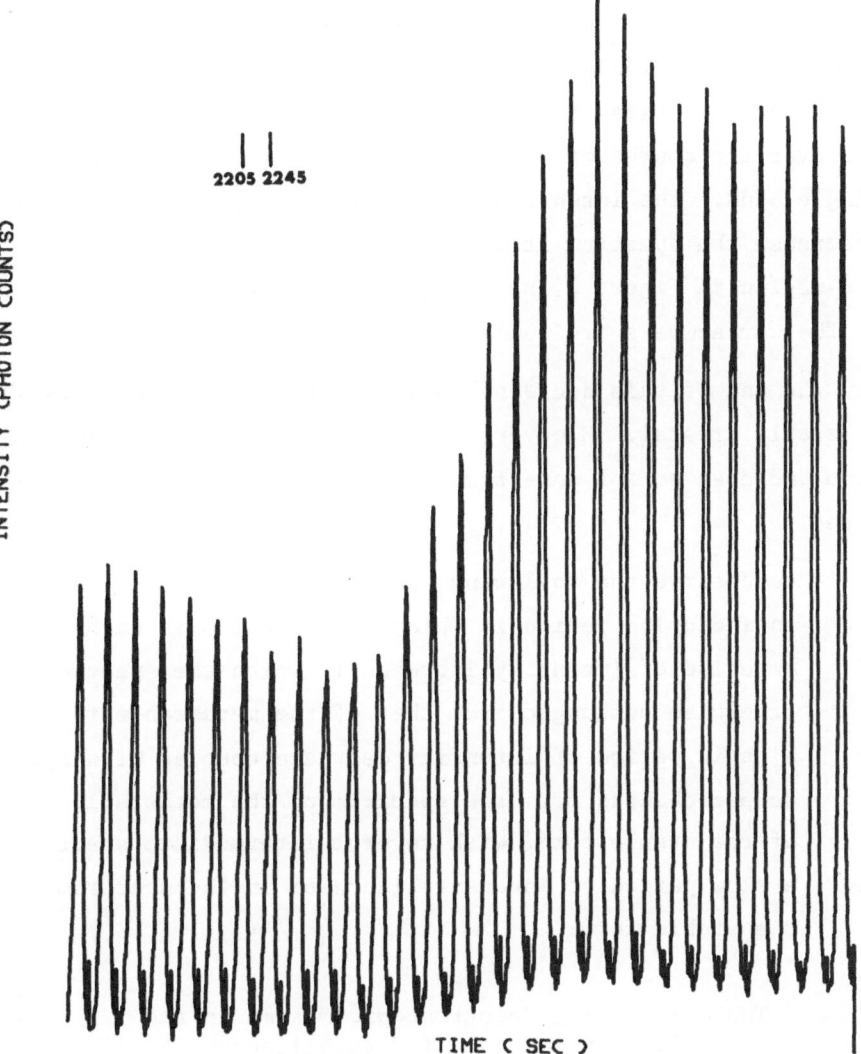

RAPID RAMAN DATA AT 44 DEG

Fig. 1   Rapid Raman spectroscopic data for the CN stretching
region displayed as continuous curves. The spectrum
is scanned repeatedly between 2205–2245 cm^{-1} in a
12 sec period, the results displayed here as a
continuous curve.

One first observes a series of spectra which are unchanged within experimental error. The number of these is truncated in the figure. At higher temperatures this period is progressively shorter. Each spectrum shows the single band characteristic of the metastable phase. After some time, the intensity begins to increase, and increases over the course of 30-40 sec to a maximum, still maintaining the single band. The intensity then decreases rapidly, and during this decrease the spectrum itself changes. As seen in an expansion of this region in Figure 2, as many as five distinct maxima can be seen. The wavenumbers are summarized in Table 1.

By the end of this decrease in intensity, the spectrum is that of the stable crystal. However, the overall intensity begins to increase and does so for some time after the "main" transition is completed.

We thus observe the following sequence:

1. An induction period in which the spectrum is unchanged.
2. A period of intensity increase in which the spectrum is otherwise unchanged from that of the metastable form.
3. A rapid period of intensity decrease coupled with spectroscopic changes from the spectrum of the metastable to that of the stable form, with several intermediate spectra seen.
4. A slow period of intensity increase in the spectrum of the stable form.

Table 1. Raman Spectroscopic Maxima in the
2200-2250 cm^{-1} Range of Crystalline 80 CB

Metastable	Intermediate	Stable
	2217	2217
	2220	
	2222	
2225	2225	
	2228	2228

Similar results have been obtained at 42, 43, 44, 45, 46, and 47°C, with the rate constants being directly dependent on temperature.

DISCUSSION

The data shown in Fig. 1 and 2 give us the beginnings of a new insight into the metastable crystalline phase of 80CB.

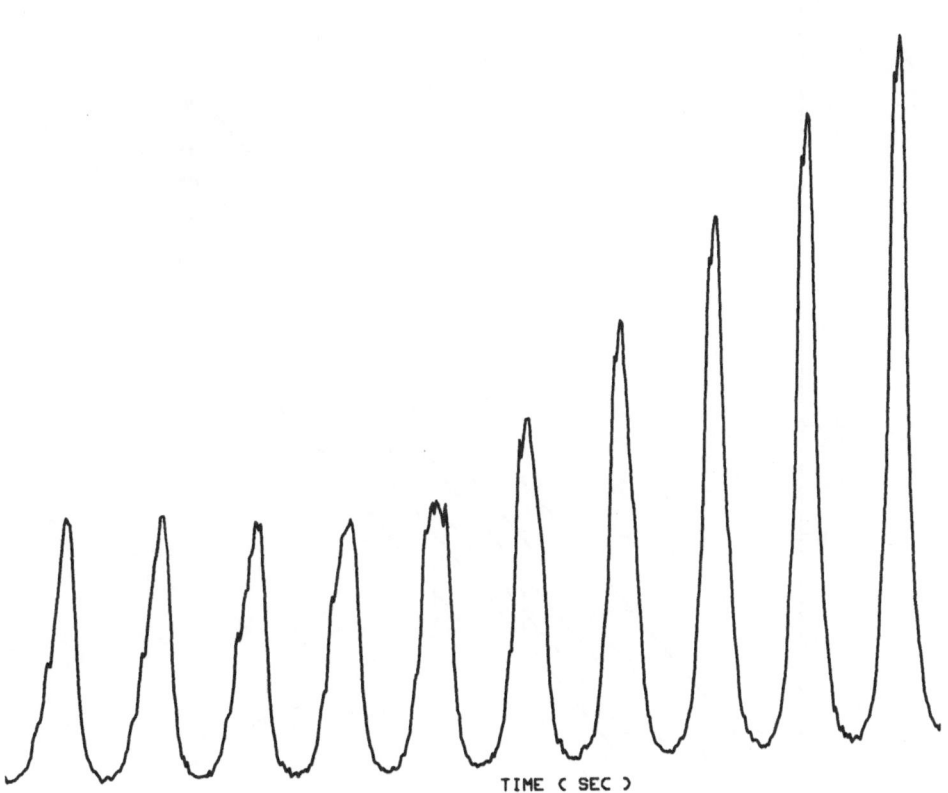

TIME (SEC)

RAPID RAMAN DATA AT 44 DEG

Fig. 2   Expanded plot of main transition region from Fig. 1. The
         evolution of bands associated with intermediate species
         can clearly be seen during the transition.

We begin with a discussion of the results in period 3. This is the main phase transition, and we postulate that initial nucleation is over before period 3 begins. The spectroscopic changes in this region are then associated with growth of nuclei and phase propagation. For such a process Avrami[8] kinetics should be obtained. This is demonstrated for the intensity at 2225 cm^{-1} in period 3 by the

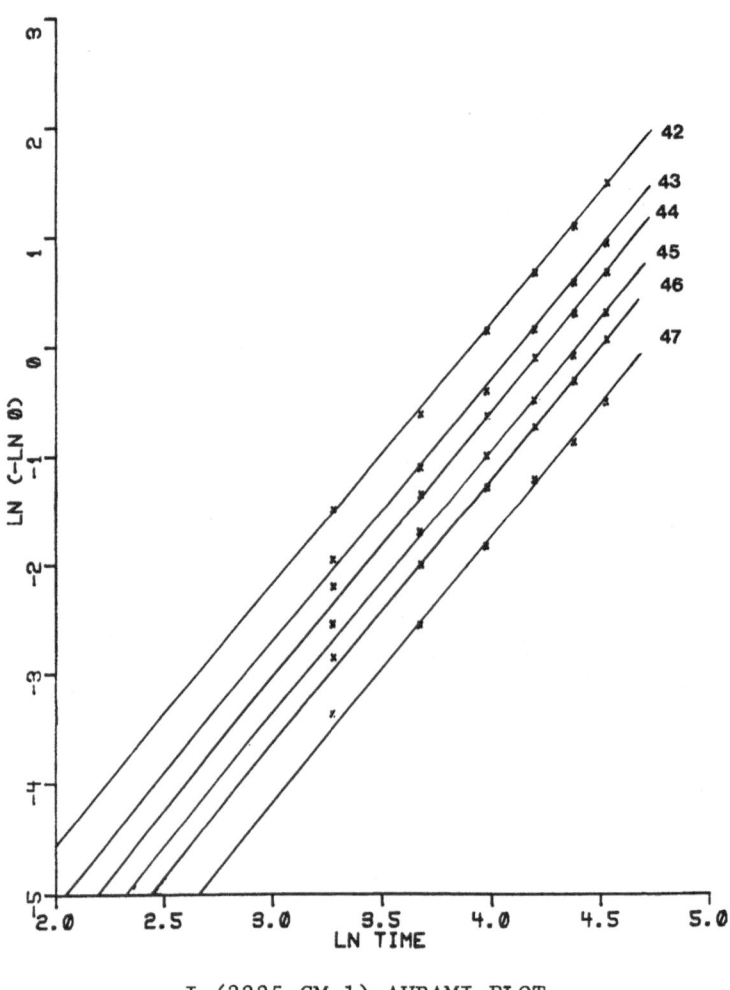

I (2225 CM-1) AVRAMI PLOT

Fig. 3   Plots of the reduced rapid Raman spectroscopic data to fit the Avrami equation for crystallization. Data at six different temperatures are shown.

plots in Fig. 3. The series of parallel plots yield an Avrami parameter n=2.27+.07.

It is by now generally agreed that simply determining n from such plots cannot give much insight into a nucleation mechanism or dimensionality of growth as was once supposed.[9] Many combinations of nucleation/dimensionality of growth yield n values near 2. It is nonetheless comforting that one of these is heterogeneous nucleation with two dimensional growth, which is quite reasonable for a crystalline phase of mesogenic material.

An activation energy for this process can be determined by well established techniques.[10] The value found from these data is 46.5 kal/mole. While this may seem high, it is similar to values found some time ago by Bulkin et al.[2] for the pretransition process.

It is significant that the number of distinct frequency maxima in the CN stretching region, and the positions of the bands, changes during the transitions. From a spectroscopic point of view there can be several causes of this change. The number of bands are primarily determined by the unit cell symmetry, or the site symmetry in a crystal. Even a reorientation of the molecules on their sites could in principle change this symmetry, but for the CN stretch to be affected this would need to be a significant reorientation about one of the axes approximately normal to the long axis of the molecule.

The number of bands may also reflect a number of distinct sites. This differs from the previous explanation as follows: In the previous paragraph, we said that a unit cell may have several molecules on sites of identical symmetry. Their CN groups may be coupled together to yield a number of modes. But as the phase transition progresses, molecules can move from one site to another. These distinct sites may have CN groups in slightly different mean fields, leading to different CN stretching frequencies. Of course, the complex spectrum observed may result from both of these mechanisms.

Given the formation of the metastable phase from a quenched nematic, and the difficulty of rotating molecules normal to the long axis, it is reasonable to conclude that crystallization occurs by progressive translations of molecules in a plane containing the director. These translations either bring the molecules through successive local potential energy minima until they are part of a growing phase, or there are several routes molecules follow to the growing phase, leading to the complex of bands observed along the way. In either case, the key idea is that the progress of the transition viewed from the vantage point of an oscillating CN group is: 1. Begins in an isotropic environment where it is uncoupled to other CN groups in the crystal. 2. Undergoes successive translations, through several distinct sites where there may be interactions between oscillators. 3. Ends up in a crystalline form which probably involves two distinct sites, one of which is more populated than the other (see ref. 7 for more details on this assignment).

Turn now to a consideration of the intensity change observed in period 4, the long time at the end of the transition. This slow tail following the main transition does not follow Avrami kinetics. During this period no frequency changes are seen, only intensity building. We believe that this is an observation of the phenomenon known as crystal perfection.[9] A crystal, once formed with molecules on sites, undergoes two processes. The molecules reorient slightly to "perfect" the crystal, actually to minimize the overall lattice energy. Crystallites may also realign slightly so as to build overall dimensions of crystalline regions. It is interesting that Raman spectroscopic intensities are sensitive to such a process.

Finally, we turn to period 2, in which, after a long induction, intensity increases by about 10% during a 30-40 second period. If one examines this as a function of temperature, the increase in intensity always occurs in the same time period relative to the main transition, i.e. we never observe the beginning of the main transition, then an increase, nor do we see the increase followed by a plateau. Always the main transition is coupled to a preceding step which manifests itself as the intensity increase. Again the spectrum during this period remains that of the metastable phase.

It is not possible to conclusively explain this spectroscopic event on the molecular level without more investigation. However, we believe that it is associated with the formation of nuclei in the sample. During the period, the concentration of nuclei increases and the effects of this on the crystal begin to propagate even before the metastable form collapses to the stable crystal. We have conducted experiments to attempt reversal of the phase transition or freezing out by quenching in period 2, but this appears to be impossible. Such experiments always yield the final crystalline phase. Thus once nuclei have formed it is not easy to stop their growth.

Further experiments on these and other systems are in progress.

REFERENCES

1. For a review, see B. J. Bulkin, Adv. in Infrared and Raman Spectroscopy 8, 151 (1980).

2. B. J. Bulkin, D. Grunbaum, A. Santoro, J. Chem. Phys. 51, 1602 (1969).

3. B. J. Bulkin, F. T. Prochaska, J. Chem. Phys. 54, 635 (1971).

4. W. J. Borer, S. S. Mitra, C. W. Brown, Phys. Rev. Lett. 27, 379 (1971).

5. M. J. Billard, M. Delhaye, J. C. Merlin, G. Vergoten, C. R. Acad. Sci. Ser. B 273, 1105 (1971).

6. J. M. Beny, B. Sombret, F. Wallart, J. Mol. Str. 45, 349 (1978).

7. B. J. Bulkin, K. Brezinsky, T. Kennelly, Mol. Cryst. Liq. Cryst. 55, 53 (1979).

8. For a good discussion see L. Mandelkern, Crystallization of Polymers, McGraw-Hill, New York, 1967, pp. 215-290.

9. B. Wunderlich, Macromolecular Physics, Volume 2, Academic Press, New York, 1976.

10. R. Becker, Ann. Physik 32, 128 (1938).

NUCLEAR SPIN-LATTICE RELAXATION DUE TO ORIENTATIONAL FLUCTUATION IN SMECTIC LIQUID CRYSTALS

Seiichi Miyajima[†], Nobuo Nakamura, and Hideaki Chihara

Department of Chemistry, Faculty of Science
Osaka University, Toyonaka, Osaka 560, Japan

ABSTRACT

Theoretical expression for nuclear spin-lattice relaxation rate is derived on the basis of thermally excited undulation mode in smectic liquid crystals. The mode is treated as essentially two-dimensional, but the effect of longest wavelength mode with respect to layer compression is also included. The dependences of relaxation rate on the Larmor frequency and on the orientational angle are deduced. Pulsed NMR experiments were done on the smectic A and B phases of HBAC (p-n-hexyloxybenzylideneamino-p'-chlorobenzene). The frequency- and temperature-dependences of proton spin-lattice relaxation rates are analyzed in the light of the theory.

INTRODUCTION

If one considers an idealized orthogonal smectic system, in which the director $\vec{n}(\vec{r})$ is perpendicular to the smectic layer in every place of the system, and the interlayer spacing does not depend on $\vec{r}$, the rotation of the director should vanish, i. e.,

$$\text{rot } \vec{n}(\vec{r}) = 0. \tag{1}$$

Thus the twist and the bend deformations are forbidden whereas the splay deformation is allowed. This kind of splay deformation causes the spatially undulating waves of the layers.[1] The focal conic appearance of the microscopic texture is related to this kind of static deformation.

The dynamics of elastic deformation in the smectic system, namely, the orientational fluctuation of the director, was first discussed by de Gennes.[2] The character of the modes depends on the direction of propagation. We take the z-axis as to lie parallel to the averaged direction of the director. If $q_{//} \simeq q_{\perp}$, two propagating modes and an overdamped viscous mode are found. If $q_{//} = 0$, a longitudinal propagating mode, an overdamped shear mode, and a very slowly damped mode (pure undulation of the layers) are found. It is the last mode,

$$\omega = i/\tau_q = i\, K_1\, q_{\perp}^2/\eta \, , \tag{2}$$

that is characteristic of the smectic system and is expected to give significant effect on NMR relaxation. The symbols η and K_1 denote the effective viscosity related to this mode, and the splay elastic constant. De Gennes discussed the dynamics of pure undulation ($q_{//} = 0$), that is, the mode purely restricted within the layer and strictly conserving the interlayer spacing. It is, however, diffi-cult to realize pure undulation in a real system.

In 1974, Ribotta et al.[3] tried to detect "impure" undulation mode by quasielastic light scattering by utilizing a homeotropically aligned monodomain smectic A system of CBOOA. By developing the theory of de Gennes, they applied the relations,

$$<|\delta \vec{n}_q|^2> = \frac{k\,T\ q_{\perp}^2}{B\,q_{//}^2 + K_1\,q_{\perp}^4} \, , \tag{3}$$

$$\tau_q^{-1} = \frac{B\,q_{//}^2 + K_1\,q_\perp^4}{\eta\,q_\perp^2}\;, \qquad (4)$$

with a boundary condition, $q_{//} = m\pi/L$ (L is the distance between glass plates, and m is an integer). Here, B is the elastic constant for layer compression. Scattered rays from the small numbers of m were observed and the validity of eqs. (3) and (4) was established by examining the $q_\perp$- and L-dependences of the intensities and the line-widths of the quasielastic lines.

The present paper gives predictions for nuclear spin-lattice relaxation rate caused by undulation mode for three different cases: (1) de Gennes' pure undulation, (2) case of monodomain smectic system under a boundary condition for $q_{//}$, (3) case of polydomain smectics. The experimental results are given for HBAC (p-n-hexyloxybenzylidene-amino-p'-chlorobenzene), and are analyzed in the light of the theory.

THEORY

Nuclear spin-lattice relaxation rate $(1/T_1)$ caused by fluctuating dipole interaction between a pair of spins is given by[4]

$$T_1^{-1} = \frac{3}{2}\,\gamma^4\,\hbar^2\,I(I+1) \sum_{p=1,2} J'(p\omega_0). \qquad (5)$$

Here, $J'(p\omega_0)$ is the Fourier transform of the time correlation function of the fluctuating part of the dipole interaction tensor, γ, I, $\hbar$ are the gyromagnetic ratio of the spin, the spin quantum number, and the Planck constant divided by 2π, respectively. We now consider, as a spin pair under question, a pair of spins the inter-spin vector of which lie parallel to the molecular long axis, and the distance r of which is kept constant. Then the coordinate transfer technique gives[5]

$$T_1^{-1} = \frac{3}{2} \, \gamma^4 \, \hbar^2 \, I(I+1) \, r^{-6} \, <p_2(\cos\Theta)>^2$$

$$\times \sum_{p=1,2} f_p(\Theta_0) \, J(p\omega_0) \qquad (6)$$

$$f_1(\Theta_0) = \frac{1}{2} \, (1 - 3 \cos^2\Theta_0 + 4 \cos^4\Theta_0), \qquad (7)$$

$$f_2(\Theta_0) = 2 \, (1 - \cos^4\Theta_0), \qquad (8)$$

$$J(p\omega_0) = \int_{-\infty}^{+\infty} <\delta\vec{n}(\vec{r}, t) \, \delta\vec{n}(\vec{r}, 0)>$$

$$\times \exp(- i \, p\omega_0 \, t) \, dt. \qquad (9)$$

Here, Θ is the angle between $\vec{n}(\vec{r})$ and an instantaneous molecular long axis, and Θ_0 is the angle between $\vec{n}_0$ and the direction of the external magnetic field. Assuming an exponential time-correlation function for $\delta\vec{n}(\vec{r}, t)$, the $1/T_1$ can be expressed by a superposition of the Lorentzians with $\vec{q}$-dependent damping rates, i. e.,

$$J(p\omega_0) = \int_{-\infty}^{+\infty} v^{-1} \sum_{\vec{q}} <|\delta\vec{n}_q|^2> \, \exp(- t/\tau_q)$$

$$\times \exp(- i \, p\omega_0 t) \, dt. \qquad (10)$$

(1) case of pure undulation

In the case of pure undulation, $q_{//}$ is taken to be zero, and $q_{\perp}$'s are integrated up to the higher cutoff wavenumber, $q_H = 2\pi/a$ (a is taken to be the molecular length). Then eq. (10) yields

$$J(p\omega_0) = \frac{k \, T}{2\pi \, L^* \, K_1 \, p\omega_0} \arctan(\frac{\omega_H}{p\omega_0}) , \qquad (11)$$

where $\omega_H = K_1 q_H^2/\eta$. Here L^*, the length of the ideal smectic system within the xy-plane, is restricted by Peierls-Landau instability.[6]

The asymptotic behavior under high and low frequency limit is given by

$$J(p\omega_o) = \frac{k\ T\ \omega_H}{2\pi\ L^*\ K_1\ (p\omega_o)^2} \tag{12}$$

when $p\omega_o \gg \omega_H$, and

$$J(p\omega_o) = \frac{k\ T}{4\ L^*\ K_1\ (p\omega_o)} \tag{13}$$

when $p\ \omega_o \ll \omega_H$.

From the experimental point of view, however, this kind of pure undulation is too much simplified one to be accepted. The real smectic system will be subject to some boundary conditions which permits long wavelength distortions parallel to the z-direction.

(2) case of monodomain smectic system under a boundary condition on $q_{//}$

When monodomain smectic liquid crystal is carefully prepared and put between glass plates distant by L in homeotropic arrangement, a boundary condition, $q_{//} = m\pi/L$, is imposed on the small displacement along the z-axis. Assuming essentially two-dimensional fluctuation, the spectral density of the director fluctuation due to undulation mode is given by

$$J(p\omega_o) = \frac{\eta\ k\ T}{8\pi\ m\ K_1\ \sqrt{K_1\ B}} \cdot \frac{1}{\sqrt{1 + \left(\dfrac{\eta\ L}{2\ m\pi\ \sqrt{B\ K_1}}\ p\omega_o\right)^2}}, \tag{14}$$

by using eqs. (3), (4), and (10). In deriving eq. (14), the cutoff effect on $q_\perp$ is neglected for simplicity. The behavior of eq. (14)

Table.   Tentative parameters
         for numerical calculations

T	350 K
B	1.0×10^{10} J m^{-3}
K_1	1.0×10^{-11} J m^{-1}
$<P_2(\cos\Theta)>$	1.0
r	2.45×10^{-10} m

is shown in Fig. 1 as a function of $\omega_o\tau_o$ where $1/\tau_o = 2m\pi \sqrt{BK_1}/\eta L$, and p is taken to be unity. The $1/T_1$ can be calculated by utilizing eqs. (6)-(10) and (14), and the parameters listed in the table. The results are summarized as follows.

(i)   the absolute value of $1/T_1$

The absolute value of $1/T_1$ in the low frequency limit depends on $<P_2>^2 \eta/K_1\sqrt{K_1 B}$. By the present numerical calculation, it was calculated to be 0.36 s^{-1} in the orientation, $\Theta_o = \pi/2$. This value is of sufficient order of magnitude to be detected experimentally under some suitable conditions.

(ii) the frequency-dependence

If the ω_o-dependence of $1/T_1$ is represented by $1/T_1 \propto \omega^\gamma$, the exponent γ varies from zero (in the low frequency limit) to -1 (in the high frequency limit) as a function of $\omega_o\tau_o$. The condition of usual NMR experiment will probably be in the high frequency region so that one can expect that $1/T_1 \propto 1/\omega_o$ in experiments.

(iii) the angular-dependence

The angular factor of $1/T_1$ is expressed by

$$T_1^{-1} \propto f(\Theta_o) = f_1(\Theta_o) + 2^\gamma f_2(\Theta_o), \tag{17}$$

and shown in Fig. 2. It has γ-dependence.

(iv) the temperature-dependence

The T-dependent factor of $1/T_1$ is $<P_2>^2 \eta kT/K_1\sqrt{K_1 B}$. So the T-dependence of $1/T_1$ is expected to be small in the temperature

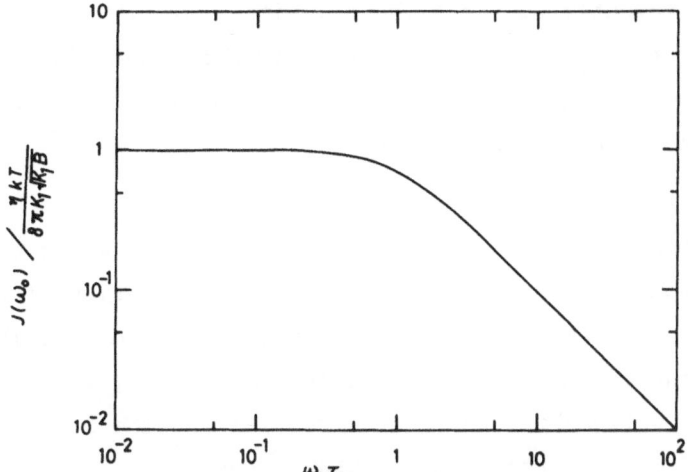

Fig. 1   The spectral density function of the director fluctuation
          due to undulation mode for monodomain smectic system,
          eq. (14) in the text.

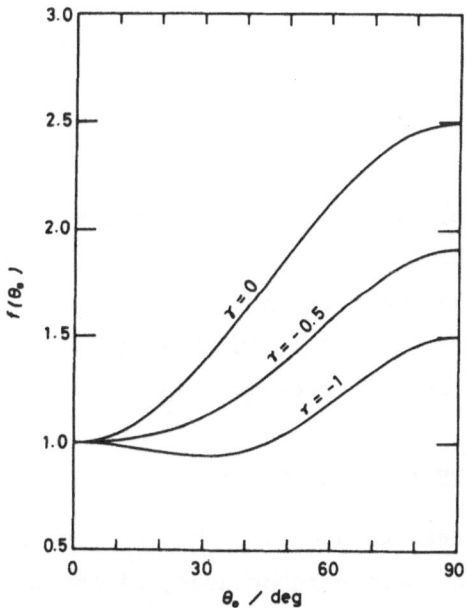

Fig. 2   Orientation dependence of the angular factor $f(\theta_0)$.
          Theoretical curves are given for three different values
          of γ.

region sufficiently far from the transition area. A strong T-
dependence of $1/T_1$ would appear when viscosity and elastic
constants would exhibit a marked change associated with critical
phenomena near the transition point.

(3) case of polydomain smectic system

When the sample consists of domains with different dimensions,
which is usually the case, the effect of distribution of L must be
introduced. We assumed, for simplicity, a uniform distribution of
L must be introduced. We assumed, for simplicity, a uniform distribution of L between L' and L_o (the smaller and the larger cutoff of
the domain size, respectively). Then we obtain

$$J(p\omega_o) = \frac{k\,T}{4(L_o - L')K_1(p\omega_o)}$$
$$\times\, \ln\left|\frac{L_o + \sqrt{L_o^2 + \dfrac{4BK_1\pi^2}{\eta^2(p\omega_o)^2}}}{L' + \sqrt{L'^2 + \dfrac{4BK_1\pi^2}{\eta^2(p\omega_o)^2}}}\right| \qquad (18)$$

In deriving eq. (18), m was fixed to be unity for simplicity. The
temperature- and the angular-dependences of $1/T_1$ are identical with
that in the case (2), whereas the crossover region ($-1 < \gamma < 0$) in
the ω_o-dependence is widened if (L_o-L') is taken large. Results of
numerical calculations are presented in Fig. 3.

EXPERIMENTAL RESULTS AND DISCUSSION

Experimental results are given on the smectic mesophases of
HBAC. HBAC has two smectic mesophases, SA and SB. The phase
relation of this compound was established by Tsuji et al.[7]
The microscopic appearance of SA and SB exhibited focal conic
texture[8] which revealed the existence of undulating deformation of

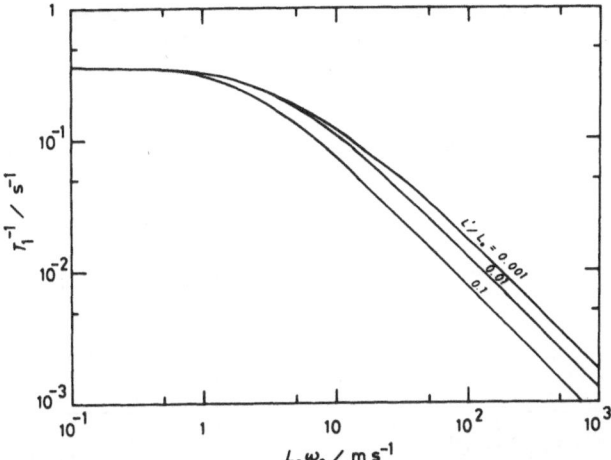

Fig. 3   Spin-lattice relaxation rate due to undulation mode:
          Case of polydomain smectics. Theoretical curves are
          calculated by eqs. (6)-(9) and (18). The angle Θ_0 was
          assumed to be randomly distributed in space to fit to
          the powder pattern. The curves corresponding to three
          different values of L'/L_0 are drawn.

the smectic layers in these mesophases.

Proton spin-lattice relaxation rates were measured by homemade
pulsed spectrometers at 10.0, 20.5, and 40.5 MHz. The molecules of
HBAC were not aligned in the whole specimen, as was evidenced by the
fact that the NMR lineshape in these phases showed powder pattern of
the dipole doublets, and that the lineshape exhibited no appreciable
change when the sample was rotated around the axis parallel to the
rf-field.

The ω_0-dependence of proton $1/T_1$ in the SA and SB mesophases are
shown in Fig. 4. These Figures show the following consequences:
(i) When one divides the observed $1/T_1$ into two contributions, the
ω_0-dependent and the ω_0-independent terms, i. e.,

$$T_1^{-1}(\omega_0, T) = A(T)\omega_0^{\gamma} + B(T), \qquad (19)$$

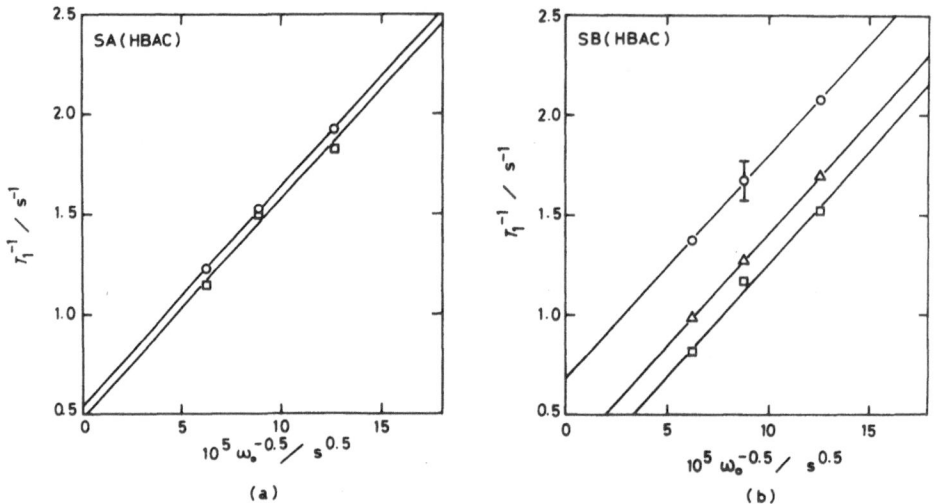

Fig. 4   Frequency-dependence of the proton spin-lattice relaxation
          rate in SA and SB mesophases of HBAC at some different
          temperatures: Fig. (a); 364.2 K (O), 368.6 K (□). Fig.
          (b); 336.4 K (O), 350.9 K (Δ), 355.2 K (□). Typical
          experimental uncertainty is indicated in Fig. (b).

$\gamma = -0.5$ holds in both of the smectic mesophases. (ii) A(T), the
ω_o-dependent contribution to $1/T_1$ is almost independent of temperature , and moreover, nearly conserved in the two different kinds of
smectic mesophases. (iii) B(T), the ω_o-independent contribution to
$1/T_1$ is strongly dependent on temperature.

     The full representation of the temperature-dependence of $1/T_1$
is shown in Fig. 5. From the Figures 4 and 5, it is difficult
to interpret the origin of the first term in eq. (19) in terms of
the rotational or translational fluctuation of molecules. It is
most reasonably related to collective excitation of orientational
fluctuation. So let us analyze the experimental results by
applying the theoretical predictions given in the previous section.
Although $\gamma = -1$ was predicted (because it was estimated that we see
the system in the high frequency region), $\gamma = -0.5$ is possible if
ω_o lies in the intermediate region (see Fig. 3).[#] Now, we assume
that $\gamma = -0.5$ results from the orientational fluctuation. Then
A(T) becomes proportional to kT if the T-dependence of B, K_1, η, and
$<p_2>$ are neglected.

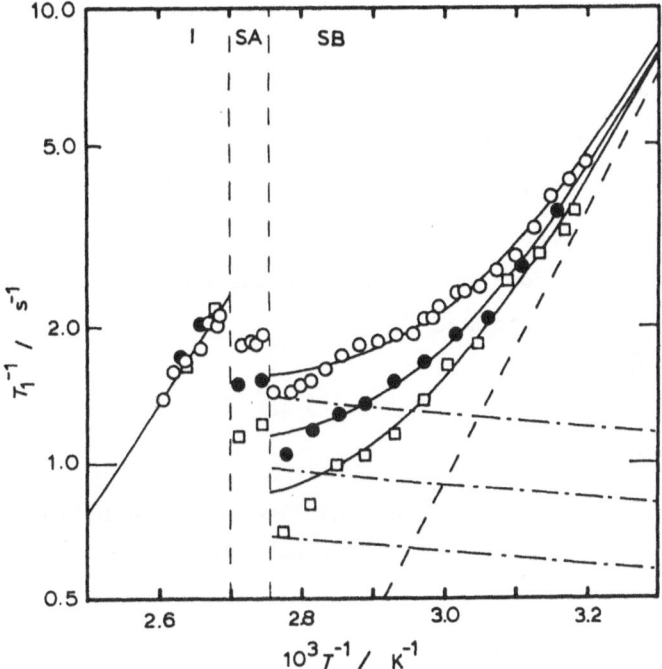

Fig. 5   Temperature-dependence of the proton spin-lattice relax-
          ation rate in SB, SA, and I (isotropic liquid) phases of
          HBAC, measured at 10.0 (O), 20.5 (●), and 40.5 (□) MHz.
          The solid lines in the SB region are the calculated ones
          using eqs. (19) and (20) with $\gamma = -1/2$ for each Larmor
          frequency. The broken line represents B(T), and the
          dashed broken lines represent A(T). The parameters are
          given in the text. The solid line in the phase I shows
          $\Delta H = 46$ kJ mol^{-1}.

        By fitting to the data in the SB region, B(T) was revealed to

exhibit Arrhenius-type behavior with the activation enthalpy (ΔH),

58 kJ mol^{-1}, which is reasonably attributed to rotational motion of

the molecule or of the chain under the extreme narrowing condition.

The solid lines in the SB region of Fig. 5 represent the calculated

curves assuming

#We cannot further examine whether this is the case or not since the
mechanical parameters, K_1, B, and η are not known to us.

$$A(T) = A_o T, \quad \text{and} \quad B(T) = B_o \exp(\Delta H/RT), \quad (20)$$

with $A_o = 31$ s$^{-3/2}$ K^{-1}, $B_o = 8.7 \times 10^{-10}$ s^{-1}, and $\Delta H = 58$ kJ mol^{-1}. Agreement with the data is satisfactory.

The experimental data of $1/T_1(\omega_o, T)$ in the smectic phases of HBAC were thus interpreted at least qualitatively by superposing the contributions from the rotational motion and the collective excitation of orientational fluctuation, the latter being characterized by undulation mode.

ACKNOWLEDGEMENT

We thank Dr. Kazuhiro Tsuji of Kwansei Gakuin University for kindly providing us with the sample of HBAC.

REFERENCES

1.   P. G. de Gennes, The Physics of Liquid Crystals, Clarendon
            Press, Oxford, 1974, Chapter 7.

2.   P. G. de Gennes, J. Phys. (Paris) 30, C4-65 (1969).

3.   R. Ribotta, D. Salin, and G. Durand, Phys. Rev. Lett. 32,
            6 (1974).

4.   R. Kubo and K. Tomita, J. Phys. Soc. Jpn. 9, 888 (1954).

5.   P. Ukleja, J. Pirs, and J. W. Doane, Phys. Rev. A14, 414 (1976).

6.   L. D. Landau and E. M. Lifshitz, Statistical Physics, 2nd Ed.,
            Pergamon Press, 1969, Chapter 13.

7.   K. Tsuji, M. Sorai, H. Suga, and S. Seki, Mol. Cryst. Liq.
            Cryst. (Letters) 41, 81 (1977).

8.   S. Miyajima, Ph.D. Thesis, Osaka University (1981).

[†]Present address: Department of Chemistry
                    College of Humanities and Sciences, Mihon
                    University, Setagaya-ku, Tokyo 156, Japan

NMR STUDIES OF MOLECULES ORIENTED IN MIXED THERMOTROPIC LIQUID CRYSTALS OF OPPOSITE DIAMAGNETIC ANISOTROPIES

C. L. Khetrapal,[*+] A. C. Kunwar[*] and
M. R. Lakshminarayana[@]

[*]Raman Research Institute, Bangalore 560080, India

[+]Bangalore NMR Facility, Indian Institute of Science
Bangalore 560012, India

[@]University of Agricultural Sciences, Bangalore 560065
India

ABSTRACT

NMR spectra of molecules oriented in a mixture of thermotropic liquid crystals of opposite diamagnetic anisotropies show novel features and provide useful information. At a critical concentration and temperature, it is observed that two discrete solute orientations corresponding to those in the two types of liquid crystals coexist. The scope and the limitations of the experiments in the determination of the chemical shift anisotropy without the use of a reference or the change of experimental conditions and the individual values of the direct and the indirect couplings between heteronuclei are examined and the results are discussed.

INTRODUCTION

A mixture of thermotropic liquid crystals of opposite diamagnetic anisotropy provides interesting features in the NMR spectra of the molecules dissolved therein[1,2]. The dipolar splittings

vary systematically with relative concentrations of the two solvents
until at a critical concentration they change abruptly such that
the values of the direct dipolar couplings between the interacting
nuclei are half (or twice) the values and have opposite signs. A
close examination of the spectra at the critical concentration
demonstrates that both the orientations corresponding to those in
the two types of solvents coexist at a particular temperature. The
temperature range at which species with both types of orientations
are present was less than 5° in the cases investigated. Possible
applications and limitations of such experiments are discussed in
the present paper.

EXPERIMENTAL

Proton NMR spectra of acetonitrile, benzene and methanol
oriented in thermotropic liquid crystals namely N-(p-ethoxy-benzyli-
dene)-p-n-butylaniline (EBBA) and the Merck phase ZLI-1167 were
studied. The solutions in EBBA and ZLI-1167 are referred to as (A)
and (B) respectively, hereafter. The conditions of the various
experiments are specified at appropriate places in the tables.

The spectra of solutions (A) and (B) were recorded on a WH-270
FT-NMR spectrometer. The accummulated free induction decays were
Fourier transformed with the help of a BNC-12 computer with a core
memory of 20K.

A known amount of the solution (A) was then mixed with the
weighed quantity of the solution (B) and the spectra were recorded.
Relative concentrations of solutions (A) and (B) were changed gradually and the spectra were recorded at each concentration until an
abrupt change in the spectral splittings was observed. Near the
concentration where the abrupt change occurs (referred to as the
'critical concentration' hereafter) variable temperature experiments
were performed such that the spectra corresponding to both the

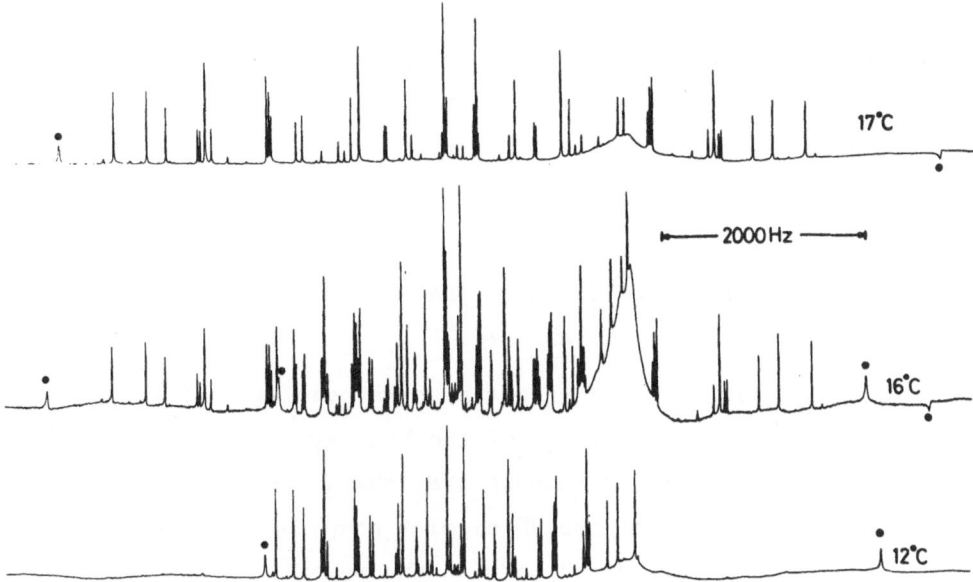

Fig. 1. Proton NMR spectra of 3 weight per cent benzene oriented in a 0.27:1 mixture of EBBA and ZLI-1167 at 270 MHz. Top trace at 17°C, central trace at 16°C and the bottom trace at 12°C. The lines (·) are due to traces of water present in the solution.

orientations were observed. Typical spectra of benzene under different orientation conditions are shown in fig. 1.

RESULTS

The proton spectra of acetonitrile without the ^{13}C-H and the ^{15}N-H satellites in general consist of 1:2:1 triplets due to the HH dipolar couplings with a total splitting of $|6D_{HH}|$ where D_{HH} is the HH direct dipolar coupling. At the critical concentration and temperature[2], two triplets with splittings in the exact ratios of 1:2 were obtained. The ^{13}C-H and the ^{15}N-H satellites were also observed and at the critical concentration and temperature there were two sets of the satellites. The determination of the dipolar couplings D_{HH}, D_{CH}, D_{CCH} and D_{NH} and the indirect spin couplings

$^{1}J_{CH}$, J_{NH} and $^{2}J_{CCH}$ was straightforward in this case[2] and the values determined are reported in table 1. The HCH bond angle and the HNC and the HCC angles were obtained from the derived dipolar couplings using the equations reported in the literature[3].

For benzene also two spectra were obtained at the critical concentration and temperature. Values of the HH dipolar couplings thus determined using the iterative LACOONOR[4], program on a DEC-10 computer are reproduced in table 2.

In methyl alcohol, the proton spectra consisted of AB_3 patterns due to relatively slow exchange of the hydroxyl protons[5,6]. At the critical concentration and temperature, the two AB_3 types of spectra were obtained. Values of the HH dipolar couplings D_{BB} and D_{AB} within the methyl protons and between the methyl and the hydroxy protons thus determined are included in table 3 together with the

TABLE 1

Parameters for 3.3 weight per cent solution of acetonitrile oriented in 0.29:1 mixture of EBBA and ZLI-1167.

Para-meters (Hz)	Temperature(°C)			
	20	21*		22
D_{HH}	−848.1±0.2	−839.1±0.2	1678.2±0.2	1665.3±0.2
D_{CH}	−585.5±0.2	−578.8±0.2	1157.5±0.2	1149.5±0.2
D_{NH}	− 26.2±0.4	− 27.0±0.4	54.1±0.4	54.1±0.4
D_{CCH}	169.8±0.2	167.8±0.2	−335.7±0.2	−334.1±0.2

$^{1}J_{CH}$ = 136.0±0.4, J_{NH} = −2.2±0.4 and $^{2}J_{CH}$ = −9.6±0.4 Hz

*The centre of the spectrum with smaller spread is 88.34±0.35 Hz up field (at 270 MHz) with respect to that of spectrum with larger spread. The scaling distance (r_{HH}) is assumed as 1.80 Å for the calculation of S_{C_3} used to obtain Δσ reported in the text.

TABLE 2

Parameters for 3 weight per cent solution of benzene oriented in
0.27:1 mixture of EBBA and ZLI-1167.

Para-meters (Hz)	Temperature(°C)			
	12	16*		17
$(D_{HH})_{ortho}$	465.39	456.08	−912.41	−902.79
$(D_{HH})_{meta}$	90.16	88.43	−176.95	−175.06
$(D_{HH})_{para}$	58.52	57.61	−115.29	−114.01
rms error	0.41	0.49	0.45	0.38

Errors of the dipolar couplings are less than 0.04 Hz.
J-values used are: $(J_{HH})_{ortho}$ = 7.549, $(J_{HH})_{meta}$ = 1.378 and
$(J_{HH})_{para}$ = 0.650 Hz.

*Separation between the centres of the two spectra is 93.52±0.06 Hz
at 270 HMz. The centre of the spectrum with the larger spread is
at higher field. The scaling distance $(r_{HH})_{ortho}$ is assumed as
2.481Å for the calculation of the S-value perpendicular to the
ring plane, for obtaining Δσ.

derived value of J_{AB}. These values were obtained by fitting the
observed and the calculated line positions using the LACOONOR
Program on the DEC-10 computer. The determined parameters, namely
the chemical shift $(\nu_A - \nu_B)$ and the direct and the indirect
spin-spin couplings are reproduced in table 3.

DISCUSSION

Table 1 shows that in acetonitrile at the critical concentration, the dipolar couplings are in the exact ratios of 1:2 with
opposite signs at 21°C. As reported earlier, the separation (Δδ)
between the centres of the two spectra provides the proton chemical
shift anisotropy (Δσ) using the relation $\Delta\sigma = \Delta\delta/S_{C_3}$ where S_{C_3} is

TABLE 3

Parameters for 4 weight per cent solution of methanol
oriented in 0.31:1 mixture of EBBA and ZLI-1167.

Para- meters (Hz)	Temperature(°C)			
	27	28*		29
D_{AB} @	− 80.9±0.3	− 79.8±0.4	161.0±0.1	160.2±0.2
D_{BB}	− 31.5±0.2	− 30.1±0.3	60.8±0.1	59.6±0.1
$\nu_A - \nu_B$ **	−294.8±0.9	−280.6±0.6	780.5±0.3	788.9±0.5
rms error	1.6	1.6	0.4	0.7

* The chemical shift difference of the methyl protons in the two
orientations is 166.4 Hz at 270 MHz. Methyl protons with larger
D_{BB} are at lower field. A value of the scaling distance (r_{HH})
within the methyl protons is assumed as 1.80 Å for the calculation of $\Delta\sigma$.

** at 270 MHz.

@ J_{AB} was derived from the spectrum of the compound oriented in
EBBA and the value (−5.0 Hz) obtained was kept fixed in these
calculations.

the larger order parameter of the C_3-axis of symmetry. A value of
−2.01±0.01 ppm is thus determined for the proton chemical shift
anisotropy.

It is well known that the spectra of oriented molecules do not,
in general, permit the separate determination of the direct and the
indirect spin-spin couplings between heteronuclei. The coexistence
of the two orientations in the exact ratios of 1:2 with opposite
signs permits the determination of ($J_{XH} + 2D_{XH}$) and ($J_{XH} - D_{XH}$)
separately between the heteronuclei where D_{XH} defines the dipolar
coupling between the heteronucleus X and the proton for the larger
orientation. These two quantities allow separate determination of
J_{XH} and D_{XH}. The values determined are given in the foot-note in
table 1.

The bond angle HCH, the angles HCC and HNC determined from the various spectra are 108.7°, 25.4° and 16.7° respectively without applying the vibrational corrections. An application of Harmonic vibrational corrections reported in the literature[7] provides the r_α-values of the angles as 109.1°, 25.0° and 16.4° respectively. These are in agreement with the microwave values[8] and those obtained by NMR studies in other solvents[7,9].

Like in acetonitrile, in benzene also two orientations are observed (table 2) at the critical concentration at 16°C. The proton chemical shift anisotropy value determined using the relation reported earlier is -1.49 ± 0.01 ppm. The magnitude of the chemical shift anisotropy is less than that (-2.59 and -3.47 ppm) obtained from studies using smectic liquid crystals[10,11]. This difference is due to different local effects on the proton chemical shift in the two methods. A quantitative calculation of the local effects is in progress. Ratios of the interproton distances in benzene are: $(r_{HH})_{para}/(r_{HH})_{ortho}$ = 1.9930 ± 0.0004 and $(r_{HH})_{meta}/(r_{HH})_{ortho}$ = 1.7280 ± 0.0004. The corresponding r_α - values obtained after applying the Harmonic vibrational corrections[11] are 1.9994 ± 0.0004 and 1.7323 ± 0.0004. The vibrationally corrected values do not show significant deviations from the regular hexagonal geometry.

For methyl alcohol, the values of the dipolar couplings at the critical concentration and at various temperatures are reported in table 3. At 28°C, two orientations are observed. The chemical shift anisotropy for the methyl protons is determined as -104.3 ppm. The magnitude of the chemical shift anisotropy is abnormally large. This is not realistic and may be understood in terms of the multiple site[12] theory analogous to that reported for the dipolar couplings and the quadrupole coupling constants. In a simplified version of the multiple site theory, only 2-sites are taken into consideration. The observed apparent order parameter in such a case is treated as a weighted average of the two order parameters

of opposite signs, in the two sites. In the present case, the multiple site theory is applicable to each of the two sites at 28°C.
Exact detailed calculations of the influence of the 'two sites' on
the proton chemical shifts in methanol are in progress.

CONCLUSION

NMR studies of acetonitrile, benzene and methanol in two thermotropic liquid crystals of opposite diamagnetic anisotropies show
the presence of two spectra with dipolar couplings in the ratios of
1:2 with opposite signs at a particular concentration and temperature. They can be used to determine the spectral parameters which
otherwise cannot be determined from the spectra of orietned molecules. The experiments also provide unique methods for the determination of chemical shift anisotropy, the local effects on the
proton chemical shifts and the evidence for the 'multiple site'
theory of orientation. The studies pose several theoretical problems which have to be investigated.

REFERENCES

1. C. L. Khetrapal and A. C. Kunwar, Mol. Cryst. Liq. Cryst. 72,
        13 (1981).

2. C. L. Khetrapal and A. C. Kunwar, Chem. Phys. Lett. 82, 170
        (1981).

3. P. Diehl and C. L. Khetrapal, NMR-Basic Principles and
        Progress, Springer-Verlag, Vol. 1, 1969.

4. P. Diehl, C. L. Khetrapal and H. P. Kellerhals, Mol. Phys. 15,
        333 (1968).

5. C. L. Khetrapal and A. C. Kunwar, J. Mag. Res. 15, 389 (1974).

6. S. B. Marks, R. Potashnik and S. D. Goren, J. Mag. Res. 17,
        132 (1975).

7. H. Bösiger and P. Diehl, J. Mag. Res., 38, 361 (1980).

8. C. C. Costain, J. Chem. Phys. 29, 864 (1968).

9. G. Englert and A. Saupe, Mol. Cryst. Liq. Cryst. $\underline{8}$, 233
        (1969).

10. J. Lounila and J. Jokisaari (private communication).

11. H. Zimmermann, Ph.D. Thesis, Basel University, Switzerland
        (1978).

12. P. Diehl, M. Reinhold, A. S. Tracey and E. Wullschleger, Mol.
        Phys. $\underline{30}$, 1781 (1975).

DISCOTIC MESOPHASE: A COMPLEMENTARY REVIEW

J. C. Dubois[*] and J. Billard[**]

[*] Laboratoire Central de Recherches, Thomson-CSF
Domaine de Corbeville, 91401 Orsay Cédex, France
[**] Laboratoire de Physique de la Matière Condensée
(équipe associée au C.N.R.S.), Collège de France
75231 Paris Cédex 05, France

ABSTRACT

Fifty four reports devoted to the discotic states of matter and related subjects which have appeared since the first review paper (January 1980) are reviewed in this article.

INTRODUCTION

Since the first review paper (1) (complementary information on the references included therein are given in the appendix) was presented in January 1980, and a second one appeared in July 1981 (2), numerous papers have been published with new data about previously reported materials and with new chemical series including new central cores and new side chains. New sequences of mesophases versus the temperature have been found, especially reentrant mesomorphism. Miscibility data is now more numerous with many examples of non ideal behavior. The structures of several discophases have been obtained along with interesting results on oriented samples. The most important new developments concern the beginning of intense experimental and theoretical physical studies. Calorimetric,

dilatometric and pressure effects have been studied. Detailed
information has been collected from N.M.R. studies on molecular
motion and order parameters. Detailed works on defects have been
achieved. Some work has also been done on hydrodynamic properties.
The importance of the discophase as a model for carbonaceous mesophases is now understood. This recent branch of mesomorphic
materials is in rapid development.

MATERIALS

Molecular Architecture

With the exception of the di-isobutylsilane diol (3) and,
accepting an hypothesis of Lydon (4) for one monotropic meosphase
of the 3'-nitro-4'-n-hexadecyloxybiphenyl-4-carboxylic acid, all
the other discogens known to date have a rigid central part and
four to eight flexible side chains.

Central Cores

With the calamitic mesogens the central part of the flat
molecules can have different symmetries. The central cores used to
date to synthesize discogens are listed in Figs. 1 and 2 according
to their thermally averaged symmetries: hexagonal 6/m m m for I
(7), tetragonal 4/m for the only lyotropic discogenic material II
(8), trigonal $\bar{3}$ m for III (9) and 3 for IV (10), binary m m m for V
(5) and VI (3), 2 m m for VII (11), 2/m for VIII (12), IX (13) and
X (14). The discogenic potentialities of nine aromatic cores with
trigonal and binary symmetries are evaluated on the basis of the
virtual transition temperatures (15). The depolarized Rayleigh
scattering method used to evaluate the mesogenic power of elongated
central parts (16) has not yet been extended to the disk-like
cores. Because of steric encumbrance the aromatic rings of the
hexaphenylbenzene are not coplanar and the hexa-p-n-alkyloxy
derivatives are not discogenic (17). Other candidate structures
tried without success are the tetra-(4-alkyloxyphenyl) tetrathiol-

fulvalenes (18) and the phloroglucinol 3-5-dialkyloxybenzoic acid
esters (15). The temperatures and the latent heats for the meltings
of eleven compounds with flat aromatic cores have been measured
(19). A large choice of central cores is offered to the chemists
to synthesize new discogens.

Side Chains

The side chains used to date are listed in Fig. 3. We use
later to note a specific compound a roman numeral for the core type
(Fig. 1 and 2) followed by a letter for the side parts type (Fig.
3) and by a number which indicates the m value, if necessary.

Recently Studied Discogens

The most numerous new data concern the triphenylene derivatives: III h 8, III i 9 (22), III d 7 to 10, III e 4 to 12 (24),
III b 11 (25) and III k 7 (15). Seven dissymmetric hexa-n-alkyloxy
triphenylenes are described: VII b 5, 9; VII b 5, 10; VII b 5, 11;
VII b 6, 8; VII b 8, 6; VII b 9, 5 and VII b 10, 5 (11), here the
first number indicates the R' radical length and the second, the R
length (Fig. 2). Two chiral triphenylene derivatives are studied:
VII g (21) and VII j (23).

Numerous recent papers are related to the truxene derivatives:
IV a 6, 7 and 8 (26), IV a 6 to 13 (27), IV a 14 (28), IV e 11
(29), IV a 15 (30) and the chiral IV g (21).

Three other series have been studied: IX f 1, 2 (13), X C 10
(14) and VIII a 6 to 9 (15).

MESOPHASES

Nomenclature

Nomenclature problems have been discussed and it has been proposed that the term discotic be reserved for the molecules themselves and the terms nematic and columnar for their mesophases

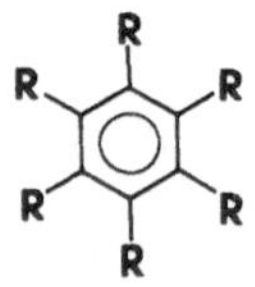

I

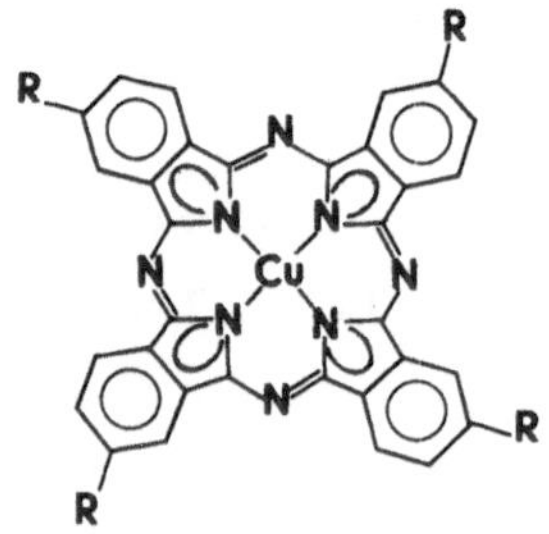

II

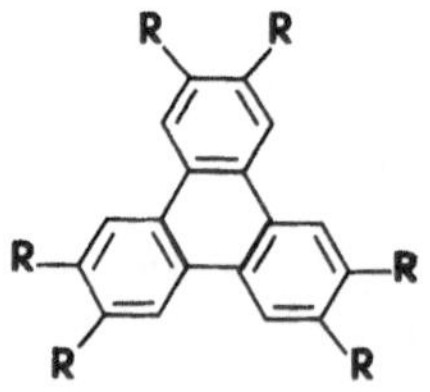

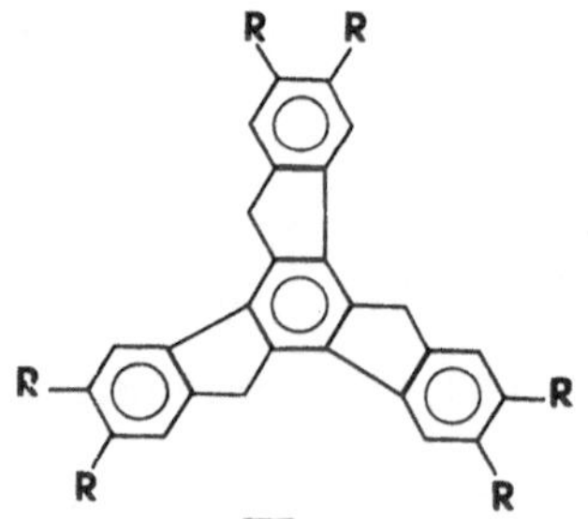

III IV

Figure 1. Discogens with hexagonal, tetragonal and trigonal thermally averaged symmetries. For the side chains R see Figure 3.

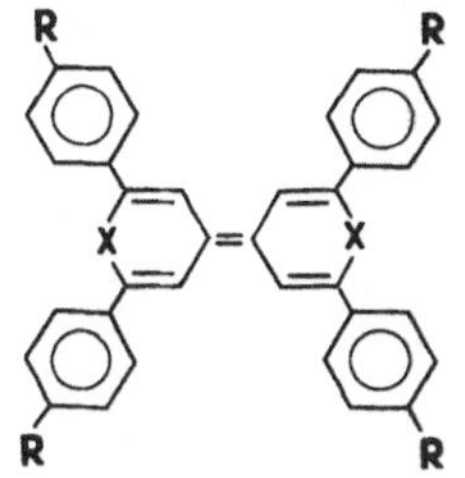

V

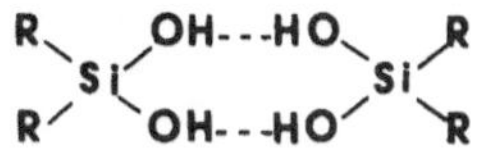

VI

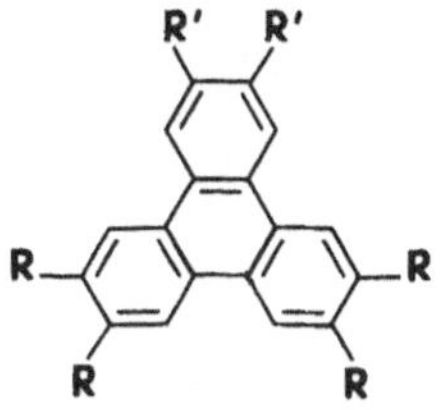

VII

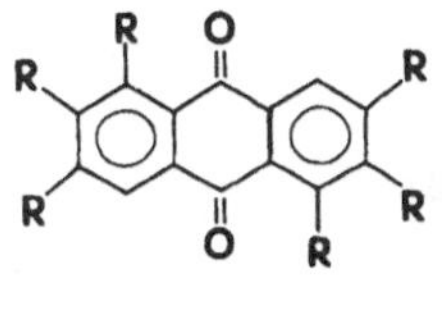

VIII

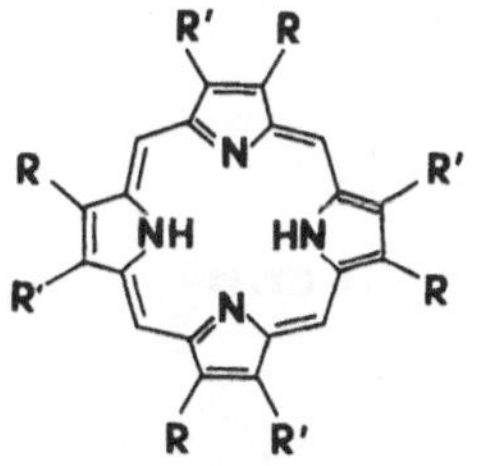

IX

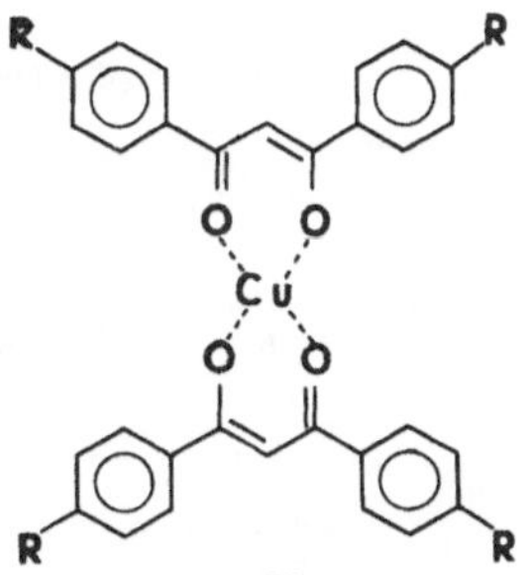

X

Figure 2. Discogens with binary thermally averaged symmetries. The hetero atom X can be oxygen (5) or sulphur (6). For the side chains R and R' see Figure 3.

(25). Georges Friedel (31) has given the names smectic (from σμηγτίκος : which cleans) and cholesteric to denote the molecules which give these mesophases (soaps for smectic). This was our reason to give, in 1978, the general name discotic (from δίσχος : quoit) to the mesophases obtained with disk-like molecules (9). We

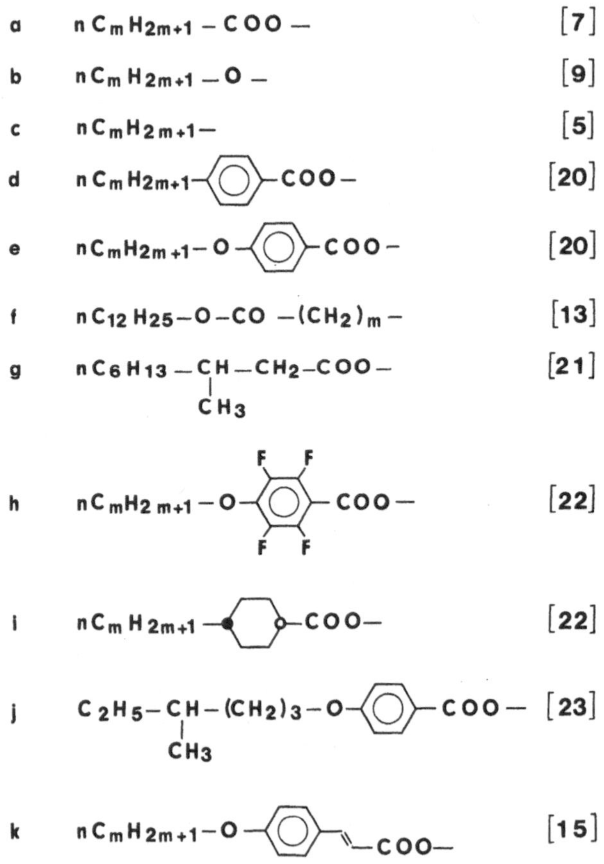

a $n\,C_m\,H_{2m+1} - COO -$ [7]

b $n\,C_m\,H_{2m+1} - O -$ [9]

c $n\,C_m\,H_{2m+1} -$ [5]

d $n\,C_m\,H_{2m+1} -$ ⬡ $- COO -$ [20]

e $n\,C_m\,H_{2m+1} - O -$ ⬡ $- COO -$ [20]

f $n\,C_{12}\,H_{25} - O - CO - (CH_2)_m -$ [13]

g $n\,C_6\,H_{13} - \underset{\underset{CH_3}{|}}{CH} - CH_2 - COO -$ [21]

h $n\,C_m\,H_{2m+1} - O -$ ⬡ $- COO -$ (F F / F F)   [22]

i $n\,C_m\,H_{2m+1} -$ ⬡ $- COO -$ [22]

j $C_2\,H_5 - \underset{\underset{CH_3}{|}}{CH} - (CH_2)_3 - O -$ ⬡ $- COO -$ [23]

k $n\,C_m\,H_{2m+1} - O -$ ⬡ $- COO -$ [15]

Figure 3.   Side Chains R (see Figs. 1 and 2) used to elaborate discogens.

think that to give the same name to the nematic mesophases obtained with elongated molecules and to the fluid discotic mesophases exhibited by quoit-like molecules is a source of confusion and we prefer discotic for all the mesophases from disk-like molecules. After the experimental evidence for discotic polymorphism (32) another problem is distinguishing the different discophases. At the second Bangalore International Liquid Crystals Conference (1979) two different notations were proposed: first one containing structural informations: D_t, D_{ho}, D_{hd}, D_{rd} and N_D with D for columnar mesophases, t for tilted, h for hexagonal, r for rectangular, o for regular, d for irregular and N_D for the fluid discophase exhibiting schlieren textures (10) and a second one similar to the one used for smectic mesophases ($\mathcal{D}_A$, $\mathcal{D}_B$ and $\mathcal{D}_C$) based on the miscibility criterion (12). One of us has suggested using $\mathcal{D}_F$ for the fluid discophase (1). The argument that $\mathcal{D}_B$ has no hexagonal symmetry contrary on smectic B (25) is not convincing. Of course the subscript index letters (given chronologically in the order of their discoveries) can't correspond to discophases having a symmetry similar to the one of the smectic phases called with the same letter, but $\mathcal{D}$ is different from S without doubt. Frequently by symmetry or miscibility studies discotic polymorphism is established before the structures are determined. Another reason is the following: the $\mathcal{D}_B$ and $\mathcal{D}_C$ discophases have different rectangular lattices of disorganized columns (33) and the notation D_{rd} is insufficient. To sum up we prefer the most simple symbols. Figure 4 gives a comparison of the two notation schemes.

Polymorphism

The $\mathcal{D}_F$ discophase was theoretically envisaged (34) prior to its observation. The geometric shape of molecules is insufficient to predict the type of discophase (35). From miscibility and X-ray studies (see later) new mesophase sequences versus the temperature were found. The first observed order versus increasing temperature is:

	$\mathscr{D}_A$ D_{hd}	$\mathscr{D}_B$ D_{rd}	$\mathscr{D}_C$?	$\mathscr{D}_E$ D_{ho}	$\mathscr{D}_F$ N_D	$\mathscr{D}_F^*$ N_D^*	$\mathscr{D}_G$ D_t
examples	III a 10 to 12	III a 7 to 12 VIII a 6 to 9	VIII a 8	III b 5 to 7	III d 8 and 9 III e 4 to 11 III k 7 IV a 9 to 13 IV e 11	III j	III e 6

Figure 4.   Correspondances between two used notations for the
discophases. For the examples chemical formulas, see
Figs. 1 to 3.

$$\mathscr{D}_C, \ \mathscr{D}_B, \ \mathscr{D}_A, \ \mathscr{D}_F \qquad\qquad (1)$$

More recently a $\mathscr{D}_F$ preceding columnar discophase was found in IV b
series (27, 28) and two $\mathscr{D}_F$ phases with one clearly reentrant in IV
b 11 (29). This result is not really surprising because it was
known that no relation exists between the symmetries and the stability temperature ranges for the polymorphic solid phases. When
the symmetry is lowered by heating a supercooling can be observed.
On the contrary for the fluid columnar transition of IV a 9 a
superheating over 2°C is observed, but not supercooling.

Discophases theoretically predicted (36, 37) have not yet been
observed.

Miscibilities

Some new isobaric binary phase diagrams have been published:
first for two members of the same series: III e 8 and 10 (24), III
a 10 and 11, III a 10 and 12 (25), IV a 9 and 10 (27), VIII a 6 and
7, VIII a 7 and 8 (15) and, second, phase diagrams for two components from two different series: III a 7 and III d 8 exhibiting a $\mathscr{D}_F$
phase between 57 and 100°C (22); III b 5 and VII b 9, 5 (11) establishing the isomorphy between the dissymmetric and symmetric triphenylene derivatives; III a 7 and III b 8 (25) proving that III b

8 has no $\mathcal{D}_B$ phase; III e 10 and IV e 11 (29) establishing clearly
the reentrant character of the low temperature $\mathcal{D}_F$ phase of IV e 11;
IX f 1, 2 and I a 6 (13); III k 7 and III e 9 (15). Two phase
diagrams concern mixtures containing a chiral component: III e 6
and III j (23); III e 7 and the non mesogenic hexa-(4-methylhexa-
noyloxy) benzene (15). Many phase diagrams establish important non
ideal behavior; specifically enhancements for the columnar discotic
ranges are observed in III a 7 - III b 8 (25); in the mixtures of
hexa-(p-n-alkyloxyphenyl) benzenes with III b 5 (17) and for different aromatic cores with III f 9 and III f 7 (15).

Structures

From microscopical observations a tilted structure is proposed
for the discophase exhibited by the I a compounds (38). Three
papers report structural information obtained from the X-ray dif-
fraction patterns for $\mathcal{D}_A$ of III a 11; $\mathcal{D}_B$ of III a 7, III a 11 (39);
$\mathcal{D}_B$ and $\mathcal{D}_C$ of VIII a 7 (33); $\mathcal{D}_F$ of III e 6, III e 11 and $\mathcal{D}_G$ of III e
6 (40). The structures of the discophases of other compounds can
be deduced from the isomorphies.

In $\mathcal{D}_A$ the molecules are irregularly stacked in columns having
an averaged cylindrical symmetry. The columns are ranged in a two-
dimensional hexagonal array. In $\mathcal{D}_B$ the columns do not have cylindrical symmetry and their array is P_{gg} rectangular. The discophase
$\mathcal{D}_C$ differs from $\mathcal{D}_B$ by the two- dimensional rectangular lattice: P_{mg}
in $\mathcal{D}_C$. The structural diversity in discophases is more rich than
we had imagined (9)! In $\mathcal{D}_E$ the lattice is rectangular, the columns
have cylindric symmetry but the molecules are more regularly stacked
in the columns than in the $\mathcal{D}_A$ phase (41). In $\mathcal{D}_F$ the centers of the
molecules have a random distribution, but the planes of the mole-
cules are, on the average, parallel. The $\mathcal{D}_G$ discophase has ordered
columns with apparently infinite-fold symmetry arranged in a rectangular array and the planes of the molecules are not perpendicular
to the column axis.

The structure of the solid phase of the non discogenic III a 3 compound has been reported (42).

Oriented Samples

Different uniform orientations can be easily obtained for the $\mathcal{D}_F$ phase. Some discophases having spontaneously a homeotropic alignment over cleaved faces of apophyllite and muscovite crystals have been described. Plates of $\mathcal{D}_F$ phases sandwiched between two glass slides coated with flat molecules possessing six polar side functions have the molecular planes parallel to the walls. With glass slides coated with silicon oxide deposited at an oblique angle the molecular planes are perpendicular to the walls (22) and parallel to the streaks (43) (parallel alignment (22)). The orientation of $\mathcal{D}_F$ can also be obtained with a 0.3 T rotating magnetic field (40) and by an electric field (44). Between two streaked glass surfaces with non parallel streaks, twisted fluid discophase ($\mathcal{D}_F$*) can be obtained. Mutual orientations of $\mathcal{D}_F$ and columnar discophases have been reported. A normal orientation (columns perpendicular to the surface and the lattice uniformly oriented by the surface) was obtained for the columnar discophase of III h 8 by cooling the liquid phase in contact with a fresh muscovite cleavage. The same normal orientation for III d 8 was given by muscovite and apophyllite cleavages. The orientation of the columns perpendicular to the surface, but with the lattice orientation uniform only in microscopic domains (normally oriented areas) is observed for $\mathcal{D}_E$ of III b 5 between glasses coated with mellitic acid. For the samples of III d 8 and III h 8 almost the whole surface has this orientation between glasses coated with hexaphenol. (Most of the areas of the columnar discophase of III d 8, obtained by cooling of $\mathcal{D}_F$ oriented with parallel alignment by streaked surfaces, have their neutral lines parallel and perpendicular to the streaks (22)). An elegant method to produce freely suspended fine strands of III a 11 has recently appeared. For $\mathcal{D}_A$ the columns are parallel to the strand axis and for $\mathcal{D}_B$ the orientation is slowly variable along the strand axis (45). By slow cooling in a rotating intense magnetic field (6.3 T) the liquid of III b 6 gives the $\mathcal{D}_E$ phase uniformly oriented (46).

PHYSICAL PROPERTIES

Many routine measurements of the molar enthalpy changes at the phase transitions have been reported in the papers devoted to the materials (see above). In accordance with theoretical analysis (47, 48) all the observed clearing transitions are first order. Precise measurements have been made, including heat capacity, on I a 6 between 13 and 393°K (49). The thermal expansions and the molar volume changes at the clearing points were measured by dilatometry for I a 6 and 7 (50) and calculated for I a 7 (47). The molar volume changes for all the transitions were obtained for III b 8 and III e 8 from the thermobarograms (51). Some tentative efforts to calculate the free energy of columnar discophases have been made (34, 52, 53). The main static magnetic susceptibilities of the liquid and $\mathcal{D}_F$ phases were measured for III e 6 and 11 (40). The principal conductivities and dielectric susceptibilities at low frequency were determined for $\mathcal{D}_F$ of III e 9 (44). The infra-red spectra from 30 to 1500 cm^{-1} of all the phases of I a 6 have been studied. In the discotic phase the side chains are highly disordered (49). Similar results were obtained by N.M.R. on the columnar discophases of III b 8 (54), III b 6 (46, 55) and III b 5 (56). These measurements also give information about translational self-diffusion, the molecular plane fluctuations and molecular rotations (57). In $\mathcal{D}_E$ of III b 6 the order parameter for the cores is elevated and weakly temperature dependent, even close to the clearing point (46, 55).

Six papers devoted to defects were recently published, the first two theoretical (58, 59) and the second three include observations on $\mathcal{D}_E$ of III b 5 (60 to 62). The rectilinear axis (9) can be virtual, $+ \pi$ and $+ 2\pi$ disclinations. Two different kinds of walls were observed and a model was proposed for the shear textures (60). The anchoring energy on the glass has been estimated and the examination of the associations of two defects established the sixfold symmetry of the $\mathcal{D}_E$ phase (61). Consequences for the topography of the free surfaces have been described (62). In a more recent paper (53) a new type of instability specific to the $\mathcal{D}_E$ columnar mesophase is theoretically treated. The textures of other discophases were sometimes described in the papers related to the materials (see above).

The discovery of the discophases has stimulated two theoretical papers on hydrodynamics of anisotropic fluids (63, 64).

CONCLUSION

We have not reviewed the phases involving disk shaped micelles (see for example (65, 66)) and other subjects related to the discotic states of matter. But the fifty-four recently published papers furnish interesting results on new materials, new polymorphic sequences, miscibilities, structures, methods to generate oriented samples. Experimental and theoretical physical studies are now in rapid development. Interest in the $\mathcal{D}_F$ phase as a model for the carbonaceous mesophases (67) (5.10^8 tons per year) is understood (68), but until now no pure discogenic hydrocarbon has been made. This void will be filled later. More general efforts are currently being made to elaborate other discogens and to elucidate the relations between the molecules and the thermodynamical stability of the mesophases in order to discover other discophases and to obtain mesophases stable at room temperature. The physical studies need to be extended; for example: analysis for the defects of other discophases as done for $\mathcal{D}_E$ and molecular motions in discophases of compounds out of the III b series. Only one dielectric measurement has been made and that at low frequency. No ultrasonic study has been performed. Another exciting problem is the following: one director is insufficient to describe the averaged molecular orientation in the optically biaxial $\mathcal{D}_B$ and $\mathcal{D}_C$ phases. To sum up it is clear that large perspectives are open for chemists and physicists in this field.

ACKNOWLEDGMENTS

We thank authors for sending preprints and reprints of their work.

REFERENCES

1. J. Billard, Liquid Crystals on One-and-Two Dimensional Order (edited by W. Helfrich and G. Heppke) Springer, Berlin 383-95 (1980).

2. S. Chandrasekhar, Mol. Cryst. Liq. Cryst., 63, 171-9 (1981).

3. J. D. Bunning, J. W. Goodby, G. W. Gray and J. E. Lydon, Liquid Crystals of One-and-Two-Dimensional Order (edited by W. Helfrich and G. Heppke) Springer, Berlin 397-402 (1980).

4. J. E. Lydon, Mol. Cryst. Liq. Cryst. Let., 72, 79-87 (1981).

5. R. Fugnitto, H. Strzelecka, A. Zann, J. C. Dubois and J. Billard, Chem. Comm., 271-2 (1980).

6. H. Strzelecka, to be published.

7. S. Chandrasekhar, B. K. Sadashiva and K. A. Suresh, Pramana, 9, 471-80 (1977).

8. S. Gaspard, A. Hochapfel and R. Viovy, C. R. Acad. Sci., Paris, 289 C, 387-90 (1979).

9. J. Billard, J. C. Dubois, Nguyen Huu Tinh and A. Zann, Nouv. J. de Chi., 2, 535-40 (1978).

10. C. Destrade, M. C. Bernaud, H. Gasparoux, A. M. Levelut and Nguyen Huu Tinh, Liquid Crystals (edited by S. Chandrasekhar) 29-32 (1980).

11. Nguyen Huu Tinh, M. C. Bernaud, G. Sigaud and C. Destrade,
     Mol. Cryst. Liq. Cryst., 65, 307-16 (1981).

12. A. Queguiner, A. Zann, J. C. Dubois and J. Billard, Liquid
     Crystals (edited by S. Chandrasekhar) 35-40 (1980).

13. J. W. Goodby, P. S. Robinson, Boom-Keng-Tao and P. E. Cladis,
     Mol. Cryst. Liq. Cryst. Let., 56, 303-9 (1980).

14. A. M. Giroud-Godquin and J. Billard, Mol. Cryst. Liq. Cryst.,
     66, 147-50 (1981).

15. P. Le Barny, J. Billard and J. C. Dubois, these Proceedings.

16. C. Destrade, Nguyen Huu Tinh and H. Gasparoux, Mol. Cryst.
     Liq. Cryst., 59, 273-88 (1980).

17. M. Dvolaitzky and J. Billard, Mol. Cryst. Liq. Cryst., Let.,
     64, 247-52 (1981).

18. A. Babeau, Nguyen Huu Tinh and H. Gasparoux, Mol. Cryst. Liq.
     Cryst. Let., 72, 171-6 (1982).

19. G. W. Smith, Mol. Cryst. Liq. Cryst. Let., 64, 15-7 (1980).

20. Nguyen Huu Tinh, C. Destrade and H. Gasparoux, Phys. Let.
     72 A, 251-4 (1979).

21. C. Destrade, Nguyen Huu Tinh, J. Malthete and J. Jacques,
     Phys. Let., 79 A, 189-92 (1980).

22. C. Vauchier, A. Zann, P. Le Barny, J. C. Dubois and J. Billard,
     Mol. Cryst. Liq. Cryst., 66, 103-14 (1981).

23. J. Malthete, C. Destrade, Nguyen Huu Tinh and J. Jacques, Mol.
     Cryst. Liq. Cryst. Let., 64, 233-8 (1981).

24. Nguyen Huu Tinh, H. Gasparoux and C. Destrade, Mol. Cryst.
     Liq. Cryst., 68, 101-11 (1981).

25. C. Destrade, Nguyen Huu Tinh and H. Gasparoux, Mol. Cryst.
     Liq. Cryst., 71, 111-35 (1981).

26. C. Destrade, J. Malthete, Nguyen Huu Tinh and H. Gasparoux,
     Phys. Let., 78 A, 82-4 (1980).

27. C. Destrade, H. Gasparoux, A. Babeau, Nguyen Huu Tinh and
     J. Malthete, Mol. Cryst. Liq. Cryst., 67, 37-48 (1981).

28. Nguyen Huu Tinh, J. Malthete and C. Destrade, Mol. Cryst. Liq. Cryst. Let., 64, 291-8 (1981).

29. Nguyen Huu Tinh, J. Malthete and C. Destrade, J. de Phys. Let., 42, 417-9 (1981).

30. C. Destrade, P. Foucher and Nguyen Huu Tinh, Fourth Inter. Liquid Cryst. Conf. Soc. Countries, Tbilissi, Abstr. I, 334-5 (1981).

31. G. Friedel, Ann. de Phys., 18, 273-474 (1922).

32. C. Destrade, M. C. Mondon and J. Malthete, J. de Phys., 40C3, 17-21 (1979).

33. J. Billard, J. C. Dubois, C. Vauchier and A. M. Levelut, Mol. Cryst. Liq. Cryst., 66, 115-22 (1981).

34. E. I. Kats, Zh. Eksp. Teor. Fiz., 75, 1819-27 (1978), Sov. Phys. J.E.T.P., 48, 916-20 (1978).

35. V. K. Pershin and Vl. K. Pershin, Fiz. Tverd. Tela, 21, 2292-7 (1979), Sov. Phys. Solid State, 21, 1319-22 (1979).

36. A. L. Tsykalo, Zh. Fiz. Khim., 54, 3014-7 (1980), Russ. J. Phys. Chem., 54, 1729-31 (1980).

37. A. L. Tsykalo, Pis'ma Zh. Tekh. Fiz., 6, 495-8 (1980), Sov. Techn. Phys. Let., 6, 213-4 (1980).

38. F. C. Frank and S. Chandrasekhar, J. de Phys. Let., 41, 1285-8 (1980).

39. A. M. Levelut, Liquid Crystals (edited by S. Chandrasekhar) Heyden, London, 21-7 (1980).

40. A. M. Levelut, F. Hardouin, H. Gasparoux, C. Destrade and Nguyen Huu Tinh, J. de Phys., 42, 147-52 (1981).

41. A. M. Levelut, J. de Phys. Let., 40, 81-4 (1979).

42. M. Pesquel, M. Cotrait, P. Marsau and V. Volpihac, J. de Phys., 41, 1039-43 (1980).

43. N. Isaert, to be published.

44. J. C. Dubois, M. Hareng, S. Le Berre, J. N. Perbet and M. Tron, Appl. Phys. Let., 38, 11-3 (1981).

45. D. H. Van Winkle and N. A. Clark, to be published.

46. D. Goldfarb, Z. Luz and H. Zimmermann, J. de Phys., <u>42</u>, 1303-11 (1981).

47. W. M. Gelbart and B. Barboy, Accounts of Chem. Res., <u>13</u>, 290-6 (1980).

48. G. E. Feldkamp, M. A. Handschy and N. A. Clark, Phys. Let., <u>85 A</u>, 359-62 (1981).

49. Michio Sorai and Hiroshi Suga, Mol. Cryst. Liq. Cryst., <u>73</u>, 47-69 (1981).

50. Th. H. Smith and G. R. Van Hecke, Mol. Cryst. Liq. Cryst., <u>68</u>, 23-8 (1981).

51. H. Gasparoux, M. F. Achard, F. Hardouin and G. Sigaud, C. R. Acad. Sci. Paris, II, to be published.

52. R. Locqueneux, Pramana, <u>16</u>, 201-9 (1981).

53. M. Kleman and P. Oswald, J. de Phys, to be published.

54. A. F. Martins and A. C. Ribeiro, Portgal. Phys., <u>11</u>, 169-81 (1980).

55. V. Rutar, R. Blinc, M. Vilfan, A. Zann and J. C. Dubois, J. de Phys., to be published.

56. M. Vilfan, G. Lahajnar, V. Rutar, R. Blinc, B. Topic, A. Zann and J. C. Dubois, J. Chem. Phys., <u>75</u>, 5250-5 (1981).

57. S. Zumer and M. Vilfan, Mol. Cryst. Liq. Cryst., <u>70</u>, 39-56 (1981).

58. M. Kleman, J. de Phys., <u>41</u>, 737-45 (1980).

59. Y. Bouligand, J. de Phys., <u>41</u>, 1297-306 (1980).

60. Y. Bouligand, J. de Phys., <u>41</u>, 1307-15 (1980).

61. P. Oswald, C. R. Acad. Sci., Paris, <u>II 292</u>, 149-52 (1981), J. de Phys. Let., <u>42</u>, 171-3 (1981).

62. P. Oswald and M. Kleman, J. de Phys., <u>42</u>, 1461-72 (1981).

63. G. E. Volovik, Pis'ma Zh. Eksp. Teor. Fiz., <u>31</u>, 297-300 (1980), J.E.T.P. Let., <u>31</u>, 273-5 (1980).

64. H. Brand and H. Pleiner, Phys. Rev., <u>24 A</u>, 2777-88 (1981).

65. M. Acimis and L. W. Reeves, Can. J. Chem., <u>58</u>, 1542-9 (1980).

66. F. Y. Fujiwara and L. W. Reeves, Can. J. Chem., $\underline{58}$, 1550-7
     (1980).

67. H. Gasparoux, C. Destrade and G. Fug, Mol. Cryst. Liq. Cryst.,
     $\underline{59}$, 109-16 (1980).

68. C. J. Atkinson, J. R. Lander and H. Marsh, Brit. Polym., $\underline{13}$,
     1-4 (1981).

APPENDIX

Listed below is complementary information on some papers
quoted in (1). The reference number of the papers given below is
that used in the first review (1).

13. D. Augie, M. Oberlin, A. Oberlin and P. Hyvernat, Carbon, $\underline{18}$,
     337-46 (1980).

17. C. Destrade, M. C. Bernaud, H. Gasparoux, A. M. Levelut and
     Nguyen Huu Tinh, Liquid Crystals (edited by S. Chandrasekhar)
     Heyden, London, 29-32 (1980).

18. R. Fugnitto, H. Strzelecka, A. Zann, J. C. Dubois and J.
     Billard, Chem., Comm., 271-2 (1980).

19. S. Chandrasekhar, B. K. Sadashiva and K. A. Suresh, Liquid
     Crystals (edited by S. Chandrasekhar) Heyden, London, 33
     (1980).

21. A. Quequiner, A. Zann, J. C. Dubois and J. Billard, Liquid
     Crystals (edited by S. Chandrasekhar) Heyden, London, 35-40
     (1980).

31. J. Prost and N. A. Clark, Liquid Crystals (edited by S.
     Chandrasekhar) Heyden, London, 53-8 (1980).

35. C. Vauchier, A. Zann, P. Le Barny, J. C. Dubois and J. Billard,
     Mol. Cryst. Liq. Cryst., $\underline{66}$, 103-14 (1981).

39. M. Sorai, K. Tsuji and S. Seki, Mol. Cryst. Liq. Cryst., $\underline{80}$,
     33-58 (1980).

40. M. Sorai, K. Tsuji, H. Suga and S. Seki, Liquid Crystals
     (edited by S. Chandrasekhar) Heyden, London, 41-51 (1980).

45.   J. Billard, J. C. Dubois, C. Vauchier and A. M. Levelut, Mol.
      Cryst. Liq. Cryst., <u>66</u>, 115-22 (1981).

We have not yet received the Proceedings of the 3rd Liquid Crystals
Conference of the Socialist Countries (Budapest, 1979): Adv. Liq.
Cryst. Res. Appl., published in 1981 and therefore did not include
any papers on discogens from this conference.

IDENTIFICATION OF THE STRUCTURE OF MESOMORPHIC POLYMERS BY MEANS
OF THEIR MISCIBILITY

G. Hardy, F. Cser, K. Nyitrai and K. Kiss

Research Institute for Plastics
H-1950 Budapest, Hungary

INTRODUCTION

Recently many polymers show mesomorphic properties.[1-5] The
conventional, low molecular mesomorphic substances have some structural classes where they can be subordinated.[6] These structures
differ in the degree of the order as it is shown in Figure 1.

The most unoriented mesomorphic structure is the nematic one.
Here the only nonstatistical symmetry is the orientation of the
molecular axis along a given direction called director. The cholesteric structure is isomorphous with the nematic one. This is
the nematic structure of molecules with chiral construction.

The smectic structures have greater order then the nematic
one. Here a reciprocal translation vector is existing, i.e. the
molecules are differentiated in some subclasses due to the different symmetry within the layers. There are two main classes,
orthogonal and tilted classes, each containing a liquid like and
some crystalline like layer symmetry. Each subclass have a symbol from A to I, what refers to the smectic structural class of
the reference substance. The miscibility law of Sackmann and
Demus[7] states, that only those structures form fully miscible

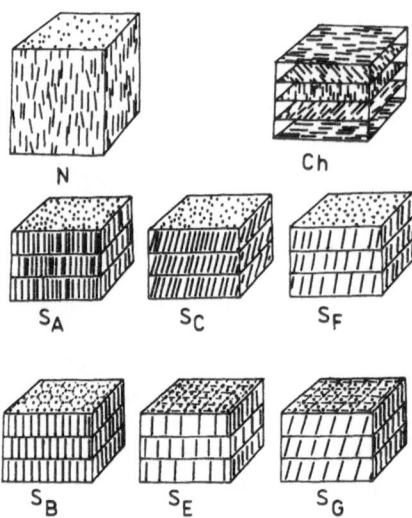

Fig. 1   General scope of liquid crystalline structures N = nema-
          tic, Ch = cholesteric, S = smectic, letters in subscript
          mean the type of the smectic structure.

series, what have the same symmetry, i.e. what belongs to the same
mesomorphic subclass. The structures found in the homologous
series of terephtaloyl-bis-p-alkyl aniline supplemented later with
those of other series is used as references for the identification.[8]   In our present paper we used terephtaloyl-bis-butyl-
aniline (TBBA) as reference compound.

MESOMORPHIC POLYMERS

     Using the miscibility test together with some other measuring
methods as x-ray diffraction, polarizing microscopy, thermomechan-
ics, NMR, we have identified the mesomorphic structure of some of
the polymers.[5,9-14]   These structures are represented in Fig. 2.

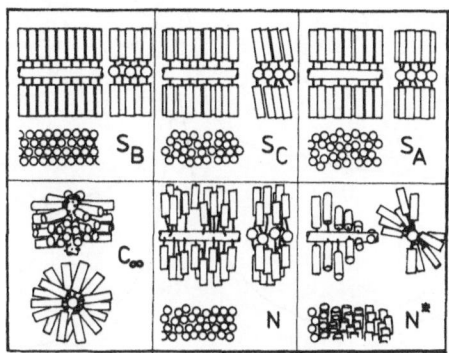

Fig. 2   Structural types of liquid crystalline polymers with meso-
          genic group in the side chains.   S = smectic, N = nematic,
          N* = cholesteric, C_∞ = aperiodic helix.

Here only the polymers with mesogenic groups in the side chain are
discussed.

The polymeric structures are basically different depending on
the chemical constitution of the monomeric units. When the mesogenic group is fixed directly to the polymeric main chain i.e.
there is no flexible chain element between them, the so called
aperiodic helical structure (C_∞) is formed.[12,13] This is a new
class of the mesomorphic substances, as structure it is structurally similar to the nematic structures but is not miscible with it
due to the great differences in the measure of the molecules. This
structure is not miscible with any of the known low molecular mesomorphic structures.

We found hexagonal layer structures (S_B) at most of the comb-
like polymers where the side chain was a long paraphinic chain.[14]
The sharp X-ray reflection peak indicates hexagonal lattice within
the layer, the magnitude of the great periodic spacing indicates
that there is one aliphatic chain in the layer (Fig. 3). We found

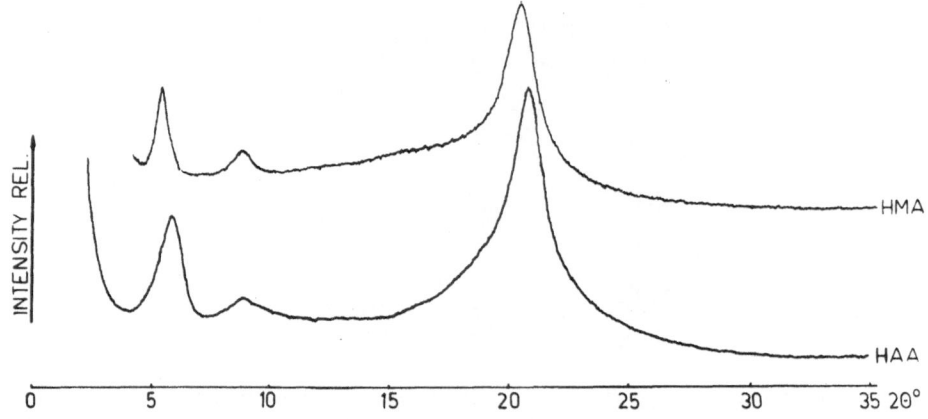

Fig. 3   Wide angle x-ray diffractogram of poly-hexadecyl acryl-
amide (HAA) and poly-hexadecyl methacrylamide (HMA).

some structures where the layers contained two aliphatic chains,
the layer period was double of the side chain length.[14] The S_B
structures formed in our experiments when the polymerization was
carried out in smectic or in solid state respectively.[15] No miscibility experiments were taken with S_B structures, as the polymers
were not miscible with the reference materials in the isotropic
state, too.

We could detect S_C structures with double layers at some polymers containing cholesteric group as mesogene. All these polymers
(cholesteryl-vinyl-succinate, VCS; cholesteryl-glycoyl-acrylate,
CSGA; cholesteryl-acryloyloxy-ethoxy-carbonate, CAEC) show amorphous structure in the wide angle X-ray scattering experiments.
They have one sharp X-ray diffraction peak in the small angle X-ray
scattering as is shown in Fig. 4 for CAEC. According to this figure, the layer period is over 4 nm, i.e. it is longer than the

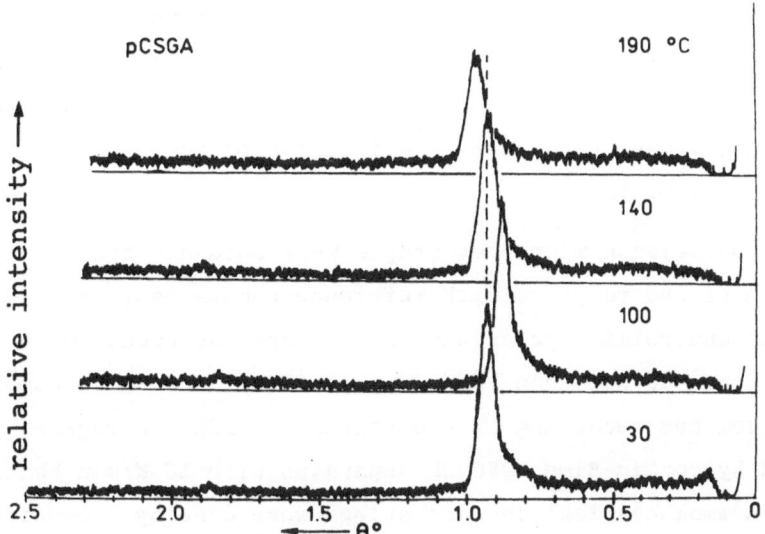

Fig. 4   Small angle X-ray diffractogram of poly-cholesteryl-
acryloyloxy-ethoxy carbonate at different temperatures.

length of the monomeric unit (cca 3 nm) and it is considerably
shorter than its twice value. The amorphous double layers can be
related to S_C structures. The miscibility experiments served to
prove these structures can not only be related to S_C, they are
even S_C structures.

In order to obtain miscibility diagrams we used polarizing
microscopy measuring the intensity of the depolarized light, DSC
thermometry and thermomechanical measurements.

EXPERIMENTAL

The preparation of the polymers (pCVS, pCSGA, pCAEC and
pMPMEB) and the reference substances (TBBA, BPPB) are given else-

where.[9,13,16,17] Small angle X-ray diffractograms were recorded
in the laboratories of Section Chemistry of the Martin-Luther University (Halle, GDR). Monocromatized Cu-Kα radiation was used.
The intensity was measured by a scintillation counter. The scanning rate was 0.2°min^{-1}.

The polarizing microphotographs were obtained from contact
preparate of the polymers with reference substances. Boetzius type
hot stage and Polmi A polarizing microscope was used. MF 6.3 mi-
croprojector and Rathenon 10/0.16 objective for a Leica camera were
applied for the recording of the textures. DSC thermograms were
obtained by Perkin-Elmer DSC 2b apparatus with 10 K/min heating
rate. Thermomechanical investigations were done by a consistome-
ter of Höppler type using a sample of 1 cm diameter and height of
1 cm with a load of 50 KN for 10 sec. Figure 5 shows a series of
the polarizing microscopical pattern of the contact preparate of
TBBA with pCSGA as the function of the temperature. The pure polymer on the right side have broken textures that are the same as the
S_C texture of TBBA. This texture remained unchanged on the whole
temperature range. The TBBA showed some other texture at the same
time. The dark area at greater temperatures indicates that the
system has a minimal clearing point at a composition near to the
pure polymer.

Figure 6 represents the DSC traces of the TBBA-pCAEC systems.
The first peak from left to right represents the melting of the
crystalline TBBA to its S_G state. An additional endotherm process
begins at lower temperatures with increasing polymer content.
Before this a step on the base line indicates that the system has
a glass temperature. In some composition above T_g an exotherm indicates a further crystallization or recrystallization of TBBA.

Next peak indicates the $S_G \rightarrow S_C$ transition.' The temperature
of their endothermic peak is decreasing with increasing polymer
content and vanishes above $X_p = 0.4$. The $S_C \rightarrow S_A$ transition can

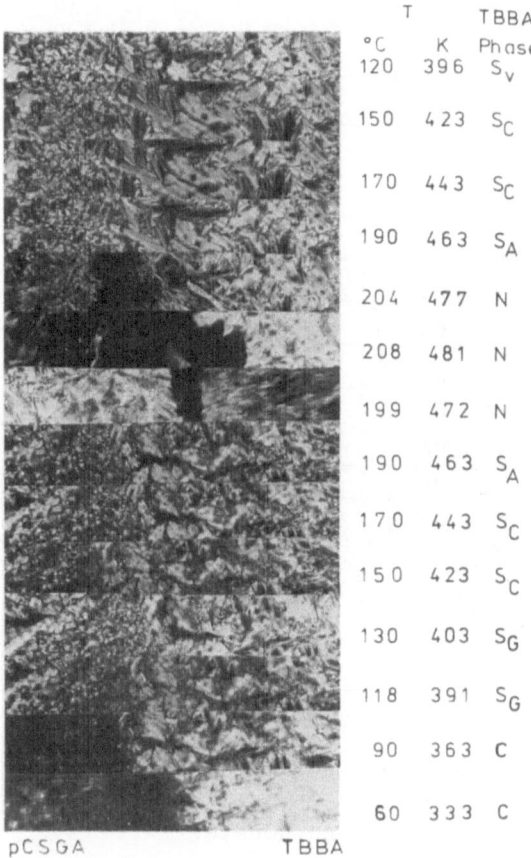

T		TBBA
°C	K	Phase
120	396	S_V
150	423	S_C
170	443	S_C
190	463	S_A
204	477	N
208	481	N
199	472	N
190	463	S_A
170	443	S_C
150	423	S_C
130	403	S_G
118	391	S_G
90	363	C
60	333	C

pCSGA   TBBA

Fig. 5   Polarizing microphotos of the contact preparatum formed
from poly-cholesteryl-succinyl-glycoyl-acrylate (pCSGA)
and terephtalyl-bis-butyl-aniline (TBBA) as the function
of the temperature.

only be observed at the pure TBBA as a step on the base line. At
samples with polymer content the indication of this process is not
sure. The phase transition is observed with polarizing microscopy
without difficulties. At higher polymer content an endothermic
process set when the system turns to S_A.

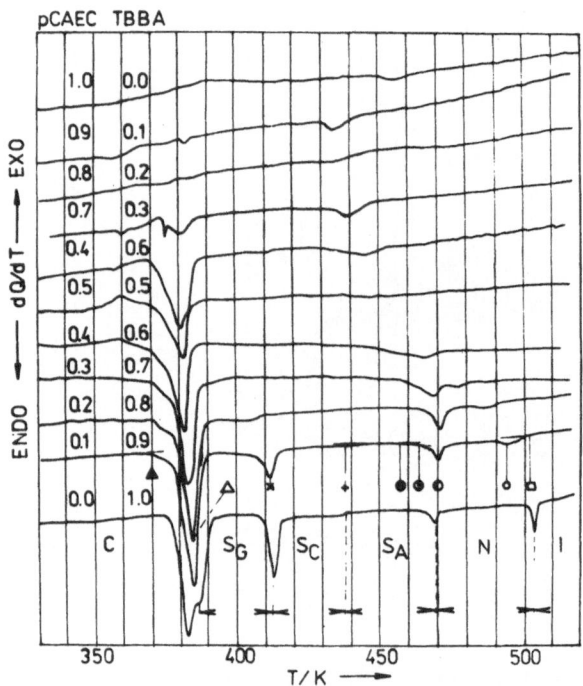

Fig. 6   DSC thermograms of systems containing different amounts
          of terephtalyl-bis-butyl aniline and poly-cholesteryl-
          acryloyloxy-ethoxy-carbonate.

Before the clearing point (last peaks) the endothermic peak
of $S_A \rightarrow N$ transition can be found. This transition temperature is
independent of the composition. The clearing point decreases with
increasing polymer content. From $X_p = 0.5$ and over the clearing
point is below the $S_A \rightarrow N$ transition, which means the nematic state
cannot be formed. A broad transition indicates the clearing process of the polymer.

The area below the endothermic peaks represent the transition
heats of the system. They are given in Figure 7. The endotherm
of C$\rightarrow$S$_G$ transition decreases with increasing polymer content and
trends to be zero at X$_p$ = 0.8. The endotherm in transition S$_G$$\rightarrow$S$_C$
behaves similarly, but it decreases linearly to zero up to X$_p$ =
0.3. These two endotherms determine the total transition heat.
The intersection of the extrapolated straightline with the abciss
indicate the amount of the TBBA which remained in the polymer as
plasticizer.

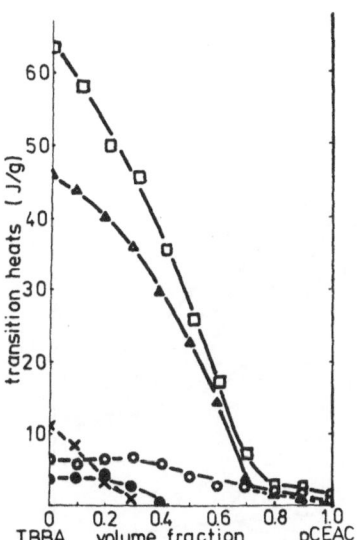

Fig. 7   Transition heats in the systems of TBBA and pCEAC as a
          function of the composition. $\bullet$ = N $\rightarrow$ I, X = S$_A$ $\rightarrow$ N,
          O = S$_G$ $\rightarrow$ S$_C$, $\triangle$ = C $\rightarrow$ S$_G$, $\square$ = total heat.

The change of the endotherm as a function of the composition
for the sum of the two transitions S$_A$ $\rightarrow$ N $\rightarrow$ I is characteristic to
that of solid solutions.

We give the miscibility diagram in Figure 8. The letters on
the diagram indicate the phases that are present. Most of the area
in the diagram are two-phase area. These are C+P$_G$, C+P$_H$, C+P$_L$,
S$_G$+P$_L$ etc. where P denotes the polymer, the subscripts G, H and L
mean glassy, high elastic or liquid states respectively. There
are only two one-phase areas that are extended between the two
coordinates. One of them is the isotropic state, the other one is
the S$_C$ state, which is equivalent to the P$_L$ state as they form
fully miscible series. The polymer melts from S$_C$ to isotropic (I)
state. The systems with increasing TBBA content first turn to S$_A$,
later, with great TBBA content, it turns in a nematic state, also.

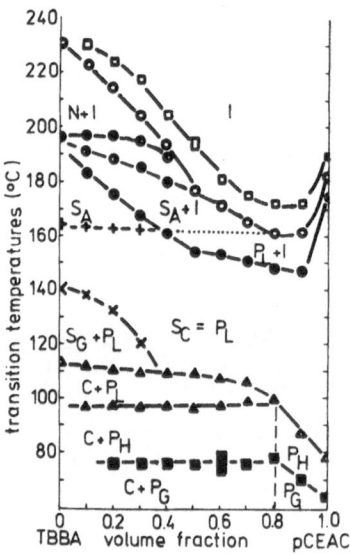

Fig. 8   State diagram of the system formed from TBBA and pCAEC.
          P = polymer phase, G, H and L in the subscript mean
          glassy, high elastic and liquid states respectively. C =
          crystalline, S = smectic, N = nematic states of TBBA
          respectively. I = isotropic solution.

It is worth noting that the glass transition temperature and the flow temperature of the polymer increases with increasing TBBA content. The latter has an endotherm which increases with increasing TBBA, too.

On the basis of similar results not shown here, the next figure (Fig. 9) represents the miscibility diagram of another polymer containing cholesteric group as mesogene; but its flexible connecting chain (spacer) is longer than that in pCAEC. This polymer is poly-cholesteryl-succinyl-glycoyl acrylate (pCSGA). Here the situation is similar to the previous system. The $S_C \rightarrow I$ tran-

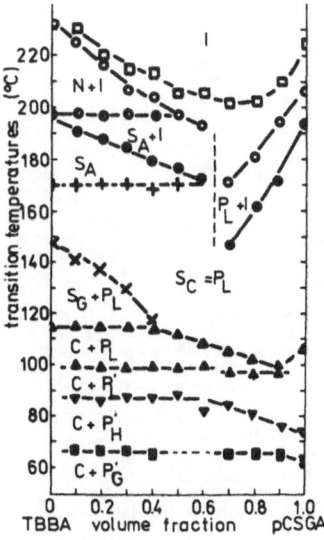

Fig. 9   State diagram of the system formed from TBBA and pCSGA. The meaning of the letters given at Fig. 8.

sition on the polymer side has a nonequilibrium character that
vanishes below X_p = 0.6. Here the transition from S_C to S_A is
appearing. Here, and in all of the investigated systems and polymers, the clearing point is a broad area; it is not a temperature
point when the systems contain polymer. The temperature dependence
of the clearing process is a function with minimal value at a
given composition, like the previous system. This is similar to
the melting of an eutectic system with limiting solubility.

Figure 10 shows the miscibility diagram of the TBBA-pCVS system. Poly-cholesteryl-vinyl-succinate has the shortest spacer
among the polymers shown here. The two phase areas are similar to
those of the previous ones. Below the melting point a true eutectic system is present here. The eutectics melt into isotropic

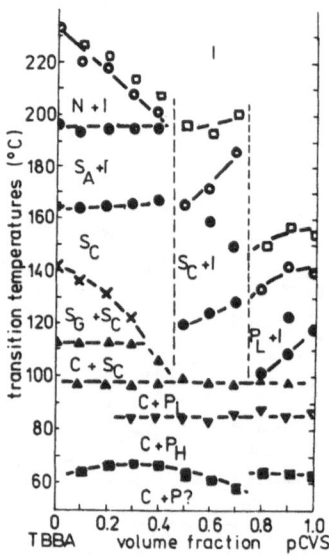

Fig. 10 State diagram of the system formed from TBBA and pCVS.
The meaning of the letters is given in Fig. 8.

states. The DSC thermograms and the polarizing microscopic investigations show strong nonequilibrium character of the systems with
medium polymer content. With a slow scanning rate we have one
phase system in the middle of the diagram, the miscibility of the
two components is a time consuming process. With a quick scanning
rate this part of the diagram seems to be two phasic. The poly-
CVS has tilted double layers, according to the X-ray experiments.
This phase does not form fully miscibilities with the similar
phase of TBBA. The short spacer should be the reason while the
complete miscibility fails.

   We have shown really miscible polymer/low molecular compound
phases with S_C structure. S_A polymers with S_A and N phases were
detected only in diluted solutions. The miscibility was never
complete in any cases. Figure 11 displays the miscibility diagram
of p-methoxy-phenyl-p-methacryloyloxy-ethoxy-benzoate (pMPMEB)
with p-buthoxyphenyl-p-propionyloxy-benzoate (BPPB). The low
molecular compound have crystalline and nematic state. The polymer has also two well defined states, but none of them is isomorphous with any of the states of BPPB. A nematic phase with great
polymer content is present, the clearing point of which is
increasing with increasing polymer content. Ringsdorf[1] stated
this polymer to be nematic below 120°C. This diagram indicates
that the polymer has no nematic state in the conventional sense.

   Our experiments show that the miscibility can well be applied
in identifying the mesomorphic state of polymers.

ACKNOWLEDGEMENT

   We are indebted to Dr. Siegmart Diele (Halle) for the small
angle X-ray diffraction experiments and the Halle group of Liquid
Crystal Research directed by Professor H. Sackman for the discussion of this topic.

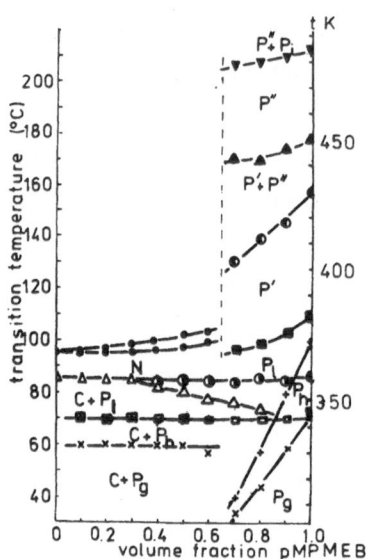

Fig. 11   State diagram of the system formed from p-butoxy-phenyl-
p-propyonyloxy benzoate (BPPB) and poly(p-methoxyphenyl-
p-acryloyloxy-ethoxy-benzoate) (pMPMEB). The meaning of
the letters is given in Fig. 8. P' and P" are liquid
polymers with different unidentified states.

REFERENCES

1.   H. Finkelmann, H. Ringsdorf, W. Siol and J. H. Wendorff: in
        Mesomorphic Order in Polymers and Polymerization in
        Liquid Crystalline Media. Ed. A. Blumstein. ACS Symp.
        Series, No. 74. Washington, D.C. (1978), p. 22.

2.   N. A. Platé and V. P. Shibaev: Comblike Polymers and Liquid
        Crystalls. (Russ) Chimiya, Moscow (1980).

3.   B. Fayolle, C. Noel and J. Billard: J. Phys. Colloque C3 40,
        C3-485 (1979).

4.   A. Rovielo and A. Sirigu: Makromol. Chem. 181, 1799 (1980).

5.   F. Cser, K. Nyitrai and G. Hardy: Acta Chim. Acad. Sci.
        Hung. 100, 463 (1979).

6.   D. Demus and L. Richter: Textures of Liquid Crystals.
            Deutsch. Vrlg. für Grundstoffindustrie, Leipzig (1978).

7.   H. Sackmann and D. Demus: Mol. Cryst. Liq. Cryst. 21, 239
            (1973).

8.   D. Demus, J. W. Goodby, G. W. Gray and H. Sackmann: Mol.
            Cryst. Liq. Cryst. 56. Lett. 311 (1980).

9.   F. Cser, K. Nyitrai and G. Hardy: Advances in Liquid Crystal
            Research and Applications. Ed. L. Bata, Pergamon.
            Akadémia, Oxford-Budapest (1980). p. 845.

10.   F. Cser, K. Nyitrai, G. Hardy, J. Menczel and J. Varga:
            J. Polym. Sci. Polym. Symp. 69, 91 (1981).

11.   F. Cser, K. Nyitrai, I. Kocsis and G. Hardy: Eur. Polym. J.
            17, 865 (1981).

12.   F. Cser: J. Phys. Colloque C3, 40, C3-459 (1979).

13.   G. Hardy, F. Cser, K. Nyitrai, G. Samay and A. Kalló: J.
            Cryst. Growth 48, 19 (1980).

14.   G. Hardy, F. Cser, K. Nyitrai and N. Fedorova: Proceedings of
            the Fourth Symposium on Radiation Chemistry. P. Hedvig
            and R. Schiller. Adadémiai Kiadó, Budapest (1976) p.
            365.

15.   F. Cser, K. Nyitrai, V. Dévényi and G. Hardy: Acta Chim.
            Acad. Sci. Hung. 96, 235 (1978).

16.   K. Nyitrai, F. Cser, G. Csermely, Bui Doc Ngoc, L. Füzes,
            G. Samay and G. Hardy: Eur. Polym. J. 14, 467 (1978).

17.   K. Nyitrai, F. Cser and G. Hardy: Acta Chim. Acad. Sci.
            Hung. 102, 361 (1979).

EFFECT OF MOLECULAR STRUCTURE ON MESOMORPHISM: 14[1]

DIFLUORINATED SIAMESE TWIN MESOGENS

Anselm C. Griffin,* Glenn A. Campbell* and
William E. Hughes[†]

Department of Chemistry* and Department of Physics and
Astronomy,[†] University of Southern Mississippi
Hattiesburg, Mississippi 39406

INTRODUCTION

Siamese Twin liquid crystalline compounds are of inherent
interest since they represent mesogenic species in which two independently liquid crystalline 'half' molecules are joined through a
chemical bond(s). Conceptually this amounts to joining together,
in pairs, near neighbor molecules in an ordinary (rod-like)
mesogenic material. Examples are shown below:

(FUSED)[2]

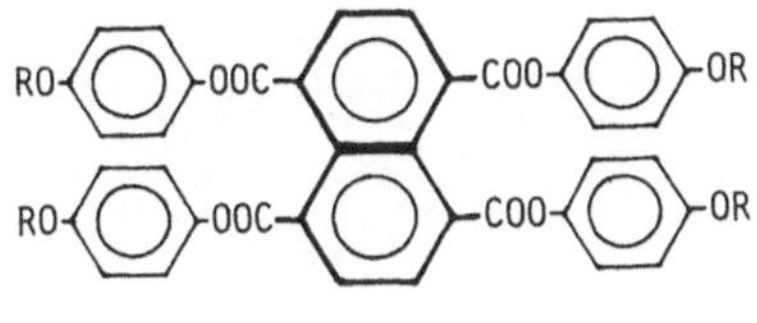

(I)

(LIGATED)[3]

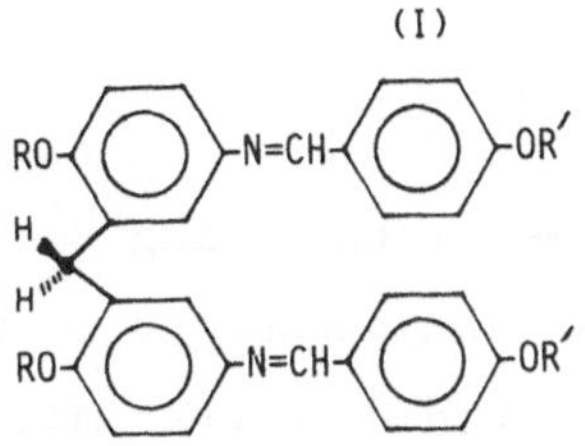

(II)

(TAIL-TO-TAIL) RO—⟨O⟩—COO—⟨O⟩—O(CH₂)₁₀—O—⟨O⟩—OOC—⟨O⟩—OR

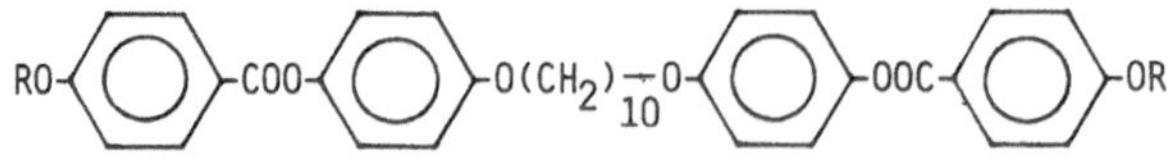

(III)

The liquid crystalline properties of such Twins are often strikingly different from those of ordinary mesogens. For example the R = R' = C_{10} homologue of series (II) above has, depending on prior thermal history, two possible smectic C layer structures differing in molecular conformation (parallel or anti-parallel) about the CH_2–Aryl bond.[3c] As part of a continuing effort to probe the limits of molecular structure compatible with liquid crystallinity, we have made new ligated twin mesogens.

We wish to report here the results of our work with Twins of the ligated type (II) in which two of the molecular 'ends' are fluorine atoms, as shown below. We will describe in the following sections the synthesis and structure determination, thermal analysis, optical microscopy, and mesophase x-ray diffraction data for these materials.

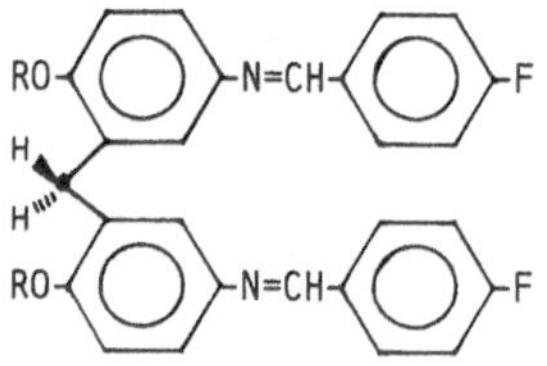

RO—⟨O⟩—N=CH—⟨O⟩—F
H
H
RO—⟨O⟩—N=CH—⟨O⟩—F

R = C_9-C_{15}

(IV)

EXPERIMENTAL

The synthetic scheme for obtaining the difluorinated Twins (IV) is shown below (Fig. 1). The scheme up to the last step has been described previously.[3a] We did however modify the original route in that the diether forming reaction (second transformation) was

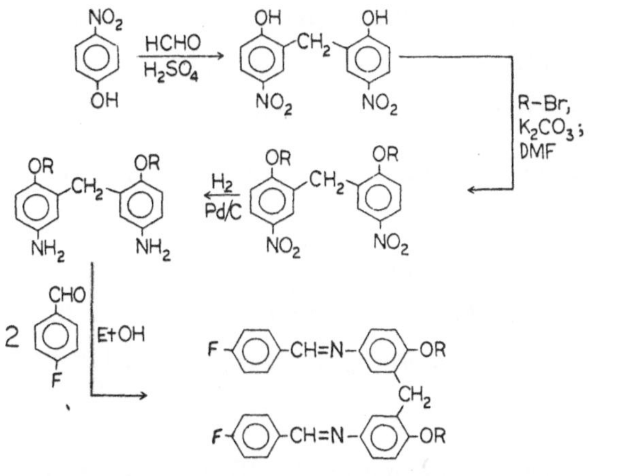

Fig. 1   Synthetic Route to Difluoro Siamese Twin Mesogens.

performed in refluxing dimethylformamide rather than in acetone
solution. This modification greatly reduced the necessary reflux
time from 100 hours to 4 hours. The final (product-forming) step
was accomplished by dissolving the diamino compound (1 molar
equivalent) in absolute ethanol at room temperature. Heating this
solution (via a hot air blower) to obtain complete dissolution of
the diamine was sometimes necessary. Two molar equivalents of
4-fluorobenzaldehyde in a minimum volume of absolute ethanol were
rapidly added to the stirred solution of diamine and the reaction
mixture was allowed to stir overnight at room temperature. The
product precipitates rather quickly from the solution. Next day
the mixture was filtered using water aspirator vacuum and the
collected solid was washed thoroughly with cold ethanol. The crude
product was vacuum filtered (aspirator) and later multiply recrystallized from a mixture of ethanol-chloroform and air dried.
Yields of pure product ranged from 60-85%. Spectral data (nmr, ir)
were obtained for each compound. Three compounds (randomly
chosen) were subjected to elemental analysis (Galbraith
Laboratories, Knoxville, TN). All results from characterization

are consistent with structures proposed. A complete set of data
is present as follows for the C-15 homologue.
Bis[2-n-pentadecyloxy-5-(4'-fluorobenzylideneamino)] methane:

<u>IR</u> (KBr pellet) cm^{-1}: 3028, 2928, 2852, 1629, 1600, 1320, 1250,
1108, 838.
^{1}H <u>NMR</u> (CDCl$_3$) δ: 0.9 - 1.9 (envelope, 58H, CH$_2$ and CH$_3$), 2.45
(singlet, 2H, Ar-CH$_2$-Ar), 4.2 (triplet, 4H, -O-CH$_2$-), 6.9 - 7.8
(envelope, 14H, Ar-H), 8.4 (broad singlet, 1H, -CH=N-).

elemental analysis:

   Calculated for C$_{57}$H$_{80}$N$_2$O$_2$F$_2$: %C 79.35; %H 9.28; %N 3.25
Found: %C 79.38 ;%H 9.48; %N 3.07

   Thermal characterization was performed using a du Pont model
990 Differential Scanning Calorimeter and optical microscopy data
were taken using a Reichert Thermovar polarizing light microscope
equipped with Mettler FP5/52 heating furnace. All mesophase x-ray
diffraction experiments were performed using a General Electric
XRD-700 diffractometer with nickel filtered copper Kα radiation
(1.54 Å). Polaroid type 157 film was used in a flat plate
geometry and samples were contained in 1.5 mm quartz capillary
tubes.

RESULTS AND DISCUSSION

   Presented in Table 1 are transition temperatures for the
difluorinated Twins. These temperatures were determined by optical
microscopy. All seven homologues show an enantiotropic smectic
A (S$_A$) phase. The optical texture of the S$_A$ phase is that of
focal conic fans and the S$_A$ assignment was made by miscibility
studies with known S$_A$ materials (contact preparations). The
column labelled K'→S$_A$ represents a transition to the S$_A$ phase
from a metastable solid (obtained by melt solidification). This
metastable solid phase for these compounds is not seen on the
first heating cycle of these materials. However once formed it

Table 1

Transition Temperatures (°C) for Difluoro Siamese Twin Mesogens

R	K→S$_A$	S$_A$→I	K'→S$_A$*
C$_9$H$_{19}$	97.8	107.2	84.2
C$_{10}$H$_{21}$	103.3	109.8	79.6
C$_{11}$H$_{23}$	99.2	109.5	78.0
C$_{12}$H$_{25}$	94.0	110.4	75.5
C$_{13}$H$_{27}$	87.8	110.1	74.4
C$_{14}$H$_{29}$	89.4	110.6	74.1
C$_{15}$H$_{31}$	87.5	109.6	73.3

* METASTABLE SOLID (OBTAINED BY MELT SOLIDIFICATION) → S$_A$ TRANSITION

transforms very slowly back to the thermodynamically more stable
solid phase, often requiring several days for complete conversion.
Figure 2 shows DSC scans of the C-9 homologue. Curve a is that of
a solvent crystallized sample with no prior thermal history. Curve
b is of the sample as in a, but was crystallized from the melt and
the curve obtained approximately 48 hours after solidification.
As can be seen from curve b, there has been some conversion K'→K
upon prolonged standing at room temperature. Heating rates are
20°C/minute. Invariably this K'→S$_A$ transition occurs at lower
temperatures than the K→S$_A$ transition. X-ray powder diffraction
photographs have shown that the thermodynamically stable solid
phase is more crystalline (more and sharper diffraction rings)
than the metastable solid phase. Figure 3 shows trends in
transition temperatures along the series. The S$_A$→I temperatures
exhibit an odd:even alternation with increasing alkoxy chain length,
albeit highly damped; only 3.4°C separate the minimum and maximum

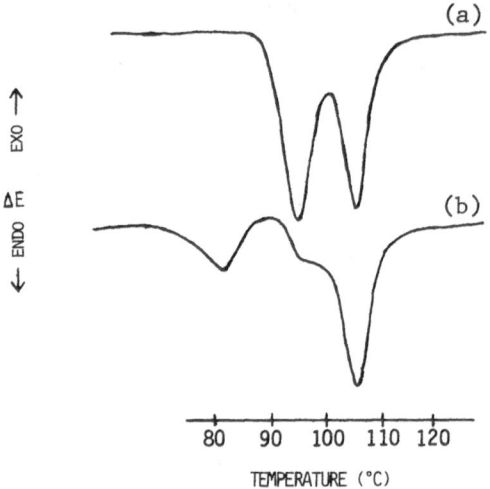

Fig. 2   DSC Curve for C-9 Homologue.

values in the homologous series.   There is for these Twins a
supercooling of the $I \rightarrow S_A$ transition, often 3-5°C below the $S_A \rightarrow I$
transition seen on heating.   The $K \rightarrow S_A$ temperatures vary irregularly
along the series, but the $K' \rightarrow S_A$ temperatures (in parentheses) show
a much more regular variation with increasing alkoxy chain length.
We feel this is reflective of the more amorphous solid phase (K')
freezing in smectic-like ordering upon solidification with little
if any differences in crystallinity or packing modes along the
series in these metastable solid phases.   We do not see in our
experiments the conversion of the metastable solid→stable solid on
heating as is usually the case for organic compounds.   Each solid
transforms rapidly to the S_A phase directly.   All transitions are,
by DSC analysis, somewhat broader than those usually found for
small molecule liquid crystals.   Since the purity is not in
question and this broadening effect is seen in all of our ligated
Twins to date, we attribute this broadening to the existence of
geometrical isomerism caused by the various arrangements of the
two trans-CH=N- linkages in each Twin molecule as described

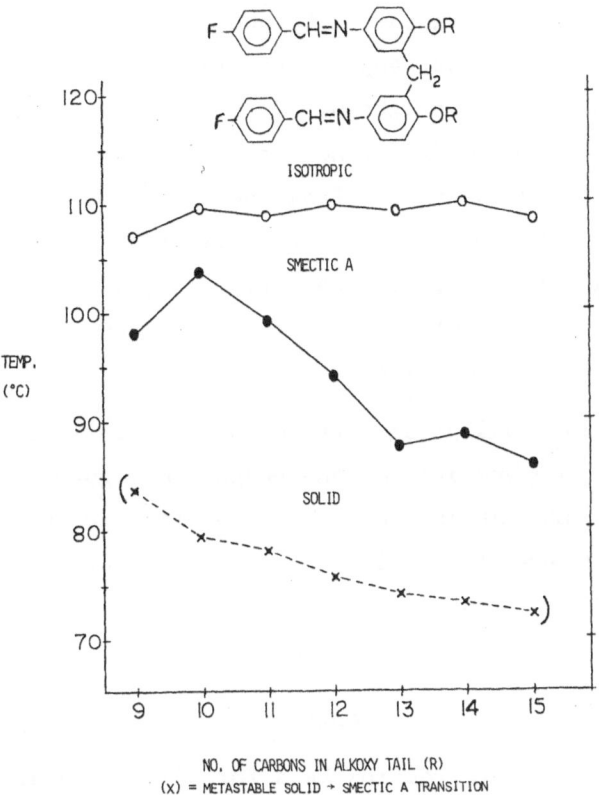

Fig. 3   Transition Temperatures (°C) for Difluoro Twins.

previously.[3b] We do not feel there is for these materials,
conformation isomerism of the type discussed earlier for other
Twin mesogens.[3c] Our x-ray evidence presented below is the basis
for this judgement.

Mesophase x-ray diffraction experiments were performed using
a heated sample chamber and all reported mesophase diffraction data
were collected about 4°C below the $S_A \to I$ transition. Additional
experiments were run using both samples which had no previous
thermal history and also samples which had been previously heated
into the mesophase before solidification. In addition we obtained
diffraction data specifically using the metastable solid phase by

heating from this phase into the S_A phase and performing the
diffraction experiment at temperatures intermediate between those
of the $K \rightarrow S_A$ and $K' \rightarrow S_A$ transition. In all cases the results were
identical indicating no difference in the S_A phase for these
compounds whether it be formed from the stable or from the
metastable solid phase. These control experiments were somewhat
time consuming but were felt necessary in light of our previous
experience with the smectic layer structure dependence on thermal
history in other Twin mesogens.[3c]

Table 2 lists mesophase diffraction data taken about 4°C
below the $S_A \rightarrow I$ temperature for the seven homologues in the series.
The d values represent the small angle diffraction maximum; in
each case there was also a wide angle maximum corresponding to
4.4 Å ($n \lambda = 2DSin\theta$, $n = 1$).

Table 2

Mesophase Diffraction Data for Difluoro Twins

$$F-\bigcirc-CH=N-\bigcirc-OR$$
$$CH_2$$
$$F-\bigcirc-CH=N-\bigcirc-OR$$

R	d(XRD), Å[A]	ℓ,Å[B]	ℓ,Å[C]
C_9H_{19}	24.3	26.3	23.9
$C_{10}H_{21}$	25.9	27.6	22.9
$C_{11}H_{23}$	27.1	28.8	22.1
$C_{12}H_{25}$	28.1	30.5	21.2
$C_{13}H_{27}$	29.5	31.5	20.6
$C_{14}H_{29}$	30.8	32.6	20.1
$C_{15}H_{31}$	32.1	34.0	19.5

A) ESTIMATED ERROR $\pm$ 0.5 Å
B) ESTIMATED ERROR $\pm$ 0.5 Å
C) ESTIMATED ERROR $\pm$ 1.2 Å

The values, l, were measured from molecular models (Fisher
Space Filling Models) in the parallel conformation (second column
from right) and the antiparallel conformation (first column from
right). In each case the measurement was made using extended
chain conformations and the length measured, l, was the minimum
distance between lines connecting molecular termini at each end
of the molecule, i.e. corresponding to the smectic layer spacing,
Fig. 4. Photographs of the two possible conformations for these
Twin compounds as shown on the following page in Fig. 5. Unfortunately, for our series the 'half Twin' analogues shown below are

not mesogenic and therefore a comparison of our d-values for the
Twins and corresponding 'half Twin' compounds is not possible. The
trend in d-values along the series indicates to us that the only
conformation in the mesophase for these materials is the 'parallel'
conformation. The constant d-l value of about 2 Å suggests random
tilt of these molecules in the S_A phase as postulated by de Vries.[5]
The fact that the d-values increase along the series rules out the
antiparallel conformation in which a decrease in d could be
expected as the alkyl chain length increases. As final evidence
for the parallel conformation we note the regular increase (about
1.4 Å) in d with increasing carbon number which is consistent with
the increase in length of the molecular models.

ACKNOWLEDGEMENT

We wish to thank the National Science Foundation
(DMR-8115703) for support of this work.

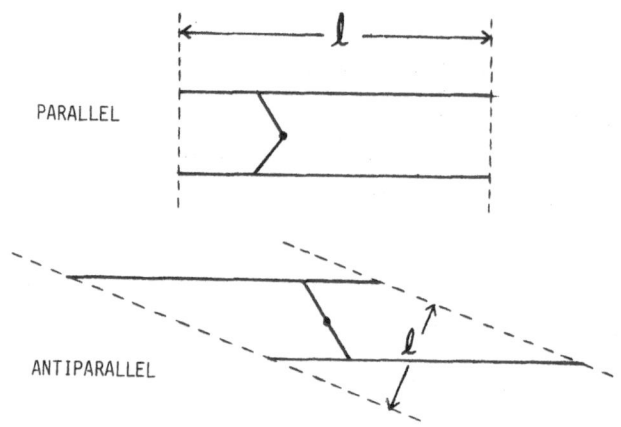

Fig. 4   Smectic Layer Model for Difluoro Twins.

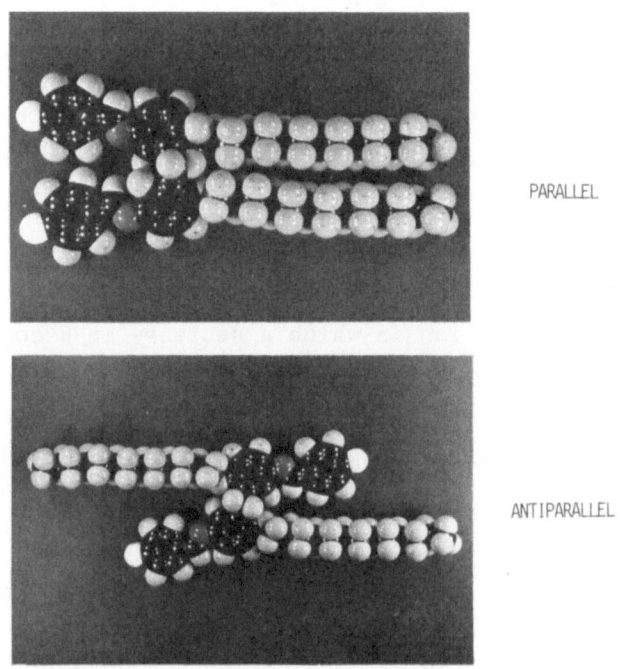

Fig. 5   Photographs of Molecular Models for Difluoro Twins.

REFERENCES

1. For part 13 in this series see A. C. Griffin, N. W. Buckley,
        J. F. Johnson, G. Bertolini, Mol. Cryst. Liq. Cryst., 73,
        35 (1981).

2. J. Malthete, J. Billard, J. Jacques, C. R. Acad. Sci. Paris,
        281, C-333 (1975).

3. (a) A. C. Griffin, S. F. Thames, M. S. Bonner, Mol. Cryst.
            Liq. Cryst. Lett., 43, 135 (1977).

    (b) A. C. Griffin, M. L. Steele, J. F. Johnson,
            G. J. Bertolini, Nouveau Journal de Chimie, 3, 697 (1979).

    (c) A. C. Griffin, N. W. Buckley, W. E. Hughes, D. L. Wertz,
            Mol. Cryst. Liq. Cryst. Lett., 64, 139 (1981).

4. A. C. Griffin, T. R. Britt, J. Am. Chem. Soc., 103, 4957 (1981).

5. A. de Vries, A. Ekachai, N. Spielberg, Mol. Cryst. Liq. Cryst.
        Lett., 49, 143 (1979).

ANISOTROPY INDUCED IN A SEMI-FLEXIBLE POLYMER BY A LIQUID-

CRYSTALLINE SOLVENT*

Yitzhak Rabin, Avinoam Ben-Shaul[1] and William M. Gelbart[2]

Department of Chemistry
University of California, Los Angeles
Los Angeles, California 90024

ABSTRACT

We outline a simple analytical scheme for treating the anisotropy of a semi-flexible polymer in a nematic field (e.g. that due to a liquid crystalline solvent). The rotational-isomer-state-model is used to describe the internal conformation energies, and the aligning field is assumed to interact separately with each of the main-chain monomers. We consider polyethylene-type molecules in which the valence angle is fixed at its tetrahedral value and the dihedral angles of successive monomers are constrained to $0°$ ("trans") and $\pm 120°$ ("gauche"). This simplification allows us to label monomer states in terms of absolute space-fixed (diamond) lattice directions and to avoid thereby the complications of the usual transformations between local coordinate frames. We calculate

*Work supported in part by NSF Grant #CHE80-24270

[1]On sabbatical leave from the Department of Physical Chemistry and the Fritz Haber Institute for Molecular Dyanmics, The Hebrew University, Jerusalem, 91904 Israel

[2]Camille and Henry Dreyfus Foundation Teacher-Scholar

monomer-monomer angular correlations and individual bond alignments
as a function of gauche energy and nematic field strength, and
compare our results with recent measurements. Finally, we comment
on the relevance of our approach to the study of long-range
orientational ordering in model membranes and polymer melts.

I.   INTRODUCTION

It is well-known that dissolving a semi-flexible polymer in a
nematic solvent leads to a lowering of the liquid crystal transition
temperature.[1] Because the angular persistence length of the solute
chain is small compared to its contour length, the polymer is only
sparingly soluble in the orientationally ordered host. Several
theoretical analyses have been offered to explain this disordering
of the liquid crystal by the polymer and to account for the nature
of the two-phase (isotropic/nematic) coexistence in these dilute
binary mixtures.[2,3]

A less studied question involves the anisotropy induced in the
solute chain by the nematic solvent. Samulski[4] has recently
determined the individual bond alignments for n-octane-d_8 dissolved
in the nematic solvent Merck Phase-5. By numerically simulating
the observed deuterium quadrupolar magnetic resonance splittings
via a model which constrains the semi-flexible chain to a hypothetical cylinder, he deduces estimates of the orientational ordering for the CD bonds on each of the four distinct carbon atoms.
The ends of the chain are found to be more disordered than the
insides. This is consistent with the expectation that an inside
bond is aligned not only by the nematic solvent but also by its
neighbors.

In this paper we present a molecular theory of the ordering
of a semi-flexible polymer by a liquid crystalline host. A
simplified version of the rotational-isomer-state model[5] is used
to describe the short range monomer-monomer interactions within

the chain. We solve this model exactly in the presence of an external field whose symmetry reflects the aligning effect of a nematic solvent. By so doing we determine all of the relevant conformational statistics. In particular we obtain the individual bond order parameters, monomer-monomer orientational correlations, angular persistence lengths, etc., all as explicit functions of the chain energies and nematic field strengths.

In Section II we describe the semi-flexible polymer in the presence of a field, within the rotational-isomer-state model, as a biased random walk on the diamond lattice. Our formulation is naturally suited to physical situations in which the isotropy of the polymer is broken by the introduction of a special space-fixed direction (due, say, to a nematic solvent[4] or flow field[6]). Specifically, we label each possible monomer "state" (orientation) in terms of a space-fixed lattice direction, rather than by a dihedral angle defined with respect to the plane of the previous two. Accordingly, we are able to side-step introduction of the transformation matrices which are commonly used[5,7,8] to propagate a rotational-isomer-state chain conformation. (This is no longer possible, of course, when one allows for valence and dihedral angles other than tetrahedral and ±120, respectively.) Instead, the only matrix which enters is an array of Boltzmann factors which is the direct generalization of the familiar "transfer matrix" from spin Ising problems.[9] n^{th}-nearest-neighbor effects are handled by dividing the chain into groups of n monomers, such that the total conformational energy can be written exactly as a sum of interactions between first-nearest-neighbor n-mers. The usual "pentane effect",[5] for example, is fully accounted for by regarding the chain as a sequence of overlapping trimers.

From the single-trimer distributions and trimer-trimer correlation functions, in the presence of a nematic field, it is straightforward to obtain all information relevant to the monomer statistics. In Section III, then, we discuss explicit results for the angular persistence lengths and for $\langle \frac{3}{2} \cos^2 \theta_{i,i+k} - \frac{1}{2} \rangle$, the correlation between the orientations of the ith and $(i+k)^{th}$ monomers. We also present there our predictions of the CC and CD bond orientational-order-parameters for arbitrary sites in the chain. Finally we comment on the relevance of the theory presented here to more complicated situations in which polymer anisotropy is coupled self-consistently to a nematic field.

II. THE MODEL: THEORY

We consider the prototype three-state rotational-isomer model in which the main-chain C-C-C angle is tetrahedral and each C-C monomer (bond) makes a dihedral angle of 0° ("trans") or ±120° (±gauche) with respect to the plane of the previous two. Accordingly, each allowed conformation of the polyethylene-type chain corresponds to a random walk on the diamond lattice. That is, each of the monomers (C-C bonds) in an arbitrary conformation is constrained to lie along one of the four (space-fixed) diamond lattice directions--these are labeled 1,2,3 and 4 in Fig. 1--we choose 1 and 2 to lie as shown in the laboratory yz-plane, and 3 and 4 to lie in the xy-plane. Each possible conformation is completely defined by specifying an ordered sequence of monomer states (lattice directions); $m_1, m_2, m_3, \ldots, m_N$, where $m_i=1,2,3,4$ and N=total number of C-C bonds.

The three-state rotational isomer model commonly includes short-range monomer-monomer interactions which involve first-, second- and third-nearest-neighbors. More explicitly, each time we add a C-C bond which is gauche (dihedral angle of ±120°) with respect to the plane of the previous two, we add g (~1/2kcal/mole for polyethylene) to the chain conformational energy. (The energy associated with a trans bond is taken to be zero). These second-nearest-neighbor interactions are trivial to handle because they

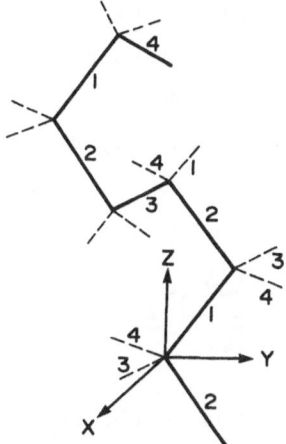

Fig. 1   The dark lines suggest part of a typical polyethylene
          conformation, the dashed line showing the CH's (or CD's).
          The nematic field lies along the space-fixed z-axis.
          "1", "2", "3", and "4" refer to the four diamond lattice
          directions along which the CC bonds can lie; "3" and "4"
          lie in the space-fixed xy-plane, while "1" and "2" are in
          the zy-plane.

lead to a total N-monomer energy which is a sum of independent
contributions (g from ±120° and 0 from 0°) from each of the bonds
i=3→N. This situation is significantly complicated by the "pentane
effect",[5] i.e. by <u>third</u>-nearest-neighbor interactions which
effectively exclude the possibility of successive gauche bonds of
opposite sign. (These local configurations involve steric hin-
rances which are large compared to kT at room temperature.) This
effect can be treated exactly by expressing the total chain energy
as a sum of nearest-neighbor interactions between overlapping
trimers.

Since there are only four lattice directions (1,2,3,4), and
since no one of these monomer states can directly repeat itself,
we need "only" allow for 4x3x3=36 trimer states. More explicitly,

$t_i=m_i,m_{i+1},m_{i+2}$ denotes the state of the ith trimer ($i=1,2\ldots,N-2$; recall that N is the total number of monomers) and m_i (=1,2,3 or 4) describes the state of the ith monomer. An arbitrary conformation of the chain is specified accordingly by an ordered sequence of overlapping trimer states: $t_1,t_2,\ldots,t_{N-2}$. The total conformation energy can then be written in the form

$$E_o(\underset{\sim}{t}) = \mathcal{E}_o(t_1) + \sum_{i=1}^{N-3} \left[\mathcal{E}_o(t_{i+1}) + I(t_i, t_{i+1}) \right] \tag{1}$$

where $\mathcal{E}_o(t)=0$ if t corresponds to the trans state of the trimer, and $\mathcal{E}_o(t)=g$ if the trimer is in the + or - gauche state; $I(t_i, t_{i+1})=p$ if the successive trimers are both gauche and of opposite sign, and I=0 otherwise. Since p>>g and g~kT,[5] it is convenient to take p→ ∞. That is, Eq. (1) is simply a sum of trans(0) and gauche (g) energies, with "pentane effect" configurations excluded: each time we add a monomer--say the $i+3^{rd}$--we add an energy

$\mathcal{E}_o(t_{i+1})=0$ or g <u>after</u> <u>first</u> <u>checking</u> that we do not have t_i, t_{i+1} = ±gauche, ∓gauche.

Consider now a nematic environment which exerts an aligning field on the above polymer chain. This field is of the "$P_2(\cos \Theta_{m_i})$", rather than the usual "$P_1(\cos \Theta_{m_i})$", type--it is afterall the monomer's axially symmetric polarizability tensor, rather than a permanent dipole moment, which is "grabbed" by the orientationally ordered solvent. Accordingly, we add to Eq. (1) a term

$$V(\underset{\sim}{t},f) = \mathcal{V}_1(t_1,f) + \sum_{i=1}^{N-3} \mathcal{v}(t_{i+1},f) \tag{2}$$

where

$$\mathcal{v}(t_{i+1},f) = -f \left[P_2(\cos \Theta_{m_{i+3}}) + \tfrac{1}{2} \right] \tag{2A}$$

and

$$\mathcal{V}_1(t_1, f) = -f\left[P_2(\cos\sigma_{m_1}) + P_2(\cos\sigma_{m_2}) + P_2(\cos\sigma_{m_3}) + \tfrac{3}{2}\right].$$ (2B)

Here $P_2(x) = \tfrac{3}{2}x^2 - \tfrac{1}{2}$ is the second Legendre polynomial and σ_{m_i} is the angle between the ith monomer (lying along the lattice direction m_i) and the nematic field (taken to lie along the laboratory Z-axis--see Fig. 1). $f = A\eta$, where A is the monomer-solvent coupling strength and η is the nematic's "P_2" order parameter. Finally, the term $\tfrac{1}{2}$ in square brackets in (2A) is introduced so that $\mathcal{V} = -f$ for trimer states whose third monomer has a Z-component while $\mathcal{V} = 0$ otherwise: note that $P_2(\cos\sigma_m) = \tfrac{1}{2}(-\tfrac{1}{2})$ for m = 1,2 (3,4). Similarly the term $\tfrac{3}{2}$ in (2B) assures that $\mathcal{V}_1 = -f, -2f$ or $-3f$ according to whether the first trimer contains one, two or three monomers in state "1" or "2".

From the above we have, for the chain energy in the presence of a field,

$$E(t, f) = \mathcal{E}_1(t_1, f) + \sum_{i=1}^{N-3} \mathcal{E}(t_i, t_{i+1})$$ (3)

where

$$\mathcal{E}_1(t_1, f) = \mathcal{E}_0(t_1) + \mathcal{V}_1(t_1, f)$$ (3A)

and

$$\mathcal{E}(t_i, t_{i+1}) = \mathcal{E}_0(t_{i+1}) + \mathcal{V}(t_{i+1}) + I(t_i, t_{i+1}).$$ (3B)

The probability of an arbitrary conformation t is then given by (here $\beta = \tfrac{1}{kT}$ as usual)

$$P(t, f) = \tfrac{1}{Q} e^{-\beta E(t, f)}$$ (4A)

where $Q = \sum_{\underset{\sim}{t}} e^{-\beta E(\underset{\sim}{t})}$ is the corresponding partition function. Defining now the <u>trimer</u> <u>transfer</u> <u>matrix</u> elements

$$W(t_i, t_{i+1}, f) = e^{-\beta E(t_i, t_{i+1}, f)} \qquad (5A)$$

and the "end"-trimer weights

$$g(t_1, f) = e^{-\beta \varepsilon_1(t_1, f)} \qquad (5B)$$

it follows trivially that the joint probability (4) can be written in the form

$$P(\underset{\sim}{t}) = \frac{1}{Q} e^{-\beta E(\underset{\sim}{t})} = \frac{1}{Q} g(t_1) \prod_{i=1}^{N-3} W(t_i, t_{i+1}) \ . \qquad (4B)$$

Note that here and henceforth we have suppressed all explicit dependences on f.

Summing the Boltzmann factor in (4B) over all states of all trimers gives the partition function ($t_1 \to u$, $t_{N-2} \to v$),

$$Q = \sum_{u,v=1}^{36} g(u) W^{N-3}(u, v) \ , \qquad (4C)$$

where $W^{N-3}(u, v)$ denotes the u, v^{th} element of the $N-3^{rd}$ power of the matrix W. Similarly, the single-trimer distribution is obtained by summing the joint probability in (4B) over all states of all trimers but the ith: for $1 < i < N-2$, say, we have

$$P_i^{(t)}(r) = \frac{1}{Q} \sum_{u,v=1}^{36} g(u) W^{i-1}(u, r) W^{N-2-i}(r, v) \qquad (6A)$$

$$= \begin{cases} \text{probability of finding the } i^{th} \\ \text{trimer in the state } r (=1 \to 36) \end{cases}$$

The corresponding single-<u>monomer</u> distribution is given by

$$P_i^{(m)}(a) = \sum_{\tau}{}' P_i^{(t)}(\tau) \qquad = \qquad \begin{array}{l}\text{probability of finding} \\ \text{the } i^{\text{th}} \text{ monomer in the} \\ \text{state } a(1 \to 4)\end{array} \qquad (6B)$$

where the prime restricts the summation to those trimer states in which the first monomer is in state a. Finally the <u>pair</u> trimer-trimer distribution is (for $1 < i < N-2$ and $0 < k < N-2-i$)

$$P_{i,i+k}^{(t)}(\tau,s) = \frac{1}{Q} \sum_{u,v=1}^{36} g(u) W^{i-1}(u,\tau) W^{k}(\tau,s) W^{N-2-i-k}(s,v) \qquad (7A)$$

from which it follows that the corresponding <u>monomer-monomer</u> correlations are given by

$$P_{i,i+k}^{(m)}(a,b) = \sum_{\tau}{}' \sum_{s}{}'' P_{i,i+k}^{(t)}(\tau,s) \qquad (7B)$$

where the double prime restricts the summation to those trimer states in which the first monomer is in state b. (Similar expressions obtain for the singlet and pair distributions involving monomers <u>at the ends</u> of the chain.)

All observables of interest can be derived from these reduced (i.e. singlet and pair) distribution functions. In particular, it is straightforward to show that the P_2-order parameter for the i^{th}C–C bond is given by

$$\eta_i^{cc} = \langle P_2(\cos\theta_{m_i}) \rangle = 2 P_i^{(m)}(m_i = 1) - \frac{1}{2} . \qquad (8)$$

The monomer-monomer correlations are simply described by

$$\eta_{i,i+k}^{cc} = \langle P_2(\cos\theta_{i,i+k}) \rangle = \frac{4}{3} \sum_{a=1}^{4} P_{i,i+k}^{(m)}(a,a) - \frac{1}{3} . \qquad (9)$$

Finally, for the p_2-order parameter associated with the CD bonds on the ith carbon atom, we have

$$\eta_i^{CD} = \sum_u {}^* P_{i-1}^{(t)}(u) - \sum_v {}^\dagger P_{i-1}^{(t)}(v) \qquad (10A)$$

for $2 \leqslant i \leqslant N$ [here *($\dagger$) restricts the summation to the three trimer states whose first two monomers are 1,2(3,4)], while for the terminal carbons we have

$$\eta_1^{CD} = \eta_{N+1}^{CD} = -\frac{2}{3} P_1^{(m)}(a{=}1) + \frac{1}{6} \qquad (10B)$$

Recall that the singlet and doublet distributions (6B) and (7B) are defined in terms of the trimer transfer matrix elements $W(t_i, t_{i+1})$ and the "end"-trimer weights $g(t_1)$. W is a 36x36 matrix, but only 84 out of its 1,296 elements are nonzero. This fact follows from Eqs. (5A), (3B) and (2A). More explicitly, because of <u>overlapping</u> of the successive trimers, a,b,c can only be succeeded by b,c,d$\neq$c. Since there are 36 a,b,c's and only 3 b,c,d$\neq$c's for each of these, there can be only 36x3=108 nonzero $W(t_i, t_{i+1})$'s. In addition, 24 of these 108 are zero because of the pentane effect, i.e. each <u>gauche</u> t_i--two-thirds of the 36 trimer states--can only be followed by <u>two</u> (rather than three) t_{i+1}. Now define $F = e^{\beta f}$ and $G = e^{-\beta g}$. Then each nonzero matrix element of W has the form $F^n G^k$, with k=0 or 1 according to whether $t_{i+1} = m_{i+1}, m_{i+2}, m_{i+3}$ is trans or gauche; and n=1 if $m_{i+3}=1$ or 2 and n=0 otherwise. Finally, for $g(t_1)$ we have $F^M G^L$ where M is the number of 1 or 2 monomers in the first trimer and L=1 for gauche states (zero otherwise).

III. RESULTS AND DISCUSSION

Fig. 2 shows η_i^{CD} for G=0.435 (see, for example, reference 5), N=7 and F=2.5 and 5.0. We have chosen this value of N in order to compare with the experimental results of Samulski for n-octane-d$_8$.[4] The F values bracket an estimate by Marcelja[10] of the monomer-

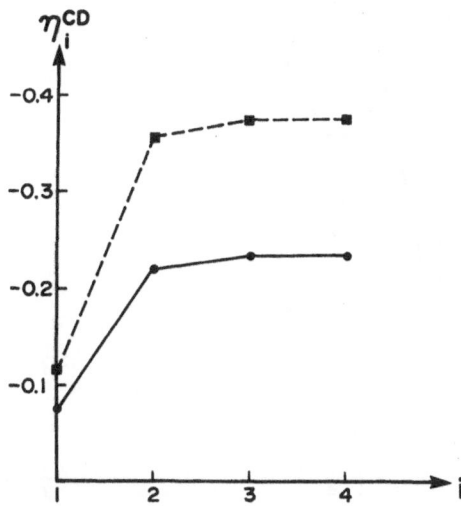

Fig. 2 $Ni^{CD} = < P_2(cos\theta_{CDi}) >$ vs i where θ_{CDi} denotes the
angle between the nematic field and the axis of the CD
bond on the ith carbon atom.
F = 2.5 (solid line) and F = 5.0 (dashed line).
G = 0.435 and N = 7 in both plots.

solvent P_2-nematic-field strength. As mentioned in the
INTRODUCTION, the order parameter increases as we head into the
middle of the chain from the ends: the P_2 alignment of the CD
bonds is greatest for the "inside" carbons (i=4,5) and least for
the "outsiders" (i=1,8). The theoretical predictions for F=2.5
and F=5.0 bracket the measured values obtained by Samulski. His
results could easily be fit more precisely by varying G and F
slightly, but no purpose would be served by doing so.

Fig. 3 shows our results for the CC bonds. Again there is an
increase in the P_2 alignment as we head into the chain from the
outside. But now there is also an "odd-even" effect, associated
with the alternating tendency of successive monomers to lie
"parallel" and "perpendicular" to the nematic field. The solid line
shows the theoretical prediction for N=7 and the dashed line for
N=20. As expected, the amplitude of the odd-even oscillations

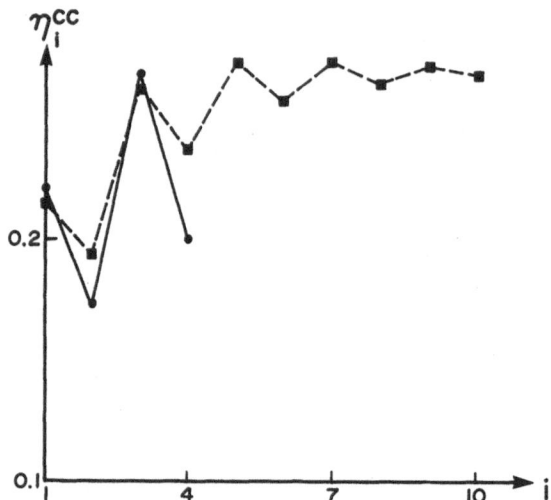

Fig. 3 $Ni^{cc} = \langle P_2(\cos\theta_{cci}) \rangle$ for N = 7 (solid line) and N = 20
          (dashed line).
          G = 0.435 and F = 2.5 in both plots.

decreases to zero as we get far enough from the terminal bonds--the
intrinsic (F=1) monomer-monomer correlations (due to the local chain-
conformational constraints) extend only over short distances.

     The explicit correlations between pairs of monomers are given
directly by Eq. (9). Fig. 4 shows our results for $\eta^{cc}_{i,i+k} = \langle P_2(\cos\theta_{i,i+k}) \rangle$
in the case of no nematic field (F=1) and for G=0.435 and N=20.
Baram and Gelbart[7] have published a similar plot, for the same
value of the gauche-trans energy difference (G). Their theory was
confined to F=1 and to the inside of a very long chain, i.e.
$i \gg 1$ and $i+k \ll N$. In this limit, our results (say for F=1, N=20,
i=7 and i+k=7$\rightarrow$15) agree with theirs. Because of the space-fixed
(diamond lattice) monomer labeling employed in the present approach,
however, it is much simpler for us to compute all of the relevant
correlations. Furthermore, it is just as easy to include "end
effects" (i.e. $i \gtrsim 1$) and nematic-field-induced corrections. Of
particular interest, for example, is the calculation of the anisot-
ropy $\langle z^2_N \rangle / \langle x^2_N \rangle$ as a function of field; this "stretching" of a

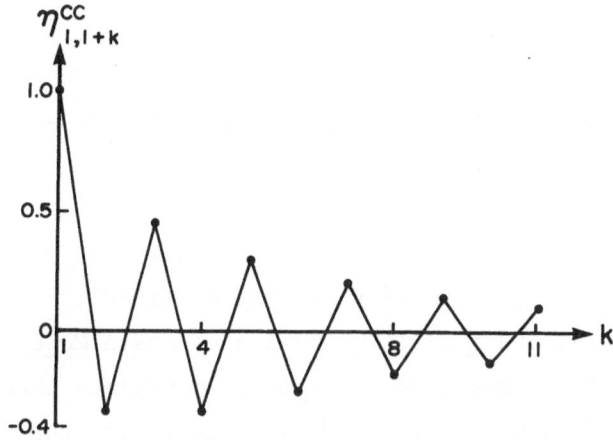

Fig. 4 $N_{1,1+k}^{cc} = \langle P_2(\cos\theta_{1,1+k}) \rangle$ vs k for N = 20, G = 0.435 and F = 1.0.

rotational-isomeric chain in a nematic environment is considered separately elsewhere.[11]

The theory outlined above can also be extended naturally to more complicated physical situations. Consider, for example, the case of a model membrane in which alkyl chains are "anchored" (via, say, ionic head groups) to a water interface.[12] Here the orientation of the first bond is essentially fixed: instead of summing over all t_1=u in (6A), say, we include only those trimer states which allow the first monomer to lie along the normal direction. This anchoring is the source of long-range orientational ordering, i.e. it breaks the isotropy of the polymer and introduces the preferred space-fixed direction. More importantly, the nematic field is now due--not to an "external" (e.g. liquid crystalline solvent) field, but--to the chain anisotropy itself. Thus the P_2 order parameter which defines V--see text following Eq. (2)--must be the same as the average of the induced bond alignments given by Eq. (8). A similar self-consistency requirement arises in the case of the nematic states of polymer melts.[13]

REFERENCES

1.   See, for example, A. Dubault, C. Casagrande, M. Veyssie and
          B. Deloche, Phys. Rev. Lett. 45, 1645 (1980).

2.   B. Kronberg, I. Bassignana and D. Patterson, J. Chem. Phys. 82,
          1714 (1978).

3.   F. Brochard, C. R. Acad. Sci. Paris, Ser. B 289, 229 (1979).

4.   E. T. Samulski, Ferroelectrics 30, 83 (1980).

5.   P. J. Flory, Statistical Mechanics of Chain Molecules, Wiley,
          New York, 1969, and references contained therein.

6.   G. Marrucci and A. Ciferi, Polymer Lett. 15, 643 (1977).

7.   A. Baram and W. M. Gelbart, J. Chem. Phys. 66, 617 and 4666
          (1977).

8.   J. Freire and M. Fixman, J. Chem. Phys. 69, 634 (1978).

9.   See, the discussion and references cited in Introduction to
          Phase Transitions and Critical Phenomena, Oxford
          University Press, New York, 1972, by H. E. Stanley.

10.   S. Marcelja, J. Chem. Phys. 60, 3599 (1974).

11.   Y. Rabin, A. Ben-Shaul and W. M. Gelbart, to be published.

12.   See, for example, S. Marcelja, Biochim. Biophys. Acta. 367,
          165 (1974).

13.   (a) P. Pincus and P. G. de Gennes, J. Poly. Sci.: Poly. Symp.
          65, 85 (1978);

      (b) G. Ronca and D. Y. Yoon, preprint entitled "Theory of
          Nematic Systems of Semiflexible Polymers: High Molecular
          Weight Limit".

ELECTROOPTICAL BEHAVIOUR OF A NOVEL STRONG-WEAK ANCHORED NEMATIC

LAYER: FIRST MEASUREMENTS OF MBBA FLEXOELECTRIC COEFFICIENT OF

SPLAY e_{1z}

H. P. Hinov and A. I. Derzhanski
Institute of Solid State Physics
Bulgarian Academy of Sciences
Boul. Lenin 72, Sofia 1184, BULGARIA

INTRODUCTION

The current state of the flexoelectric theory now clearly points out two essential types of the flexoelectricity in the nematic liquid crystals - the dipolar and the quadrupolar ones introduced and developed by Meyer (1) and by Prost and Marcerou (2) respectively. Of great interest is the experimental determination of the flexoelectric coefficients of bend e_{3x} (or e_{33}) and splay e_{1z} (or e_{11}) respectively involved by Meyer (1) and their comparison with the theoretical estimations given by Helfrich (3) and by Derzhanski and Petrov (4).

The bending of a dielectrically stable MBBA layer in a transversal electric field observed by Flannery et al (5) has been interpreted by Helfrich (6) in terms of the flexoelectricity who calculated for the first time the value of the flexoelectric coefficient of bend e_{3x}. Later Schmidt et al (7) reevaluated the magnitude of this flexocoefficient and clarified its sign on the basis of a very carefully performed experiment.

The total flexoelectric coefficient $(e_{1z} + e_{3x})$ has been measured in a gradient electric field by different authors (8, 9,

10). On the other hand the difference between the flexoelectric coefficients of splay and bend $(e_{1z} - e_{3x})$ respectively has been obtained from the threshold voltage measurements indicating the
appearance of the early-discovered by Vistin' domains of second
kind (11) which have been interpreted by Barnik et al (12,13) in
terms of the two-dimensional flexoelectric theory developed by
Bobylev and Pikin (14,15). In addition, the experiments performed
gave different values of the flexoelectric coefficient of bend
e_{3x} on account of a possible surface polarization (9,16) essential
for homeotropic orientation of the liquid crystal layer.

Unfortunately up to now the flexoelectric coefficient of splay
e_{1z} had not been measured due to the complications arising with
the formation of weakly-anchored planar nematic layers. Recently
Derzhanski et al (17) suggested a promising method for the measurement of the splay flexoelectric coefficient e_{1z} based on the creation of possible flexoelectric deformations in asymmetrically
strong-weak anchored nematic layers in a transversal d.c. electric
field.

PREPARATION OF ASYMMETRICALLY STRONG-WEAK ANCHORED MBBA LAYERS

It is well-known from the experiments performed during the
last ten years that the planar nematic layers usually are strongly-
anchored whereas the homeotropic nematic layers can be either weakly-
anchored or strongly-anchored depending on the nature, thickness
and density of the surfactants utilized. At this point it is
important to note that in the most cases only one molecular surfactant layer had been deposited by means of the well-known
Blodgett-Langmuir's technique. For instance, the orientation of
nematic liquid crystals by one molecular soap or lecithin layer has
been investigated in detail by the group of Ter-Minassian-
Saraga (18-21) in France and by Hiltrop and Stegemeyer (22-24) in
West Germany respectively. All these experimental results point

out that strong-weak anchored nematic layers can be prepared when
the orientation of the molecules at the one of the boundaries of
the cells under study is planar (strong anchoring) and homeotropic
(weak anchoring) at the other boundary. Indeed the idea for the
construction of such kind of cells has been proposed by Matsumoto
et al (25) in 1976 and many hybrid-aligned liquid crystal cells
have been investigated. However, it has been accepted in most of
the cases strong anchoring of the liquid crystal molecules at both
the boundaries. In the real experiments this condition, in effect,
is very hard to be obtained and usually one investigates hybrid-
aligned cells with strong-weak anchoring. This fact has been
correctly recognized by Chigrinov (26) who proposed a simple
method for the measurement of the second-order elastic coefficient
K_{13} (27) on the basis of strong-weak anchored hybrid-aligned
nematic layers. The theoretical results obtained by this author
reveal the possibility for the creation of strong-weak (or planar-
homeotropic) nematic layers when the following important inequality
holds:

$$W_{s\theta} \, d \leq K_{11}$$

where $W_{s\theta}$ is the θ strength-coupling surface constant, d is the
thickness of the cells under study and K_{11} is the splay elastic
constant (it has been assumed that the liquid crystal molecules at
the other boundary are fixed by strong surface forces in planar
position).

Let us stress here that this is the UNIQUE condition for the
creation of PLANAR STRONG-WEAK anchored nematic layers. It is NOT
POSSIBLE, in general, to create strong-weak PLANAR layers when the
equilibrium condition for the surface orientation of the liquid
crystal is different from the homeotropic one i.e. when the
minimum of the surface energy of the liquid crystal is given by
the well-known expression $W_{s\theta}\sin^2(\theta - \theta_0)$ with θ_0 being the
initial tilt.

Furthermore, the surfactants can be deposited on smooth,
rubbed or SiO treated under vacuum evaporation glass plates (28-30)
and the liquid crystal orientation in these cases is determined by
the competition between the complex physico-chemical forces and the
pure elastic forces. In addition, some surfactants can give tilted
orientation of the liquid crystal molecules (31) which is strongly affected by the temperature (32,33). Note that under strong-
weak anchoring usually one means θ-polar strong-weak anchoring
(Fig. 1). The case is more complicated for bulk liquid crystal
deformations including θ-polar and ϕ-asimuthal realignment of the
director and θ and ϕ coupling are different (Fig. 1). Weakly-
anchored θ, ϕ nematic layers can be prepared by means of thick
soap deposition on the glass plates (34). The great difference in
the anchoring of the liquid crystal at the deposition of one molecular soap layer or many soap layers can be illustrated for instance with the different kind of the electrohydrodynamic generated

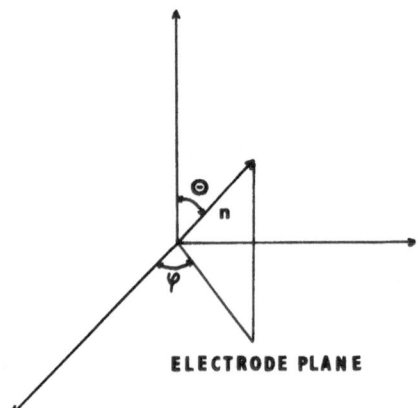

ELECTRODE PLANE

Fig. 1 The general surface energy of a nematic layer can be
        expressed with the aid of two independent deformation
        angles designated by θ-polar and ϕ-asimuthal angles.

after the application of d.c. electric field across differently
prepared MBBA layers. In the case of deposition of one soap molecular layer only θ-electrohydrodynamic domains have been generated
(Fig. 2) whereas the thick soap deposition gives a way for the
development of complex θ,ϕ electrohydrodynamic domains started
from the flexoelectric ones (35).

During the experimental measurements of the flexoelectric coefficient of splay e_{1z} strong-weak anchored MBBA layers are prepared by means of rubbing of the two glass plates with diamond
paste followed by thin or thick deposition of common soap (Na salt
of fatty acids) on one of them respectively (Fig. 2) (34). We
noticed that the efficiency of the soap deposition was higher when

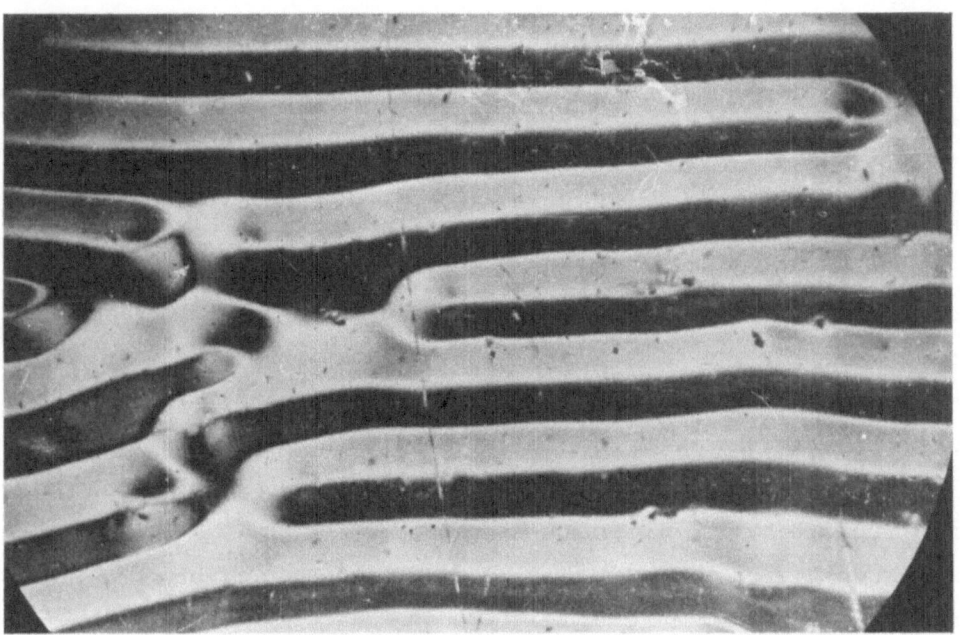

Fig. 2 θ,ϕ Electrohydrodynamic generated in a strong-weak
        anchored MBBA layer under the influence of d.c.
        voltage when one of the glass plates is coated
        with thick soap layer.

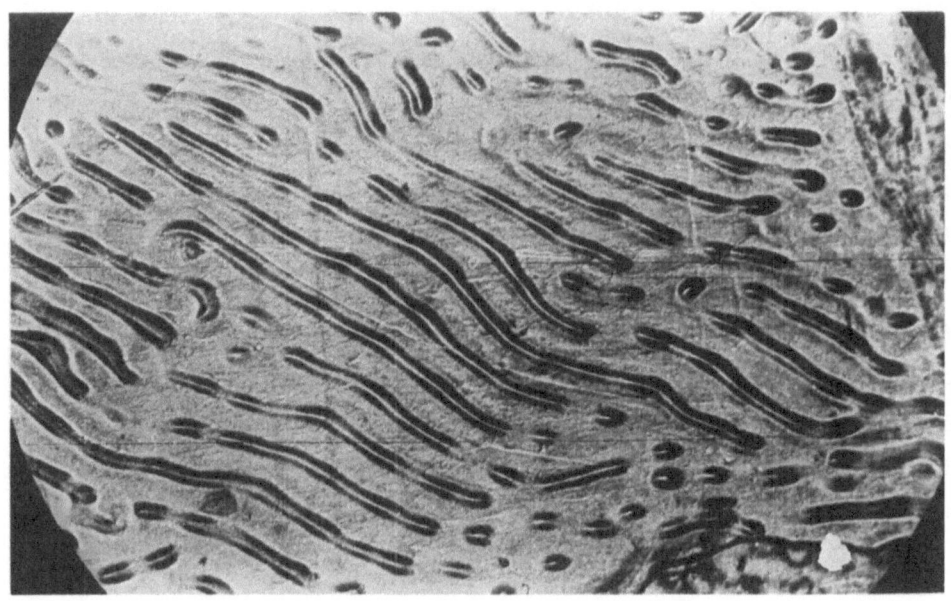

Fig. 3   θ-Electrohydrodynamic generated in a strong-weak
         anchored MBBA layer under the influence of d.c.
         voltage when one of the glass plates is coated only
         with one molecular soap layer.

the soap was rubbed on with a cloth as in Chatelain's method. The
soap films were very thin and not observable with the naked eye.
The coupling of the liquid crystal molecules with the surfaces
treated in this way was weak and the orientation of the molecules
was tilted, however without any degeneration (36-38). Furthermore, in accordance with the recent results obtained by Nakamura
(39) it was remarked that the sign of the tilt angle θ_0 can be
predetermined by the direction of rubbing (Fig. 4). This important
experimental fact was confirmed by direct conoscopic observations
demonstrated either uniform tilt when the deformation angles at the
two boundaries are with equal sign or nearly planar weak-anchored
MBBA layers with the slight splay of the liquid crystal molecules
when the tilt at the two boundaries is with different sign. The
nearly planar alignment of the liquid crystal layers was with very

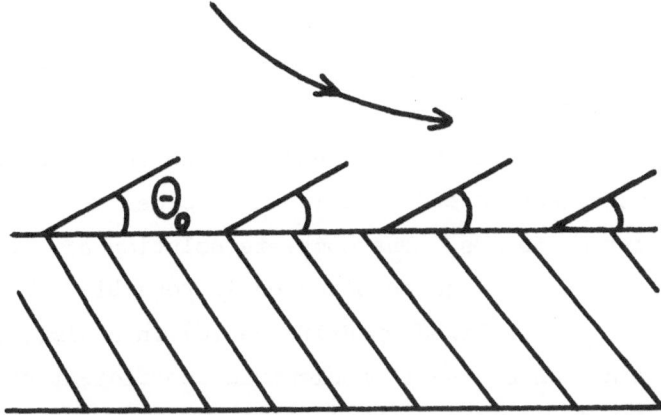

Fig. 4 The rubbing direction can determine the sign of the
initial tilt angle θ_0.

weak θ-polar anchoring which was estimated to be in the range of
10^{-4} erg/cm^2 [40] (On the other hand the asimuthal ϕ-surface anchoring was stronger and able to form a well-aligned twist layer).

The conoscopic images and fringes (41-46) have shown that the
thinner soap deposition determines a tilt angle below 45° whereas
the utilization of thicker soap deposition leads to tilt angles
above 80°. It should be noted that the strength of the anchoring
and the value of the tilt angle depend strongly on the thickness of
the soap deposition. It is easy to understand that the experimental results for obtaining of weakly-anchored MBBA layers are not
well reproducible and further elaboration is needed. However, a
great deal of these uncertainties can be removed with the use of
strong-weak anchlored MBBA layers. The strong coupling at one of
the boundaries has a crucial role for the realiztion of a relatively homogeneous liquid crystal orientation with a small tilt at the
other boundary with soap treatment depending strongly on the value
of the surface-coupling constant $W_{s\theta}$. On the other hand the strong-
weak anchoring of the liquid crystal molecules can prevent the
possible full degeneration of the liquid crystal orientation which
is typical for weak-anchored nematic layers (36-38).

THEORY

    Consider a strong-weak anchored MBBA layer in a transversal
d.c. electric field (Figs. 5a and 5b). The equilibrium position
of the liquid crystal molecules depends on the influence of the
dielectric, flexoelectric and elastic bulk forces as well as on
the elastic surface forces. The complete solution of the problem
under consideration although complicated is possible. However, we
shall follow the way utilized by Helfrich (6) in a similar problem
for the determination of the flexoelectric coefficient of bend e_{3x}
since first, it gives a clear physical picture for the competition
between the different forces and second, it gives a nearly correct
solution of the problem under consideration for small tilting of
the liquid crystal layers.

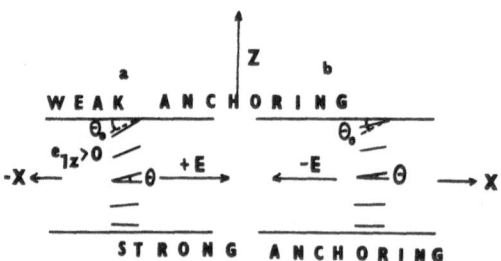

Fig. 5 Flexoelectric splaying of a strong-weak anchored ne-
        matic layer with a slight tilt at the upper boundary:
    a) the flexoelectric deformations increase the initial
       splaying of the layer
    b) the flexoelectric deformations decrease the initial
       splaying of the layer.

The free flexoelectric energy H_E per cm^2 at constant and uniform electric field E (the space charge effects were not taken into account) according to Mayer (1) and Helfrich (6) has the following form:

$$H_E = (1/2)K_{11} \int_0^d (\frac{d\theta}{dz} - \frac{e_{1z}E}{K_{11}})^2 \, dz + (1/2(W_{s\theta}(\theta_d-\theta_0)^2 \quad (1)$$

$$- \frac{|\Delta\varepsilon| E^2}{8\pi} \int_0^d \theta^2 \, dz$$

where Z was choosen to be perpendicular to the glass plates and the electric field E can be directed along X or -X respectively (Figs. 5a and 5b), K_{11} is the splay elastic coefficient, e_{1z} is the splay flexoelectric coefficient, $\theta_0 < 45°$ is the initial tilt, θ_d is the value of the surface angle given by the solution of the problem, $W_{s\theta}$ is the θ-surface coupling constant and $\Delta\varepsilon$ is the dielectric anisotropy which for the MBBA case is negative.

The validity of (1) is restricted to small changes in the deformation angle: $|\theta|<<\pi/2$. Positive $d\theta/dz$ means that the orientation lines splay toward the direction of the field. Furthermore it is assumed that the liquid crystal orientation is fixed in planar position by strong-anchoring forces i.e. $\theta(0) = 0$.

Let us in agreement with Helfrich's (6) results take $d\theta/dz$ to be independent of Z. Minimizing H_E then leads to the following equation connecting the flexoelectric deformation with all the constants entering in the problem under consideration:

$$\frac{d\theta}{dz} = \frac{(e_{1z}E/K_{11}) + (w_{s\theta}\theta_0/K_{11})}{1 + \frac{W_{s\theta}d}{K_{11}} - \frac{|\Delta\varepsilon| E^2 d^2}{12\pi K_{11}}} \quad (2)$$

The optical path difference l_1 produced by the initial splay deformations of the nematic layer under consideration has the following form:

$$l_1 = l(0) - l(\theta_0) = (1/2)n_e \left(\frac{n_e^2}{n_0^2} - 1\right) \int_0^d \theta^2(z)\,dz$$

$$= (1/2)n_e \left(\frac{n_e^2}{n_0^2} - 1\right)(d^3/3) \frac{(W_{s\theta}\theta_0/K_{11})^2}{(1 + \frac{W_{s\theta}d}{K_{11}})^2} \qquad (3)$$

when the electric field is removed. In addition, the optical path difference l_2 produced by a deformed nematic layer under the combined action of the surface and flexoelectric torques can be expressed with the following formula:

$$l_2 = l(0) - l(\theta_0, E) = (1/2)n_e \left(\frac{n_e^2}{n_0^2} - 1\right) \int_0^d \theta^2(z)dz$$

$$= (1/2)n_e \left(\frac{n_e^2}{n_0^2} - 1\right) (d^3/3) \frac{\left(\frac{e_{1z}E}{K_{11}} + \frac{W_{s\theta}\theta_0}{K_{11}}\right)^2}{(1 + \frac{W_{s\theta}d}{K_{11}} - \frac{|\Delta\varepsilon|}{12\pi}\frac{E^2 d^2}{K_{11}})^2} \qquad (4)$$

Note that $l(\theta_0, E = 0) = l(\theta_0)$.

Finally one calculates the optical path difference due to the flexoelectric deformations:

$$l_2 - l_1 = l(\theta_0, E) - l(\theta_0)$$

$$\approx (1/2(n_e \left(\frac{n_e^2}{n_0^2} - 1\right)(d^3/3) \frac{\left(\frac{e_{1z}E}{K_{11}}\right)^2 + 2\left(e_{1z}E/K_{11}\right)\frac{W_{s\theta}\theta_0}{K_{11}}}{(1 + \frac{W_{s\theta}d}{K_{11}} - \frac{|\Delta\varepsilon|}{12\pi}\frac{E^2 d^2}{K_{11}})^2}$$

$$\qquad (5)$$

It is easy to understand that the weak flexoelectric deformations due to the splaying of the nematic layer can be optically detected when the dielectric effects are negligible:

$$\frac{|\Delta\epsilon|}{12\pi} \quad \frac{E^2 d^2}{K_{11}} \quad \ll 1 \tag{6}$$

as well as when the strength of the surface anchoring of the liquid crystal molecules at the weak-anchored boundary is with such a magnitude to ensure the validity of the following two inequalities:

$$\frac{W_{s\theta} d}{K_{11}} \ll 1 \;, \qquad W_{s\theta}\theta_0 \;\ll\; e_{1z}E, \qquad \theta_0 < 45° \tag{7}$$

The upper limit of a rather weakly θ-anchored nematic layer has been estimated by Derzhanski and Hinov (47) to be:

$$W_{s\theta} \leq \frac{3 \times 10^{-7}}{d} \quad erg/cm^2 \tag{8}$$

Similary, the upper limit of a rather weakly ϕ-anchored nematic layer has been assessed by Cognard (48) to be:

$$W_{s\theta} \leq 2 \times 10^{-4} \; erg/cm^2 \tag{9}$$

It should be pointed out that both the elastic coefficients of splay K_{11} and θ-surface coupling parameter $W_{s\theta}$ decrease with the temperature. Taking the upper limit of the splay constant which at room temperature is about $7,5 \times 10^{-7}$ dyne (49) the following important inequality is obtained:

$$W_{s\theta} d \;\ll\; 7,5 \times 10^{-7} \; dyne \tag{10}$$

which in effect is crucial for the measurement of the MBBA flexoelectric coefficient of splay e_{1z} on one hand as well as for the
preparation of novel strong-weak anchored MBBA layers. For thin
liquid crystal cells with a typical thickness about 10 microns the
surface soupling parameter $W_s\theta$ should be in the range of 10^{-4}
erg/cm^2. In addition, this value of the surface coupling parameter
is sufficient to ensure the second of the inequalities (7):

$$W_{s\theta}\theta_0 \ll e_{1z}E$$

for small initial tilt angles: $\theta_0 < 20°$, e_{1z} in the range of 10^{-4}
dyne$^{1/2}$ and a value of the electric field in the range of 1-2
stat. V/cm (it is not possible to increase further the electric
field due to the electrohydrodynamic movement of the fluid which
starts at about 750 V/cm (see the experimental results).

EXPERIMENTAL RESULTS

I. Determining of $|e_{1z}|$

A number of strong-weak anchored MBBA layers 10 μm thick were
prepared by means of rubbing of the two glass plates with diamond
paste followed by thin or thick soap deposition on one of them
respectively. The MBBA layers obtained in this way were homogeneously oriented with a small tilt and weak anchoring at the
soap-treated glass plate. This fact was confirmed by the strong
flickering effect due to the thermal fluctuations as well as by
the flow relaxation of a realigned liquid crystal. In addition,
the sharp Frederiks transition has occurred at about 2000V/cm
(which is about a half of the required threshold field in conventional sandwich cells 10 microns thick) indicated the nearly good
planar orientation of the strong-weak anchored MBBA layer.

The measurements of the splay flexoelectric coefficient were
performed in transmitted white light under an interferometric-
polarizing polish microscope MPI 5 employing the sensitive diffe-

rential regime. Of interest is to note that in this regime is possible the investigation of homogeneously oriented samples with small changes in the liquid crystal orientation (the optical path interval is 3λ and the accuracy of the measurements is $\lambda/250$).

The flexoelectric splay deformations were caused in a transversal d.c. electric field (Figs. 5a and 5b). The flexoelectric splaying of the MBBA liquid crystal layer under study started at about 300V/cm (the distance between the electrode strips was around 0,8 mm). At above 750 V/cm a strong electrohydrodynamic developed chiefly around the electrodes (50-52) has disturbed the flexoelectric splaying of the liquid crystal under investigation. Very often the flexoelectric deformations started in different regions of the MBBA layer due to the inhomogeneity in the surface anchoring at the weak-anchored boundary which make difficult the performance of the optical path measurements.

The correct measurements of the flexoelectric coefficient of splay e_{1z} requires the utilization of the formula (5) provided to know the value of $(W_{s\theta}\theta_0/K_{11})$ $(\frac{|\Delta\epsilon|}{12\pi}\frac{E^2d^2}{K_{11}}$ $< 0,1$ for the MBBA cases (53)):

$$e_{1z} = (K_{11}/E)\{ - \frac{W_{s\theta}\,\theta_0}{K_{11}} + [(\frac{W_{s\theta}\theta_0}{K_{11}})^2 + \frac{6(1_2 - 1_1)}{n_e(\frac{n_e}{n_0^2} - 1)d^3}]^{1/2} \}\tag{11}$$

Since the value of $(W_{s\theta}\,\theta_0/K_{11})$ is known only approximately and varies from sample to sample it is more convenient to utilize the following more simple formula:

$$e_{1z}^2 = \frac{1(E)\, -1(0)}{(1/6)n_e\, (\frac{n_e^2}{n_0^2} - 1)E^2\, d^3}\, K_{11}^2\tag{12}$$

when $(W_{s\theta} \theta_0)$ is smaller than e_{1z} E. Note that this formula is more correct for hybrid-aligned nematic layers with planar-homeotropic orientation and strong-weak anchoring respectively.

The measurements were performed for four values of the applied voltage which were 35 V, 40 V, 50 V and 55 V (the distance between the electrodes as noted was 0,8 mm). The corresponding phase difference 1(E) −1(0) between the two rays was calculated according to the relation:

$$1(E) - 1(0) \quad = \quad \frac{\Delta p \lambda}{h} \quad = \quad \frac{\Delta p \; 0,550}{2530} \tag{13}$$

where Δp is the difference between the positions of the micrometric screw of the microscope to indicate the phase difference $p(E)$ − $p(0)$ respectively at each value of the electric field applied across the liquid crystal layer under study, $\lambda = 0,550$ μm is the wavelength of the white light which was utilized in the experiment and h = 2530 microns is the distance between each two interference bands of the white light.

In our case the corresponding values of Δp were measured to be 60, 170, 270 and 370 respectively. Since the measurements were performed at room temperature (T $\backsimeq$ 20°C) for the other parameters were utilized the following numerical values:

$n_e \simeq 1,80$, $n_0 \simeq 1,55$ [54], $K_{11} \simeq 7,5 \times 10^{-7}$ dyne [49]

d = 10 μm . (T $_{MBBA}$ $\simeq$ 44°C)

In this way the following four values of the splay flexoelectric coefficient $|e_{1z}|$ were calculated:

$0,6 \times 10^{-4}$ dyne$^{1/2}$, $\quad 0,85 \times 10^{-4}$ dyne$^{1/2}$, $\quad 0,85 \times 10^{-4}$ dyne$^{1/2}$

$0,90 \times 10^{-4}$ dyne$^{1/2}$

These values clearly show that the difference between the last three measurements of the flexoelectric coefficient of splay $|e_{1z}|$ is negligible. On the contrary the first measurement gives lower value of the flexoelectric coefficient of splay due to the influence of the initial tilt angle θ_0.

The accuracy of the measurements of the modulus of the flexo-electric coefficient $|e_{1z}|$ can be evaluated according to us from the following relation:

$$\frac{\Delta|e_{1z}|}{|e_{1z}|} \simeq \frac{|\Delta E|}{|E|} + \frac{\Delta K_{11}}{K_{11}} + (3/2) \frac{\Delta d}{d} \tag{14}$$

(in our opinion the uncertainities in the value of the optical indices should influence slightly the accuracy of the measurements).

It is reasonable to accept the following uncertainities in the measurements of the parameters entering in the problem under consideration:

$$\frac{|\Delta E|}{|E|} \sim 5\%, \qquad \frac{\Delta K_{11}}{K_{11}} \sim 10\%, \quad \frac{\Delta d}{d} \sim 10\% \tag{15}$$

Finally one can accept that the most probable value of the modulus of the splay flexoelectric coefficient is in the range of

$$|e_{1z}| = (0,9 \pm 0,2) \, 10^{-4} \, dyne^{1/2} \tag{16}$$

II. Determination of the Sign

The optical path difference due to the flexoelectric deformations in presence of initial tilt depends strongly on the sign of the electric field (see the relation (5)). It is clear that at positive sign of both the initial tilt θ_0 and the flexoelectric coefficient $|e_{1z}|$ when the electric field points out from anode to

cathode the flexoelectric splaying leads to the increase of the
initial splay (Fig. 5a). Contrary, at the reverse of the sign of
the electric field the flexoelectric splaying should leads to decrease of the initial tilt (Fig. 5a) (the same effect can be obtained with the change of the sign of the flexoelectric coefficient e_{1z}). The optical path expressed by the relation (5) clearly shows that in the first case ($\theta_0 > 0$, $E > 0$, $e_{1z} > 0$) one should
observe strong variations in $l_2 - l_1$ whereas in the other case the
variations in $l_2 - l_1$ should be smaller for the range of the electric
field utilized in our experiment: $E < 750$ V/cm.

This situation was really observed in our experiment. The
variations in the optical path at positive sign of the electric
field and positive value of the initial tilt θ_0 which was measured
conoscopically to be smaller than $40°$ were always larger in comparison with those measured at the reverse of the sign of the
electric field (let us stress here that it is very hard to obtain
weakly-anchored tilted layers when the liquid crystal orientation
is nearly to the planar ones).

The sign of the flexoelectric coefficient e_{1z} can be easy determined from the experimental observations taken into account
the sign of the electric field and the sign of the initial tilt
θ_0 which according to Nakamura (39) can be predetermined with the
rubbing direction.

Our experimental results clearly indicate the positive sign
of the MBBA flexoelectric coefficient of splay: $e_{1z} > 0$

DISCUSSION

The value of the flexoelectric coefficient of splay e_{1z} is
from two to three times larger than that calculated by Helfrich
(3) on the basis of the dipolar flexoelectric theory (1). In
our opinion this is due to quadrupolar contributions according

to the quadrupolar flexoelectric theory developed by Prost and Marcerou (2).

The uncertainities in the value of e_{1z} calculated in our experiment are due also to the soap which contaminates the liquid crystal, to the small tilt at the soap-treated wall, to the possible hydrodynamic effects generated near to the electrodes, to second-order elasticity effects which can change the effective value of the splay elastic constant K_{11}, to the possible surface polarization (16) at higher tilts, to the rapid deterioration of the liquid crystal, etc.

For more precision determination of e_{1z} including the thickness variations are necessary experimental measurements performed on different MBBA layers with thicknesses below 10 microns to prevent the possible electrohydrodynamic. Of interest is the utilization of surfactants which can induce homeotropic orientation of the nematic layer under study with very low anchoring (say in the range of 10^{-4} erg/cm^2). Such kinds of surfactants can facilitate very much the construction of good planar strong-weak anchored nematic layers with unique electro-optical properties: low threshold voltages, slow relaxation of the liquid crystal deformations which can be done very fast with the application of additional a.c. electric field etc.

REFERENCES

1. R. B. Meyer, Phys. Rev. Lett. 22, 918 (1969).

2. J. Prost and J. P. Marcerou, J. Physique 38, 315 (1977).

3. W. Helfrich, Z. Nat. 26a, 833 (1971).

4. A. I. Derzhanski and A. G. Petrov, Phys. Lett 36A, 483 (1971).

5. W. Haas, J. Adams and J. B. Flannery, Phys. Rev. Lett. 25, 1326 (1970).

6. W. Helfrich, Phys. Lett. 35A, 393 (1971).

7. D. Schmidt, M. Schadt and W. Helfrich, Z. Nat. 27a, 277 (1972).

8. A. I. Derzhanski and M. D. Mitov, C. R. Acad. Bulg. Sci. <u>28</u>, 1331 (1975).

9. J. Prost and P. S. Pershan, J. Appl. Phys. <u>47</u>, 2298 (1976).

10. J. P. Marcerou and J. Prost, Mol. Cryst. Liq. Cryst. <u>58</u>, 259 (1980).

11. L. K. Vistin', Kristaloggr. <u>15</u>, 594 (1970).

12. M. I. Barnik, L. M. Blinov, A. N. Trufanov and B. A. Umanski, Zh. Eksp. Teor. Fiz. <u>73</u>, 19 (1977).

13. M. I. Barnik, L. M. Blinov, A. N. Trufanov and B. A. Umanski, J. Physique <u>3</u>, 417 (1978).

14. Yu. P. Bobylev, V. G. Chigrinov and S. A. Pikin, J. Physique Colloq. <u>40-C</u>, C3-331 (1979).

15. Yu. P. Bobylev and S. A. Pikin, Zh. Eksp. Teor. Fiz. <u>72</u>, 369 (1977).

16. A. G. Petrov and A. I. Derzhanski, Mol. Cryst. Liq. Cryst. Lett. <u>41</u>, 41 (1977).

17. A. I. Derzhanski, A. G. Petrov and M. D. Mitov, J. Physique 39, 273 (1978).

18. J. E. Proust, L. Ter-Minassian-Saraga and E. Guyon, Sol. St. Comm. 11, 1227 (1972).

19. J. E. Proust and L. Ter-Minassian-Saraga, C. R. Acad. Sci. Paris <u>276</u>, C-1731 (1973).

20. J. E. Proust and L. Ter-Minassian-Saraga, J. Physique Colloq. <u>36</u>, C1-77 (1975).

21. E. Perz and J. E. Proust, C. R. Acad. Sci. Paris <u>282</u>, C-559 (1976).

22. K. Hiltrop and H. Stegemeyer, Ber. Bunsenges. Phys. Chem. <u>82</u>, 884 (1978).

23. K. Hiltrop and H. Stegemeyer, Mol. Cryst. Liq. Cryst. Lett. <u>49</u>, 61 (1978).

24. K. Hiltrop and H. Stegemeyer, Ber. Bunsenges. Phys. Chem. <u>85</u>, 582 (1981).

25.   Sh. Matsumoto, M. Kawamoto and K. Mizunoya, J. Appl. Phys. 47, 3842 (1976).

26.   V. G. Chigrinov, The Fourth Liquid Crystal Conference of Socialist Countries, Tbilisi 5-8 X (1981).

27.   J. Nehring and A. Saupe, J. Chem. Phys. 54, 337 (1971).

28.   W. R. Heffner, D. W. Berreman, M. Sammon and S. Meiboom, Appl. Phys. Lett. 36, 144 (1980).

29.   T. Uchida, M. Ohgawara and M. Wada, Jpn. J. Appl. Phys. 19, 2127 (1980).

30.   M. Nakamura and M. Ura, J. Appl. Phys. 52, 210 (1981).

31.   G. Porte, J. Physique 37, 1245 (1976).

32.   A. Toda, H. Mada and Sh. Kobayashi, Jpn. J. Appl. Phys. 17, 261 (1978).

33.   H. Mada, Mol. Cryst. Liq. Cryst. 51, 43 (1979).

34.   H. P. Hinov, Mol. Cryst. Liq. Cryst. 74, 1789 (1981).

35.   H. P. Hinov, Z. Nat. 37a, 334 (1982).

36.   G. Ryschenkow and M. Kleman, J. Che. Phys. 64, 404 (1976).

37.   M. Kléman and G. Ryschenkow, J. Chem. Phys. 64, 413 (1976).

38.   E. Guyon and W. Urbach, "Nonemissive Electrooptic Displays", Eds. A. R. Kmetz and F. K. VonWillisen, 121 (1976).

39.   N. Nakamura, J. Appl. Phys. 52, 456 (1981).

40.   H. P. Hinov, to be published.

41.   K. Fahrenschon, H. Gruler and M. F. Schiekel, Appl. Phys. 11, 67 (1976).

42.   K. Fahrenschon and M. F. Schiekel, J. Electrochem. Soc. 124, 953 (1977).

43.   R. Vilanove, E. Guyon, C. Mitescu and P. Pieranski, J. Physique 35, 153 (1974).

44.   E. P. Raynes, D. K. Rowell and I. A. Shanks, Mol. Cryst. Liq. Cryst. Lett. 34, 105 (1976).

45.   W. A. Crossland, J. H. Morrisy and B. Needham, J. Phys. D: Appl. Phys. 9, 2001 (1976).

46. M. R. Johnson and P. A. Penz, IEEE Tran. El. Dev. <u>ED-24</u>, 805
        (1977).

47. A. I. Derzhanski and H. P. Hinov, J. Physique <u>38</u>, 1013 (1977).

48. J. Cognard, Mol. Cryst. Liq. Cryst. Lett. <u>64</u>, 331 (1981).

49. W. H. de Jeu, W. A. P. Claassen and A. M. J. Spruijt, Mol. Cryst.
        Liq. Cryst. <u>37</u>, 269 (1976).

50. R. Chang, Mol. Cryst. Liq. Cryst. <u>20</u>, 267 (1973).

51. Sh. Kai, K. Yamaguchi and K. Hirakawa, Jpn. J. Appl. Phys. <u>14</u>,
        1653 (1975).

52. Sh. Kai, K. Hirakawa, Memoirs Fac. Engn. Kyushu University <u>36</u>,
        269 (1977).

53. D. Diguet, F. Rondelez and G. Durand, C. R. Acad. Sc. Paris
        <u>271</u>, B-954 (1970).

54. M. Laurent and R. Journeaux, Mol. Cryst. Liq. Cryst. <u>36</u>, 171
        (1976).

ON THERMOTROPIC DISC-LIKE MESOGENS

Nguyen Huu Tinh, J. Malthête[*], H. Gasparoux, and
C. Destrade

Centre de Recherche Paul Pascal
Université de Bordeaux I
Domaine Universitaire
33405 Talence Cédex - France

* Laboratoire de Chimie des Interactions Moleculaires
Collège de France
75231 Paris Cedex - France

INTRODUCTION

The existence of thermotropic disc-like mesogens has been
suggested for years. This idea originates first on some simple
geometrical considerations: between unidimensional rigid cores
(thermotropic rod-like mesogens) and tridimensional systems (plastic liquid crystals) it was easy to imagine that bidimensional
aromatic flat molecules may lead to mesomorphism.

   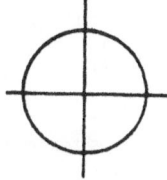   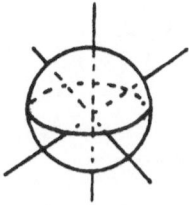

rod like molecules   disc-like molecules   globular molecules

Another motivation lay in the discovery around 1961[1,2], of the so called carbonaceous mesophase. We know that this anisotropic phase occurs at high temperature during the carbonization of graphitizable substances such as heavy petroleum residues[3,6], the obtained fluid mesophase presents an optical texture very similar to the nematic one. The chemical nature of this mesophase is not yet well known, but it could be composed of a complex mixture of flat polyaromatic plate-like or disc-like molecules, so there was also a need to get pure simple component models to study this mesophase.

This project was entirely successful around 1977-1978[7-10], almost simultaneously in several laboratories. Some hexasubstituted benzene and triphenylene derivatives exhibit between the solid and isotropic phases a highly viscous and birefringent mesophase corresponding to tubular or columnar arrangement of discs. Later on the discovery of a complex columnar polymorphism[11], then of a nematic phase built up with discs[12-15] (and of course of its chiral variant[16]) at last of several rentrant sequences[17-18] open new field of reflexion.

DISC-LIKE MOLECULES PHASES

Up to now only a few disc-like cores lead to mesomorphic phases. They are for example hexasubstituted derivatives of benzene[7], triphenylene[8-13], anthraquinone[19-20], truxene[14-18].

benzene   triphenylene   anthraquinone   truxene

with R = $C_nH_{2n+1}O-$, $C_nH_{2n+1}COO-$

C_nH_{2n+1}⟨O⟩COO , $C_nH_{2n+1}O-$⟨O⟩COO

Columnar phases

 - Several columnar phases (D phase) have been definitively
identified: they consist in regular two dimensional periodic array
and the different phases can differ either by the symmetry group of
the columns lattice or by the stacking of molecules in each column.
The columns can be upright or tilted with respect to the molecular
reference plan.

 - Hexagonal columnar phases - Two D columnar phases with an
hexagonal bidimensional lattice have been described as an ordered
(D_{ho})[21] or disordered arrangement of molecules in the columns
(D_{hd})[7], these phases are present in all the published series.

 - Rectangular columnar phases - With a rectangular lattice of
columns we know only disordered arrangement of molecules, we call
them D_{rd} phases. They are generally of the pgg symmetry[22,23] but
one example of pmg symmetry has been described in one rufigallol
derivative[19,20] (hexasubstituted anthraquinone).

 - Tilted phases - A single example of strongly tilted D columnar phase (D_t)[23] has been founded in the hexa(p-alkoxybenzoyloxy)tri-
phenylene series (tilted angle $\alpha = 55°$).

N_D nematic phase

 We have given the description of several hexa(p-alkyl or p-
alkoxybenzoyloxy)triphenylenes which exhibit a very fluid thermotropic mesophase with optical schlieren textures in any point
similar to the nematic one. In fact this new phase corresponds to
a structure in which the director is perpendicular to the average
plane in which the flat molecules are aligned. We call this nematic
phase "N_D". The nematic nature of this phase has been checked by
some X-ray measurements[23]. Another fine confirmation has been given
by the fact that mixture of these substances with some chiral triphenylene derivatives leads to very typical cholesteric textures[24].

Recently a pure disc-like substance with cholesteric properties has been described[16].

PHASES SEQUENCES

In most cases, the disc-like mesogenes exhibit a "normal" sequence but with truxene derivatives we have found an "inverted" sequence and later on a reentrant sequence.

Normal sequence:

- Hexaalkanoyloxybenzenes (HAB)[7]. Only one phase is observed in this series. This phase is D_{hd}

        K D_{hd} I

- Hexaalkoxytriphenylenes (HET)[8-10]. The only phase observed with these derivatives is D_{ho}

        K D_{ho} I

- Hexaalkanoyloxytriphenylenes (HAT)[10-11]. The short alkyl chain derivatives (from C_7 to C_{10}) exhibit only a D_{rd} phase

        K D_{rd} I

With long alkyl chain (from C_{11} to C_{13}) we have found a mesomorphic polymorphism.

        K D_x D_{rd} D_{hd} I

The last columnar D_x phase was not carried out on X-ray measurements.

- Hexaalkanoates of rufigallol (HAR)[19-20] (or hexaalkanoyloxy-anthaquinones) the only one described derivative of this series, the octanoate, exhibits a $D_{rd}{}^+$ enantiotropic phase with Pgg symmetry and an another $D_{rd}{}^{++}$ monotropic phase with Pmg symmetry

        K $D_{rd}{}^{++}$ $D_{rd}{}^+$ I

- Hexa(p-alkyl or p-alkoxybenzoyloxy)triphenylenes (HBT)[12-13]. This is the first series in which we have obtained the N_D nematic

phase with optical schlieren texture and we have described three
different normal sequences

$$K \qquad N_D \qquad I$$

$$K \quad D_{rd} \qquad N_D \qquad I$$

$$K \quad D_t \qquad N_D \qquad I$$

Inverted sequence

These normal sequences observed in these various series show
the great analogies between this new condensed state of matter and
rod like liquid crystals[*]: But in the series of hexaalkanoyloxytruxenes (HATX)[14-15] with short alkyl chains we have found a new
phenomenon: an inverted sequence in which a fluid N_D nematic phase
is observed below a higher viscous D_{rd} columnar one. This phenomenon is never observed with pure rod like mesogens. This inverted
sequence is:

$$K \quad N_D \quad D_{rd} \quad D_{ho} \quad I$$

Reentrant sequence

The discovery of such a N_D nematic phase at lower temperature
than a viscous D_{rd} columnar phase suggests us at first that it
could be a reentrant N_D nematic phase. In fact we are sure now
that it is not a real one. One cannot find for example the nematic
phase again at higher temperature. But it provides an exciting
observation leading with the reentrant phenomenon research.

In fact the study of highly purified hexaalkanoyloxytruxenes
with very long alkyl chains provides us the required phenomenon: an

---

[*] highly ordered phases at low temperatures and the less ones at
high temperatures.

hexagonal reentrant columnar phase with the sequence[17].

$$K \qquad (D_{ho}) \qquad N_D \qquad D_{rd} \qquad D_{ho} \qquad I$$

If it is noted that hexa(p-alkoxybenzoyloxy)triphenylenes (HBT) exhibit a N_D nematic phase at high temperature and HATX derivatives at low temperature it is easy to understand why we tried to build up hexa(p-alkoxybenzoyloxy)truxenes with the hope to obtain a reentrant N_D nematic phase. This project is entirely successful and we have obtained two new reentrant sequency with long alkoxy chains (HBTX)[18]

$$K \qquad D_{rd} \qquad N_D \qquad D_{rd} \qquad N_D \qquad I$$

$$K \qquad D_{rd} \qquad N_D \qquad D_{rd} \qquad D_{ho} \qquad I$$

In brief, all the observed phases sequences of disc-like liquid crystals are:

Monomorphism: D_{hd} (HAB)
                  D_{ho} (HET)
                  $D_{rd}{}^+$ (HAT)
                  N_D (HBT)

Dimorphism: $D_{rd}{}^{++}$ $D_{rd}{}^+$ (HAR)
                  $D_{rd}{}^+$ N_D (HBT)
                  D_t N_D (HBT)

Trimorphism: D_x D_{rd} D_{hd} (HAT)
                  N_D $D_{rd}{}^+$ D_{ho} (HATX)

Tetramorphism: D_{ho} N_D $D_{rd}{}^+$ D_{ho} (HATX)
                  $D_{rd}{}^+$ N_D $D_{rd}{}^+$ N_D (HBTX)
                  D_{rd} N_D $D_{rd}{}^+$ D_{ho} (HBTX)

\+ Pgg symmetry

\++ Pmg symmetry

CONCLUSION

With only a few disc-like cores and about seventy derivatives one can find the N_D nematic, N_D^* cholesteric, six different columnar phases and three reentrant sequences. This remark will surely stimulate new synthetic work in the future.

REFERENCES

1. J. D. Brooks and G. H. Taylor, Carbon 3, 185 (1965).

2. J. D. Brooks and G. H. Taylor, Chemistry and Physics on
      Carbon, Eds. P. L. Walker and E. Arnold, New York, 1968,
      vol. 4, 243.

3. J. E. Zimmer and J. L. White, Mol. Cryst. Liq. Cryst. 38, 177
      (1977).

4. J. L. White and J. E. Zimmer, Carbon 16, 469 (1978).

5. H. Gasparoux, C. Destrade and G. Fug, Mol. Cryst. Liq. Cryst.
      59, 109 (1980).

6. H. Gasparoux, Liq. Cryst. of One and Two Dimensional Order
      Springer-Verlag Berlin, Heidenberg, New York, 1980, 373.

7. S. Chandrasekhar, B. K. Shadashiva and K. A. Suresh, Pramana
      9, 471 (1977).

8. Nguyen Huu Tinh, J. C. Dubois, J. Malthête, and C. Destrade,
      C. R. Acad. Sc. 286C, 463 (1978).

9. J. Billard, J. C. Dubois, Nguyen Huu Tinh and A. Zann, Nouv.
      J. de Chim. 2, 535 (1978).

10. C. Destrade, M. C. Mondon and J. Malthête, J. de Phys. C3, 40,
      17 (1979).

11. C. Destrade, M. C. Mondon Bernaud and Nguyen Huu Tinh, Mol.
      Cryst. Liq. Cryst. Lett 49, 169 (1979).

12. Nguyen Huu Tinh, C. Destrade and H. Gasparoux, Phys. Lett. A.
      72, 251 (1979).

13. Nguyen Huu Tinh, H. Gasparoux and C. Destrade, Mol. Cryst.
      Liq. Cryst. 68, 101 (1981).

14. C. Destrade, J. Malthête, Nguyen Huu Tinh and H. Gasparoux,
        Phys. Lett. A. <u>78</u>, 82 (1980).

15. C. Destrade, J. Malthête, H. Gasparoux, A. Babeau and Nguyen
        Huu Tinh, Mol. Cryst. Liq. Cryst. <u>67</u>, 37 (1981).

16. J. Malthête, C. Destrade, Nguyen Huu Tinh and J. Jacques, Mol.
        Cryst. Liq. Cryst. Lett. <u>64</u>, 233 (1981).

17. Nguyen Huu Tinh, J. Malthête and C. Destrade, Mol. Cryst. Liq.
        Cryst. Lett. <u>64</u>, 291 (1981).

18. Nguyen Huu Tinh, J. Malthête and C. Destrade, J. de Phys.
        Lett. <u>42</u>, 417 (1981).

19. A. Queguiner, A. Zann, J. C. Dubois and J. Billard, Proc. Int.
        Liq. Cryst. Conf. Bangalore, Heydon and Son London, 1980,
        35.

20. J. Billard, J. C. Dubois, C. Vaucher and A. M. Levelut, Mol.
        Cryst. Liq. Cryst. <u>66</u>, 115 (1981).

21. A. M. Levelut, J. de Phys. Lett. <u>40</u>, 81 (1979).
        It would not be fair to pass over the first observed
        thermotropic columnar phases in silence obtained in this
        cases with rod like molecules (neat soaps) Spegt P. A.
        and Skoulios, A. E. Acta-Cryst. (1963) 16, 301 ibidem
        (1966) 21, 892.

22. A. M. Levelut, Proc. Int. Liq. Cryst. Conf. Bangalore, Heyden
        and Son London, 1980, 21.

23. A. M. Levelut, F. Hardouin, H. Gasparoux, C. Destrade and
        Nguyen Huu Tinh, J. de Phys. <u>42</u>, 147 (1981).

24. C. Destrade, Nguyen Huu Tinh, J. Malthête and J. Jacques,
        Phys. Lett. A. 189 (1980).

ON THE NATURE OF THE SMECTIC-C PHASE

Satyendra Kumar* and Thomas Lee Polgreen**

*Department of Physics and Center for Materials Science
and Engineering, 13-2025, Massachusetts Institute of
Technology, Cambridge, MA 02139

**Department of Physics and Materials Research Laboratory
University of Illinois at Urbana-Champaign
1110 West Green, Urbana, IL 61801

ABSTRACT

Temperature dependence of the smectic layer thickness of the
Sm-C phase of HOAB has been measured by high resolution x-ray diffraction. Layer undulations appear to be an intrinsic property of
the Sm-C phase. The positional-order correlation length parallel to
the smectic plane is found to be larger in the Sm-C than in the Sm-A
phase. These results successfully explain several experimental
observations found in the literature.

INTRODUCTION

In the smectic phases, the long axes of the molecules are nearly
parallel to a preferred direction and the molecular centers of mass
are arranged in equidistant layers. The orientational-order
parameter [$S = 1/2 \langle 3 \cos^2\theta - 1 \rangle$, where θ is the angle between long axis
of a molecule and the preferred direction of alignment] is saturated
(i.e. $S \approx 1$) and changes very little in a smectic phase. In the

smectic-A (Sm-A) and -C (Sm-C) phases the arrangement of molecules inside the smectic layers is random (liquid-like) and free molecular rotation about their long axes is assumed. In the Sm-A phase, the molecules are aligned perpendicular to the layers; while in the Sm-C phase, they are tilted relative to the layers. The tilt is strongly dependent upon temperature in most compounds. There are instances, however, where no temperature dependence of the molecular tilt has so far been observed.[1,2] One such compound, 4-4'-bis(heptyloxy) azoxybenzene (HOAB), has been very widely studied. The absence of temperature dependence of the molecular tilt has led to the belief that two different Sm-C phases exist.

In this paper we present results of high-resolution x-ray experiments on the Sm-C phase of HOAB and some other compounds. We find that (a) the layer thickness, d, (and hence molecular tilt) in the Sm-C phase is temperature dependent; (b) the Sm-C phase, in both bulk and thin samples, has undulations; and (c) the Sm-A and -C phases have different positional-order correlation length, $\xi_{||}$ parallel to the smectic layer. We also point out several experimental observations previously not fully understood. These observations are explained on the basis of results (b) and (c).

EXPERIMENT AND RESULTS

The compound HOAB was purified by recrystallization from ethyl alcohol and degassed by repeated melting and cooling (thaw cycle) in vacuum. It has a first order Sm-C to nematic transition at 94.75°C and a nematic to isotropic transition at 124°C and has been a subject of numerous x-ray studies[2-5] in the past. We made high resolution x-ray measurements using a three crystal diffractometer and CuK_{α} (λ=1.54178Å) radiation, as described in reference 6. The temperature precision was better than 0.05°C. The profile of Bragg reflections was recorded and a Lorentzian curve was fit to determine the peak position and the scattering angle θ. The value of d cal-

culated, using the value of θ thus obtained, has an accuracy of
0.04% and is shown in Fig. 1. For completeness the value of d for
cybotactic clusters in the nematic phase, taken from reference 5,
is also plotted. Evidently, d in the Sm-C phase has temperature
dependence comparable to that in cybotactic groups. The behaviour
of d is similar to that in the Sm-C phase of TBBA, away from the
Sm-C to -A transition. It is reasonable to assume that the temperature dependence of the Sm-C layer thickness of such compounds has
not been measured with enough precision and that d in the Sm-C
phase is always temperature dependent.

During a similar x-ray study[6] of the Smectic-C phase of TBBA
and its homologs, it was found that the Bragg peak becomes asym-

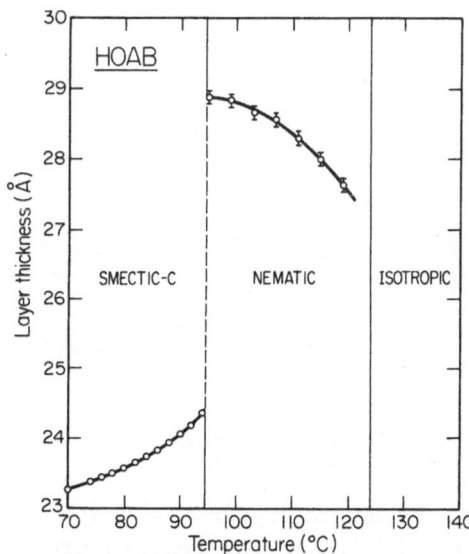

Fig. 1   Layer thickness of the Sm-C phase and cybotactic clusters
          in the nematic phase of HOAB as a function of temperature.
          The nematic phase data taken from reference 5.

metric, as shown in Fig. 2 for TB5A. The asymmetry increases as
the sample is cooled in the Sm-C phase and disappears at the
transition to the Sm-F or Sm-H phase. The rate of cooling in the
Sm-C phase appears to have a significant effect on this asymmetry.
On the other hand, it is stable in time over a period of days.
This effect can be explained by assuming that the Sm-C layers are
bent as shown in Fig. 3. This bending, called undulations, could
be caused by mechanical stress arising out of the requirement that
the smectic layers fill the same volume even after a decrease in
their thickness. This gives rise to local variation of the magnitude
and direction of the molecular tilt. The variation of local tilt
causes a spread in the values of d which is observed as an asymmetry
of Bragg reflections. This explanation predicts two qualitative

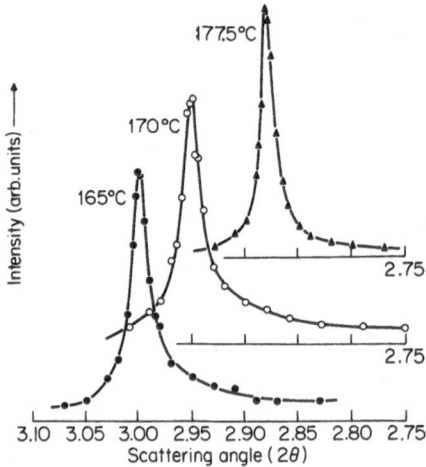

Fig. 2   The growth of asymmetry in the Bragg reflections from the
          Sm-C phase of TB5A as the temperature is lowered. The
          height of each peak is normalized to the same value.

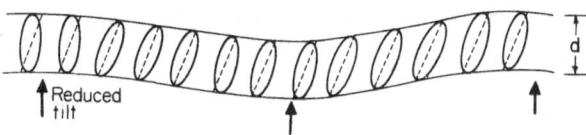

Fig. 3   Undulations in the Sm-C layers cause the local molecular
            tilt to change. The variations of tilt causes the vari-
            ation of d and asymmetry of Bragg reflections. The arrows
            point toward the regions of decreased molecular tilt.

features of the Bragg profiles for the Sm-C phase. (i) The wings
of the peaks should broaden only on the lower angle (larger d) side
but never beyond the Sm-A peak position. (ii) Since undulations
cause only a local decompression of layers, the minimum d will
correspond to the undistorted Sm-C phase. The spread on the smaller
d (larger angle) side should be very small. The experimentally
obtained profiles of Bragg reflections do indeed possess features
(i) and (ii).

Johnson and Saupe, in a previous study,[7] had created undulations in thin Sm-C films between two glass plates by cooling
while keeping the thickness fixed. The increased tilt upon cooling causes mechanical stress and thus the undulations are induced.
They found the period of these undulations to be about 7 microns.
We have observed undulations microscopically in thin slides without
any constraints on their thickness.

X-ray measurements of the positional-order correlation
length,[8] $\xi_{||}$ parallel to the smectic plane of N-(4-heptyloxy-
benzylidene)-4-n-hexylaniline (HBHA) yield unusually large value
($\sim$50-55Å) for the Sm-C phase. It is in agreement with the width
(3°) of the outer ring reported by Benattar[9] for the Sm-C phase
of TB5A. For the Sm-A phase, a value of $\xi_{||}$ $\sim$22Å can be estimated

from reference 10 for 70.5. An approximately equal value of $\xi_{||}$
has been measured by Kortan et al.[11] for the Sm-A phase of a
mixture of 60CB and 80CB. It is interesting to note that even in
free-standing films of 40.8,[12] the value of $\xi_{||}$ is the same (21Å)
as in bulk samples. Obviously the value of $\xi_{||}$ in the Sm-C phase
is more than double of that in the Sm-A phase. This indicates that
the local field experienced by the molecules in the Sm-C phase is
different (stronger) than in the Sm-A phase.

DISCUSSION

During several experimental studies, some effects have been
observed that could not be explained on the basis of the conventional
picture of the Sm-C phase. Unusual assumptions were used to explain
these effects. In this section, we will discuss some of these
observations and show that they can be explained by: (i) existence
of undulations and (ii) stronger local field than the Sm-A phase.

While measuring molecular tilt in the smectic phases of TBBA
by conoscopy, Niessen and denOuden[13] observed bird-like isogyres
in the Sm-C phase. Similar patterns were later observed by us in
the Sm-C phase of TB8A.[6] To explain this unexpected shape, an
assumption[13] was made that the sample is a pile of two or more
slices and that the optic axis in each slice was oriented differently.
The difference in the orientation of optic axes in neighbouring
slices was taken to be as large as 5°. Such an artificial division
of sample into slices is prohibited by homogeneity of the system.
However, existence of a pair of undulations perpendicular to each
other, as has been observed experimentally,[7] can give rise to a
variation of the direction of molecular tilt inside a layer as well
as from one layer to another. This variation, clearly, can be
responsible for the bird-like isogyres.

Leadbetter and Wrighton[14] have reported x-ray measurements of
the orientational-order parameter, S, at different temperatures in

the nematic, Sm-A and Sm-C phases of TBBA. It was found that the value of S in the Sm-C phase is significantly smaller than in the Sm-A phase. On the basis of the conventional picture of the Sm-C phase, S should be equal to its value in the Sm-A phase, which is not in agreement with the experimental results. The presence of undulations would change local tilt of the molecules and result in a de-enhancement of the orientational-order.

In several NMR experiments[15,16] it was found necessary to assume a Gaussian distribution of the molecular tilt with 5° variance to explain experimental results. Taylor et al.[4] also noted that the intensities in their optical experiment did not agree with predicted values. It appears that undulations were, at least in part, responsible for these effects.

A study of the Sm-C phase by Galerne[18] provides substantial support to our conclusion that the local field in the Sm-C phase is different (stronger) than in the Sm-A phase. In his study, measurements were made of the three indices of refraction, n_x, n_y and n_z, where n_z was along the long molecular axis. n_x was perpendicular to n_z and the direction of molecular tilt, and parallel to the smectic plane. It was found that below the Sm-A to Sm-C transition, $(n_z - n_x)$ and $(n_y - n_x)$ increase as expected. However, n_x was also found to increase at a considerably faster rate than in the Sm-A phase. A similar increase in refractive indices has been observed by Lockhart[19] et al. for $\bar{8}$S5 and $\bar{9}$S5 compounds. n_x should not vary with temperature because it is perpendicular to both the director and the direction of molecular tilt. The variation of density[20] with temperature in the Sm-C phase is at the same rate as in the Sm-A phase and can be ruled out as a factor responsible for the increase in n_x. These measurements suggest that the local field in the Sm-C phase is not only different from the Sm-A phase but also that it is temperature dependent. This difference of local field is also responsible for increased $\xi_{||}$ in the Sm-C phase. Our x-ray measurements were not precise enough to measure any temperature dependence of $\xi_{||}$. The change in local field at the Sm-A to Sm-C transition lends some support to the 'dipole freeze-out' model of McMillan.[21]

CONCLUSION

Through x-ray studies we have determined temperature dependence of layer thickness in the Sm-C phase of HOAB, which was believed to have temperature independent tilt. The layer undulations are found to occur in bulk as well as thin samples. It would not be unreasonable to assume the undulations to be an intrinsic property of the Sm-C phase. We have also found the positional-order correlation length parallel to the smectic-C layers to be larger than in the Sm-A phase, indicating a difference in local field in these two phases. On the basis of these results we have been able to provide an explanation of several experimental observations not in conformity with the conventional Sm-C model.

ACKNOWLEDGEMENTS

The authors would like to thank W. L. McMillan, J. D. Litster, J. Berlinsky, and L. J. Yu for helpful discussions and suggestions. This research was funded by the National Science Foundation under grant Nos. NSF-DMR-77-23999 (MRL, UIUC) and NSF-DMR-78-23555 (MIT).

REFERENCES

1. T. R. Taylor, J. L. Fergason and S. L. Arora, Phys. Rev. Lett. 24, 259 (1970).
2. A. deVries, J. Phys. (Paris) Colloq. 36, C1-1 (1975).
3. W. L. McMillan, Phys. Rev. 8A, 328 (1973).
4. I. G. Chistyakov, L. S. Schabischev, R. I. Jarenov, and L. A. Gusakova, Mol. Cryst. Liq. Cryst. 7, 279 (1969).
5. M. Teruchi, J. Phys. (Paris) Colloq. 40, C3-401 (1979).

6.   Satyendra Kumar, Phys. Rev. 23A, 3207 (1981).

7.   D. L. Johnson and A. Saupe, Phys. Rev. 15A, 2079 (1977).

8.   Satyendra Kumar, to be published.

9.   J. J. Benattar, F. Moussa, and M. Lambert, J. Phys. (Paris)
     Lett. 42, L-67 (1981).

10. W. M. DeJeu and J. A. DePoorter, Phys. Lett. 61A, 114 (1977).

11. A. R. Kortan, unpublished work.

12. D. E. Moncton and R. Pindak, Phys. Rev. Lett. 43, 701 (1979).

13. A. K. Niessen and A. denOuden, Philips Res. Repts. 29, 119
     (1974).

14. A. J. Leadbetter and P. G. Wrighton, J. Phys. (Paris) Colloq.
     40, C3-234 (1979).

15. A. Wulf, J. Chem. Phys. 63, 1564 (1975).

16. S. Meiboom and Zeev Luz, Mol. Cryst. Liq. Cryst. 22, 143 (1973).

17. T. R. Taylor, S. L. Arora, and J. L. Fergason, Phys. Lett. 25,
     722 (1972).

18. Y. Galerne, J. Phys. (Paris) 39, 1311 (1978).

19. T. E. Lockhart, D. W. Allender, E. Gelerinter, and
     D. L. Johnson, Phys. Rev. 20A, 1655 (1979).

20. D. Guillon and A. Skoulios, J. Phys. (Paris) 38, 79 (1977).

21. W. L. McMillan, Phys. Rev. 8A, 1921 (1973).

CONTRIBUTORS

M. F. Achard, Centre de Recherche Paul Pascal, Universite de Bordeaux I, 33405 Talence Cedex - France

S. M. Aharoni, Corporate Research and Development, Allied Corporation, Morristown, NJ 07960

F. V. Allen, EM Chemicals, Hawthorne, NY 10532

V. Amirthalingam, Reactor Chemistry Section, Bhabha Atomic Research Centre, Trombay, Bombay 400 085

J. Asrar, Polymer Science Program, Department of Chemistry, University of Lowell, Lowell, MA 01854

B. L. Bales, Department of Physics and Astronomy, California State University, Northridge, Northridge, CA 91330

D. A. Balzarini, Department of Physics, University of British Columbia, Vancouver, B. C. Canada

J. Billard, Laboratoire De Physique De La Matiere Condensee, (Equipe Associee Au C.N.R.S.), College De France, 75231 Paris Cedex 05, France

B. C. Benicewicz, (1) Department of Chemistry and Institute of Materials Science, University of Connecticut, Storrs, CT 06268 (2) Celanese Research Company, Summit, New Jersey 07901

A. Ben-Shaul, Department of Chemistry, University of California, Los Angeles, Los Angeles, California 90024

D. W. Berreman, Bell Laboratories, Murray Hill, NJ 07974

J. Billard, Laboratoire de Physique de la Matiere Condensee, (equipe associee au C.N.R.S.), College de France, 75231 Paris Cedex 05, France

1141

H. Birecki, Hewlett-Packard Company, Solid State Laboratory, 1501
    Page Mill Road, Palo Alto, CA   94304

A. Blumstein, Polymer Science Program, Department of Chemistry,
    University of Lowell, Lowell, MA   01854

R. B. Blumstein, Polymer Science Program, Department of Chemistry,
    University of Lowell, Lowell, MA   01854

M. F. Bone, Edward Davies Chemical Laboratories, University College
    of Wales, Aberystwyth SY23 1NE, UK

L. Bosio, Laboratoire de Physicochimie Structurale et Macromole-
    culaire, et Laboratoire de Physique des Liquides et Electro-
    chimie, ESPCI, 10 rue Vauquelin, 75231 Paris Cedex 05, France

B. J. Bulkin, Department of Chemistry, Polytechnic Institute of New
    York, Brooklyn, N.Y.   11201

D. J. Byron, Department of Physical Sciences, Trent Polytechnic,
    Burton Street, Nottingham NG1 4BU, England

G. A. Campbell, Department of Chemistry, University of Southern
    Mississippi, Hattiesburg, Mississippi   39406

E. F. Carr, Physics Department, University of Maine, Orono, ME
    04469

J. A. Castellano, Stanford Resources, Inc., P. O. Box 20324, San
    Jose, CA   95160

P. M. Chaikin, Department of Physics, University of California, Los
    Angeles, CA   90024

S. Chandrasekhar, Raman Research Institute, Bangalore 560 080,
    India

H. Chihara, Department of Chemistry, Faculty of Science, Osaka
    University, Toyonaka, Osaka 560, Japan

N. J. Chou, IBM Research Laboratory, Yorktown, NY   10598

P. E. Cladis, Bell Laboratories, Murray Hill, NJ   07974

M. G. Clark, Royal Signals and Radar Establishment, Malvern,
    Worcestershire WR14 3PS, UK

P. J. Collings, Department of Physics, Kenyon College, Gambier,
    Ohio 43022

R. J. Cox, IBM Research Laboratory, 5600 Cottle Road, San Jose, CA 95193

P. P. Crooker, Department of Physics and Astronomy, University of
     Hawaii, Honolulu, Hawaii 96822

F. Cser, Research Institute for Plastics, H-1950 Budapest, Hungary

B. L. Dawson, IBM Research Laboratory, 5600 Cottle Road, San Jose,
     CA 95193

A. de Vries, Kent State University, Liquid Crystal Institute, Kent,
     Ohio 44242

K. U. Deniz, Nuclear Physics Division, Bhabha Atomic Research
     Centre, Trombay, Bombay 400 085

A. I. Derzhanski, Institute of Solid State Physics, Bulgarian
     Academy of Sciences, Boul. Lenin 72, Sofia 1184, Bulgaria

D. Destrade, Centre de Recherche Paul Pascal, Universite de Bordeaux
     I, Domaine Universitaire, 33405 Talence Cedex - France

J. W. Doane, Liquid Crystal Institute, Kent State University, Kent,
     Ohio 44242

R. A. Dolin, Department of Physics and Astronomy, California State
     University, Northridge, Northridge, CA 91330

B. Dozier, Department of Physics, University of California, Los
     Angeles, CA 90024

J. C. Dubois, Laboratoire Central de Recherches, Thomson-CSF,
     Domaine de Corbeville, 91401 Orsay Cedex, France

D. A. Dunmur, Department of Chemistry, University of Sheffield,
     Sheffield, U.K.

D. B. DuPre, Department of Chemistry, University of Louisville,
     Louisville, Kentucky 40292

A. J. East, Celanese Research Company, Summit, New Jersey 07901

J. Economy, IBM San Jose Research Center, San Jose, CA 95193

R. Eidenschink, E. Merck, Darmstadt, West Germany

M. El-Nokaly, Chemistry Department, University of Missouri-Rolla,
     Rolla, Missouri 65401

B. Fayolle, Rhone-Poulenc, Centre de Recherches des Carrieres,
     69190 Saint-Fons, France

K. Feng, Department of Physics, University of California, San
     Diego, La Jolla, CA 92093

J. R. Fernandes, Department of Chemistry, University of Louisville, Louisville, Kentucky 40292

R. W. Filas, Bell Laboratories, Murray Hill, NJ 07974

H. Finkelmann, Institut fur Physikalische Chemie, Technische Universitat Clausthal, 3392 Clausthal-Zellerfeld, FRG

P. L. Finn, Bell Laboratories, Murray Hill, NJ 07974

J. H. Flack, Department of Physics and Astronomy, University of Hawaii, Honolulu, Hawaii 96822

S. E. Friberg, Chemistry Department, University of Missouri-Rolla, Rolla, Missouri 65401

C. Friedrich, Laboratoire de Physicochimie Structurale et Macromoleculaire, et Laboratoire de Physique des Liquides et Electrochimie, ESPCI, 10 rue Vauquelin, 75231 Paris Cedex 05, France

H. Gasparoux, Centre de Recherche Paul Pascal, Universite de Bordeaux I, Domaine Universitaire, 33405 Talence Cedex - France

W. M. Gelbart, Department of Chemistry, University of California, Los Angeles, Los Angeles, California 90024

P. R. Gerber, Central Research Units, F Hoffmann-La Roche & Co., Limited Company, 4002 Basel, Switzerland

J. W. Goodby, Bell Laboratories, Murray Hill, NJ 07974

A. C. Griffin, Department of Chemistry, University of Southern Mississippi, Hattiesburg, Mississippi 39406

S. Gurnani, Biochemistry and Food Technology Division, Bhabha Atomic Research Centre, Trombay, Bombay 400 085

F. Hardouin, Centre de Recherche Paul Pascal, Universite de Bordeaux I, 33405 Talence Cedex - France

G. Hardy, Research Institute for Plastics, H-1950 Budapest, Hungary

M. Helene, Chemistry Department, University of Waterloo, Waterloo, Ontario, Canada N2L 3G1

J. Her, Department of Physics and Astronomy, State University of New York at Buffalo, Amherst, NY 14260

H. Hervet, College de France, Paris, France

K. Hiltrop, Department of Physical Chemistry, University of
Paderborn, 4790 Paderborn, Warburger StraBse 100, Gebaude J.,
Postfach 1621

H. P. Hinov, Institute of Solid State Physics, Bulgarian Academy of
Sciences, Boul. Lenin 72, Sofia 1184, Bulgaria

J. T. Ho, Department of Physics and Astronomy, State University of
New York at Buffalo, Amherst, NY 14260

A. W. Horton, Section Chemical Biology and Oncology, Department of
Public Health and Preventive Medicine, Department of Biochem-
istry, School of Medicine, Oregon Health Sciences University,
Portland, Oregon 97201

C. C. Huang, School of Physics and Astronomy, University of
Minnesota, Minneapolis, MN 55455

S. J. Huang, Department of Chemistry and Institute of Materials
Science, University of Connecticut, Storrs, CT 06268

R. L. Hubbard, Tektronix 50/426, P. O. Box 500, Beaverton, OR
97075

W. E. Hughes, Department of Physics and Astronomy, University of
Southern Mississippi, Hattiesburg, Mississippi 39406

A. Isenberg, Martin-Luther-Universitat Halle-Wittenberg, Sektion
Chemie, DDR-4020 Halle/Saale, Weinbergweg 16

M. K. Jain, Department of Chemistry, University of Delaware,
Newark, Delaware 19711

J. E. Jensen, Hughes Research Laboratories, 3011 Malibu Canyon
Road, Malibu, California 90265

J.-II Jin, Chemistry Department, College of Science, Korea University, 1-Anam Dong, Seoul 132, Korea

D. L. Johnson, Department of Physics and Astronomy, University of
Hawaii, Honolulu, Hawaii 96822

J. F. Johnson, Department of Chemistry and Institute of Materials
Science, University of Connecticut, Storrs, CT 06268

T. J. Jones, Department of Chemistry, Harvey Mudd College, Claremont, CA 91711

K. Kiss, Research Institute for Plastics, H-1950 Budapest, Hungary

C. L. Khetrapal, Raman Research Institute, Bangalore 560080, India;
     Bangalore NMR Facility, Indian Institute of Science,
     Bangalore 560012, India

K. R. Koehler/Beran, Tektronix 50/426, P. O. Box 500, Beaverton, OR
     97075

R. W. H. Kozlowski, Physics Department, University of Maine, Orono,
     ME 04469

S. Kumar, Department of Physics and Center for Materials Science
     and Engineering, 13-2025, Massachusetts Institute of Techno-
     logy, Cambridge, MA 02139

A. C. Kunwar, Raman Research Institute, Bangalore 560080, India

D. Lacey, Department of Physical Sciences, Trent Polytechnic,
     Burton Street, Nottingham NG1 4BU, England

M. R. Lakshminarayana, University of Agricultural Sciences,
     Bangalore 560065, India

D. W. Larsen, Chemistry Department, University of Missouri-St.
     Louis, St. Louis, Missouri 63121

F. Laupretre, Laboratoire de Physicochimie Structurale et
     Macromoleculaire, et Laboratoire de Physique des Liquides et
     Electrochimie, ESPCI, 10 rue Vauquelin, 75231 Paris Cedex 05,
     France

P. LeBarny, Laboratoire Central de Recherches, Thomson-CSF, Domaine
     de Corbeville, 91401 Orsay Cedex, France

L. Leger, College de France, Paris, France

R. W. Lenz, Chemical Engineering Department, University of
     Massachusetts, Amherst, MA 01003

T. M. Leslie, Bell Laboratories, Murray Hill, NJ 07974

A. M. Levelut, Laboratoire de Physique des Solides, Bat. 510,
     Universite Paris-Sud, 91405 Orsay - France

J. C. H. Liang, Tektronix 50/426, P. O. Box 500, Beaverton, OR
     97075

H. M. Lindsay, Department of Physics, University of California, Los Angeles, CA 90024

S. Long, Department of Physics and Astronomy, University of Hawaii, Honolulu, Hawaii 96822

B. Luhmann, Institut fur Physikalische Chemie, Technische Universitat Clausthal, 3392 Clausthal-Zellerfeld, FRG

N. V. Madhusudana, Raman Research Institute, Bangalore 560 080, India

L. Makaruk, Institute of Organic Chemistry and Technology, Technical University (Politechnika), Koszykowa 75, Warsaw, Poland

J. Maltheete, Laboratoire de Chimie des Interactions Moleculaires, College de France, 75231 Paris Cedex - France

J. D. Margerum, Hughes Research Laboratories, 3011 Malibu Canyon Road, Malibu, California 90265

D. G. McDonnel, Royal Signals and Radar Establishment, Malvern, Worcestershire WR14 3PS, UK

P. Meurisse, Laboratoire de Physicochimie Structurale et Macromole- culaire, et Laboratoire de Physique des Liquides et Electro- chimie, ESPCI, 10 rue Vauquelin, 75231 Paris Cedex 05, France

W. H. Miller, Department of Chemistry, University of Sheffield, Sheffield, U.K.

E. B. Mirza, Reactor Chemistry Section, Bhabha Atomic Research Centre, Trombay, Bombay 400 085

P. A. Mitchel, Department of Physics and Astronomy, University of Maine at Orono, Orono, ME 04469

S. Miyajima, Department of Chemistry, Faculty of Science, Osaka University, Toyonaka, Osaka 560, Japan

N. Nakamura, Department of Chemistry, Faculty of Science, Osaka University, Toyonaka, Osaka 560, Japan

M. E. Neubert, Liquid Crystal Institute, Kent State University, Kent, Ohio 44242

C. Noel, Laboratoire de Physicochimie Structurale et Macromole- culaire, et Laboratoire de Physique des Liquides et Electro- chimie, ESPCI, 10 rue Vauquelin, 75231 Paris Cedex 05, France

K. Nyitrai, Research Institute for Plastics, H-1950 Budapest, Hungary

P. Palffy-Muhoray, Department of Physics, University of British Columbia, Vancouver, B.C., Canada

P. S. Parvathanathan, Nuclear Physics Division, Bhabha Atomic Research Centre, Trombay, Bombay 400 085

J. A. Pavlisko, Research Triangle Institute, Research Triangle Park, NC

G. R. Penk, Section Chemical Biology and Oncology, Department of Public Health and Preventive Medicine, School of Medicine, Oregon Health Sciences University, Portland, Oregon 97201

P. Photinos, Liquid Crystal Institute, Kent State University, Kent, Ohio 44242

P. A. Pincus, Department of Physics, University of California, Los Angeles, CA 90024

A. C. Pineda, Department of Chemistry, Harvey Mudd College, Claremont, CA 91711

H. Polanska, Institute of Organic Chemistry and Technology, Technical University (Politechnika), Koszykowa 75, Warsaw, Poland

T. L. Polgreen, Department of Physics and Materials Research Lab., University of Illinois at Urbana-Champaign, 1110 West Green, Urbana, IL 61801

J. S. Prasad, Department of Physics, University of Mysore, Mysore 570 006, India

A. H. Price, Edward Davies Chemical Laboratories, University College of Wales, Aberystwyth SY23 1NE, UK

Y. Rabin, Department of Chemistry, University of California, Los Angeles, Los Angeles, California 90024

F. Ramirez, Department of Chemistry, State University of New York at Stony Brook, Stony Brook, NY 11794

L. W. Reeves, Chemistry Department, University of Waterloo, Waterloo, Ontario, Canada N2L 3G1

G. Rehage, Institut fur Physikalische Chemie, Technische Universitat Clausthal, 3392 Clausthal-Zellerfeld, FRG

C. Robinson, Chemistry Department, University of Waterloo,
    Waterloo, Ontario, Canada N2L 3G1

M. Roemer, E. Merck, Darmstadt, West Germany

J. Rogers, Department of Chemistry, University of Delaware, Newark,
    Delaware 19711

N. D. Russell, Department of Physics, Kenyon College, Gambier, Ohio
    43022

E. T. Samulski, Department of Chemistry, Institute of Materials
    Science, University of Connecticut, Storrs, CT 06268

F. C. Saunders, Royal Signals and Radar Establishment, Malvern,
    Worcestershire WR14 3PS, UK

A. Saupe, Liquid Crystal Institute, Kent State University, Kent,
    Ohio 44242

K. L. Savithramma, Raman Research Institute, Bangalore 560 080,
    India

R. N. Schwartz, Hughes Research Labs., 3011 Malibu Canyon Blvd.,
    Malibu, CA 90265

M. Shamsai, Physics Department, University of Maine, Orono, ME
    04469

P. Sheng, Corporate Research Science Labs., Exxon Research &
    Engineering Co., P. O. Box 45, Linden, NJ 07036

N. C. Shivaprakash, Department of Physics, University of Mysore,
    Mysore 570 006, India

J. M. Sloan, Department of Chemistry, Polytechnic Institute of New
    York, Brooklyn, N.Y. 11201

M. Sorai, Chemical Thermodynamics Laboratory and Department of
    Chemistry, Faculty of Science, Osaka University, Toyonaka,
    Osaka 560, Japan

G. J. Sprokel, IBM Research Laboratory, San Jose, CA 95193

H. Stegemeyer, Department of Physical Chemistry, University of
    Paderborn, 4790 Paderborn, Warburger StraBse 100, Gebaude J.,
    Postfach 1621

H. Stevens, Institut fur Physikalische Chemie, Technische Universitat Clausthal, 3392 Clausthal-Zellerfeld, FRG

H. Suga, Chemical Thermodynamics Laboratory and Department of Chemistry, Faculty of Science, Osaka University, Toyonaka, Osaka 560, Japan

C. E. Tarr, Department of Physics and Astronomy, University of Maine at Orono, Orono, ME   04469

N. H. Tinh, Centre de Recherche Paul Pascal, Universite de Bordeaux I, 33405 Talence Cedex - France

H. Toriumi, Department of Chemistry, Institute of Materials Science, University of Connecticut, Storrs, CT   06268

J. Tsay, IBM San Jose Research Center, San Jose, CA   95193

C. van Ast, Hughes Research Laboratories, 3011 Malibu Canyon Road, Malibu, California   90265

G. R. Van Hecke, Department of Chemistry, Harvey Mudd College, Claremont, CA   91711

J. M. Viner, School of Physics and Astronomy, University of Minnesota, Minneapolis, MN   55455

J. Virlet, Department de Physicochimie, CEN-Saclay, B.P. n°2, 91191 Gif Sur Yvette

R. L. Vold, Department of Chemistry, University of California, San Diego, La Jolla, CA   92093

R. R. Vold, Department of Chemistry, University of California, San Diego, La Jolla, CA   92093

W. Volksen, IBM San Jose Research Center, San Jose, CA   95193

H.-M. Vorbrodt, Martin-Luther-Universitat Halle-Wittenberg, Sektion Chemie, DDR-4020 Halle/Saale, Weinbergweg 16

R. A. Wheeler, Department of Chemistry, Harvey Mudd College, Claremont, CA   91711

R. C. Wilson, Department of Physical Sciences, Trent Polytechnic, Burton Street, Nottingham NG1 4BU, England

C. W. Woo, Department of Physics, University of California, San Diego, La Jolla, CA   92093

S.-M. Wong, Hughes Research Laboratories, 3011 Malibu Canyon Road,
    Malibu, California 90265

L. Wright, Royal Signals and Radar Establishment, Malvern, Worces-
    tershire WR14 3PS, UK

Z. Yaniv, Liquid Crystal Institute, Kent State University, Kent,
    Ohio 44242

H. Yoshioka, Chemical Thermodynamics Laboratory and Department of
    Chemistry, Faculty of Science, Osaka University, Toyonaka,
    Osaka 560, Japan

L.-P. Yu, Department of Chemistry and Institute of Materials
    Science, University of Connecticut, Storrs, CT 06268

H. Zaschke, Martin-Luther-Universitat Halle-Wittenberg, Sektion
    Chemie, DDR-4020 Halle/Saale, Weinbergweg 16

INDEX